Light Scattering, Size Exclusion Chromatography and Asymmetric Flow Field Flow Fractionation

Light Scattering, Size Exclusion Chromatography and Asymmetric Flow Field Flow Fractionation

Powerful Tools for the Characterization of Polymers, Proteins and Nanoparticles

Stepan Podzimek

WILEY

A John Wiley & Sons, Inc., Publication

Published by John Wiley & Sons, Inc., Hoboken, New Jersey.
Published simultaneously in Canada.

Library of Congress Cataloging-in-Publication Data:

Podzimek, Stepan, 1955–
    Light scattering, size exclusion chromatography and asymmetric flow
field flow fractionation : powerful tools for the characterization of
polymers, proteins and nanoparticles / Stepan Podzimek.
        p. cm.
    Includes bibliographical references and index.
    ISBN 978-0-470-38617-0 (cloth)
    1. Polymers–Analysis. 2. Polymers–Separation. 3. Light–Scattering.
4. Chromatographic analysis. I. Title.
    QD139.P6P625 2010
    543'.8–dc22
                                                    2010010798

Printed in Singapore.

10 9 8 7 6 5 4 3 2 1

Contents

Preface   ix

1 Polymers   1

1.1   Introduction   1

1.2   Molecular Structure of Polymers   2

    1.2.1   Macromolecules in Dilute Solution   4

1.3   Molar Mass Distribution   10

    1.3.1   Description of Molar Mass Distribution   13

        1.3.1.1   Distribution Functions   17
        1.3.1.2   Molar Mass Averages   21

1.4   Methods for the Determination of Molar Mass   23

    1.4.1   Method of End Groups   23
    1.4.2   Osmometry   24

        1.4.2.1   Vapor Pressure Osmometry   24
        1.4.2.2   Membrane Osmometry   25

    1.4.3   Dilute Solution Viscometry   26

        1.4.3.1   Properties of Mark-Houwink Exponent   30
        1.4.3.2   Molecular Size from Intrinsic Viscosity   31
        1.4.3.3   Dependence of Intrinsic Viscosity on Polymer Structure, Temperature, and Solvent   33

    1.4.4   Matrix-Assisted Laser Desorption Ionization Time-of-Flight Mass Spectrometry   34
    1.4.5   Analytical Ultracentrifugation   35

1.5   Keynotes   36

1.6   References   36

2 Light Scattering   37

2.1   Theory and Basic Principles   37

2.2   Types of Light Scattering   39

    2.2.1   Static Light Scattering   40

        2.2.1.1   Particle Scattering Functions   47
        2.2.1.2   Light Scattering Formalisms   54
        2.2.1.3   Processing the Experimental Data   54

    2.2.2   Dynamic Light Scattering   59

2.3   Light Scattering Instrumentation   63
2.4   Specific Refractive Index Increment   65
2.5   Light Scattering in Batch and Chromatography Mode   72
2.6   Parameters Affecting Accuracy of Molar Mass Determined by Light Scattering   78
2.7   Examples of Light Scattering Measurement in Batch Mode   84
2.8   Keynotes   96
2.9   References   97

3 Size Exclusion Chromatography   **99**

3.1   Introduction   99
3.2   Separation Mechanisms   102

    3.2.1   Steric Exclusion   102
    3.2.2   Restricted Diffusion   103
    3.2.3   Separation by Flow   103
    3.2.4   Peak Broadening and Separation Efficiency   105
    3.2.5   Secondary Separation Mechanisms   113

3.3   Instrumentation   114

    3.3.1   Solvents   118
    3.3.2   Columns and Column Packing   122
    3.3.3   Detectors   127

        3.3.3.1   UV Detector   130
        3.3.3.2   Refractive Index Detector   131
        3.3.3.3   Infrared Detector   133
        3.3.3.4   Evaporative Light Scattering Detector   134
        3.3.3.5   Viscosity Detector   135
        3.3.3.6   Light Scattering Detector   140
        3.3.3.7   Other Types of Detectors   142

3.4   Column Calibration   143

    3.4.1   Universal Calibration   149
    3.4.2   Flow Marker   152

3.5   SEC Measurements and Data Processing   154

    3.5.1   Sample Preparation   154

3.5.1.1   Sample Derivatization   159

3.5.2   Determination of Molar Mass and Molar Mass Distribution   159
3.5.3   Reporting Results   173
3.5.4   Characterization of Chemical Composition of Copolymers and
Polymer Blends   174
3.5.5   Characterization of Oligomers   175
3.5.6   Influence of Separation Conditions   184
3.5.7   Accuracy, Repeatability, and Reproducibility of SEC
Measurements   192

3.6   Applications of SEC   198
3.7   Keynotes   204
3.8   References   205

4   Combination of SEC and Light Scattering   207

4.1   Introduction   207
4.2   Data Collection and Processing   208

4.2.1   Processing MALS Data   219

4.2.1.1   Debye Fit Method   220
4.2.1.2   Zimm Fit Method   220
4.2.1.3   Berry Fit Method   221
4.2.1.4   Random Coil Fit Method   221
4.2.1.5   Influence of Light Scattering Formalism on Molar Mass and
RMS Radius   221

4.2.2   Determination of Molar Mass and RMS Radius Averages and
Distributions   232
4.2.3   Chromatogram Processing   235
4.2.4   Influence of Concentration and Second Virial Coefficient   240
4.2.5   Repeatability and Reproducibility   240
4.2.6   Accuracy of Results   242

4.3   Applications of SEC-MALS   243

4.3.1   Determination of Molar Mass Distribution   243
4.3.2   Fast Determination of Molar Mass   247
4.3.3   Characterization of Complex Polymers   251

4.3.3.1   Branched Polymers   251
4.3.3.2   Copolymers and Polymer Blends   252

4.3.4   Conformation Plots   254
4.3.5   Mark-Houwink Plots   255

4.4   Keynotes   257
4.5   References   257

5   Asymmetric Flow Field Flow Fractionation   **259**

5.1   Introduction   259
5.2   Theory and Basic Principles   261

   5.2.1   Separation Mechanisms   271
   5.2.2   Resolution and Band Broadening   273

5.3   Instrumentation   277
5.4   Measurements and Data Processing   281

   5.4.1   Influence of Separation Conditions   285

      5.4.1.1   Isocratic and Gradient Experiments   287
      5.4.1.2   Overloading   288

   5.4.2   Practical Measurements   289

5.5   A4F Applications   291
5.6   Keynotes   301
5.7   References   303

6   Characterization of Branched Polymers   **307**

6.1   Introduction   307
6.2   Detection and Characterization of Branching   311

   6.2.1   SEC Elution Behavior of Branched Polymers   318
   6.2.2   Distribution of Branching   321
   6.2.3   Average Branching Ratios   330
   6.2.4   Other Methods for the Identification and Characterization of
            Branching   333

6.3   Examples of Characterization of Branching   337
6.4   Keynotes   344
6.5   References   345

Symbols   347

Abbreviations   353

Index   355

Preface

This book brings together three powerful methods of polymer analysis and characterization, namely light scattering and two analytical separation techniques, size exclusion chromatography and asymmetric flow field flow fractionation. Each of these methods has been known and used in polymer research for several decades, and each of them has its specific advantages and limitations. Many of the limitations can be overcome by combination of light scattering with one of the separation methods. Bringing together three different techniques into a single book, showing their advantages and limitations and explaining how they complement each other and how their combinations overcome the limitations, should be the main benefit for readers, who might include university students, analysts in manufacturing quality control, and scientists in academic and industrial research laboratories. The application area of the methods that are presented includes various synthetic and natural polymers, proteins, and nanoparticles. The ability of these methods to characterize and study biomacromolecules makes them particularly attractive, because detailed knowledge of structure and structure–properties relationships is a pathway to new materials capable of replacing traditional crude-oil-based raw materials. The importance of these methods is evident in their numerous applications in medical and pharmaceutical research, including drugs, drug delivery systems, and materials for medical devices.

Molar mass is a characteristic that distinguishes polymers from low-molar-mass organic compounds. Unlike organic compounds, which have a single molar mass corresponding to their chemical formula, polymers typically consist of molecules covering a specific molar mass range. The molar mass distribution of a given polymer sample is related to many important properties and also yields information about the production process or the changes brought about during polymer application or degradation. In protein chemistry, the ability of proteins to form various oligomers affects their capability to crystallize and their possible therapeutic applications, and such demonstration of the absence of oligomers is of vital importance. The size distribution of nanoparticles, which have a wide variety of potential applications in material and biomedical fields, is crucial for their applicability and properties of final products.

Light scattering is one of the few physical techniques that provide absolute molar mass. The term *absolute* means that the molar mass is determined on the basis of fundamental physical principles using an exactly derived relationship between the intensity of light scattered by a dilute polymer solution and the molar mass of scattering molecules. In addition to molar mass, the light scattering measurements yield valuable information about the molecular size and intensity of interactions of polymer molecules with solvent. Light scattering technique is also able to provide information about branching of polymer molecules, which is another important type of nonuniformity of synthetic and natural polymers having significant impact on their various properties. The most serious limitation of the classical light scattering of nonfractionated polydisperse polymers is that it yields average quantities that are often unable to distinguish among different polymer samples or describe structure–properties relationships.

Size exclusion chromatography (SEC) has been used in polymer research since the mid-sixties and since that time has found great popularity among polymer chemists. The instrumental development of the method has been driven by the development of other types of liquid chromatography. Today's SEC instruments are highly reliable and relatively easy to use. However, it may be just the relative simplicity that often results in poor reproducibility of the method, especially in the sense of long-term reproducibility within a laboratory or reproducibility among various laboratories. The poor reproducibility is a consequence of high sensitivity of the results to numerous operational parameters, which is often overlooked by inexperienced users. In addition, the SEC results are often misinterpreted in the sense of the absolute correctness of the obtained molar masses. The most serious limitation of SEC is that the method does not measure any physical quantity directly related to molar mass. The method solely separates polymer molecules according to their hydrodynamic volume; to transfer the obtained chromatograms to molar mass information the SEC columns must be calibrated, that is, one has to establish the relation between the elution volume and molar mass. This procedure, called *calibration*, has several pitfalls, and finding true calibration for many synthetic and natural polymers is uncertain or even impossible. It has become common practice that a calibration curve established with polymer standards of a given chemical composition is used for processing the data of other polymers of significantly different chemical composition or molecular architecture. As a consequence of that, the resulting molar masses may differ from the true ones quite significantly. This fact is often not understood by SEC users. Although the apparent values obtained by incorrect calibration may be useful for a simple comparison of polymer samples and finding the effect of polymerization conditions on molar mass distribution, the molar masses obtained by this approach cannot be used for detailed polymer characterization. Thus the calibration of SEC columns remains the most serious limitation of the SEC method.

The most effective solution of the calibration issue is the combination of SEC with a method capable of direct measurement of molar mass. Light scattering, especially in the form of multi-angle light scattering (MALS), has been proved to be the most suitable method for this purpose. The MALS detector not only effectively solves the calibration problem, but also significantly improves the reproducibility and repeatability of the measurements. In addition, the combination of SEC and MALS allows very detailed characterization of branching and detection of even minute amounts of aggregates. The latter ability makes SEC-MALS highly attractive for protein characterization. However, even in the case of SEC-MALS there are still several potential limitations resulting from the nature of SEC separation that is achieved in columns packed by porous materials. The passage of polymer molecules through the porous column bed is a possible source of several problems, namely degradation of polymer molecules by shearing forces, interaction of polymer molecules with column packing, and anchoring of branched molecules in pores of column packing.

Asymmetric flow field flow fractionation (A4F) is one of the field flow fractionation techniques. The method has coexisted with SEC for several decades. However, until now it has not achieved the popularity or as wide an application range as SEC. The reason for that has been mainly more complicated instrumentation and even more uncertain determination of molar mass calibration. Recent developments in A4F instrumentation have brought a new generation of commercially available instruments that are as easy to use as SEC. The modern A4F instruments even allow easy switching from A4F to SEC mode and vice versa. The combination of A4F with a MALS detector allows efficient determination of molar mass and size distribution, identification of aggregates, and characterization of branching. The separation in A4F is achieved by a flow of polymer molecules or particles in an empty channel, which strongly reduces or even completely eliminates SEC limitations such as shearing degradation, interactions with column packing, or anchoring in pores. The A4F-MALS hyphenated method has been recently finding its way into many pharmaceutical, polymer, and nano-related research and quality control laboratories.

This book minimizes theory to the explanation of basic principles and emphasizes the practical approach of achieving reproducible and correct results. The focus is on giving guidelines for using the instruments properly, planning the experiments, acquiring reliable data, data processing, and the proper interpretation of the obtained results. The book draws from my long experience based on my own work in the laboratories of industrial research, academia, and an instrument manufacturer, as well as experience gained by my visits to many laboratories and interactions with users of light scattering, SEC, and A4F. This book presents a selection of interesting and informative examples from thousands of experimental data files collected during my experimental work. The book targets novices who are about to perform their first experiments and need to learn basic principles and methodology, as well as experienced users who may need to confirm their own understanding or help in interpreting their results.

This book would have been impossible without my 20-months' stay with Wyatt Technology Corporation in Santa Barbara and without long cooperation and support from this company. All MALS and A4F results presented in the book were acquired using instruments from Wyatt Technology Corporation. My special thanks go to Dr. Philip Wyatt, CEO and founder of Wyatt Technology Corporation, his sons, Geofrey and Clifford, president and vice president of the company, and Dr. Christoph Johann, director of Wyatt Technology Europe.

STEPAN PODZIMEK

November 2010, Pardubice, Czech Republic

Chapter 1

Polymers

1.1 INTRODUCTION

Polymers can be characterized by many methods that find applications in organic chemistry, such as, for example, nuclear magnetic resonance, infrared spectroscopy, or liquid chromatography. On the other hand, there are several methods that find utilization almost exclusively in the field of polymer chemistry. Examples include light scattering, dilute solution viscometry, size exclusion chromatography, and flow field flow fractionation.

Polymer is a substance composed of macromolecules, that is, molecules built of a big number of small molecules linked together by covalent bonds. The entirely manmade polymers (synthetic polymers) are relatively new materials that did not exist a hundred years ago. The first synthetic polymer, phenol-formaldehyde resin, Bakelite, appeared shortly before World War I. Further synthetic polymers, developed before World War II, were neoprene, nylon, poly(vinyl chloride), polystyrene, polyacrylonitrile, and poly(vinyl butyral); poly(vinyl butyral) was first used in automotive safety glass to prevent flying glass during car accidents and continues to be used for this important application. World War II encouraged further development of polymers as a result of war shortages and demands for new materials with enhanced properties. Other important polymers included polytetrafluoroethylene (Teflon), polysiloxanes (silicones), polyester fibers and plastics such as poly(ethylene terephthalate) (PET), aromatic polyamides (Kevlar), and polyetheretherketone (PEEK). Nowadays, the synthetic polymers are used in a variety of applications covering, for example, electronics, medical uses, communications, food, printing inks, aerospace, packaging, and automobiles.

Synthetic polymers can be classified as *thermoplasts*, which soften under heat and can be reversibly melted and dissolved, and *thermosets*, which, by the action of heat or chemical substances, undergo chemical reaction and form insoluble materials that cannot be melted or dissolved. Mixtures of molecules of

Light Scattering, Size Exclusion Chromatography and Asymmetric Flow Field Flow Fractionation: Powerful Tools for the Characterization of Polymers, Proteins and Nanoparticles, by Stepan Podzimek
Copyright © 2011 John Wiley & Sons, Inc.

relatively low molar mass (hundreds to thousands g/mol) that are able to react mutually or with other compounds and form cross-linked materials are often called *synthetic resins*. The term *oligomer* refers to a polymer molecule with relatively low molar mass (roughly below 10,000 g/mol) whose properties vary significantly with the removal of one or a few of the units. Besides synthetic polymers, many polymers can be found in the nature. Various polysaccharides (e.g., cellulose, starch, dextran, hyaluronic acid) represent an important group of biopolymers (natural polymers); some of them are an essential part of food or have other important applications. Proteins are other examples of biopolymers, which represent a specific and tremendously rising field of research, where the use of efficient analytical tools is necessary for the characterization and process development of protein therapeutics.

1.2   MOLECULAR STRUCTURE OF POLYMERS

The terms configuration and conformation are used to describe the geometric structure of a polymer and are often confused. *Configuration* refers to the molecular structure that is determined by chemical bonds. The configuration of a polymer cannot be altered unless chemical bonds are broken and reformed. *Conformation* refers to the order that arises from the rotation of molecules about the single bonds. If two atoms are joined by a single bond, then rotation about that bond is possible since it does not require breaking the bond. However, a rotation about a double bond is impossible. The term conformation refers to spatial structure of a macromolecule in dilute solution. Depending on the thermodynamic quality of solvent and properties of a polymer chain, the polymer may adopt a random coil, compact sphere-like shape or highly extended rod-like conformation. The terms *topology* or *architecture* often refer to the polymer chain arrangement with respect to branching.

The part of a macromolecule from which the macromolecule is built is called a *monomer unit* while the smallest part of a macromolecule that repeats periodically is called a *structural repeating unit*. Polymers can consist of one or more kinds of monomer unit. The former are called *homopolymers*, the latter *copolymers*. Synthetic polymers are usually varied mixtures of molecules of different molar mass (M) and often also of different chemical composition and/or molecular architecture. That is, they are nonuniform (polydisperse) materials. *Polydispersity* means that a given property, such as molar mass, spans a continuous range. Various possible nonuniformities are outlined in the following:

- Molar mass.
- Chemical composition: A *random copolymer* contains a random arrangement of the monomers and can be denoted schematically as -A-B-A-B-A-A-B-B-B-B-A-B-B-A-B-. The particular macromolecules can differ in their overall chemical composition as well as in the sequential arrangement of monomers in the polymer chain. A *block copolymer* contains

linear blocks of monomers of the same type -A-A-A-A-A-A-A-A-A-B-B-B-B-B-B- and the possible heterogeneity includes various block length or existence of homopolymer fractions. A *graft copolymer* contains a linear main chain consisting of one type of monomer with branches made up of other monomers, when the molecules may differ in the number, position, and length of the branches. An *alternating copolymer* consists of regularly alternating units -A-B-A-B-A-B-A-B-A-B- such as, for example, in the well-known Nylon 66 $(-CO-(CH_2)_4-CO-NH-(CH_2)_6-NH-)_n$ and the heterogeneity is limited to the molar mass and end groups. The characterization of a copolymer is always much more complex than that of a homopolymer.

- End groups: X-A-A-A-A-A-A-A-X, Y-A-A-A-A-A-A-A-Y, X-A-A-A-A-A-A-A-Y.

- *Cis* and *trans* isomerization: The *cis* configuration arises when substituent groups are on the same side of a carbon–carbon double bond. *Trans* refers to the substituents on opposite sides of the double bond. These structures cannot be changed by rotation. Technically important examples include polybutadiene or unsaturated polyesters based on maleic acid.

- Branching: A branched polymer is formed when there are side chains attached to a main polymer chain. There are many ways in which a branched polymer can be arranged. Possible branching topology includes randomly branched polymers, stars, combs, hyperbranched polymers, and dendrimers.

- Tacticity: spatial arrangement on chiral centers within a macromolecule (atactic, isotactic, syndiotactic polymers, such as polypropylene).

- Head-to-tail or head-to-head (tail-to-tail) configuration of vinyl polymers: $-CH_2-CHR-CH_2-CHR-$, $-CH_2-CHR-CHR-CH_2-$.

Polymers can be nonuniform in one or more properties. It is worth mentioning that monodisperse polymers (i.e., uniform with respect to all properties) are exceptional for synthetic polymers and most of the natural polymers. A polystyrene sample prepared by anionic polymerization that has a very narrow molar mass distribution is the most common example of an almost monodisperse polymer in the field of synthetic polymers. Examples of polymers that are heterogeneous in more than one distributed property are copolymers and branched polymers. Although the term *polydisperse* can apply to various heterogeneities, it is often understood only with respect to polydispersity of molar mass. The importance of a given heterogeneity may depend on molar mass and application. For example, the end groups are of primary importance for synthetic resins, like, for example, epoxies, where the end epoxy groups are essential for curing process. However, the influence of end groups diminishes with increasing molar mass and for most of the polymers the effect of end groups on their properties is negligible. Besides molar mass, chemical composition is another important characteristic governing polymer properties and applications.

The two most important sources of chemical heterogeneity are: (1) *statistical heterogeneity*, when compositional variation arises from random combinations of comonomers in polymer chains, and (2) *conversion heterogeneity*, when differences in the reactivity of the comonomers cause the change of monomer mixture composition with conversion and such molecules with different composition are formed at different conversion. While the former type of heterogeneity is almost negligible, the latter is usually the main source of the compositional heterogeneity in polymers.

It is of utmost importance for polymer chemists and analysts to be aware of all possible nonuniformities of polymers in order to choose a suitable experimental method for the characterization, interpret the experimental data, and understand the polymer properties and behavior. Two polymer samples may be identical in one or more properties but differ in others. Although the polymer properties are generally distributed, solely average values can be often obtained by the analysis. Two polymer samples can be identical in an average property but the property distributions can be different. However, average properties are often used instead of distributions in order to simplify the description of a polymer sample or because the distribution cannot be determined due to time or instrumental limitations. In addition to nonuniformity resulting from the randomness of the polymerization process, many commercially important polymer-based materials are polymer blends, that is, mixtures of two or more polymeric components; also various low-molar-mass compounds are added to polymers to modify their properties and protect them against degradation.

1.2.1   Macromolecules in Dilute Solution

Understanding the shape, size, and hydrodynamic behavior of polymer molecules in dilute solutions is essential not only for understanding the property–structure relationships, but also for understanding the principles of polymer characterization, such as column calibration in size exclusion chromatography or the characterization of branching. In a dilute solution the polymer molecules are isolated from each other so that the interactions of polymer–solvent prevail over the intermolecular interactions of polymer–polymer. The macromolecules take the most statistically probable conformations and usually form so-called *random coils* (coiled polymeric domains swollen with the solvent). The polymer coil must not be assumed to be a rigid, motionless object, but due to rotation about single bonds the coil can create a large number of various conformations. That means a polymer chain shows a dynamic behavior with fast and randomly changing conformations. It is impossible to study the number of various conformers and their corresponding conformations, but the experimental measurements always provide statistical averages of macromolecular dimensions. The polymers that can easily transform from one conformation to another and that can form a large number of various conformations are flexible, while those polymers for which the transition from one conformation to another is restricted by high potential barrier and the number of possible conformations is limited are rigid. The flexible polymers typically consist of only single C-C bonds in the main chain and no or small chain substituents, while double bonds or cyclic structures in the main chain as well as large chain substituents increase chain rigidity.

The conformation of a real chain is defined by valence angles and restricted torsion due to different potential energy associated with different torsion angles (*trans* position being at minimum potential energy). In addition, two segments cannot occupy the same space element at the same time, and the chain expands due to the *excluded volume* effect. The excluded volume is a result of materiality of the polymer chain. It refers to the fact that one part of a long-chain molecule cannot occupy space that is already occupied by another part of the same molecule. Excluded volume causes the ends of a polymer chain in a solution to be further apart than they would be were there no excluded volume. The effect of excluded volume decreases with increasing chain rigidity and decreasing chain length, because the bonds in a polymer chain are to a certain extent stiff and such a collision of two segments of the same chain can only occur when the chain between the two segments can create a sufficiently large loop. In thermodynamically good solvents, the interactions between polymer segments and solvent molecules are energetically favorable and the solvent creates a solvating envelope around the polymer chain, which results in further expansion of the polymer coil. In a thermodynamically poor solvent, the intramolecular interactions between polymer segments are intensive and under specific conditions can precisely compensate the effect of excluded volume. Such conditions (solvent and temperature) are called *theta conditions* and polymer coil dimensions under these conditions *unperturbed dimensions* (zero subscript is used to indicate unperturbed dimensions). In theta conditions, the long-range interactions arising from excluded volume are eliminated and the chain conformation is defined solely by bond angles and short-range interactions given by the hindrances to rotation about bonds (i.e., steric or other interactions involving neighboring groups). A characteristic feature of theta conditions is that the second virial coefficient is zero. Commonly all theoretical calculations are done under the assumption of unperturbed chain dimensions, while the real experiments are mostly carried out far from theta point. This fact must be considered when the experimental results are being compared with the theoretical predictions.

The dimensions of a linear chain can be described by the *mean square end-to-end distance* $\langle r^2 \rangle$ or the square root of this quantity $\langle r^2 \rangle^{\frac{1}{2}}$. The angle brackets denote the average over all conformations. In a three-dimensional space, the distance between the two ends is a vector, which fluctuates with regard to the dimension and direction. The scalar product of the vector with itself is a quantity fluctuating only with respect to the dimension. Note that squares of vector quantities are usually used in theoretical calculations to eliminate the directional part of the vectors. However, the end-to-end distance becomes completely meaningless in the case of branched polymers that have more than two ends. Another parameter describing the size of the polymer chain, which can be effectively used for the characterization of branched molecules, is the *mean square* (MS) *radius* $\langle R^2 \rangle$ and the *root mean square* (RMS) *radius* $\langle R^2 \rangle^{\frac{1}{2}}$. The RMS radius can be used generally for the size description of a particle of any shape. The RMS radius is frequently called *radius of gyration* and symbols $\langle r_g^2 \rangle^{\frac{1}{2}}$ or $\langle s^2 \rangle^{\frac{1}{2}}$ are also used in scientific literature. For the sake of simplicity, the symbol R is mostly used for the RMS radius in this book. The RMS radius is often mistakenly associated with the term *gyration*, although there is no gyration involved in the RMS radius definition. The integration is over the mass elements of the molecule with respect to the center of gravity of the molecule (i.e., the subscript g refers to the center of gravity and not to gyration). For the RMS radius definition and the determination, see Chapter 2.

The mean square end-to-end distance of a real chain in solution is expressed as:

$$\langle r^2 \rangle = \alpha^2 \langle r^2 \rangle_0 \tag{1.1}$$

where α is the *expansion factor*, which represents the effect of long-range interactions, that is, the effect of excluded volume, and swelling of the chain by the polymer–solvent interactions. The effect of bond angle restriction and steric hindrances to rotation about single bonds is represented by unperturbed dimension $\langle r^2 \rangle_0$. The expansion factor expresses the deviation of a polymer chain from theta state. Besides the expansion factor based on the end-to-end distance there are other expansion factors defined by other dimensional characteristics, namely by the RMS radius and intrinsic viscosity:

$$\alpha_R = \sqrt{\frac{\langle R^2 \rangle}{\langle R^2 \rangle_0}} \tag{1.2}$$

$$\alpha_\eta = \left(\frac{[\eta]}{[\eta]_0} \right)^{\frac{1}{3}} \tag{1.3}$$

It is worth noting that expansion factors defined by Equations 1.1–1.3 are not expected to be exactly equal.

The simplest model of a polymer in solution is a *freely jointed*, or *random flight*, chain[1] (Figure 1.1). It is a hypothetical model based on the assumptions that (1) chain consists of n immaterial segments of identical length l; and (2) $(i + 1)$th segment freely moves around its joint with the ith segment. The angles at the segment junctions are all of equal probability and the rotations about segments are free. That means a polymer molecule is formed by a random walk of fixed-length, linearly connected segments that occupy zero volume and have all bond and torsion angles equiprobable. Since the segments are assumed to be of zero volume, two or more segments can occupy the same volume element in the space.

For a sufficiently long chain, the value of $\langle r^2 \rangle_0$ for a freely jointed chain is directly proportional to the number of segments:

$$\langle r^2 \rangle_{0,j} = nl^2 \tag{1.4}$$

$$\langle r^2 \rangle_{0,j}^{1/2} = \sqrt{n} \times l \tag{1.5}$$

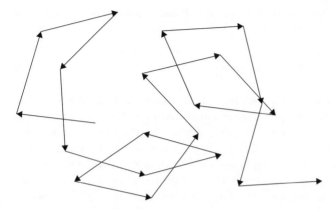

Figure 1.1   Schematic representation of freely jointed model of polymer chain formed by a random walk of 20 segments in two-dimensional space.

Here n is the number of segments (rigid sections) of the length l, and the subscripts zero and j are used to indicate unperturbed dimensions and freely jointed model, respectively. In simple single-strand chains, bonds are taken as the rigid sections. The MS radius is in a simple relation to the mean square end-to-end distance:

$$\langle R^2 \rangle_0 = \frac{\langle r^2 \rangle_0}{6} \tag{1.6}$$

A *freely rotating* chain is a hypothetical model consisting of n segments of fixed length l jointed at fixed angles. It assumes free internal rotation under fixed bond angles (i.e., all torsion angles are equally likely). For a chain consisting of only one kind of bond of length l and for $n \rightarrow \infty$, the mean square end-to-end distance is:

$$\langle r^2 \rangle_{0,r} = nl^2 \times \frac{(1 + \cos\theta)}{(1 - \cos\theta)} \tag{1.7}$$

where θ is the supplement of the valence bond angle and subscript r indicates a freely rotating chain. For carbon polymer chains the valence angle is $109.5°$ (i.e., $\cos\theta = 1/3$) and thus the mean square end-to-end distance is a double that of the freely jointed chain. Although the freely rotating chain represents a more realistic model of polymer chains, the state of entirely free rotation is rare. The freely rotating behavior diminishes with increasing size of the main chain substituents. The ratio of the root mean square end-to-end distance of a real polymer chain with unperturbed dimensions to that of a freely rotating chain with the same structure:

$$\sigma = \left(\frac{\langle r^2 \rangle_0}{\langle r^2 \rangle_{0,r}} \right)^{\frac{1}{2}} \tag{1.8}$$

is called the *steric factor*, which reflects the effect of hindrance to free rotation.

The unperturbed dimensions of a flexible polymer chain can be characterized by the so-called *characteristic ratio*:

$$C_n = \frac{\langle r^2 \rangle_0}{nl^2} \tag{1.9}$$

where n is the number of rigid sections in the chain, each of length l. The characteristic ratio is the ratio of the mean square end-to-end distance in the theta state divided by the value expected from the freely jointed chain. C_n approaches an asymptotic value as n increases (i.e., $C_n = C_\infty$ for $n \to \infty$). In simple chains, the bonds can be taken as the segments and the number of segments can be calculated from the degree of polymerization P or the molar mass M and the molar mass of the monomer unit M_0. For vinyl polymers, $n = 2P$ or $2M/M_0$ and $l = 0.154$ nm. If all of the segments are not of equal length, the mean square value of l is used:

$$l^2 = \frac{1}{n} \sum_i l_i^2 \tag{1.10}$$

For the freely jointed chain, $C_\infty = 1$, and for real polymer chains, $C_\infty > 1$. The increasing value of C_∞ indicates greater deviation from freely jointed behavior. For a polymer with N' chain bonds per monomer unit, Equation 1.9 can be rearranged as:

$$C_\infty = \frac{\langle r^2 \rangle_0}{M} \frac{M_0}{N'l^2} \tag{1.11}$$

Unperturbed chain dimensions of polystyrene can be used as concrete examples of the previous characteristics: $K_0 = (82 \pm 5) \times 10^{-3}$ mL/g, $\langle r^2 \rangle_0^{\frac{1}{2}} / M^{\frac{1}{2}} = (670 \pm 15) \times 10^{-4}$ nm, $\sigma = 2.22 \pm 0.05$, $C_\infty = 9.85$. The data were determined in various solvents at a temperature around $30°C$.

The determination of unperturbed dimensions of polymer chains can be achieved by means of the Flory-Fox equation:[3]

$$[\eta] = K_0 M^{0.5} \alpha_\eta^3 \tag{1.12}$$

where

$$K_0 = \Phi_0 \left(\frac{\langle r^2 \rangle_0}{M} \right)^{1.5} \tag{1.13}$$

and $[\eta]$ is the intrinsic viscosity.

Under theta conditions, there is no excluded volume effect, $\alpha_\eta = 1$, and Equation 1.12 can be written as:

$$[\eta]_0 = K_0 M^{0.5} \tag{1.14}$$

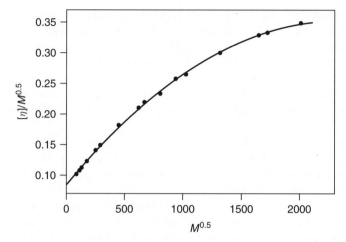

Figure 1.2   Plot of $[\eta]/M^{0.5}$ versus $M^{0.5}$ for polystyrene in THF (intrinsic viscosity expressed in mL/g). The intercept $(K_0) = 84.3 \times 10^{-3}$ mL/g.

Equations 1.12 and 1.14 correspond to the Mark-Houwink equation, but the exponent a has a constant value of 0.5. Flory constant Φ_0 is a universal constant for linear flexible chain molecules under theta conditions. In fact, Φ_0 is not a real constant, because different values were reported in the literature, the range being somewhere in the limits of 2.1×10^{21} to 2.87×10^{21} (for end-to-end distance expressed in centimeters and intrinsic viscosity in deciliters per gram). The Flory-Fox equation can be used outside θ conditions (see Section 1.4.3.2). The measurement of intrinsic viscosity under θ conditions yields K_0. The ratio $\langle r^2 \rangle_0 / M$ is obtained from Equation 1.13, which then yields C_∞ from Equation 1.11. Measurements of intrinsic viscosity at theta conditions can encounter experimental difficulties and various procedures that allow determination of K_0 from the intrinsic viscosities determined in thermodynamically good solvents (i.e., $T \neq \theta$) were proposed. The estimation of K_0 can be obtained by measurements in thermodynamically good solvents using, for example, the Burchard-Stockmayer-Fixman method:[4,5]

$$[\eta] = K_0 M^{0.5} + 0.51\Phi_0 BM \qquad (1.15)$$

where B is a constant. Linear extrapolation of the relation $[\eta]/M^{0.5}$ versus $M^{0.5}$ yields K_0 as the intercept. However, in very good solvents, especially if the molar mass range is broad, the $[\eta]/M^{0.5}$ versus $M^{0.5}$ plot is markedly curved, which makes extrapolation rather uncertain. An example of a Burchard-Stockmayer-Fixman plot for a polymer in thermodynamically good solvent is shown in Figure 1.2. The obtained constant K_0 of 84.3×10^{-3} mL/g yields ratios $\langle r^2 \rangle_0^{1/2} / M^{\frac{1}{2}} = 696 \times 10^{-4}$ nm and $C_\infty = 10.6$, which are in good agreement with literature values.[2]

It must be emphasized that the extrapolation procedures for the estimation of K_0 are valid for flexible chains and should not be applied to polymers with semi-flexible chains (Mark-Houwink exponent $a > 0.85$). A wormlike chain model[6] is used to describe the behavior of semiflexible polymers such as some types of polysaccharides, aromatic polyesters, aromatic polyamides, and polypeptides in helical conformation. The value of $\langle r^2 \rangle_0 / M$ and other characteristics of the semiflexible polymers can be obtained, for example, by a procedure developed by Bohdanecky.[7]

Contour length is another term that can be used to describe chain molecules. It is the maximum end-to-end distance of a linear polymer chain, which for a single-strand polymer molecule usually means the end-to-end distance of the chain extended to the all-*trans* conformation.

A real polymer chain consisting of n segments of the length l can be approximated by a freely jointed chain consisting of n' segments of the length l' under the condition that the values of $\langle r^2 \rangle_0$ and the totally expanded chain lengths for the freely jointed chain and the real chain are identical:

$$\langle r^2 \rangle_0 = n' l'^2 \tag{1.16}$$

$$nl = n' l' \tag{1.17}$$

Such a model chain is called an *equivalent chain* and its segment a *statistical segment* (Kuhn segment). The number of segments (bonds) in a statistical segment is proportional to chain rigidity.

1.3   MOLAR MASS DISTRIBUTION

It is the high molar mass that distinguishes polymers from organic low-molar-mass compounds. The molar mass and molar mass distribution of synthetic and natural polymers are their most important characteristics with a strong relation to various properties and industrial applications. The polymer properties influenced by molar mass include melt and solution viscosity, tensile strength, toughness, impact strength, adhesive strength, elasticity, brittleness, abrasion resistance, flex life, softening temperature, solubility, chemical resistance, cure time, diffusion coefficient, film and fiber forming ability, ability to be fabricated, and processing temperature. The ability of a polymer to form fibers and films is possible from a certain molar mass and the film and fiber properties are related to molar mass.

A polymer containing high-molar-mass fractions shows greater elastic effect. However, the relation of some properties to molar mass may not be straightforward. A certain property may be related more to a certain molar mass average, and polydispersity usually plays an important role. Different molar mass averages can be related to different polymer properties since either high-molar-mass or low-molar-mass fractions can primarily influence specific properties. For example, the tensile strength is particularly related to the weight-average molar mass (M_w) since it is most influenced by the large molecules in the material. The flex life (ability of a polymer material to bend many times before breaking) is more related to z-average molar mass (M_z), because extremely large molecules are most important for this property. The number-average molar mass (M_n) is needed for kinetics studies and stoichiometric calculations. Relatively narrow molar mass distribution and high molar mass are beneficial for fiber-forming polymers, where molecules with high molar mass increase the tensile strength, while polymers for pressure-sensitive adhesives benefit from broad polydispersity since the high-molar-mass fractions enhance the material strength and the lower-molar-mass fractions have a desirable plasticizing effect. Resistance of plastics to the surface-initiated failure of stressed polymers in the presence of surface active substances such as alcohols or soaps (environmental stress cracking) increases with increasing molar mass, and is considerably decreased by the presence of low-molar-mass chains. The molar mass distribution is also important for polymers used as plasma expanders (e.g., hydroxyethyl starch, dextran), because the circulation time in blood depends on it, and the adverse effects are caused by too high levels of the low-molar-mass fractions. Many times the positive influence of increasing molar mass must be balanced with the ability of a polymer to be processed (e.g., tensile strength versus melt or solution viscosity). Solubility of polymers decreases with increasing molar mass because of the decrease of the second virial coefficient (see Equation 2.4). It is important to note that there are no commonly good molar mass averages or molar mass distributions for a polymer sample. The optimum values depend on the nature of the polymer, the way of processing, and especially on the required end-use properties. A molar mass distribution of a polymer sample that is known as a good one for a given application can serve as a reference to which other samples are compared.

The viscosity of polymer melts is proportional to the 3.4-power of M_w:

$$\eta = k \times M_w^{3.4} \tag{1.18}$$

where k is a proportionality constant. For some polymers, the melt viscosity may become related to an average somewhere between the M_w and M_z. Polymer melts typically show non-Newtonian behavior (i.e., their viscosity decreases with increasing shear stress). The rate of viscosity reduction with shear is related to molar mass and polydispersity; generally it is enhanced by the presence of high-molar-mass components. The glass transition temperature (T_g) is related to the M_n according to the relation:

$$T_g = T_g(\infty) - \frac{K}{M_n} \tag{1.19}$$

where $T_g(\infty)$ is a glass transition temperature of a polymer with indefinite molar mass and K is a constant. In a solution of macromolecules, the diffusion rate decreases with increasing molar mass according to relation:

$$D = K_D M^{-\beta} \tag{1.20}$$

where D is the translational diffusion coefficient characterizing the ability of molecules to move in solution and K_D and β are constants for a given polymer, solvent, and temperature.

The exponent β generally lies in the range of $0.33 < \beta < 1.0$, the value depending on the molecular conformation and thermodynamic quality of solvent. Example of molar mass dependence of diffusion coefficient is shown in Figure 1.3.

Although the molar mass and molar mass distribution affect many polymer properties in bulk, the determination of molar mass is possible only in the form of dilute polymer solution. That means the solubility of a polymer is a necessary condition for the determination of its molar mass. The requirement of solubility may not be fulfilled for some important technical polymers such as Teflon or PEEK. In addition, a polymer in a certain solvent may not form a true solution where all molecules are dispersed in the form of individual molecules, but the solution may also contain supermolecular structures that have a significant effect on the average value of molar mass. Obviously, in the case of thermosets, the characterization of molar mass is possible only before their crosslinking.

Several notations are used in polymer science to express molar mass, namely *molar mass* or *molecular mass* (mass of polymer divided by the amount of polymer expressed in moles, dimension g/mol), and *relative molecular mass* or *molecular weight* (mass of the polymer related to 1/12 of the mass of the ^{12}C atom, dimensionless). A relative unit, *Dalton* (Da), is also sometimes used. The molar mass expressed in different units is numerically identical, which means a polymer having molar mass 10^5 g/mol has relative molecular mass 10^5 or 10^5 Da. The number of monomeric units in a macromolecule or oligomer molecule is called *polymerization degree* (P). The relation between the molar mass and polymerization degree is:

$$M = P \times M_0 \tag{1.21}$$

where M_0 is the molar mass of a monomeric unit.

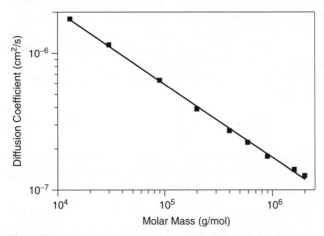

Figure 1.3 Molar mass dependence of diffusion coefficient of polystyrene in THF at room temperature. Data obtained by DLS of narrow polystyrene standards.
$D = 0.00027 \times M^{-0.533}$ (cm^2/s).

1.3.1 Description of Molar Mass Distribution

In contrast to low-molar-mass compounds a polymer does not have a single molar mass and almost all synthetic and natural polymers are *polydisperse* in molar mass. Frequencies of particular molar masses in a given polymer sample are described by the *molar mass distribution*. The *differential*, $f(M)$, and *cumulative* (*integral*), $I(M)$, molar mass distribution functions are used for the description of molar mass distribution. The differential distribution is often preferred in the polymer community over the cumulative molar mass distribution. However, the cumulative molar mass distribution allows easy determination of the weight fraction of polymer below or above a certain molar mass limit or the weight fraction of polymer in a specific molar mass range. Such information cannot be readily deduced from the differential distribution, which is more convenient for the assessment of the molar mass distribution symmetry, revealing multimodal distribution or the determination of the molar mass minimum and maximum and the molar mass of the most abundant fractions.

The molar mass distribution can be described graphically by a distribution curve or mathematically by a distribution function. Although the molar mass is a discrete quantity, the discreteness of molar mass is usually ignored[i] and the distribution is treated as a continuous one. The distribution function expressing the amount of material as weight fraction is called the *weight molar mass distribution*. It is also possible to use mole fraction (number fraction) and the distribution is called the *number molar mass distribution*. The subscripts w and n are used to differentiate between the weight and number distributions, respectively. The weight distributions are mostly used in routine practice. It is usual practice to express the molar mass axis in logarithmic scale or as a logarithm of M. The notations $f_w(M)$ and $F_w(\log M)$ can be used in order to distinguish the distributions based on normal and logarithmic scale, respectively. The function $F_w(\log M)$ is typically more symmetrical and better descriptive than $f_w(M)$. *Note:* Expression of the axis in $\log M$ is significantly less convenient for reading than using the molar mass axis in logarithmic scale (compare, for example, Figures 1.4 and 3.20).

The following equations describe various distribution functions and relations between them:

$$f_w(M)dM = F_w(\log M)d \log M \tag{1.22}$$

$$f_w(M) = \frac{F_w(\log M)}{2.303 \times M} \tag{1.23}$$

$$\int_0^\infty f_w(M)dM = 1 \tag{1.24}$$

---

[i]Oligomers that can be at least partly separated by a chromatographic technique or mass spectroscopy may be exceptions from the continuous expression of molar mass distribution.

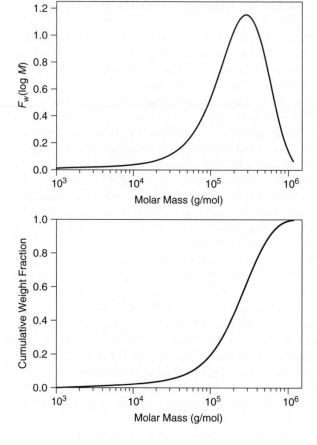

Figure 1.4   Differential weight distribution (top) and cumulative weight distribution (bottom) of NIST SRM 706 polystyrene. $M_n = 66{,}000$ g/mol, $M_w = 279{,}000$ g/mol, $M_z = 442{,}000$ g/mol, $M_{z+1} = 588{,}000$ g/mol.

$$\int_0^\infty f_n(M)\,dM = 1 \tag{1.25}$$

$$f(M) = \frac{dI(M)}{dM} \tag{1.26}$$

$$I_w(M) = \int_0^M f_w(M)\,dM \tag{1.27}$$

$$I_n(M) = \int_0^M f_n(M)\,dM \tag{1.28}$$

The weight fraction w of polymer material having molar mass between M and $M + dM$ is $f_w(M)\,dM$, and alternatively for the distribution expressed in the logarithmic scale $F_w(\log M)\,d(\log M)$. The weight fraction of molecules with molar mass smaller than or equal to M_1 is given by $\int_0^{M_1} f_w(M)\,dM$ or $I_w(M_1)$. The weight fraction of molecules with molar mass between M_1 and M_2 is given by: $\int_{M_1}^{M_2} f_w(M)\,dM$ or $I_w(M_2) - I_w(M_1)$. In the graphical representation by differential distribution curve, this fraction equals the area bounded by the curve, abscissa, and two vertical lines at the position of M_1 and M_2. Similarly, using the number distributions yields the same information expressed in the number fractions: for example, the number (mole) fraction of molecules with molar mass $\leq M_1$ is $\int_0^{M_1} f_n(M)\,dM$ or $I_n(M_1)$.

The relation between the weight and number distribution can be derived using the definition of the weight fraction:

$$w_i = \frac{m_i}{\sum\limits_i m_i} = \frac{n_i M_i}{\sum\limits_i n_i M_i} \tag{1.29}$$

After extension of the above expression by $1/\sum\limits_i n_i$ we get:

$$w_i = \frac{M_i}{M_n} x_i \tag{1.30}$$

which allows transformation of the weight fraction w to the number fraction x and vice versa. In the above equations, m_i, n_i, w_i, and x_i are the weight (mass), number of moles, weight fraction, and number (mole) fraction of molecules having molar mass M_i, respectively. The number fraction is defined as:

$$x_i = \frac{n_i}{\sum n_i} \tag{1.31}$$

Alternatively to Equation 1.30, the relation between the weight and number distribution can be expressed as:

$$f_w(M) = \frac{M}{M_n} f_n(M) \tag{1.32}$$

An example of the cumulative and differential distribution curves is shown in Figure 1.4 and Table 1.1. Using the data in Figure 1.4 and Table 1.1, one can easily read the following information: (1) The sample contains about 1.6% wt molecules with molar mass below 10,000 g/mol; (2) the most abundant fractions around the apex of the distribution have molar mass around 287,000 g/mol; for example, the weight fraction of the molecules having the molar mass in the range of about $285,200 - 287,500$ g/mol is $1.1609 \times (\log 287454 - \log 285154) = 0.0041$, that is, about 0.41% wt.; (3) the minimum molar mass is about 2,000, the maximum is about 1.3×10^6 g/mol; and (4) the weight fraction of molecules with molar mass in the range of 10,000 g/mol to 100,000 g/mol is about $0.1888 - 0.0156 = 0.173$, i.e., 17.3% wt.

Graphical overlay of distribution curves can serve for comparison of a series of samples and finding even subtle differences among them. The distribution

Table 1.1   Tabular Data for Differential and Cumulative Distribution Curves Shown in Figure 1.4

Molar Mass (g/mol)	Differential Weight Distribution	Cumulative Weight Distribution
9688	0.0382	0.0152
9766	0.0384	0.0153
9845	0.0387	0.0150
9924	0.0389	0.0155
10005	0.0391	0.0156
10085	0.0393	0.0158
10167	0.0396	0.0159
~	~	~
97961	0.5231	0.1830
98751	0.5283	0.1849
99547	0.5335	0.1868
100350	0.5387	0.1888
101160	0.5440	0.1907
101976	0.5494	0.1927
102798	0.5547	0.1947
~	~	~
278364	1.1598	0.5943
280609	1.1603	0.5983
282872	1.1606	0.6022
285154	1.1608	0.6062
287454	1.1609	0.6102
289773	1.1608	0.6142
292110	1.1606	0.6181
294466	1.1603	0.6221
296841	1.1598	0.6261
299235	1.1592	0.6301
301649	1.1585	0.6340
304082	1.1576	0.6380
306535	1.1566	0.6419
309007	1.1554	0.6459
~	~	~

functions can also provide information about the polymer composition or polymerization process. Two or more peaks on the differential distribution indicate a blend of two or more polymers with different molar mass distributions. Multimodal distribution may also indicate the presence of a single polymer prepared under polymerization conditions (temperature, concentration of monomers, or initiator) that were changed in the course of polymerization. Tails or shoulders at the high-molar-mass regions of molar mass distribution are signs of appreciable amounts of species with very high molar masses. It is important to note that these species can significantly affect the polymer properties. A high-molar-mass tail on the distribution curve often indicates the presence of branched molecules. A substantial level of oligomers in a polymer sample is indicated by a tail running to low molar masses (see, for example, Figure 1.4).

1.3.1.1 *Distribution Functions*

The molar mass distribution of polymers can be described by an analytical function of two or more parameters. Several molar mass distribution functions were derived from the kinetics of polymerization or estimated empirically to fit the experimentally determined distribution curves. At the beginning of polymer science, the empirical distribution functions were often used to estimate the entire distribution from the knowledge of two molar mass averages. With the development of size exclusion chromatography, the importance of the distribution functions as the means of the estimation of molar mass distribution dropped significantly, but they still can find utilization in theoretical modeling and calculations. In addition, the discrepancy between the experimental and theoretical distribution functions can indicate side reactions. Several examples of various distribution functions are given here:

Schulz-Zimm distribution:

$$f_w(M) = \frac{a^{b+1}}{\Gamma(b+1)} M^b e^{-aM} \tag{1.33}$$

where a and b are two positive parameters, and $\Gamma(b+1)$ is the gamma function of $(b+1)$. The parameters a and b can be adjusted to fit the real distribution of a polymer sample. The average molar masses of the Schulz-Zimm distribution can be expressed in terms of the two parameters:

$$M_n = \frac{b}{a} \tag{1.34}$$

$$M_w = \frac{b+1}{a} \tag{1.35}$$

The distribution function proposed by *Tung* is expressed as:

$$f_w(M) = npM^{n-1} e^{-pM^n} \tag{1.36}$$

The main advantage of the distribution function expressed by Equation 1.36 is that it can be integrated analytically to the cumulative distribution form:

$$I_w(M) = 1 - e^{-pM^n} \tag{1.37}$$

The average molar masses for those polymers that follow Equation 1.36 are:

$$M_n = \frac{p^{-1/n}}{\Gamma(1 - 1/n)} \tag{1.38}$$

$$M_w = \Gamma(1 + 1/n)p^{-1/n} \tag{1.39}$$

Another distribution function of interest is the *log-normal distribution* with two parameters M_0 and β:

$$f_w(M) = \frac{1}{\sqrt[\beta]{\pi}} \frac{1}{M} \exp\left(-\frac{1}{\beta^2} \ln^2 \frac{M}{M_0}\right) \tag{1.40}$$

The average molar masses expressed in terms of M_0 and β are:

$$M_n = M_0 e^{\frac{-\beta^2}{4}} \tag{1.41}$$

$$M_w = M_0 e^{\frac{\beta^2}{4}} \tag{1.42}$$

It is obvious that in the above equations the molar mass can be replaced by the polymerization degree.

Molar mass distribution can be calculated theoretically for various polymerization reactions. The distribution of molar mass in polymers prepared by polycondensation has been derived by Flory, assuming that the functional group reactivities are independent of the chain length. For the polycondensation of monomer type A-B (e.g., hydroxy acid) or the equimolar mixture of monomers A-A and B-B (e.g., dicarboxylic acid with glycol), the probability that a randomly selected functional group has reacted is equal to the reaction conversion q:

$$q = \frac{n_o - n}{n_o} \tag{1.43}$$

where n and n_0 are the number of moles of functional groups at the reaction time t and at the beginning of reaction, respectively. The conversion q ranges within 0 and 1 and can be equaled with the probability that a randomly selected functional group has reacted (i.e., it is incorporated in the polymer chain). Similarly, the probability that a randomly selected functional group has not reacted is $(1 - q)$. The fraction of molecules having the polymerization degree P equals the probability that a randomly selected molecule has the polymerization degree P, that is, contains $(P - 1)$ reacted functional groups and one unreacted functional group at the end of polymer chain. The total probability is given as the product of partial probabilities, and for the mole fraction of molecules having the polymerization degree P one gets:

$$x_P = q^{P-1}(1 - q) \tag{1.44}$$

Expression 1.44 fulfils the condition $\sum_{P=1}^{\infty} x_P = 1$ (sum of infinite series). The number average polymerization degree is obtained as:

$$P_n = \sum_{P=1}^{\infty} x_P P = \sum_{P=1}^{\infty} (1-q)Pq^{P-1} = \frac{1}{1-q} \tag{1.45}$$

The weight fraction of molecules having the polymerization degree P is obtained using Equation 1.30 as:

$$w_P = \frac{P}{P_n} x_P = (1-q)^2 P q^{P-1} \tag{1.46}$$

Equation 1.46 leads to the expression for the weight-average polymerization degree:

$$P_w = (1-q)^2 \sum_{P=1}^{\infty} P^2 q^{P-1} = \frac{1+q}{1-q} \tag{1.47}$$

and polydispersity:

$$\frac{P_w}{P_n} = 1+q \tag{1.48}$$

The obtained distribution is usually called the *Flory distribution* or *most probable distribution*. The following arrangement of the above relations can be beneficial for practical applications: For large P it is true that $P = (P-1)$, $(1-q)^2 = \frac{1}{P_n^2}$, $\ln q = q - 1 = -\frac{1}{P_n}$, and $q = e^{\ln q}$, which leads to:

$$f_w(P) = \frac{P}{P_n^2} e^{-\frac{P}{P_n}} \tag{1.49}$$

$$f_w(M) = \frac{M}{M_n^2} e^{-\frac{M}{M_n}} \tag{1.50}$$

Using the definition of the cumulative distribution and Equations 1.49 and 1.50, and integration by parts, one gets:

$$I_w(P) = 1 - e^{-\frac{P}{P_n}} \left(1 + \frac{P}{P_n}\right) \tag{1.51}$$

$$I_w(M) = 1 - e^{-\frac{M}{M_n}} \left(1 + \frac{M}{M_n}\right) \tag{1.52}$$

Once the value of M_n is determined experimentally, it can be used for the calculation of the theoretical distribution function of a polymer prepared by polycondensation. In the case of polycondensation of monomers with average functionality larger than 2, the average polymerization degrees are:

$$P_n = \frac{1}{1 - \frac{qf}{2}} \tag{1.53}$$

$$P_w = \frac{1+q}{1-q(f-1)} \tag{1.54}$$

where f is an average monomer functionality

$$f = \frac{\sum n_i f_i}{\sum n_i} = \sum_i x_i f_i \tag{1.55}$$

where n_i and x_i are the number of moles and number fraction of f functional monomer, respectively.

An addition of a small amount of monomer B+B[i] to monomer A-B or equimolar mixture of monomers A-A and B-B results in the decrease of the polymerization degree:

$$P_n = \frac{1+r}{r-2qr+1} \tag{1.56}$$

where $r = n_A/n_B$ is the ratio of the number of moles of functional groups A and B and q is the conversion of functional group A. For example, 1% excess of monomer B-B (i.e., $r = 0.99$) decreases the maximum polymerization degree at 100% conversion to $P = 199$.

The above equations lead to following conclusions concerning the molar mass of polymers prepared by polycondensation:[ii]

- One of the most critical factors to achieve a high degree of polymerization is a strict stoichiometric balance of the two functional groups. An imbalanced ratio of the two functional groups leads to a lower degree of polymerization.
- The maximum polydispersity M_w/M_n of the polymers prepared by polycondensation of monomer A-B or an equimolar mixture of monomers A-A and B-B is two. This polydispersity is achieved at 100% conversion.
- In the case of functionality larger than 2.0, the polycondensation can reach the gel point (i.e., the conversion when P_w achieves indefinite value).

The molar mass distribution of polymers prepared by free radical polymerization is governed by the termination process. Polymers formed by free radical polymerization with termination by disproportionation have the molar mass distribution expressed as:

$$f_w(P) = \frac{P}{P_n^2} e^{-\frac{P}{P_n}} \tag{1.57}$$

And for products with termination by combination (coupling):

$$f_w(P) = \frac{4P^2}{P_n^3} e^{-\frac{2P}{P_n}} \tag{1.58}$$

[i]It can be identical with monomer B-B, but not necessarily.
[ii]The same conclusions can be applied to the polymers prepared by polyaddition.

The polydispersities of polymers prepared by free radical polymerization are either 2.0 or 1.5 for the termination by disproportionation or combination, respectively. Termination by combination results in narrower polydispersity, because radicals of various lengths (i.e., also short and long) are combined. The values of polydispersity and distributions according to Equations 1.57 and 1.58 apply to products formed during a short time period when the concentrations of monomer and initiator are constant. Commercial polymers, which are always polymerized to high conversions, have distributions different from those described by Equations 1.57 and 1.58, because in the course of free radical polymerization the monomer and initiator concentrations usually vary with conversion and consequently the polydispersity of final product is broader. The final distribution function is superposition of partial distribution functions having arisen at various conversions. Indication of whether termination occurs by combination or disproportionation can be obtained from the measurements of polydispersity of polymers prepared during a short time period for which the monomer-to-initiator ratio remains constant (in practice, this requirement is achieved by measurement at low conversion when the polymerization goes to only 5 or 10% conversion). However, both termination reactions usually occur concurrently.

Molar mass distribution of polymers prepared by free radical polymerization can be influenced by chain transfer to polymer, which results in branching of polymer chains and markedly increased polydispersity. Also the change of reaction conditions (temperature, concentration of monomer, initiator or chain transfer agent) increases the broadness of the molar mass distribution. Let us also remember that the molar mass of polymers made by free radical polymerization decreases with increasing temperature and increasing initiator concentration.

1.3.1.2 Molar Mass Averages

Several average molar masses are used in polymer science. Different instrumental techniques provide different kinds of molar mass averages. They are defined as follows:

Number-average:

$$M_n = \int_0^\infty M f_n(M)\, dM = \sum_i x_i M_i = \frac{\sum_i n_i M_i}{\sum_i n_i} = \frac{\sum_i m_i}{\sum_i m_i / M_i} = \frac{1}{\sum_i w_i / M_i} \quad (1.59)$$

Weight-average:

$$M_w = \int_0^\infty M f_w(M)\, dM = \sum_i w_i M_i = \frac{\sum_i m_i M_i}{\sum_i m_i} = \frac{\sum_i n_i M_i^2}{\sum_i n_i M_i} \quad (1.60)$$

Z-average:

$$M_z = \frac{\int\limits_0^\infty M^2 f_w(M)\, dM}{\int\limits_0^\infty M f_w(M)\, dM} = \frac{\sum\limits_i m_i M_i^2}{\sum\limits_i m_i M_i} = \frac{\sum\limits_i n_i M_i^3}{\sum\limits_i n_i M_i^2} \qquad (1.61)$$

Z+1-average:

$$M_{z+1} = \frac{\int\limits_0^\infty M^3 f_w(M)\, dM}{\int\limits_0^\infty M^2 f_w(M)\, dM} = \frac{\sum\limits_i m_i M_i^3}{\sum\limits_i m_i M_i^2} = \frac{\sum\limits_i n_i M_i^4}{\sum\limits_i n_i M_i^3} \qquad (1.62)$$

Viscosity-average:

$$M_v = \left[\int\limits_0^\infty M^a f_w(M)\, dM \right]^{\frac{1}{a}} = \left[\sum_i w_i M_i^a \right]^{\frac{1}{a}} \qquad (1.63)$$

In the above equations, n_i is the number of moles (or the number of molecules N_i), m_i is the mass, x_i is the mole fraction, and w_i is the weight fraction of molecules with molar mass M_i. The exponent a is the exponent of the Mark-Houwink equation. If the exponent a becomes unity, the viscosity-average becomes identical to the weight-average. Commonly the M_v value lies between M_n and M_w (closer to M_w). The different averages are differently sensitive to different molar masses. Namely, M_n is sensitive mainly to the fractions with low molar masses while M_w and particularly M_z and even more M_{z+1} are sensitive to high-molar-mass fractions. For monodisperse polymers all molar averages are identical. The mutual relation of molar mass averages for polydisperse polymers is $M_n < M_v < M_w < M_z < M_{z+1}$. To illustrate the sensitivity of particular averages to low and high molar masses, let us consider a hypothetical sample consisting of 1% by weight of molecules with molar mass $M = 10^7$ g/mol, 98% wt molecules with $M = 10^5$ g/mol, and 1% molecules with $M = 10^3$ g/mol. The molar mass averages for this sample are: $M_n = 50,500$ g/mol, $M_w = 198,000$ g/mol, $M_z = 5,100,000$ g/mol, and $M_{z+1} = 9,904,000$ g/mol. Note that from the viewpoint of number of molecules the most abundant fraction is that with $M = 10^3$ g/mol (50.5 molar %).

The ratio of M_w/M_n is a measure of the broadness of the molar mass distribution and is often called *polydispersity* or *polydispersity index*. If a polymer is truly monodisperse, the ratio M_w/M_n is unity because all molar mass averages are identical. The ratio M_w/M_n increases as the polymer polydispersity increases. The ratio M_z/M_w can be used as an additional parameter or can be alternatively applied instead of M_w/M_n if the M_n value cannot be reliably determined. Other indices of polydispersity are the ratio M_z/M_n and index $U = M_w/M_n - 1$.

1.4 METHODS FOR THE DETERMINATION OF MOLAR MASS

Molar mass of polymers can be determined by various methods that differ in the type of molar mass average and applicable molar mass range. With the development of size exclusion chromatography, the importance of some traditional techniques, such as, for example, osmometry or analytical ultracentrifugation, decreased markedly. SEC gives not only molar mass averages, but also a complete description of molar mass distribution. The disadvantage of SEC in its conventional form is that it is a relative technique that requires a careful calibration; that is, SEC is not an absolute technique for measuring polymer molar mass. The term *absolute* means that molar mass is related to an exactly determinable physical quantity such as vapor pressure lowering, osmotic pressure, or intensity of scattered light. Ebullioscopy (measuring the boiling point elevation of a solution) and cryoscopy (measuring the freezing point depression of a solution), once widely used and still listed in some polymer textbooks, nowadays have no real applicability and their meaning is solely in the sense of understanding basic physical principles and historical perspectives.

1.4.1 Method of End Groups

The end-group method, which provides M_n, is applicable solely to linear polymers with easily determinable end groups such as epoxy resins, polyesters, or polyamides. The end groups are determined by titration or using UV, IR, or NMR spectroscopy. The concentration of end groups in a polymer sample decreases with increasing molar mass, which limits the applicability of the method to polymers with M_n below about 10,000 g/mol. The obtained results may be influenced by the presence of chain defects or cyclization (overestimation of M_n) and by the branching (underestimation of M_n). On the other hand, the discrepancies of M_n obtained by the method of end groups and by another suitable method may identify branching or chain end defects. In the case of linear polyesters prepared by the polycondensation of hydroxy carboxylic acids, that is, polyesters with the equal number of hydroxyl and carboxyl ends, the determination of M_n is according to the equation:

$$M_n = \frac{56 \times 1000}{AN} \qquad (1.64)$$

For polyesters prepared by polycondensation of dicarboxylic acids and diols, the relation is:

$$M_n = \frac{2 \times 56 \times 1000}{AN + HN} \qquad (1.65)$$

where AN and HN are acid number and hydroxyl number, respectively. They are determined by titration and expressed in mg KOH per gram of sample. Factor 56 is the molar mass of potassium hydroxide that is used for titration. Despite several limitations, the method of end groups is still relatively frequently used for the characterization of various oligomeric materials. The main advantage is given by simplicity and low demand for instrumental equipment.

1.4.2 Osmometry

Osmometry yields values of M_n. There are two types of osmometry applicable to polymers of different molar mass range. They are *vapor pressure osmometry* (vapor phase osmometry) (VPO) and *membrane osmometry* (MO). Vapor pressure and osmotic pressure are colligative properties, that is, properties that depend on the number of molecules in a given volume of solvent and not on the properties (e.g., size or mass) of the molecules. Unlike MO, VPO requires calibration with a standard of known molar mass, and it may not be considered an absolute method of molar mass determination in a precise sense of the term. However, in contrast to SEC, a low-molar-mass compound of accurately known molar mass is used for the calibration of VPO apparatus. A general advantage of osmometry is independency of the chemical nonuniformity; VPO and MO are very suitable for the characterization of heterogeneous copolymers. However, the practical meaning of both methods has declined tremendously with the development of SEC and nowadays they find only limited applications.

1.4.2.1 Vapor Pressure Osmometry

The principle of VPO is based on the fact that the vapor pressure of a solution is lower than that of the pure solvent at the same temperature and pressure. At sufficiently low concentrations, the magnitude of the vapor pressure decrease is directly proportional to the molar concentration of dissolved compound. Vapor pressure is not measured directly, but is measured indirectly by using thermistors (resistors whose resistance varies with temperature) that are connected in a Wheatstone bridge. The thermistors are placed in a measuring chamber that contains a reservoir of solvent and porous paper wicks to provide a saturated solvent atmosphere around the thermistors. If drops of solvent are placed on both thermistors, the thermistors will be at the same temperature. When a drop of solution is placed on one thermistor and solvent is placed on the other, the temperature difference is created. Due to lower solvent pressure above the solution, condensation of solvent into the solution from the saturated solvent atmosphere occurs. Solvent condensation releases heat that increases the solution temperature. Condensation continues until the solution temperature rises enough to compensate the decrease of solvent pressure. In the system equilibrium, the decrease of solvent pressure by the presence of dissolved compound is compensated by the increase of pressure due to increased temperature. The measurements are made at multiple concentrations and M_n is obtained by the extrapolation of the relation $\Delta R/c$ versus c to zero concentration:

$$M_n = \frac{K}{\left(\frac{\Delta R}{c}\right)_{c \to 0}} \tag{1.66}$$

where c is the concentration, ΔR is the difference in resistance of the thermistors that is measured instead of temperature difference, and K is a calibration constant of the instrument obtained by measuring a compound of known molar mass (e.g., sucrose octaacetate for organic solutions or sucrose for aqueous solutions).

A serious disadvantage of VPO is excessive sensitivity to the presence of low-molar-mass compounds, such as residual monomers, solvents, or moisture, which can result in serious underestimation of M_n. For example, presence of solely 0.1% wt of water in a polymer with $M_n = 2,000$ g/mol decreases the experimental value of M_n to about 1,800 g/mol. VPO is applicable to M_n of about 20,000 g/mol (some instruments claim applicability up to 10^5 g/mol). In the past, VPO replaced ebullioscopy and cryoscopy and enjoyed widespread use, but nowadays the VPO method is relatively rarely used and it has been replaced by SEC. Besides polymer chemistry, osmometers are used for determining the concentration of dissolved salts or sugar in blood or urine samples.

1.4.2.2 Membrane Osmometry

Membrane osmometry is an absolute method for the determination of M_n. In an MO instrument, a solution of a polymer is separated from pure solvent by a semipermeable membrane, which allows the solvent to pass through while being impermeable for the polymer. Due to the difference of chemical potential of pure solvent and solvent in solution, there is a solvent flow from the pure solvent side to the solution side. This solvent flow can be eliminated by applying pressure to the solution side. The applied pressure equals the osmotic pressure of the solution. A schematic illustration of the osmometer is shown in Figure 1.5. Solvent diffuses to the solution until the osmotic pressure is balanced by the hydrostatic pressure:

$$p = \Delta h d g \tag{1.67}$$

where Δh is the difference in the heights of the columns of solvent and solution, d is the density, and g is the gravitational acceleration.

The osmotic pressure π is related to the molar mass by the equation:

$$\frac{\pi}{c} = RT \left(\frac{1}{M_n} + A_2 c + A_3 c^2 + \cdots \right) \tag{1.68}$$

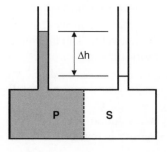

Figure 1.5   Schematic representation of membrane osmometer. P = polymer solution, S = pure solvent, Δh = height difference. Dashed line represents semipermeable membrane.

where A_2 and A_3 are the second and third virial coefficients, c is the concentration of polymer (g/mL), R is the gas constant, and T is the absolute temperature. At low concentrations the higher terms in Equation 1.68 are negligible and a plot of π/c versus c is linear with the intercept yielding M_n. The ratio of π/c is called the *reduced osmotic pressure*. The second virial coefficient is an additional result from the measurement of osmotic pressure that provides information about the polymer–solvent interactions and thus about the thermodynamic quality of solvent. Upward curvature on the π/c versus c plot indicates contributions from the third and higher virial coefficients. Membrane osmometry has a potential of the determination of theta conditions for a given solvent–polymer pair, but this potential is limited by solubility problems.

The apparatus outlined in Figure 1.5 may require several hours or even days to achieve equilibrium. The automated osmometers introduced in the early 1960s allowed measurements of osmotic pressure in only a few minutes.

The low-molar-mass limit of MO is given by membrane permeability, which is around 5,000 g/mol. Consequently, if a sample contains oligomeric species that can permeate through the membrane, the osmotic pressure is too low and the obtained M_n is overestimated. Generally, polydisperse polymers with pronounced oligomeric tail are not suitable for the MO method. The upper limit is about 5×10^5 g/mol.

1.4.3 Dilute Solution Viscometry

The dimensions of the polymer coils in a dilute solution affect the viscous properties of the solution. As a matter of fact, viscometry is not a method of the determination of molar mass, because the molar mass is not measured directly. Using the viscometric data, the molar mass can be obtained indirectly using the previously established Mark-Houwink relation. Viscometry of dilute solutions is an important method of the characterization of polymers with significant advantage given by simple and cheap instrumentation. In addition to the determination of molar mass, measurements of dilute solution viscosities can provide information about polymer branching, polymer size and its temperature dependence, and chain flexibility. Besides classical measurements of polymer solutions by a capillary viscometer, viscometry has become an important type of detection for SEC. Therefore, the viscometric characterization of dilute polymer solutions requires special attention. In order to characterize polymer molecules, the increase of viscosity brought about by the polymer molecules is measured instead of the absolute viscosity of the polymer solution. Classical viscometric measurements are performed using a capillary viscometer (Ubbelohde, Ostwald). The measured quantity is the *specific viscosity* (η_{sp}) or *relative viscosity* (η_{rel}):

$$\eta_{sp} = \frac{\eta - \eta_0}{\eta_0} = \frac{t - t_0}{t_0} = \eta_{rel} - 1 \tag{1.69}$$

where η and η_0 are the viscosities of dilute polymer solution and pure solvent, respectively. In practice, the measurement of viscosity is replaced with the measurement of time (t) needed for a certain volume of the solvent or solution to flow from one mark to the other. The specific viscosity is a measure of the increase of viscosity due to the addition of polymer, and according to Einstein it is proportional to the volume fraction (φ) of a polymer in solution:

$$\eta_{sp} = 2.5 \times \varphi \tag{1.70}$$

where $\varphi = \frac{V_{polymer}}{V_{solution}}$ (i.e., fraction of volume occupied by polymer molecules to the entire solution volume). The factor 2.5 assumes that polymer molecules behave like hard spheres.

Another quantity of interest is the *reduced viscosity* (*viscosity number*):

$$\eta_{red} = \frac{\eta_{sp}}{c} \tag{1.71}$$

or the *inherent viscosity* (*logarithmic viscosity number*):

$$\eta_{inh} = \frac{\ln \eta_{rel}}{c} \tag{1.72}$$

In the limit of infinite dilution, the reduced viscosity is known as the *intrinsic viscosity* (*limiting viscosity number*):

$$[\eta] = \lim_{c \to 0} \frac{\eta_{sp}}{c} \tag{1.73}$$

or alternatively

$$[\eta] = \lim_{c \to 0} \frac{\ln \eta_{rel}}{c} \tag{1.74}$$

The intrinsic viscosity reflects the ability of a polymer molecule to enhance the viscosity and depends on the size and shape of the polymer molecule. It is an important quantity describing the behavior of a polymer in solution and the polymer molecular structure. The concentration dependence of the specific viscosity can be described by the Huggins equation:

$$\frac{\eta_{sp}}{c} = [\eta] + k_H [\eta]^2 c + \cdots \tag{1.75}$$

where c is the concentration (g/mL or g/dL), k_H is the Huggins constant for a given polymer, solvent, and temperature, and $[\eta]$ is the intrinsic viscosity. For a sufficiently low concentration, the plot of η_{sp}/c versus c is linear with intercept equal to the intrinsic viscosity. The term k_H is obtained from the slope of the plot. For polymers in thermodynamically good solvents, k_H usually has a value around 1/3, while larger values of 0.5−1 are typical of poor solvents. Thus the Huggins constant can be used as a measure of solvent quality for a given polymer. An alternative expression was proposed by Kraemer:

$$\frac{\ln \eta_{rel}}{c} = [\eta] + k_K [\eta]^2 c + \cdots \tag{1.76}$$

where k_K is the Kraemer constant. For polymers in good solvents, k_K is negative and the mutual relation of the two constants is $k_H - k_K = 0.5$. The most accurate procedure of the determination of the intrinsic viscosity is to plot according to both Equations 1.75 and 1.76 and to take the mutual intercept as $[\eta]$. A small discrepancy of the intercept can be eliminated by taking an average value. An example of the determination of the intrinsic viscosity according to the Huggins and Kraemer procedures is depicted in Figure 1.6. Although various other extrapolation procedures were proposed, those according to Huggins and Kraemer represent the most often used and most reliable ones. When constant k_H or k_K is known, Equations 1.75 and 1.76 allow the calculation of intrinsic viscosity from a single point measurement. However, the measurements at multiple concentrations and extrapolation procedures should be preferred for precise determination of intrinsic viscosity. For the measurement the sample is typically prepared as a stock solution that is diluted by addition of solvent to prepare low concentrations. A more accurate procedure, which eliminates random errors during the preparation of stock solution, involves separate preparation of each solution. The initial concentration should provide $\eta_{sp} \approx 0.8$ and the specific viscosity for the most dilute solution should be about 0.1. However, the most important parameter is the linearity of the dependence of η_{sp}/c versus c or $\ln \eta_{rel}/c$ versus c with no indication of curvature at higher concentrations. It can be seen from Figure 1.6 that even the points that do not fulfill the requirement of $\eta_{sp} < 0.8$ show no deviation from the linearity. Another important requirement is that the solution can be considered dilute (i.e., consisting of mutually isolated polymeric coils

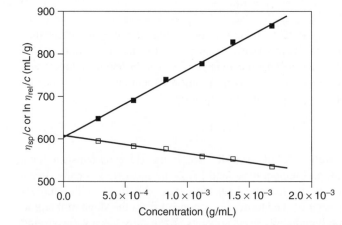

Figure 1.6   Plots of η_{sp}/c (■) and $\ln \eta_{rel}/c$ (□) versus c for a sample of hyaluronic acid sodium salt of $M_w = 335,000$ g/mol in aqueous 0.1 M sodium phosphate buffer pH 7 at 25°C. The mutual intercept is $[\eta]$. The intercepts are 604 mL/g and 608 mL/g for Huggins and Kraemer plots, respectively. The slopes allow calculation of Huggins and Kraemer constants: $k_H = 0.43$, $k_K = -0.11$.

Source: Courtesy Martina Hermannova, Contipro C, Czech Republic.

without significant polymer–polymer intermolecular interactions). The concentration when the solution is completely filled with the polymer coils and the coils start to penetrate each other is called *critical concentration* (c^*). Critical concentration is reached when the volume fraction of polymer $\varphi = 1$. Then, using Equation 1.70, the critical concentration can be estimated as:

$$c^* \approx \frac{2.5}{[\eta]} \tag{1.77}$$

The critical concentration is indirectly proportional to the intrinsic viscosity, and the determination of the intrinsic viscosity should be performed below c^* to assure exact linear extrapolation. The two highest concentrations in Figure 1.6 do not fulfill the requirement of $\eta_{sp} < 0.8$, but they are still well below the critical concentrations, which appears a more relevant requirement since the two points do not show any deviation from linear fit.

The viscometer should be selected so that $t_0 \approx 100$ sec. The solutions must be filtered to remove dust and other insoluble matters. Precise temperature control is very important for the determination of intrinsic viscosity. However, maintaining the constant temperature is of greater importance than absolute temperature itself.

The relation between the intrinsic viscosity and molar mass[i] is given by the Mark-Houwink equation (also called Mark-Houwink-Kuhn or Mark-Houwink-Kuhn-Sakurada):

$$[\eta] = KM^a \tag{1.78}$$

where K and a are the constants for a given polymer, solvent, and temperature. The intrinsic viscosity of a polydisperse polymer is related to the molar mass distribution and the obtained intrinsic viscosity is the weight-average:

$$[\eta] = \sum_i w_i [\eta]_i \tag{1.79}$$

where w_i is the weight fraction of polymer with the intrinsic viscosity $[\eta]_i$. Replacing the intrinsic viscosity in Equation 1.79 with Equation 1.78 and calculating M from Equation 1.78 shows that the molar mass obtained from the intrinsic viscosity is the viscosity average:

$$M = \left(\sum_i w_i M_i^a \right)^{\frac{1}{a}} = M_v \tag{1.80}$$

Once K and a values are reliably established, the molar mass can be determined from the value of $[\eta]$ for a polymer under investigation. For many polymers, the

---

[i]The use of this equation with the relative molecular weight is recommended by IUPAC Compendium of Chemical Terminology to avoid awkward and variable dimensions of constant K in the case that molar mass in g/mol is used.

parameters K and a can be found in the literature.[2] The experimental determination of K and a requires a series of samples with narrow molar mass distribution and known values of the intrinsic viscosity and molar mass. The intrinsic viscosity is plotted in a double logarithmic scale against molar mass, and the Mark-Houwink constant K and exponent a are obtained from the intercept and slope, respectively. The traditional method of determination of the parameters of the Mark-Houwink equation included fractionation of a polydisperse sample into a series of narrow fractions and the characterization of the obtained fractions by viscometry and light scattering and/or membrane osmometry. The requirement of narrow polydispersity was often not fulfilled. Equation 1.78 yields the viscosity average molar mass only if the constants K and a were determined by the measurements of monodisperse fractions. For all other cases, when the Mark-Houwink relationship was determined using polymer samples of polydispersity $M_w/M_n > 1$ and the average molar mass M_n or M_w was used, the Mark-Houwink equation should be corrected with a correction factor (correction factors are listed, for example, in *Polymer Handbook*[2]). However, the influence of polydispersity is relatively small when $[\eta]$ is correlated with M_w and $M_w/M_n < 1.5$. Although the Mark-Houwink relation of linear polymers is linear over a broad range of molar masses, one should be cautious when the parameters K and a are used beyond the molar mass range over which they were determined. The molar mass range of polymer samples used for the determination of the $[\eta]-M$ relationship should cover at least one order of magnitude. The parameters K and a should always be quoted not only with the solvent and temperature, but also with the number of samples and their molar mass range and polydispersity, and the method of the molar mass determination. As a rule of thumb, the values obtained with narrow polymers using light scattering should be preferred.

1.4.3.1   Properties of Mark-Houwink Exponent

The literature shows relatively significant differences of the values K and a even for polymers measured under identical conditions. The low K values are associated with the high a values and vice versa, which means the differences are to a certain level mutually compensated when the constants are used to calculate the molar mass averages.

An important property of the exponent a is that it bears information about the polymer conformation. Generally, for flexible polymer molecules in thermodynamically good solvents the exponent a ranges from about 0.65 to about 0.75. In theta solvent, $a = 0.5$, while higher values of a, equaling or even exceeding unity, can be found for less flexible rodlike macromolecules, such as some polysaccharides or polyelectrolytes. Especially polyelectrolytes are able to create expanded rodlike macromolecules due to the repulsive electrostatic forces of charges within the polymer chain. The maximum value of a is two, which corresponds to rigid rods. On the other hand, the increasing compactness of the macromolecules results in decreasing a, which in the ultimate case of compact spheres equals zero and so the intrinsic viscosity becomes independent of the molar mass. Although compact sphere-like macromolecules are rare, the decrease of the exponent *a* is observed for branched polymers that often show an evenly decreasing slope of the Mark-Houwink plot with increasing molar mass (see examples and further discussion in Chapter 6).

The Mark-Houwink relations established for high-molar-mass polymers are not valid for oligomers. In the region of low molar masses, the Mark-Houwink relations show a break where the slope drops to about 0.5, no matter the polymer composition and thermodynamic quality of the solvent. This is a consequence of short polymer chains that have no excluded volume effect, because the situation where a polymer chain segment cannot take the exact place of another segment cannot happen for too-short chains. Different Mark-Houwink exponents of oligomers and polymers are illustrated in Figure 1.7, which shows log – log plot of [*η*] against *M* for polystyrene in THF. The exponent *a* obtained for lower-molar-mass polystyrene is markedly lower than that for polystyrene of high molar mass, yet it is a larger-than-expected value of 0.5, which may reflect the fact that the chains consisting of 13–125 monomeric units are not short enough to completely eliminate the effect of excluded volume.

1.4.3.2   *Molecular Size from Intrinsic Viscosity*

The dimension of intrinsic viscosity, mL/g, suggests this quantity to be related to the volume occupied by the polymer molecules in solution. The hydrodynamic volume of a polymer coil is expressed as:

$$V_h = \frac{[\eta]M}{\gamma N_A} \tag{1.81}$$

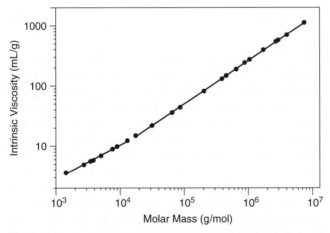

Figure 1.7 Mark-Houwink plot for polystyrene in THF at 35°C. Mark-Houwink parameters $a = 0.565$, $K = 5.66 \times 10^{-2}$ mL/g (*M* from 1400 to 13,000 g/mol); $a = 0.719, K = 1.27 \times 10^{-2}$ mL/g (*M* from 17,500 to 7.5×10^6 g/mol). The data were obtained by SEC-MALS-VIS analysis of narrow polystyrene standards.

where M is the molar mass, N_A is Avogadro's number, and γ is Simha's parameter related to the shape of a particle, which equals 2.5 for a hard sphere. Using $\gamma = 2.5$, the hydrodynamic volume is the volume of a hydrodynamically equivalent sphere (i.e., a hypothetical sphere that would have the same intrinsic viscosity as the actual polymer molecule). The hydrodynamically equivalent sphere can be similarly defined on the basis of diffusion coefficient. The size of a hydrodynamically equivalent sphere may be different for different types of motion of the macromolecules (i.e., viscous flow or diffusion), and thus the hydrodynamic radii calculated on the basis of intrinsic viscosity and diffusion coefficient need not be absolutely identical. The hydrodynamic radius (radius of hydrodynamically equivalent sphere) is then calculated as:

$$R_h = \left(\frac{3[\eta]M}{10\pi N_A} \right)^{\frac{1}{3}} \tag{1.82}$$

The hydrodynamic radius calculated from the intrinsic viscosity is often called the *viscometric radius* (R_η) and the term *hydrodynamic radius* is reserved for the radius based on the diffusion coefficient. However, the difference between the two radii is usually negligible. The intrinsic viscosity also can yield the RMS radius using the Flory-Fox and Ptitsyn-Eizner equations:[3,8]

$$R = \frac{1}{\sqrt{6}} \left(\frac{[\eta]M}{\Phi} \right)^{\frac{1}{3}} \tag{1.83}$$

$$\Phi = 2.86 \times 10^{21}(1 - 2.63\varepsilon + 2.86\varepsilon^2) \tag{1.84}$$

$$\varepsilon = \frac{2a - 1}{3} \tag{1.85}$$

where a is the exponent of Mark-Houwink equation. Equation 1.84 expresses Flory's universal constant Φ_0 after correction for non-theta conditions. The correction parameter $\varepsilon = 0$ at theta conditions, whereas at thermodynamically good solvents it is related to the exponent of the Mark-Houwink equation. Equation 1.83 yields the RMS radius in cm using the intrinsic viscosity expressed in dL/g. For example, for polystyrene molecules of molar mass 10^6 g/mol, using the parameters of the Mark-Houwink equation $a = 0.717$ and $K = 1.17 \times 10^{-4}$ dL/g, one gets $\varepsilon = 0.145$, $\Phi = 1.94 \times 10^{21}$, and $R = 4.4 \times 10^{-6}$ cm (i.e., 44 nm). The practical meaning of Equation 1.83 is that it can provide RMS radii for smaller polymers that are not accessible directly by the light scattering measurements. Equation 1.83 also can be used in reversed direction for the calculation of intrinsic viscosity from the RMS radius and molar mass. Since both RMS radius and molar mass are obtained by light scattering, a light scattering photometer can yield the intrinsic viscosity without a viscometer. Equation 1.83 also shows the relation between the exponents of the Mark-Houwink equation and the relationship RMS radius versus molar mass:

$$R = k \times M^b \tag{1.86}$$

$$b = \frac{1+a}{3} \tag{1.87}$$

The exponent a can range from zero for compact spheres to two for rigid rods and thus the values of exponent b are in a narrower interval of $0.33-1.0$. This fact explains why the conformation plots of branched polymers are usually significantly less curved than the corresponding Mark-Houwink plots. A typical value for the exponent $a \approx 0.7$ corresponds to $b \approx 0.57$. In contrast to the Mark-Houwink parameters K and a, which were published for many polymers in various solvents, the literature values of k and b are significantly less frequent. Since these parameters are needed for the characterization of branching, the possibility of calculating them on the basis of the Mark-Houwink parameters using Equation 1.83 is of practical importance.

Using Equation 1.82 and the Stokes-Einstein equation, one can get the relation for the exponent of the molar mass dependence of the diffusion coefficient (Equation 1.20):

$$\beta = \frac{1+a}{3} \tag{1.88}$$

1.4.3.3 Dependence of Intrinsic Viscosity on Polymer Structure, Temperature, and Solvent

The influence of the chemical composition of a polymer on the intrinsic viscosity is not straightforward. Double bonds and cycles in the main chain and large side groups decrease the chain flexibility and affect the ability of the chain to form a random coil and so result in increased intrinsic viscosity. However, the chemical structure of the polymer also affects the interactions of the polymer with the solvent, and it may not be easy to distinguish between the influence of solvation and the steric hindrance of different chain structures. The solvation of a polymer chain has a significant impact on the expansion of the polymer coil. The solvent interacting with a polymer chain creates a solvating envelope that increases with increasing thermodynamic quality of the solvent and thus the intrinsic viscosity increases with increasing thermodynamic quality of the solvent.

Ionic groups along the polymer chain have a strong effect on the intrinsic viscosity. A polymer coil of polyelectrolyte can expand due to the electrostatic repulsive forces between the ionic groups compared to a neutral polymer of a similar chemical structure (e.g., polyacrylic acid sodium salt versus polyacrylamide, hyaluronic acid sodium salt versus pullulan). The coil expansion of the polyelectrolyte is determined by the degree of dissociation of the ionic groups, which depends on pH. The expansion of the polyelectrolyte chain is strongly affected by addition of salt, which shields charged groups. The coil expansion and thus the intrinsic viscosity decreases with an increasing salt concentration.

The viscosity of a polymer solution decreases exponentially with temperature according to the equation:

$$\eta = A \times e^{\frac{E}{RT}} \tag{1.89}$$

where E is the flow activation energy, T is the absolute temperature, R is the gas universal constant, and A is a constant. The behavior of the intrinsic viscosity is different, and the temperature coefficient $d[\eta]/dT$ can be positive, negative, or zero. Increased temperature can increase chain flexibility due to the lower potential barrier of the transition from one conformation to another, which decreases the intrinsic viscosity due to a more coiled macromolecular structure. On the other hand, the rising temperature increases the thermodynamic quality of the solvent. The intrinsic viscosity is more sensitive to temperature in the vicinity of theta temperature. The increase of the intrinsic viscosity with temperature is used for an interesting and important application of some copolymers as additives for motor oils. These polymers dissolved in the oil at room temperature are closed to theta conditions and their molecules are highly coiled. Due to the relatively low hydrodynamic volume of the macromolecules, their contribution to solution viscosity is low. During the oil application at high temperatures, the thermodynamic quality of the oil increases, leading to high solvation of the macromolecules, their strong expansion, and thus a partial compensation for the overall decrease of the oil viscosity and thus the lubrication efficiency.

1.4.4   Matrix-Assisted Laser Desorption Ionization Time-of-Flight Mass Spectrometry

Matrix-assisted laser desorption ionization time-of-flight mass spectrometry (MALDI-TOF MS) is a relatively new technique of the determination of molar mass with accuracy and resolution unachievable with other methods. A mixture of a UV-light-absorbing matrix and a polymer sample is irradiated by a nanosecond laser pulse. Most of the laser energy is absorbed by the matrix, which prevents unwanted fragmentation of the polymer. Another purpose of the matrix is to isolate the analyte molecules from each other and thus to prevent clustering of the analyte into high-mass complexes that would be too large for desorption and analysis. The ionized macromolecules are accelerated in an electric field and enter the TOF tube. During the flight in the tube, the molecules are separated according to their mass-to-charge ratio and reach the detector at different times. In this way, each molar mass gives a distinct signal in the mass spectrum. In addition to the molar mass distribution MALDI-TOF MS can also provide compositional information concerning end groups and chemical composition. However, in the case of polydisperse polymers the method commonly fails due to the mass-dependent desorption ionization process and/or mass-dependent detection efficiency due to the signal intensity discrimination against higher-molar-mass components. As a consequence of that, the most abundant ions in the obtained spectrum may not coincide with the apex of the molar mass distribution and the fractions with the highest molar mass can be completely missing. The inability of MALDI to provide reliable molar mass distribution for polydisperse polymers can be overcome by combination with SEC by collecting fractions and performing MALDI-TOF MS analysis offline. In a SEC-MALDI-TOF MS combination, the refractive index detector of the SEC instrument determines the polymer concentration in each fraction and the corresponding molar masses are measured by MALDI-TOF MS. The determination of the average molar mass of each fraction allows establishing the SEC calibration, which can then be used to compute molar mass distribution by conventional SEC. However, the speed of the analysis and the expensive instrumentation are the limitations of the SEC-MALDI methodology. For more information about the MALDI-TOF technique, refer to reference 9.

1.4.5 Analytical Ultracentrifugation

Analytical ultracentrifugation (UC) represents a traditional technique of polymer characterization that played an important role at the beginning of polymer science in the 1920s. The method was invented by Theodor Svedberg, who introduced both the *sedimentation velocity method* and the *sedimentation equilibrium method*. The method of sedimentation velocity is used to measure the sedimentation coefficient that can be used for the calculation of molar mass. The method is based on monitoring the movement of concentration boundary in dilute polymer solution placed in the ultracentrifuge cell under high centrifugal field (e.g., 70,000 rotations per minute). The determination of sedimentation coefficient is performed at multiple concentrations to allow the extrapolation of sedimentation coefficient to zero concentration. The molar mass can be calculated from the Svedberg equation:

$$M = \frac{RTs}{D(1 - \bar{v}d)} \tag{1.90}$$

where R is the gas constant, T is the temperature (K), s is the sedimentation coefficient (second), D is the diffusion coefficient, $\bar{v}$ is the partial specific volume (mL/g), and d is the density.

The sedimentation equilibrium experiment yields both weight-average M_w and the centrifuge-average M_z. The centrifugal field is significantly lower compared to the former method (e.g., 15,000 rpm). When the sedimentation equilibrium is reached, the molecules are distributed in the centrifuge cell according to their molar mass. Although the method can provide valuable information about sample purity, characterize assembly and disassembly mechanisms of biomacromolecular complexes, characterize macromolecular conformational changes, or measure equilibrium constants for self- and hetero-associating systems, in the area of synthetic polymers it has been completely replaced with size exclusion chromatography. Major limitations for routine applications are analysis time needed for the experiments (time needed to reach equilibrium can be up to several days) and expensive instrumentation.

1.5 KEYNOTES

- There are several types of nonuniformities that can occur in most synthetic and natural polymers. The most frequently studied nonunifiormity is that of molar mass, but the others should be kept in mind when polymers are being characterized.
- Freely jointed chain represents the basic model of polymer chain in dilute solution.
- Excluded volume refers to the fact that one part of a long chain molecule cannot occupy space that is already occupied by another part of the same molecule. Excluded volume causes the polymer chain to expand to a larger size compared to what it would be were there no excluded volume effect.
- The effect of excluded volume can be neutralized by specific conditions (solvent and temperature) that are called theta conditions.
- Besides excluded volume, the size and shape of a polymer chain is affected by chain structure (i.e., bond angles, bond length, ability to rotate about the single bonds, and presence of double bonds or cycles in the main chain).
- The size and properties of polymer chains must be evaluated by averaging over all possible conformations given by all angles of rotation about the bonds.
- Molar mass distribution can be described by cumulative and differential distribution functions. Both distributions can be expressed in term of number fraction and weight fraction.
- The molar mass moments used in polymer science are the number-average, weight-average, z-average, $(z+1)$-average, and viscosity average. The type of average should be stated whenever talking about the polymer molar mass.

1.6 REFERENCES

1. Flory, P. J., *Statistical Mechanics of Chain Molecules*, John Wiley & Sons, New York (1969).
2. Brandrup, J., Immergut, E. H., Grulke, E. A. (editors), *Polymer Handbook*, 4th Edition, John Wiley & Sons, New York (1999).
3. Flory, P. J. and Fox, T. G., *J. Am. Chem. Soc.*, **73**, 1904 (1951).
4. Burchard, W., *Makromol. Chem.*, **50**, 20 (1960).
5. Stockmayer, W. H. and Fixman, M., *J. Polym. Sci., Part C*, **1**, 137 (1963).
6. Kratky, O. and Porod, G., *Rec. Trav. Chim.*, **68**, 1106 (1949).
7. Bohdanecky, M., *Macromolecules*, **16**, 1483 (1983).
8. Ptitsyn, O. B. and Eizner, Yu. E., *Sov. Phys. Tech. Phys.*, **4**, 1020 (1960).
9. Pasch, H. and Schrepp, W., *MALDI-TOF Mass Spectrometry of Synthetic Polymers*, Springer, Berlin (2003).

Chapter 2

Light Scattering

2.1 THEORY AND BASIC PRINCIPLES

This chapter describes basic principles of scattering of light by dilute solutions of macromolecules or dispersions of colloidal particles and presents various examples of the application of light scattering for the characterization of unfractionated polymer solutions. Principles and applications of the light scattering hyphenated with a separation technique are presented in Chapters 4 and 5. Generally, it is not necessary to distinguish between the solutions of macromolecules or dispersions of particles. In fact, many real systems may contain dissolved macromolecules along with highly compact branched macromolecules or supermolecular structures and the borderline between *dissolved* and *dispersed* may not always be clear. The importance of light scattering is obvious because it is one of the few absolute methods available for the determination of molar mass of polymers. In addition, light scattering can provide information about the macromolecular size and structure and interactions of macromolecules with solvent and with each other.

An important advantage of light scattering is its applicability over an extremely broad range of molar masses. The first experiments focused on the application of light scattering for polymer characterization were carried out in the 1940s and 1950s, that is, before availability of personal computers, lasers, membrane filters, and size exclusion chromatography. Several important technological developments, including lasers and advances in personal computers and software, converted light scattering from an originally tedious and uncertain method of limited applicability to a routine and easy-to-use modern analytical technique. The advent of size exclusion chromatography and other separation techniques and development of light scattering instruments capable of working as chromatography detectors tremendously extended obtainable information as well as application areas.

Light Scattering, Size Exclusion Chromatography and Asymmetric Flow Field Flow Fractionation: Powerful Tools for the Characterization of Polymers, Proteins and Nanoparticles, by Stepan Podzimek
Copyright © 2011 John Wiley & Sons, Inc.

Light scattering can be simply defined as a natural phenomenon, namely a result of the interaction of light with matter. The light is the oscillating field consisting of electric and magnetic parts. The oscillating electric field interacting with a neutral molecule creates a dipole, which, due to the oscillation of the incident radiation, oscillates as well. The oscillating dipole becomes a source of new radiation. The schematic illustration is shown in Figure 2.1. It is obvious that the ability of a molecule to scatter light is in relation to the tendency of the electron cloud of the molecule to be displaced from its normal shape by an external electric field. This tendency is called *polarizability*, which is directly proportional to the specific refractive index increment (*dn/dc*).

A perfect crystal that is much larger than the wavelength of the incident light can be considered as composed of identical small elements that are much smaller compared to the wavelength. Every small element can be considered to be a point source of scattered light. In any direction of observation it is possible to find two identical volume elements distant from each other such that the light beams scattered by these two elements reach the detector phase shifted by exactly $\lambda/2$. In a perfect crystal, identical volume elements contain an identical number of molecules and therefore the intensities of light scattered by the two volume elements are identical. Two light beams of identical intensity, shifted by $\lambda/2$, cancel each other completely due to interference. Consequently, large perfect crystals do not scatter light. Real crystals scatter light due to local inhomogeneities and surface layers. Pure liquids scatter light much more intensively than the same number of molecules in the form of crystal. The explanation for this fact is that the number of molecules in a volume element of a liquid fluctuates with time. Consequently, the intensities scattered by the same volume elements are not identical and are not completely canceled by interference. Although the macroscopic density of a liquid is constant with time, the microscopic density fluctuations explain the non-zero intensity of light scattered by pure liquids.

The intensity of light scattered by a dilute polymer solution consists of two contributions: (1) intensity scattered by a solvent itself, and (2) intensity scattered by macromolecules. The difference between the two scattered intensities is called *excess scattering*, which bears information about the dissolved macromolecules. Similarly to the case of pure liquids, small volume elements of solution contain different numbers of macromolecules and thus the fluctuation of solute concentration accounts for the excess scattering.

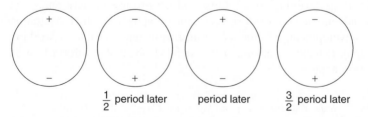

Figure 2.1 Oscillation of induced dipole due to the oscillation of incident light.

2.2 TYPES OF LIGHT SCATTERING

There are three different types of light scattering:

1. *Static light scattering* (also called *elastic*, *Rayleigh*, or *classical* light scattering) occurs at the same wavelength as the incident light. Static light scattering refers to an experiment in which the scattered light intensity is determined at a given scattering angle by averaging the fluctuating intensity over a long time scale compared with the time scale of the intensity fluctuation. The information about the scattering macromolecules is obtained from the measurement of angular and concentration dependence of the intensity of scattered light. This type of experiment yields the molar mass, molecular size expressed as the root mean square (RMS) radius, and the second virial coefficient. This book deals mostly with the static type of light scattering.

2. *Dynamic light scattering* (*quasielastic* light scattering) also occurs at the same wavelength as that of the incident light, but the fluctuations of the scattered light intensity are studied instead of the average intensity. Due to the Brownian motion of the scattering particles, the intensity of scattered light fluctuates. The fluctuation is over extremely short time intervals. In fact, static and dynamic light scattering apply to the same phenomena and the only difference is in the way of collecting and processing the experimental data. In static light scattering we measure the time-averaged intensity of scattered light while in dynamic light scattering we analyze how the scattered light intensity fluctuates with time. Today's light scattering instruments allow performing both types of light scattering experiments simultaneously. *Note:* The photodiodes used to detect the scattering light intensity at the static experiments are not able to follow extremely rapid changes of the intensity and thus monitor an average intensity of the scattered light. The difference between the static and dynamic light scattering measurements is illustrated in Figure 2.2. The dynamic light scattering experiment yields information about the diffusion coefficient of scattering particles and consequently about the hydrodynamic radius.

3. *Raman scattering* occurs at a wavelength different from that of incident light. Raman spectroscopy can provide structural information and is not a topic of this book.

Several examples of light scattering can be found in nature. The most visible example of light scattering is the blue color of sky. As will be shown later, the intensity of scattered light strongly depends on the light wavelength. The blue color is the most intensively scattered light from the incident sunlight, which is scattered by dust and ice particles and molecules at the upper layers of atmosphere. Since the blue part of sunlight is scattered with the highest intensity, the sky ultimately appears as blue. Similar effects are responsible for red sunset or

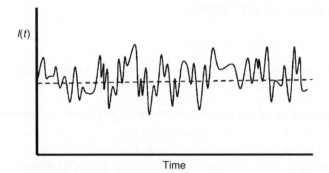

Figure 2.2 Sketch of fluctuation of scattered light intensity with time. For a static light scattering experiment the average scattered intensity (indicated by the dotted line) is measured, while for dynamic light scattering the intensity fluctuations are analyzed.

sunrise. The only difference is that the incident light comes to the atmosphere from a different angle and travels a longer path in the atmosphere compared to the sunlight that one can see during the day. In the longer path, the scattering actually has a filtering effect and one can see the least scattered light as red.

2.2.1 Static Light Scattering

The fundamental description of the theory, instrumentation including assembly of a flow cell in multi-angle light scattering (MALS) photometers, and applications of light scattering were reviewed in an extensive article by Wyatt.[1] A book by Pavel Kratochvil can become a source of valuable information about light scattering from polymer solutions.[2]

The basic equation that relates the intensity of scattered light with the properties of the macromolecules in solution is:[3]

$$\frac{R_\theta}{K^*c} = MP(\theta) - 2A_2cM^2P^2(\theta) + \cdots \tag{2.1}$$

where R_θ is the excess Rayleigh ratio, c is the concentration of polymer in solution (g/mL), M is the molar mass, A_2 is the second virial coefficient, K^* is the optical constant, and $P(\theta)$ is the *particle scattering function* (also called *particle scattering factor*). The properties of some of the quantities are discussed in the following text.

The excess Rayleigh ratio describes the angle-dependent intensity of light scattered by a sample. As a matter of fact, the light scattering intensity is measured as voltage yielded by photodiodes:

$$R_\theta = \frac{(I_\theta - I_{\theta,solvent})r^2}{I_0V} = f\frac{V_\theta - V_{\theta,solvent}}{V_{laser}} \tag{2.2}$$

where I_θ is the scattered light intensity of the solution; $I_{\theta,solvent}$ is the scattered light intensity of the solvent; I_0 is the intensity of the incident radiation; V is the volume of the scattering solution; r is the distance between the scattering volume and detector; V_θ, $V_{\theta,solvent}$, and V_{laser} are detector signal voltages of the solution, solvent, and laser, respectively; and f is an instrumental constant related to the geometry of the apparatus, structure of the scattering cell, and the refractive indices of the solvent and scattering cell. The constant f is usually determined by a solvent of the well-known Rayleigh ratio. Note that dividing by the laser signal compensates for any change of laser intensity due to power supply instability and temperature fluctuations or laser aging. The subscript θ implies the angle between the scattering direction and the incident light beam ($\theta = 0°$ being straightforward from the perspective of the incident light). The word *excess* means that one measures the contribution of dissolved molecules to the Rayleigh ratio of the entire solution, that is, the difference between the Rayleigh ratios of solution and pure solvent. The dimension of R_θ is length^{-1}, mostly cm^{-1}. The advantage of the Rayleigh ratio is that it is independent of the incident light intensity and the geometrical arrangement of the apparatus. *Note:* Equation 2.2 assumes a vertically polarized light source. The constant K^* is an optical constant that is defined for the vertically polarized incident light as:

$$K^* = \frac{4\pi^2 n_0^2}{\lambda_0^4 N_A}(dn/dc)^2 \tag{2.3}$$

where n_0 is the refractive index of the solvent at the incident wavelength, λ_0 is the incident radiation wavelength at vacuum, N_A is Avogadro's number, and dn/dc is the specific refractive index increment of scattering macromolecules. The second virial coefficient A_2 is related to the thermodynamic quality of solvent. Note that Equation 2.1 has further terms with the third and higher virial coefficients that can be neglected at very low concentrations. Generally, the R_θ/K^*c versus c plot is linear at the region of low concentrations with the slope directly proportional to the second virial coefficient and deviating from linearity at higher concentrations. Modern light scattering photometers are sufficiently sensitive to allow working at low concentrations where the effect of the term with the third virial coefficient is negligible.

The A_2 is a quantity that is sometimes overlooked as a result of polymer characterization by light scattering. It characterizes the intensity of interactions between the solvent and dissolved polymer. The positive values of the order of magnitude of $\geq 10^{-4}$ mol mL/g^2 are typical for so-called thermodynamically good solvents, where the polymer–solvent interactions are stronger than mutual interactions among individual polymer molecules or intramolecular interactions among different segments of a macromolecule. In other words, in thermodynamically good solvents the dissolved macromolecules prefer to interact with the solvent over interacting with themselves. In so-called *theta* solvents, the A_2 equals zero and the strength of polymer–solvent interactions is identical with that of polymer–polymer interactions.

Slightly negative values indicate the tendency of polymer to precipitate from solution, because the interactions of polymer–polymer are stronger than those of polymer–solvent, or we can say that the molecules prefer themselves to the solvent. Solutions with larger negative A_2 do not exist, because such thermodynamically poor solvents do not dissolve polymers at all. The value of A_2 depends on temperature and polymer molar mass. With the increasing temperature the A_2 increases, which explains the well-known fact that polymers dissolve more rapidly at elevated temperatures. The increasing temperature, as a matter of fact, increases the thermodynamic quality of the solvent. The molar mass dependence of A_2 is given by a simple expression:

$$A_2 = kM^{-\vartheta} \tag{2.4}$$

where k is a constant and exponent ϑ is mostly in the range of 0.15 to 0.35. For instance, using the data from reference 4 for polystyrene in THF at room temperature, the following relationship is obtained:

$$A_2 = 0.01 \times M^{-0.25} (\text{mol mL/g}^2) \tag{2.5}$$

Decreasing value of A_2 with increasing molar mass explains why polymer solubility decreases as molar mass increases. In contrast to solutions of monodisperse polymers, where only identical macromolecules interact with each other, in solutions of polydisperse polymers all different macromolecules interact with each other with the frequencies given by the molar mass distribution of a given polymer. Therefore, in the case of polydisperse samples the A_2 obtained by a light scattering experiment is an average that is not as clearly defined as the averages of molar mass and RMS radius.

For a polydisperse polymer, the total Raileigh ratio can be expressed as a sum of the particular Rayleigh ratios corresponding to scattering by each molar mass:

$$R_\theta = K^* \sum_i c_i M_i \tag{2.6}$$

which can be rearranged as

$$\frac{R_\theta}{K^* c} = \frac{\sum c_i M_i}{c} = \sum_i w_i M_i = M_w \tag{2.7}$$

where c_i and w_i are the concentration and the weight fraction of molecules with molar mass M_i, respectively. That means the Rayleight ratio of a dilute solution of a polydisperse polymer is directly proportional to the weight-average molar mass (M_w), and thus the batch light scattering measurements of polydisperse polymers yield the values of M_w.

The particle scattering function $P(\theta)$ describes the decrease of the scattered light intensity with increasing angle of observation. It is defined as the ratio of the intensity of radiation scattered at an angle of observation θ to the intensity of radiation scattered at zero angle:

$$P(\theta) = \left(\frac{R_\theta}{R_0}\right)_{c=0} \tag{2.8}$$

The decline of light scattering intensity with increasing angle is due to intramolecular interference of light beams scattered by different points of the same particle. A small scattering particle with the maximum distance of two points of the particle below about $\lambda/20$ behaves as a point source of scattered light for which all scattered beams are in phase. In contrast, the light beams scattered by different parts of a large particle reach the photodetector phase shifted. Phase-shifted light beams interfere destructively and thus the intensity is lower than would correspond to the intensities of individual light beams. The destructive interference depends on the scattering angle, as demonstrated in Figure 2.3. At zero scattering angle, the path lengths of the scattered beams are identical and there is no destructive interference, while due to the interference of the two phase-shifted beams the intensity of the resulting radiation at an angle θ is smaller than at zero angle with decreasing trend toward higher angles. For very large colloidal particles, the particle scattering function can even show maxima and minima.

For small angles, the particle scattering function can be approximated as:

$$\lim_{\theta \to 0} P(\theta) = 1 - \frac{16\pi^2}{3\lambda^2}R^2\sin^2(\theta/2) = 1 - \frac{\mu^2}{3}R^2 \tag{2.9}$$

where $\lambda = \lambda_o/n_o$ is the wavelength of the incident light in a given solvent, R^2 is the mean square radius, and μ is the scattering vector:

$$\mu = \frac{4\pi}{\lambda}\sin(\theta/2) \tag{2.10}$$

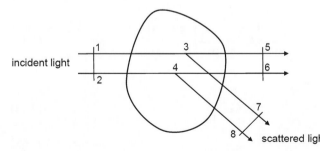

Figure 2.3   Scattering of light by a large particle. The path length 1–3–5 equals the path 2–4–6, but path 1–3–7 is longer than 2–4–8. The different path lengths of light scattered by different parts of a large particle cause reduction of scattering intensity due to intramolecular interference.

As for $x \to 0 \lim[1/(1-x)] = 1 + x$, Equation 2.9 can be for $\theta \to 0$ written as

$$\lim_{\theta \to 0} \frac{1}{P(\theta)} = 1 + \frac{16\pi^2}{3\lambda^2} R^2 \sin^2(\theta/2) = 1 + \frac{\mu^2}{3} R^2 \qquad (2.11)$$

According to Equations 2.9 and 2.11, the plots of $P(\theta)$ or $P^{-1}(\theta)$ against $\sin^2(\theta/2)$ are straight lines with the slopes proportional to the R^2. This very important fact tells us that the slope of the angular variation of the scattered light intensity at angle zero yields the root mean square (RMS) radius (R). It is also worth noting that the information about the size of scattering particles can be revealed solely from the measurement of the angular variation of scattered light intensity. The RMS radius describes the distribution of mass around the center of gravity. A particle can be assumed to consist of mass elements of mass m_i and the RMS radius is defined as the square root of the weight-average of r_i^2 for all the mass elements:

$$R = \left(\frac{\sum m_i r_i^2}{\sum m_i} \right)^{1/2} \qquad (2.12)$$

where r_i is the distance of the ith mass element of mass m_i from the center of gravity. In contrast to other size parameters, the RMS radius makes no assumption of particle shape. As already mentioned, the RMS radius is a quantity purely derived from the light scattering data, which can be determined even if the concentration and dn/dc of the scattering particles are unknown. Equation 2.12 is valid for a rigid particle. A random coil, which is by far the most frequent shape for synthetic and natural polymers, is a flexible structure that can create a large number of various conformations, each of them having its particular RMS radius. The quantity measurable experimentally is an average value—that is, for a non-rigid macromolecule the distance r_i is an average over all possible conformations.

The RMS radius is in simple relationships with other size parameters characteristic for various particle shapes:

For a random coil:

$$R^2 = \frac{\langle r^2 \rangle}{6} \qquad (2.13)$$

For a rod:

$$R^2 = \frac{L^2}{12} \qquad (2.14)$$

For a compact sphere:

$$R^2 = \frac{3D^2}{20} = \frac{3}{5}a^2 \qquad (2.15)$$

For a hollow sphere (spherical shell):

$$R^2 = a^2 \tag{2.16}$$

where $\langle r^2 \rangle$, L, D, and a are the mean square end-to-end distance of random coil, rod length, sphere diameter, and sphere radius, respectively.

Equations 2.9 and 2.11 describe how the intensity of scattered light decreases with the increasing angle of observation at the region of very low angles. This angular dependence becomes more pronounced with increasing size of scattering molecules. On the other hand, small molecules with the RMS radius roughly below 10 nm scatter light with the same intensity in all scattering angles. That means no size information can be obtained from elastic light scattering measurement for small molecules. It must be emphasized at this point that the 10 nm RMS radius is not a magical borderline below which the molecules scatter with equal intensity in all directions and above which the measurement of RMS radius is feasible. The diminution of the angular dependence is gradual and the limit for which the RMS radius can be or cannot be measured depends also on the signal-to-noise ratio of the light scattering signal. Scattering intensities for small molecules that scatter evenly in all angles and for larger molecules that scatter with declining intensity are compared in Figure 2.4. Note that although the scattering intensity of larger molecules decreases with increasing angle, its absolute value is markedly larger than that of small molecules.

An important property of the RMS radius is that for polydisperse polymers the experimental value is the z-average as shown in the following: For an infinitely dilute solution of a polydisperse polymer, the intensity of scattered light is given as the sum of the contributions of particular polymer molecules:

$$R_\theta = \sum_i R_{\theta,i} = K^* \sum_i c_i M_i P_i(\theta) = K^* c M_w P(\theta)_{average} \tag{2.17}$$

where $P(\theta)_{average}$ is an average particle scattering function. The comparison of the last two terms of Equation 2.17 defines the average particle scattering function of polydisperse polymer to be the z-average:

$$P(\theta)_{average} = \frac{\sum c_i M_i P_i(\theta)}{c M_w} = \frac{1}{M_w} \sum_i w_i M_i P_i(\theta)$$

$$= \frac{1}{M_w} \int_0^\infty f_w(M) M P_M(\theta) \, dM = P_z(\theta) \tag{2.18}$$

The average particle scattering function for a polydisperse polymer of known distribution function $f_w(M)$ can be derived using the particle scattering function for a given particle shape on condition that the integral in Equation 2.18 can be solved analytically.

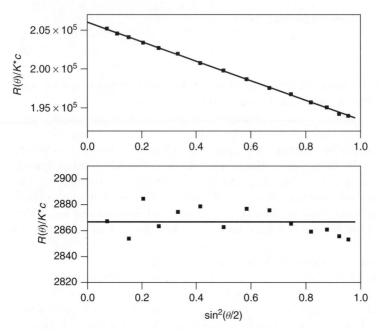

Figure 2.4   Angular variation of intensity of scattered light for polystyrene with nominal molar mass of 200,000 g/mol (top) and 2500 g/mol (bottom). Note the scattering of experimental data points for the bottom plot corresponds to the molar mass of less than 50 g/mol.

For small angles, the particle scattering function can be expressed using Equation 2.9:

$$P_z(\theta) = \frac{1}{M_w} \sum_i w_i M_i \left(1 - \frac{\mu^2}{3} R_i^2\right) = 1 - \frac{\mu^2}{3} \frac{1}{M_w} \sum_i w_i M_i R_i^2 \qquad (2.19)$$

$$\frac{1}{M_w} \sum_i w_i M_i R_i^2 = R_z^2 \qquad (2.20)$$

That means the initial slope of the scattered light intensity on $\sin^2(\theta/2)$ is proportional to the z-average RMS radius. The fact that light scattering yields the molar mass and the RMS radius of different moments, that is, the weight-average molar mass against the z-average RMS radius, is very important for the interpretation of the light scattering results. It was shown in Section 1.3.1.2 that the two averages are sensitive to the presence of small amounts of fractions with very high molar mass, and that this effect is more pronounced for the z-average.

The particle scattering function describes the decrease of light scattering intensity with the observation angle for large particles due to destructive interference of phase-shifted light beams and must not be confused with the angular dependence due to the use of unpolarized light. The scattering of unpolarized light can be considered as scattering of light polarized in two perpendicular directions:

vertically polarized light for which the scattering intensity is independent of angle, and horizontally polarized light for which the scattering intensity depends on scattering angle. The intensity of a wave is proportional to the square of its amplitude, but in the case of light polarized in a horizontal plane the intensity is not given by the whole amplitude, but solely by its projection to the direction of observation as illustrated in Figure 2.5. In the case of unpolarized light, the scattering intensity depends on scattering angle by the term $(1 + \cos^2 \theta)$, where unity corresponds to the vertical component, which is angularly independent, and $\cos^2\theta$ reflects the angular dependence of the horizontal component. The maximum scattering intensity is at $\theta = 0°$, while the minimum scattering intensity is at $\theta = 90°$. The optical constant as described by Equation 2.3 is valid for vertically polarized incident light for which the angular dependence is solely due to the intramolecular interference. The measurements are typically carried out using a vertically polarized light source.

2.2.1.1 Particle Scattering Functions

The particle scattering functions were calculated for various particle shapes, such as random coils, rods, and spheres. A general form of the particle scattering function for a particle consisting of N identical scattering elements can be expressed as:[5]

$$P(\theta) = \frac{1}{N^2} \sum_i \sum_j \frac{\sin(\mu r_{ij})}{\mu r_{ij}} \qquad (2.21)$$

where μ is the scattering vector expressed by Equation 2.10, and r_{ij} is the distance between the ith and jth scattering element. For small products of $\mu r_{i,j}$ (i.e., when the scattering molecule is small or the angle is low), Equation 2.21 can be approximated by Equation 2.9. From Equation 2.21 the particle scattering functions can be calculated for a specific particle shape. For linear random coils, the particle scattering function is:[6]

$$P(\theta) = \frac{2}{x^2}(e^{-x} - 1 + x) \qquad (2.22)$$

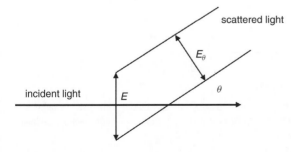

Figure 2.5 Schematic depiction of scattering of light polarized in horizontal plane (top view). E is the amplitude of scattered light, E_θ is the projection of E to the direction of observation θ : $E_\theta = E \sin\left(\frac{\pi}{2} - \theta\right) = E \cos\theta$.

where

$$x = \frac{8\pi^2}{3\lambda^2} \langle r^2 \rangle \sin^2(\theta/2) \tag{2.23}$$

where λ is the wavelength of the light in a given solvent and $\langle r^2 \rangle$ is the mean square end-to-end distance of the chain related to the RMS radius according to the Equation 2.13. Figure 2.6 compares particle scattering functions for linear random coils of various size calculated by means of Equations 2.22 and 2.9. It is obvious that the difference of the two particle scattering functions for relatively small molecules is negligible even for scattering angles far from zero. For large molecules, the two scattering functions overlap as the angle approaches zero. It should be noted that Equation 2.22 is valid for polymers dissolved in a theta solvent, in which the second and higher virial coefficients are zero. In thermo-dynamically good solvents ($A_2 > 0$), the expansion of the polymer coils affects the particle scattering function.

For homogeneous spheres, the particle scattering function is:

$$P(\theta) = \left[\frac{3}{x^3} (\sin x - x \cos x) \right]^2 \tag{2.24}$$

where

$$x = a\mu = \frac{2\pi D}{\lambda} \sin(\theta/2) \tag{2.25}$$

where μ is the scattering vector, D is the diameter, and a is the radius of the sphere. In the case of real samples, spheres can be represented by solid poly-mer particles prepared by emulsion polymerization, various kinds of organic and

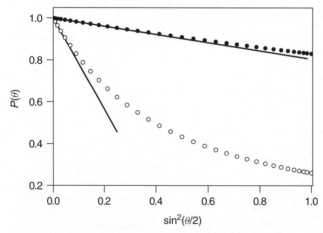

Figure 2.6 Particle scattering functions for monodisperse linear random coils of RMS radius 30 nm ($\bullet$) and 100 nm ($\circ$) calculated by means of Equation 2.22. Solid lines represent approximate function represented by Equation 2.9.

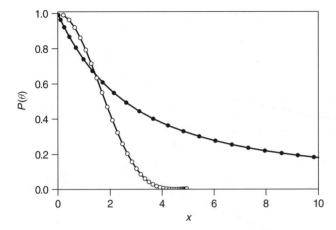

Figure 2.7 Particle scattering functions for monodisperse linear random coils (•) and homogeneous spheres (○) according to Equations 2.22 and 2.24.

inorganic nanoparticles, incompletely soluble polymer fractions (usually called *microgels*), or highly branched molecules such as stars with a high number of arms or dendritic and hyperbranched polymers. *Note:* Equations 2.22 and 2.24 are valid for monodisperse systems of given R or a.

The particle scattering functions according to Equations 2.22 and 2.24 are functions of only one variable x, which includes a dimensional parameter characteristic for a given particle shape and angle of observation. The dimension of a particle is related to the wavelength of light in a given solvent and the angular parameter is expressed as $sin(\theta/2)$. Note that the same value of x, and thus also of $P(\theta)$, can be obtained by an infinite number of combinations of dimensional and angular factors. Graphical representations of particle scattering functions for the two most common particle shapes (random coils and solid spheres) are given in Figure 2.7. Plots of $P^{-1}(\theta)$ versus $sin^2(\theta/2)$ for random coils and solid spheres of different size are contrasted in Figure 2.8. A significant difference between the particle scattering functions of the two most common shapes can be found for large particles while for smaller particles the plots almost overlap. The RMS radius of 30 nm for linear random coils corresponds to molar mass of roughly 500,000 g/mol. Note the strong curvature of the particle scattering function for large solid spheres, which in some special cases may provide further information about the sample under analysis.

Various coordinates can be used to describe the angular variation of the scattered light intensity, for instance, $P^{-1}(\theta)$ versus $\mu^2 R^2$, $P^{-1}(\theta)$ versus $sin^2(\theta/2)$, or $P(\theta)$ versus x. Alternatively, instead of $P^{-1}(\theta)$, the ratio K^*c/R_θ is plotted against $sin^2(\theta/2)$. Both dependences are completely equivalent, because K^*c/R_θ is directly proportional to $P^{-1}(\theta)$ with the proportionality constant $1/M$ or $1/M_w$. The plot K^*c/R_θ or R_θ/K^*c versus $sin^2(\theta/2)$ is usually used for simultaneous determination of molar mass from the intercept and R from the slope, while

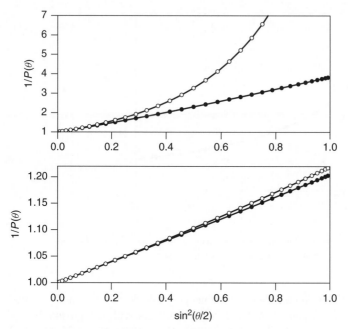

Figure 2.8   Reciprocal particle scattering function $P^{-1}(\theta)$ for monodisperse linear random coils (•) and homogeneous spheres (○) of RMS radius 100 nm (top) and 30 nm (bottom).

$P^{-1}(\theta)$ or $P(\theta)$ versus $\sin^2(\theta/2)$ can be beneficial for the comparison of angular dependences of samples of different molar mass.

Besides random coils and solid spheres, the particle scattering functions were derived for other particle shapes, such as infinitely thin rods, hollow spheres, infinitely thin discs, ellipsoids, cylinders, polydisperse coils, coils outside theta temperature, regular stars, polydisperse stars, monodisperse regular combs, and so on. A review of particle scattering functions for different particle models can be found, for example, in references 7 and 8. In principle, the comparison of an experimentally determined particle scattering function with $P(\theta)$ calculated theoretically allows the determination of particle shape. However, the practical meaning of such determination is limited due to the following reasons: The great majority of natural and synthetic polymers form random coils, and other shapes, such as rods or ellipsoids, are rare. In other words, polymer chemists mostly characterize random coils, whose diversity is given by their contraction or expansion due to polymer–solvent interactions or size reduction due to the presence of branches. However, these parameters cannot be simply inferred from the shape of particle scattering function. Another limitation of the estimation of the particle shape from the $P(\theta)$ function is given by the fact that the comparison of experimental and theoretical particle scattering functions requires a wide range of $P(\theta)$ values, which are accessible solely for very large particles. Let us consider the wavelength of 690 nm, tetrahydrofuran of $n = 1.401$, maximum angle of observation of about $155°$, and a polymer molecule of molar mass of 10^6 g/mol, which corresponds to the RMS radius of about 45 nm.

The maximum value of $x \approx 1.3$ can be calculated for this molecule. That means the experimentally accessible x values for small and mid-sized molecules are too small to get into the range where different particle shapes can be reliably distinguished from each other (see Figure 2.7). It can be concluded that information about the particle shape can be obtained only for polymers of very large molar mass (order of magnitude several millions g/mol). Moreover, the investigated polymer must be monodisperse in order to eliminate the influence of polydispersity on $P(\theta)$, because the effect of polydispersity may overwhelm the effect of particle shape. It can be concluded that the ability of classical light scattering to characterize polymer shape and conformation is limited and can be enhanced only by combination with a separation technique that reduces the effect of polydispersity and provides a relation between the RMS radius and molar mass.

The most typical shapes of the particle scattering function are as follows: (1) For small particles, with the RMS radius <10 nm, the Rayleigh ratio does not depend on the angle of observation; (2) linear or slightly curved particle scattering functions are typical for molecules with the RMS radii of several tens of nm; and (3) strongly curved particle scattering functions are characteristic of very large polymers (RMS radius >100 nm), very broad polymers, or systems containing macromolecules of a common molar mass of the order of 10^5 g/mol plus small amounts of very large particles. These large particles can be macromolecules of ultra-high molar mass, supermolecular aggregates consisting of a large number of individual macromolecules, or crosslinked swollen particles (microgels). Even a trace weight fraction of species with molar mass and molecular dimensions much larger than the rest of the molecules strongly affects the intensity and angular variation of the scattered light. This phenomenon occurs because the large macromolecules scatter light very intensely mainly at low scattering angles, due to very high molar mass, but the scattering intensity diminishes quickly with increasing angle of observation due to the very large size (i.e., intensive intramolecular interference).

On the other hand, smaller macromolecules with lower molar mass scatter light less intensely than the large molecules at low angles, but the decrease of the scattering intensity toward high angles is not that steep. That means in the case of a mixture of smaller and large macromolecules, the light scattering intensities at low angles are governed mainly by large particles, while the contribution of smaller molecules becomes more important at high angles. The superposition of angular dependencies of smaller and very large particles yields a strongly curved angular dependence as shown in Figure 2.9. As seen in this figure, the presence of solely 0.02% wt of a component with very high molar mass results in strongly curved plots regardless of the light scattering formalism used for data processing. The slopes at low angles are significantly higher than those at high angles. This is because very large macromolecules (10^7 g/mol in this particular example) contribute to scattering intensity relatively more at low angles, while

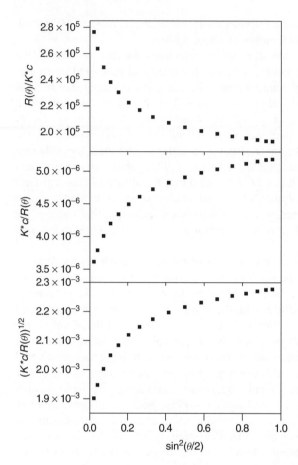

Figure 2.9   Debye plots (from top to bottom: Debye, Zimm, and Berry formalism) for a mixture of PS standards of nominal molar mass 2×10^5 and 10×10^6 g/mol prepared at concentrations of approximately 500 μg/mL and 0.1 μg/mL, respectively.

the smaller molecules (200,000 g/mol in this example) contribute more at higher angles. The data in Figure 2.9 were obtained by a model mixture, but a similar situation occurs in the case of very broad continuous distribution and the angular dependences of extremely polydisperse polymers are strongly curved. In batch measurements it may be difficult to remove the large fractions from solutions, and in fact it is often not obvious what exactly should be removed, since it may be difficult to distinguish what is still an integral part of the sample and what is dust and other impurities.

The large structures may have a strong effect on the sample properties and their identification and characterization may bring important information about the polymerization process. On the other hand, they have strong effect on the molar mass and especially on the RMS radius and their presence in solution can completely disturb the information inferred about the major polymeric part of the sample. It must be emphasized that supermolecular aggregates are often found in polymers of very high molar mass (10^6 g/mol and more), in highly branched polymers, crystalline polymers, or stereolegular polymers capable of crystallization. The aggregates often can be destroyed by a long period of dissolution (several days, a week, or even longer) or at elevated temperatures. However, intensive mechanical agitation should be avoided because of possible shearing degradation.

The light scattering theory outlined in the preceding text is based on the Rayleigh-Gans-Debye (RGD) approximation:

$$|m - 1| \ll 1 \text{ and } 2ka|m - 1| \ll 1 \tag{2.26}$$

where a is a characteristic radius of the molecules, m is the relative refractive index of scattering molecules n/n_0, n is the refractive index of the solvated molecules, n_0 is the refractive index of the solvent, λ_0 is the vacuum wavelength of the light, $k = 2\pi n_0/\lambda_0$. Inequalities 2.26 mean that the relative refractive index of scattering particles is close to unity, that is, the particle refractive index is almost indistinguishable from that of solvent, and the total phase shift of the incident light wave is negligible.

These conditions are fulfilled for many common samples, but they will not be true for some very large, dense molecules and particles. As a matter of fact, the RGD theory is often applied to large particles where its validity is uncertain. For instance, the aqueous polymer latex (refractive index $\approx 1.6, a \approx 100$ nm), $m = 1.6/1.33 \approx 1.2$ does not exactly satisfy the RGD conditions. However, the comparison of results obtained by MALS with those determined by dynamic light scattering or electron microscopy suggests that the theory works even when RGD requirements are not strictly satisfied. It was also shown that RGD approximation becomes more valid in the limit of $\theta \rightarrow 0$.[9] Finally, the more general and more complex Mie theory, valid for spheres of any size and refractive index, can be applied to the particles that do not comply with RGD approximation.

The basic light scattering Equation 2.1 can be summarized into two fundamental light scattering principles:

1. The intensity of light scattered by a dilute polymer solution is directly proportional to the concentration and molar mass of polymer molecules.
2. The angular variation of the scattered light intensity is related to the size of scattering molecules.

These principles are important not only for proper understanding and interpretation of the experimental data, but also for planning and performing the measurements. Even trace amounts of particles with enormously high molar mass (aggregates, highly compact particles, or dust) generate very intensive light scattering signals. On the other hand, investigation of oligomers requires preparation of solutions with sufficiently high concentrations. It is sometimes not well understood that appropriate concentrations for batch light scattering measurement of dissolved polymers can vary over a very broad concentration range, such as 10^{-5} to 10^{-2} g/mL, depending only on the molar mass of the polymer under investigation. The concentration needed for the characterization of dispersed particles can be several orders of magnitude lower compared to that needed for the measurement of macromolecular solutions.

Incipient users of light scattering sometimes ask about the detection limit of a MALS detector. The first light scattering principle explains that a MALS detector has no detection limit in the sense of minimum detectable concentration, because the response of a light scattering photometer is proportional not only to the concentration, but also to the molar mass. Each molar mass has its own detection limit and the detection limits can differ by many orders of magnitude. As a matter of fact, a MALS detector is rather used to characterize than to detect. That means the concentration of analyzed sample should be appropriate in order to get an intensive signal-to-noise ratio and thus proper characterization.

2.2.1.2 Light Scattering Formalisms

Equation 2.1 can be rearranged into the following alternative forms:[3,10]

$$\frac{K^*c}{R_\theta} = \frac{1}{MP(\theta)} + 2A_2c + \cdots \qquad (2.27)$$

$$\sqrt{\frac{K^*c}{R_\theta}} = \frac{1}{\sqrt{MP(\theta)}} + A_2c\sqrt{MP(\theta)} + \cdots \qquad (2.28)$$

Similarly to the case of Equation 2.1, the above equations have higher concentration terms. The particular formalisms describe the same phenomenon (i.e., the concentration and angular variation of light scattered by a polymer solution or a colloidal dispersion). Equations 2.1, 2.27, and 2.28 represent different ways of the processing of the experimental data and are usually called *Debye, Zimm,* and *Berry formalisms,* respectively.

Frequently used light scattering terminology can be sort of confusing since the term *Debye* is used for the plot of the light scattering intensity versus angle of observation at a given concentration and also for one of the possible light scattering formalisms. The light scattering intensities acquired at a single concentration can be extrapolated to zero angle, neglecting the concentration dependence. This procedure, called a *Debye plot,* can be done using Zimm, Debye, or Berry formalism. The Debye plot is always used for processing the data acquired in an online mode, but can be also applied for a batch measurement. A *Zimm plot* means processing the light scattering data that were collected at multiple angles and multiple concentrations. In contrast to the Debye plot, the concentration dependence is taken into account and the obtained data are extrapolated not only to zero angle, but also to zero concentration. This processing can be performed using Zimm, Debye, or Berry formalisms. For discussion of the influence of the light scattering formalism on the molar mass and RMS radius, see also Section 4.2.1.5.

2.2.1.3 Processing the Experimental Data

Raw data generated by a light scattering photometer are voltages yielded by photodiodes at various angles. However, the requested physical quantity is the intensity of scattered light, which is expressed as the *Rayleigh ratio*. The conversion of voltages into Rayleigh ratios is done by means of the instrumental constant f, which in addition takes account of the scattering volume and the distance of the scattering volume from the detector. In routine practice, the constant f is determined by a standard liquid, a solution of standard polymer of known molar mass, or a dispersion of standard colloids. To convert voltage recorded by a light scattering photometer to the Rayleigh ratio, the instrument must by calibrated by a standard under identical conditions (wavelength, temperature, and geometrical arrangement of the cell).

Calibration of the light scattering photometer involves the measurement of the voltage of the standard of known Rayleigh ratio and the determination of the constant f from the measured voltage and the Rayleigh ratio. The voltages generated by a sample are then converted to the Rayleigh ratios using the constant f. It is worth mentioning that the instrumental constant f is independent of solvent that is used for the measurements of real samples, but it slightly depends on temperature. The absolute scattering power of the standard must be determined by an absolute method. The calibration by pure liquids has the advantage that the scattering power depends only on the temperature and wavelength of the incident light and there is no other compound involved. Toluene is an example of a liquid that was carefully characterized with respect to its absolute scattering power and that is easy to purify and stable at room temperature. The formerly often-used benzene cannot be recommended, due to high toxicity. 1,2,4-Trichlorobenzene, a frequently used solvent for high-temperature SEC, can be used as a standard for calibration of high-temperature MALS photometers. Calibration using solution of a polymer standard may be uncertain in the sense of absolute correctness of the molar mass of standard and the method of its determination. The M_w of standard used for the calibration of a light scattering photometer is often measured by another light scattering photometer. An even worse situation might be the case where the standard would be characterized by conventional SEC with column calibration. The calibration constant of a light scattering photometer, once properly determined, usually remains constant for several months or even years and there is no need for frequent recalibration. That is one of the advantages of light scattering. However, the calibration constant may be affected by a stray light scattered by dirt in the flow cell or caused by a misaligned laser beam. The stray light causes the calibration constant to be too small, resulting in calculated molar masses that are also too small (the percentage errors in the calibration constant and molar mass are identical).

Note that the molar masses are affected only through the calibration constant, because the stray light added to the sample measurement is included also in the baselines and thus it is eliminated during data processing. Using solvent with a high Rayleigh ratio, the possible effect of stray light is minimized and thus low-Rayleigh-ratio solvents, such as water, methanol, or THF, cannot be recommended for the calibration of MALS photometers.

Normalization is another procedure that is needed in the case of a MALS photometer. The normalization provides a set of coefficients, one for each of the detectors placed at various angles around the cell of a MALS detector. Unlike calibration constant f, the normalization coefficients are valid for a given solvent and a change to a solvent of different refractive index requires determination of another set of normalization coefficients. The principle of normalization is explained in the following.

Assuming vertical polarization of incident light, small particles scatter with the same intensity at all angles. That means the particular photodiodes placed at various angles around the cell of a MALS photometer should yield identical voltages when the flow cell is filled with a dilute solution of a small polymer. However, in reality, the voltages are not identical for the following reasons: (1) The photodiodes are not identical and they can produce slightly different voltages for identical light intensity, and (2) the photodiodes monitor different scattering volumes. This effect is sketched in Figure 2.10, which shows that each photodiode views a different illuminated volume. The photodiodes at low and high angles look along the beam and see a larger illuminated volume, while the intermediate photodiodes look across the beam and see a smaller illuminated volume (the smallest observed volume is at $90°$). Therefore, a set of normalization coefficients relating each detector to the $90°$ detector must be determined.

The normalization coefficient for a given angle (N_θ) becomes a part of Equation 2.2, where the right side becomes equal to $N_\theta f (V_\theta - V_{\theta,solvent})/V_{laser}$. The normalization coefficient of the $90°$degree detector is always unity and only this detector is calibrated. The normalization involves measurement of a solution of a small polymer that scatters equally in all angles. The light scattering software records voltages corresponding to particular photodetectors and calculates a series of normalization coefficients. The voltages yielded by particular photodiodes become identical and equal to $90°$ voltage after multiplication with corresponding normalization coefficients. The obtained normalization coefficients are solvent related, because the scattered light beam is refracted when passing from solvent into the flow cell glass. Due to the refraction of scattered light passing from the solvent into the glass, the photodiode monitors light scattered at a different angle than the fixed angle at which the photodiode is placed (Figure 2.11). As a matter of fact, the refraction explains why the MALS photometers can measure at relatively small scattering angles with minimized effect of transmitted light. Polymers suitable for normalization are those with molar

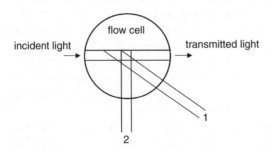

Figure 2.10   Schematic illustration of different scattering volumes monitored by two photodetectors (1, 2) placed at different positions around the flow cell (top view). The arrows indicate inlet and outlet from the flow cell. Larger scattering volume is viewed by detector 1.

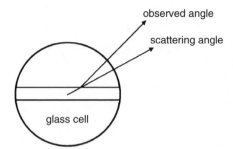

observed angle

scattering angle

glass cell

Figure 2.11   Sketch of flow cell refraction (top view).

mass high enough to scatter intensively, but small enough to scatter equally in all directions. To make the normalization more accurate the light scattering software can correct for a non–infinitely small molecular size of the polymer used for normalization using a theoretical particle scattering function and known RMS radius of the polymer. Polystyrene of molar mass 20,000–30,000 g/mol is a good example of sufficiently small polymer for organic solvents. Bovine serum albumin (BSA) or about 30,000 g/mol pullulan, dextran, or poly(ethylene glycol) can be used in aqueous solvents. The polymer used for normalization should be preferentially of narrow molar mass distribution, because fractions with very low molar mass undesirably decrease the light scattering intensity while fractions with very high molar mass do not fulfill the requirement for small, evenly scattering molecules. The requirement of narrow molar mass distribution is of primary importance in the case of batch measurements, while in the case of combination of a MALS detector with a separation technique a part of a chromatogram corresponding to small molecules can be selected for the normalization. An example of using a proper part of a chromatogram for normalization is shown in Figure 2.12 for BSA that fulfills the requirement of small molecules, but typically contains a certain amount of larger aggregates. The peaks of dimer and trimer as well as lower-intensity data points at the descending part of the peak are not used for the normalization as indicated by the two vertical lines.

Processing the light scattering data requires correction for two effects: (1) Due to the interference of light beams scattered by different mass points of a large particle, the intensity of the resulting radiation is smaller than the sum of particular intensities of light scattered by all the individual mass points of that particle. This phenomenon is called *intramolecular (intraparticle) interference of scattered light*. The decrease of the scattered light intensity due to the intramolecular interference is described by the particle scattering function. (2) In a polymer solution of a finite concentration, the light scattered by different macromolecules also interferes. This effect, which is called *intermolecular interference*, causes the intensity of light scattered by a solution to be smaller than the sum of scattered intensities by the individual macromolecules. The extent of the intermolecular interference is related to thermodynamic properties of the polymer–solvent system and is characterized by the second and higher

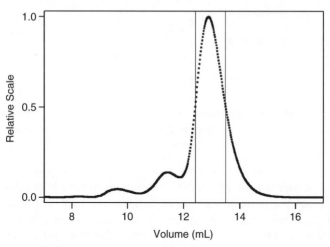

Figure 2.12 Normalization of a MALS photometer in an SEC mode using BSA. Only the data points between the vertical lines are used for normalization to avoid influence of aggregates and data points with low intensities.

virial coefficients. To process the data acquired by a light scattering experiment and to determine true characteristics of the dissolved polymer, the two non-ideality effects (the effect of terms with A_2 and higher virial coefficients and the effect of intramolecular interference) have to be eliminated.

To correct for the non-ideal solution behavior, we can neglect terms beyond the A_2 and plot K^*c/R_θ as a function of c. For low concentrations the relation is linear, giving the second virial coefficient from the slope (slope $= 2A_2$) and the weight-average molar mass from the intercept (intercept $= 1/M_w$). However, the obtained results are valid solely for small polymer molecules since the concentration extrapolation ignores the effect of intramolecular interference. To eliminate the scattered light intensity reduction due to the intramolecular interference, the light scattering experiment would have to be carried out at zero angle. However, it is impossible to measure the intensity of scattered light at zero angle, because the intensity of transmitted light is several orders of magnitude higher than that of scattered light. Instead, the light scattering intensity is measured at multiple angles and then extrapolated to zero angle, for which the particle scattering function equals unity.

The concentration and angular variation of the intensity of scattered light can be simultaneously processed using the so-called Zimm plot, which is a relation between K^*c/R_θ and $\sin^2(\theta/2) + kc$. The constant k is chosen to spread out the experimental data points and affects solely the visual appearance of the plot, but has no influence on the obtained results. Details of the processing and examples of a Zimm plot for various polymers are presented further in Section 2.7.

2.2.2   Dynamic Light Scattering

Macromolecules or colloidal particles dissolved or dispersed in solvent undergo Brownian motion. The light beams scattered at a given time by different particles are to a certain extent phase shifted and mutually interfere. The interference can be either positive or negative depending on the mutual position of the scattering particles. After a time delay the scattering particles move to another position and the intensity of light receipted by the detector is different, because the mutual position of the particles is different and the phase shift of scattered beams is different as well (see Figure 2.13). The distance among the scattering particles in solution is therefore constantly changing with time, which results in the fluctuation of the intensity of scattered light. The fluctuation of intensity reflects the motion of the scattering particles. In the case of large particles, which move slowly, the intensity fluctuates slowly too, while for small particles moving rapidly the intensity fluctuation is rapid too. That means the fluctuation of the intensity bears information about the moving particles.

The information about the moving particles is obtained by the analysis of the intensity fluctuation, which is done by transformation into the intensity autocorrelation function using a piece of hardware called a *correlator*. The time scale of the scattered light intensity fluctuations is analyzed by a mathematical process called *autocorrelation*. The autocorrelation function expresses the mutual relationship of the signal with itself (i.e., it reports how quickly on average the light intensity changes with time). The autocorrelation function expresses the probability that after time delay τ the intensity of the scattered light will be identical

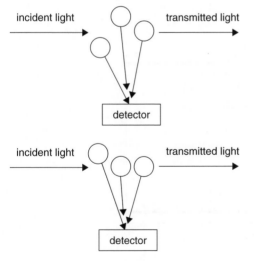

Figure 2.13   Illustration of scattering of light by particles undergoing Brownian motion. Position of particles at time t (top) and $t + \tau$ (bottom). Light beams scattered by particles at two different times travel different distances to the detector and thus their mutual phase shift is different.

to that in the initial time. At short time delays, the correlation is high because the scattering particles do not have enough time to change their mutual position to a great extent compared to the original state. The two signals are almost identical when compared after a very short time interval. With the increasing time delay, the correlation starts to exponentially decrease. After a certain time delay, there is no correlation between the scattering intensity at the initial time t and the final time $t + \tau$. Figure 2.14 shows an example of different profiles of autocorrelation functions for particles of different hydrodynamic radius. A faster decay corresponds to smaller molecules, while a slower decay indicates larger molecules. The normalized intensity autocorrelation function is defined as:

$$g(\tau) = \frac{\langle I(t)I(t+\tau) \rangle}{\langle I(t) \rangle^2} \tag{2.29}$$

where $I(t)$ is the detected intensity as a function of time t, $\langle I(t) \rangle^2$ is the average scattered intensity squared, τ is a delay time, and the brackets indicate averaging over all t. The autocorrelation function is established by multiplying the scattered intensity as a function of time with itself after shifting by a delay time τ, and the obtained products are averaged over a sufficiently long time period. The autocorrelation function is calculated for various values of τ, ranging typically from about 1 μs to several seconds, and plotted against τ (usually on a log-scale time axis).

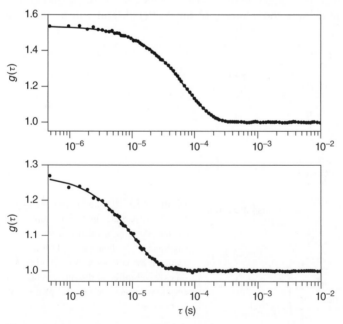

Figure 2.14 Autocorrelation function for polystyrene of $R_h = 30$ nm (top) and 4 nm (bottom). Tetrahydrofuran at ambient temperature.

For a monodisperse sample, the normalized autocorrelation function is described by an exponential function:

$$g(\tau) = 1 + \beta e^{-2D\mu^2\tau} \tag{2.30}$$

where $\mu = \frac{4\pi}{\lambda}\sin(\theta/2)$, D is the translational diffusion coefficient, β is the signal amplitude of the autocorrelation function, λ is the wavelength of light in a given solvent, and θ is the detection angle with respect to the direction of the incident beam. The analysis of the autocorrelation function allows determination of the translation diffusion coefficient of the macromolecules or particles. Using the diffusion coefficient and assuming the spherical shape of the molecules, one can calculate the hydrodynamic radius using the Stokes-Einstein relation:

$$R_h = \frac{kT}{6\pi\eta D} \tag{2.31}$$

where k is Boltzman's constant, T is the absolute temperature, and η is the solvent viscosity.

The hydrodynamic radius R_h is the radius of the so-called hydrodynamically equivalent sphere that would have the same diffusion coefficient as the molecules or particles under investigation. It must be stressed that the hydrodynamic radius is by definition a different size parameter from the RMS radius. The mutual relation of the two radii depends on the particle shape and thus the comparison of R_h and R allows, at least in principle, the estimation of particle shape. For a compact sphere, the hydrodynamic radius equals the geometric radius and can be related to the RMS radius using the following equation (compare Equation 2.15):

$$R^2 = \frac{3}{5}R_h^2 \tag{2.32}$$

It can be mentioned that there is another method for the determination of hydrodynamic radius based on the intrinsic viscosity as shown in Section 1.4.3.2. A significant advantage of R_h compared to R is that it can be measured also for small particles down to about 1 nm. On the other hand, the characterization of larger particles requires longer acquisition time, which limits the applicability of the method in the online mode, because in SEC or A4F of polydisperse samples the size of particles flowing through the flow cell changes with time. That means the autocorrelation function is not acquired for a rigorously monodisperse fraction and this effect becomes more pronounced with increasing acquisition time. If the relation between the diffusion coefficient and molar mass is known, the diffusion coefficient can be used for the estimation of molar mass. However, such calculation is meaningful only for polymers of given chemical composition under given experimental conditions (solvent, temperature) and dynamic light scattering cannot be considered to be the absolute method of molar mass determination.

The diffusion coefficient and the hydrodynamic radius can be accurately determined only for dilute solutions or dispersions, because in the case of more concentrated samples the particle mobility is influenced by interactions with neighboring particles. However, in the case of concentrated dispersions such as those prepared by emulsion polymerization the necessity to dilute the sample may question the obtained results, because both aggregation and disaggregation can accompany the dilution.

Although dynamic light scattering represents an absolute method for the determination of hydrodynamic radius, the measurement of polydisperse systems in batch mode offers only limited possibility to characterize the size distribution. For a mixture of small and large particles, the autocorrelation function may show two decays—a faster one representing smaller particles and a slower one representing larger particles. The autocorrelation function of light scattered from a polydisperse population of particles is described by the sum of the autocorrelation functions of all particles, weighed by their normalized intensities. Information about the size distribution is obtained by the analysis of the autocorrelation function under a few assumptions.

Generally two methods are used to extract the size distribution data from the batch DLS measurement: The method of *cumulants* assumes a Gaussian distribution of diffusion rates and uses the first and second cumulant only to calculate a mean and Gaussian width of the diffusion rates. The cumulants of a distribution are closely related to distribution's moments and they are used to calculate the corresponding values of hydrodynamic radius. The *regularization* method assumes a smooth distribution of hydrodynamic radius. The software determines a number of R_h distributions that all fit the experimental data equally well, and chooses between them based on the smoothness of the distribution, favoring smooth distributions over spiked distributions. The obtained size distribution is weighted by scattered light intensity and can be converted to mass or number distribution as follows:

According to the basic light scattering equation, the scattering intensity of the ith particle (I_i) can be written in a simplified form as:

$$I_i = KM_ic_i = KM_i^2N_i \qquad (2.33)$$

where K is the proportionality constant, and M_i, c_i, and N_i are the molar mass, the concentration (g/mL), and the number concentration (particle number/mL) of scattering particles, respectively. The relation between the molar mass M and the radius a for a spherical particle is:

$$M_i = a_i^3 N_A \frac{4\pi}{3\bar{v}} \qquad (2.34)$$

where N_A is Avogadro's number, $\bar{v}$ the partial specific volume (mL/g). After substitution of M_i Equation 2.33 can be rearranged into the following form:

$$I_i = K'a_i^3c_i = K''a_i^6N_i \qquad (2.35)$$

which relates the scattering intensity to the mass concentration or the number concentration of the ith particle. To determine the mass fraction, the intensity fraction $I_i/\sum I_i$ is divided by $R_{h,i}^3$, and to determine the number fraction, by

$R_{h,i}^6$. It is worth mentioning that different types of distribution are measured by different analytical methods and thus the type of distribution (intensity %, mass % or number %) must be taken into account when the results from different analytical techniques are compared. The mass % and number % distributions are shifted toward lower hydrodynamic radii compared to the intensity distributions. The results from the batch DLS are often expressed using a bar diagram. The number of bars and their heights can serve as an estimation of the size distribution, but they are not the true distribution—for instance, three bars do not mean a mixture of three species of distinct size.

2.3   LIGHT SCATTERING INSTRUMENTATION

A classical light scattering instrument is called *goniometer*. In the goniometer the sample is placed in a glass or quartz cell and the detector photodiode is mounted on a bearing allowing measurements of light scattering intensity at various angles. That means each scattering angle is measured separately in a sequence. Modern instruments mostly allow performing both static and dynamic light scattering experiments. The most serious disadvantage of the goniometer is the relatively long time required for a single measurement of angular intensity dependence. For this reason, goniometers cannot be used for online measurements in combination with analytical separation techniques.

Another approach offers instruments that allow simultaneous measurements of the scattered light intensity at multiple angles. Simultaneous measurement can cover the whole range of angles at once, which is a principal difference from the goniometer-based technique. Such instruments are usually called *multi-angle light scattering (MALS) detectors* or *photometers* (see scheme in Figure 2.15) and they have achieved great popularity due to their ability to perform rapid batch measurements of unfractionated samples, but more frequently they are used as online detectors for size exclusion chromatography or other separation techniques. Unlike the goniometer, where a single photodetector moves around the cell, the cell of a MALS photometer is surrounded by an array of photodiodes. Each photodiode is placed at a fixed angle, but depending on the solvent refractive index the observed angles change as shown in Figure 2.11. Alternatively, a set of optical fibers can be used to collect the scattered light and transfer it for the detection and processing. MALS photometers can measure static and dynamic light scattering simultaneously. To perform the dynamic light scattering experiment the light scattered in the flow cell is collected by an optical fiber, detected by an avalanche photodiode and analyzed by a digital correlator that measures the autocorrelation function of the intensity signal carried by the optical fiber.

A Brookhaven Instruments BI-200SM system allowing studies of both static and dynamic light scattering is an example of goniometer. A Wyatt Technology Corporation DAWN® HELEOS™ is an example of a MALS photometer with currently the highest available number (18) of photodetectors. Wyatt Technology Corporation is also the manufacturer of MALS photometers operating at eight or

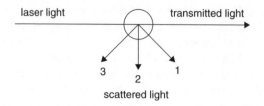

scattered light

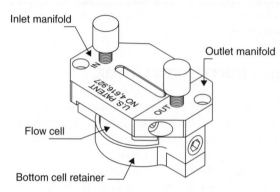

Inlet manifold

Outlet manifold

Flow cell

Bottom cell retainer

Figure 2.15   Top: Scheme of a MALS photometer. Flow cell is surrounded by photodetectors (1, 2, 3) placed at various angles. Bottom: The flow cell assembly consists of a glass cylinder with a bored channel assembled in a channel holder allowing continuous inlet and outlet flow. *Source:* Courtesy of Wyatt Technology Corporation.

three angles, namely HELEOS™ 8 and miniDAWN™ TREOS. The light scattering instrumental portfolio of this company is completed by various dynamic light scattering instruments capable of measurement in chromatography or batch mode. Brookhaven Instruments manufactures a fiber-optic seven-angle MALS photometer, BI-MwA. A two-angle light scattering photometer PD2020 with scattering angles of 15° and 90° is available from Precision Detectors (acquired by Varian). A serious disadvantage of the two-angle system is that the extrapolation using solely two angles is highly uncertain.

Single-angle photometers operating at 90° or a very low angle close to zero (e.g., 7°) are other types of light scattering instruments. The former instruments are limited to the characterization of small polymers such as proteins with negligible angular variation of the scattered light intensity. The latter instruments, often called *low-angle laser light scattering (LALLS)*, do not require extrapolation of scattered light intensity to zero angle, because the particle scattering function at a very low angle approaches unity. The LALLS approach completely ignores the particle scattering function, which can be seen as an advantage since the reduction of measurement to a single angle simplifies the processing of data compared to multi-angle detectors. However, very high sensitivity to dust particles present in solvent or shedding from SEC columns is a serious limitation of these instruments, which typically suffer from a significantly lower signal-to-noise ratio compared to the MALS photometers. In addition, the size information is completely missing, because of ignoring the angular variation of scattered light intensity, which represents a serious obstacle in branching studies.

2.4 SPECIFIC REFRACTIVE INDEX INCREMENT

The specific refractive index increment is an important parameter necessary for processing the light scattering data. The *dn/dc* appears in the optical constant K^*, and it is also needed to calculate the absolute concentration of polymer molecules in online experiments using MALS or viscometric detectors. The specific refractive index increment is also associated with the sensitivity of the light scattering measurement, because at a given molar mass and concentration the intensity of scattered light increases with *dn/dc* squared. That means the same polymer will scatter light with different intensities at different solvents. In some cases it may be necessary to change solvent to get *dn/dc* high enough to be able to perform light scattering measurements. According to my experience, *dn/dc* as low as 0.05 mL/g allows reliable measurement of polymers with molar masses down to about 1,000 g/mol. However, measurements at very low molar masses and low *dn/dc* require solutions of sufficiently high concentration. The optical constant K^* is proportional to the square of *dn/dc* and thus the error of $N\%$ in the *dn/dc* results in about $2N\%$ error of molar mass. Consequently, accurate *dn/dc* is essential for accurate determination of molar mass. This is especially valid in the case of batch measurements. In online experiments, where the *dn/dc* is used also for the determination of concentration, the $N\%$ error in *dn/dc* results in the same percentage error in molar mass, because of partial compensation (see Equations 2.1 and 2.37).

The *dn/dc* is a constant for a given polymer in a given solvent and depends on the wavelength and temperature. It decreases with increasing wavelength and this dependence becomes less pronounced toward higher wavelengths. The differences of *dn/dc* between most recently employed wavelengths of 633 nm, 658 nm, and 690 nm are negligible. The *dn/dc* can be defined as the slope of the dependence of refractive index of polymer solution on its concentration. The determination of *dn/dc* is accurate to about ± 0.001 mL/g when properly performed. Experimentally it can be measured with a differential refractometer. The measurement is relatively easy; nevertheless, several requirements must be fulfilled in order to obtain correct results. The most important requirement is purity of the polymer sample that is used for the measurement. The sample must be free of solvents, residual monomers, moisture, or other impurities. The percentage of impurities results in the same percentage error in *dn/dc*, and consequently the same error in molar mass from online experiment, and in double error in molar mass determined using the batch technique. If the purity of the sample is questionable, the sample should be purified by a suitable purification process such as drying or precipitation. Another requirement is that the solvent used for the preparation of polymer solutions must be identical with the solvent that is injected into the reference cell of the differential refractometer. That means the content of the impurities must be the same in the polymer solutions and the solvent used for their preparation.

Note: Absolute purity of the solvents used for the *dn/dc* determination is not an imperative and regularly a purity >99% is sufficient. The really crucial requirement is solvent identity and thus the solvent used for the preparation of polymer solutions should be from the same bottle as the solvent in the reference cell.

Great caution must be taken in the case of solvents that are chemically unstable and/or highly hygroscopic. Tetrahydrofuran (THF), which is widely used in organic SEC due to its relatively low toxicity, low viscosity, and ability to dissolve many synthetic polymers, is also a typical example of solvent vulnerable to chemical changes due to exposure to light even if it is stabilized. In addition, it is highly hygroscopic. However, the determination of *dn/dc* in THF is quite feasible using the following procedure:

1. Use volumetric flasks of the same type. However, there is no need for colored glass (i.e., regular clear laboratory glass is sufficient).

2. Take the needed volume of THF into a large flask (e.g., 250 mL). Do not degas the solvent.

3. Prepare stock solution in a 50- or 100-mL flask; use analytical balance with the precision of 0.1 mg. Fill the flask carefully to the mark. *Note:* The error in concentration of the stock solution will result in the same error of *dn/dc*, so pay close attention to this step.

4. Immediately after sample dissolution use serial dilution to prepare at least four more samples with concentration covering about one order of magnitude. Weigh the stock solution and THF rather than using the volumetric method. However, it is possible to use, for example, a 10-mL syringe to facilitate dosing the stock. The example procedure is as follows: add 1 mL of stock into a 10-mL volumetric flask, take a weight using analytical balance, fill the flask to the mark with THF, and take the total weight. The diluted concentration equals the concentration of stock solution times the weight of stock solution divided by the total weight. *Note:* Although the time between the sample preparation and the measurement should be as short as possible, some polymer samples may need several hours or even days to dissolve properly. If the solution contains insoluble impurities, filter it with about 1 μm filter (filtration can be easily done with a 10-mL syringe and a syringe filter). Be aware that a significant amount of insolubles affects the accuracy of the measurement.

5. Keep all solutions and THF in the same place with the same light conditions. Seal all the flasks properly.

6. Inject the solutions into the cell of RI detector starting and ending with pure THF. The injection can be made with a syringe pump (e.g., from Razel Scientific, the same arrangement as that used for the microbatch MALS measurements). Use different syringes for different solutions. It is possible to use disposable plastic syringes, but without a rubber tip. Since injections with the syringe pump require 5 to 10 min for each solution, it

is possible to make injections and dilutions simultaneously. Alternatively, THF and particular solutions can be injected using an HPLC pump and injector with a sample loop large enough (≈ 1 mL) to create flat plateaus for each concentration.

The determination should be repeated at least twice to assure accuracy of the result. It is important to emphasize that once the reliable *dn/dc* is obtained, its measurement need not be repeated. An example of the measurement in THF is depicted in Figure 2.16. The decent appearance of the plot proves the feasibility of the measurement in THF. *Note:* The determination of *dn/dc* in water and aqueous salt solutions is typically easier than measurements in THF.

An alternative determination of *dn/dc* can be performed in an online mode when the sample is injected into the SEC columns and the *dn/dc* is calculated

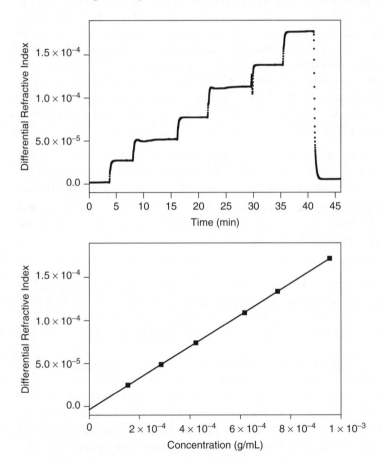

Figure 2.16 Determination of *dn/dc* of polystyrene in THF at 633 nm and 25°C. Plots of differential refractive index versus time (top) and differential refractive index versus concentration (bottom): *dn/dc* = 0.183 ± 0.001 mL/g; RI detector calibrated by means of NaCl aqueous solutions using *dn/dc* = 0.174 mL/g.

from the injected mass and the response of an RI detector. In this case, the determination of dn/dc and sample characterization with respect to the molar mass and size distribution can be done simultaneously within one analysis. However, it can be recommended to focus first on the determination of dn/dc. Besides the requirement of sample purity the sample must elute from SEC columns with 100% mass recovery. Although this requirement is fulfilled for many polymers, adsorption of a part of sample in the columns may be a source of significant error. Therefore, multiple injections shall provide identical results and deviations larger than ± 0.002 mL/g may indicate adsorption issue. In the online mode, the accuracy of the dn/dc measurement is also affected by the accuracy of the injection device, because the injected mass is calculated from the sample concentration and the injected volume. It can be noted that some manual injectors add about 10 μL volume to the sample loop volume. Since typical injection volume in SEC is 100 μL, 10% erroneous dn/dc are determined using such injectors. Therefore, the injection of samples with an autosampler should be preferred. The accuracy of the injected volume by an autosampler can be easily checked by mass difference after multiple injections of a specified volume (e.g., 5×100 μL injection of pure water in several-minute time intervals). The dn/dc from SEC data can be calculated using an appropriate template of light scattering software (e.g., Wyatt Technology Corporation ASTRA® V offers such calculation), and it can be also rapidly determined by the following procedure:

1. Inject known volume of exactly known concentration into SEC columns (e.g., 100 μL of 0.2% w/v solution, which yields 0.2 mg total injected mass). Enter apparent dn/dc of 0.1 mL/g into the light scattering software.

2. Compare injected mass (calculated from the sample concentration and injected volume, i.e., 0.2 mg in our example) with the mass calculated using the apparent dn/dc of 0.1 mL/g.

3. Calculate sample $dn/dc = 0.1 \times$ (calculated mass)/(injected mass).

Although the offline method of injecting solutions directly into the cell of an RI detector may be considered more accurate, it must be stressed that both procedures yield identical results. This fact is illustrated in Table 2.1, which lists dn/dc of three polysiloxanes of different composition determined by online and offline methods.

The determination of dn/dc in multicomponent solvents (solvent blends or salt solutions) requires dialysis of solution against the solvent. The dialysis results in the osmotic equilibrium and thus even distribution of low-molar-mass components between the solution and solvent. The dialysis may be of particular importance in the case of polyelectrolytes that are mostly measured in aqueous salt solutions to suppress the electrostatic repulsive forces. The aqueous salt solutions are mixed solvents and thus the dialysis becomes a necessary step in the accurate determination of dn/dc. The need for dialysis is given by the selective sorption of one of the components of a multicomponent solvent. A dilute polymer solution is in fact a two-phase system, one phase being the polymer

Table 2.1   Comparison of *dn/dc* of Polysiloxanes in THF Determined in SEC Mode
Assuming 100% Mass Recovery and Using Optilab® RI Detector in Offline Mode

Sample	*dn/dc* (mL/g)	
	100% Mass Recovery	Offline Mode
1	0.050	0.0495 ± 0.0006
	0.049	
2	0.153	0.1533 ± 0.0004
	0.154	
3	0.061	0.0616 ± 0.0005
	0.062	

coils (i.e., the polymer domains highly swollen with the solvent), and the second phase being the bulk solvent occupying the space among polymer coils. In the case of two-component (multicomponent) solvents, the ratio of the two solvents is generally different in each phase, because of different thermodynamic quality of the particular components with regard to a polymer. The solvent that is thermodynamically better for a given polymer is preferentially solvated in the polymer domains and thus the solvent phase becomes enriched with the thermodynamically poorer solvent, which means the polymer is in fact dissolved in a different solvent than that used for sample preparation. Consequently, the differential refractive index measurement of the solution against the mixed solvent of the original composition yields incorrect *dn/dc*.

The compositional difference between the solvent in the polymer domains and the solvent in the surrounding space increases as the polymer concentration increases, because more thermodynamically better solvent is sorbed by polymer coils. To reestablish the original composition of the solvent outside the polymer domains, the polymer solution and the mixed solvent must be brought into contact through a semipermeable membrane. The necessary condition is that the volume of solution is finite while the volume of solvent is infinite. The driving force of the solvent transport is the concentration gradient. When the equilibrium is reached, the solvent in the polymer coils is still enriched with the thermodynamically better component, but the composition of the bulk solvent among the polymer molecules is identical with that used for sample preparation. Since the volume of pure solvent used for the dialysis is infinite, or in practice significantly larger than the volume of solution, the redistribution of molecules does not affect the solvent composition.

Dialysis can be performed using a dialysis cassette (e.g., Pierce) or seamless dialysis tubing (e.g., Sigma), which allow dialysis of polymer solution in mixed solvent against significantly larger volume of pure mixed solvent in a suitable vessel. Dialysis eliminates the change of the bulk solvent composition caused by the preferential sorption. In the absence of selective sorption, the dialysis is not necessary because the overall composition of the solvent in solution is not changed. Dialysis is also not needed in SEC, because the equilibrium is achieved during the elution of sample through SEC columns. That means the dn/dc determined in the online mode is equivalent to the dn/dc value measured in the offline mode with the dialysis step.

Note that the preferential sorption affects also the light scattering intensity, because the polymer–solvent complex with selectively sorbed solvent molecules has different dn/dc than the polymer itself. In mixed solvents, the intensity of scattered light is given by the dn/dc of the complex polymer–solvent, whereas the intensity corresponding to the dn/dc of the polymer alone is needed in order to get true molar mass. That means the light scattering measurement of a polymer in a multicomponent solvent yields an apparent molar mass instead of the true molar mass. The ratio of the correct molar mass (determined by the measurement in a single component solvent) to the apparent molar mass (determined in the mixed solvent) can provide a certain measure of the preferential sorption.

For polymer blends and copolymers, the average dn/dc can be calculated from the dn/dc values of particular homopolymers and the composition:

$$dn/dc = (dn/dc)_A w_A + (dn/dc)_B w_B \qquad (2.36)$$

where w is the weight fraction of homopolymers in the blend or monomers in the copolymer.

In the case of heterogeneous copolymers, the dn/dc according to Equation 2.36 is correct only for polymer molecules having the average chemical composition. Other molecules of the chemical composition different from the bulk composition have dn/dc corresponding to their actual chemical composition. Note that the solution of a chemically heterogeneous copolymer scatters light even if the average dn/dc is zero, because due to the polydispersity of chemical composition most of the molecules have non-zero dn/dc. In contrast to homopolymers, the intensity of light scattered by the solution of a chemically heterogeneous copolymer depends not only on the molar mass, but also on the distribution of chemical composition. That means the molar mass of heterogeneous copolymer determined by the light scattering measurement is generally incorrect, and only an apparent value is obtained. On the other hand, the measurements of molar mass in several solvents of different refractive index can provide correct molar mass and two parameters characteristic of the chemical heterogeneity.[11]

Selected dn/dc values for various polymers are listed in Table 2.2. In general, the dn/dc depends on both the polymer composition and solvent. It increases with decreasing refractive index of solvent. If reliable dn/dc cannot be found in the literature or determined experimentally, it can be estimated from the value in one solvent and difference of the refractive indices of given solvent and the solvent for which the dn/dc is to be estimated. Let us illustrate the procedure for polystyrene and tetrahydrofuran and toluene: dn/dc of polystyrene in THF at 633 nm is 0.185 mL/g, $n_{THF} = 1.401$, $n_{toluene} = 1.488$, and estimated dn/dc of polystyrene in toluene $= 0.185 - (1.488 - 1.401) = 0.098$ mL/g is close to the real value of 0.105 mL/g. A review of dn/dc for different polymers and solvents can be found in references 12 and 13.

Table 2.2   Selected dn/dc Values (Room Temperature, 633–690 nm)

Polymer	Solvent	dn/dc (mL/g)
Polystyrene	THF	0.185
Polystyrene	Toluene	0.105
Polystyrene	TCB$^{135°C}$	0.047
Poly(methyl acrylate)	THF	0.068
Poly(methyl methacrylate)	THF	0.084
Poly(butyl acrylate)	THF	0.064
Poly(butyl methacrylate)	THF	0.076
Poly(isobutyl methacrylate)	THF	0.075
Poly(cyclohexyl acrylate)	THF	0.095
Poly(benzyl acrylate)	THF	0.138
Poly(benzyl methacrylate)	THF	0.144
Poly(lauryl methacrylate)	THF	0.079
Poly(methoxyethyl methacrylate)	THF	0.077
Polyisoprene	THF	0.127
Polybutadiene	THF	0.130
Polyisobutylene	THF	0.112
Polyvinylacetate	THF	0.059
Polyethylene	TCB$^{135°C}$	−0.104
Poly(1-hexene)	THF	0.076
Polycarbonate	THF	0.186
Bisphenol A epoxy resin	THF	0.183 ($M_w > 9000$ g/mol)
		0.178 ($M_w \approx 3000$ g/mol)
Phenol-formaldehyde novolac	THF	0.220
Poly(phenyl acetylene)	THF	0.286
Poly(DL-lactic acid)	THF	0.049
Polybutandiol	THF	0.069
Poly(ethylene glycol)	Water	0.135
Bovine serum albumin (BSA)	Water, aqueous buffers	0.185
Dextran	Water, aqueous buffers	0.145
Pullulan	Water, aqueous buffers	0.145
Hyaluronic acid sodium salt	Water, aqueous buffers	0.155

In the case of lack of literature and experimental data, the dn/dc can be estimated from the value for a polymer of similar chemical composition and the difference in chemical composition between the reference and polymer under investigation. General rules are as follows: The aromatic segments in the polymer chain increase dn/dc while long aliphatic chains have the opposite effect, double bonds slightly increase the dn/dc, but the effect is moderate compared to aromatic rings. Probably the maximum value of dn/dc in THF of 0.286 mL/g was reported for poly(phenyl acetylene),[14] which is due to the highly aromatic structure and double bonds in the polymer chain. On the other hand, aliphatic polymers with no double bonds and no aromatic rings have *dn/dc* several times lower. The *dn/dc* depends on the polymer chemical composition, but slight changes of chemical composition typically have minor effects on *dn/dc*. In the range of low molar masses, the *dn/dc* increases with increasing molar mass. However, from molar masses of several thousands the *dn/dc* becomes constant and the molar mass dependence of *dn/dc* can be safely neglected. The *dn/dc* of a polymer in certain solvents can be close to zero (e.g., polydimethyl siloxane in THF). Also negative values of *dn/dc* are possible (e.g., polyethylene in TCB). The negative *dn/dc* has no impact on the light scattering signal, but the RI peak appears negative and requires changing the RI detector polarity.

2.5 LIGHT SCATTERING IN BATCH AND CHROMATOGRAPHY MODE

There are two types of light scattering experiments: *batch* measurement and *chromatography* (online) measurement. In the batch mode, the MALS detector is used as a standalone instrument to characterize an unfractionated polymer sample. The advantage of the batch mode is that it can eliminate possible problems arising from chromatographic separations (e.g., interactions of sample with column packing or shearing degradation of large molecules in SEC columns). It is also possible to work in solvents that may be difficult for chromatography, such as dimethylsulfoxide or concentrated acids. Batch mode is also suitable for solvent studies, because in the SEC mode it takes a relatively long time to switch from one solvent to another. However, substantially more information is typically obtained when the MALS detector is connected to a separation system (mostly SEC, but other types of chromatography or field flow fractionation are applicable as well) and used as an online detector having the capability of determining the molar mass and RMS radius distributions. A significant advantage of the MALS detectors over goniometers is that they allow easy switching from batch to chromatography mode and vice versa, while goniometers can be used for batch measurements only.

A typical example of the batch measurement is shown in Figure 2.17. For the sake of simplicity, Figure 2.17 shows only the signal recorded by a photodiode at position 90°, but signals of light scattered at other angles were recorded simultaneously as well. The sample was prepared at multiple concentrations covering about one order of magnitude. The sample preparation can be done by the serial dilutions of a stock solution or by the separate preparations of each concentration. The procedure is similar to that used for the determination of *dn/dc*, but the small difference of composition of solvent used for sample preparation and solvent injected for the determination of solvent offset does not affect the obtained results.

The sample solutions can be injected directly into the flow cell of the MALS detector or they can be measured in scintillation vials. To distinguish

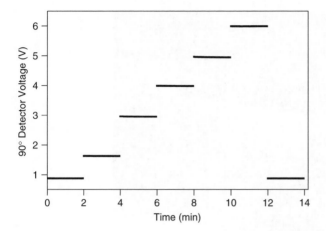

Figure 2.17   Detector voltages for 90° photodiode obtained by microbatch measurement of polystyrene NIST SRM 706. For corresponding Zimm plot, see Figure 2.21. Sample concentrations: 5.152e-5, 1.450e-4, 2.207e-4, 2.938e-4, 3.760e-4 g/mL.

the two types of measurements, the former is usually called a microbatch, the latter a batch. Both experiments have specific advantages and disadvantages. The scintillation vials are easy to use and they are well suitable for long-term time-dependent experiments. In the period between the two measurements the sample can be stored directly in the scintillation vials under requested conditions and re-measured.

Another advantage of the scintillation vials is that they are cheap and disposable and therefore especially suitable for samples that have a tendency to deposit in the flow cell (e.g., some proteins or nanoparticles). A microcuvette may become an option for samples available only in small quantities. The microbatch experiment takes full advantage of the perfect optical quality of the flow cell and thus more consistent Zimm plots are usually obtained. A syringe pump and single-use syringes with attached filters represents the most efficient way of sample injection in the microbatch mode. A typical example of a syringe pump setup is shown in Figure 2.18. An alternative method of measurement of a Zimm plot in the microbatch mode uses an HPLC pump and sample injections with a large sample loop.

Another rarely used method employs a binary HPLC pump connected to two solvent reservoirs. One reservoir is filled with a sample solution of exactly known concentration and the second reservoir is filled with a solvent. The pump gradient is programmed stepwise to generate different sample concentrations by mixing solvent and sample solution in different ratios, typically from 100% solvent to 100% sample solution. A sample and solvent filtration must be performed by an inline filter connected before the MALS detector inlet.

The batch or microbatch experiment provides three important quantities: M_w, R_z, and A_2. The molar masses and RMS radii measured in chromatography

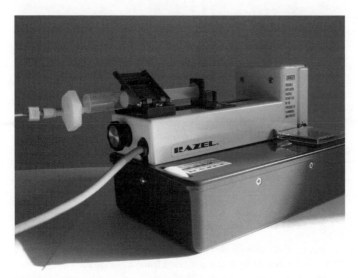

Figure 2.18 Photograph of a syringe pump with a syringe attached to a membrane filter and a luer adaptor. Connection to a detector inlet is with finger-tight PEEK fittings and PEEK tubing.

and batch mode are typically well comparable. In principle, it is also possible to measure the A_2 in chromatography mode, but the upper concentration range in the SEC-MALS experiment is limited regarding the possible overloading effect of columns, whereas at very low concentrations the effect of A_2 is negligible. Thus significantly more reliable values of A_2 are determined using a traditional batch arrangement.

It is of ultimate importance to realize that any light scattering experiment can be completely discarded by submicrometer dust particles scattering more intensely than the polymer molecules themselves. The clarification of polymer solutions is therefore of primary importance for successful batch/microbatch measurements. The clarification includes either centrifugation or filtration. Filtration using a disposable syringe filter unit is an easy and fast way of clarification with mostly satisfactory results. Several companies manufacture membrane filters (e.g., Millipore, Whatman) that are available in various membrane materials and pore sizes. The available pore sizes include 0.02 μm, 0.1 μm, 0.2 μm, 0.45 μm, 1 μm, and 5 μm.

The signal noise decreases with decreasing pore size of the filter used for sample clarification and thus the pore size of the filter should be as low as possible for a given sample. However, a polymer sample can contain a fraction of molecules or supermolecular structures that can be removed by filtration when a too-small filter size is used.

Special attention must be paid to the filtration of highly polydisperse polymers containing molar masses spanning several orders of magnitude. The fractions of very high molar mass can be completely removed from solution by filtration, which has significant effect on the obtained molar mass and especially on the RMS radius. Filters of 0.45 μm pore size usually work well for samples containing molecules up to molar mass of order of magnitude 10^7 g/mol. *Note:* For typical random coils in thermodynamically good solvents, the RMS radius of the molar mass of 10^7 g/mol is about 170 nm. Filters of pore size 1 μm or even 5 μm should be used for polymers that are expected to contain fractions with molar masses up to order of magnitude several tens or even hundreds of millions g/mol. The previous numbers represent only rough guidelines and for unknown samples that are suspected to contain ultra-high-molar-mass fractions test measurements using filters of various pore size should be performed.

The upper limit of molar mass for which a given filter pore size can be safely used for sample clarification depends also on the thermodynamic quality of solvent, because polymer molecules are more expanded and thus larger in thermodynamically good solvents. That means the same macromolecules can be removed when filtered in a thermodynamically good solvent and may pass using the same filter in a thermodynamically poor solvent. Sample filtration is usually performed using a disposable syringe attached to a disposable filter unit.

Clogging of the filter and increased syringe backpressure are obvious evidence of removing a part of sample. Besides removing a part of sample due to the presence of particles larger than filter size, sample adsorption to a filter membrane can occur. The change of concentration due to the adsorption on filter membrane has a strongly negative effect in the case of batch experiments, while in the online experiments the actual concentration of eluting molecules is measured with a concentration-sensitive detector. In the case of adsorption, using different filter material or conditioning the filter with several mL of sample solution before collecting the sample solution for measurement can eliminate this effect. An important recommendation for batch experiments is discarding the first few drops of the sample solution because of possible elution of dust particles from the filter itself during the early stage of filtration. Another aspect to consider is that during hand filtration with a syringe a relatively high pressure may occur, which can result in shearing degradation of large molecules during their passage through the filter membrane.

The sample clarification procedure is quite different in the case of online SEC experiments. As a matter of fact, the filtration of polymer solutions is carried out only to protect the columns from mechanical impurities that may be present in samples. The columns themselves function as very efficient filters that clarify the injected sample before it reaches the cell of the light scattering detector. However, the SEC columns are not only filters, but a potential source of dust particles as well. The dust particles eluting from SEC columns have two sources. They are residues from the manufacturing process or particles that were filtered out from the mobile phase during the previous measurements.

Especially new columns bleed submicrometer particles. The bleeding effect can diminish almost completely after several days of continuous flushing, but it usually appears again when the columns are exposed to a flow rate change. The source of particles from the column structure can be eliminated by innovative techniques yielding packing materials that exhibit very low particle shedding.

SEC columns specifically designed for use with light scattering detectors are available, for example, from Polymer Laboratories, part of Varian (PLgel LS), or Wyatt Technology Corporation (Wyatt MALS columns).

Since excessive baseline noise arising from small particles in the eluent can significantly deteriorate the quality of the light scattering data it is imperative to flush the columns properly before measurements. Usually the SEC columns must be flushed for several hours or even several days before the measurement can be performed in order to completely minimize the noise generated by particles. An important requirement is flushing the columns at a constant flow rate that is used for real measurements. Some experimentalists have a tendency to reduce the flow rate or even completely stop the pump when the SEC-MALS setup is not in use. However, it must be stressed that any change of the flow rate creates a pressure pulse that may release particles from columns and it may take again several hours to achieve a stable MALS signal. It is not widely known that when the pump is stopped the created pressure drop releases particles from the column that reach the MALS detector. If left for a long period of time the particles may adhere to the flow cell walls and may not be flushed out when the flow is switched on again. Therefore, for the sake of keeping the flow cell clean, it is highly recommended to disconnect the MALS photometer before switching off the pump, and when the pump is started again to bypass the MALS for at least several hours to flush the columns first. After several hours of flushing, the column outlet should be connected to the MALS detector inlet without stopping the pump. The use of PEEK finger-tight fittings can facilitate the connections of a column outlet and a MALS detector inlet. It may be noted that finger-tight PEEK fittings work well also with stainless-steel tubings.

In connection with the spurious effect of dust particles in chromatography eluent it is necessary to emphasize the importance of perfect cleanliness of the flow cell of a MALS detector. Bright red light spots visible in the flow cell are a clear indication of the particles trapped in the flow cell. Also increase of the solvent background voltage indicates that there is an additional source of scattered light in the flow cell. However, a slight increase of the light scattering intensity during the day can be due to the warmup of the light scattering instrument or due to the change of laboratory temperature. A rigorous test of the flow cell can be easily performed in the chromatography mode by measurement of a small polymer. It can be the same polymer as used for the normalization and both normalization and test of the flow cell cleanliness can be performed from one data file. The principle of the test is that a small polymer scatters light with equal intensities in all directions and thus, assuming proper normalization, all chromatograms recorded by particular photodetectors are identical. That means, viewing the chromatograms in a three-dimensional plot with rotation and elevation angles set to $0°$, all signals from particular photodetectors should superimpose on each other and appear as a single chromatogram. Any stray light in the flow cell causes the signals at different angles to not perfectly overlay on either the leading or trailing edge of a peak. The procedure is illustrated in Figure 2.19,

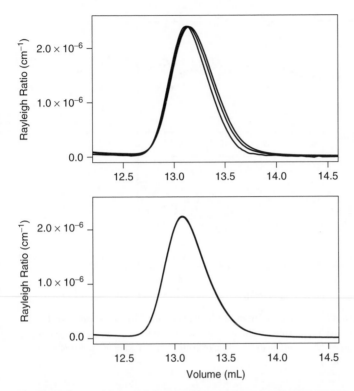

Figure 2.19 Detecting stray light in the flow cell of a MALS detector: three-dimensional plot of MALS chromatograms recorded at 45°, 90°, and 135° viewed with rotation and elevation angles set to 0°. Top: Symptoms of stray light in the flow cell. Bottom: Perfect overlay of all three signals proves clean flow cell. Sample: narrow polystyrene standard of $M = 30,000$ g/mol.

and since it is very beneficial, every light scattering software should offer this kind of test (available in ASTRA® software of Wyatt Technology Corporation).

The negative effect of particles shedding from the SEC columns is illustrated in Figure 2.20, which compares signals corresponding to a clean filtered aqueous buffer injected into the flow cell with a syringe filter and the same buffer eluting from an SEC column. The buffer was pre-filtered with a 0.2-μm filter and online filtered using a 0.1-μm filter, which means the noise can be attributed mostly to the particles generated from the columns. Note that the particles eluting from the SEC column not only increase the signal noise but also markedly increase the absolute signal offset.

Pre-filtration of SEC solvents and connection of an inline filter between the HPLC pump and injector may significantly reduce the level of particles in the mobile phase. Pre-filtration is especially important in the case of aqueous solvents containing inorganic salts, because the salts are typically a significant source of dust particles. In addition, highly polar water is a great absorber of dust. On the other hand, the pre-filtration of organic solvents such as THF or

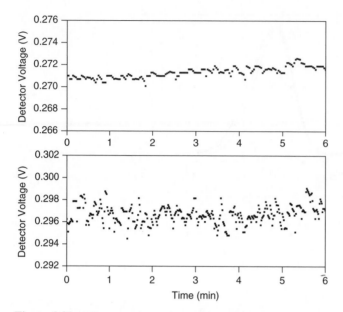

Figure 2.20   Effect of particles bleeding from SEC column on the MALS detector signal.
MALS 90° signal corresponding to solvent filtered with 0.02 μm filter (top) and to solvent eluting
from SEC column (bottom). The continuous bleeding of particles from column increases not only
the noise level, but also the baseline voltage.

toluene is not necessary with the exception of organic solvents containing salts
(e.g., dimethylformamide with LiBr). Generally, from the viewpoint of dust it is
easier to work with organic solvents than with water and aqueous salt solutions.
It is also necessary to stress that the electronic noise generated by a modern light
scattering detector itself is significantly below the noise level generated by the
chromatography system.

It must be pointed out that the filtering effect of columns can also remove
a part of ultra-high-molar-mass fractions from the analyzed sample and thus
significantly affect the obtained results. Crosschecks with batch measurement or
the use of field flow fractionation can reveal a column filtration effect and/or
shearing degradation. Another recommendation is that samples that may contain
fractions with very high molar mass should be characterized with SEC columns
packed with larger particles (10 μm or 20 μm instead of 5 μm) in order to
minimize shearing degradation.

2.6 PARAMETERS AFFECTING ACCURACY OF MOLAR MASS DETERMINED BY LIGHT SCATTERING

Equations 2.1, 2.27, and 2.28 relate the molar mass and the intensity of light
scattered by a polymer solution. They were rigorously derived and from this
viewpoint the light scattering is undoubtedly an absolute method of molar mass

determination, absolute in the sense that the molar mass is directly related to a measurable physical quantity. However, the method is as absolute as absolute constants and other parameters are used for the calculation. The following three quantities influence the absolute accuracy of the light scattering experiment: (1) the instrumental constant f, (2) the specific refractive index increment dn/dc, and (3) the concentration of polymer molecules in solution.

The instrumental constant f is related to the method employed for the calibration of a light scattering instrument. As already mentioned, light scattering instruments are often calibrated with toluene and thus the accuracy of the Rayleigh ratio of toluene is in direct relation to the accuracy of the molar mass. The Rayleigh ratio of 1.404×10^{-5} cm^{-1} at wavelength 633 nm and 25°C is used by Wyatt Technology Corporation software ASTRA. The origin of this number is reference 15. According my experience, this value provides reliable molar masses well comparable with other methods of molar mass determination. A very similar value of toluene Rayleigh ratio of 1.39×10^{-5} cm^{-1} (633 nm, 20°C) is given in Huglin's monograph.[11] The Rayleigh ratios of toluene for other recently used wavelengths 690 nm and 658 nm are 9.780×10^{-6} cm^{-1} and 1.193×10^{-5} cm^{-1}, respectively.

For batch experiments, polymer solutions can be prepared at highly accurate concentrations using analytical balance. However, the impurities present in polymer such as residual monomers, solvents, or moisture result in incorrect concentrations. Another crucial requirement for the preparation of sample solutions is 100% solubility of a polymer sample. It must be stressed that polymers generally dissolve slowly and it usually takes several hours for polymer to dissolve completely. Fractions with very high molar mass may require even several days. Solutions containing a measurable level of incompletely dissolved supermolecular structures even after a week in solution are not exceptional. The dissolution can be promoted by elevated temperature, but intensive shaking or sonication cannot be recommended due to possible degradation of large polymer chains by shearing forces. Even in the case of thermodynamically good solvents industrial polymers can contain insoluble fractions created by various side reactions.

In the case of online experiments the situation is quite different. The exact concentration of injected polymer need not be known, because it is calculated from the response of a concentration-sensitive detector. However, in contrast to the conventional SEC, where only relative slice areas are needed, the absolute concentrations of polymer molecules eluting at particular elution volume slices must be determined by a concentration-sensitive detector. An RI detector is the most widely used concentration detector for the SEC-MALS experiments. The signal of the RI detector in volts is converted to refractive index units by means of the RI detector calibration constant and the obtained refractive index difference is further used with dn/dc to calculate the concentration according to the following equation:

$$c_i = \frac{\alpha(V_i - V_{i,baseline})}{dn/dc} \qquad (2.37)$$

where α is the RI detector calibration constant (in RI units per volt), and V_i and $V_{i,baseline}$ are the RI detector sample and baseline voltages, respectively. The calibration constant α is determined by injections of solutions of exactly known concentration of a compound of accurately known dn/dc. Aqueous sodium chloride can be recommended for the RI detector calibration using the procedure exactly equivalent to that used for the determination of dn/dc. The online RI detector calibration is carried out by injection of an exactly known mass of a polymer of known dn/dc into SEC columns. The RI detector calibration constant is calculated from the RI peak area, injected mass, and polymer dn/dc. This procedure is equivalent to the online determination of dn/dc. Polystyrene is the most suitable polymer for THF due to its perfect solubility and no tendency to interactions with styrene-divinylbenzene column packing. In addition, it is available in the form of narrow standards that elute as sharp peaks that are easy to integrate. Pullulan or dextran can be used for online RI detector calibration in aqueous solvents. The accuracy of the obtained RI calibration constant is directly related to the accuracy of dn/dc of the compound used for the calibration.

Most of the RI detectors operate at a wavelength different from that of MALS photometer. Widely used Waters RI detectors 2414 operating at a wavelength of 880 nm or the previous model 410 operating at 930 nm can be given as examples. However, most of the literature dn/dc values are listed for wavelength 633 nm or lower. The following equation shows the wavelength dependence of aqueous sodium chloride solution calculated using the data from reference 16:

$$dn/dc = 1.673 \times 10^{-7}\lambda_0^2 - 2.237 \times 10^{-4}\lambda_0 + 0.2487 \qquad (2.38)$$

The above equation yields dn/dc of aqueous sodium chloride of 0.174 mL/g for the two commonly used wavelengths of 633 nm and 690 nm. An important fact is that the difference between the two wavelengths is negligible. A slightly different dn/dc of aqueous sodium chloride of 0.179 mL/g can be found in reference 13. A difference less than 3% between the dn/dc values of 0.174 and 0.179 mL/g indicates possible uncertainty associated with the calibration of an RI detector. Since the wavelength dependence of dn/dc diminishes with the increasing wavelength, the RI detectors working at high wavelengths should be preferred over those operating with polychromatic light. Unfortunately, there is a lack of extensive study of the effect of the difference between the wavelength of the RI detector and the wavelength at which the dn/dc of the compound used for the RI detector calibration is valid. It can be hypothesized that the error of RI detector calibration constant α caused by the erroneous dn/dc of the compound used for the detector calibration is compensated by the same error of dn/dc of polymer under investigation.

To give a concrete example for the previous sentence: an RI detector operating at 930 nm is calibrated by polystyrene in THF using dn/dc of 0.185 mL/g valid at 633 nm; the RI detector is then used for the analysis of poly(methyl methacrylate) with dn/dc of 0.084 mL/g that is valid at 633 nm as well. The idea is that the dn/dc difference due to the wavelength difference is identical for both polymers and thus the effect on concentration is mutually canceled. Although this concept sounds speculative, results of a short study published in reference 17 did not indicate significant error when a Waters 410 RI detector operating at a wavelength of 930 nm was calibrated using polystyrene dn/dc valid for 633 nm, and then used for SEC-MALS characterization of various polymers. Anyway, the use of an RI detector with a wavelength identical with that of a MALS detector yields more trustworthy results. A Wyatt Technology Corporation Optilab® rEX is an example of an RI detector whose operating wavelength matches the wavelength of the MALS detectors manufactured by the company.

Another possible source of inaccuracy is the difference between the RI detector temperature and the reference temperature at which the dn/dc of a polymer under investigation or a compound used for the RI detector calibration was determined. Many RI detectors have heating, but no cooling ability. In order to stabilize the detector flow cell temperature it is necessary to set the operating temperature of the RI detector $5-10°C$ above the room temperature. That means the RI detector is typically working at $35°C$, while the literature dn/dc was determined at $25°C$. The temperature coefficient of the specific refractive index increment, $d(dn/dc)/dt$ is 1×10^{-4} to 5×10^{-4} mL/g°C. That means that a change of temperature by $1°C$ results in dn/dc change of $0.0001-0.0005$ mL/g. Results published in reference 17 did not indicate significant errors due to the temperature of RI detector. The temperature dependence of dn/dc affects not only the concentration determined by the RI detector, but also the molar mass calculated from the intensity of scattered light, because the dn/dc is in the optical constant K^*. In a less-favorable case of the temperature coefficient the change of dn/dc due to a temperature difference of $10°C$ may be around 0.005 mL/g. For dn/dc of 0.185 mL/g (e.g., polystyrene in THF or BSA in aqueous salt solution), the error of dn/dc due to a $10°$ temperature difference is likely less than 3%.

A UV detector is another concentration detector that can be used in combination with the MALS detector in the online mode. To determine the concentration by the UV detector the following parameters must be known: (1) the UV response factor in absorbance units per volt (typically 1 AU/volt), (2) the cell length (typically 1 cm), and (3) the extinction coefficient in mLg^{-1} cm^{-1}. In the ideal case, the concentrations determined by RI and UV detectors can be crosschecked. If the extinction coefficient of the analyzed polymer is known, it is possible to calculate the mass from the UV peak area and then to use this mass to calculate the unknown dn/dc of the polymer from the RI peak area (the reverse procedure, when the dn/dc is known and the extinction coefficient unknown, is the same). The obtained dn/dc can be then used for the determination of molar mass from the light scattering intensity. The advantage of this procedure is that it does not require 100% sample purity or 100% mass recovery from the SEC columns. However, most of the synthetic and natural polymers have no UV response, styrene-containing polymers and proteins being important exceptions.

Possible uncertainty of results associated with the dn/dc and RI detector calibration constant is depicted in Table 2.3. In this experiment, a Waters 2414

Table 2.3 M_w of Polystyrene NIST SRM 706 Measured by SEC-MALS with UV and RI Detection and by Batch Measurement

Determination of polystyrene UV extinction coefficient @ 254 nm and determination of RI calibration constant using 90,000 and 170,000 polystyrene narrow standards		
1 No. of injections	Extinction coefficient $(mLg^{-1}\ cm^{-1})$	RI calibration constant relative uncertainty(%)
8	1595 ± 14	0.7
Analysis of broad polystyrene NIST 706 by SEC-MALS (10 injections)		
M_w $(10^3$ g/mol)		Mass recovery (%)
2 RI detection UV detection		RI detection UV detection
283.4 ± 0.2 276.0 ± 0.3 $(287.5 \pm 0.1)^*$		100 102.6 ± 0.1
3 Batch mode (average from three measurements) $M_w = (288.1 \pm 2.2) \times 10^3$ g/mol		
4 Nominal $M_w = (285 \pm 23) \times 10^3$ g/mol		

Notes:

Row 1: Waters 2414 RI detector working at 880 nm was calibrated online by PS standards using *dn/dc* of 0.185 mL/g; Waters 2487 UV detector was used online to determine extinction coefficient of polystyrene.

Row 2: The obtained extinction coefficient and RI calibration constant were used to determine M_w by SEC-MALS using a three-angle miniDAWN™ photometer. *The result obtained by triplicate measurements using an SEC-MALS-RI setup consisting of an 18-angle DAWN® EOS photometer and Optilab® DSP RI detector with wavelength matching that of MALS; RI calibration offline by means of NaCl.

Row 3: Batch mode using an 18-angle DAWN® EOS photometer; measurements using serial dilutions of stock solution; each stock solution was made up independently.

Row 4: According to NIST certificate.

RI detector with operating wavelength of 880 nm was carefully calibrated in SEC mode using eight injections of narrow polystyrene standards and *dn/dc* of 0.185 mL/g valid for 633 nm and THF. The instrumental setup also included a Waters 2487 UV detector operating at 254 nm, which was used to determine the extinction coefficient of polystyrene simultaneously with the calibration of the RI detector. The same UV-MALS-RI setup was then used for the characterization of broad polystyrene NIST SRM 706 that was processed using both UV and RI detector signals. The results show about 2.6% difference between the M_w and the calculated mass determined from the signals of UV and RI detectors. The difference may reflect possible errors given by the use of an RI detector of different wavelength than that for which the *dn/dc* is valid. However, the difference may be also due to accuracy of volume delay between the UV and MALS and MALS and RI detectors, and interdetector peak broadening.

An important finding is that the M_w obtained by the online experiment agrees with that determined in the batch mode. This fact verifies both the calibration constant of the RI detector and the UV extinction coefficient.

Another important finding is that the obtained M_w agrees with that determined with another SEC-MALS setup consisting of different MALS and RI detectors. All the obtained M_w values agree well with the reference value no matter whether in batch or chromatography mode or what concentration detector is used. National Institute of Standards and Technology (NIST) Standard Reference Material (SRM) 706a polystyrene is still available and thus a few notes concerning its characterization may be worthwhile. The polystyrene sample (originally labeled as SRM 706) was prepared by thermal polymerization of styrene at 140°C to 37% conversion. The originally reported values (in 1967) were as follows:

Measured by light scattering	$M_w = 257,800 \pm 930$ g/mol
Measured by sedimentation equilibrium	$M_w = 288,100 \pm 9600$ g/mol
Osmotic pressure measurements	$M_n = 136,500$ g/mol
Intrinsic viscosity[i] (benzene at 25°C)	$[\eta] = 93.7 \pm 0.19$ mL/g
(cyclohexane at 35°C)	$[\eta] = 39.5 \pm 0.10$ mL/g

The light scattering and sedimentation measurements were carried out in cyclohexane at 35°C and the data were processed using 0.1705 mL/g for the refractive index increment and 0.930 mL/g for the partial specific volume. Ratios of molar masses $M_z : M_w : M_n = 2.9 : 2.1 : 1$ were reported based on a viscometric analysis of 41 fractions. The recertification was carried out by light scattering in 1998 and yielded the following results:

$$M_w = 285,000 \pm 23,000 \text{ g/mol}$$

The recertification measurement was carried out in toluene at 25°C. The specific refractive index was determined as 0.1089 ± 0.0009 mL/g. The dn/dc was determined using a Chromatix KMX-16 differential refractometer calibrated with sodium chloride solutions. The light scattering measurements of the solutions were made on a Brookhaven Instrument Model BI-200 goniometer, which was calibrated by benzene. The recertification report also provides $A_2 = 0.000411 \pm 0.00003$ mol mL/g^2 and $R_z = 27.8 \pm 1.0$ nm. Another important finding of the NIST SRM 706 recertification is that the SEC study did not indicate any difference between SRM 706 and SRM 706a.

Light absorption and fluorescence represent other potential sources of errors in light scattering measurements. Fortunately, most of the synthetic and natural polymers do not absorb visible light or show fluorescence. However, for some specific polymers both effects must be considered. If polymer yields colored solutions, they must be checked by a spectrometer for absorption at the wavelength

---

[i]Note the significant influence of the thermodynamic quality of solvent on the intrinsic viscosity.

of the light scattering photometer. Light scattering photometers that monitor the intensity of transmitted light allow easy detection of absorption, and also the light transmitted through the flow cell can be used by the light scattering software to compensate for the absorption. To eliminate the effect of absorption the intensity of scattered light is related to the intensity of transmitted light monitored by the forward photodetector. ASTRA® V (Wyatt Technology Corporation) is an example of the light scattering software that can eliminate the effect of absorption. Fluorescence (i.e., the emission of light after irradiation by visible light) is another source of errors. If it is caused by impurities, the problem can be solved by sample purification or measurement in the online mode when the polymer under investigation is separated from the fluorescing impurities. If the fluorescence is caused by the polymer itself, the emitted light superimposes with the scattered light, which leads to the overestimation of molar mass. The intensity of the fluorescent light may be significantly higher than that of the scattered light and thus the obtained molar masses may be enormously high.

Less intense fluorescence can be eliminated by fluorescence filters, that is, monochromatic filters that are placed between the scattering solution and photodetectors. These filters transmit only the true scattered light that has the wavelength of the incident light. Concerning the absolute accuracy of light scattering, it is necessary to realize that small errors in particular parameters—concentration, calibration constant of light scattering instrument, dn/dc of the polymer under investigation or the compound used for the RI detector calibration, and RI detector calibration constant—can mutually partly or even completely compensate. On the other hand, they can work synergistically.

2.7 EXAMPLES OF LIGHT SCATTERING MEASUREMENT IN BATCH MODE

Processing the light scattering data obtained by the batch experiment is demonstrated on well-known polystyrene NIST SRM 706. Figure 2.21 shows a Zimm plot of NIST 706 polystyrene (the corresponding raw data are depicted in Figure 2.17). The data were obtained by preparation of sample in THF at five different concentrations. The solutions were prepared by dilution of the stock solution. The dilution was done on the basis of weight. The solutions were injected directly into the flow cell of a MALS photometer by means of a syringe pump using disposable syringes attached to 0.45 μm filter units. The Zimm plot allows simultaneous extrapolation of the concentration and angular dependence of the light scattering intensities to zero angle and zero concentration.

It is worth mentioning that the Zimm plot processes the three-dimensional function R_θ versus c and θ using a two-dimensional plot. The first-order polynomials fit both angular and concentration dependencies for the data in Figure 2.21, but higher-order fits may be necessary, especially for angular dependence. As for the concentration dependence, the experiments should be carried out at concentrations low enough that the first-order polynomial is sufficient to fit the data.

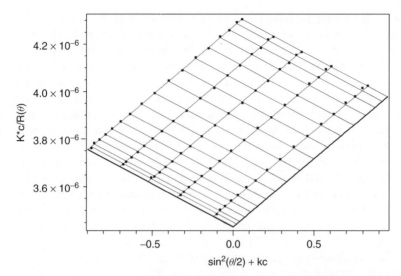

Figure 2.21   Zimm plot for NIST SRM 706 polystyrene. Each of five lines consisting of 17 data points represents angular variation acquired for a particular concentration; thick lines are concentration dependence at zero angle and angular dependence at zero concentration. Formalisms according to Zimm, THF as solvent, room temperature, vertically polarized light 690 nm, negative scale factor k, concentration: 1st-order fit, angle: 1st-order fit. $M_w = 291,600 \pm 200$ g/mol; $R_z = 27.8 \pm 0.1$ nm; $A_2 = (4.33 \pm 0.04) \times 10^{-4}$ mol mL/g^2.

The resulting extrapolated lines are the angular dependence of the scattered light intensity at zero concentration and the concentration dependence of scattered light intensity at zero angle. Both lines cross with the K^*c/R_θ axis at the intercept equal to the reciprocal weight-average molar mass $1/M_w$. The slope of the concentration dependence at zero angle yields the second virial coefficient: slope $= 2A_2$.

The angular dependence of scattered light intensity at zero concentration can be expressed using Equations 2.27 and 2.11 as:

$$\left(\frac{K^*c}{R_\theta}\right)_{c=0} = \frac{1}{M_w} + \frac{16\pi^2}{3\lambda^2}\frac{1}{M_w}R_z^2 \sin^2\left(\frac{\theta}{2}\right) \tag{2.39}$$

The slope of this relation at zero angle (m_0) equals:

$$m_0 = \frac{16\pi^2}{3\lambda^2}\frac{1}{M_w}R_z^2 \tag{2.40}$$

and the intercept of this line is equal to $1/M_w$. The z-average RMS radius is calculated from the slope of the angular variation at zero concentration:

$$R_z = \sqrt{\frac{3m_0\lambda^2 M_w}{16\pi^2}} \tag{2.41}$$

Although according to Equation 2.41 the calculation of R_z involves M_w, incorrect M_w due to incorrect dn/dc does not affect the obtained R_z, because the error in M_w generated by the error in dn/dc generates proportional error in the slope m_0 (see Figure 2.22). Using the relation between R and M for linear polystyrene ($R = 0.014 \times M^{0.585}$), the RMS radius of 22 nm corresponds to the M_w obtained by the Zimm plot, which is markedly lower than the value determined experimentally from the angular variation. This fact illustrates that for polydisperse polymers the molar mass and RMS radius cannot be directly compared, because they are of different type and different sensitivity to the polydispersity. Assuming linear topology, the value of M_z of 434,000 g/mol can be estimated from R_z and the $R-M$ relationship. This value is quite comparable to that determined by SEC-MALS or A4F-MALS measurements (see Figure 5.25). The second virial coefficient of order of magnitude 10^{-4} mol.mL/g^2 confirms THF as a thermodynamically good solvent for polystyrene. Note that the R_z determined for NIST 706 in THF equals the reference value obtained in toluene, which indicates similar thermodynamical quality of the two solvents and consequently similar expansion of the polymer chain.

Figure 2.23 depicts corresponding Debye plots for the lowest and highest concentrations. In this case the information about the second virial coefficient is missing and the M_w and R_z are calculated from the intercept and slope of the angular dependence, respectively. The M_w determined at the lowest and highest concentrations are 98.7% and 91.4% of the value obtained by means of the

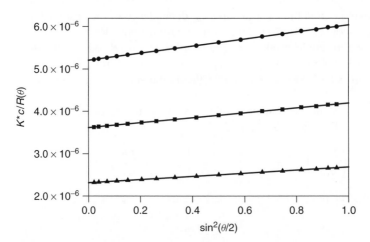

Figure 2.22 Influence of the accuracy of dn/dc on the Debye plot (Zimm formalism) of polystyrene: (■) correct $dn/dc = 0.185$ mL/g, intercept = 3.616E-6, slope = 5.795E-7, (▲) incorrect $dn/dc = 0.148$ mL/g, intercept = 2.314E-6, slope = 3.709E-7, (●) incorrect $dn/dc = 0.222$ mL/g, intercept = 5.207E-6, slope = 8.345E-7. The molar mass (reciprocal intercept) calculated by erroneous dn/dc is incorrect, while the RMS radius calculated from the slope according to Equation 2.41 is independent of dn/dc, because the ratio slope/intercept remains constant.

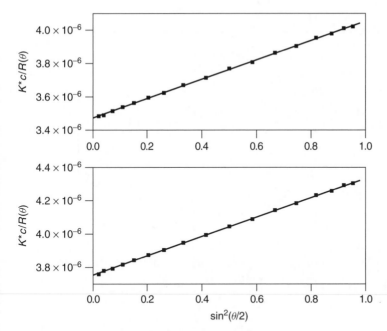

Figure 2.23 Deby plots obtained from data presented in Figure 2.21 for the lowest (top) and the highest (bottom) concentration: $c = 5.152e-5$ g/mL: $M_w = 287,900 \pm 100$ g/mol, $R_z = 27.7 \pm 0.1$ nm; $c = 3.760e-4$ g/mL: $M_w = 266,400 \pm 100$ g/mol, $R_z = 26.7 \pm 0.1$ nm.

Zimm plot. In the case of the Debye plot, the R_z is slightly concentration dependent due to the concentration dependence of M_w. The Debye plot at a single concentration has practical meaning in the case of a limited sample amount or when the sample throughput is to be increased. The error due to neglecting the concentration dependence increases with increasing concentration and thus the Debye experiments should be performed at the lowest possible concentrations.

Figure 2.24 shows the same data for NIST SRM 706 polystyrene processed using the so-called Debye and Berry formalisms (i.e., Equations 2.1 and 2.28, respectively). It can be concluded that for this polymer various formalisms yield almost identical results. However, this is not a generally valid conclusion as shown further.

The effect of processing the data with various formalisms is further demonstrated in Table 2.4, which compares results determined by particular mechanisms for several polymers of different molar mass and size. In conclusion, one can say that the results for small polymers with no or moderate angular dependency of scattered light intensity are independent of the formalism.

For larger polymers, the differences of molar masses and RMS radii determined by particular mechanisms are quite significant and can reach several tens of percent. The quantity that matters is the size and not the molar mass as evident from the results obtained for poly(methyl methacrylate) with significantly larger M_w but smaller R_z compared to hyaluronic acid.

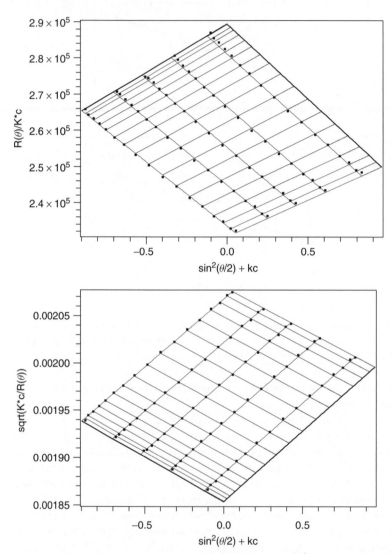

Figure 2.24 Zimm plot for NIST SRM 706 polystyrene obtained using Debye formalism (top) and Berry formalism (bottom). THF, room temperature, vertically polarized light 690 nm; negative scale factor, concentration: 1st-order fit, angle: 1st-order fit. Debye: $M_w = 289,400 \pm 400$ g/mol; $R_z = 25.6 \pm 0.2$ nm; $A_2 = (3.82 \pm 0.07) \times 10^{-4}$ mol mL/g^2. Berry: $M_w = 291,000 \pm 200$ g/mol; $R_z = 27.2 \pm 0.1$ nm; $A_2 = (4.19 \pm 0.04) \times 10^{-4}$ mol mL/g^2. (Zimm formalism, see Figure 2.21.)

Table 2.4 Characteristics of Polymers Determined in Batch Mode Using Various Light Scattering Formalisms

Sample	M_w (10^3 g/mol)			R_z (nm)			A_2 (10^{-4} mol mL/g^2)		
	Z	B	D	Z	B	D	Z	B	D
PS	2.9	2.9	2.8	—	—	—	17.8	16.0	12.3
EP (2)	8.4	8.3	8.1	—	—	—	14.5	13.2	10.4
PS (A)	337	336	336$^{(2)}$	33.8	32.9	34.4$^{(2)}$	4.1	3.9	3.6
HA	437	382	315$^{(3)}$	72.0	61	58$^{(3)}$	27.4	19.8	10.6
PMMA (Y)	612	609	609$^{(2)}$	41.6	39.9	41.4$^{(2)}$	2.6	2.5	2.2
PS	4340	3953$^{(2)}$	3237$^{(3)}$	133	122$^{(2)}$	101$^{(3)}$	2.8	2.1	1.1

PS = linear polystyrene, EP = epoxy resin based on bisphenol A, HA = hyaluronic acid sodium salt, PMMA = poly(methyl methacrylate), Z = Zimm, B = Berry, D = Debye, [2],[3] 2nd respective 3rd-order polynomial used to fit angular data; otherwise linear fit.

Choosing the proper formalisms may not be always obvious. A fitness of the formalism to the experimental data points and uncertainty calculated by the software can be used as guidance. Generally, the Zimm formalism is linear even for large molecules, but compared to other formalisms it yields larger M_w and R_z. The effect of the light scattering formalism is presented in more detail in Section 4.2.1.

A Zimm plot for sodium salt of hyaluronic acid is depicted in Figure 2.25. Compared to M_w the sample has high R_z, which in this particular case indicates extended chain conformation. However, high R_z can be also due to very high polydispersity and presence of fractions with very high molar mass. The A_2 of order of magnitude 10^{-3} mol mL/g^2 suggests intensive polymer–solvent interactions. The high A_2 accounts for significant inaccuracy of the molar mass determined by means of the Debye plot method, which gives errors in M_w of about 10% and 62% for the lowest and highest concentration, respectively.

Figure 2.26 shows an example of a Zimm plot for highly branched polystyrene with M_w substantially higher compared to NIST polystyrene. The second-order polynomial is necessary to fit the angular data points. The slope of the concentration dependence is close to zero although the A_2 of 2.7×10^{-4} mol.mL/g^2 corresponds to the M_w according to Equation 2.5, which can be attributed to the presence of highly branched molecules and the fact that A_2 decreases with increasing degree of branching. The decrease of A_2 due to branching is given by a more compact structure of branched macromolecules where polymer segments are forced to intramolecular interactions. The A_2 equal to zero allows accurate determination of molar mass using the Debye plot technique at a single concentration. In this case, even the Debye plot at the highest concentration yields M_w virtually identical with that determined by means of the Zimm plot. The experimental R_z is larger than the value of

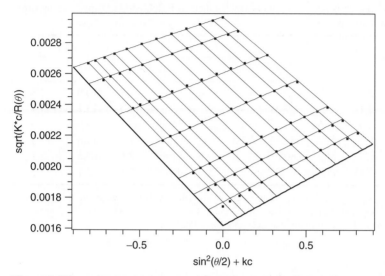

Figure 2.25 Zimm plot for hyaluronic acid sodium salt. Berry formalism, phosphate buffer, room temperature, vertically polarized light 690 nm; negative scale factor k, concentration: 1st-order fit, angle: 1st-order fit. Concentrations: 8.345e-5, 1.656e-4, 2.497e-4, 4.176e-4, 5.833e-4, 7.478e-4, 8.334e-4 g/mL. $M_w = 382,000 \pm 2000$ g/mol; $R_z = 61.1 \pm 0.4$ nm; $A_2 = (1.98 \pm 0.01) \times 10^{-3}$ mol mL/g^2.

Source: Courtesy Martina Hermannova, Contipro C, Czech Republic.

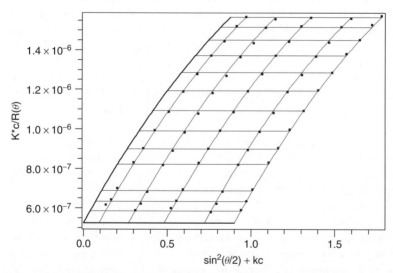

Figure 2.26 Zimm plot for randomly branched polystyrene. Zimm formalism, THF, room temperature, vertically polarized light 690 nm, positive scale factor k, concentration: zero-order fit, angle: 2nd-order fit. Concentrations: 9.866e-6, 2.830e-5, 5.054e-5, 7.555e-5, 9.448e-5 g/mL. $M_w = 1,908,000 \pm 7000$ g/mol; $R_z = 116.2 \pm 0.6$ nm; $A_2 \rightarrow 0$ mol mL/g^2.

66 nm calculated for the given M_w from the R-versus-M relationship for linear polystyrene.

Branching reduces the molecular size and the information about branching can be generally revealed from the comparison of RMS radii of linear and branched polymers of the same molar mass. However, randomly branched polymers with a high degree of branching are characterized by very broad molar mass distribution with a high-molar-mass tail. Since the high-molar-mass fractions affect the z-average more than the weight-average, the reducing effect of branching on R_z is overcompensated by the presence of fractions with very high molar mass, which accounts for the R_z larger than would correspond to the linear polymer of the same M_w.

Figure 2.27 shows another example of a Zimm plot for branched polymer. In this case the branched polymer was prepared by GTP and has a starlike topology with numerous arms. Highly compact structure corresponds to the high molar mass and relatively small size. The RMS radius of corresponding linear polymer of the same M_w is 63 nm (i.e., markedly larger that the experimental value). In this particular case of highly compact structure and absence of a high-molar-mass tail (confirmed by A4F-MALS), the value of R_z is lower than that of corresponding linear polymer of the same M_w. The data in Figures 2.26 and 2.27 show that the comparison of experimental value of R_z with the theoretical R calculated for linear polymer of the same M_w yields very uncertain and limited information

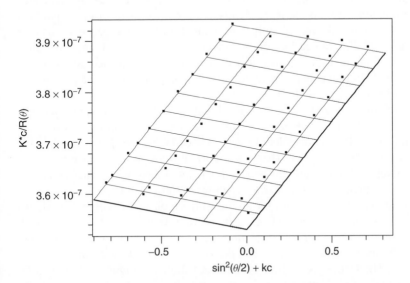

Figure 2.27 Zimm plot for star-branched poly(benzyl methacrylate) with high number of arms. Zimm formalism, THF, room temperature, vertically polarized light 690 nm, negative scale factor k, concentration: 1st-order fit, angle: 1st-order fit. Concentrations: 8.939e-6, 2.213e-5, 3.983e-5, 5.915e-5, 7.843e-5 g/mL. $M_w = 2{,}833{,}000 \pm 6000$ g/mol; $R_z = 23.5 \pm 0.4$ nm; $A_2 = (3.8 \pm 0.6) \times 10^{-5}$ mol mL/g^2.

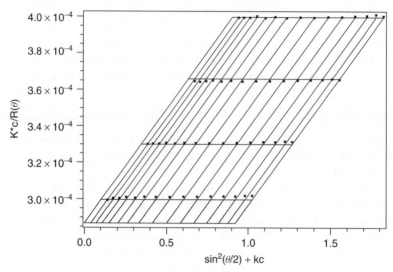

Figure 2.28 Zimm plot for bisphenol A–based epoxy resin. Zimm formalism, THF, room temperature, vertically polarized light 690 nm, positive scale factor k, concentration: 1st-order fit, angle: zero-order fit. Concentrations: 3.678e-3, 1.238e-2, 2.263e-2, 3.226e-2 g/mL. $M_w = 3,490 \pm 10$ g/mol; R_z N/A; $A_2 = (1.75 \pm 0.01) \times 10^{-3}$ mol mL/g^2.

about branching. Only if the experimental R_z is smaller than the R calculated from the relation between the RMS radius and molar mass for a linear polymer can one safely conclude that the polymer is branched. However, even in this case there is no possibility of getting information about the degree of branching unless the polymer is monodisperse. For most of the randomly branched polymers the reduction of RMS radius is partly or completely compensated by the presence of fractions with very high molar mass.

An example of a Zimm plot for an oligomer is shown in Figure 2.28. The sample shows no angular variation of the scattered light intensity due to very small molecules. High A_2 of order of magnitude of 10^{-3} corresponds to relatively low molar mass. The excellent shape of the Zimm plot presented in Figure 2.28 also disproves the sometimes-still-believed perception that light scattering is suitable only for polymers with molar mass above 10,000 g/mol.

An example of a Zimm plot for an ultra-high-molar-mass polymer is shown in Figure 2.29. Despite very high molar mass, the A_2 is at the 10^{-4} mol.mL/g^2 level, roughly corresponding to Equation 2.5. The data show high uncertainty and also significant difference between the results obtained by Zimm and Berry formalisms. The Debye formalism, which is not shown in Figure 2.29, was completely unable to fit the experimental data points. It is worth mentioning that the Zimm formalism yields markedly larger values of M_w and R_z than the Berry formalism. In the case of polymers with very high molar mass, the difference between the highest and lowest value of K^*c/R_θ is large and thus a small inaccuracy of extrapolation to zero angle significantly affects the value of M_w, which

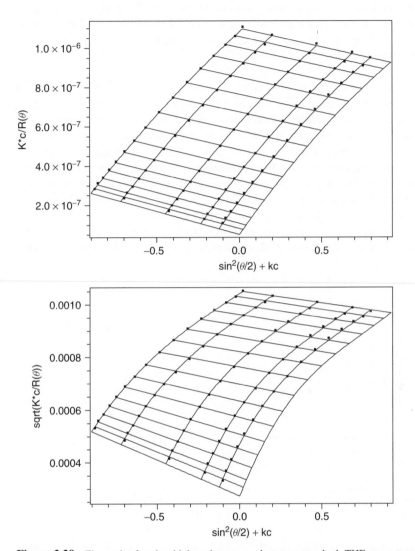

Figure 2.29 Zimm plot for ultra-high-molar-mass polystyrene standard. THF, room temperature, vertically polarized light 690 nm, negative scale factor k. Concentrations: 6.614e-5, 1.250e-4, 2.388e-4, 3.835e-4, 4.795e-4 g/mL. Zimm formalism (top): concentration: 1st-order fit, angle: 2nd-order fit, $M_w = (19.3 \pm 3.6) \times 10^6$ g/mol; $R_z = 325 \pm 33$; $A_2 = (2.20 \pm 0.15) \times 10^{-4}$ mol mL/g^2. Berry formalism (bottom): concentration: 1st-order fit, angle: 3rd-order fit. $M_w = (13.4 \pm 1.4) \times 10^6$ g/mol; $R_z = 230 \pm 14$; $A_2 = (1.40 \pm 0.13) \times 10^{-4}$ mol mL/g^2.

accounts for high M_w uncertainty. The uncertainty of R_z is also much higher compared to smaller polymers. Although very large polymer molecules have steep initial slope of angular dependence that can be measured precisely, the uncertainty and accuracy of R_z are influenced by the value of M_w, as seen from Equation 2.41.

Another application of the batch light scattering method is shown in Figure 2.30. The measurement was carried out using a heated/cooled MALS photometer allowing measurements below room temperature. The sample was a polymer creating supermolecular structures at temperatures close to $0°C$. The measurement was carried out in a scintillation vial. The aggregation of individual molecules is indicated by significant increase of scattered light intensity. The data also allow the determination of the size of arisen aggregates. Very high RMS radius of the aggregates yields highly curved angular dependency, and thus only the lower angles are used for data processing.

Light scattering can be used as a fast method of studying protein temperature induced aggregation, as demonstrated in Figure 2.31. In this example,

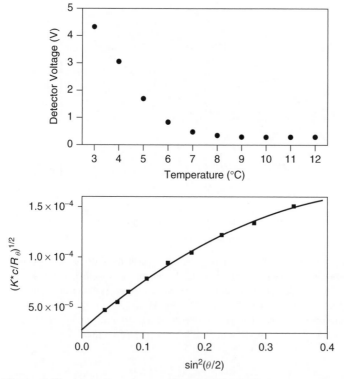

Figure 2.30 Temperature dependence of light scattering intensity of a polymer creating aggregates (top) and Debye plot at $3°C$ (bottom). Second-order fit using Berry formalism: $R_z = 385 \pm 4$ nm.

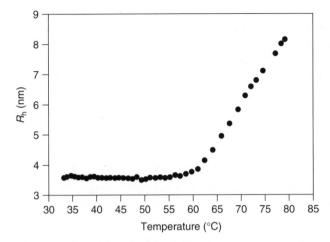

Figure 2.31 Melting study for BSA: hydrodynamic radius of BSA measured as a function of temperature. Onset temperature $= 59.0 \pm 0.1°C$.
Source: Courtesy of Roger Scherrers, Wyatt Technology Europe. Measured using photometer Dynapro™ NanoStar.

DLS employing thermal ramping was used to measure changes of hydrodynamic radius. A folded protein can exhibit unfolding with increase in temperature, resulting in the exposure of hydrophobic regions and eventual protein aggregation.

Generally, the light scattering technique can be used for sensitive detection of processes that are associated with the change of molar mass or molecular size. An interesting application is the quantitative characterization of reversible protein–protein association. The method employs static light scattering under different compositions and concentrations of protein and buffer. As protein complexes are formed, scattered light intensity and M_w increase. The method was reported as composition gradient multi-angle light scattering (CG-MALS).[18,19] The procedure of preparing solutions of different compositions and delivering them into a MALS photometer and a concentration detector can be fully automatized by a commercially available instrument called Calypso (Wyatt Technology Corporation). The instrument is based on three computer-controlled syringe pumps that yield different compositions by varying the relative flow rates of the pumps connected to the two sample vials and a buffer reservoir. The specified compositions are delivered via a static mixer into the flow cells of MALS and concentration (UV, RI) detectors.

The previous examples of light scattering measurements in batch/microbatch mode showed several Zimm plots of neat appearance, and demonstrated that the technique is quite applicable over a broad range of molar masses and polymer structure. The batch measurement is relatively fast and does not require more than two hours for one sample, including sample preparation. The particular solutions can be prepared individually or by dilution of a stock solution. Since the appearance of the Zimm plot strongly depends on the accuracy of the concentrations, usually better-looking Zimm plots are obtained using serial dilutions of a stock solution. The particular dilutions must be made with great care, keeping four significant figures of accuracy. Dilutions on the basis of weight should be used instead of dilutions on the basis of volume.

2.8 KEYNOTES

- Light scattering is the result of interaction of light with matter.
- In static (elastic, Rayleigh) light scattering the time-averaged intensity of the scattered light is measured, whereas in dynamic (quasielastic) light scattering the time-dependent fluctuations are of interest.
- The intensity of scattered light is expressed by the quantity called Rayleigh ratio, which is independent of the intensity of incident light and light scattering apparatus.
- The absolute intensity of scattered light is directly proportional to the molar mass and concentration of scattering molecules. The size of scattering molecules is reflected in the decrease of the scattered light intensity with the angle of detection.
- The angular variation of the scattered light intensity is described by the particle scattering function. The particle scattering function reflects the intramolecular interference, which is eliminated by extrapolation to zero angle.
- In solutions of finite concentration, the light scattered from different molecules interferes. This intermolecular interference is eliminated by extrapolation to zero concentration or by measurement at very low concentrations.
- A light scattering experiment yields three pieces of information: (1) the weight-average molar mass, (2) the z-average RMS radius, and (3) the second virial coefficient. The RMS radius can be determined without knowledge of concentration and/or dn/dc solely from the angular variation of scattered light intensity.
- A light scattering experiment can be performed in batch mode on an unfractionated sample or in chromatography mode when a light scattering detector is placed online after a separation analytical instrument.
- The traditional batch light scattering experiment involves the measurement of a series of solutions of various concentrations at different angles. The obtained angular and concentration dependence of the intensity of scattered light is simultaneously extrapolated to zero angle and zero concentration by means of a Zimm plot. The concentration range should cover about one order of magnitude. Samples with different molar mass require different

concentrations of measured solutions: Figures 2.21–2.29 show appropriate concentrations for various molar masses.

- A simpler way of processing the light scattering data is extrapolation of solely angular intensity at a single concentration. The method is called a Debye plot and usually yields underestimated M_w with error proportional to the concentration.

- The experimental data can be processed using various light scattering formalisms. They are: Debye, Zimm and Berry, which plot R_θ/K^*c, K^*c/R_θ, and $(K^*c/R_\theta)^{1/2}$ versus $\sin^2(\theta/2)$, respectively.

- Light scattering is an absolute method for determination of molar mass whose absolute accuracy is given by the accuracy of Rayleigh ratio of standard used for the calibration of a light scattering instrument, accuracy of the dn/dc, and accuracy of concentration. In the case of SEC-MALS experiments, the calibration constant of the RI detector is another important parameter affecting the accuracy of the molar mass. Concerning the determination of molar mass, the RI detector plays as important a role as the MALS detector.

- Light scattering experiments are very sensitive to submicrometer particles in solutions measured in batch mode or mobile phase eluting from a separation device. The optical clarification of the sample is of crucial importance for the batch experiments. The clarification should remove dust and other scattering matters not belonging to the polymer under analysis. On the other hand, the clarification should not remove a relevant part of the sample. Clarification of the sample itself is not crucial in the case of online experiment, but possible particle shedding from SEC columns is another source of unwanted signal noise.

- A perfectly clean flow cell of the MALS photometer is crucial for accurate measurements. Bright spots visible in the low cell and increase of the baseline signal are indications of impurities in the cell.

- Although light scattering yields both molar mass and molecular size, the ability to characterize branching of polydisperse polymers by measurements of unfractionated samples is highly limited due to different sensitivity of M_w and R_z to the high-molar-mass fractions. For polydisperse polymers, the reduction of RMS radius due to the branching is compensated by the increase of R_z due to the presence of high-molar-mass fractions.

2.9 REFERENCES

1. Wyatt, P. J., *Analytica Chimica Acta*, **272**, 1 (1993).
2. Kratochvil, P., *Classical Light Scattering from Polymer Solutions, Polymer Science Library*, Jenkins, A. D. (editor), Elsevier, Amsterdam (1987).
3. Zimm, B. H., *J. Chem. Phys.*, **16**, 1093 (1948).

4. Schulz, G. V. and Baumann, H., *Makromol. Chem.*, **114**, 122 (1968).
5. Debye, P., *Ann. Phys.*, **46**, 809 (1915).
6. Debye, P., *J. Phys. Coll. Chem.*, **51**, 18 (1947).
7. Burchard, W., *Advances in Polymer Science*, **48**, Springer-Verlag, Berlin Heidelberg (1983), p. 58.
8. Burchard, W., *Macromolecules*, **10**, 919 (1977).
9. Kerker, M., Farone, W. A., and Matijevic, E., *J. Opt. Soc. Am.*, **53**, 758 (1963).
10. Berry, G. C., *J. Chem. Phys.*, **44**, 4550 (1966).
11. Huglin, M. B. (editor), *Light Scattering from Polymer Solutions*, Academic Press, London (1972).
12. Mori, S. and Barth, H. G., *Size Exclusion Chromatography*, Springer, Berlin (1999).
13. Brandrup, J., Immergut, E. H., Grulke, E. A. (editors), *Polymer Handbook*, 4th Edition, John Wiley & Sons New York (1999).
14. Sedlacek, J., Vohlidal, J., and Grubisic-Gallot, Z., *Makromol. Chem. Rapid Commun.*, **14**, 51 (1993).
15. Kaye, W. and McDaniel, J. B., *Applied Optics*, **13**, 1934 (1974).
16. Becker, A., Kohler, W., and Muller, B., *Berichte der Bunsengesellschaft fur Physikalische Chemie*, **99**, 600 (1995).
17. Podzimek, S., in *Multiple Detection in Size-Exclusion Chromatography*, Striegel, A. M. (ed.), ACS Symposium Series **893**, Washington, D.C. (2004), p. 109.
18. Some, D., Hanlon, A., and Kamron, S., *American Biotechnology Laboratory*, **26**, March, 18 (2008).
19. Some, D., Hitchner, E., and Ferullo, J., *American Biotechnology Laboratory*, **27**, February, 16 (2009).

Chapter 3

Size Exclusion Chromatography

3.1 INTRODUCTION

Polymer analysis can be performed in three different modes of column liquid chromatography: (1) size exclusion chromatography (SEC), (2) liquid chromatography (LC)[i] at critical conditions, and (3) various kinds of LC based on enthalpic interactions. Enthalpic interactions represent attractive and repulsive forces between solute, solvent, and stationary phase and include adsorption, partition, ion interactions, and specific biochemical interactions (bioaffinity). The dissolution–precipitation process is another mechanism playing an important role in the separation of polymers and oligomers. The term *adsorption* LC applies to the distribution of solute between the solution and a solid surface; that is, the sample components are separated due to their varying degree of adsorption onto the solid surfaces. The stationary phase in adsorption chromatography is usually silica gel.[ii] The term *partition (liquid–liquid)* chromatography describes the distribution of solute between two chemically different liquid phases. In partition chromatography, the stationary phase is usually a liquid, which is mechanically coated or chemically bonded on an inert solid support. The term *normal-phase* HPLC is used for nonpolar solvent and polar column (silica gel), while the term *reversed-phase* HPLC is used for polar solvent/nonpolar column (e.g., silica gel modified with C_{18} or C_8 hydrocarbon).

Besides SEC, reversed-phase HPLC, typically performed in gradient elution mode, is often used in polymer and especially oligomer analysis. SEC is, at

---

[i] With the development of high-performance columns and advanced instrumentation, the term *high-performance liquid chromatography (HPLC)* has been frequently used.

[ii] The properties and polarity of silica gel can be further modified with polar functional groups such as cyano—C_2H_4CN, diol—$C_3H_6OCH_2CHOHCH_2OH$, or amino—$C_3H_6\ NH_2$.

Light Scattering, Size Exclusion Chromatography and Asymmetric Flow Field Flow Fractionation: Powerful Tools for the Characterization of Polymers, Proteins and Nanoparticles, by Stepan Podzimek
Copyright © 2011 John Wiley & Sons, Inc.

99

least in its ideal state, governed purely by entropy, whereas the interaction LC is controlled by enthalpic effects, and the balance between enthalpic and entropic effects controls the separation by LC at critical conditions. Interaction LC separates according to both molar mass and chemical composition; the increase in molar mass results in the increase of retention time, while the presence of polar functional groups retention time decreases (reversed-phase LC) or increases (normal-phase LC). LC at critical conditions separates molecules independently of their molar mass according to their chemical composition and the type of end groups. Although this type of separation is attractive for many samples, its wide application is limited by the high sensitivity to experimental conditions.

In contrast to interaction LC and especially LC at critical conditions, where each polymer under analysis requires very specific separation conditions (column, mobile phase, temperature), SEC analysis can be performed using generic conditions for various polymers with the only requirements those of solubility of a polymer in an SEC solvent and an appropriate separation range of the SEC columns. The terms *LC* and *HPLC* include all chromatographic techniques in which the mobile phase is a liquid. Strictly speaking, SEC is one type of liquid chromatography and it is not appropriate to speak of SEC and HPLC as if they were two different methods. However, in common practice the two terms are used in order to distinguish an HPLC method based on interactions and typically applied to oligomeric and low-molar-mass compounds from the SEC method based on entropic separation and typically used for the analysis of polymers.

SEC, which is also well known under the name *gel permeation chromatography* (GPC) or *gel filtration*, represents one of the most important methods of polymer analysis that is widely used for polymer characterization and understanding and predicting polymer performance. SEC has almost completely replaced traditional methods of molar mass determination such as osmometry or ultracentrifugation, and nowadays even light scattering and viscometry are mostly carried out in combination with SEC. The advantages of SEC include relative simplicity, versatility, and ability to determine the complete distribution of molar masses as opposed to other methods providing solely an average molar mass, speed of measurement, and low sample demand. In addition, SEC benefits from intensive development of instrumentation driven by other types of liquid chromatography and from the fact that the instrumentation is available from many manufacturers.

SEC with solely a concentration detector and the calculation of molar mass distribution based on the column calibration can be called *conventional* SEC if there is a need of distinguishing from SEC combined with a light scattering or a viscometric detector. SEC can be defined as one method of molar mass determination even though the molar mass is not measured directly. SEC can be also defined as a special type of column chromatography—special in the sense of the nonexistence of interactions between the analyte and stationary phase. It must be emphasized that the absence of enthalpic interactions makes SEC unique among other types of liquid chromatography. As a consequence of the absence of interactions the method separates compounds purely according to their size in solution. However, the interactions are completely absent under ideal SEC conditions, and in real SEC, various types of interactions often appear as undesirable side phenomena.

SEC has become the most popular method for determination of molar mass distribution and molar mass moments, and, in combination with light scattering and viscometric detectors, for determination of the distribution of root mean square (RMS) radius and intrinsic viscosity, detections of aggregation, and characterization of the molecular conformation and branching. The requested information is typically obtained in a timeframe of about 30 min. Just for curiosity, obtaining such information by traditional fractionation methods and characterization of the obtained fractions with batch light scattering and viscometry required several weeks of intensive work. SEC is often used to study polymerization kinetics, and for investigation of polymer degradation and ageing, determination of low-molar-mass additives in polymers, and characterization of oligomers. One of the advantages of the method is that it is suitable for use in research laboratories as well as for industrial plant applications.

The method dates back to the 1950s, when Porath and Flodin successfully separated water-soluble compounds using crosslinked dextrane gels.[1] They called the method *gel filtration* and proposed the idea of tailoring preparation of gels for different molecular size ranges. The milestone in the separation of synthetic polymers in nonaqueous solvent was the work by Moore,[2] who prepared a series of gels of different pore sizes based on crosslinked styrene-divinyl benzene copolymers and described efficient separation of polystyrene in the molar mass range of $700-10^6$ g/mol. He suggested the name *gel permeation chromatography*.

The first experiments were carried out using simple devices where the eluent flowed through the gel bed solely by gravitation forces and the fractions were collected for further characterization. The second generation of GPC instruments appeared in the 1960s (Waters GPC 100 and Waters GPC 200). These utilized low-pressure pumps operating at pressure below 10 bars. The 4-ft (120-cm) × 7.8-mm stainless-steel columns were packed with relatively soft gel packing of particle size of several tens of μm. The columns were assembled into sets consisting of three to six columns packed with gel of different pore sizes to cover sufficient molar mass range. The elution volume was determined by a siphon where each pour-out was registered by a count on a chromatogram. A significant improvement of the method was the development of online refractive index (RI) and UV detectors. A chart recorder recorded the detector signals and the obtained chromatograms were manually processed using a pencil and a ruler by drawing several tens of equidistant vertical lines from the baseline to a point on the chromatogram. The calculation of the calibration curves, the molar mass averages, and distribution plots was performed using large computers. The overlay of several chromatograms was typically done manually, using a pen and a sheet of transparent paper. The instruments also offered fraction collectors and simple autoinjectors. The third generation of instruments utilized high-pressure pumps, significantly shorter columns (typically 1 ft (30 cm) × 7.8 mm), smaller column packing of about 10 μm of mostly spherical shape, electronic integrators, and later the first personal computers. Current development is mostly aimed at new columns with increased separation efficiency and reduced secondary separation mechanisms, advanced detection systems capable of direct measurements of molar mass and other molecular characteristics, increased sample throughput and methodology of processing, and interpretation of the experimental data.

3.2 SEPARATION MECHANISMS

The basic principle of separation by SEC can be described as follows. The chromatographic columns are packed with small particles of porous material. The space among the particles and pores are filled with a mobile phase. The sample is injected in the form of a dilute solution in the same solvent as used in the mobile phase into a series of columns that are continuously flowed with the mobile phase. The concentration of molecules eluting from the column outlet is monitored online by a concentration detector, usually an RI detector. The molecules permeate into the pores; the smaller molecules can permeate deeper into pores and they can permeate into smaller pores while large molecules are excluded from the pores with effective size smaller than the size of the molecules. Consequently, the large molecules elute from the columns first, followed by molecules with decreasing molecular size. This principle is generally known and accepted as the major separation mechanism called *steric exclusion*. However, there are other separation mechanisms that may play a role under specific conditions. The main SEC separation mechanisms are (1) steric exclusion, (2) restricted diffusion, and (3) separation by flow.

3.2.1 Steric Exclusion

The concept of steric exclusion is based on the idea that different a volume of pores in the SEC column is available for polymer molecules of different size. The basic idea is that the molecules have enough time to diffuse into the pores and back. In other words, the diffusion coefficients are large enough that the time necessary for polymer molecules to diffuse in and out of pores is significantly shorter than the residence time in which molecules stay in a given section of the column. Typically, the experimental conditions allow the establishment of the diffusion equilibrium and thus the steric exclusion is the primary separation mechanism in SEC.

The elution volume V_e can be expressed as

$$V_e = V_0 + K_d V_i \tag{3.1}$$

where K_d is the distribution coefficient, V_0 is the total volume of the solvent outside the pores, and V_i is the total volume of the solvent inside the pores. Molecules that are larger than the largest pores elute at elution volume V_0 (*limit of total exclusion*) and small molecules that can permeate into all pores elute at volume $(V_0 + V_i)$ at the *limit of total permeation*. Under ideal conditions without interactions the distribution coefficient lies in the interval $0 \leq K_d \leq 1$. The distribution coefficient represents a volume fraction of pores that are available for molecules of a given molecular size. In contrast to other types of liquid chromatography, where the retention volumes can achieve very high values, the elution volume in SEC is limited by the interval $V_0 - V_t$, where $V_t = V_0 + V_i$ is the total volume of the mobile phase in the column. The effective volume of an SEC column in which the molecules can elute is approximately in the range of $0.4V_c - 0.8V_c$, where V_c is the volume of the column:

$$V_c = V_0 + V_i + V_g \tag{3.2}$$

where V_g is the volume occupied by the solid matrix (gel). For instance, for a 300×8-mm SEC column, the effective volume is within the range of about 6–12 mL. As a consequence of the limited range of the elution volume, the selectivity of SEC is lower than that of other types of liquid chromatography.

The distribution coefficient is independent of the length and inner diameter of the SEC column, but depends on the pore size distribution of the material used as the column packing. Although this property of the distribution coefficient allows good comparison of the results obtained using different packings under different experimental conditions, the distribution coefficient is rarely used and the elution volume itself is mostly used to plot SEC chromatograms.

Much attention was paid to finding a direct relation between the distribution coefficient and the size or even molar mass of eluting macromolecules. This is a very attractive idea, because it would make SEC an absolute method of molar mass determination. However, the obtained theoretical results never found real applications.

3.2.2 Restricted Diffusion

The restricted diffusion separation mechanism is based on the idea that the time needed for macromolecules to diffuse in and out of the pores is comparable with the time that they stay in a given column zone. In such a case, the permeation depth is governed by the diffusion coefficient, which is indirectly related to molecular size. The large molecules penetrate slowly and thus do not stay in a given chromatographic zone long enough to penetrate into the entire available volume. The idea of restricted diffusion implies that the elution volume should depend on the flow rate, which is typically not true. However, separation by restricted diffusion may partially take place at the separation of high-molar-mass polymers or at high flow rates.

3.2.3 Separation by Flow

Separation by flow (hydrodynamic chromatography) is based on the idea of flow through a narrow capillary in which there is a parabolic velocity profile of the liquid flow. A column packed with small, solid nonporous particles creates a system of narrow capillaries. For each molecule there is an excluded volume in the proximity of the channel walls given by its geometrical dimensions. The large molecules statistically more frequently occur closer to the capillary center and therefore flow faster than the smaller molecules, which can be situated close to the wall where the flow is slow. Although hydrodynamic separation is in principle different from size exclusion, it separates according to the particles' size in the same order as SEC and thus does not destroy steric separation. Separation by flow may take place at the region of very high molar masses. This is probably one of the reasons why the experimental determination of the limit of the total exclusion is not as straightforward as one might expect, because separation by flow can occur beyond the molar mass corresponding to the total exclusion limit.

Although over a certain molar mass all molecules should elute together at elution volume V_0, molecules larger than the column exclusion limit are separated by flow and elute at elution volumes smaller than V_0. An example of concurrent SEC and hydrodynamic separation of polystyrene standards is shown in Figure 3.1. The nominal exclusion limit of the applied column set is about 30,000 g/mol. That means the standards with molar mass 34,500 g/mol and less are separated by SEC, whereas the standards over this limit are separated by flow. The most significant parameters affecting the separation by flow are the ratio of the polymer hydrodynamic diameter to the diameter of the packing particles and the pore diameter related to the diameter of the packing particles.

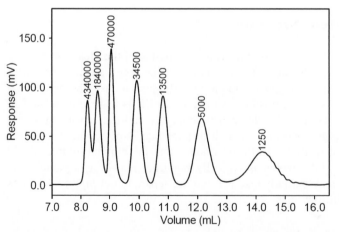

Figure 3.1   Separation of polystyrene standards using $2 \times$ PLgel Mixed-E 300×7.5-mm 3-μm columns. THF, 1 mL/min, 40°C. Standards 1250 g/mol, 5000 g/mol, 13,500 g/mol, and 34,500 g/mol are separated by SEC separation, whereas standards 470×10^3 g/mol, 1.84×10^6 g/mol, and 4.34×10^6 g/mol are separated by hydrodynamic chromatography.

3.2.4 Peak Broadening and Separation Efficiency

A monodisperse compound injected into an SEC column should theoretically elute as a narrow rectangular peak of width equal to the injected volume and the area proportional to the detector sensitivity and injected mass. However, in real chromatography the span between the elution volume of the first eluting molecules and the elution volume of the last eluting molecules is much greater than the volume of injected sample and the real peak is of different shape and height compared to the ideal rectangular peak. The explanation is that the peak of the monodisperse solute is spread during the flow through the column. The major sources of peak broadening are diffusion along the column axis, eddy diffusion, limited velocity of the mass transfer between the stationary phase and the mobile phase, and the void volume of the chromatographic system.

A peak of a monodisperse compound can be approximated by a Gaussian function:

$$G(V) = \frac{A}{\sigma\sqrt{2\pi}} e^{-\frac{(V-V_e)^2}{2\sigma^2}} \tag{3.3}$$

where $G(V)$ is the detector response as a function of elution volume, σ is the standard deviation of the Gaussian function, V is the elution volume, V_e is the elution volume of the apex of the chromatographic peak, and A is a parameter related to the sensitivity of the detection system and directly proportional to the injected mass (i.e., it can be understood as the area of the peak). The standard deviation describes the broadness of the peak. Besides σ the broadness of the peak can be characterized by a width at the baseline W or at half height $W_{1/2}$ related to the standard deviation as $W = 4\sigma$ and $W_{1/2} = 2.355\sigma$.

A chromatogram of a polydisperse polymer can be assumed to be a superposition of chromatograms of individual macromolecules with different molar masses, each of them being approximated by a Gaussian function. A real chromatogram of a polydisperse polymer is broadened due to the real broadness of the molar mass distribution and due to the peak broadening effects. Because of peak broadening, the real peak width is broader than would correspond to the molar mass distribution and thus processing an SEC chromatogram without peak broadening correction yields broader molar mass distribution. A general relation between the experimental chromatogram $G(V)$ and the true chromatogram $W(y)$ is expressed by Tung's equation:[3]

$$G(V) = \int_0^\infty W(y)G(V,y)\,dy \tag{3.4}$$

where V and y are elution volumes and $G(V,y)$ is the instrumental spreading function. The spreading function can be understood as a response of a given chromatographic column to the infinitesimally short injection of a perfectly monodisperse polymer with molar mass corresponding to the elution volume y.

The chromatogram $W(y)$ would be recorded by means of a hypothetical column with unlimited resolution power.

Peak broadening (band broadening; zone spreading; axial, longitudinal, column, or instrumental dispersion) is an unwanted effect that decreases separation efficiency. In SEC the available elution volumes are limited by the exclusion limit and limit of total permeation and thus the minimization of the sources of peak broadening is of primary importance. Knowledge and understanding of the factors leading to peak broadening are important for the optimization of separation conditions. The most frequent peak broadening is the symmetrical or Gaussian type that broadens the molar mass distribution, decreases M_n, and increases M_w and M_z. In a rigorous manner, the compensation of the peak broadening is necessary for the calculation of the true molar mass distribution and true molar mass averages from the chromatogram obtained by SEC experiment.

To compensate for peak broadening it is necessary to transfer the experimental chromatogram into an ideal chromatogram that would be recorded without peak broadening. The peak width and shape of the corrected chromatogram reflect only the width and shape of the molar mass distribution. Numerous methods were proposed to correct for the effect of peak broadening. Peak broadening correction used to be a very popular subject of scientific papers dedicated to SEC theory. A comprehensive and critical overview of the different methods used to correct band broadening can be found in reference 4. However, none of the published methods is readily adaptable for routine measurements. With the development of high-performance columns, correction for peak broadening became less important and usually is not applied. The effect of peak broadening can be significant in the case of SEC analysis of very narrow polymers where determination of the true polydispersity may be of interest. An interesting question related to the correction for peak broadening may be, "How narrow are narrow polymer standards?" In the case of broad polymer samples, the effect of peak broadening usually can be ignored without significant impact on the accuracy of results.

A simple method of peak broadening correction assumes Gaussian SEC chromatograms and nominal polydispersity of polystyrene standards to be correct. Several standards are measured and a correction factor

$$R = \left[\frac{(M_w/M_n)_{\text{nom}}}{(M_w/M_n)_{\text{exp}}} \right]^{\frac{1}{2}} \tag{3.5}$$

is calculated by simply comparing the nominal (nom) and experimental (exp) polydispersity. The true molar mass averages of samples under analysis are then calculated from the experimental results and R:

$$M_{n,\text{true}} = \frac{M_{n,\text{exp}}}{R} \quad \text{and} \quad M_{w,\text{true}} = R M_{w,\text{exp}} \tag{3.6}$$

Although this procedure is rather simplified, and the correctness of polydispersity of the calibration standards may be questionable, one can see that even in the case of a relatively large difference between nominal and experimental polydispersities the correction of molar mass averages lies in the range of only a few percent.

The extent of peak broadening and the separation efficiency of the chromatographic column can be characterized by the *number of theoretical plates* and the *height equivalent to a theoretical plate* (*H, plate height*). The plate height is defined as the variance (σ^2) of the eluting peak divided by the length of the column (*L*):

$$H = \frac{\sigma^2}{L} \tag{3.7}$$

The relation between the separation efficiency and the parameters characterizing the chromatographic separation process was derived by van Deemter et al.[5] and expressed as:

$$H = A + \frac{B}{u} + Cu \tag{3.8}$$

where *u* is the linear velocity of the mobile phase (cm/s), *A* is the contribution of eddy diffusion, *B* is the contribution of axial diffusion and *C* is related to the mass transfer between the stationary and mobile phases. Equation 3.8 describes three basic processes that contribute to plate height: (1) eddy diffusion, caused by the flow of solute at unequal rates through the tortuous pathways of the bed of the packing particles (which means the packed column functions as some kind of a static mixer); (2) axial diffusion, in which solute molecules diffuse axially from the center of the zone; and (3) nonequilibrium or mass transfer, in which the limited speed of diffusion in and out of the stationary phase causes broadening of the solute zone, because the molecules being at a given time inside the pores are delayed behind those that are already carried by the flow stream.

Linear velocity is easily calculated from the volume flow rate in mL/min dividing by πr^2, where *r* is the inner radius of the column. Equation 3.8 shows that the axial diffusion becomes less pronounced with increasing flow rate while the effect of the mass transfer between the phases increases. In contrast to low-molar-mass compounds, the polymers have generally low diffusion coefficient and thus the effect of axial diffusion is negligible as shown by the interrupted flow experiment when the polymer was kept in the columns for many hours before the elution continued.[6] The relation *H* versus *u* touches minimum at a certain *u* that represents the most optimum flow rate for a given column. In routine practice the maximum pressure limit for the columns and the analysis time must be considered and a flow rate in the range of 0.5 to 1 mL/min is mostly used with standard 300 × 8-mm SEC columns.

The number of theoretical plates (*N*) can be calculated from the chromatogram of a monodisperse low-molar-mass compound:

$$N = 16 \left(\frac{V_e}{W} \right)^2 = 5.54 \left(\frac{V_e}{W_{1/2}} \right)^2 \tag{3.9}$$

where W and $W_{1/2}$ are the peak width in the baseline and in half height, respectively. The peak width W is measured between two crosspoints of the baseline and two tangents drawn from the inflection points of the ascending and descending sides of the peak. The height equivalent to a theoretical plate is calculated as L/N. The peak width is indirectly proportional to the square root of the plate number (Equation 3.9), that is, smaller plate numbers mean broader peaks and less resolution. It may be important to note that the plates do not really exist; they are created by the imagination, which helps to understand the separation process and serves as a way of measuring the column efficiency. The plate height decreases with decreasing particle size and thus columns packed with smaller particles are more efficient, but they are also less suitable for the analysis of polymers with high molar mass owing to high backpressure and possibility of shearing degradation. As shown by semiempirical prediction[7] and confirmed by experimental data,[8] the plate height increases with decreasing distribution coefficient K_d. For example, the H-versus-K_d plot published in reference 7 shows about a triplicate increase of H with the decrease of K_d from 0.4 to 0.2. At very low K_d close to the total exclusion limit the trend is reversed and H starts to decrease. That means the H-versus-K_d plot shows a maximum at low K_d, but the plate height values at low K_d are still significantly higher compared to the region of larger K_d.

Peak symmetry is another parameter that characterizes the quality of column:

$$s = \frac{a}{b} \tag{3.10}$$

where a and b are the peak widths measured at 10% of peak height on either side of the perpendicular from the baseline to the peak apex (a is the distance from the center line to the descending part of peak).

Peak symmetry is a measure of peak tailing that has negative effect on the resolution. A little tailing (roughly $s < 1.2$) is normal, but excessive tailing indicates a damaged column bed, enthalpic interactions of the analyzed molecules with column packing, or other non-SEC effects. Various compounds can be used to measure the number of theoretical plates and symmetry, for example, propyl benzene, acetone, toluene, dicyclohexyl phthalate, ethyleneglycol, glucose, and orthodichlorobenzene. Modern 300×7.5-mm SEC columns provide per column more than 24,000 plates, 15,000 plates, 10,000 plates, and 5,000 plates for 3, 5, 10, and 20 μm packing size, respectively. However, the plate number usually decreases with column use.

The separation of two components can be characterized by *resolution*:

$$R_S = \frac{2(V_2 - V_1)}{W_1 + W_2} \tag{3.11}$$

where V and W are elution volume and baseline width of the two components. Resolution is the difference in elution volumes divided by the average peak width. The difference of elution volumes of the two components ($V_2 - V_1$) is a reflection of the selectivity of the separation process, whereas W represents the zone spreading that is related to the efficiency of the process. At a resolution of 1 the two peaks are not completely separated, but the peak area overlap is only 2%. The complete separation is achieved at a resolution of 1.25.

Equation 3.11 tells us that the resolution is controlled by both selectivity and efficiency, and that with columns of low efficiency (e.g., worn out by excessive or improper use), which provide very broad peaks, one can barely achieve satisfactory resolution; on the other hand, even highly efficient columns cannot provide satisfactory resolution if the separation process is not sufficiently selective. Resolution increases with the square root of column length or number of theoretical plates. If the column length (number of theoretical plates) is doubled, the resolution increases by a factor of $\sqrt{2}$. In SEC the quantity called *specific resolution* (R_{SP}), which takes molar mass of the two compounds into account, provides better characterization of the separation efficiency of SEC columns:

$$R_{SP} = \frac{2(V_2 - V_1)}{(W_1 + W_2)(\log M_1 - \log M_2)} \tag{3.12}$$

The specific resolution can be determined by two monodisperse polymer standards. If the polymer standards are not monodisperse, the widths W_1 and W_2 in the above equation are divided by the polydispersity M_w/M_n.

The resolution of two compounds in the interaction types of liquid chromatography is defined by the equation:

$$R_S = \frac{2(V'_{R2} - V'_{R1})}{W_1 + W_2} = \frac{1}{4} \frac{r_{1,2} - 1}{r_{1,2}} \frac{k_2}{1 + k_2} \sqrt{N} \tag{3.13}$$

where $V'_R = V_R - V_t$ is the net retention volume (the term *retention volume* is used as the equivalent to *elution volume* to differentiate between the interaction and non-interaction separation),

$$r_{1,2} = \frac{V'_{R2}}{V'_{R1}} \tag{3.14}$$

is the relative retention (selectivity), and

$$k = \frac{V_R - V_t}{V_t} \tag{3.15}$$

is the capacity factor.

Equation 3.13 shows that the resolution in interaction liquid chromatography is controlled by selectivity ($r_{1,2}$), capacity (k), and column efficiency (N). In SEC, contrary to other types of liquid chromatography, capacity does not play a role and resolution can be controlled solely by selectivity and column efficiency. The number of theoretical plates is related to the quality of column packing and column length. The usual way of increasing the number of theoretical plates in SEC is connection of two or more columns in series. However, it is necessary to stress that the resolution is proportional only to the square root of the number of theoretical plates and doubling the number of theoretical plates by using, for example, four columns instead of two improves resolution by a factor of 1.4, but the analysis time increases by a factor of 2. Selectivity in SEC is related to the slope of the calibration curve. That means SEC columns of high plate numbers do not necessarily provide good resolution unless the slope of the calibration curve is sufficiently low. Although the calibration curves are usually slightly curved, they can be approximated by a linear fit at least for a certain range of the elution volume:

$$\log M = a + bV \tag{3.16}$$

Then the slope equals:

$$b = \frac{\log M_1 - \log M_2}{V_2 - V_1} \tag{3.17}$$

where M_1 and M_2 are molar masses eluting in the elution volumes V_1 and V_2. The slope of the linear fit shows how well a column can separate peak apexes of polymers of different molar mass, while the peak widths of monodisperse polymers are related to the number of theoretical plates. The separation increases with decreasing slope, and to achieve good separation the slope should be of minimum value. The slope of the calibration curve is related to the pore volume, which should be as large as possible. Selectivity further depends on the pore size of the column packing material and the selection of SEC columns with separation range appropriate for a given polymer sample is of primary importance. For example, if the pore sizes are too small for the macromolecules to be separated, all molecules elute at the limit of total exclusion. On the other hand, packing material with large pore sizes cannot separate oligomeric mixtures. For a given column set, the slope of the calibration curve and consequently the selectivity can be different for different molar mass regions.

Figure 3.2 compares three column sets of different separation range and selectivity. Note that the column sets (□) and (■) have similar separation range, but their slopes in the region of lower and higher molar masses are different. Figure 3.3 compares the calibration curves established by the same type but a different number of SEC columns. In this case the use of more columns not only improves the selectivity due to lower slope, but also increases the number of theoretical plates. If the columns are properly calibrated and of appropriate separation range, the molar mass results should not depend on the columns employed for the measurements. However, in reality one can often find differences between the distribution curves determined using different column sets even though other separation conditions are identical (e.g., see Figure 3.4).

The reciprocal of the slope of the calibration curve over one molar mass decade (i.e., logarithm of molar mass interval of one) related to the column cross-sectional area is another parameter that can be used to characterize the

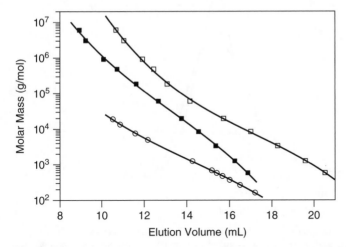

Figure 3.2 Calibration curves of three column sets 2 × 300 × 7.5 mm with different selectivity in the region of high and low molar masses. Set (□) compared to set (■) shows lower selectivity in the region of high molar masses and higher selectivity in the lower-molar-mass region. Set (□): R_{SP} = 2.86 and 3.83 for pairs of standards 915,000/60,450 and 60,450/3370, respectively. Set (■) : R_{SP} = 4.22 and 3.37 for pairs of standards 915,000/60,450 and 60,450/3370, respectively. Set (○) has the highest selectivity in the lower-molar-mass region, but the separation range is limited.

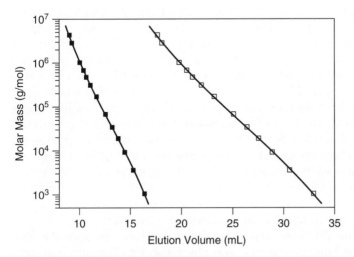

Figure 3.3 Calibration curves obtained by sets of two and four PLgel Mixed-C 300 × 7.5 columns. Two columns (■) : R_{SP} = 3.88, SP = 4.94; four columns (□): R_{SP} = 4.39, SP = 10.30. R_{SP} calculated for standards 675,000 and 68,000 g/mol.

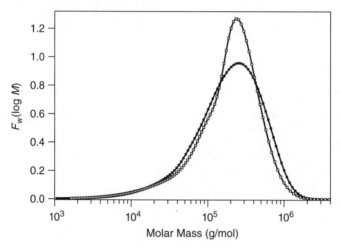

Figure 3.4   Differential distribution plots determined using column sets (■) and (□) from Figure 3.2. Molar mass averages: M_n (g/mol) = 89,900 ± 1200 (Set □), 79,900 ± 900 (Set ■), M_w (g/mol) = 262,800 ± 1000 (Set □), 277,600 ± 500 (Set ■), M_z (g/mol) = 444,000 ± 2000 (Set □), 526,900 ± 2600 (Set ■).

separation performance (*SP*) of SEC columns:[9]

$$SP = \frac{V_M - V_{10M}}{\pi r^2} \tag{3.18}$$

where r is the inner radius of the column (in cm), and V_M and V_{10M} are the elution volumes (in mL) of calibration standards with molar mass M and ten times the value of M, respectively. According to reference 9, the minimum SP is six, while the value of R_{SP} should be greater than 2.5.[10] The two parameters defined by Equations 3.12 and 3.18 are not entirely equivalent: R_{SP} reflects both selectivity in the sense of difference of elution volumes and efficiency in the sense of the peak width; SP reflects only the column selectivity in terms of the slope of the calibration curve. The limit of separation performance of six was suggested by experience and may not be fulfilled by two mixed columns. For example, separation of two polystyrene standards of molar masses of 675,000 g/mol and 68,000 g/mol using two PLgel Mixed-C 300 × 7.5 columns results in SP of about 4.9 and R_{SP} of about 3.9 (i.e., SP is below and R_{SP} well above the recommended values). As shown in Section 3.5.6, the increased number of columns need not necessarily have significant effect on the molar mass averages.

SEC resolution decreases with increasing molar mass, as is clearly illustrated by the separation of oligomers, where the resolution of neighboring members of the oligomeric series rapidly diminishes with the increasing polymerization degree (see chromatograms in Section 3.5.5). In the high-molar-mass region large differences in molar mass are reflected in small changes in elution volume.

3.2.5 Secondary Separation Mechanisms

All mechanisms describing the principle of SEC separation assume that the stationary phase acts only as an inert matrix that does not interact with the solute. In real SEC highly skewed chromatograms and even polymer molecules eluting behind the limit of total permeation can occur. However, severe tailing can also result from a possible void at the column inlet, which can be created during long-term usage of the column. Tailing from poor column packing can be easily distinguished from tailing due to column interactions by measuring a few narrow calibration standards. If the peaks of standards are symmetrical, the peak tailing can be explained by interactions. The characteristic feature of the tailing due to damaged column packing is that all peaks in a chromatogram show similar shape. The measurement of molar mass or RMS radius across the peak by a MALS detector often reveals that molar mass and RMS radius decrease only over a part of the peak and then reach a plateau parallel with the volume axis or even show the increase from certain elution volume. Such behavior indicates enthalpic interactions of solute with stationary phase, or a specific anchoring behavior of branched polymers (described in Section 6.2.1).

The term *adsorption* is generally used to describe SEC separation in which enthalpic interactions play a role. The enthalpic interactions can affect the SEC separation to different extents depending on particular properties of SEC column packing, sample, and solvent. In the ultimate case the sample can be irreversibly retained in the SEC columns. The type of interactions may not be always known. They can be dipole–dipole interactions, dispersion forces, $\pi-\pi$ interactions (noncovalent interactions between organic compounds containing aromatic rings), hydrogen bonding, ionic exclusion and inclusion, hydrophobic interactions, or interactions of polar groups with the polar groups of the column packing. Ionic interactions are common in separation of various water-soluble polymers in aqueous solvents. Ion exclusion is a consequence of the presence of anionic groups in the SEC packing, when negatively charged polymers are exluded from the pores due to electrostatic repulsive forces. This effect results in earlier elution than would correspond to hydrodynamic volume and thus in the overestimation of the molar mass calculated on the basis of column calibration. Ion inclusion is interaction of anionic and cationic functionalities, resulting in delayed elution and thus underestimation of molar mass. Reduction of mobile phase pH to prevent dissociation of carboxyl groups and increase of the ionic strength of the mobile phase in order to shield electrostatic interactions are possible ways to eliminate ionic interactions.

Besides solute–column packing interactions the elution volume in SEC can be influenced by incompatibility between solute and packing, which results in early elution. Strong interactions are very common in the case of polymers bearing amino groups analyzed in tertahydrofuran using columns packed with styrene-divinylbenzene gels. Amine-containing polymers have been recently applied for the production of environmentally friendly waterborne paints and thus problems with their SEC characterization may be frequent. The interaction of amine-containing polymers with styrene-divinylbenzene gels is sort of surprising, because the gels are hydrophobic and they are not supposed to contain polar functional groups capable of interacting with polymer. It appears that the styrene-divinylbenzene gels may contain polar groups from irreversibly adsorbed or chemically bonded auxiliary compounds of the manufacturing process or polar groups that are created during column usage and aging. THF has a tendency to form peroxides and contains a certain level of oxygen even if degassed. Both oxygen and peroxides can contribute to the oxidative formation of carbonyl, carboxyl, or peroxy functional groups. The functional groups in SEC stationary phase can also occur due to irreversibly retained previously analyzed materials. My experience shows that column properties depend on its history and previous applications (i.e., columns have some type of memory effect). Deterioration of separation efficiency is usually faster if the columns are applied to the analysis of polar polymers as compared to when they are used for the characterization of well-soluble neutral polymers. Besides amino groups, carboxyl groups represent another type of group increasing probability of interactions.

Hydrophobic interactions are frequent in the case of water-soluble polymers analyzed on hydrophylic gel packing based on poly(hydroxyethyl methacrylate) (HEMA) crosslinked with ethylenglycol dimethacrylate. The polymer molecules of water-soluble polymers typically consist of hydrophylic polar groups giving water solubility and hydrophobic chains. The HEMA packing shows similar composition and such hydrophobic chains of polymer interact with the hydrophobic backbone of HEMA. The analysis of sulphonated polystyrene is a typical example.

Enthalpic interactions can result in the adsorption of a part of the injected samples on the surface and in the pores of column packing. This effect is often indicated by poor repeatability and by different peak areas obtained by repeated injections. Sometimes the columns can show some kind of saturation effect; that is, after column saturation by several consecutive injections the peak areas become constant and the repeatability of molar mass averages improves.

Various mobile-phase additives can be used to suppress the interactions. Diethanol amine in the amount of 0.1% can be used for the analysis of amine-containing polymers in THF to prevent adsorption. An example is presented in Figure 3.5. Another way of suppressing the interactions is saturation of the column by multiple injections of the problematic polymer or more effectively by the addition of the polymer directly into the mobile phase. Adsorption also can be suppressed by sample derivatization (see Section 3.5.1.1).

3.3   INSTRUMENTATION

There is no principal difference between the instrumental setup used for SEC measurements and other types of liquid chromatography, and a "liquid chromatograph" can be easily converted to an "SEC chromatograph" by a simple

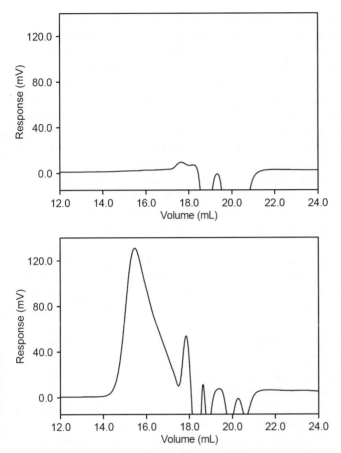

Figure 3.5 RI chromatograms of copolymer 2-ethylhexyl methacrylate (25%) and dimethylaminopropyl methacrylamide (75%) measured in THF (top) and THF containing 0.1% diethanol amine (bottom). In pure THF, the sample was completely retained in the columns due to enthalpic interactions that were suppressed by the addition of diethanol amine.

swapover of columns. In contrast to other types of liquid chromatography, SEC does not use gradient elution. Also some types of detectors are rarely used while other detectors, such as light scattering or viscometry, are almost exclusively used in SEC.

A chromatographic setup for SEC consists of a solvent reservoir, a degassing device, a solvent delivery pump, an injector, columns, a detector system, a waste reservoir, and a PC with SEC software allowing data acquisition and processing. A pulse dampener is another possible part of the solvent delivery system, which may be needed in some applications. The pulse dampener, which is placed between the pump and injector, is usually not needed for RI detectors in the case of a modern HPLC pump, but it may improve the signal of an online viscometer that is extremely sensitive to flow pulsation.

Mobile-phase degassing is necessary to avoid malfunction of the solvent delivery pump. An air bubble in the HPLC pump causes a momentary drop in the backpressure and decrease of the flow rate. The air bubbles that get into the column system stay in solution because of the system pressure, but when they arrive in a detector cell, where the system pressure is low, the bubbles can be released and cause spikes on the chromatogram. The elimination of spikes by a pressure restrictor after the detection system is not the best solution, especially in the case of RI detectors, which are generally not designed to withstand high pressures. Helium sparging, a traditional and effective way of degassing, has been widely replaced with inline vacuum degassers, which have become a part of most of today's solvent delivery systems. In contrast to inline degassing, helium sparging removes the air already in the solvent reservoir, which may be advantageous in the case of oxygen-sensitive solvents such as THF.

The solvent reservoir is typically a glass bottle of 1–2-L volume. Glass solvent containers are usually delivered with the HPLC pump. They are equipped with plastic caps containing several holes for polytetrafluorethylene (PTFE) tubing that delivers solvent to the pump. The end of the PTEF tubing is connected to a glass or stainless-steel frit that protects the pump from mechanical impurities. An original bottle, in which the solvent is supplied, can be used as a solvent container. The holes for the PTFE tubings can be directly drilled into the caps. A brown bottle should be used for THF to prevent degradation caused by light. The solvent reservoir must hold a sufficient amount of solvent to allow measurements of number of runs without the need for solvent refill. This is especially important in the case of an RI detector when the refill of solvent typically results in a significant change of baseline signal. The reservoir should isolate the solvent from contact with the laboratory environment in order to prevent solvent evaporation, reaction with oxygen, light-induced degradation, and absorption of moisture.

These requirements obviously do not apply to all solvents; for example, toluene has almost no tendency to degradation or moisture absorption, whereas THF is a typical example of solvent that requires special attention. Some experimenters recommend some means of agitation (e.g., magnetic stirring) to assure solvent homogeneity. Nevertheless, agitation is usually not needed for most solvents, including aqueous buffers.

A high-quality pump that is able to deliver solvent at a constant, pulse-free and reproducible flow rate is a crucial part of an SEC instrumental setup. Today's HPLC pumps are typically dual-head, having two independently driven pistons for optimal flow control. The pulse-free flow is achieved by a synchronized action of two pistons, where one piston fills the solvent into the chamber while the other provides flow to the columns. Inlet and outlet check valves are an important part of the HPLC pump. The most common check valve is the ball-type check valve, usually consisting of a ruby ball and a saphire seat. The check valve operates in a way such that when the pressure below the ball is higher than above the ball, the ball is lifted off the seat and liquid can flow through the valve. When the pressure above the ball is higher, it closes the valve by pushing the ball back to the seat, preventing reverse liquid flow. It is worth mentioning that many pump-related problems are caused by improperly working check valves.

The sample solution can be injected by a manual six-port sample injection valve, or by means of an autoinjector (autosampler). The loop of the manual valve is filled with sample solution in the sample load position, and then the valve rotor is manually turned into the injection position to flush the loop into the eluent stream. The volume of the loop is usually 20–200 μL, but smaller or large volumes are also available. An excess volume of sample solution is used to flush the solvent completely out and to fill the loop properly. A typical feature of the manual injection valves is that they typically add about 10 μL of an extra volume corresponding to the inside channels. This fact is not important for conventional SEC or HPLC, but it should be taken into account for the online calibration of RI detectors or the determination of a specific refractive index increment in the case of a light scattering detection.

Automatic injectors called *autosamplers* or *autoinjectors* allow automatic analysis of large sample series and overnight operation. The samples are placed into 1–4 mL vials sealed with septa to avoid solvent evaporation. A significant advantage of most autosamplers over manual injectors is the possibility of changing the injection volume without changing the sample loop. The required volume of sample solution is withdrawn from a vial by a stepper motor–controlled syringe into a sample loop and injected into the eluent flow similarly as in the case of manual injectors. In the case of high-temperature SEC instruments, the autosampler compartment must be kept at high temperature and the sample dissolution is usually performed directly in the autosampler compartment. The autosampler can be programmed to agitate the sample periodically by rotation of a sample tray to promote dissolution, and the sample is then automatically filtered prior to injection.

A temperature-controlled column oven is another optional part of the SEC instrumental setup, which can contribute to results accuracy and repeatability because it maintains the column temperature constant. The column oven may not be necessary in the case of constant room temperature and especially when the molar mass is determined by a light scattering detector instead of column calibration.

An SEC setup can be purchased separately as a modular system or as an integrated apparatus incorporating a pump, an injector, a column oven, and detectors into one system. The integrated form is necessary in the case of high-temperature systems that are used for polymers that are not soluble in any solvent at room temperature. Polyolefins, such as polyethylene and polypropylene, are typical examples of polymers of great industrial importance, which require analysis at high temperatures. High-temperature SEC typically means that the columns are kept at a temperature over 100°C, mostly around 150°C. The temperature in the high-temperature SEC system must be maintained from the injector over columns to the detection system in order to prevent sample precipitation. The detectors are typically placed in the column oven. Should the sample leave the oven to reach another detector, such as a high-temperature light scattering photometer, the connecting lines must be heated as well. In high-temperature SEC the actual flow rate is higher than the nominal flow rate of the chromatographic pump because of thermal expansion of the mobile phase. In trichlorobenzene at 150°C the flow rate increase is about 10%.

Some solvents may require temperatures of 60–80°C in order to decrease their viscosity. If the sample dissolves at room temperature, or remains in solution after dissolving at elevated temperature, then an ambient SEC system can be used for the measurements with column oven maintaining the requested temperature.

3.3.1 Solvents

A solvent is an important part of the SEC analysis. The solvent for SEC must be compatible with the column packing material and must dissolve the polymer requiring analysis. The solvent must form stable solutions to prevent precipitation of polymer from solution prior to or during analysis. From this viewpoint, thermodynamically good solvents should be preferred. The thermodynamic quality of the solvent increases with increasing temperature and thus analysis above room temperature is necessary for some polymers. The solvent viscosity preferably should be low, because with increasing viscosity the separation efficiency decreases and the backpressure increases. However, some polymers are difficult to dissolve and the only choice may be highly viscous solvent such as dimethyl sulfoxide. Additionally, the solvents for SEC should be of low toxicity, nonflammable, stable, noncorrosive with respect to the chromatography system and columns, cheap, and easily purifyable. An important parameter is the specific refractive index increment of a polymer requiring analysis that is directly proportional to the RI detector response. For UV detection the solvents must have almost no absorption at the wavelength that is to be used. Typically, it is impossible to meet all the previous requirements and mostly the requirement of good solubility prevails over others. Unlike gradient liquid chromatography, solvents of HPLC purity are not necessary and purity over 99% is mostly appropriate for SEC measurements.

Tetrahydrofuran (THF) represents the most common solvent for organic SEC, because it dissolves a wide range of synthetic polymers. THF does not dissolve biopolymers such as proteins or polysaccharides, and some important synthetic polymers such as polyamides, polyolefins, poly(ethylen terephthalate), and poly(vinyl alcohol). Since THF is highly hygroscopic it should be kept in closed bottles to prevent absorption of moisture from the environment. Besides ability to dissolve many polymers, other advantages of THF include low UV cutoff, low viscosity, and acceptable price. Disadvantages include relatively high toxicity, high volatility, high tendency to absorb moisture and air, formation of explosive mixtures with air, and formation of explosive peroxides and other compounds with oxygen. The reactivity of THF with oxygen is potentially dangerous, because the peroxides can concentrate by THF evaporation and they can explode at temperatures around 120°C. Despite potential danger THF can be safely utilized with no special laboratory precautions. Oxidation is efficiently prevented by stabilization with 0.025% 2,6-ditert-butyl-4-hydroxy toluene (BHT).

Eluent from SEC columns containing small amounts of polymer can be easily distilled and reused in order to reduce analysis costs and environmental burden. The distillation is performed with the addition of copper oxide (about 1 g/L) to remove possibly present peroxides. Water can be removed by molecular sieves 3 Å or by potassium hydroxide prior to distillation, but the distillation itself is mostly sufficient for the purification of waste THF. The distillation is performed using a filled glass distillation column of about 1 m in length placed in a fume hood covered for safety reasons with fencing. Freshly distilled THF can be used for SEC measurements with UV detection or stabilized with BHT. THF forms strong hydrogen bonds with hydroxyl groups and thus creates associates containing one THF molecule per hydroxyl group. The associates have larger dimensions compared to the original compounds and elute at lower elution volumes than would correspond to the dimensions of molecules without association with THF.

Association between sample molecules and THF must be taken into account for the interpretation of chromatograms of some oligomers. A typical feature of RI chromatograms recorded from the analysis in THF is the existence of three solvent peaks. These peaks (Figure 3.6) belong to water, nitrogen, and oxygen. A good resolution of the three peaks is an indication of good separation efficiency of the columns and thus the peaks can be used as a quick check of column separation performance (Figure 3.7). One of the three peaks can be utilized as a flow marker.

Toluene is a traditional SEC solvent that dissolves nonpolar and medium-polar polymers such as polystyrene, polybutadiene, and polyisoprene. It is used

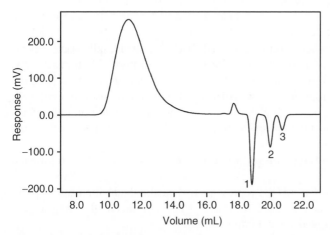

Figure 3.6 RI chromatogram of a polydisperse polymer in THF showing three typical negative peaks: 1 = water, 2 = nitrogen, 3 = oxygen.

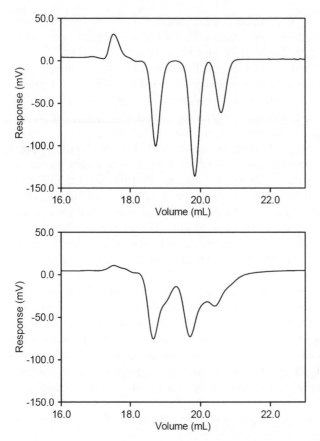

Figure 3.7 Solvent peaks in THF recorded with good columns (top) and the same columns after long-term use (bottom).

instead of THF for the analysis of poly(dimethyl siloxane), which has very low RI response in THF. Advantages include relatively low toxicity and excellent stability, but it cannot be used with UV detection.

Chloroform is another traditionally used SEC solvent. Despite some advantages such as very good ability to dissolve various polymers, UV transparency, and nonflammability, $CHCl_3$ has several serious disadvantages. It is very toxic, which is especially serious due to high volatility. The oxidation leads to extremely dangerous phosgene. The waste disposal cost is high and in addition chloroform is highly corrosive and thus its application is justified only if there is no other appropriate alternative.

THF, toluene, and chloroform represent good solvents for crosslinked polystyrene gels.

1,2,4-Trichlorobenzene (TCB) represents a typical solvent for SEC of polyolefins at temperatures above 135°C. It dissolves polystyrene and thus polystyrene standards can be employed for column calibration.

1,1,1,3,3,3-Hexafluoro-2-propanol (hexafluoroisopropanol, HFIP) is an interesting solvent due to its properties. It exhibits strong hydrogen-bonding properties and ability to dissolve many polymers, including those insoluble in other organic solvents such as polyamides (e.g., nylon 6), polyacrylonitrile, and poly(ethylene terephthalate). HFIP is corrosive, and UV transparent up to about 190 nm with low refractive index ($n_D = 1.276$), which results in significantly higher polymer dn/dc compared to other solvents. The applicability of HFIP is limited by high price, which can be reduced by redistillation and/or using narrow-bore SEC columns. Addition of about 0.1% sodium trifluoroacetate is used to suppress polyelectrolyte effects of the analyzed polymers. Poly(methyl methacrylate) narrow standards must be used for column calibration instead of the typically used polystyrene, which is insoluble in HFIP. Very low refractive index and consequently dn/dc significantly higher than in other solvents make HFIP an ideal solvent for SEC measurements with a light scattering detector.

Dimethylformamide (DMF) is typically used with the addition of about 0.1% lithium bromide. It is a good solvent for polyacrylonitrile or poly(vinyl alcohol). Polystyrene standards are soluble in DMF, but interaction with polystyrene-based packing can influence the elution and thus poly(ethylene glycol) and poly(ethyle oxide) standards can be used instead.

N-methylpyrrolidone (NMP) has desirable properties such as low volatility, low flammability, relatively low toxicity, and ability to dissolve many polymers. It may represent a good alternative for polymers that are not soluble in THF.

Dimethylsulfoxide (DMSO) is another organic solvent that can be used for polymers that are insoluble in THF or other organic solvents. DMSO represents the best choice for urea formaldehyde resins that are insoluble in most other solvents. It can be also used for the analysis of starch.

o-Chloronaphthalene is an infrequently used solvent that can be used at very high temperatures of about 220°C for the characterization of polyphenylene sulfide and possibly other polymers with limited solubility.

Water, usually with the addition of various salts, is used for the analysis of proteins and other water-soluble polymers such as dextran, pullulan, hyaluronic acid, or poly(acrylic acid). Typical salts are $NaNO_3$ or Na_2SO_4 in concentration of about 0.1 M or phosphate buffer. Other possible additives include surfactants such as sodium dodecyl sulphate (SDS) or urea. The salt concentration and pH of aqueous solvents may alter solubility of biopolymers. The ionic strength affects the conformation and size of polyeclectrolytes that expand at low salt concentrations and create random coil conformation when repulsive electrostatic forces are sufficiently shielded by salt ions.

Mixed eluents can be used in some special cases in order to achieve solubility, and suppress aggregation or interactions of solute with column packing. For example, addition of 10–20% organic solvent (methanol, acetonitrile, THF) to water can suppress hydrophobic interactions of water-soluble polymers with column material. Addition of a small level of methanol to THF can promote solubility of polar compounds. However, the separation process and obtained results can be affected by preferential solvation of the analyzed polymer with one component of the mobile phase. Other possible issues are higher tendency of RI signal to drift due to preferential evaporation of one of the components and extremely intensive solvent peaks at the end of the chromatogram that can affect the determination of peak limits in the case of imperfect separation from the main polymer peak.

All solvents should be of purity above 99% and free of mechanical impurities. Pre-filtration of solvents such as toluene or THF is not necessary, but it should be always performed in the case of solvents containing salts, because the salts, even if of high chemical purity, often contain a substantial level of particulate matter, such as mechanical impurities from the manufacturing process or bits of plastic generated by rubbing the plastic container during transport and handling.

3.3.2 Columns and Column Packing

SEC columns are the most important part of an SEC instrumental setup that is critical for efficient separation. Today's columns are typically 300 mm in length and 7.5, 7.8, or 8 mm in inner diameter. Stainless steel is the most frequent material for SEC columns. Biocompatible glass columns are used for the separation of biomaterials that can degrade in contact with metals. Narrow-bore columns of 4.6-mm inner diameter allow significant reduction of solvent usage and thus solvent purchasing and disposal costs. To maintain the same linear velocity through the columns, the volume flow rate must be reduced by a factor of about three, which results in significantly lower solvent consumption. A disadvantage of narrow-bore columns is that the band-broadening effects are more severe because the contribution of the outside column volume in connecting tubings and detectors is related to smaller column volume and thus it is relatively more significant. To use the potential of narrow-bore columns it is necessary to minimize the system dispersion and therefore the columns are not well suitable for multiple detection systems. Preparative columns of dimension, for example, 250 × 25 mm, are generally packed with the same packing materials as those used for the analytical columns. The columns allow fractionation of the sample into several fractions by collecting the eluent and subsequent solvent evaporation. Compared to regular 300 × 7.5-mm analytical columns, the 250 × 25-mm columns offer more than 10 times scaleup. The applicability of semipreparative SEC is limited by high column price and significantly smaller load and resolution compared to interaction types of liquid chromatography.

Rapid columns of dimensions 150 × 7.5 mm or 100 × 10 mm can be used when significantly reduced run time is requested. They can find utilization in direct manufacturing control when the results are needed in a very short time or for the high-throughput screening measurements of large sample series. The fast SEC columns bring several benefits, such as analysis time less than 7 minutes, increased sample throughput, reduced solvent consumption and consequently reduction of purchase and waste disposal costs, and lower column pressure and consequently reduced possibility of shearing degradation. In addition, the purchase price of a single fast column is markedly below that of several conventional columns, and low solvent consumption allows use of expensive solvents such as

HFIP. However, the selectivity and efficiency of a fast column are significantly lower compared to a system of two or three regular columns and thus their applications are limited.

Guard columns are used to protect the main analytical columns from materials that can irreversibly adhere to column packing. They are generally 50 mm in length of inner diameter as the analytical columns, packed with the same packing material as the main analytical columns. The packing material in the guard column should trap materials that otherwise would bind irreversibly to the top of the analytical column, thus prolonging the analytical column lifetime. Although a guard column can in some cases protect and extend the lifetime of the analytical columns, not all contaminants are commonly trapped by flow through a 50-mm-long guard column. Although one might think that the extra column length added by the guard column should improve the separation, in fact the guard columns do not contribute to or even slightly reduce efficiency and separation. The reason for that may be extra volume added by the additional fittings and tubing. It may also not be unambiguous to decide when the guard column needs to be replaced. Since the guard columns are relatively expensive, it may not be acceptable to change them on regular basis. A substantially cheaper online filter with replaceable stainless-steel frit can be alternatively used to protect the analytical columns from blockage with insoluble impurities. It must be emphasized that blocked frit of the inline filter can significantly increase the shearing degradation of the high-molar-mass polymers, as demonstrated in Figure 3.8. An inline filter can be installed after the injector even if the guard column is used.

Styrene-divinylbenzene gels are the most widely used packing materials for organic applications. HEMA-based gels are frequent materials for aqueous applications. Other packing materials include, for example, polyacrylamide gels, sulfonated polystyrene, poly(vinyl alcohol) gels, and gels based on dextran crosslinked with epichlorohydrin. Silica-based packing materials are highly rigid and allow application in a wide range of solvents including water and organic solvents. The silanol groups on the surface of silica gel can be deactivated by chemical modification using, for example, glycidoxypropyltrimethoxysilane, which is bonded to silanol groups, and consequently the epoxy ring is opened to form a hydrophilic surface covered with dihydroxy groups. The silica-based columns with organic hydroxyl modification are often used for the characterization of proteins. Modification with alkyl chlorosilane can create a hydrophobic surface usable for applications in organic solvents.

In order to analyze polydisperse polymers it is necessary to cover sufficiently broad molar mass range. The traditional approach is to combine several individual pore size columns in series. Individual pore size columns are available in several grades with different separation range. Traditionally, the packing materials are designated as 100, 500, 1000, 10^4, 10^5, and 10^6 Å with corresponding separation limits up to 4×10^3, $500 - 3 \times 10^4$, $500 - 6 \times 10^4$, $10^4 - 6 \times 10^5$, $6 \times 10^4 - 2 \times 10^6$, and $6 \times 10^5 - 10^7$ g/mol. The exclusion limits are approximately valid for polystyrene in THF and can vary depending on polymer chemical and molecular structure, temperature, and solvent. The symbol Å

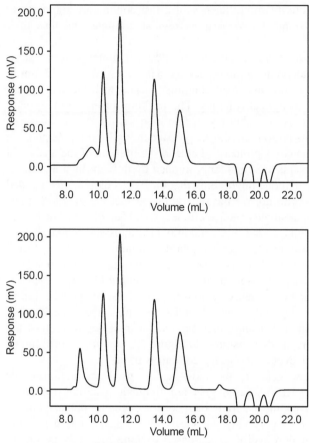

Figure 3.8 Chromatograms of PS standards ($2.85 \times 10^6, 4.7 \times 10^5, 1.7 \times 10^5, 1.9 \times 10^4$, and 3600 g/mol) with blocked (top) and clean (bottom) frit of inline filter.

does not apply to the real pore diameter, but it was historically introduced as the exclusion limit expressed by fully extended polystyrene chain.

The individual pore size columns must be combined so that the separation ranges of the columns overlay, which means there are no gaps in the pore volume distribution. An improper combination of the individual pore size columns may result in false peak shoulders that are due to the mismatch of the pore size and volume. That means the unusual peak shapes are artifacts and not reflections of the true pattern of molar mass distribution. A more recent approach for covering a broad separation range is using mixed columns packed with a mixture of individual pore size materials that are blended to cover a specific molar mass range.

Although the columns are often blended in order to get linear calibration curves, the real calibration curves are slightly curved as in the case of individual pore size columns. The mixed bed columns are typically available in several types covering different molar mass range. For example, PLgel mixed columns are labeled as E, D, C, B, A and cover the molar mass range up to 3×10^4, 200 − 4×10^5, 200 − 2×10^6, 500 − 10^7, 2000 − 4×10^7 g/mol, respectively. Packing materials with multipore structure instead of the combination of individual pore size materials is the most recent development in column technology.

Generally, mixed and multipore columns simplify the column selection and reduce the possibility of artifacts in the peak shape. A significant advantage is that very different molar masses and polydispersity can be analyzed on the same column set. To increase the number of theoretical plates the mixed and multipore columns are usually connected in series. For most practical cases the combination of two 5- or 10-μm columns is adequate and using three or four columns usually does not bring more information about the analyzed polymers. Three to four columns should be used in the case of 20-μm packing designed for the analysis of polymers containing fractions with molar mass up to several tens of millions g/mol. The calibration curves for individual and mixed columns are usually available in the manufacturers' brochures and they can be used to choose an appropriate type of columns. The manufacturers of SEC columns include Waters (Ultrastyragel, Ultrahydrogel columns), Polymer Laboratories—a part of Varian (PLgel, PL aquagel-OH), Shodex (KF-800, K-800 series), Tosoh (TSK-GEL), Phenomenex (Phenogel), Jordi, Polymer Standards Service, and MZ Analysentechnik.

Besides pore volume and pore size, the particle size of column packing is another important parameter related to both the number of theoretical plates and backpressure. The two quantities increase with decreasing particle size. Available particle sizes are 3, 5, 10, and 20 μm. The higher the column backpressure the higher the possibility of degradation of high-molar-mass fractions of the sample by shearing forces. Small particle size is used for the analysis of oligomers and lower-molar-mass polymers, whereas 20-μm packing is used for the characterization of high-molar-mass polymers to reduce shearing degradation, which can happen for molar masses in the range of several millions and more.

SEC columns are not only important, but also a relatively expensive part of an SEC setup and proper care and handling is necessary to prolong their lifetime. Columns that are frequently used will not last forever. Experience shows that columns that are in everyday service can show appropriate separation efficiency even after several years although the same type of columns may deteriorate within a few months or even faster. Besides the column handling, the chemical nature of the samples that are analyzed is the most important for column duration. Samples containing microgels and other insoluble fractions and highly polar functional groups are more likely to reduce column lifetime compared to well-soluble neutral polymers.

Excessive peak broadening and tailing due to the use of columns that have already lost efficiency result in erroneous results, usually underestimated M_n and overestimated M_w and polydispersity. SEC column cleaning or regeneration is mostly impossible and columns that have lost resolution should be replaced.

Columns do not lose separation efficiency even when stored for several years assuming they are stored at room temperature with both ends tightly plugged to prevent solvent evaporation. In the case of aqueous columns, the solvent in the columns should be replaced with about 0.02% sodium azide or alternatively with 5% aqueous methanol to prevent growth of microorganisms. Unlike many HPLC columns, SEC columns packed with swollen organic gels must be prevented from drying since deterioration of column performance as a result of drying is irreversible. Maximum flow rate and column pressure are specified by the column manufacturers.

To avoid column damage the maximum operating pressure recommended by a manufacturer should not be exceeded even for a short period of time. The flow direction is labeled directly on the columns. A typical flow rate for regular 7.5-mm inner diameter columns in THF is 1 mL/min. Lower flow rates of 0.5 or 0.8 mL/min are mostly used in aqueous solvents. When using columns with increased or decreased inner diameter, the volumetric flow rate should be adjusted accordingly to maintain an equivalent linear velocity through the column. Higher viscosity eluents such as DMSO should be used at reduced flow rates or elevated temperatures. Columns packed with styrene-divinylbenzene–based gels must not be used in non-solvents for polystyrene such as water, alcohols, or hydrocarbons. Although modern styrene-divinylbenzene packing materials are highly rigid and my own experience indicates that the columns can survive even a short accidental flushing with methanol, the use of the previously stated solvents should be avoided.

Using a stepwise change of the flow rate in several increments or continuous flow rate gradient over several minutes to reach a full flow rate or to stop it was traditionally recommended to protect the columns from mechanical damage. This practice can be still obeyed even though it may not be absolutely necessary with today's rigid packing materials. It must be stressed that even microscopic shrinkage of the gel bed due to the shrinkage by chemical or mechanical exposure or solvent evaporation results in voids that cause severe peak tailing and reduction of plate numbers. One defective column typically causes peak spreading that cannot be overcome by any number of good columns and thus one defective column in a series causes poor separation efficiency of the entire column set.

The columns can be transferred from one solvent to another assuming the solvents are miscible. The transfer should be performed at a low flow rate of about 0.2 mL/min. However, to maintain high column efficiency the solvent replacement should not be done too often. Instead, it is preferable to have several column sets in different solvents. Some manufacturers offer columns directly packed in special solvents such as DMF or HFIP. Styrene-divinylbenzene–packed columns allow working at elevated temperatures around 150°C. The high temperature is necessary for the analysis of polyolefins and other polymers of limited solubility. To avoid an excessive temperature shock the column temperature should be raised or lowered at a reduced rate, such as 1°C per minute, and if the pump is intended to stop it should be stopped after room temperature is achieved. The pH of the mobile phase is important in aqueous silica–based columns, which should be usually in the pH range of 3–7.5. The pH range of aqueous columns is a part of the manufacturer's recommendations. Although UV detection may demand working in nonstabilized THF, it should be replaced with stabilized THF before storage or when only RI detection is needed.

To eliminate the effect of possible partial solvent evaporation at the column inlet and outlet during long storage, the column can be connected in a reversed direction; that is, the outlet of the column is connected to the solvent delivery system and the pump is activated at a low flow rate of about 0.2 mL/min until a few drops appear at the column inlet. Then the column is connected in a normal direction and the flow rate is increased.

Caution should be taken when connecting columns into series and to an injector and a detector. Leaks in the connections can result in detector noise, drifting baseline, and elution volume irregularities, and leakage inside detectors can even result in their complete damage. It is necessary to keep in mind that improperly tightened fittings may cause leaks, but overtightening should be avoided as well, especially when the fitting is reused. Testing the connections for leaks can be most effectively performed by dabbing with a piece of folded filter paper and identifying the leak as either a dark spot or smell of organic solvent on the paper. A new fitting should be used rather than overtightening when slightly tightening the fitting does not solve the leakage problem.

3.3.3 Detectors

Detectors are used to monitor the molecules of solute eluting from the separation system. The refractive index (RI) detector, the ultraviolet (UV) detector, the light scattering photometer, the viscometer, the infrared (IR) photometer, and the evaporative light scattering detector (ELSD) represent the most important detectors in SEC. The detector monitors the change of the mobile-phase composition and converts it into an electrical signal, which is further acquired and processed by computer and SEC software.

The RI and UV detectors represent concentration-sensitive detectors with the signal proportional solely to the concentration of sample in the eluate. The signal of a light scattering detector is proportional to the product of molar mass and concentration, while the viscometer signal is proportional to the product of concentration and molar mass to the power of the Mark-Houwink exponent. That means that for most polydisperse polymers the peak of the viscometer is after the peak of the light scattering detector, with the exception of polymers with the Mark-Houwink exponent ≥ 1, where the peaks overlay or the viscometer peak can even forego the light scattering one.

The most important parameter for the concentration detectors is the *detection limit*, which represents the minimum injected amount detectable under given chromatographic conditions. The smallest detectable signal is usually considered to be double the height of the largest noise spike. When the baseline signal is zoomed enough, one can see that the baseline is irregular even if no peak elutes

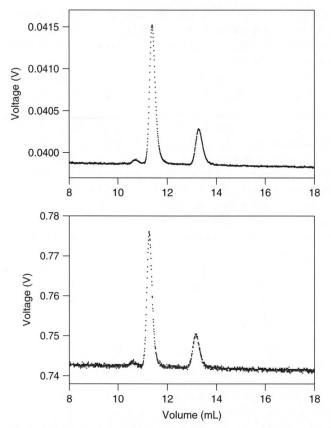

Figure 3.9 Comparison of signal-to-noise ratio for two generations of light scattering detectors. Data measured under identical conditions with the two detectors connected in series.

from the columns. Although the noise should be minimized, its absolute level is not as important as the ratio of the signal compared to the baseline noise (i.e., the *signal-to-noise ratio*). For quantitative analysis the signal-to-noise ratio should be at least ten. If the signal-to-noise ratios of different detectors are to be compared, they should be determined under identical experimental conditions; preferably the two detectors should be connected in series. An example of such a comparison is shown in Figure 3.9 for two generations of light scattering detectors.

Drift is an even increase or decrease of the baseline signal that appears as the slope. It can be normal for a certain period of time due to warming up of the lamps and electronics and column flushing and stabilization. After warmup, the baseline signal should show negligible change with time. The most important sources of drift are elution of the old mobile phase after the refill or change of the eluent, change of the mobile-phase composition due to evaporation of one component in the case of mixed solvents or buffers or chemical instability such as in the case of THF, change of temperature, and elution of compounds adsorbed in columns from previous injections. The signal noise and drift specified by the detector manufacturers are usually far below the noise and drift that originate from the instability of the chromatographic system.

The *linearity* of detector response means that a detector yields signal with a peak area directly proportional to the injected amount. The plot of peak area versus injected mass should be linear in a wide range with the slope giving the detector *response*. The detector response is different for different compounds depending on the detector type and properties of a specific compound. The extinction coefficient, the specific refractive index increment dn/dc, and the product of $(dn/dc)^2$ and molar mass are the relevant properties for UV, RI, and light scattering detectors, respectively. That means different compounds can show significantly different detector responses when injected in the same amounts.

The *time constant* characterizes how quickly a detector can record a peak. It can be defined as the time needed to reach a certain percentage of full scale (e.g., 98%). It is expressed in seconds and generally the detector noise decreases with increasing time constant, but it should not be too high for detecting very narrow peaks. Most detectors offer a choice of the time constant. In SEC, the polymer peaks are usually broad and there is no need to use small time constants.

Volume of the flow cell and cell path length are other important detector characteristics. The volume of the detector cell contributes to the peak broadening. On the other hand, too short pass length has a negative impact on the sensitivity, whereas detectors with short path lengths can be used for semi-preparative analyses. A standard cell volume of RI or UV detector is $8-10~\mu L$. Regarding peak broadening, the cell volume must be related to peak width of the narrowest peak in the chromatogram. Since SEC chromatograms are usually broad, covering elution volume of several milliliters, detector cell volume is not a serious issue.

Different types of detectors can show significantly different sensitivity to the baseline noise and drift. An RI detector is more sensitive to mobile-phase composition and temperature changes compared to a UV detector. A light scattering detector is very sensitive to bleeding of particles from the column packing, which are completely unseen by RI and UV detectors, whereas its sensitivity to temperature or flow rate fluctuations is negligible. A randomly noisy detector baseline can arise from several sources. These include bubbles caused by improper solvent degassing, pump malfunction, faulty detector lamp, electric power fluctuation, temperature instability, disturbance from the environment (e.g., nearby improperly shielded electrical appliances or air conditioning), or sometimes even slight pulses from a dripping waste tube.

Noise originating in the pump is usually indicated by pressure pulsation and it can be demonstrated by turning off the pump and obtaining a noise-free (yet usually drifting) baseline. The correction of the pump problem usually requires changing the pump seals, check valves, and pistons. If the noise persists when the pump is off, the detector lamp and other possibilities should be checked.

3.3.3.1 UV Detector

A *UV detector* is the most common type of detector used in liquid chromatography and can be used for compounds containing a double bond, conjugated double bonds, an aromatic ring, a carbonyl group C=O, or a nitro group NO_2. Its applicability in SEC is limited by the fact that many important polymers have weak or no absorption of UV light. The applicability of the UV detector is also limited by the fact that THF has a higher UV cutoff of about 230 nm compared to other common HPLC solvents such as methanol or acetonitrile. The applicability of a UV detector includes polystyrene and styrene-containing copolymers, nitrocellulose, bisphenol A–based epoxy resins, phenol-formaldehyde resins, unsaturated polyesters, alkyds, poly(ethylene terephthalate), and proteins. The UV detector measures the absorbance (*A*) of the eluate, which is defined as the logarithm of the ratio of the intensities of the incident light (I_0) and the transmitted light (*I*). According to the Lambert-Beer law the absorbance is related to the absorption (extinction) coefficient a (mLg^{-1} cm^{-1}), the concentration c (g/mL), and the cell length L (cm):

$$A = \log\left(\frac{I_0}{I}\right) = acL \tag{3.19}$$

For most samples, namely those with UV absorbance given by the monomer unit, the response of the UV detector is proportional to the injected mass of the analyzed sample. However, in the case of polymer molecules bearing a single absorbing unit, the UV response is proportional to the number of eluting molecules. In such a specific case the molar mass of molecules at a given elution volume can be determined from the ratio of the signal of a mass-sensitive detector (usually a refractometer) to that of a number-sensitive detector.

The sensitivity of a UV detector is expressed as the response factor in absorbance units (AU) per volt. For example, the response factor one AU per volt means that the absorbance one AU causes the signal output one volt. The path length of a typical UV detector is 10 mm, but the thickness is small in order to keep the total cell volume small. The family of UV detectors includes a fixed-wavelength detector, a variable-wavelength detector, and a photodiode array (PDA) detector. Technical details can differ for the particular models of different manufacturers. Modern UV detectors do not use a reference cell and zero baseline is adjusted electronically using the signal from a reference photodiode.

Some detectors allow simultaneous data acquisition at dual wavelength. A programmable-wavelength UV detector consists of a deuterium lamp, an optical system of mirrors that collects light from the lamp and directs it toward the grating. The operation wavelength is determined by the position of grating. The light of a particular wavelength is focused onto the entrance of the flow cell and the transmitted light is detected by a photodiode. A beamsplitter located in front of the flow cell diverts a portion of the light to a reference photodiode. In the case of a PDA detector the light from the deuterium lamp passes directly through the flow cell. The sample in the flow cell absorbs at specific wavelengths and the light exiting the flow cell is directed onto the grating. The light from the grating is dispersed into 1.2-nm wavelength beams that are recorded by the array of 512 photodiodes.

Usually, the response from several photodiodes is cumulated, so that the spectral band is larger than 1.2 nm (e.g., 2.4 nm, 4.8 nm). The reference diode receives the light through the beamsplitter assembly. The PDA detector acquires all data in a specified wavelength range and thus UV spectra of compounds eluting at particular elution volumes can be obtained and the chromatograms can be extracted from the collected data at any suitable wavelength.

The UV detectors offer a highly sensitive detection of UV-absorbing compounds, but they cannot be applied for the detection of a wide range of nonabsorbing compounds. For interpretation of the experimental results it is necessary to keep in mind that two peaks of about the same area can correspond to compounds that are present in concentrations different by several orders of magnitude. Similarly, a minor peak in a chromatogram can correspond to a weakly absorbing compound that may represent a substantial part of the analyzed sample.

If the UV detector is used in combination with a light scattering detector or an online viscometer, the absolute concentration c_i of the molecules eluting at the ith elution volume increment i can be determined according to Equation 3.20:

$$c_i = \frac{RF_{UV}(V_i - V_{i,baseline})}{aL} = \frac{\alpha_{UV}(V_i - V_{i,baseline})}{a} \tag{3.20}$$

where RF_{UV} is the UV detector response factor in absorbance units per volt, L is the flow cell length, a is the extinction coefficient ($mLg^{-1} cm^{-1}$), and V_i and $V_{i,baseline}$ are the detector signals in volts for sample and baseline, respectively. The ratio of RF_{UV}/L represents the calibration constant of the UV detector (α_{UV}). The UV response factor supplied by the manufacturer is generally accurate and can be easily verified. The procedure for determining the UV detector calibration constant is similar to that for determining the RI calibration constant. The only difference is that the extinction coefficient is used instead of the dn/dc. *Note:* SEC-MALS analysis with a UV detector requires not only dn/dc, but the extinction coefficient as well.

3.3.3.2 Refractive Index Detector

Refractive index (RI) (differential refractive index, DRI) detectors are the most common detectors in SEC. They can be used universally for all compounds with a non-zero specific refractive index increment. Although dn/dc of different polymers can differ substantially and also negative values are possible, dn/dc values close to zero are rare. In the case of zero dn/dc the detector response must be improved by the choice of a different solvent of different refractive index. For compounds with high absorption of UV light, such as those containing aromatic rings, the sensitivity of an RI detector may be significantly lower compared to a UV detector. On the other hand, for many nonabsorbing compounds the RI detector is the only choice.

Compared to UV detectors the RI detectors are highly sensitive to temperature and flow rate fluctuation. To eliminate the influence of the temperature, the cell is incorporated in a massive metal thermostated block. Most RI detectors have no cooling capability and thus the operating temperature must be set about $10°C$ above room temperature. The eluent passes through relatively long stainless-steel tubing before it reaches the flow cell to equilibrate the temperature with that of the cell. This fact typically results in significantly larger interdetector volume compared to UV detectors. Instability of the flow rate causes a significant increase of noise level. A malfunction of pump seals or check valves results in a regular short-term pulsation of a baseline. Therefore, a properly working pump is essential for keeping the noise level low. The RI detectors are highly sensitive to any changes of the composition of eluate and it typically takes several hours to flush the SEC columns and stabilize the signal. A solvent refill during the pump operation usually results in a significant change of the baseline signal especially in the case of THF. Regular fluctuation with a several-minute period may be due to an online vacuum degasser, which keeps the vacuum level within certain limits. In the case where pressure gets over a maximum limit the pump of the degasser starts working until the pressure drops to the desired limit. Due to the vacuum fluctuation the level of degassing fluctuates as well and the RI detector signal regularly fluctuates because of different levels of solvent degassing. This problem does not appear with degassers working in continuous mode, that is, keeping the chamber pressure at a constant value, or when degassing is achieved with helium sparging.

A typical feature of RI detectors is a low-pressure resistance of the flow cell, which usually does not allow connection of other detectors after the RI detector.

The principle of an RI detector is based on the refraction (bending) of a light beam when passing from one medium into another at an angle that is not perpendicular to the interface surface. The extent of the light refraction is related to the difference of the refractive indices of the two media.

A *deflection*-type RI detector has a flow cell separated into two parts. During the purge period both parts of the flow cell, that is, the reference cell and the measuring cell, are filled with a solvent eluting from the columns. When both parts of the cell are properly flushed the reference part is closed and the eluent passes only through the measuring cell. Shortly after the purge the refractive indices of solvent in the reference cell and measuring cell are identical. The light from the lamp is focused and passes both parts of the flow cell and the deflection of the light beam, which is proportional to the difference of the refractive indices in both cells, is monitored by a photodiode. The zero detector output can be adjusted optically or electronically. As the sample components eluting from the column pass through the measuring cell, the refractive index of solution changes and the light beam passing though the cell is refracted.

Conventionally, a dual-element photodiode is used to measure the light beam angular deflection. The dual-element photodiode contains two photodiode light detectors placed side by side. As the light beam bends, it moves away from one photodetector onto the other. The actual light beam position is characterized by the voltage difference between the two elements of the photodiode, which is proportional to the refractive index difference between the measuring cell and the reference cell, which is proportional to the concentration of the component in the measuring cell. Once the light beam moves entirely off of one photodetector and onto the other, there is no way to determine the beam position and the detector signal saturates. An innovative way to measure light beam deflection is based on an array of 512 light-measuring elements instead of using two elements (Optilab® rEX of Wyatt Technology Corporation). Each element of the photodiode array, called a pixel, precisely measures the intensity of light impacting it. The data from the 512 pixels are analyzed using mathematical algorithms to determine the position of the light beam on the array. Using a photodiode array and advanced mathematical analysis techniques, the light beam position is measured with very high accuracy. The photodiode array is 1.3 cm long, and so the beam may move more than a centimeter before it slides off the end of the photodiode array.

A significant advantage of this type of RI detector is that it does not require any change of sensitivity setting. In contrast to conventional RI detectors, for which high sensitivity is associated with a limited dynamic range or a great dynamic range is associated with reduced sensitivity, full sensitivity is achieved over the entire range and thus small peaks can be detected alongside huge peaks within a single run with no signal saturation.

Another type of RI detector is an *interferometer*, where the light beam is split into two beams of equal intensity by a beam splitter. One beam passes through a reference cell, the other through a measuring cell. As the beams enter the cells they are in phase with one another. After passage through both cells the two beams are recombined by a second beam splitter. If the refractive indices of the liquids in the reference and measuring cell are different, the two light beams are phase shifted, which results in an attenuation of the light intensity due to interference.

In contrast to conventional SEC, the use of an RI detector with a light scattering photometer or a viscometer requires determination of concentration in absolute units (g/mL). To determine the absolute concentration one must know the RI detector calibration constant (in refractive units per volt). Unlike the response factor of UV detectors, the calibration constant provided by an RI detector manufacturer (with the exception of Optilab® rEX) is not accurate and must be determined (see Section 2.6).

3.3.3.3 Infrared Detector

Infrared (IR) detectors work on a principle similar to UV detectors. The major limitation of the IR detector in SEC is the fact that the solvents used as the mobile phase also have a strong absorption at wavelengths that could be potentially used to monitor eluting polymers. The IR detector currently finds application mainly in the case of SEC of polyolefins in trichlorobenzene, where it can be used not only for monitoring the concentration but also for the determination of chemical heterogeneity of polyethylene copolymers and branching in polyethylene. The determination of short-chain and long-chain branching distribution across the elution volume axis is based on the simultaneous measurements of a selective absorbance in the CH_3 region and a broad-band absorbance of CH_2 and CH_3. The ratio of these two signals is directly proportional to the number of branches. In the high-temperature SEC of polyolefins, the IR detector can be used as an alternative to an RI detector with the advantage given by high sensitivity and signal stability. An IR detector suitable for a high-temperature SEC analysis of polyolefins is available, for example, from the Spanish company PolymerChar.

In the case of combination with a light scattering detector, the IR detector must be calibrated in order to provide absolute concentration. The calibration is equivalent to the determination of the extinction coefficient by the UV detector in online mode. A significant advantage of the IR detector over the differential refractometer in the combination of SEC with light scattering and viscosity detectors is that the IR detector can be connected as the first detector in the series and thus the measurement of concentration is not affected by the dilution effect of the viscometer.

A possible solution of the limitation caused by solvent absorption is using an interface that permits removing the solvent from the eluent. In such an application the IR detector is not used as the concentration-sensitive detector, but provides information about chemical composition of sample eluting from the SEC columns. The sample can be deposited continuously onto a rotating germanium disk that can be subsequently scanned to provide the polymer composition as a function of molar mass, or strictly speaking, as a function of elution volume. The eluent from the column is sprayed with a nitrogen stream through a nozzle to the disc. The nonvolatile solute from the eluent is deposited on the disc, from which the IR spectra are measured. Alternatively, the sample can be deposited as a series of spots on the surface of a moving stainless-steel belt. The belt continuously transfers the spots into the diffuse reflectance accessory of the FTIR spectrometer, enabling identification of the deposited solutes by measurement of the diffuse reflectance IR spectra.[11] Although the principle appears promising for the characterization of chemically heterogeneous samples, it has not found a wide application.

3.3.3.4 Evaporative Light Scattering Detector

The *evaporative light scattering detector* (ELSD) is another type of universal detector representing a possible alternative to the RI detector. In contrast to the RI detector, the ELSD allows the use of mobile-phase gradient, which may be of interest for other types of liquid chromatography. The ELSD offers some potential advantages, such as no requirement for the sample to have a strong absorbing chromophore, universal response with no effect of absorption coefficient or *dn/dc*, and high sensitivity. The eluent is sprayed into a stream of gas, the solvent is vaporized, and the solute that is less volatile than the solvent creates an aerosol of small particles. The particles scatter the light from the light source. The scattered light intensity is proportional to the concentration of the eluting polymer. The response of the detector can be optimized by the temperature of the evaporator and the flow rate of the nebulizing gas (typically nitrogen). The temperature of the evaporator is selected according to the type of sample and mobile phase. Low-molar-mass compounds and lower oligomers can be lost with the nebulizing gas. Compared to UV and RI detectors, the ELSD is destructive and cannot be used for preparative applications. The ELSD is sometimes confused with light scattering detectors.

3.3.3.5 Viscosity Detector

A *viscosity detector* (*online viscometer*, *viscometric detector*) is exclusively used in combination with SEC. It is often considered to be a molar mass–sensitive detector, even though the primary quantity measured by a viscometer is not molar mass, but the specific viscosity, which is, using the concentration from a concentration-sensitive detector, converted into the intrinsic viscosity. Having the intrinsic viscosity, one can read the molar mass from the universal calibration dependence. However, the method suffers from drawbacks similar to conventional SEC, namely sensitivity to the flow rate variations, temperature fluctuations, non-size-exclusion separation mechanisms, and efficiency of the columns. The failure of the universal calibration for polyelectrolytes in aqueous solvents was reported.[12]

An online viscometer measures a pressure drop between a line containing the polymer solution and one containing pure eluent. According to Poiseuille's law for laminar flow, the measurement of viscosity can be replaced by the measurement of pressure difference:

$$\dot{V} = \frac{\Delta P \pi R^4}{8 \eta L} \tag{3.21}$$

where $\dot{V}$ is the volumetric flow rate through the tube (mL/min), ΔP is the pressure difference across the tube (i.e., the inlet pressure minus the outlet pressure), R is the radius of the tube, η is the viscosity, and L is the length of the tube.

Online viscometers are available in several different designs. The simplest type, measuring the pressure drop across a capillary, is highly sensitive to flow rate and temperature fluctuations. A significant reduction of signal noise is obtained with the bridge type of viscometer, outlined in Figure 3.10. The fluid stream splits at the top of the bridge, and half of the solvent flows through each arm. Since the bridge is symmetric, the differential pressure transducer in the center of the bridge measures zero. When a sample elutes, it is also split evenly. In the left arm of the bridge is a delay volume, where the eluting sample is retained. At the time of sample elution the sample enters the delay volume, but solvent still exits, causing a pressure imbalance in the bridge. This imbalance pressure, combined with the inlet pressure, gives the specific viscosity (η_{sp}) according to the equation:

$$\eta_{sp} = \frac{\eta}{\eta_0} - 1 = \frac{4 \Delta P}{IP - 2 \Delta P} \tag{3.22}$$

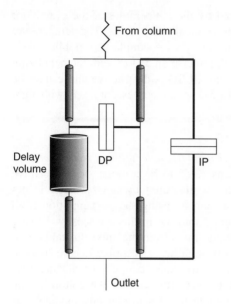

From column

Delay volume

DP

IP

Outlet

Figure 3.10 Scheme of four-capillary bridge viscometer ViscoStar™. DP = differential pressure transducer, IP = inlet pressure transducer. *Source:* Courtesy of Wyatt Technology Corporation.

where η is the viscosity of the sample solution, η_0 is the viscosity of the solvent, ΔP is the imbalance pressure across the bridge, and IP is the inlet pressure (i.e., pressure difference between the top and the bottom of the bridge).

According to Equation 3.22, the specific viscosity measurement is performed by the two pressure measurements ΔP and IP. That means the accuracy of the specific viscosity is based only on the calibrated transducers. The measurement is independent of the flow rate, but both ΔP and IP are directly proportional to the applied flow rate, which can be in some cases used to improve the signal-to-noise ratio, because most of the HPLC pumps operate better at higher flow rates. However, typical flow rates used for the measurements with an online viscometer do not differ from those used for other types of SEC measurements.

At the end of the run, the delay volume is flushed with new solvent, which causes a negative signal of the viscometer and a positive signal of the RI detector (see Figure 3.11). The volume delay (a set of three empty columns) can be changed by a disconnection of one or two delay columns to match the configuration of the viscometer with the number of SEC columns and type of analyzed samples. The need to flush the delay volume reservoir prolongs the run time and thus the use of narrow-bore 4.6-mm columns requiring lower flow rates is not advisable.

When combined with a concentration detector, the online viscometer can be used to determine the intrinsic viscosity using the well-known relation:

$$[\eta] = \lim_{c \to 0} \frac{\eta_{sp}}{c} \tag{3.23}$$

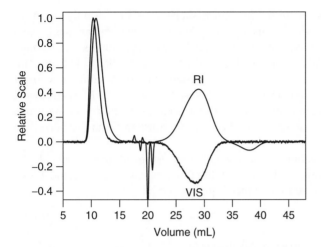

Figure 3.11   Positive and negative signals of RI detector and viscometer caused by flushing the delay volume of the viscometer.

where c is the concentration of polymer molecules eluting from the columns. As in the case of light scattering detectors, the concentration measured by a concentration detector must be expressed in g/mL, which requires the absolute calibration constant of an applied detector to be known. In fact, without correcting for band broadening, the intrinsic viscosity is the weight-average of the molecules eluting at a given elution volume. However, band broadening is mostly neglected and the values of intrinsic viscosity are assumed to be measured for monodisperse elution volume slices.

Since the RI detector is usually connected as the last detector in the series, the concentration of the eluting molecules is decreased by the split ratio between the two sides of the viscometer bridge. That means the sample that exits the viscometer is diluted by a factor of about two. Therefore the RI detector does not measure the same concentrations that flow through the detectors connected before the viscometer. If the two arms were absolutely identical, the dilution factor would be exactly 50%. In practice, the two arms are never exactly identical.

To correct for the change in concentration, it is necessary to determine the exact dilution ratio. The determination of the dilution factor is simple and involves injection of a sample giving an easily processable chromatogram (e.g., narrow standard well separated from the solvent peaks). The exact sample concentration need not be known. The sample is injected with and without the viscometer and the dilution factor is simply computed as the ratio of the two peak areas.

The data from the viscometer and concentration detector allow not only the determination of the molar mass using the universal calibration, but also the computation of the distribution of intrinsic viscosity and intrinsic viscosity averages. Some scientists have suggested using the particular averages of intrinsic viscosity instead of molar mass averages for polymer characterization and relation of the molecular structure with application properties.

Intrinsic viscosity is one of the fundamental characteristics of polymer materials. Its practical application is mainly in combination with molar mass, especially for the characterization of macromolecular size and polymer branching. That means the most efficient application of the viscometric detector is in combination with a light scattering detector that measures molar mass directly without need for universal calibration. The data can be used to generate a Mark-Houwink plot, that is, log–log relation between the intrinsic viscosity and molar mass, and to calculate the hydrodynamic radius through the Einstein-Simha relation (see Equation 1.82).

An interesting application of the viscosity detector is an alternative calculation of the number-average molar mass without a concentration detector. Using the definition of hydrodynamic volume (Equation 1.81):

$$V_h \approx [\eta]M \tag{3.24}$$

the number-average molar mass is calculated as

$$M_n = \frac{\sum_i c_i}{\sum_i \frac{c_i}{M_i}} = \frac{\sum_i c_i}{\sum_i \frac{c_i}{V_{h,i}/[\eta]_i}} \tag{3.25}$$

At low concentrations the intrinsic viscosity at the ith elution volume increment can be replaced by $\eta_{sp,i}/c_i$ and Equation 3.25 can be rearranged as

$$M_n = \frac{\sum_i c_i}{\sum_i \frac{\eta_{sp,i}}{V_{h,i}}} \tag{3.26}$$

The concentration (in g/mL) of molecules eluting at the ith elution volume slice is

$$c_i = \frac{mH_i}{\Delta V \sum_i H_i} \tag{3.27}$$

where m is the injected mass, H_i is the height of the RI chromatogram at volume slice i, and ΔV is the elution volume between two data points. Then

$$\sum_i c_i = m/\Delta V \tag{3.28}$$

and Equation 3.26 can be expressed as

$$M_n = \frac{m}{\Delta V \sum_i \frac{\eta_{sp,i}}{V_{h,i}}} \tag{3.29}$$

The hydrodynamic volumes at particular elution volume slices $V_{h,i}$ are directly determined from the universal calibration curve and the values $\eta_{sp,i}$ are measured by the viscometer. A significant advantage of Equation 3.29 is that the value of M_n is not influenced by the calibration constant of the RI detector or by the specific refractive index increment and thus the method represents an interesting way of determining M_n or verifying results obtained by other methods. Of course, one has to assume 100% mass recovery of the injected polymer from the columns. According to Equation 3.26, in the case of imperfect SEC separation, that is, when band broadening is not negligible or in the case of complex polymers, the molar mass determined at a given elution volume is the number-average.

For shear-sensitive polymers (i.e., those showing decrease of solution viscosity with increasing shear rate), the results can be affected by the *shear-thinning* effect. For a radial velocity profile $v(r)$ the shear rate $\gamma = dv(r)/dr$. The shear rate in a capillary in the distance r from the center of the capillary is:

$$\gamma = \frac{-4\dot{V}r}{\pi R^4} \tag{3.30}$$

where $\dot{V}$ is the volumetric flow rate and R is the radius of capillary. The negative sign is because the fluid velocity $v(r)$ is at maximum at the capillary center $v(0)$, and zero at the capillary wall $v(R)$. The maximum shear rate at the capillary wall is:

$$\gamma_{max} = \frac{4\dot{V}}{\pi R^3} \tag{3.31}$$

and the average value is:

$$\gamma_{average} = \frac{8\dot{V}}{3\pi R^3} \tag{3.32}$$

For a typical flow rate of 1 mL/min (i.e., 0.0167 mL/s) and a capillary of 0.025 cm, the average shear rate is about 900 cm^{-1}, which can be for very large molecules over the limit of Newtonian flow behavior. The limit of Newtonian behavior increases as concentration and molar mass decrease, and as ionic strength increases for the case of polyelectrolytes. At concentrations typically used in SEC experiments even shear-thinning polymers can be still on their Newtonian plateau. Behind the limit of Newtonian behavior the specific viscosity becomes underestimated. Note that the shear rate effect on the shear-sensitive polymers is the opposite in the case of an online viscometer compared to a capillary viscometer. In the case of the online viscometer, the concentration dependence of the specific viscosity is neglected, and the intrinsic viscosity is equaled with the reduced viscosity ($\frac{\eta_{sp}}{c}$). On the other hand, when the measurement is performed by a capillary viscometer, the intrinsic viscosity is determined from the y-axis intercept of the concentration dependence of reduced viscosity, and the non-Newtonian behavior, being more prominent at higher concentrations, in fact results in the decrease of the slope of the concentration dependence of the reduced viscosity and thus overestimation of the intrinsic viscosity.

3.3.3.6 Light Scattering Detector

Light scattering detectors are almost exclusively used in SEC and also in FFF measurements. In principle, it is possible to use a light scattering detector in combination with other types of liquid chromatography, but the majority of applications lie in the area of SEC. Light scattering detectors represent the most powerful detection in SEC since they eliminate the necessity of column calibration and provide information unavailable by conventional SEC. Three different types of light scattering detectors are used that differ in the method of the determination of molar mass. *Multi-angle* light scattering (MALS) detectors permit extrapolating the light scattering intensities measured at various angles to zero angle where the particle scattering function is unity. *Low-angle* light scattering detectors use an angle low enough to assume that the particle scattering function is equal to one. However, for very large molecules $P(\theta)$ can decrease considerably below one even at low angles. *Right-angle* light scattering (RALS) can be used either for small polymers assuming particle scattering function $P(\theta) = 1$, or as part of a so-called *triple detection* system (sometimes termed SEC^3) consisting of a RALS detector, a viscometer, and an RI detector. The $90°$ light intensity is used to estimate the molar mass considering the particle scattering function is unity. The approximate molar mass is used together with the intrinsic viscosity obtained from the online viscometer to estimate the RMS radius from Equations 1.83–1.85. Then the particle scattering function is calculated using the estimated R and the equation for a given particle shape. Increasingly accurate approximations of the particle scattering function and the molar mass can be calculated by iterations until the values obtained can be considered constant. The limitation of this procedure is that it requires using the particle scattering function appropriate to the analyzed molecules. Probably the most serious limitation of the triple detection approach is caused by the generally unknown Flory's constant Φ for a given polymer under given experimental conditions. For example, in the case of branched polymers Φ will vary with the degree of branching.

The theory of light scattering is explained in Chapter 2. The principles and methodology of the combination of SEC with multi-angle light scattering detectors are described in Chapter 4. An interesting application of a light scattering detector was described in reference 13, which suggests improved methodology of the determination of the link between SEC results and polymer rheological properties. The method may be useful for samples with a molar mass distribution skewed toward high molar masses, which show at low elution volumes an intensive signal of a light scattering detector, but a weak RI signal due to very low concentration of the molecules eluting at the beginning of the chromatogram (see Figure 3.12 for a sample of polyethylene). For such samples conventional SEC and SEC combined with light scattering and RI detectors may suffer from weak RI signal at the region of low elution volumes, which can result in poor precision of the higher-order molar mass averages (M_z, M_{z+1}).

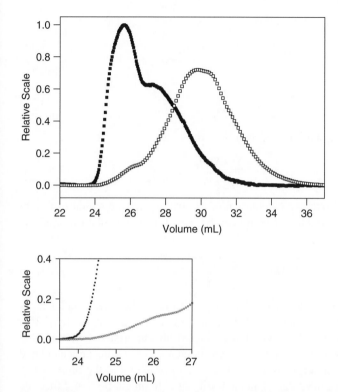

Figure 3.12   RI ($\square$) and $90°$ MALS ($\blacksquare$) signals for a sample of polyethylene illustrating the advantage of the calculation of M_z using solely light scattering signal (Equation 3.36).

The alternative calculation of higher-order molar mass averages requires a simple change of a computational algorithm. The light scattering signal is used instead of the RI signal to calculate the molar mass averages from the conventional SEC calibration curve determined by means of any of the available procedures. The advantage of the method is that it uses the high sensitivity of a light scattering detector to high-molar-mass fractions, but avoids the problem of diminishing RI signal at the region of low elution volumes. The calculation is explained in the following: Using the conventional calibration and RI detection approach, M_z is calculated as:

$$M_z = \frac{\sum c_i M_i^2}{\sum c_i M_i} = \frac{\sum RI_i M_i^2}{\sum RI_i M_i} \tag{3.33}$$

where M_i its the molar mass from the calibration curve at the ith elution volume slice, and c_i is the concentration of the molecules at the ith slice that can be replaced by the RI detector response RI_i. The light scattering signal LS_i is proportional to the concentration and molar mass of molecules eluting at the

SEC slice i:

$$LS_i \approx c_i M_i \approx RI_i M_i \tag{3.34}$$

and thus

$$RI_i \approx \frac{LS_i}{M_i} \tag{3.35}$$

Using Equation 3.35, Equation 3.33 can be written as:

$$M_z = \frac{\sum LS_i M_i}{\sum LS_i} \tag{3.36}$$

The above equation allows the calculation of M_z using only the signal of a light scattering detector. A similar procedure can be used for the calculation of M_{z+1}:

$$M_{z+1} = \frac{\sum LS_i M_i^2}{\sum LS_i M_i} \tag{3.37}$$

Equations 3.36 and 3.37 permit the calculation of higher-order molar mass averages from only the light scattering signal and the SEC calibration curve with no need for an RI detector. A potential advantage of this approach is the fact that the molar mass averages sensitive to the high-molar-mass fractions are calculated solely from the light scattering signal, which is also sensitive to the high-molar-mass fractions. The disadvantage is the necessity of column calibration. Since the method focuses on high-molar-mass fractions that are often branched, the molar masses according to the SEC calibration curve can be significantly erroneous. For many samples the errors given by the incorrect calibration can be larger than those given by a low RI signal when the light scattering detector is used together with an RI detector to calculate the molar mass directly without a calibration curve. A possible solution to this dilemma can be self-calibration, that is, using the data points across the chromatogram where both light scattering and RI signals are sufficiently intensive to establish the calibration curve that is extrapolated toward low elution volumes and used with the light scattering signal for the determination of higher-order molar mass averages.

3.3.3.7   Other Types of Detectors

Online mass spectrometers (MS) are widely used in other types of LC because of high sensitivity and ability to provide qualitative information. Of the many different mass spectrometric methods, matrix-assisted laser desorption ionization time-of-flight (MALDI-TOF) and electrospray ionization (ESI) are the most promising as SEC detectors. Both MALDI and ESI represent soft ionization methods that do not generally lead to fragmentation. Although application of an MS detector in SEC is possible and has been reported in the literature, real applications, especially in the area of synthetic organic soluble polymers, are rare. *Inductively coupled plasma mass spectrometry* (ICP-MS) has been reported for the detection of various metals along the elution volume profile of biological and environmental samples.[14,15]

Detectors based on *nuclear magnetic resonance* using flowthrough microprobes can provide structural information or the determination of molar mass in the oligomeric region based on the determination of end groups. Besides the substantially high price of such instruments the disadvantages include generally low sensitivity of the online NMR spectrometer, which can be partly solved by using a stop-and-go technique. Note that due to generally low diffusion coefficients of polymers the spreading of the elution zone during the stop period is negligible, and thus the stop-and-go technique is very applicable in SEC.

Various kinds of detectors, specifically designed for combination with SEC, have been reported in the literature, but none of them has found regular applications. A density detector and an osmometric detector have been used in combination with SEC. The density detector, which may be an alternative to an RI detector, is based on the increase of density due to polymer elution. The measurement is based on the change of the period of oscillation of a measuring cell in the form of a U-shaped oscillating tube. Lehmann and Kohler[16] reported the use of membrane osmometry as an online detector in combination with SEC. They designed an osmometer containing cylindrical semipermeable membrane and an outer glass tube. The polymer solution flows through the bore of the membrane capillary. The reference cell filled with solvent is the volume between the membrane and the outer glass tube. The osmometric detector has several limitations such as relatively long response time, impossibility of detecting lower molar mass due to permeation of oligomers through the semipermeable membrane, and decreasing sensitivity toward high molar masses. These limitations are missing in the case of light scattering detectors and thus routine use of online osmometers is unlikely.

3.4 COLUMN CALIBRATION

SEC has become the most intensively used method for the determination of molar mass of synthetic and natural polymers. However, in the conventional form with a concentration-sensitive detector only, SEC is not an absolute method of molar mass determination since there is no direct relation between the measured quantities (i.e., elution volume and detector response, and molar mass). SEC is a method that can easily provide numbers, but the numbers may not always be meaningful. An appropriate column calibration must be established in order to get true information about the molar mass distribution of the analyzed polymers. Since there is no exact theoretical relation between the elution volume and molar mass or size of the eluting molecules, the SEC columns used for the analysis must be calibrated. The calibration relation is then used for the samples requiring characterization to convert the elution volume into the molar mass. The calibration of SEC columns means establishing the relation between the logarithm of molar mass and elution volume V. The calibration curve can be mostly described by the polynomial:

$$\log M = a + bV + cV^2 + dV^3 \qquad (3.38)$$

In the simplest case, the log M is related linearly with the elution volume (i.e., $c = d = 0$), but for most of the available SEC columns the third-order polynomial represents the most appropriate fit. Some SEC softwares offer higher-order polynomials or other types of fits (e.g., point-to-point). However, higher-order polynomials may produce unrealistic maxima or minima on the calibration curve. The applied SEC software should permit the evaluation of the fit in the sense of correlation coefficient and differences between the nominal molar masses of calibration standards and molar masses calculated from the calibration curve.

A straightforward method of calibration of SEC columns is the analysis of a series of standards of accurately known molar mass and very narrow polydispersity under the same experimental conditions as those used for the analysis of real polymer samples. The identical conditions involve mainly the solvent, the flow rate, and the temperature. The obtained calibration curve is valid for a given SEC column set and cannot be used for other columns even in the case of identical column type and manufacturer. Small temperature deviations in the range of about $1°C$ have no significant impact on the accuracy of the measurement, while the flow rate is the absolutely key parameter, for which even minor fluctuations result in serious errors in molar mass.

The molar mass of the standards used for the column calibration should be determined by a suitable absolute method. The primary methods are light scattering, membrane osmometry, and vapor-phase osmometry. However, it seems that only SEC is used to characterize some of the commercially available standards. The polydispersity of standards should be characterized by SEC, because for narrow standards the values obtained by the combination of light scattering and osmometry are not reliable because of the small differences between M_w and M_n. The requirement of narrow polydispersity is important, because only if this requirement is fulfilled one can assume the validity of the equation $M_n = M_w = M_{PEAK}$. The quantity M_{PEAK} is the molar mass corresponding exactly to the maximum of the chromatogram.

The procedure of the establishment of the calibration curve using narrow standards is illustrated in Figure 3.13. The standards can be analyzed as mixtures of 3–5 standards of molar mass different enough to obtain almost baseline separation. The chromatograms are obtained for each standard mixture, the elution volumes are calculated from the time of injection to the maximum of each peak, and the calibration is established as a plot of log M_{PEAK} versus peak elution volumes.

Establishing a good calibration curve requires standards covering the entire molar mass range of the polymer samples to be analyzed; the analyzed polymers should elute only within the range of the calibration data. Although this requirement may not be always fulfilled, caution should be taken when the polynomial

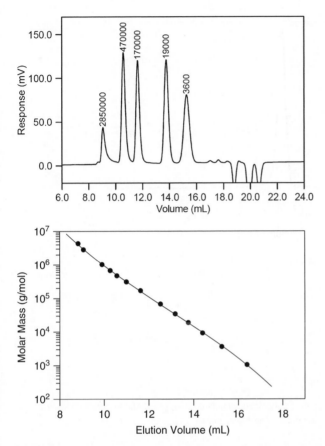

Figure 3.13 Chromatogram of a mixture of polystyrene standards (top) and corresponding calibration curve obtained by the analysis of three standard mixtures (bottom). Molar masses of standards are labeled at the peaks. Calibration curve: $\log M = 16.38664 - 1.86284 \times V + 0.11088 \times V^2 - 0.00287 \times V^3$.

is extrapolated outside the first and last data point of the calibration curve. This is especially true in the proximity of the exclusion limit and limit of total permeation. In addition, the standards should be approximately equidistant on the elution volume axis. Typically 10–15 standards are sufficient to establish the calibration curve; using more standards does not increase accuracy.

Calibration standards for polystyrene covering molar mass range of about $600-10^7$ g/mol are available from several suppliers (e.g., Waters, Polymer Laboratories, Pressure Chemical Company, PSS Polymer Standards Service). The polystyrene standards are prepared by anionic polymerization using stringently purified reagents and synthesis conditions. Some standards prepared by anionic polymerization can contain a small amount of polymer with double molar mass, which appears on a chromatogram as a small satellite peak with lower elution

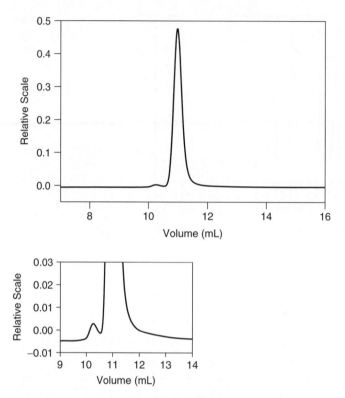

Figure 3.14 RI chromatogram of polystyrene standard of M_{PEAK} 200,000 g/mol showing a small peak of an impurity with about double molar mass.

volume (Figure 3.14). This polymer impurity does not necessarily disturb the determination of a calibration data point, especially when it is well resolved from the main peak, but it influences M_w measured by batch light scattering. The standards are typically characterized with M_{PEAK} and polydispersity M_w/M_n. The nominal polydispersity is mostly less than 1.1. In order to properly characterize the standards, at least one molar mass average should be determined by an absolute method. Unfortunately, this requirement is not always fulfilled. The standards are available as kits or individual molar masses or in the form of mixtures in vials or deposited on small spatulas. With the pre-prepared vial or spatula the calibration solution is easily prepared by addition of an appropriate volume of eluent into a vial or placing a spatula into a vial upon addition of solvent. The pre-prepared kits are available, for example, from Polymer Laboratories (part of Varian), under the trade names EasiVial™ and EasiCal™.

The narrow standards are also available for other homopolymers, for example, poly(methyl methacrylate), poly(ethylene glycol), poly(ethylene oxide), polytetrahydrofuran, dextran, pullulan, sulfonated polystyrene, and poly(acrylic acid). Some standards may not cover a wide molar mass range, and for many homopolymers and especially copolymers the narrow standards are unavailable. Preparation of standards by precipitation fractionation of a polydisperse polymer and subsequent characterization of the obtained fractions by absolute methods is a very laborious and time-consuming process, which typically yields fractions with polydispersity >1.2.

Polydisperse standards for column calibration can be obtained either by preparation of broad polymer samples covering a broad molar mass range (using, for example, free radical polymerization with different initiator concentration), or by fractionation of a broad polymer. In the latter case the polydispersity of the fractions is lower than that of broad polymers. In either way, at least two molar mass averages must be determined by absolute methods for each polydisperse standard. Alternatively, one molar mass average and the polydispersity estimated by SEC using column calibration established for another polymer can be used. For polydisperse standards, taking M_n, M_v or M_w as M_{PEAK} leads to errors that increase with increasing polydispersity. The correct M_{PEAK} lies somewhere between M_n and M_w and usually the viscosity average M_v is the closest average to M_{PEAK}, but not identical. The relation between the M_{PEAK} and other molar mass averages depends on the type of molar mass distribution. For the Schulz-Zimm type of distribution M_{PEAK} is approximately equal to M_w.

For the logarithmic normal distribution M_{PEAK} is approximately equal to $(M_n \times M_w)^{1/2}$ and there are the following relations between M_n, M_v or M_w and M_{PEAK}:

$$M_n = M_{PEAK}\ e^{-\frac{\beta^2}{4}} \tag{3.39}$$

$$M_v = M_{PEAK}\ e^{\frac{a\beta^2}{4}} \tag{3.40}$$

$$M_w = M_{PEAK}\ e^{\frac{\beta^2}{4}} \tag{3.41}$$

where a is the exponent of the Mark-Houwink equation and β is a parameter related to the polydispersity. The determination of M_{PEAK} of broad standards requires knowledge of not only an average molar mass for each standard, but of distribution type and polydispersity as well.

A single polydisperse polymer with at least two known molar mass averages can be used for the determination of the calibration curve using a trial-and-error search method consisting of comparison of the experimental molar mass averages with those calculated from the SEC chromatogram. A linear calibration curve can be established if only M_n and M_w are known. The procedure may be a part of commercially available SEC software (e.g., Waters Empower™). The broad standards of known molar mass averages can be obtained by SEC-MALS measurements.

Another calibration procedure based on a broad standard uses a polymer sample with a known molar mass distribution curve. The procedure involves comparison of the normalized peak area with the cumulative fraction taken from the distribution curve. The procedure is outlined in Figure 3.15. The particular

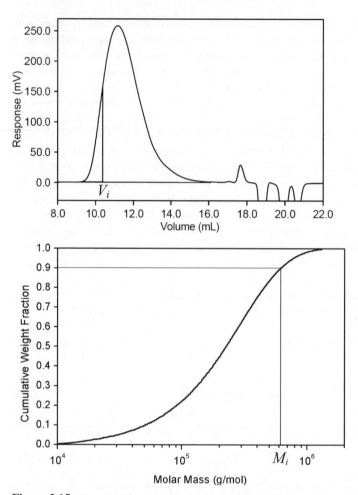

Figure 3.15 Procedure for the determination of calibration curve using a broad polymer standard with known molar mass distribution. RI chromatogram with baseline and a cut at the elution volume corresponding to 10% of total peak area (top) and cumulative distribution curve obtained by SEC-MALS analysis (bottom). The molar mass M_i corresponds to the elution volume V_i, that is, the coordinates of the calibration data point are M_i, V_i. Further data points are obtained by equivalent procedures.

elution volumes V_i are matched with the molar masses M_i taken from the molar mass distribution curve so that the peak area fraction and the weight fraction are identical. The molar mass distribution curve of polydisperse standard is determined experimentally, preferably by means of SEC-MALS, or using two molar mass averages and assuming validity of a theoretical distribution function.

Unlike calibration based on the narrow standards, the calibration curves obtained by broad standards are affected by instrumental peak broadening and thus are more prone to errors when applied to the molar mass characterization of polymer samples. Nevertheless, although calibration by means of a broad standard is less accurate than that established by a series of narrow standards, it may yield more reliable results than those based on the calibration established by narrow standards of different chemical composition than that of polymer under investigation.

3.4.1 Universal Calibration

In view of the fact that SEC separation is governed by the size of polymer molecules in dilute solution, it can be assumed that the hydrodynamic volumes (Equation 1.81) of all species eluting at the same elution volume are identical and that the product $[\eta]M$ represents a universal calibration parameter. The *universal calibration* is a function of $\log([\eta]M) = f(V)$ that can be, as in the case of conventional calibration, described by the first- to third-order polynomial. The intrinsic viscosity of narrow standards can be determined experimentally or calculated from the Mark-Houwink equation. The idea of universal calibration was proved for the first time by Benoit and co-workers,[17] who plotted $\log([\eta]M)$ versus elution volume and obtained a common plot for polymers of various chemical composition and architecture, namely linear polystyrene, star polystyrene, comb polystyrene, poly(methyl methacrylate), poly(vinyl chloride), copolymer styrene–methyl methacrylate, poly(phenyl siloxane), and polybutadiene. The concept of universal calibration appears to be widely valid with the exception of polymers encountering secondary separation mechanisms that are common in aqueous solvents for ionic polymers, highly polar polymers, or polymers measured in theta solvents. Universal calibration was also shown to fail in the case of low-molar-mass polymers.[18]

The concept of universal calibration allows the transformation of the calibration based on well-defined narrow standards (usually polystyrene in organic solvents or pullulan in aqueous solvents) into a calibration valid for a polymer requiring analysis. The hydrodynamic volumes of the standard and polymer under investigation are identical at each elution volume, which, using the definition of the hydrodynamic volume, gives:

$$\log([\eta]_P M_P) = \log([\eta]_S M_S) \tag{3.42}$$

where S and P refer to standard polymer and polymer under analysis, respectively. Entering the Mark-Houwink equation into Equation 3.42 leads to:

$$\log M_P = \frac{1}{1+a_P} \log \frac{K_S}{K_P} + \frac{1+a_S}{1+a_P} \log M_S \tag{3.43}$$

where K and a are the constants of the Mark-Houwink equation for the standard and the polymer requiring analysis. The molar mass of each standard can be recalculated to the molar mass of the polymer requiring analysis using Equation 3.43 and the constants of the Mark-Houwink equation for the standard and the polymer. The literature often shows several combinations of Mark-Houwink parameters for a polymer in the same solvent at the same temperature. Even for such common polymers as polystyrene or poly(methyl methacrylate) the literature values may differ noticeably, and the situation is typically worse for other polymers. Generally, small K is related to large a, and vice versa. Any combination of K and a usually can be used to calculate the viscosity-average molar mass from the intrinsic viscosity and the obtained results are similar. However, when the Mark-Houwink constants are to be used for universal calibration, their careful selection for the calibration standard and the polymer under investigation is very important. The parameters to consider are: number of samples used to get the Mark-Houwink plot, molar mass range (at least one order of magnitude), polydispersity of samples $(M_w/M_n < 1.5)$, type of molar mass average $(M_w$ preferred), and linearity of the obtained plot.

An alternative universal calibration procedure assumes good thermodynamic quality of the SEC solvent for the standard and polymer under analysis, that is, the state when the two polymers will have similar polymer–solvent interactions. Then the unperturbed root mean square end-to-end distance $\langle r \rangle_0^{1/2}$ can be used as an alternative universal calibration parameter.

The relation between $\langle r \rangle_0^{1/2}$ and $[\eta]$ describes the equation derived by Flory and Fox (Equations 1.12 and 1.13). In thermodynamically good solvents the two polymers have similar polymer–solvent interactions and their expansion factors are almost identical. The calibration curves of a standard and a polymer are then related by:

$$\log M_P = \log M_S + \log \frac{(\langle r^2 \rangle_0/M)_S}{(\langle r^2 \rangle_0/M)_P} \tag{3.44}$$

where $(\langle r^2 \rangle_0/M)^{1/2}$ is a constant for a polymer independent of molar mass, which can be found in the literature.[19] Note that according to Equation 3.44 the calibration curves of the standard and polymer under analysis are assumed to be parallel, which is often not fulfilled.

Besides the transformation of the calibration curve of a standard to the calibration curve of a polymer, the concept of universal calibration can be also used for the determination of the constants of the Mark-Houwink equation using the chromatogram of a broad polymer sample for which at least two molar mass averages or one molar mass average and the intrinsic viscosity are known. The procedure uses two of the following equations:

$$[\eta] = \sum_i w_i [\eta]_i \tag{3.45}$$

$$M_n = \left[\sum_i (w_i/M_i) \right]^{-1} \tag{3.46}$$

$$M_w = \sum_i w_i M_i \tag{3.47}$$

where w_i is the weight fraction of polymer eluting at the elution volume V_i and $[\eta]_i$ and M_i are the corresponding intrinsic viscosity and molar mass. Identifying $J_i = [\eta]_i M_i$, the above equations can be rearranged as:

$$[\eta] = K^{\frac{1}{1+a}} \sum_i w_i J_i^{a/(1+a)} \tag{3.48}$$

$$M_n = \left[K^{\frac{1}{1+a}} \sum_i \frac{w_i}{J_i^{1/(1+a)}} \right]^{-1} \tag{3.49}$$

$$M_w = \left(\frac{1}{K} \right)^{\frac{1}{1+a}} \sum_i w_i J_i^{1/(1+a)} \tag{3.50}$$

The values of w_i can be determined from the experimental chromatogram and J_i from the universal calibration curve. The procedure is theoretically capable of yielding the Mark-Houwink constants, but the obtained values may differ for different samples of the same polymer and the method represents a generally less reliable way of determining the constants K and a.

The most effective application of universal calibration is the use of $\log([\eta]M)$-versus-V dependence with an online viscometer, which continuously measures the intrinsic viscosity as a function of elution volume. Using the intrinsic viscosities determined at particular elution volumes one can calculate the corresponding molar masses from the universal calibration curve. This application of universal calibration generally provides more reliable results than calculation based on Equations 3.43 or 3.44, because it is independent of the accuracy of the universal calibration parameters.

For many real polymer samples reliable universal calibration parameters or well-defined standards are unavailable. The application of a calibration determined by available standards, mostly polystyrene in the case of organic soluble polymers, for processing the chromatograms of other polymers became a routine practice in many laboratories of polymer research and quality control. This practice can be called a bad SEC habit. It must be stressed that the molar masses obtained by the use of calibration prepared with narrow standards of one polymer to polymers of other types are often in error by a factor of several tens percent, and by as much as an order of magnitude in the case of branched polymers (for example, see Figure 3.16).

Although the application of polystyrene calibration to other polymers can be acceptable for a rough estimation of the molar masses and especially for the mutual comparison of samples of identical chemical composition, it should not be used for a fundamental description of polymer samples, stoichiometric calculations, or kinetic studies. The procedure is especially erroneous and even misleading in the case of branched polymers and polymers interacting with column packing. If the polystyrene calibration is applied, this fact should be reported together with the obtained results and they should not be treated as the absolute values. The molar mass averages and the molar mass distribution obtained by

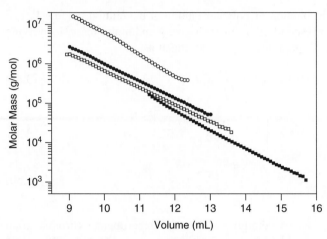

Figure 3.16 Calibration curves of polystyrene (□), epoxy resin based on bisphenol A (■), linear (•) and star-branched (○) poly(benzyl methacrylate) determined by SEC-MALS. Conditions: 2 × PLgel Mixed-C 300 × 7.5 mm, THF, 30°C.

means of polystyrene calibration represent the values for a hypothetical sample of polystyrene that would have the same distribution of hydrodynamic volume as the polymer under analysis.

3.4.2 Flow Marker

The calibration curve is valid for a given column set and separation conditions used for the chromatography runs. These conditions include solvent, temperature, flow rate, and also current status of the columns, because the separation efficiency of the columns can change with time. An absolutely constant flow rate is essential for further use of calibration curve. Since the calibration curve is a log-M-versus-V relation, small deviations of the elution volume have significant influence on the molar mass calculated from the calibration equation. Even modern HPLC pumps may show slight flow rate fluctuations that are typically insignificant for other types of liquid chromatography but may be significant from the viewpoint of the molar mass determination by SEC.

Flow marker (internal standard) is a low-molar-mass compound that is added to the solutions of standards used to establish the calibration relation and to a solution of a polymer requiring analysis. A reference elution volume and a tolerance window (in %) is entered into the SEC software used to acquire and process the data. The calibration relation is valid for a certain elution volume of the flow marker measured at the time when the calibration was established. If the elution volume of the flow marker varies between the runs, the elution volumes of particular volume slices are shifted by the ratio of $\frac{V_{FM,ref}}{V_{FM}}$, where V_{FM} and $V_{FM,ref}$ are the elution volume of the flow marker and the reference flow marker position, respectively. Then the molar masses M_i are calculated for elution volumes $V_i \times \frac{V_{FM,ref}}{V_{FM}}$. That means that if the flow marker appears, for example, at lower elution volume compared to the reference value, the entire polymer peak is shifted slice by slice along the elution volume axis toward higher elution volumes to compensate for the faster pump rate. Toluene, benzene, and sulphur have been used as a flow marker in the case of THF, glucose can be an option in aqueous solvents, and many other compounds can be employed as well.

A good flow marker should be readily soluble in SEC solvent, nonreactive with the solvent and polymer sample, low-toxic, stable, and easily available, and it should elute after the polymer peak at elution volume different from those of impurity peaks. The application of flow marker can undoubtedly improve the accuracy and repeatability of the SEC measurements, but it has several potential drawbacks. First, the flow marker can compensate for a deviation of the flow rate from nominal value that is constant during the entire SEC run, but cannot eliminate the influence of short-term flow rate fluctuations. For many polymer samples the flow marker can overlay with the oligomers, residual solvents, and monomers or additives present in the sample. In addition, the flow markers may co-elute with the solvent peaks from the SEC eluent itself. That is especially the case with THF, which usually shows three negative peaks at the end of chromatogram, or aqueous buffer solutions where an intensive salt peak usually appears at the end of chromatogram.

The mutual interference of positive and negative peaks can change the position of the flow marker and thus its application can be counterproductive. In THF, one of the three negative peaks can be used as the flow marker, assuming the sample does not contain a significant amount of impurities co-eluting with the solvent peaks. The effect of the flow marker on the elimination of flow rate deviations is demonstrated in Table 3.1. The data show that a flow rate deviation as low as 0.2% results in a reckonable deviation of molar mass, and 1% deviation causes significant difference of about 10%. Besides flow rate variations the flow marker can partly compensate for the change of SEC column properties. However, the use of flow marker is not always straightforward and alternatively daily establishment of the calibration can be used. Laboratories equipped with an autosampler can implement this approach, because assuming a typical sample run time of 30–45 min and injection of three standard mixtures there is still sufficient instrumental time for numerous analyses. Polystyrene standards in THF are stable for several months with no sign of degradation. The only requirement is to keep the solutions sealed properly to avoid THF evaporation. The samples can be stored directly in the autosampler vials using new nonpunctured seals. Storage in a refrigerator can reduce THF evaporation. Also solutions of water-soluble standards can be kept for several months in a refrigerator especially when microbiological degradation is minimized by addition of 0.02% sodium azide.

Measurement of flow rate can be performed to identify possible sources of errors due to flow rate deviations. The procedure is simple and requires collection of the effluent from the detector outlet into a 5- or 10-mL volumetric flask and

Table 3.1 Influence of Flow Rate on M_w of Polydisperse Polystyrene and Correction of Flow Rate Deviations with Flow Marker

Flow (mL/min)	M_w (g/mol)		Flow Marker Elution Volume (mL)
	No Flow Marker	Flow Marker	
1	$273,200 \pm 1,000$	$277,500 \pm 500$	19.93–19.94
0.998	$268,300 \pm 300$	$277,500 \pm 300$	19.97
0.995	$262,800 \pm 100$	$279,800 \pm 300$	20.02
0.99	$249,800 \pm 500$	$279,600 \pm 400$	20.10
0.98	$226,200 \pm 1300$	$279,800 \pm 900$	20.25–20.28
0.97	$199,500 \pm 2700$	$280,800 \pm 400$	20.46–20.52
0.95	$153,900 \pm 300$	$281,300 \pm 300$	20.95

Conditions: THF at 1 mL/min, column temperature 40°C, 2 × PLgel Mixed-C 300 × 7.5 mm columns, flow marker elution volume for calibration standards = 19.91 mL.

the measurement of time by a stopwatch. The flow rate is then calculated with a precision of ±0.01 mL/min.

Regular analysis of a reference polymer should be included in the analysis of each sample series irrespective of the use of flow marker. The reference polymer should be stable in the SEC solvent and show no interactions with column packing. It need not be necessarily of well-known molar mass averages, because its primary application is to check repeatability of the measurements. The reference sample preferably should have molar mass range similar to typical samples that are to be analyzed, but the chemical composition can be different. A broad polystyrene in THF and dextran in aqueous solvents can be recommended as good reference polymers, but other types according to particular columns, solvent, and application area can be used as well. The reference sample should be available in sufficient amounts so that it can be used over many years and possibly shared with other laboratories when interlaboratory comparability of the results becomes of interest. The reference polymer can be measured repeatedly during long sample series. A deviation of the molar mass averages over an acceptable limit results in the reestablishment of the calibration curve and check of the pump flow rate, temperature stability, and column performance.

3.5 SEC MEASUREMENTS AND DATA PROCESSING

3.5.1 Sample Preparation

Appropriate preparation of sample is an important part of successful SEC measurement. It is not always recognized that an inappropriate sample preparation may result in significant errors in molar masses or in poor repeatability. The ultimate condition for the analysis of a polymer by SEC is its solubility in a suitable solvent. The solubility of a polymer sample in a given solvent can be predicted according to the solubility parameter δ that is defined as:

$$\delta = \left(\frac{\Delta E}{V}\right)^{\frac{1}{2}} \qquad (3.51)$$

where $\Delta E / V$ is the energy of vaporization per unit volume. A polymer dissolves in a solvent if the solubility parameters of the solvent and polymer are identical, whereas the probability of dissolution decreases with increasing difference of solubility parameters. In contrast to solvents, the solubility parameters of polymers cannot be measured directly, because polymers do not vaporize, but they must be measured indirectly from swelling experiments with solvents of known δ. Solubility parameters are available for various polymers and solvents in the literature, but the prediction of solubility on the basis of solubility parameters is limited and a typical approach is empirical testing of a polymer requesting analysis in the solvents that are usually used in a given laboratory.

Commonly, the sample requiring analysis is prepared in the mobile phase and only exceptionally it is dissolved in a different solvent than that used as the mobile phase. Such a procedure can be used if the sample solubility is significantly better in a different solvent than in the mobile phase. To protect SEC columns it is necessary to verify that the sample remains in solution after injection into the SEC columns. Tests can be performed by dilution of a sample solution by SEC solvent. A sample solution at higher concentration than that used for SEC analysis must remain clear with no precipitation after significant dilution with SEC solvent. However, the injection of a sample in a solvent different from the mobile phase should be used only in justified cases. This procedure also results in a very intensive solvent peak at the end of the chromatogram.

It must be stressed that dissolving a polymer often requires several hours or even several days and exceptionally a full dissolution may take even longer. The obtained polymer solution may contain insoluble parts if the dissolution time is not sufficient. The insolubles consist of the high-molar-mass part of the sample and improper sample preparation may result in a loss of part of a sample and consequently an incorrect picture of the molar mass distribution. Generally, the time needed to dissolve a polymer sample completely depends on the thermodynamic quality of the solvent, molar mass of the polymers, polymer branching, and polymer crystallinity. The dissolution rate increases with decreasing molar mass and increasing thermodynamic quality of the solvent, and decreases with increasing branching and crystallinity.

A thermodynamically good solvent can dissolve a polymer up to high concentrations in a wide temperature range, while polymer solubility is limited in thermodynamically poor solvents. The sample solubility increases with dissolution temperature and some polymers are soluble only at elevated temperatures. The samples soluble only at elevated temperatures may either remain in solution even after cooling down to the room temperature and so can be analyzed using ambient SEC, or remain in solution only at high temperature and so must be analyzed at that temperature by means of a high-temperature SEC system.

Although gentle shaking and warming at about 60°C can decrease the dissolution time and help dissolve polymers with limited solubility, intensive manual shaking, high-speed agitation, or dissolution in an ultrasonic bath should be avoided, because polymers can contain fractions with high molar mass that may undergo shearing degradation when exposed to intensive shaking or agitation. The probability of shearing degradation increases with increasing molar mass. The shearing degradation is also more probable in thermodynamically good solvents and at elevated temperatures, because of the expansion of polymer chains.

To prepare a sample for SEC analysis a known amount of polymer is placed in a volumetric flask or other suitable container and filled with the solvent. Although knowledge of the exact sample concentration is not needed for data processing, the sample concentration should be appropriate. The appropriate concentration is a result of counteracting requirements such as detector response, loading capacity of the columns, solution viscosity of the sample, and concentration dependence of the elution volume. The sample should be injected at a concentration and volume that results in sufficiently intensive detector response while column overloading and viscosity effects are minimal. Typical sample concentrations range from 0.1 to 1% w/v. Sample concentrations recommended by column manufacturers are listed in Table 3.2. As a rule of thumb, the optimum sample concentration decreases with increasing molar mass and samples containing fractions with molar mass of several millions g/mol should be injected at the lowest possible concentration. Broad polymers usually can be injected at higher concentrations than narrow polymers, because the sample zone in SEC columns is spread over larger elution volume. The sample concentration of oligomers is typically not a serious issue.

Injected volume is another important operating parameter influencing the accuracy of the obtained results. Polymers creating highly viscous solutions may require injection of larger, more dilute solutions, but the injection of too-large

Table 3.2   Sample Concentration Recommended by Column Manufacturers

Manufacturer	Molar Mass (g/mol)	Sample Concentration (% w/v)
1	Up to 25,000	<0.25
	25,000–200,000	<0.1
	200,000–2,000,000	<0.05
	>2,000,000	<0.02
2	Up to 5,000	<1.0
	5,000–25,000	<0.5
	25,000–200,000	<0.25
	200,000–2,000,000	<0.1
	>2,000,000	<0.05
3	Not specified	0.05–0.5

volumes contributes to peak broadening and reduces separation performance, which is especially true in the case of 4.6-mm narrow-bore columns.

The injection volume in SEC is usually several times larger than in other types of liquid chromatography. A typical injection volume per regular 300×8-mm column is 50 to 100 μL, and should be decreased accordingly for narrow-bore 4.6-mm columns. The injection volume for narrow standards can be about half of that used for polydisperse polymers to improve peak shape and resolution. For example, using two 300×7.5-mm columns, 50 μL can be used for the injection of standards and 100 μL for the injection of broad polymer samples.

The optimum concentration for narrow polystyrene standards is approximately 0.03% w/v for molar masses up to about 1 million g/mol, 0.015% w/v for standards with molar mass of 1–5 million g/mol, and 0.01% w/v for standards with molar mass around 10^7 g/mol. A polydisperse polymer can be analyzed using a concentration of about 0.2–0.3% w/v and the concentration can be adjusted according to the detector response, peak shape, average molar mass, and sample polydispersity.

Excessive injected mass can result in column overloading, which does not damage the columns, but affects the peak shape and position and thus the molar mass distribution. As shown further in Section 3.5.6, the molar mass averages decrease with the increasing injected mass. Consequently, random changes of injected mass due to improper sample preparations can negatively influence the accuracy and repeatability of the measurements. This can happen when a small piece of sample is dissolved without a proper weighing or if samples containing different amounts of solvents or fillers are prepared at the same amount without taking into account the real concentration of polymer in the sample. Paints, which besides polymer binder contain solvents, fillers, and other additives, are typical materials for which the polymer content should be determined before the sample is prepared for SEC analysis.

Filtration is an important step in sample preparation for SEC measurement. Many industrial polymers contain insoluble materials such as fillers or species created during polymerization such as highly crosslinked microgels. Filtration protects column frits from blockage. The blockage increases the system back pressure and can even distort the chromatographic peaks. In addition, the blockage of column frits increases the probability of shearing degradation. Filtration of the sample through a 0.45-μm disposable filter is an efficient way of removing insoluble species. To reduce the possibility of removing part of a sample by filter membrane, the filter should be flushed with a small amount of sample solution before transferring into the injection device. Although sample filtration protects SEC columns, it should be always performed with caution, because it may remove or degrade high-molar-mass fractions in the sample. Samples containing ultra-high-molar-mass fractions may require filters with larger porosity (e.g., 1 μm). Samples such as calibration standards or pure polymers that show complete dissolution with no indication of particulate matter do not need to be filtered. An alternative procedure of sample clarification is to centrifuge the sample with a benchtop centrifuge for a few minutes and then inject the supernatant.

Most synthetic and also natural polymers show good stability in solutions. Keeping the solutions well sealed at room temperature or in a refrigerator can enhance their stability. Nevertheless, sample stability after dissolution is a potential issue that may have a negative impact on the accuracy of the experimental results. If a polymer sample is suspected of degradation in solution, the measurements should be repeated in suitable time intervals to see whether the molar mass is constant or decreases with time. If the degradation is eminent, then the sample should be prepared in another solvent where the degradation does not occur, or at least measured immediately after dissolution. Water, which is typically present in THF, can promote hydrolysis of polyesters as demonstrated in Figure 3.17. If needed, THF can be effectively dried with molecular sieves. Poly(phenyl acetylene) is a typical polymer that undergoes fast degradation in THF.

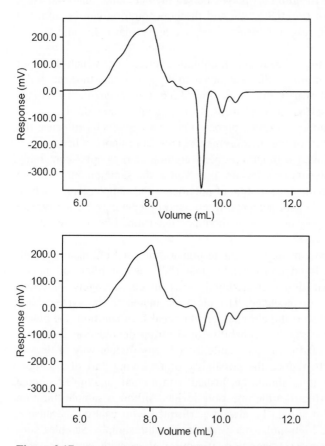

Figure 3.17   RI chromatograms of polyester resin: fresh solution (top) and the same solution after three days (bottom). Note the significant decrease of negative water peak as a result of hydrolysis of the polyester chain. M_n (g/mol) = 2020 (fresh solution), 1900 (after 3 days); M_w (g/mol) = 6480 (fresh solution), 5520 (after 3 days); M_z (g/mol) = 16,100 (fresh solution), 12,700 (after 3 days).

3.5.1.1 *Sample Derivatization*

Derivatization represents a possible part of sample preparation. In other types of liquid chromatography, derivatization is often used to increase the detection limit, and in gas chromatography the purpose of derivatization is to increase volatility and improve thermal stability. In SEC, derivatization is used significantly less frequently and may improve sample solubility or suppress enthalpic interactions of polymer with the column packing. In principle, the derivatization reaction can be carried out before the analysis or between the column and detector, the former being the most applicable in SEC. The derivatization reactions must not be associated with degradation of the polymer chain. The derivatization changes the molar mass, and the number and mass of bonded molecules should be taken into account at the interpretation of the obtained results. The derivatization involves the reaction of a specific functional group of a polymer with a suitable chemical agent. The chemical composition of a polymer to be analyzed and its solubility govern the choice of the derivatization procedure. Most of the polymers are by their nature not capable of easy derivatization. Fortunately, there is also no need of derivatization of such polymers, because they can be easily analyzed in their net form. Various derivatization reactions and derivatives can be found in an excellent book by Blau and Halket.[20] The following reactions can serve as examples of suitable derivatizations:

- *Silylation* with N,O-Bis(trimethylsilyl)trifluoroacetamide of polymers containing a high level of carboxyl groups can significantly improve the peak shape for analysis in THF on styrene-divinylbenzene gels as demonstrated in Figure 3.18. The silylation can be easily done by addition of 0.3 mL of silylation agent to 3 mL of about 0.3% w/v solution of polymer in THF. The solution is heated up to about 60°C for three hours, filtered, and injected into SEC columns.

- *Acetylation* with acetic anhydride under catalysis with N-methylimidazole can improve THF solubility of polymers containing hydroxyl groups or amine groups. The reaction results in the formation of esters or amides of acetic acid. Acetic anhydride and N-methylimidazole are added in the amount of 20 μL per 1 mL of about 0.3% w/v THF solution, and after one hour at room temperature the solution can be analyzed. The reaction has been successfully applied to the analysis of hyperbranched polyesters based on dimethylol propionic acid, which showed limited solubility in THF with tendency to aggregation.

3.5.2 Determination of Molar Mass and Molar Mass Distribution

A primary result of an SEC experiment is a *chromatogram*, which is a record of a detector signal against elution volume. The chromatogram bears information about the concentration of molecules eluting at various elution volumes, which

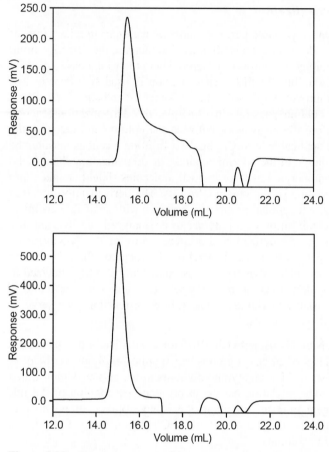

Figure 3.18 Chromatograms of acrylic copolymer containing a high level of carboxyl groups before (top) and after (bottom) silylation with N,O-Bis(trimethylsilyl)trifluoroacetamide. The silylation of the carboxyl groups eliminates the interactions with column packing. Conditions: 2 × PLgel Mixed-C 300 × 7.5 columns, THF, 40°C, 1 mL/min.

is related to the intensity of detector signal, and the molar mass, which is related to the elution volume. Although the chromatogram reflects the molar mass distribution of the sample under analysis, chromatograms of the same sample measured using different columns, and possibly other parameters such as temperature or flow rate, are different. Assuming proper calibration and data processing, the molar mass distributions obtained by processing chromatograms acquired under different chromatographic conditions should be identical within the experimental uncertainties. That means the transformation of the chromatograms into molar mass distribution curves eliminates the differences given by different instrumentation.

The simplest way to determine molar mass from the SEC chromatogram is the determination of M_{PEAK}, which is the molar mass corresponding to the maximum of the peak. This method is applicable only for monodisperse or very narrow polymers; for polydisperse samples it cannot be recommended. Not only is the method applicable solely to unimodal chromatograms, but it results in the determination of a molar mass average of unknown type for which the relation $M_n < M_{PEAK} < M_w$ is valid. In addition, a single value of M_{PEAK} does not provide any information about sample polydispersity. Consequently, calculation of the entire molar mass distribution and all molar mass averages should be performed for proper description of a polymer under analysis.

A fundamentally correct procedure of the calculation of molar mass distribution from an SEC chromatogram involves correction for peak broadening. Correction for peak broadening is the most complicated and uncertain step in processing SEC data and there are no simple and completely adequate methods. With the development of modern high-performance columns, the need for peak broadening correction has faded for most practical cases. The necessity of peak broadening correction also decreases with the increasing sample polydispersity. In the case of narrow polymers, a possible error is not in the sense of molar mass, but mainly in the sense of sample polydispersity, which is undoubtedly overestimated due to peak broadening. The need for peak broadening correction is also further eliminated in the case of light scattering detection when molar mass is directly determined after the sample elutes from the SEC columns.

The following procedure for processing chromatograms assumes that band broadening can be neglected. The procedure consists of determination of baseline and peak integration interval (i.e., beginning and end of the peak). The chromatogram is separated into regular intervals by vertical lines that are sufficiently close to each other. The relative height of each slice represents the weight fraction of molecules eluting at a given elution volume (the slice area can be used instead of the slice height):

$$w_i = \frac{H_i}{\sum_i H_i} = \frac{A_i}{\sum_i A_i} = \frac{A_i}{A} \tag{3.52}$$

where w_i is the weight fraction of molecules eluting at elution volume V_i, H_i and A_i are the height and the area of the ith slice, respectively, and A is the total peak area. The sampling frequency is typically 0.5–2 seconds and further decrease of the slice width does not benefit the obtained results. The molar mass corresponding to each elution volume slice is calculated from the calibration curve expressed by Equation 3.38. The obtained values of H_i or A_i and M_i are used to calculate the molar mass averages:

$$M_n = \frac{\sum H_i}{\sum \frac{H_i}{M_i}} = \frac{\sum A_i}{\sum \frac{A_i}{M_i}} \tag{3.53}$$

$$M_w = \frac{\sum H_i M_i}{\sum H_i} = \frac{\sum A_i M_i}{\sum A_i} \tag{3.54}$$

$$M_z = \frac{\sum H_i M_i^2}{\sum H_i M_i} = \frac{\sum A_i M_i^2}{\sum A_i M_i} \tag{3.55}$$

$$M_{z+1} = \frac{\sum H_i M_i^3}{\sum H_i M_i^2} = \frac{\sum A_i M_i^3}{\sum A_i M_i^2} \tag{3.56}$$

$$M_v = \left(\frac{\sum H_i M_i^a}{\sum H_i} \right)^{\frac{1}{a}} = \left(\frac{\sum A_i M_i^a}{\sum A_i} \right)^{\frac{1}{a}} \tag{3.57}$$

where a is the exponent of the Mark-Houwink equation. If the number of slices is sufficiently large, the use of area slices or slice heights provides equivalent results. Besides the values of molar mass averages, most of the commercially available SEC softwares report also the value of M_{PEAK}. Although M_{PEAK} does not have an exact definition as do other molar mass averages, it can be understood as the molar mass of the most abundant fractions in a given polymer sample and thus it can provide certain information about the analyzed sample. However, caution should be taken in the case of bimodal samples, where using M_{PEAK} can lead to false conclusions.

The SEC analysis yields not only all molar mass averages, but also the differential and cumulative distribution curves. The normalized chromatogram $g(V)$ is obtained according to the equation:

$$g(V) = \frac{G(V)}{\int\limits_0^\infty G(V) dV} \tag{3.58}$$

where $G(V)$ is the unnormalized detector signal. The relation between the normalized chromatogram and the differential distribution curve is:

$$g(V) dV = -f_w(M) dM \tag{3.59}$$

where $f_w(M)$ is the weight differential distribution function. The negative sign is because the molar mass decreases with increasing elution volume. In a rigorous manner, Equation 3.59 should use a normalized chromatogram corrected for band broadening $w(y)$ instead of $g(V)$.

Using Equation 3.59, the differential distribution function equals:

$$f_w(M) = \frac{-g(V)}{\frac{dM}{dV}} = \frac{-g(V)}{\frac{d(\log M)}{dV}} \frac{\log e}{M} = \frac{-g(V)}{2.303 M \frac{d(\log M)}{dV}} \tag{3.60}$$

The weight fraction of polymer in the molar mass interval M to $(M + dM)$ is $f_w(M)dM$ or in the molar mass interval $\log M$ to $(\log M + d\log M)$ is $F_w(\log M)d\log M$. The differential function $F_w(\log M)$ is calculated according to the equation:

$$F_w(\log M) = \frac{-g(V)}{\frac{d(\log M)}{dV}} \tag{3.61}$$

The differential distribution curve expressed in log M is sometimes labeled as $dw/d(\log M)$ or $dwt/d(\log M)$, where w or wt indicates the weight fraction. The derivatives in Equations 3.60 and 3.61 are obtained analytically by differentiating the calibration curve. If the calibration curve is fitted by the third order polynomial (Equation 3.38), the derivative to be used in Equations 3.60 and 3.61 is:

$$\frac{d(\log M)}{dV} = b + 2cV + 3dV^2 \tag{3.62}$$

In the case where the calibration curve is linear, the derivative is constant over all elution volumes and equals the slope of the calibration curve.

The cumulative distribution expressing the weight fraction less than or equal to M_i is given by the equation:

$$I_w(M_i) = \int_{V_b}^{V_i} g(V)\, dV = \frac{\sum_1^i A_i}{A} \tag{3.63}$$

where M_i is the molar mass of molecules eluting from the columns at elution volume V_i, V_b is the beginning elution volume of the peak processing, A_i is the slice area, and A is the total peak area. The procedure for processing the chromatogram and calculating the molar mass distribution and molar mass averages is illustrated in Figure 3.19. The corresponding distribution curves calculated by Waters Empower™ software are depicted in Figure 3.20 and the numerical data are listed in Table 3.3. Note that in this particular example the two distribution curves are drawn from high to low molar masses in accordance with the SEC elution order, and the molar mass axis is expressed in log M. Taking, for example, the data at slice number 130, one can read that about 41.1% of molecules have molar mass larger than 251,000 g/mol or alternatively that 58.9% molecules in the analyzed sample have molar mass below this value. The slice fraction $A_{130} = 2.58 \times 10^5 / 3.56 \times 10^7 = 0.00725$, that is, 0.725% molecules have molar mass corresponding to this slice. The same value is obtained from the differential distribution function: $F_w(\log M) = 0.94632$, $d(\log M) = (\log 251{,}300 - \log 246{,}900) = 0.00767$, weight fraction of molecules with molar mass in the range of 246,900 g/mol $-251{,}300$ g/mol $= 0.94632 \times 0.00767 = 0.00726$, that is, 0.726%.

The most important steps in processing the experimental data are the construction of the baseline and the selection of the peak limits. The baseline construction is the easiest part of chromatogram processing, assuming a stable detector signal. The procedure for constructing the baseline is to draw a linear line from a point before the beginning of the polymer peak to the point after the solvent and impurity peaks. If the baseline is stable and properly established, moving the baseline beginning and end points along the elution volume axis should not markedly change the results, as demonstrated in Figure 3.21. Especially in the case of samples containing oligomeric tail, positioning the baseline

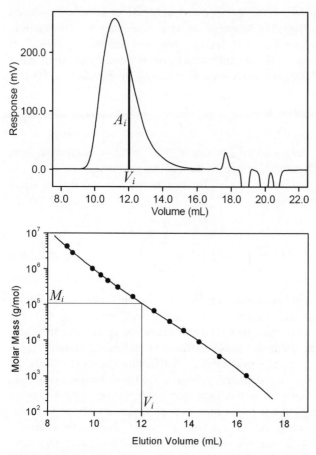

Figure 3.19 Procedure for the calculation of molar mass averages and molar mass distribution curves from an SEC chromatogram and a calibration curve.

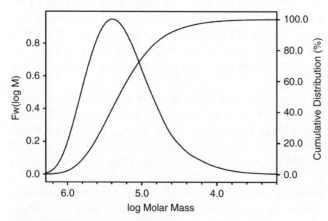

Figure 3.20 Differential and cumulative molar mass distribution plots corresponding to Figure 3.19. Waters Empower™ style.

Table 3.3   Example of Slice Data Corresponding to Figure 3.19

Slice #	V (mL)	M (g/mol)	Area (μVs)	$I_w(M)(\%)$	$F_w(\log M)$
5	9.117	2.787E+06	2.19E+02	0.002	0.00066
6	9.133	2.728E+06	2.48E+02	0.003	0.00075
7	9.150	2.670E+06	2.76E+02	0.004	0.00084
8	9.167	2.614E+06	3.19E+02	0.005	0.00097
9	9.183	2.559E+06	3.62E+02	0.006	0.00110
10	9.200	2.506E+06	4.09E+02	0.007	0.00125
11	9.217	2.453E+06	4.80E+02	0.008	0.00147
12	9.233	2.402E+06	5.51E+02	0.010	0.00169
13	9.250	2.352E+06	6.41E+02	0.012	0.00197
14	9.267	2.303E+06	7.40E+02	0.014	0.00228
15	9.283	2.255E+06	8.62E+02	0.016	0.00266
16	9.300	2.209E+06	9.98E+02	0.019	0.00308
17	9.317	2.163E+06	1.15E+03	0.022	0.00356
18	9.333	2.118E+06	1.33E+03	0.026	0.00413
19	9.350	2.075E+06	1.54E+03	0.030	0.00479
20	9.367	2.032E+06	1.78E+03	0.035	0.00553
21	9.383	1.990E+06	2.04E+03	0.041	0.00636
22	9.400	1.950E+06	2.34E+03	0.048	0.00731
23	9.417	1.910E+06	2.68E+03	0.055	0.00840
24	9.433	1.871E+06	3.05E+03	0.064	0.00957
...	...	...	...	...	...
120	11.033	3.000E+05	2.55E+05	33.849	0.92419
121	11.050	2.947E+05	2.56E+05	34.565	0.92788
122	11.067	2.895E+05	2.56E+05	35.284	0.93129
123	11.083	2.844E+05	2.57E+05	36.004	0.93433
124	11.100	2.794E+05	2.57E+05	36.726	0.93704
125	11.117	2.745E+05	2.58E+05	37.448	0.93937
126	11.133	2.697E+05	2.58E+05	38.172	0.94140
127	11.150	2.649E+05	2.58E+05	38.895	0.94313
128	11.167	2.603E+05	2.58E+05	39.620	0.94447
129	11.183	2.557E+05	2.58E+05	40.344	0.94554
130	11.200	2.513E+05	2.58E+05	41.068	0.94632
131	11.217	2.469E+05	2.58E+05	41.791	0.94667
132	11.233	2.426E+05	2.58E+05	42.514	0.94680
133	11.250	2.384E+05	2.57E+05	43.236	0.94658
134	11.267	2.342E+05	2.57E+05	43.957	0.94606
135	11.283	2.301E+05	2.56E+05	44.676	0.94526
136	11.300	2.261E+05	2.56E+05	45.394	0.94413
137	11.317	2.222E+05	2.55E+05	46.110	0.94268
138	11.333	2.184E+05	2.55E+05	46.824	0.94093
139	11.350	2.146E+05	2.54E+05	47.536	0.93891
...	...	...	...	...	...

Total area= 3.56E+07.

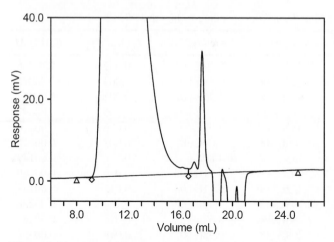

Figure 3.21 Example of stable baseline. Baseline position is marked with triangles; diamonds show integration limits. Baseline is drawn from 8 to 25 mL; moving the baseline 1 mL backward or forward has no impact on the molar mass averages: M_n (g/mol) $= 75,600$; $76,000$; $75,200$; M_w (g/mol) $= 272,500$; $272,500$; $272,600$; M_z (g/mol) $= 499,100$; $498,400$; $499,800$ for the baseline start/end points at 7/26 mL, 8/25 mL, and 9/24 mL, respectively.

end after the solvent peaks should be preferred over drawing the baseline from the peak beginning to the peak end (see Figure 3.22). It is worth mentioning that the automatic data processing algorithms have a strong tendency to position the baseline end point on the oligomeric tail of the chromatogram. Positioning the baseline end on the oligomeric tail has a significant influence on the M_n while the influence on the higher-order molar mass averages is negligible.

To obtain a stable detector signal, the columns must be flushed for several hours before the first injection can be performed. Other necessary conditions for a stable baseline are sufficient warming-up and stabilization of RI detector and constant temperature of RI detector and SEC columns. Most RI detectors are temperature controlled in the sense of heating and stable column temperature can be easily achieved with a column oven, which, however, is not a necessary part of an SEC experimental setup because a stable room environment is mostly sufficient. Keeping the RI detector on continuously even if it is not used for a couple of days improves the signal stability and shortens the time needed to get the SEC setup ready for the measurements. In fact, the entire SEC setup can be kept running even if not used overnight or over the weekend. HPLC pumps are designed to work and they work best when kept running. If an SEC instrument is not to be used for a period of time, the pump can be set to a lower flow rate. To prevent excessive solvent consumption, the column outlet can be recycled back to the solvent reservoir. Some RI detectors enable easy switching between the waste and recycle outlets. The time-life limited components such as the laser of a light scattering detector or the lamp of an UV detector can be switched off since their warmup time is relatively short (sufficient time is usually about 30

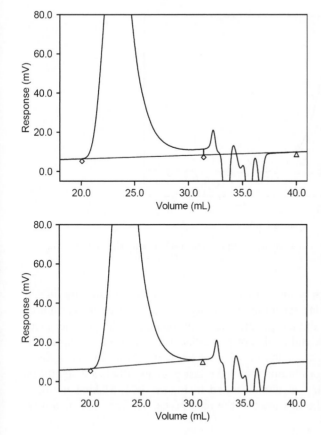

Figure 3.22 Correct (top) and incorrect (bottom) position of the baseline end point for a polymer sample containing oligomeric fractions. Baseline end at 40.0 mL: $M_n = 60,000$ g/mol, $M_w = 290,000$ g/mol, $M_z = 469,000$ g/mol; baseline end at 31.4 mL: $M_n = 134,000$ g/mol, $M_w = 298,000$ g/mol, $M_z = 463,000$ g/mol. Conditions: 3 × Styragel HR5E 300 × 7.8 mm, THF, 25°C, 1 mL/min, 75 μL 0.25% w/v. Baseline end is marked with a triangle.

minutes). Evenly drifting baseline usually does not represent a serious obstacle in data processing, but the baseline fluctuation makes the baseline construction uncertain and it should be eliminated.

Selection of the peak interval, although it appears to be a simple procedure, is not always straightforward. As mentioned, many synthetic and natural polymers contain fractions with molar masses down to several hundreds g/mol. In such a case the end of the polymer peak is flat and parallel to the baseline. Chromatograms of polymer with and without oligomeric tail are shown in Figures 3.22 and 3.23. For many samples the detector response from oligomeric fractions overlays with the onset of the response from the impurities. The position of the integration end has significant impact on M_n and the inclusion of the oligomeric part of the peak results in significantly lower values.

The issue of baseline construction and selection of the integration end was investigated by Mori et al.,[21] who studied NIST SRM 706 polystyrene and verified that the flat parallel baseline after the polymer peak is really caused by oligomers and that it is not an artifact. They also suggested the use of a special Solvent-Peak Separation column (available from Shodex) that is connected before

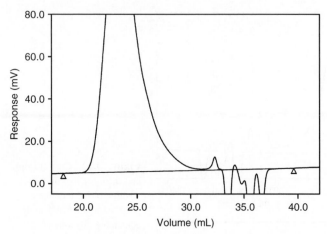

Figure 3.23 RI chromatogram of a polymer sample without oligomeric tail (compare with Figure 3.22).

the regular columns. Such a column can improve separation in the low-molar-mass range without affecting separation in the high-molar-mass range, but it does not solve the problem of the flat parallel signal at the end of the polymer peak. Determination of the peak end is also of particular importance in processing the data of oligomers. Chromatograms of oligomers usually show several peaks at their ends. The insertion or elimination of the last peak usually has a significant impact on the obtained M_n, as demonstrated in Figure 3.24. Making the decision whether to include the last peak in the data processing may not be always

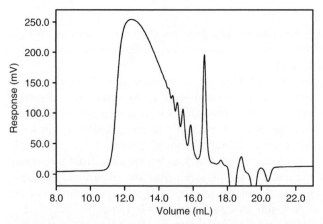

Figure 3.24 RI chromatogram of polyester based on dimethylol propionic acid (DMP) and p-hydroxy benzoic acid (HB). The peak at 16.7 mL has relative area of 6.3% and probably belongs to DMP-HB ester. The molar mass averages with this peak (1) excluded from integration: $M_n = 1960$ g/mol, $M_w = 3490$ g/mol, $M_z = 5170$ g/mol; (2) included in integration: $M_n = 1430$ g/mol, $M_w = 3290$ g/mol, $M_z = 5140$ g/mol.

easy especially when the identity of the peak is unknown, because it can be a member of an oligomeric series, but it can also belong to additives, monomers, or impurities.

There is no magic rule for solving the problem of determination of the peak end. The following recommendations should help get reliable M_n and improve reproducibility of the measurements: (1) The peak end set by a skilled operator is usually more reliable than using automatic peak processing by SEC software. However, intuitive assignment of the peak end by different individuals may result in noticeably different values of M_n. (2) Oligomers, if present, are an integral part of the sample with possibly significant impact on the application properties and shall be included in the determination of molar mass. On the other hand, residual monomers, solvents, and additives are not a part of polymer itself and should be excluded from the calculation of polymer molar mass distribution. However, their presence can be mentioned as a part of the SEC results and illustrated by a chromatogram. They can be characterized and quantified by other more appropriate techniques such as other types of liquid chromatography or by gas chromatography, or simply by the determination of solids. (3) Peak end should be assigned consistently for a series of similar samples to facilitate their mutual comparison. For many polymers the oligomeric tail overlays with the solvent and impurity peaks and the peak end can be positioned at the valley, as shown in Figure 3.25. (4) Zooming the area around the peak baseline is important for proper peak selection. (5) Preparation of sample in the solvent taken from the pump solvent reservoir can minimize the solvent peak on an RI chromatogram. Especially in the case of THF, the use of solvent from different bottles can result in an intensive peak of stabilizer that can overlay with the lowest oligomers and result in significantly lower M_n when included in chromatogram processing.

The influence of the peak end (cutoff elution volume) on the molar mass averages is demonstrated on NIST SRM 706 polystyrene in Figure 3.26 and Table 3.4 In contrast to the determination of the polymer peak end, the determination of the peak beginning is usually straightforward. However, some samples may contain low amounts of high-molar-mass fractions, which cause very low increase of the ascending part of the polymer peak. For such samples it may not be easy to distinguish between the drifting baseline and the rise of the peak due to the elution of small amounts of higher-molar-mass fractions. These fractions can be easily detected by means of SEC columns of lower exclusion limit, as shown in Figure 3.27. Although the columns with low exclusion limit do not provide true molar mass distribution, they can be used for sensitive detection and quantification of higher-molar-mass impurities in oligomers.

An important assumption for the calculation of the true molar mass distribution from SEC data is that the detector response is proportional only to the concentration and that it is independent of elution volume. Only if this assumption is fulfilled the ratio $A_i / \sum A_i$ equals the weight fraction w_i of molecules eluting at the elution volume V_i. This requirement is typically well fulfilled in the case of homopolymers, but it may not be fulfilled in the case of copolymers or polymer blends when the chemical composition may change along the elution volume axis.

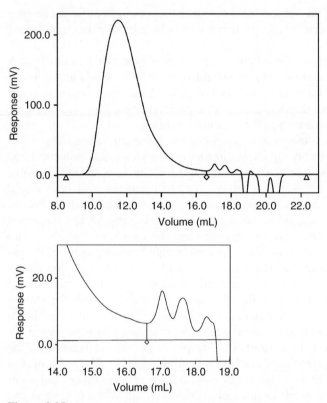

Figure 3.25   Example of positioning the peak end (indicated by diamond) at the valley between the oligomeric tail and solvent and impurity peaks.

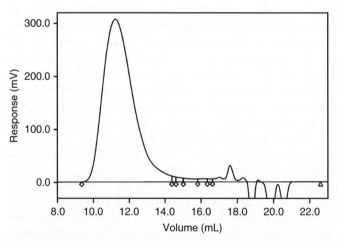

Figure 3.26   RI chromatogram of NIST SRM 706 polystyrene with different peak ends. Molar mass averages corresponding to various peak end positions are listed in Table 3.4. Conditions: columns 2 × PLgel Mixed-C 300 × 7.5 mm, THF at 1 mL/min.

Table 3.4   Effect of Peak End on Molar Mass Averages of Polystyrene NIST SRM 706
See chromatogram in Figure 3.26

Peak End (mL)	M at Peak End (g/mol)	M_n (10^3 g/mol)	M_w (10^3 g/mol)	M_z (10^3 g/mol)
16.64	700	55.3	270.3	464.5
16.35	1000	66.1	271.0	464.5
15.81	2000	85.8	272.4	464.5
15.0	5000	108.5	274.6	464.6
14.62	8000	118.0	275.9	464.6
14.38	10,000	123.9	276.9	464.7

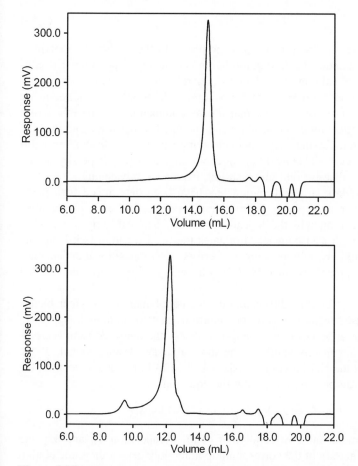

Figure 3.27   Chromatograms of oligomeric sample containing a small amount of higher-molar-mass fractions analyzed on columns with exclusion limit of 2×10^6 g/mol (top) and 30,000 g/mol (bottom). The columns with low exclusion limit visualize fractions with higher molar mass. Columns: $2 \times$ PLgel Mixed-C (top) and $2 \times$ PLgel Mixed-E (bottom), THF at 1 mL/min.

In the case of similar specific refractive index increments of particular homopolymers, neglecting the chemical heterogeneity does not result in significant errors in w_i. However, similar refractive index increments of homopolymers are not always the case: For instance, polystyrene in THF has dn/dc of 0.185 mL/g while most acrylates and methacrylates that are often used for copolymerization with styrene have dn/dc in the range of 0.6–0.84 mL/g. Another possible error in the determination of molar mass of copolymers arises from the column calibration. If the parent homopolymers have significantly different calibration curves, then each elution volume requires different calibration appropriate to the given chemical composition. The rigorous calibration would be the logarithm of molar mass as a function of not only elution volume, but chemical composition as well:

$$\log M = f(V, w_A) \tag{3.64}$$

where w_A is the weight fraction of co-monomer A in the analyzed copolymer. The determination of such a calibration relation is typically impossible. The concept of the universal calibration is theoretically reliable, because the copolymer molecules separate on the basis of hydrodynamic volume as do homopolymers. The limitation is caused by the fact that the hydrodynamic volume of copolymer is generally a function of molar mass and chemical composition and thus the molecules of varied molar mass and chemical composition co-elute at a given elution volume. That means chemical heterogeneity increases polydispersity within the elution volume slice beyond that resulting from peak broadening.

Although the Mark-Houwink parameters have been determined for several copolymers, they are unavailable for most real samples, and even if available, their applicability is limited by the dependence on the chemical composition. Universal calibration with a viscometer eliminates the need for the Mark-Houwink constants and the only remaining sources of errors are compositional dependence of the detector response and increased polydispersity of the particular elution volume slices.

In conventional SEC the chromatograms of copolymers or polymer blends are mostly processed using the same procedure as that for homopolymers with the calibration that appears the most appropriate for the analyzed samples, typically the calibration that is valid for the most abundant homopolymer. If the composition along the elution volume axis is known from the dual detection, the calibration curve can be estimated using the equation:

$$\log M_c = w_B \log M_B + (1 - w_B) \log M_A \tag{3.65}$$

where M_c is the molar mass of copolymer of the composition w_B and M_A and M_B are the molar masses of the corresponding homopolymers at the same elution volume. It has been shown that Equation 3.65 provides suitable results for block copolymers, but does not apply for statistical and alternating copolymers. Nevertheless, chemical composition as a function of elution volume is often unknown and thus the application of Equation 3.65 is rare. Due to the complexity of most copolymers, the obtained results should be expected to be less correct compared to the characterization of corresponding homopolymers.

Conventional calibration also provides incorrect results for branched polymers, because branched molecules are generally more compact compared to corresponding linear molecules of the same molar mass. Consequently, branched molecules elute later than their linear counterparts and the molar mass determined from column calibration is underestimated.

3.5.3   Reporting Results

Proper reporting of results is an important part of data processing. The primary goal of the SEC experiment is usually the characterization of molar mass distribution. The report should contain not only the characteristics of the sample, but also information that would permit the data to be easily found and, if needed, further reprocessing. The information in the report should include (1) sample information such as sample unanimous name, information about research project and customer, date of data acquisition and processing, and injection volume; (2) RI and/or UV chromatogram showing the baseline and peak limits; (3) table with molar mass averages and possibly other characteristics of the chromatogram useful for the interpretation of results and comparison with other data; and (4) cumulative and differential distribution curves, either in separate plots or overlaid in a single plot.

The raw chromatogram is useful to illustrate the stability of the baseline, to show integration limits, and eventually to indicate presence of other compounds, such as residual monomers, solvents, or oligomers. Molar mass averages are reported by software with accuracy to the last unit. With respect to the repeatability of the measurement, the molar mass averages should be rounded off to hundreds or thousands or even tens of thousands at the molar mass order of 10^6 g/mol. Similarly, the polydispersity M_w/M_n should be rounded off to one or two decimal figures.

The report should also state the total peak area that can reveal concentration inconsistency resulting from improper sample preparation or from different levels of polymer in sample due to solvents, fillers, or other nonpolymeric compounds. The peak area can also indicate adsorption of part of the sample in column packing. For samples of identical chemical composition prepared at the same concentration the peak area must be identical.

End time of peak processing may be important for samples showing oligomeric tail where the determination of peak end is uncertain. This information may improve consistency of results when similar samples are analyzed after some time. Besides molar mass averages and distribution curves it is possible to report other characteristics that may be helpful for the characterization of samples. These may be percentage of fractions below a certain molar mass limit (e.g., fractions below 1,000 g/mol), molar masses at 10 and 90% of the cumulative distribution, or M_w of the lowest 10% fractions and

M_w of the highest 10% fractions. For example, the M_w of the 10% lowest and highest fractions are required for the characterization of dextran samples by the *Pharmacopoeia*.

3.5.4 Characterization of Chemical Composition of Copolymers and Polymer Blends

A complete description of a broad chemically heterogeneous binary copolymer requires two-dimensional function $f_w(M, w_A)$, which provides the weight fraction of molecules with molar mass M and chemical composition w_A. The situation is much more complicated in the case of ternary or even more complex copolymers. Using SEC the only obtainable information about the chemical composition of copolymers and polymer blends is chemical composition as a function of molar mass. The determination of the distribution of chemical composition, that is, function $f_w(w_A)$, requires separation according to the chemical composition by means of other types of liquid chromatography, critical mode of liquid chromatography or combination of various chromatographic modes in two-dimensional chromatography. Details and other literature references can be found, for instance, in a book by Pasch and Trathnigg.[22]

Chemical composition as a function of molar mass can be determined either (1) by fraction collection, using semipreparative/preparative SEC columns and subsequent analysis of the obtained fractions by analytical SEC to determine the molar mass distribution, and spectral or chemical methods to characterize chemical composition, or (2) online, using multiple detection. The former approach is laborious, but generally applicable; the latter is mostly limited to copolymers and polymer blends where one component is UV active while the other is not or where the two components have significantly different UV response at two different wavelengths.

Unsaturated polyesters composed of maleic (fumaric) unsaturated units along with units of aromatic saturated acid are an example of copolymers with different UV response at different wavelengths. The chemical composition at particular elution volumes can be determined from the relation between the absorbance ratio at two wavelengths and the weight fraction of one of the absorbing components:[23]

$$\frac{A_{\lambda_1}}{A_{\lambda_2}} = f(x_A) \tag{3.66}$$

where A_λ is the absorbance at a given wavelength and x_A is the mole fraction of one of the two absorbing components. Equation 3.66 is obtained in SEC mode from the total peak areas recorded at two different wavelengths by injecting copolymers of known overall composition and/or mixtures of two homopolymers, or in offline mode using a UV spectrometer.

In the case of various styrene copolymers (e.g., styrene-acrylate, styrene-methacrylate, styrene-vinylacetate) the UV detector records only UV-absorbing units while the RI detector monitors both absorbing and non-absorbing units. The weight fraction of monomer units B (w_B) for each elution volume can be calculated using the following equation:[24]

$$w_B = \frac{(A_{RI} - A_{UV} K_2) K_1}{A_{RI} K_1 + A_{UV} K_2 - A_{UV} K_1 K_2} \tag{3.67}$$

where A_{RI} and A_{UV} is the area of the ith volume slice of the chromatogram recorded by the RI and UV detectors, respectively, K_1 is the ratio of the RI detector response for homopolymer A to that for homopolymer B, and K_2 is the ratio of the RI detector response to the UV detector response for UV-absorbing homopolymer A.

Another procedure for calculating the composition across the chromatogram requires the knowledge of the extinction coefficient of the absorbing homopolymer and the specific refractive index increments of both homopolymers. The method finds an important application in the characterization of protein conjugates. Poly(ethylene glycol) (PEG) or carbohydrate are frequently used to modify the pharmacokinetic properties of protein therapeutics. By adding PEG or carbohydrate the half-life of protein can be significantly increased. If the *dn/dc* for protein and modifier and the protein extinction coefficient are known, it is possible to calculate the total molar mass of the conjugate and molar masses of the particular constituents and protein fraction on a slice-by-slice basis. Similar calculation can be performed for synthetic copolymers. The calculation is included in ASTRA® software (Wyatt Technology Corporation).

An online or offline infrared detector can be used to monitor chemical composition as a function of elution volume. The most important application is the determination of short-chain branching in polyethylene, which has direct impact on the end-use properties.

3.5.5 Characterization of Oligomers

Several typical examples of chromatograms of oligomers are depicted in Figures 3.28 and 3.29. Various synthetic resins (e.g., epoxies, unsaturated polyesters, urea-formaldehyde or phenol-formaldehyde condensates), are typical oligomeric materials. SEC chromatograms of some resins are that typical that the chromatogram itself can identify the type of the resin. Column efficiency and resolution increases with decreasing particle size of column packing. This fact can be fully utilized in the analysis of oligomers where flow-induced shear degradation is unlikely, and therefore 3-μm packing size should be used. Chromatograms of oligomers typically consist of several overlapping peaks corresponding to individual oligomers of low polymerization degrees. With increasing polymerization degree the resolution of particular oligomers decreases until the individual peaks become a smooth curve similar to that of polymers. Note that chromatograms of oligomers clearly demonstrate decreasing resolution of SEC with increasing molar mass. Suitable pore sizes of the column packing

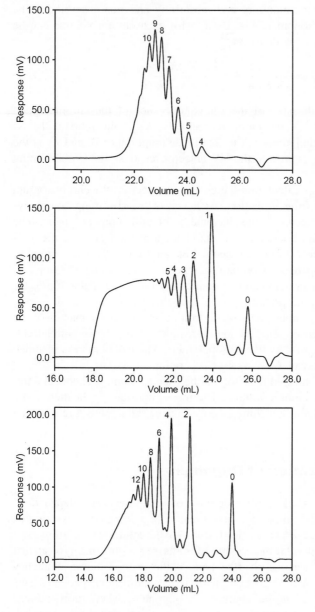

Figure 3.28 Examples of chromatograms of various oligomers: from top to bottom, polyethylene glycol (PEG) of nominal molar mass 440 g/mol, phenolic novolac, and epoxy resin based on bisphenol A prepared by reaction of low-molar-mass epoxy resin with bisphenol A. Separation conditions 3 × PLgel Mixed-E 3 μm 300 × 7.5 mm columns, THF at 1 mL/min, column temperature 40°C, inject 30 μL 0.6% w/v. PEG: $n = 4, 5, 6, \ldots$ oligomers with molar mass according to equation $M = 18 + n \times 44$. Novolac: $n = 0, 1, 2, 3, \ldots$ oligomers with molar mass according to equation $M = 94 + n \times 106; 0 \approx$ phenol, $1 \approx$ dihydroxydiphenyl methane. Epoxy resin: $n = 0, 2, 4, \ldots$ oligomers with molar mass according to equation $M = 340 + n \times 284; 0 \approx$ diglycidyl ether of bisphenol A.

material and high column efficiency are the most important aspects of the high resolution of oligomers.

Figure 3.30 shows the influence of column selectivity and efficiency on the resolution of oligomers. In this example, column efficiency and selectivity were controlled by number and type of SEC columns, namely PLgel Mixed-E, which

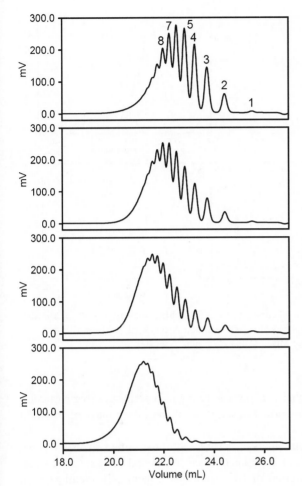

Figure 3.29 Chromatograms of oligostyrene of nominal molar mass (from top to bottom): 680 g/mol, 900 g/mol, 1050 g/mol, and 1250 g/mol, $n = 1, 2, 3, \ldots$ oligomers with molar mass $M = 58 + n \times 104$. Conditions as in Figure 3.28.

are specifically designed for the separation of oligomers, versus PLgel Mixed-C, which cover a broad molar mass range, but their selectivity in the region of lower molar masses is lower compared to the Mixed-E columns. Besides efficiency and selectivity of the SEC columns, the resolution depends on the difference of the molar masses of the two neighboring oligomers in the oligomeric series. Typically, the maximum polymerization degree for which at least partial resolution can be achieved is around ten. It must be emphasized that the resolution power of SEC concerning the ability to separate oligomers is below that of other types of HPLC. Gradient-reversed-phase HPLC can separate oligomers up to a polymerization degree of around 25.

The advantages of SEC compared to HPLC include simpler operating conditions (isocratic elution), lower demand on the purity of the mobile phase, possibility of using an RI detector, and easier identification of particular oligomers because of their explicit separation according to the hydrodynamic volume.

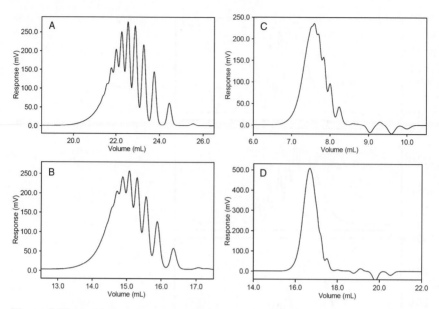

Figure 3.30   Resolution of styrene oligomers using different SEC columns: 3 × PLgel Mixed-E 300 × 7.5 mm 3 μm (A), 2 × PLgel Mixed-E 300 × 7.5 mm 3 μm (B), 1 × PLgel Mixed-E 300 × 7.5 mm 3 μm (C), 2 × PLgel Mixed-C 300 × 7.5 mm 5 μm (D).

It is also worth mentioning that it is not necessary to baseline-resolve particular oligomers in order to get the correct description of the molar mass distribution as seen in Figure 3.31.

The chromatograms of oligomers are processed in the same way as the chromatograms of polymers to get the molar mass averages and distribution curves. The procedure for the calibration of SEC columns is generally identical with that of polymers. For some oligomeric materials it is possible to separate several oligomers with low polymerization degrees and identify their peaks in a chromatogram. Then the calibration curve can be established by relating the elution volumes of oligomeric peaks with the molar masses of the corresponding oligomers. However, the calibration curve based on the oligomeric peaks is valid up to relatively low molar masses, and extrapolation toward the region of higher molar masses, where the experimental data points are missing, may be uncertain and can lead to significant errors. This is especially true if the higher oligomers are branched, such as in the case of phenol-formaldehyde polycondensates. The molar mass range of the calibration curve can be extended by the isolation of particular oligomers by semipreparative HPLC and their subsequent application for the calibration of SEC columns. The obtained calibration can be related to easily obtainable polystyrene calibration. Such a procedure is outlined in Figure 3.32, which compares the calibration curve of polystyrene with that of bisphenol A–based epoxy resin. The two calibrations are in the following

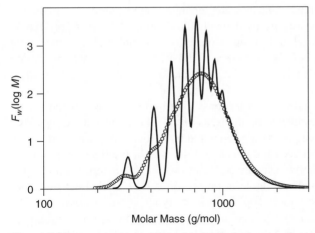

Figure 3.31 Comparison of differential distribution curves of polystyrene of nominal molar mass 680 g/mol determined using columns of different resolution: 3 × PLgel Mixed-E (solid line) versus 2 × PLgel Mixed-C (○). M_n (g/mol) = 680 (3 × Mixed-E), 670 (2 × Mixed-C), M_w (g/mol) = 790 (3 × Mixed-E), 790 (2 × Mixed-C), M_z (g/mol) = 900 (3 × Mixed-E), 920 (2 × Mixed-C).

relation:

$$M_{EP} = 3.895 \times M_{PS}^{0.783} \, (25°C, \text{THF}) \tag{3.68}$$

where subscripts *PS* and *EP* refer to polystyrene and epoxy resins, respectively. Equation 3.68 permits the transformation of the calibration obtained by polystyrene standards to the calibration valid for bisphenol A–based epoxy resins. Using SEC columns with a sufficient resolution in the oligomeric region, the

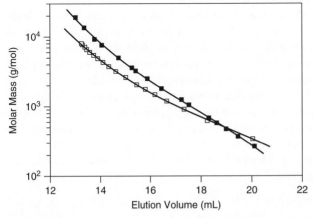

Figure 3.32 Calibration curves of polystyrene (■) and bisphenol A–based epoxy resins (□). The epoxy oligomers were isolated by semipreparative reversed-phase HPLC. Columns Microgel 100 Å, 500 Å, 1000 Å 250 × 7.8 mm, THF at 1 mL/min, room temperature.

low-molar-mass polystyrene standards are separated into several oligomers and their individual molar masses are used for the calibration. Note that polystyrene standards contain one butyl end group (from butyl lithium) and thus the molar mass of styrene oligomers is $M = 58 + n \times 104$ where n is the polymerization degree. That means the analogy for the monomer is hexylbenezene instead of ethylbenzene.

Besides the determination of molar mass distribution, content of the lowest oligomers can be easily estimated on the basis of the relative peak areas of particular peaks. The overlapping peaks are separated by vertical lines arising from the baseline to the valleys between the peaks and the weight fractions of oligomers are estimated as their relative peak areas. Exact quantitative determination of the low-molar-mass compounds available in pure form can be carried out by a procedure identical with that widely applied in other types of chromatography. The procedure is demonstrated by the determination of the lowest epoxy oligomer (diglycidyl ether of bisphenol A) in epoxy resin (Figure 3.33). Quantitative analysis by means of the *external standard* procedure involves the preparation of several solutions of the compound to be quantified at exactly known concentrations and their injections into SEC columns. The calibration graph is then established by plotting the peak area versus injected mass or concentration of the injected solutions. The calibration graph should be established using at least three data points. The sample is analyzed under the same conditions as those used for the calibration standards and the mass corresponding to the peak area is determined from the calibration graph. The procedure can be used to quantify the residual monomers, stabilizers, plasticizers, or other additives in polymers. Alternatively, the *internal standard* procedure or the procedure of *standard addition* can be applied for quantitative analysis of low-molar-mass compounds.

A potential source of errors associated with processing the chromatograms of oligomers is the dependence of the specific refractive index increment (dn/dc) or the extinction coefficient (a) on molar mass, which is typically significant at low polymerization degrees. Correction of the relative peak areas can be made if the relation between dn/dc or a and polymerization degree is known. In most practical cases the polymerization degree dependence of dn/dc or a is unknown and therefore must be neglected. The experimental results indicate that the increase of dn/dc or a with polymerization degree is negligible over molar mass of about 5,000 g/mol, and even for lower molar masses the errors caused by neglecting the molar mass dependence of dn/dc and a are mostly below 10%. Examples of the molar mass dependence of dn/dc and a can be found in the literature for polystyrene (dn/dc),[18,25] polyisobutene (dn/dc),[18] and bisphenol A–based epoxy resins (dn/dc,[25] extinction coefficient[26]). The following examples illustrate possible differences of dn/dc and a:[i]

Bispenol A–based epoxy resin, $M = 340$ g/mol: $dn/dc = 0.152$ mL/g

Bispenol A–based epoxy resin, $M_n = 1490$ g/mol: $dn/dc = 0.178$ mL/g

[i]The values are valid for THF and the wavelength of 633–690 nm (dn/dc) and 278 nm (a).

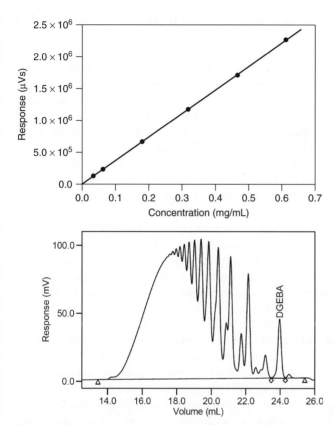

Figure 3.33   Calibration curve for diglycidyl ether of bisphenol A (DGEBA) (top) and DGEBA quantification in epoxy resin (bottom). DGEBA peak area = 613, 800 μVs; using the calibration curve the area corresponds to 0.166 mg/mL of DGEBA, concentration of injected solution = 6.06 mg/mL; that is, the concentration of DGEBA in the resin = 2.7% wt. Conditions as in Figure 3.28.

Bispenol A–based epoxy resin, M_n = 2780 g/mol: dn/dc = 0.183 mL/g
Bispenol A–based epoxy resin, M = 340 g/mol: a = 10.8 Lg^{-1} cm^{-1}
Bispenol A–based epoxy resin, M = 1760 g/mol: a = 12.0 Lg^{-1} cm^{-1}
Bispenol A–based epoxy resin, M = 3180 g/mol: a = 12.1 Lg^{-1} cm^{-1}
Bispenol A–based epoxy resin, M > 30, 000 g/mol: a = 12.2 Lg^{-1} cm^{-1}
Polystyrene, M = 580 g/mol: dn/dc = 0.161 mL/g
Polystyrene, M = 2500 g/mol: dn/dc = 0.182 mL/g
Polystyrene, M > 10^4 g/mol: dn/dc = 0.185 mL/g

The differences of dn/dc and a between particular members of the oligomeric series can be assumed to be small only if their chemical composition is identical. This assumption is not always fulfilled, such as in the case of polyesters based on a blend of aliphatic and aromatic acids, where a and dn/dc of the individual oligomers depend on the number of aromatic units.

Identification of the peaks in chromatograms of oligomers can be performed on the basis of expected chemical composition. Knowledge of the elution volumes of monomers or compounds of similar chemical composition as that of the analyzed oligomers can help assign particular peaks. Model oligomers synthesized using either excess or elimination of one component can be used to interpret the chromatograms of real materials. Measurements of samples taken from the batch during the course of the polyreaction can facilitate the identification of the peaks in chromatograms on the basis of changes of chromatogram pattern with reaction conversion. The expected oligomeric structure should provide good correlation of the calibration curve obtained by relating the logarithms of molar mass to the elution volumes of the peaks assigned to particular oligomers. The elution volumes of oligomers containing hydroxyl groups are influenced by the association with THF. If some of the analyzed oligomers contain hydroxyl groups and some do not, the experimental data points fit on the same curve only if multiples of THF molar mass corresponding to the number of hydroxyl groups are added to the molar masses of oligomers.[27] Another procedure for identification of chemical structure involves isolation of several oligomers and their identification by a spectral technique. A direct online application of an ESI MS detector is limited by poor ionization in THF, which is commonly used as SEC mobile phase. Applications of SEC in combination with MALDI-TOF MS were also reported.

A potentially important issue related to the analysis of oligomers arises from regulatory requirements such as REACH (Registration, Evaluation, and Authorization of Chemicals). A polymer under definition of REACH is a substance consisting of molecules characterized by the sequence of one or more types of monomer unit. Such molecules must be distributed over a range of molecular weights. In accordance with REACH, a polymer is defined as a substance meeting the following two criteria:

1. Over 50% of the weight for that substance consists of polymer molecules. A polymer molecule is a molecule that contains a sequence of at least three monomer units, which are covalently bound to at least one other monomer unit or other reactant. *Note:* In terms of polymer chemistry the requested polymerization degree is four or more.

2. The amount of polymer molecules presenting the same molecular weight must be less than 50 weight percent of the substance.

The method preferred by REACH to determine whether a substance falls under the definition of a polymer is SEC. However, performing the yes/no analysis under the REACH definition requires identification of particular oligomeric peaks in an SEC chromatogram to correctly identify their polymerization degrees. This may not be always a straightforward procedure and the proper characterization may require combination of SEC results with those obtained by other techniques such as MS or HPLC.

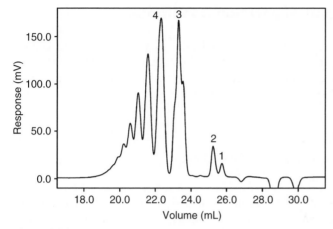

Figure 3.34 Chromatogram of oligomer prepared by glycolysis of poly(ethylene terephthalate) (PET). Composition of reaction mixture: 65% PET, 28% diethylene glycol, 7% propylene glycol. Polystyrene equivalent molar masses/relative peak areas for peaks 1, 2, 3, and 4: 146 g/mol/1.4%, 194 g/mol/3.0%, 513 g/mol/27.2%, and 793 g/mol/25.0%, respectively. Conditions as in Figure 3.28.

Let us examine whether the compound depicted in Figure 3.34 is a polymer according to the REACH classification. The investigated material is a product of glycolysis of polyethylene terephthalate (PET). The decomposition of PET in a glycolic environment is based on the transesterification of PET with glycol(s), resulting in a mixture of aromatic polyols, which can be further processed in the production of polyurethanes or unsaturated polyester resins. The obtained polyols are oligomers consisting of various numbers of terephthalate (T) units, ethylene glycol (EG) units, and units of glycol used for glycolysis (propylene glycol, PG, and diethylene glycol, DiEG, in the case of the sample shown in Figure 3.34). The procedure allows recycling of the waste PET and thus is environmentally important.

The classification of the analyzed PET alcoholysate is based on two assumptions: (1) The RI detector response of the particular oligomers is approximately identical and thus the relative peak areas can be used for the quantification, and (2) polystyrene calibration provides approximately correct determination of molar masses of terephthalic-based oligomers. An oligomer fulfilling the requirement of REACH is the reaction product PG-T-EG-T-PG with molar mass of 474 g/mol or similar oligomers with different end glycol units. The chromatogram proves that there is no oligomer in the analyzed mixture in an amount over 50%. The shape of peak #3 indicates a mixture of three compounds, most likely oligomers with different end glycol units. The polystyrene equivalent molar mass corresponding to peak #3 allows us to identify the compounds as PG-T-EG-T-PG and similar oligomers with different end units. It can be concluded that under given assumptions the compound in Figure 3.34 is a polymer according to the REACH classification.

3.5.6   Influence of Separation Conditions

Molar masses determined with conventional SEC using the traditional column calibration approach strongly depend on the calibration used for data processing. In addition, the obtained results are sensitive to several operating parameters of the SEC instrumental setup. The most important parameters are:

- Flow rate
- Column temperature
- Sample concentration and injected volume
- Column performance

It has been demonstrated in Table 3.1 that flow rate is the most essential parameter for accurate and repeatable determination of molar mass. The elution volume of polymer molecules may slightly depend on flow rate, but this dependence is not important for practical measurements because the calibration standards and polymer samples are measured at the same flow rate. SEC experiments are carried out at a fixed flow rate selected according to solvent viscosity and column specification, keeping in mind the productivity of the measurement. Using standard 8-mm-inner-diameter columns, a typical flow rate is 1 mL/min for THF and 0.5–0.8 mL/min for aqueous solvents. Lower flow rates may slightly increase efficiency. In addition, lower flow rates of about 0.3 mL/min may be beneficial from the viewpoint of possible shearing degradation of high-molar-mass fractions.

The determination of molar mass distribution is based on the assumption of efficient polymer separation according to the hydrodynamic volume. It has been stated that the resolution R_{SP} should be at least 2.5 and the parameter SP more than 6. The effect of resolution is shown in Table 3.5, which compares molar mass averages of three polydisperse polymers determined by different numbers of SEC columns. Although the molar mass averages are not entirely identical, the differences are relatively small. It is obvious that increasing the resolution by using too many columns does not benefit the obtained results. Taking into account the accuracy of measurement, run time, and solvent consumption, it can be concluded that two standard $300 \times 7.5–8$-mm columns packed with mixed-pore packing should be adequate for most practical applications, and even a single column would provide sufficient results when time per analysis becomes an important factor.

Column temperature is another important parameter influencing the results generated by column calibration. Unless required by limited sample solubility, SEC experiments are typically carried out at room temperature or slightly above room temperature. Figure 3.35 shows chromatograms of a mixture of polystyrene standards and corresponding calibration curves acquired at three different temperatures. The shift of the calibration curves is a result of two concurrently acting factors: a temperature expansion of the mobile phase that results in an increased flow rate, and an expansion of the hydrodynamic volume of standards

Table 3.5   Effect of Resolution on the Molar Mass Averages of Polydisperse
Polystyrene

Polystyrene Sample	M_n (10^3 g/mol)		M_w (10^3 g/mol)		M_z (10^3 g/mol)	
	$1 \times$ HR5E	$3 \times$ HR5E	$1 \times$ HR5E	$3 \times$ HR5E	$1 \times$ HR5E	$3 \times$ HR5E
M1	56.9 ± 0.2	60.3 ± 0.4	283.2 ± 0.9	289.7 ± 0.7	482.3 ± 1.7	468.8 ± 0.9
M3	80.8 ± 0.8	89.3 ± 0.7	289.3 ± 1.3	296.5 ± 0.5	535.8 ± 2.4	521.0 ± 1.7
M4	42.7 ± 0.2	45.7 ± 0.2	235.4 ± 0.3	241.5 ± 0.5	483.7 ± 5.6	468.4 ± 1.6

SEC columns: $1 \times$ versus $3 \times$ Waters Styragel HR5E 300×7.8 mm, $R_{SP} = 2.4$ versus 3.3,
$SP = 2.6$ versus 7.6 calculated for standards with M_{PEAK} 675,000 and 68,000 g/mol. *Conditions:*
THF at 1 mL/min, calibration with 13 polystyrene standards $10^3 - 4.3 \times 10^6$ g/mol (concentration
0.03% w/v of each standard), sample concentration 0.25% w/v, sample injection 25 μL per column.

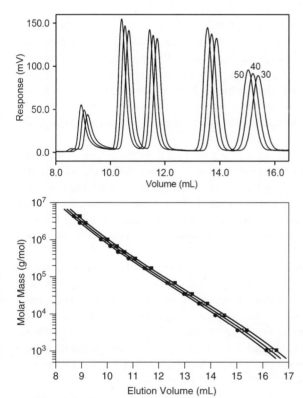

Figure 3.35   Chromatograms
of polystyrene standards (top)
and corresponding calibration
curves (bottom) obtained at
$30°C$ (■), $40°C$ (□), and $50°C$
(●). Columns $2 \times$ PLgel
Mixed-C 300×7.5 mm, THF
at 1 mL/min. Standards
2,850,000 g/mol,
470,000 g/mol, 170,000 g/mol,
19,000 g/mol, and 3600 g/mol.

due to increased thermodynamic quality of the solvent. Both effects result in
faster elution.

Table 3.6 shows M_w values of polydisperse polystyrene measured at temperatures of 30, 40, and $50°C$ using the calibration relation determined at $30°C$.
The decrease of elution volume due to increasing temperature results in markedly
increasing M_w when the chromatograms collected at 40 and $50°C$ are processed

Table 3.6   Effect of Column Temperature on M_w of Polydisperse Polystyrene and the Compensation by Flow Marker

| Temperature (°C) | M_w (g/mol) | | Flow Marker Elution Volume (mL) |
	No Flow Marker	Flow Marker	
30	273,000 ± 200	276,500 ± 600	20.25 − 20.26
40	319,700 ± 400	251,100 ± 200	19.82
50	370,900 ± 400	234,800 ± 100	19.47

Conditions: Column calibration at 30°C, THF at 1 mL/min, 2 × PLgel Mixed-C 300 × 7.5 mm columns, flow marker elution volume of calibration standards = 20.23 mL.

with the calibration curve established at 30°C. The use of a flow marker over-compensates for the effect of the increased temperature and the resulting M_w are too low. The results emphasize the necessity of using the identical temperature for column calibration and the sample measurements. The elution volume at higher temperatures can be also influenced by expansion of gel pores and decreased viscosity of the mobile phase, which consequently increases the diffusion coefficients. The temperature may also affect the secondary separation mechanisms.

Table 3.7 demonstrates the influence of injected mass on the molar mass moments of broad polystyrene. Different injected masses were achieved by different injection volumes of a sample of given concentration or by injection of the same volume of solutions of different concentration. Glancing at Table 3.7 one can see a significant decrease of molar mass with increasing injected mass. A relatively small difference in the injected mass results in a measurable difference of the molar mass.

The influence of the concentration of calibration standards on the obtained molar masses is shown in Table 3.8. Three calibration solutions were prepared as mixtures of four or five narrow polystyrene standards covering the molar mass range of $10^3 − 4.3 × 10^6$ g/mol. The regular concentration of the calibration standards was 0.03% w/v of each of the standard. The two standards with the highest molar mass ($4.34 × 10^6$ and $2.85 × 10^6$ g/mol) were prepared at half concentration. The calibration solutions were then prepared at concentrations lower by a factor of two and four and two and three times higher. Table 3.8 lists individual concentrations per standard and the total concentrations of all standards in the calibration solutions. The molar mass averages were calculated for sample prepared at a concentration of 0.25% w/v using calibration curves obtained for the particular concentrations of standards. The results show a negligible effect of the concentration of calibration solutions on M_n, and clearly evident impact on the higher molar mass averages. However, the influence of the concentration of polystyrene standards appears to be low over relatively broad concentration range.

Table 3.7 Effect of Injected Mass on Molar Mass Averages of Broad Polystyrene Determined by Conventional SEC

Injected Mass (mg)	Volume (μL)	Concentration (% w/v)	M_n (g/mol)	M_w (g/mol)	M_z (g/mol)
0.05	100	0.05	89,900	292,100	533,000
0.10	100	0.10	91,900	290,200	528,400
0.15	100	0.15	87,100	285,300	520,800
0.20	100	0.20	87,800	283,500	518,300
0.25	100	0.25	85,300	276,700	508,400
0.50	100	0.5	84,200	259,400	492,900
0.75	100	0.75	78,300	231,400	475,500
1.0	100	1.0	68,400	201,400	459,800
1.5	100	1.5	45,800	164,000	421,000
0.025	10	0.25	91,200	305,800	558,100
0.05	20	0.25	91,100	300,900	548,800
0.125	50	0.25	88,300	290,700	531,300
0.188	75	0.25	87,000	285,500	522,300
0.31	125	0.25	86,200	275,600	506,600
0.38	150	0.25	85,300	272,100	501,300
0.50	200	0.25	83,800	265,800	492,500
0.1	10	1.0	91,800	295,800	543,000
0.15	10	1.5	88,400	285,400	530,300
0.2	10	2.0	86,700	273,300	517,900

The concentration dependence of the elution behavior was studied by Janca et al.[28–33] Three contributions of concentration dependence of the elution volume were considered: change of the hydrodynamic size of the analyzed macromolecules, viscosity of the polymer solution in the pores of column packing, and secondary exclusion due to occupancy of a pore by another polymer molecule. The hydrodynamic volume of polymer random coils decreases with increasing concentration, which leads to the increase of elution volume. The dependence is more pronounced in thermodynamically good solvents, and the polymer dimensions become independent of concentration in theta conditions. The secondary exclusion resulting from the occupation of a part of pore volume by other polymer molecules becomes more pronounced at higher concentrations and the elution volume decreases. The viscosity effect is related to the mobility of macromolecules. The effect of the viscosity is more dominant at the region of higher molar masses.

At a certain viscosity of polymer solutions the chromatograms may become more complicated with several irregular maxima not corresponding to the true molar mass distribution. This phenomenon is called *viscous fingering* and may lead to a complete failure of SEC separation. Viscous fingering appears if the viscosity of the injected solution is markedly greater than that of the mobile phase. The high viscosity of the injected solution may result in the decreased

Table 3.8   Effect of the Concentration of Polystyrene Standards on the Molar Mass Averages of Broad Polystyrene

Concentration per Standard (% v/w)	Concentration total (% v/w)	M_n (g/mol)	M_w (g/mol)	M_z (g/mol)
0.004 or 0.008	0.026–0.034	85,500	273,300	498,400
0.008 or 0.015	0.053–0.068	85,300	274,500	501,100
0.015 or 0.030	0.105–0.135	85,600	276,900	507,000
0.030 or 0.060	0.210–0.270	85,800	282,000	520,900
0.045 or 0.090	0.315–0.405	86,300	288,300	539,200

Calibration with 13 polystyrene standards ranging from 10^3 to 4.3×10^6 g/mol, injection in three mixtures containing 4 or 5 standards.

velocity of the injected zone, whereas the faster mobile phase has a tendency to finger its way through the injected sample zone, which distorts the peak shape.

Investigation of the concentration dependence of the elution volume is facilitated by a light scattering detector capable of direct measurement of molar mass and size of the molecules eluting from SEC columns. An example of such a study is presented in Figure 3.36, which shows results obtained for different injected masses of polydisperse polystyrene. With the increasing injected mass the chromatograms are broadened and shifted to lower elution volumes. The shift of the molar mass plots toward higher elution volumes appears from a certain injected mass, below which the effect of sample concentration is negligible. The corresponding molar mass averages are listed in Table 3.9.

Comparison of Tables 3.7 and 3.9 shows significantly lower sensitivity of M_w determined by the light scattering detector to injected mass compared to M_w obtained by column calibration. In the case of conventional SEC the decrease of M_w with increasing injected mass is caused by the shift of the chromatograms toward higher elution volumes. In the case of SEC-MALS the decrease of M_w is due to neglecting the concentration dependence of the intensity of scattered light, but the effect is relatively small within the range of concentrations usually used for SEC measurements. With the MALS detector the averages M_n and M_z are determined on the assumption of the monodisperse slices, and thus the increased polydispersity within the elution volume slices due to column overloading increases M_n and decreases M_z.

It is obvious that the accuracy of the molar mass determined by SEC with column calibration depends on the injected mass of the sample and somewhat on the injected mass of the calibration solutions. To find out an optimum concentration and injected volume in order to reduce the errors caused by the concentration dependence of elution volume may not be straightforward. The concentration of the injected solution begins to change immediately after the injection as a result of sample dilution with a solvent and separation according to molar mass distribution. This effect depends on the width of the molar mass distribution,

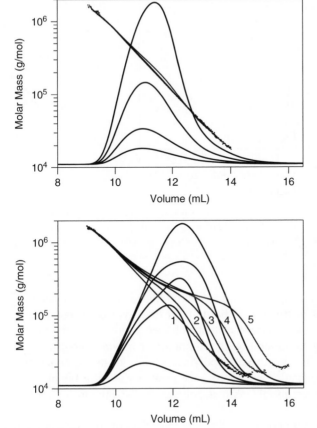

Figure 3.36 Effect of injected mass: Molar mass versus elution volume plots and RI chromatograms of polydisperse polystyrene analyzed under different injected mass. Injected mass: 0.05 mg, 0.11 mg, 0.24 mg, and 0.49 mg (top), 1 = 0.19 mg, 2 = 0.73 mg, 3 = 1.06 mg, 4 = 1.45 mg, 5 = 1.93 mg (bottom). Columns 2 × PLgel Mixed-C 300 × 7.5 mm, THF at 1 mL/min, 40°C, injected volume 100 μL. Molar mass averages are listed in Table 3.9.

Table 3.9 Effect of Injected Mass on Molar Mass Averages of Broad Polystyrene Determined by SEC-MALS

Injected mass (mg)	M_n (g/mol)	M_w (g/mol)	M_z (g/mol)
0.05	130,000	294,000	480,000
0.11	124,000	293,000	477,000
0.15	118,000	288,000	470,000
0.19	121,000	288,000	471,000
0.24	118,000	287,000	469,000
0.49	121,000	283,000	458,000
0.73	127,000	280,000	442,000
1.06	144,000	276,000	419,000
1.45	159,000	273,000	400,000
1.93	184,000	269,000	380,000

Data obtained by conventional SEC are in Table 3.7; graphical representation is in Figure 3.36.

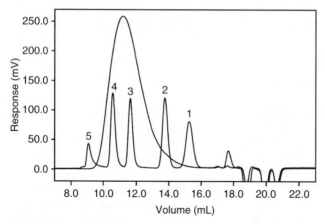

Figure 3.37 Overlay of chromatograms of a mixture of polystyrene standards and broad polystyrene. Injection: 100 μL 0.2% w/v for broad polymer, 50 μL 0.03% w/v for each of standards 1–4, and 0.015% w/v for standard 5. Molar masses of standards 1–5: 3600 g/mol, 19,000 g/mol, 170,000 g/mol, 470,000 g/mol, and 2,850,000 g/mol, respectively.

which means that broader samples are always more diluted compared to samples with narrow molar mass distribution. The dilution of polydisperse polymer in SEC columns is responsible for the fact that polymer sample concentration is mostly different from the concentration of calibration standards. The difference of the concentration of the polydisperse polymer sample from the concentration of the calibration standards varies across the peak as shown in Figure 3.37, which overlays chromatograms of a broad polymer and a mixture of narrow standards.

Several methods were proposed to eliminate the effect of the concentration of sample and calibration solutions on the molar masses determined by SEC. One of the proposed methods of reducing the concentration effect involves calibration by means of standards of different concentrations, which yields a series of calibration curves valid for different concentrations and an extrapolated curve for zero concentration. The chromatogram of a polydisperse polymer is then separated into several sections and each of them is processed using the calibration dependence valid for the concentration most appropriate to the given section. Other methods of reducing the concentration effect are an extrapolation procedure where reciprocal molar mass averages determined at various concentrations are plotted against concentration and extrapolated to zero concentration, or processing the chromatogram with the use of a calibration curve extrapolated to zero concentration. A significant disadvantage of the latter method is that it does not eliminate the concentration effect of the sample itself, which is obviously more serious than the concentration effect of the calibration standards.

It is evident that the procedures for correcting the concentration effect are not suitable for routine measurements. For example, to eliminate the concentration effect completely by using calibration curves valid for different concentrations would require the separation of a chromatogram into a relatively high number of sections. Although the concentration effect represents a potentially serious source of inaccuracy, its reduction should not eliminate the speed and simplicity of SEC measurements. Keeping the concentration of sample and calibration standards at an appropriate level can minimize the effect of concentration. As a rule of thumb, the concentrations of the analyzed samples and calibration standards should be kept as low as possible, but achieve sufficient detector signal-to-noise ratio. The injected mass can be easily reduced by dilution of the injected sample or decreasing the injected volume. Remeasuring samples showing peak shoulders at lower injected mass can reveal multicomponent polymer blends, as demonstrated in Figure 3.38.

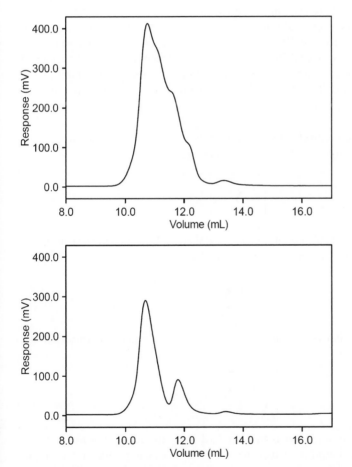

Figure 3.38 Chromatograms of polymer blend analyzed at the concentrations of 0.3% w/v (top) and 0.15% w/v (bottom). Columns 2 × PLgel Mixed-C 300 × 7.5 mm, THF at 1 mL/min, 100-μL injection.

3.5.7   Accuracy, Repeatability, and Reproducibility of SEC Measurements

The term *repeatability* (precision) refers to the variation of the experimental results generated by a single person or instrument on the same sample under identical conditions. A measurement may be considered repeatable when the variation of results is smaller than a certain agreed limit. The repeatability conditions also include measurements over a short period of time. *Reproducibility* refers to the ability of the experiment to be reproduced by someone else working independently. In contrast to repeatability, which is related to successive experiments, reproducibility is related to the agreement of experiments performed by different operators, with different apparatus and laboratory locations, or of measurements repeated in the same laboratory after a relatively long time period. Reproducibility is often confused with repeatability, and results generated in a single laboratory by a single person by multiple injections within one day are often used to demonstrate good reproducibility achievable with a given instrumentation.

Repeatability is of crucial importance for the interpretation of experimental results. When the results are at the verge of the limit of possible repeatability, the conclusions concerning the effect of synthetic conditions on molar mass, relation of molar mass with sample properties, or differences between samples might be false.

Repeatability and reproducibility are usually evaluated by the *standard deviation* defined for a large number of measurements as:

$$\sigma = \sqrt{\frac{1}{N} \sum_{i=1}^{N} (x_i - \bar{x})^2} \tag{3.69}$$

where N is the number of measurements, x_i is the result of the ith measurement, and $\bar{x}$ is the arithmetic mean (average) defined as:

$$\bar{x} = \frac{1}{N} \sum_{i=1}^{N} x_i \tag{3.70}$$

For a small number of measurements the standard deviation is estimated as:

$$\sigma = \sqrt{\frac{1}{N-1} \sum_{i=1}^{N} (x_i - \bar{x})^2} \tag{3.71}$$

Relative standard deviation is the standard deviation relative to the average. Besides standard deviation the experimental differences can be illustrated graphically as a plot of molar mass average versus laboratory number or measurement number. To estimate repeatability, standard deviation from multiple SEC runs can be calculated.

Accuracy (correctness) can be defined as agreement with the truth and can be estimated by comparison with results obtained by measurements with an absolute method or literature values. Questions related to accuracy, reproducibility, and repeatability are:

1. Do the obtained results agree with the true values?
2. Are the results comparable with those generated in other laboratories?
3. Are the results identical when the measurements are repeated after a certain period of time?

A well-characterized polymer with reliably determined molar mass averages is needed to address the first question. Molar mass averages determined by conventional SEC are compared with the results determined by light scattering and membrane osmometry for several samples in Table 3.10, which shows a tendency of SEC to yield M_n values lower than those from membrane osmometry. Note that SEC and MO have a counteracting tendency to errors in M_n. In conventional SEC the band-broadening M_n decreases, while in the case of MO the M_n may be increased due to the permeation of oligomeric fractions through the membrane. For samples containing oligomeric fractions, SEC can be a more reliable source of M_n than absolute membrane osmometry. The agreement of M_w obtained by SEC and light scattering is satisfactory and indicates that neglecting the peak-broadening correction does not affect the experimental results significantly.

Reproducibility of the measurements can be evaluated by means of *round-robin tests*, where the same samples are sent to different laboratories where they are measured using available apparatus, columns, and standards and the methodology routinely used in the laboratory. Results of ten round-robin tests[9] carried out for various polymers in the late 1980s and the 1990s showed repeatability

Table 3.10 Comparison of Molar Mass Averages Determined by Conventional SEC with Column Calibration* and Absolute Methods

	M_n (10^3 g/mol)		M_w (10^3 g/mol)		
Sample	SEC	MO	SEC	Batch MALS	SEC-MALS
PS (NIST 706)	89 ± 1	137	273 ± 1	288 ± 2	283 ± 1
PS (A)	—	—	327 ± 5	344 ± 8	326 ± 4
PS (AN)	82 ± 5	120	273 ± 7	303 ± 1	294 ± 1
PS (1)	70 ± 1	139	355 ± 1	386 ± 1	370 ± 1
PS (2)	105 ± 1	153	362 ± 1	398 ± 6	375 ± 1
PS (S)	39 ± 1	66	217 ± 1	236 ± 1	225 ± 1
PS (K)	84 ± 1	98	230 ± 1	239 ± 3	238 ± 1
PMMA (Y)	145 ± 3	185	498 ± 1	604 ± 10	558 ± 7
PMMA (J)	39 ± 1	33	84 ± 1	96 ± 1	89 ± 1

*Calibration is based on polystyrene or poly(methyl methacrylate) narrow standards.

Table 3.11 Results of Two Round-Robin Tests on Synthetic Resins (S) and Epoxy Resins (EP): Molar Mass Averages and Relative Standard Deviations

Sample	No. of Participants	M_n (10^3 g/mol)	RSD (%)	M_w (10^3 g/mol)	RSD (%)	M_z (10^3 g/mol)	RSD (%)
S (1)	6	4.0	37	9.0	9	14.2	9
S (2)	6	3.9	49	52.2	25	320	34
EP (1)	16	1.8	16	5.0	11	—	—
EP (2)	16	4.0	13	12.4	9	—	—

and reproducibility of M_n in the ranges of 0.7–5.6% and 3.5–33.4%, respectively; and the repeatability and reproducibility of M_w in the ranges of 0.4–5.9% and 4.3–17.7%, respectively. Results of two round-robin tests organized by the author are listed in Table 3.11. Experience and round-robin tests show that:

- Repeatability is significantly better than reproducibility.
- The molar mass averages generated in two laboratories can often differ by a factor of two or more.
- The reproducibility of M_w is markedly better compared to that of M_n, which can be explained by typically higher uncertainty in the selection of the peak end.
- Reproducibility is better for well-resolved samples compared to those with high-molar-mass or low-molar-mass tails.
- Samples containing functional groups along the polymer chain show larger uncertainty of results due to interactions with the surface of the column packing.

Significantly better repeatability compared to reproducibility is true also in the same laboratory, as demonstrated in Tables 3.12 and 3.13, which compare repeatability with a long-term reproducibility generated in a single laboratory using the same SEC setup, column type, and calibration standards. It is necessary to emphasize that good repeatability may lead to self-satisfaction and overestimation of accuracy of SEC and its ability to generate comparable results when the same sample is characterized in different laboratories. An isolated operator without a confrontation with other laboratories can generate systematically shifted results.

The results of a Japanese round-robin test[34] showed significantly more favorable results when the data generated only in experienced laboratories, namely those belonging to manufacturers of SEC instrumentation and columns, were selected for calculation of standard deviations. This fact corresponds to my own experience, which is demonstrated in Table 3.14. Although Table 3.14 compares results from only two laboratories, the very good agreement is most likely not accidental, because it was obtained for five different samples. It appears that

Table 3.12   Repeatability of Molar Mass Determination of Broad Polystyrene from Ten Consecutive Runs

Run #	M_n (g/mol)	M_w (g/mol)	M_z (g/mol)
1	85,200	274,300	503,300
2	84,800	274,600	504,800
3	85,200	275,700	506,800
4	84,000	276,000	507,300
5	85,300	277,000	508,100
6	86,000	277,400	509,900
7	85,000	277,800	511,700
8	86,300	277,800	510,100
9	85,900	278,100	510,300
10	85,400	278,400	511,400
Mean	85,300	276,700	508,400
SD (g/mol)	700	1500	2800
RSD (%)	0.8	0.5	0.6

For comparison of reproducibility, see Table 3.13.
Conditions: THF at 1 mL/min, 2 × PLgel Mixed-C 300 × 7.5 mm columns, no flow marker.

SEC reproducibility is markedly better when experienced operators carry out the measurements despite using different instrumentation, columns, and calibration standards.

Relatively poor reproducibility of conventional SEC can be explained by sensitivity to many experimental and processing parameters. These are: type of columns, injection volume, concentration of injected samples and standards, type and manufacturer of calibration standards and their number, universal calibration yes/no, universal calibration parameters, fitting the calibration data, flow rate, whether flow marker is used, column temperature, sample preparation (incomplete dissolution, improper filtration), and the way of baseline and peak selection.

Regular measurements of a reference sample can address repeatability and reproducibility within a single laboratory and facilitate comparison of results acquired with a long time gap. Long-term reproducibility within a single laboratory is influenced mainly by the change in separation efficiency of the columns. Column performance usually deteriorates with time, which is indicated by decreasing plate number. Column properties may also change with respect to column interactivity (i.e., tendency to interact with polymers). Irreversibly adsorbed molecules from previous injections can occupy a part of the packing pores and thus decrease the pore volume available for sample separation. The change of the interactivity may not be revealed even by regular analysis of a control polymer, because if the chemical composition of the control polymer differs from that of samples being analyzed, then the interaction behavior is most likely different. Deviations of the pump flow rate from the nominal value are another important source of poor reproducibility within a laboratory. One

Table 3.13 Reproducibility of Molar Mass Averages of Broad Polystyrene* Measured Over a Period of One Year

Date Acquired	M_n (g/mol)	M_w (g/mol)	M_z (g/mol)
01/03	88,100	282,300	530,000
01/08	86,100	282,500	530,200
01/11	87,100	282,800	527,300
01/22	87,100	282,300	526,200
01/29	85,000	267,000	496,800
01/31	71,000	267,000	511,300
02/07	76,300	266,200	497,700
02/12	81,000	267,300	497,700
. . .	. . .	. . .	. . .
06/05	87,400	276,100	507,300
06/07	85,800	275,200	509,400
06/13	87,100	275,600	504,100
06/14	87,600	276,200	502,700
06/18	75,600	270,400	496,400
06/21	86,700	274,500	501,600
06/22	85,200	275,400	504,600
06/27	84,853	282,000	517,900
07/11	82,900	272,300	501,700
. . .	. . .	. . .	. . .
11/01	80,200	279,400	510,700
11/02	85,400	286,500	526,800
11/07	81,700	268,800	496,100
11/08	85,100	284,100	524,100
11/22	85,600	274,140	505,800
11/26	82,700	273,640	505,400
11/27	89,000	275,700	505,200
12/11	87,000	270,100	498,700
Total no. of measurements	73	73	73
Mean	82,200	272,800	506,200
SD (g/mol)	4,800	6,500	10,400
RSD (%)	5.8	2.4	2.1

*The same sample as in Table 3.12.
Conditions: THF at 1 mL/min, 2 × PLgel Mixed-C 300 × 7.5 mm columns, no flow marker, calibration before each measurement.

should not forget temperature, which may become important in laboratories without air conditioning and without a column oven.

Various round-robin tests usually did not reveal any major source of poor reproducibly. That means the variation of SEC results is due to a combination of several different factors, including the different experience and individual

Table 3.14 Molar Mass Averages of Acrylic Copolymers Determined in Two
Independent, Very Experienced Laboratories: Example of Very Good Reproducibility

| | M_n (g/mol) | | M_w (g/mol) | |
Sample No.	Laboratory #1	Laboratory #2	Laboratory #1	Laboratory #2
1	10,500	10,300	29,300	32,900
2	12,300	12,300	33,800	34,600
3	10,900	10,700	32,200	32,800
4	13,500	13,800	40,200	41,500
5	3,300	2,500	6,300	6,300

approach of operators. Reproducibility can be markedly improved by imple-
menting proper rules and procedures. On the other hand, SEC is certainly a truly
routine technique, and simplicity and ease-of-use should not be sacrificed in the
name of reproducibility. It is necessary to balance between bad laboratory practices and procedures, leading to poor reproducibility, and far-too-strict rules that
are difficult to fulfill and that reduce the full potential of the SEC method.

The most important measures leading to improvement of the interlaboratory
reproducibility of conventional SEC results are:

- Selection of columns appropriate for samples to be analyzed, using mixed
  columns if possible, and using the same type of columns
- The same calibration standards from the same vendor
- The same concentration and injection volume of the calibration standards
- The same number of calibration data points and the same type of data fit,
  usually third-order polynomial
- The same constants of the Mark-Houwink equation if the universal cali-
  bration is applied
- The same injection volume and concentration of samples
- Recalibration before each sample series and using a flow marker if possible
- Measurement under stabilized conditions giving minimum baseline drift
- Unified methodology for finding the baseline and peak integration limits,
  rejecting results acquired with an unstable baseline
- Regular analysis of a common reference sample
- Using the same type of detector (mostly RI)
- Regular maintenance of the pump with respect to flow rate accuracy and
  repeatability

In the case of light scattering detection, some of the above parameters become
far less important and the most important points are:

- A common method for the calibration of light scattering and concentration
  detectors

- Using identical dn/dc for the polymers being analyzed

In both conventional SEC and SEC combined with a light scattering detector, the report should contain information necessary for mutual comparison of results generated in different laboratories or within a given laboratory over a long time period. The information in the report should address the points stated above.

3.6   APPLICATIONS OF SEC

An important question to ask before characterization of polymers is what sort of information is needed, for what purpose, and in what time period. Useful information about polymers usually comes from molar mass averages, typically M_n, M_w, and M_z. However, more information can be revealed from distribution functions that yield information about the synthetic process and help us understand the structure–property relations. Therefore, the most important major application of SEC is determination of the molar mass distribution and molar mass averages. This goal can be achieved using column calibration or most effectively in combination with a light scattering detector that measures the molar mass directly.

Information about molar mass distribution is important in all areas of polymer science covering the theory, synthesis, properties, and final applications. The combination of SEC with a light scattering detector and also a viscometer can yield further information about the molecular structure. The molar mass averages are the characteristics that are easy to arrange into tables and relate with polymer properties. However, polymer properties can depend not only on the average molar masses, but also on the entire molar mass distribution and its details. Especially fractions with molar mass significantly lower or higher than the majority of polymer molecules can considerably affect the application properties. An overlay of distribution curves is especially suitable for the comparison of properties of similar samples and can explain possible differences among them. Molar mass characteristics can be used for structure–property studies, and they can be also valuable for evaluation of the reproducibility of manufacturing process and to study the influence of polymerization conditions on the molecular structure of the obtained products. In contrast to SEC, the traditional quantities such as solution or melt viscosity or content of functional groups (e.g., acid number, epoxy content) provide only an average characteristic.

Another advantage of SEC in the area of manufacturing process and quality control compared to other methods are speed and simplicity of sample preparation. The results of SEC measurement are available in very short time, which allows for taking measures before the next manufacturing batch is started.

The typical analysis time of about 30 min can be further reduced by using special columns for high-throughput analysis, which, however, provide only limited resolution. Due to the speed of the measurements, SEC can be also used for direct manufacturing process monitoring, especially in the case of polycondensation reactions. The obtained chromatograms need not be necessarily converted

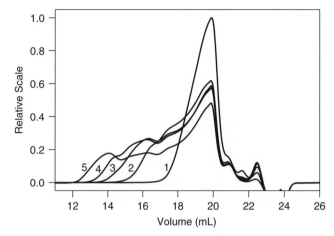

Figure 3.39 Monitoring the course of preparation of alkyd resin by a series of chromatograms of samples taken during the reaction (reaction conversion increases from sample #1 to #5).

into molar mass distribution; a simple fingerprint comparison can be sufficient for sample comparison or illustration of the growth of a polymer chain during the polyreaction or degradation process due to chemical or physical exposition. An example of the monitoring of the course of a polycondensation reaction by overlay of chromatograms is shown in Figure 3.39.

A simple fingerprint comparison of two polyols for polyurethane foams is depicted in Figure 3.40. It is evident that a simple chromatogram overlay can reveal differences between the two polyols, one containing a significant fraction of lower oligomers with the main fraction shifted to higher molar masses as compared to the other sample. The molar mass averages, although polystyrene equivalent values, correspond to the chromatograms: M_n is lower for the polyol containing the oligomeric fractions, while its M_w is higher, corresponding to the shift of the molar mass distribution to higher molar masses.

Figure 3.41 shows another example of simple interpretation of SEC data that allows us to differentiate between samples with different end-use properties.

In the case of oligomeric materials, SEC can determine the content of lower oligomers. High repeatability of SEC allows a sensitive fingerprint comparison of chromatograms of oligomeric materials such as various types of synthetic resins. Another application is the quantification of low-molar-mass additives that are added to polymers to improve their properties and to protect them against degradation. The additives include plasticizers, dyes, antioxidants, and UV stabilizers, and these can be easily quantified after the calibration of detector response.

Other low-molar-mass compounds that can be present in polymers are residual monomers, initiators, catalysts, solvents, and various side products. SEC can be used for the quantification of additives and other low-molar-mass compounds only if their molecular sizes are sufficiently different from each other. If this requirement is not fulfilled, other types of liquid chromatography must be used.

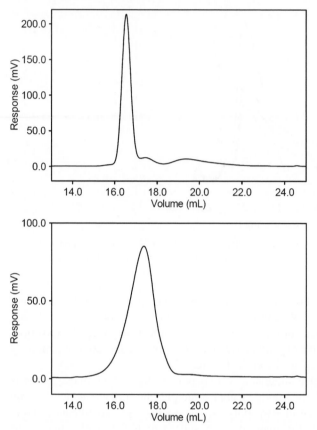

Figure 3.40 Chromatograms of polypropylene glycol–based polyols for polyurethane foams showing different molar mass distribution. Top: $M_n = 6500$ g/mol, $M_w = 9900$ g/mol, $M_w/M_n = 1.51$. Bottom: $M_n = 7100$ g/mol, $M_w = 8600$ g/mol, $M_w/M_n = 1.21$. Conditions as in Figure 3.28.

There are two different methods of determination of low-molar-mass compounds in polymers. One is the extraction of low-molar-mass compounds with a suitable solvent and subsequent SEC or HPLC analysis of the obtained extract. The extraction can be carried out by immersing a polymer film or a thin sheet in a solvent for many hours or several days or using a Soxhlet extractor. The other method is to inject the polymer solution directly into an SEC instrument with columns capable of separation of the additive from polymer. A typical example of a chromatogram of a polymer containing a low-molar-mass additive is shown in Figure 3.42.

There are also other, less-usual applications; for instance, the high repeatability of SEC has been used to identify sources of crude oil contamination on the basis of fingerprint comparison. SEC can be also used for the characterization of various low-mol-mass compounds, but the separation ability is limited compared

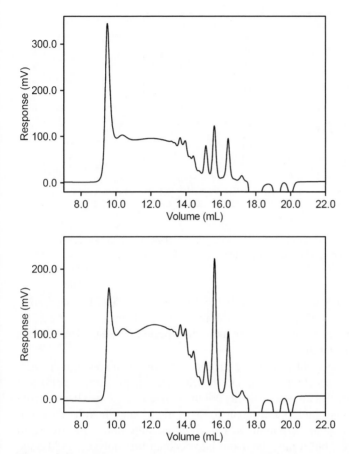

Figure 3.41 Chromatograms of two resin samples with good (top) and bad (bottom) properties obtained using columns with exclusion limit of about 30,000 g/mol. Chromatograms show presence of low-molar-mass compounds with molar mass of several hundreds as well as compounds with molar mass exceeding the exclusion limit of SEC columns. The two batches differ in the content of low-molar-mass compounds and also in the fractions with high molar mass. Columns 2 × PLgel Mixed-E 300 × 7.5 mm 3 μm, THF at 1 mL/min, injection 50 μL 0.3% w/v.

to other types of liquid chromatography. Possible applications include the characterization of crude oil, natural oil, asphalt, and tar. The obtained chromatograms may be complicated overlays of several peaks, because the complete resolution of all compounds is virtually impossible.

Special techniques include preparative, recycle, inverse, and differential SEC. Some techniques have found only limited applications, while preparative SEC may be relatively frequently used.

SEC is an important part of two-dimensional separations (2D-LC). The sequence is usually LC in the first dimension and SEC in the second dimension. The fractions separated by LC separation are transferred into the second

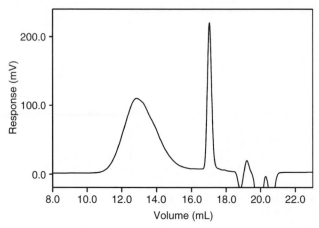

Figure 3.42   RI chromatogram of polymer containing low-molar-mass additive (copolymer styrene-methylmethacrylate-butylmethacrylate containing nonylesters of phthalic acid).

dimensions by an automatically controlled switching valve. The LC separation permits separation according to chemical composition, which is further accomplished with the determination of molar mass distribution by SEC. The reviews can be found in the literature.[35-37]

SEC plays an important role in the characterization of biomacromolecules, typically various polysaccharides (starch, dextran, pectin, hyaluronic acid) and proteins. Polysaccharides are polydisperse polymers that need to be characterized by the molar mass distribution and molar mass averages, and thus the data processing is similar to synthetic polymers. The column calibration is mostly based on pullulan standards, but it must be taken into account that polysaccharides are often branched. The use of calibration established by linear pullulan standards for branched polysaccharides provides strongly erroneous results. Pullulan calibration also yields significant errors when applied to polysaccharides bearing anionic groups (e.g., hyaluronic acid) that have expanded coil structure due to electrostatic repulsive forces. Calibration based on dextran standards may yield erroneous results even when applied to the characterization of dextran samples, because of possibly differing degrees of branching of standards and the analyzed samples.

A very important application of SEC in the area of protein characterization is the determination of oligomers and aggregates. The aggregates reduce the ability of protein to crystallize (crystals being necessary for x-ray structural studies). The quantification is usually based on the relative peak areas in RI or UV chromatograms. An example of the separation of protein oligomers is shown in Figure 3.43. Note that in protein terminology, in contrast to that used in macromolecular chemistry, *oligomer* does not refer to a lower-molar-mass polymer, but to a higher-molar-mass compound containing multiple protein molecules bonded by covalent or other chemical bonds. The term *monomer* refers

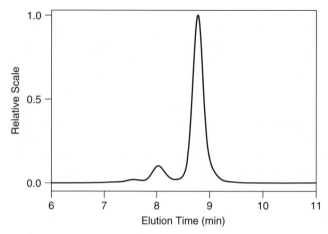

Figure 3.43 UV chromatogram of BSA showing separation of monomer, dimer, and trimer. Column Shodex Protein KW 803 300 × 8 mm, phosphate buffer, flow 0.5 mL/min, UV at 280 nm.

to a single protein molecule. That means *monomer bovine serum albumine* means the protein molecules of molar mass of 66.4×10^4 g/mol. A distinctive feature of proteins is that they are strictly monodisperse (i.e., the molar mass within each chromatographic peak is identical). The determination of molar mass of proteins and protein oligomers based on the calibration curve established by standard proteins is very inaccurate because of possibly differing protein conformation. The elution volume of proteins is also frequently affected by secondary interactions with the stationary phase. To determine the true molar mass of proteins, a light scattering detector is needed even more than in the case of other synthetic and natural polymers. For more information on SEC of proteins see the review in reference 38.

Preparative and semipreparative SEC can serve to isolate polymer fractions for their further characterization. The scale of regular analytical 300 × 8-mm columns in combination with repeated injections can provide sufficient amounts of sample for spectral analysis. To increase the loadability of SEC columns, which is proportional to the cross-sectional area of the columns, it is necessary to use wider bore semipreparative or preparative columns. The flow rate for the preparative columns must be increased proportionally to keep the same linear flow velocity and thus the run time comparable with the analytical columns. The increase of sample load can be achieved by increasing the sample concentration and injection volume. The increase of sample concentration is preferred over the increase of the injection volume, but is limited by viscosity effects of polymers containing high-molar-mass fractions. Therefore, both sample concentration and injection volume are usually increased to achieve sufficient mass injection. Peak saturation is often observed in preparative SEC, which can be addressed by decreasing detector sensitivity or using wavelengths of UV detector at which the sample absorption is minimal. The fractions are obtained in the form of highly diluted solutions. The solvent must be evaporated using vacuum and/or elevated temperature. In the case of stabilized THF, the stabilizer (usually BHT) is concentrated and this must be taken into account for the interpretation of spectral results; otherwise, nonstabilized THF must be used. In aqueous SEC, the mobile phase typically contains salts that are concentrated in the isolated fractions. A possible application of preparative SEC is the isolation of organic contaminants from matrices for environmental analyses. The sample for environmental analysis prepared by extraction can contain compounds with higher molar mass such as fats or dyes that can be efficiently removed by SEC. That means SEC is used as a cleaning tool for agricultural products, animal fat, and soil for the determination of pesticides, polychlorobiphenyls, and polycyclic aromatic hydrocarbons.

3.7 KEYNOTES

- Size exclusion chromatography separates polymer molecules according to their hydrodynamic volume and not according to the molar mass. That means molecules of different molar mass, different chemical composition, and different degree of branching can co-elute at a common elution volume. A unimodal SEC peak does not prove the homogeneity of the chemical composition.

- SEC belongs to the group of liquid chromatography techniques. A distinctive feature of SEC is absence of enthalpic interactions between the analyzed molecules and the stationary phase. However, in real SEC, various types of interactions often play a significant role and it is necessary to minimize them by selection of suitable column packing, mobile phase, temperature, additives, or sample derivatization.

- Peak broadening in SEC columns and other parts of the instrumental setup is another undesirable phenomenon in SEC, which affects the experimentally determined molar mass distribution. However, using columns with adequate efficiency it can be neglected without serious errors in molar mass distribution.

- In order to analyze broad polymers over a wide range of molar masses it is necessary to use SEC columns with appropriate separation range. A traditional option is to use multiple columns of different pore sizes linked together in series. Another possibility is to use columns packed with a mixed-bed packing of different pore sizes at an optimized mix ratio, or a packing material composed of monodisperse particles in which a wide range of pore sizes is contained within a single particle. False molar mass distribution is obtained if the analyzed sample contains fractions with molar masses outside the separation range of the applied columns.

- In conventional mode with only a concentration detector, the SEC columns must be calibrated in order to transform elution volume to molar mass. When an adequate calibration curve is established, SEC is capable of providing correct molar mass distribution. The most accurate calibration is achieved by a series of well-characterized narrow standards of the same chemical composition as that of the polymer under analysis. These standards are mostly unavailable and alternatively broad polymer standards or universal calibration must be used.

- For many polymers it is impossible to establish correct calibration and the true molar masses can be determined only by using the universal calibration in combination with a viscometer or by means of a light scattering detector.

- If the SEC data are evaluated by means of calibration established for a different polymer than that requiring analysis, the obtained molar mass distribution and molar mass averages are only apparent values that would have a polymer of the same chemical composition as the calibration standards and the same distribution of hydrodynamic volume as the polymer under analysis.

- A typical feature of SEC measurements is mostly very good short-term repeatability, whereas reproducibility is often significantly worse, especially when results generated in different laboratories using different instrumental conditions are compared.

- Analysis of a reference polymer should be performed along with the sample analysis to control for repeatability and reproducibility of the measurements.

- SEC results are sensitive to several experimental variables. The most important variables are flow rate, temperature, and sample concentration. Constant flow rate and column temperature can be achieved with suitable instrumentation, whereas sample concentration requires the careful attention of the operator.

- The following generic conditions can be recommended for most routine applications: two 300×8-mm mixed columns covering at least a molar mass range of $500 - 2 \times 10^6$ g/mol, sample concentration for polymers with M_w of order of magnitude of 10^5 g/mol around 0.2%, sample injection volume of 100 μL, standard concentration of 0.015–0.03% w/v, standard injection volume of 50 μL, and number of standards 10–15.

- *Modern Size-Exclusion Liquid Chromatography*[39] can be recommended to all who are interested in this technique.

3.8 REFERENCES

1. Porath, J. and Flodin, P., *Nature (London)*, **183**, 1657 (1959).
2. Moore, J. C., *J. Polym. Sci. A*, **2**, 835 (1964).
3. Tung, L. H., *J. Appl. Polym. Sci.*, **10**, 375 (1966).
4. Baumgarten, J. L., Busnel, J. P., and Meira, G. R., *J. Liq. Chromatogr. Relat. Technol.*, **25**, 1967 (2002).
5. van Deemter, J. J., Zuiderweg, F. J., and Klinkenberg, A., *Chem. Eng. Sci.*, **5**, 271 (1956).
6. Striegel, A. M., *J. Chromatogr.*, **932**, 21 (2001).
7. Kubin, M., *J. Chromatogr.*, **108**, 1 (1975).

8. Tung, L. H. and Runyon, J. R., *J. Appl. Polym. Sci.*, **13**, 2397 (1969).
9. Bruessau, R. J., *Macromol. Symp.*, **110**, 15 (1996).
10. ISO Standard 13885-1 (1998).
11. Mottaleb, M. A., Cooksey, B. G., and Littlejohn, D., *Fresenius' J. Anal. Chem.*, **358**, 536 (1997).
12. Reed, W. F., *Macromol. Chem. Phys.*, **196**, 1539 (1995).
13. Yau, W. W., Gillespie, D., and Hammons, K., *International GPC Symposium 2000*, Las Vegas, October 22–25, 2000.
14. Wrobel, K., Sadi, B. B. M., Wrobel, K., Castillo, J. R., and Caruso, J. A., *Anal. Chem.*, **75**, 761 (2003).
15. Szpunar, J., Pellerin, P., Makarov, A., Doco, T., Williams, P., and Lobinski, J., *J. Anal. At. Spectrom.*, **14**, 639 (1999).
16. Lehmann, U. and Kohler, W. *Macromolecules*, **29**, 3212 (1996).
17. Grubisic, Z., Rempp, P., and Benoit, H., *J. Polym. Sci. B*, **5**, 753 (1967).
18. Chance, R. R., Baniukiewicz, S. P., Mintz, D., Strate, G. V., and Hadjichristidis, N., *Int. J. Polym. Anal. Character.*, **1**, 3 (1995).
19. Brandrup, J., Immergut, E. H., Grulke, E. A. (editors), *Polymer Handbook*, 4th Edition, John Wiley & Sons, New York (1999).
20. Blau, K. and Halket, J. (editors), *Handbook of Derivatives for Chromatography*, John Wiley & Sons, Chichester (1993).
21. Mori, S., Kato, H., and Nishimura, Y., *J. Liq. Chrom. & Rel. Technol.*, **19**, 2077 (1996).
22. Pasch, H. and Trathnigg, B., *HPLC of Polymers*, Springer-Verlag, Berlin (1999).
23. Podzimek, S., Hanus, J., Klaban, J., and Kitzler, J., *J. Liq. Chromatogr.*, **13**, 1809 (1990).
24. Runyon, J. R., Barnes, D. A., Rund, J. F., and Tung, L. H., *J. Appl. Polym. Sci.*, **13**, 2359 (1969).
25. Podzimek, S., in *Multiple Detection in Size-Exclusion Chromatography*, Striegel, A. M. (editor), ACS Symposium Series **893**, Washington, D.C. (2004), p. 107.
26. Braun, D. and Lee, D. W., *Angew. Makromol. Chem.*, **48**, 161 (1975).
27. Podzimek, S., Dobas, I., Svestka, S., Horalek, J., Tkaczyk, M., and Kubin, M., *J. Appl. Polym. Sci.*, **41**, 1161 (1990).
28. Janca, J., *J. Chromatogr.*, **134**, 263 (1977).
29. Janca, J. and Pokorny, S., *J. Chromatogr.*, **148**, 31 (1978).
30. Janca, J. and Pokorny, S., *J. Chromatogr.*, **156**, 27 (1978).
31. Janca, J., *J. Chromatogr.*, **187**, 21 (1980).
32. Janca, J., *Polym. J.*, **12**, 405 (1980).
33. Janca, J., Pokorny, S., Vilenchik, L. Z., and Belenkii, B. G., *J. Chromatogr.*, **211**, 39 (1981).
34. Aida, H., Matsuo, T., Hashiya, S., and Urushisaki, M., *Kobunshi Ronbunshu*, **48**, 507 (1991).
35. Berek, D., in *Handbook of Size Exclusion Chromatography and Related Techniques*, 2nd Edition, Wu, C.-S. (editor), Marcel Dekker, New York (2004), p. 501.
36. Pasch, H., in *Multiple Detection in Size-Exclusion Chromatography*, Striegel, A. M. (editor), ACS Symposium Series **893**, Washington, D.C. (2004), p. 230.
37. Teraoka, I., in *Multiple Detection in Size-Exclusion Chromatography*, Striegel, A. M. (editor), ACS Symposium Series **893**, Washington, D.C. (2004), p. 246.
38. Baker, J. O., William, S, A., Himmel, M. E., and Chen, M., in *Handbook of Size Exclusion Chromatography and Related Techniques*, 2nd Edition, Wu, C.-S. (editor), Marcel Dekker, New York (2004), p. 439.
39. Striegel, A. M., Yau, W. W., Kirkland, J. J., and Bly, D. D., *Modern Size-Exclusion Liquid Chromatography*, 2nd Edition, John Wiley & Sons, Hoboken, NJ (2009).

Chapter 4

Combination of SEC
and Light Scattering

4.1 INTRODUCTION

Without separation by SEC, the light scattering measurements yield only the weight-average molar mass, the z-average RMS radius, and the second virial coefficient. Although these quantities provide valuable information about the polymer samples requiring characterization, the combination of light scattering with SEC separation brings both methods to a new qualitative level. There is an evident synergy effect of the two experimental techniques combining separation of molecules according to their hydrodynamic size with an immediate online characterization of their molecular structure. A light scattering detector connected online after the outlet from SEC columns measures molar mass from the intensity of light scattered by eluting molecules. The multi-angle light scattering (MALS) detector provides molecular size as another important piece of information that appears especially useful in relation with molar mass.[i] Since the RMS radius is determined from the angular variation of the scattered light intensity, a MALS-type light scattering detector is necessary for the determination of this quantity. In fact, the light scattering detector completely eliminates the necessity of column calibration and various SEC calibration methods may appear completely useless. However, knowledge of the calibration procedures and principles of universal calibration is necessary for understanding the basic principles of SEC itself. In addition, the universal calibration procedure can be reversed and used for the determination of some polymer characteristics such as constants of the Mark-Houwink equation or unperturbed dimensions.

The basic theory and instrumentation of light scattering are described in Chapter 2. In this chapter, the focus is on the light scattering instrument used as

[i]This chapter focuses on MALS detectors with at least three detection angles.

Light Scattering, Size Exclusion Chromatography and Asymmetric Flow Field Flow Fractionation: Powerful Tools for the Characterization of Polymers, Proteins and Nanoparticles, by Stepan Podzimek
Copyright © 2011 John Wiley & Sons, Inc.

an SEC detector and on aspects specific to the combination of light scattering and SEC. The principles that apply to the combination of light scattering with SEC are generally valid for the combination of light scattering with various techniques of flow field flow fractionation or other analytical separation techniques. The basic principles of light scattering and its combination with SEC were reviewed by Wyatt.[1,2] Despite the significant advancement in instrumentation since the date of publishing of reference 1, the paper is still a valuable source of information.

The Debye equation:

$$\frac{R_{\theta,i}}{K^*c_i} = M_i P_i(\theta) - 2A_{2,i}c_i M_i^2 P_i^2(\theta) \tag{4.1}$$

is the basis of the calculations performed in SEC-MALS. In the above equation $R_{\theta,i}$ is the excess Rayleigh ratio (cm^{-1}), c_i is the concentration of the macromolecules in mL/g, M_i is the molar mass, $A_{2,i}$ is the second virial coefficient, $P_i(\theta)$ is the particle scattering function that involves the RMS radius of scattering particles (Equation 2.12), and K^* is the optical constant including the solvent refractive index, the vacuum wavelength of the incident light, and the specific refractive index increment dn/dc (for K^* definition see Equation 2.3). The subscript i implies that the quantities are valid at the ith elution volume slice (for the sake of clarity, the subscript i is not further used). Similarly as in the case of conventional SEC the obtained chromatograms are processed at regular elution volume intervals (usually one-half, one, or two seconds). *Note:* Equation 4.1 has higher concentration terms that become significant at higher concentrations. However, these terms can be mostly safely ignored at the low concentrations that are in SEC columns.

4.2 DATA COLLECTION AND PROCESSING

For successful SEC-MALS measurements it is necessary to determine several parameters that allow proper processing of the acquired light scattering data and the calculation of correct molar masses and RMS radii. These are (1) calibration constant of the MALS photometer, (2) normalization coefficients of the MALS photometer, (3) calibration constant of the concentration detector (RI, UV), (4) volume delay between the MALS photometer and other detectors, and (5) correction for the band broadening. The meaning and the determination of the MALS and RI (UV) calibration constants and normalization coefficients are explained in Sections 2.2.1.3 and 2.6. Recall that the *MALS calibration constant* is used to transfer the voltages yielded by the photodiodes to the values of Rayleigh ratio, the *RI calibration constant* is used to calculate the absolute concentration in g/mL from the RI signal expressed in volts, and *normalization* is the process by which the various detectors' signals are related to the 90° detector signal and the MALS calibration constant. The calibration constants of MALS and RI detectors are independent of SEC mobile phase, whereas the change of mobile phase requires new normalization.

Volume delay between the MALS and other detectors consists of the volume of the tubings inside the detectors, volume of the cells, and volume of the interdetector connecting tubings and fittings. Only the volume of connecting tubings is partly under the control of the operator. Due to the volume between the detectors, the concentration, which is measured at a given time by a concentration-sensitive detector, does not match the intensity of light that is at the same time measured by the light scattering detector. The volume delay is used by the software to correct for the time that it takes fluid to transfer between the instruments so that a given light scattering intensity is matched with the appropriate concentration. Once the volume delay has been determined, there is no need to determine it again until the tubings between the instruments are changed for tubings of different length and/or inner diameter.

Note that the volume delay between UV and MALS is significantly lower than between MALS and RI. If the UV and RI detectors are used simultaneously, the sequence is UV-MALS-RI instead of MALS-UV-RI in order to minimize the delay between MALS and RI.

The delay volume is usually established by measurement of a monodisperse polymer and aligning the peaks obtained by MALS and concentration-sensitive detectors. Only in the case of monodisperse polymers the peak elution volumes of chromatograms recorded by the MALS and RI, and possibly UV and/or viscosity (VIS), detectors are identical. This is due to the different responses of various detectors to the molar mass and concentration. If the samples used to determine the volume delay are not monodisperse, the peaks will not be in alignment because the peak of a concentration detector will be at the maximum of polymer concentration, c, while the peak of a MALS detector will be at the maximum of the product of $M \times c$, and the peak of a VIS detector will correspond to the maximum of $M^a \times c$.

Proteins such as bovine serum albumin (BSA) can be used for the determination of volume delay in aqueous SEC. Polyethyleneoxide (PEO) can be alternatively used for SEC columns that retain proteins. Polydispersity of pullulan and dextran standards is usually higher than that of PEO, which results in overestimation of volume delay. Polystyrene standards are mostly used for determination of volume delay in THF and other organic solvents.

The most accurate results are obtained using a standard with molar mass approximately in the range of 200,000–400,000 g/mol. Standards with too-high molar mass are usually of higher polydispersity while low standards with molar mass in the range of several thousands can be partly separated by SEC columns and consequently the volume delay is overestimated. Standards with high molar mass can partly undergo shearing degradation, which further increases the polydispersity.

The effect of molar mass on volume delay determined by polystyrene standards is shown in Figure 4.1. The lower-molar-mass standard is separated by efficient columns and the volume difference between the RI and MALS peaks is given not only by the actual volume between the two detectors, but by the volume difference caused by standard polydispersity as well. The experience shows

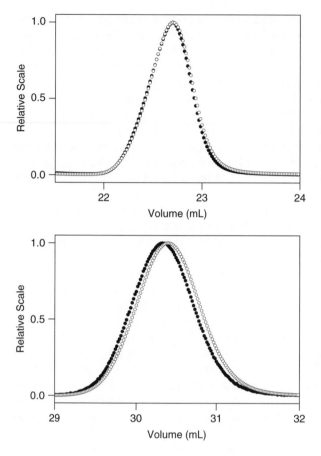

Figure 4.1 Determination of interdetector volume delay: effect of molar mass on the experimentally determined volume delay. MALS (•) and RI (○) signals of polystyrene standard with nominal molar mass of 200,000 g/mol (top) overlap while there is an additional volume difference of 0.067 mL for the peaks of 3,600 g/mol standard (bottom). Columns: 4 × PLgel Mixed-C 300 × 7.5 mm, THF at 1 mL/min. Volume delay determined by 200,000 g/mol standard = 0.217 mL, nominal $M_w/M_n = 1.05$ (3,600 g/mol standard), and 1.06 (200,000 g/mol standard). The signals are scaled to the same intensity.

that the difference in volume delays determined by 200,000 and 30,000 g/mol polystyrene standards is negligible even in the case of columns with very high resolution. For a monodisperse standard, the accurate volume delay should give an identical molar mass across the peak.

Band broadening in SEC-MALS means different phenomena than in conventional SEC. Band broadening in SEC-MALS terminology is a result of peak movement between the multiple detectors connected in series. When a peak moves along the flow path created by several detectors connected in series, each detector flow cell acts as a small mixing chamber that causes, together with axial broadening, the peak to broaden with a slight tail toward higher elution volumes. That means the peak recorded by the subsequent detector in series is broader compared with the peak recorded by the previous detector. Although peak broadening is present for all analyzed samples, it is particularly evident in the case of monodisperse samples for which the molar mass across the peak is constant. Due to peak broadening the molar mass across the peak of a monodisperse polymer is not flat, as one would expect, but has a "sad grimace" pattern.

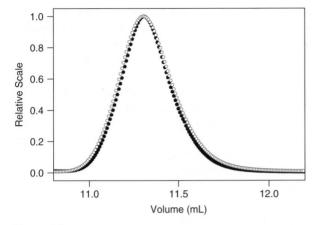

Figure 4.2 Interdetector peak broadening: signals of RI (○) detector and MALS photometer at 90° (●) for narrow polystyrene standard of nominal molar mass 200,000 g/mol. The signals are scaled to the same intensity. Columns: 2 × PLgel Mixed-C 300 × 7.5 mm columns, THF at 1 mL/min.

The peak of a monodisperse polymer recorded by the RI detector, which is always connected after the light scattering detector, is slightly broadened and tailed compared to the peak from the MALS detector, as shown in Figure 4.2. Consequently, the concentrations at particular elution volume slices do not correspond precisely to the light scattering intensities and the molar masses at the peak center are overestimated while those at the beginning and end of the peak are underestimated, which results in the "sad grimace" profile of the molar mass–versus–elution volume plot. The band-broadening effect is proportional to the interdetector volume delay and thus it is significantly less pronounced for the combination of UV-MALS as compared to the combination of MALS-RI. The peak-broadening effect is more severe for such applications where three detectors in series are used, especially in the case of combination MALS-VIS-RI. Although an average molar mass for the entire peak is usually not significantly affected, the band broadening represents at least a potential source of errors and it is also a certain "cosmetic" issue when the results for narrow or monodisperse polymers are presented. Correction for peak broadening has become a part of commercially available light scattering software ASTRA® (Wyatt Technology Corporation). To correct for the peak-broadening effect, the software expands the light scattering trace to fit the broadened RI chromatogram. The procedure may result in a slight loss of resolution, but mostly this effect is negligible. To compensate for peak broadening correctly, it is necessary to measure strictly monodisperse polymer. The requirement of monodispersity is entirely fulfilled only in the case of proteins. The measurement of narrow polymers such as polystyrene standards may result in a slight overcompensation of the broadening effect. To compensate for peak broadening by means of narrow, yet not completely monodisperse polymers, the molar mass should be roughly in the range of 200,000–400,000 g/mol.

Once the parameters for the correction of band broadening are determined utilizing a narrow polymer, they can be used for data processing of other samples. The effect of band broadening on the molar mass–versus–elution volume profile of narrow polymer standards is shown in Figure 4.3. Due to significantly lower volume delay between UV and MALS, the band-broadening effect is minimal compared with the molar mass–versus–elution volume profile obtained by the RI detector. Figure 4.3 also shows that the band-broadening effect can be eliminated by software correction. Although band broadening may have a strong effect on the molar mass–versus–elution volume plots of narrow and monodipserse samples, the effect on the molar mass–versus–elution volume plots and distribution curves of polydisperse polymers is entirely imperceptible (Figure 4.4).

As a matter of fact, lower separation efficiency of SEC columns becomes a virtue when the determination of volume delay and/or peak-broadening correction is of interest. Since SEC resolution decreases with increasing molar mass, SEC is practically incapable of resolving slightly different molecules in a higher molar mass range. Due to the limited SEC resolution power, a narrow polymer standard of sufficiently high molar mass virtually behaves as monodisperse even if the polydispersity index is not exactly equal to unity. For this reason, a single short column (e.g., 150×7.5 mm) can be used to minimize the separation according to molecular size in order to increase the accuracy of the determination of the volume delay and band-broadening correction.

A typical example of chromatograms obtained by SEC-MALS analysis of a polydisperse polymer is shown in Figure 4.5, which depicts signals from the light scattering detector at $90°$ and the refractive index detector. Notice that the peaks given by the two detectors are of different elution volumes and shapes. As already explained, this is due to different sensitivity of particular detectors to molar mass and concentration. A typical feature of polydisperse polymers is that the peak of the MALS detector is always before the peak of the RI detector. The same elution volume of the two detectors occurs only for monodisperse or very narrow polymers; on the other hand, a large volume difference between the peaks of MALS and RI detectors is indicative of very high polydispersity. Note that the chromatograms in Figure 4.5 are scaled to the same relative intensity, but the true voltage outputs from the two detectors are not identical.

Different sensitivities of the MALS and RI detectors to molar mass and concentration is very evident from the chromatograms of narrow standards of different molar mass, as shown in Figure 4.6. Although the injected mass of 30,000 g/mol standard is double that of 200,000 g/mol standard, the signal of the MALS detector is significantly lower and only the RI detector provides true information about the concentration.

In Figures 4.5 and 4.6, only the MALS chromatogram from the $90°$ light scattering detector is shown. However, there are other chromatograms recorded by the MALS detector at various angles. The signals from all available light scattering detectors that are placed at different angular positions are shown in a three-dimensional plot in Figure 4.7. For molecules with RMS radius greater than about 10 nm, the light scattering signals at lower angles are greater than

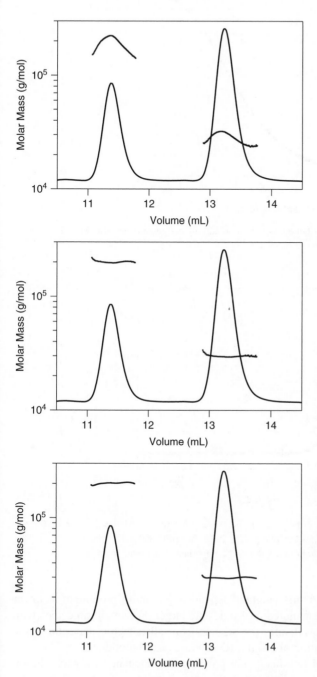

Figure 4.3 Molar mass–versus–elution volume plots for 200,000 and 30,000 g/mol polystyrene standards processed using as concentration detector RI detector (top), UV detector (center), and RI detector applying ASTRA® band-broadening correction (bottom). RI chromatogram is superimposed. $M_w = 203{,}100$ g/mol, 29,900 g/mol (top); 199,000 g/mol, 29,800 g/mol (center); 201,400 g/mol, 29,800 g/mol (bottom). Volume delays: UV to MALS = 0.017 mL, MALS to RI = 0.233 mL.

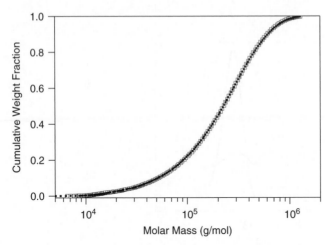

Figure 4.4 Cumulative distribution plots of a broad polymer obtained by SEC-MALS without (■) and with (□) band-broadening correction.

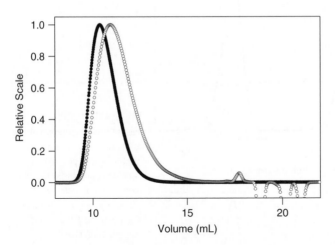

Figure 4.5 Chromatograms of polydisperse polystyrene recorded by MALS detector at 90° (•) and refractive index detector (○). Conditions: 2 × PLgel Mixed-C 300 × 7.5 columns, 100 μL 0.25% w/v, THF at 1 mL/min. The signals are scaled to the same intensity.

those at higher ones, and such angular dependence is more prominent at lower elution volumes, where the eluted molecules are larger. For small molecules with RMS radius below about 10 nm, the signals from all detectors have the same intensity and no information about the RMS radius can be obtained.

Molar mass and RMS radius of the polymer are calculated for each elution volume slice of the selected chromatographic peak by means of a so-called *Debye plot*, as depicted in Figure 4.8. The data points in the Debye plot correspond to

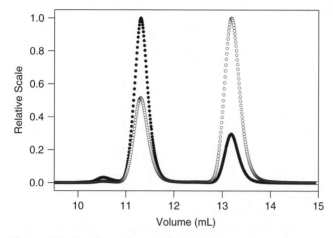

Figure 4.6 Superimposition of 90° MALS (●) and RI (○) signals of narrow polystyrene standards of nominal molar mass 200,000 g/mol and 30,000 g/mol. Conditions: 2 × PLgel Mixed-C 300 × 7.5 columns, THF at 1 mL/min injection 100 μL 0.15% w/v (200,000 polystyrene) and 0.3% w/v (30,000 polystyrene).

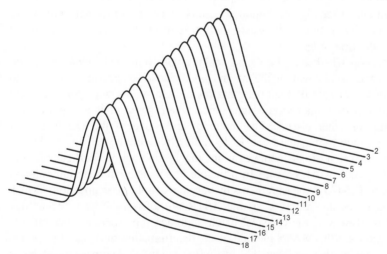

Figure 4.7 Three-dimensional view (rotation angle 30°, elevation angle 45°) of multiple MALS signals for polydisperse polystyrene: 17 scattering angles in the range of 17°–155°.

the light scattering intensities at various angles at a particular elution volume. A Debye plot, which is an arrangement of R_θ/K^*c plotted against $\sin^2(\theta/2)$, is used to extrapolate the light scattering intensities to zero scattering angle. There is no strict need to use all available data points of the MALS photometer, but some data points can be dropped and not included in the least-squares fit. Especially the lowest and partly also the highest angles are often affected by particles eluting

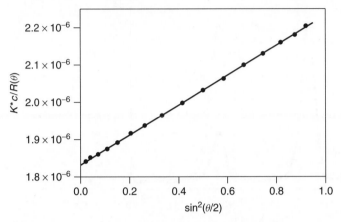

Figure 4.8 Debye plot (Zimm light scattering formalism) for one elution volume slice of Figure 4.7. $M = 546,000$ g/mol, $R = 31.7$ nm, $c = 5.6 \times 10^{-5}$ g/mL.

from SEC columns, incorrect normalization, or impurities in the flow cell of the MALS detector. In any event, the improper angular data points can be safely dropped without affecting the obtained results since the other data points are sufficient to derive results. This is especially true when the angular variation can be fitted by the first-order fit.

The decision whether to drop a particular angle from the calculation can be based on visual assessment of the fit and on the results uncertainty calculated by the software. A rule of thumb is that fewer data points with good fit yield more accurate results than more data points, where some of them are affected by impurities or other effects. The intercept and slope of the Debye plot at zero angle provide molar mass and RMS radius, respectively. Since the Debye plot is obtained at a single concentration, the second virial coefficient is not determined by SEC-MALS. *Note:* In principle it is possible to determine the A_2 by multiple SEC-MALS measurements by varying sample mass injections, which is supported by commercially available ASTRA® software (Wyatt Technology Corporation). However, the classical batch approach represents a more reliable way to A_2 and the SEC-MALS method of the determination of A_2 can be applied mainly for samples that are available in limited amounts. Another advantage of the online determination of A_2 is that the exact concentration of the analyzed solution need not be known, because the injected mass is determined from the signal of a concentration-sensitive detector. Variation of injected amount can be easily achieved by different injection volumes and thus a single solution can be used for the measurements. Note that for the online determination of A_2 an efficient SEC separation is not needed and the measurements can be performed using a single mixed-bed short SEC column (e.g., 150×7.5 mm).

Figure 4.9 plots molar mass across the peak of polydisperse polymer together with corresponding RMS radius. An example of slice data is shown in Table 4.1.

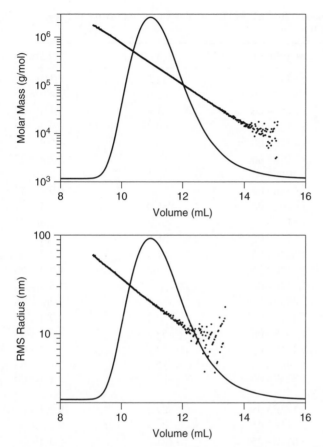

Figure 4.9   Molar mass (top) and RMS radius (bottom)–versus–elution volume plots of polydisperse polymer from Figure 4.5. RI chromatogram is superimposed in the plots.

Evenly decreasing molar mass and RMS radius across the entire chromatogram is indicative of good separation by pure size exclusion mechanism, whereas the molar mass or radius plots parallel with the elution volume axis or the rising of molar mass and radius with increasing volume indicate non–size exclusion separation, which can be due to some kind of enthalpic interactions of polymer molecules with the column packing or anchoring of branched molecules in the pores. The molar mass–versus–elution volume plot can also easily detect the presence of fractions with molar mass above the exclusion limit of the columns employed for the analysis or reveal poor column efficiency.

Figure 4.9 also demonstrates different detection limits of MALS for molar mass and RMS radius. The data points for the RMS radius become scattered as the radius approaches values of about 10 nm, but the molar mass, which is determined from the light scattering intensity, can be determined for smaller molecules. Although the SEC-MALS combined method does not require the

Table 4.1 Slice Data Obtained by SEC-MALS Analysis of Polydisperse Polymer Corresponding to Figure 4.9

Volume (mL)	M (g/mol)	R (nm)	c (g/mL)
9.050	1.875E+06	62.5	2.47E−07
9.067	1.845E+06	62.0	2.87E−07
9.083	1.815E+06	61.4	3.33E−07
9.100	1.786E+06	60.8	3.92E−07
9.117	1.757E+06	60.2	4.50E−07
9.133	1.729E+06	59.6	5.23E−07
9.150	1.701E+06	59.1	6.01E−07
9.167	1.673E+06	58.5	6.96E−07
9.183	1.646E+06	58.0	8.01E−07
9.200	1.619E+06	57.4	9.18E−07
9.217	1.593E+06	56.9	1.05E−06
9.233	1.568E+06	56.3	1.20E−06
9.250	1.542E+06	55.8	1.36E−06
9.267	1.517E+06	55.3	1.55E−06
9.283	1.493E+06	54.7	1.76E−06
9.300	1.469E+06	54.2	1.97E−06
9.317	1.445E+06	53.7	2.21E−06
9.333	1.421E+06	53.2	2.49E−06
9.350	1.398E+06	52.7	2.80E−06
9.367	1.376E+06	52.2	3.13E−06
. . .	. . .	. . .	. . .
10.867	3.173E+05	22.2	9.31E−05
10.883	3.122E+05	22.0	9.33E−05
10.900	3.072E+05	21.7	9.34E−05
10.917	3.022E+05	21.5	9.35E−05
10.933	2.973E+05	21.3	9.36E−05
10.950	2.925E+05	21.1	9.36E−05
10.967	2.878E+05	20.9	9.36E−05
10.983	2.831E+05	20.7	9.35E−05
11.000	2.786E+05	20.5	9.35E−05
11.017	2.740E+05	20.3	9.34E−05
11.033	2.696E+05	20.1	9.33E−05
11.050	2.653E+05	20.0	9.31E−05
11.067	2.610E+05	19.8	9.29E−05
11.083	2.568E+05	19.6	9.27E−05
11.100	2.526E+05	19.4	9.25E−05
11.117	2.485E+05	19.2	9.22E−05
11.133	2.445E+05	19.0	9.19E−05
11.150	2.405E+05	18.8	9.16E−05
11.167	2.367E+05	18.7	9.13E−05
11.183	2.328E+05	18.5	9.09E−05
. . .	. . .	. . .	. . .

determination of the calibration curve as in the case of conventional SEC, the calibration curve can be easily established from the acquired data and used as an additional source of information about the analyzed polymer and its SEC elution behavior. It is worth mentioning that the molar mass–versus–elution volume plot can be used for the characterization of separation efficiency on the basis of the slope of the molar mass–versus–elution volume plot.

4.2.1   Processing MALS Data

The light scattering intensity at zero angle is needed to obtain molar mass by light scattering measurements. However, the scattered intensity at zero scattering angle is not experimentally measurable. One possible solution of this problem is to measure at a very low angle by so-called low-angle light scattering, which assumes the scattering intensity at a low angle to be equal to that at zero angle. The disadvantage of the low-angle approach is that measurements at very low angles suffer from high noise generated by dust particles and other mechanical impurities in the mobile phase. Because of very high mass, the scattering intensity of dust particles is many times higher than that of significantly less dense random coils of comparable size. The effect of disturbing particles decreases with increasing angle of observation, the decrease being much steeper for compact spheres compared to random coils (compare Figures 2.7 and 2.8). In consequence, dust particles disturb light scattering signals even at trace concentrations, which is particularly true at low angles. To avoid the disturbing effect of particles in the mobile phase, the measurement is performed at multiple angles far from zero degree and the zero intensity is estimated by the extrapolation of data points from higher angles. The multi-angle approach efficiently solves the problem with noise and in addition allows the determination of the RMS radius, which can be obtained from the angular dependence close to zero angle.

Molar mass and RMS radius can be also obtained by fitting the light scattering intensities to a given particle scattering function, but a particular disadvantage of this approach is that it requires a priori knowledge of the polymer conformation (i.e., whether the molecules under investigation are random coils or another shape). In fact, this is not a serious limitation for most synthetic and natural polymers, because they generally create random coils. One possible disadvantage of the multi-angle approach is that different extrapolation methods may lead to different results, which means that careful selection and optimization of the extrapolation method is an important part of processing the experimental data.

To determine molar mass and RMS radius the experimental data can be processed in various ways that are described Sections 4.2.1.1–4.2.1.4. Note that the primary experimental results are always identical (i.e., the angular variations of light scattering intensities and concentrations collected at regular elution volume intervals across the chromatographic peak), and only the method of subsequent processing is different.

4.2.1.1 Debye Fit Method

To obtain the molar mass and the RMS radius, a Debye plot, that is, a plot of R_θ/K^*c versus $\sin^2(\theta/2)$, is constructed, and the obtained plot is fitted by a polynomial to get the intercept at zero angle R_0/K^*c and the slope at zero angle $m_0 = d(R_\theta/K^*c)/d(\sin^2(\theta/2))_{\theta\to0}$. At angle approaching zero, the particle scattering function approaches unity and the equation becomes

$$\frac{R_{\theta\to0}}{K^*c} = \frac{R_0}{K^*c} = M - 2A_2cM^2 \tag{4.2}$$

$$M = \frac{2\left(1 - \sqrt{1 - 8A_2c\left(\frac{R_0}{K^*c}\right)}\right)}{8A_2c} \tag{4.3}$$

or an equivalent form to avoid errors when $A_2 \to 0$:

$$M = \frac{2\left(\frac{R_0}{K^*c}\right)}{1 + \sqrt{1 - 8A_2c\left(\frac{R_0}{K^*c}\right)}} \tag{4.4}$$

and if $A_2 = 0$, then

$$M = \frac{R_0}{K^*c} \tag{4.5}$$

The mean square radius is calculated according to the equation:

$$R^2 = \frac{-3m_0\lambda^2}{16\pi^2 M(1 - 4A_2Mc)} \tag{4.6}$$

and for $A_2 = 0$:

$$R^2 = \frac{-3m_0\lambda^2}{16\pi^2 M} \tag{4.7}$$

4.2.1.2 Zimm Fit Method

The calculation according to the Zimm formalism, which is K^*c/R_θ versus $\sin^2(\theta/2)$, leads to:

$$M = \left(\frac{K^*c}{R_0} - 2A_2c\right)^{-1} \tag{4.8}$$

and

$$R^2 = \frac{3m_0\lambda^2 M}{16\pi^2} \tag{4.9}$$

where m_0 is the slope of the polynomial at angles approaching zero, $m_0 = d(K^*c/R_\theta)/d(\sin^2(\theta/2))_{\theta\to0}$. For negligible A_2, Equation 4.8 gives molar mass as reciprocal intercept $\left(\frac{K^*c}{R_0}\right)^{-1}$.

4.2.1.3 Berry Fit Method

The molar mass and RMS radius are calculated using the Berry light scattering method, that is, a plot of $\sqrt{K^*c/R_\theta}$ versus $\sin^2(\theta/2)$, from the following equations:

$$M = \frac{4}{(\sqrt{K^*c/R_0} + \sqrt{K^*c/R_0 - 4A_2c})^2} \tag{4.10}$$

for $A_2 = 0$:

$$M = \frac{1}{(\sqrt{K^*c/R_0})^2} \tag{4.11}$$

and

$$R^2 = \frac{3\lambda^2 m_0}{8\pi^2 \sqrt{M}(1/M - A_2c)} \tag{4.12}$$

for $A_2 = 0$:

$$R^2 = \frac{3\lambda^2 m_0}{8\pi^2 1/\sqrt{M}} \tag{4.13}$$

where $m_0 = d(\sqrt{K^*c/R_\theta})/d(\sin^2(\theta/2))_{\theta \to 0}$.

Note that all three methods of constructing the Debye plot differ in the quantity used on the ordinate, but the abscissa is always the same—$\sin^2(\theta/2)$. Depending on the applied method, the Debye plot appears differently and may yield different values of M and R, although it is constructed from the same data.

Note that the error of dn/dc changes both the slope and the intercept of the Debye plot, which affects the molar mass, but the RMS radius remains unchanged, because m_0 and M in Equations 4.7, 4.9 and 4.13 mutually compensate (see Figure 2.22).

4.2.1.4 Random Coil Fit Method

The random coil fit method does not fit angular variation of scattered light intensity as do the previous methods. Instead, it uses the theoretical particle scattering function for random coils (Equation 2.22), and an iterative nonlinear least-squares fit. In contrast to the previous methods, the random coil fit method assumes that the analyzed molecules are random coils. This assumption is fulfilled for most synthetic and natural polymers and thus the random coil fit method represents a suitable alternative to other methods.

4.2.1.5 Influence of Light Scattering Formalism on Molar Mass and RMS Radius

The particular light scattering formalisms for constructing the Debye plot were investigated for two different polymer shapes (homogeneous spheres and random coils) with radii from 25 to 250 nm in reference 3. The essential conclusions obtained in reference 3 by model calculations are as follows:

1. Extrapolation to zero angle requires only the lowest angles and the data points at higher angles do not contribute significantly to the results. For a given polynomial degree more accurate results are obtained using only low angles for the extrapolation. If higher angles are used, it is necessary to use a higher-order polynomial to reach the same accuracy as for the low angles. Use of all angles is meaningful only when the experimental light scattering data are fitted to a theoretically derived particle scattering function.

2. For random coils the Berry method is superior in terms of accuracy and robustness, whereas for spheres the Debye method is superior. The Berry method should be preferred especially for large random coils. The model calculation suggested that a linear fit according to the Berry using the lowest reliable angular data is the most general procedure.

3. For molar mass determination, the extrapolation error is below 1% for all three methods in the case of small molecules (RMS radius <50 nm). With increasing size, the particular plots become more curved, which requires higher-order polynomials to fit the light scattering intensities. For large random coils ($\approx$250 nm), the error in molar mass is <5% for the Berry and Zimm method using the third- and fourth-order fit, but a fifth-order polynomial is required for the Debye method. The relative errors in RMS radius follow the same trend as for molar mass but are larger.

The above conclusions were brought about mainly by theoretical calculations. The experimental investigations published in reference 4 lead to the following conclusions:

1. The Debye fit method requires a strongly increasing polynomial degree with increasing molecular size.

2. The Zimm plot is linear up to very high molar masses, but compared to other methods it slightly overestimates molar masses and more considerably RMS radii of large molecules.

3. The Berry method is linear over a broad molar mass range.

4. The random coil method gives results quite comparable to the other methods and appears to work well even for highly compact branched molecules.

To explain the differences between particular extrapolation methods and help users make the most appropriate choice of methods in a particular situation, the data obtained for polystyrene of various molar masses are shown in Figures 4.10–4.14 and the corresponding molar masses and RMS radii are compared in Table 4.2. The plots were obtained at different elution volumes of a broad polystyrene sample or at peaks of narrow polystyrene standards. Polystyrene is well known to create random coil conformation in THF and thus the random coil

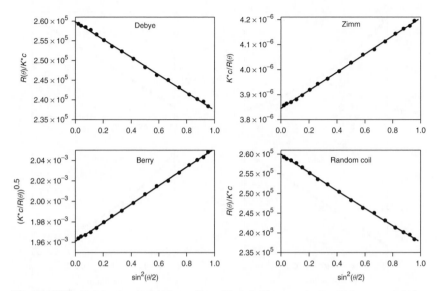

Figure 4.10 Debye plots for ≈20-nm slice of broad polystyrene using various extrapolation methods.

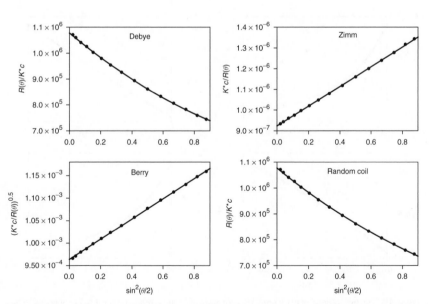

Figure 4.11 Debye plots for ≈46-nm slice of broad polystyrene using various extrapolation methods. Second-order polynomial for Debye method; first-order fits for Zimm and Berry methods.

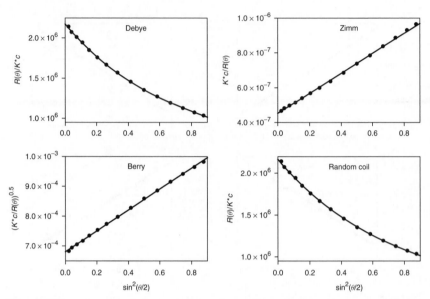

Figure 4.12   Debye plots for ≈70-nm slice of broad polystyrene using various extrapolation methods. Third-order polynomial for Debye method; first-order fits for Zimm and Berry methods.

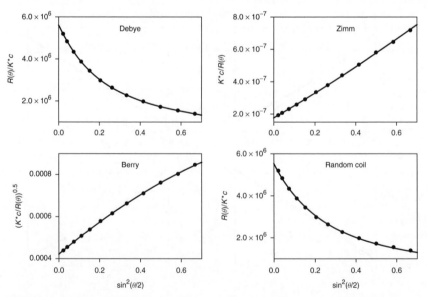

Figure 4.13   Debye plots for ≈130-nm slice of narrow polystyrene standard using various extrapolation methods. Fifth-order polynomial for Debye method; second-order fits for Zimm and Berry methods.

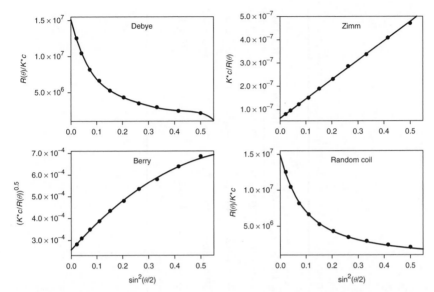

Figure 4.14   Debye plots for ≈200-nm slice of narrow polystyrene standard using various extrapolation methods. Fifth-order polynomial for Debye method; second-order fit for Berry method; first-order fit for Zimm method.

method should theoretically provide the most accurate results (assuming validity of particle scattering function outside theta temperature).

The first example (Figure 4.10) is the case where all methods work equally well and yield almost-identical molar mass and RMS radius. All available light scattering data points are used and no curvature is present in either plot. All results are associated with very low uncertainty based on the statistical consistency of the data.

The second example is for a slice with a higher molar mass, about 1 million g/mol, and RMS radius around 46 nm (Figure 4.11). In this case, the Debye method requires a second-order polynomial while the Zimm and Berry methods are still linear. All methods provide almost-identical molar masses, but the RMS radius determined by Zimm extrapolation is slightly higher than the values obtained by other methods.

This is even more evident for larger random coils of the RMS radius of about 70 nm (Figure 4.12), where Debye extrapolation needs a third-order polynomial and the overestimation of the RMS radius by the Zimm method compared to the random coil method is almost 10%. However, the agreement of molar masses is still very good for all methods. The curvature of the Debye plot further increases with increasing molecular size and a fifth-order polynomial becomes necessary for very high molecular sizes over about 100 nm (Figure 4.13 and 4.14). A second-order fit is usually needed for processing the light scattering data of such large molecules with the Berry and Zimm methods,

Table 4.2 Molar Masses and RMS Radii Corresponding to Figures 4.10–4.14 Determined Using Various Light Scattering Formalisms

Figure	M (10^3 g/mol)				R (nm)			
	Debye	Zimm	Berry	Random	Debye	Zimm	Berry	Random
4.10	259.8 ± 0.1	260.1 ± 0.1	260.0 ± 0.1	260.0 ± 0.1	19.9 ± 0.1	20.8 ± 0.1	20.6 ± 0.1	20.6 ± 0.1
4.11	1078 ± 1	1083 ± 1	1076 ± 1	1077 ± 1	46.3 ± 0.3	48.7 ± 0.1	46.3 ± 0.1	46.5 ± 0.1
4.12	2174 ± 6	2208 ± 8	2158 ± 6	2167 ± 23	70.4 ± 0.9	76.4 ± 0.3	68.9 ± 0.3	69.9 ± 0.2
4.13	5739 ± 26	5807 ± 24	5745 ± 15	5646 ± 44	133 ± 2	141 ± 1	135 ± 1	128 ± 1
4.14	$15{,}090 \pm 330$	$16{,}200 \pm 130$	$15{,}160 \pm 200$	$14{,}830 \pm 390$	208 ± 9	249 ± 1	216 ± 2	207 ± 3

but the Zimm method can still be approximated by a linear fit if the light scattering intensities acquired at high angles are dropped.

Overestimation of the RMS radii of very large molecules by the Zimm method can reach about 20%, but significantly lower error is obtained for molar mass. It is necessary to emphasize that the molecular size, not molar mass, is the relevant parameter for the curvature of the fit as illustrated in Figure 4.15 for highly compact star-branched molecules. The Debye plot is perfectly linear despite high molar mass of the analyzed molecules, which are, however, small due to the high degree of branching. The obtained results suggest that good fit of the experimental light scattering intensities can be obtained with the random coil method, which actually works well not only for linear random coils, but for highly branched molecules as well (Figure 4.16).

For processing the data of polydisperse polymers it is necessary to select the method that will be appropriate for the entire molar mass range, because the light scattering software may not allow for using different processing methods or polynomial orders for various parts of the chromatogram. In fact, this would be a simple modification of the software, but changing the extrapolation method or fit order would result in irregularities on the distribution plots. The Debye method appears the least suitable for polymers covering a broad molar mass range, because the first-order fit will not be suitable for the high molar masses, while the higher-order fits may be too high for lower molar masses.

The angular dependence of the scattering intensity can become even more complicated when smaller molecules co-elute with significantly larger molecules or even supermolecular structures. Such a situation can happen at the beginning of a chromatogram if the analyzed sample contains molecules exceeding the exclusion limit of the applied columns. In such a case the large molecules are pushed forward and mixed with smaller molecules. Another possibility is when slices at higher elution volumes are contaminated by large molecules that

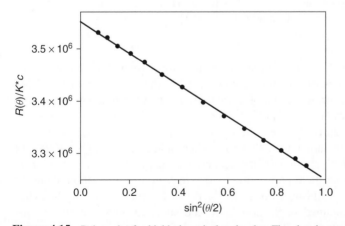

Figure 4.15 Debye plot for highly branched molecules. The plot shows no curvature despite high molar mass of 3.55×10^6 g/mol. RMS radius = 19.8 nm.

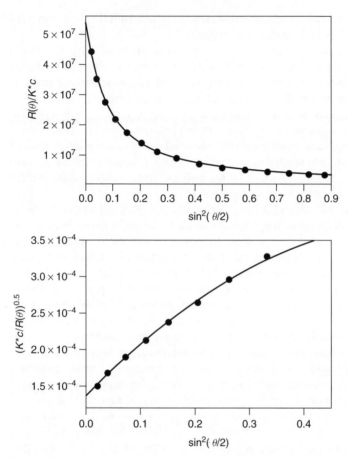

Figure 4.16 Debye plots for a slice of randomly branched polystyrene obtained by random coil (top) and Berry (bottom) formalisms: $M = 5.37 \times 10^7$ g/mol, $R = 222$ nm and $M = 5.42 \times 10^7$ g/ mol, $R = 228$ nm, respectively. Branching ratio $g = 0.23$.

are delayed by enthalpic interactions or, in the case of branched molecules, by anchoring in the pores of column packing. In FFF, the co-elution of small and large species can occur at the end of a fractogram when the field is switched off before complete sample elution, or due to steric separation of large species.

As shown in Figure 2.9, even a trace amount by mass of species with molar mass and size much larger than those of other molecules strongly affects the shape of the angular variation of the scattered light intensity. Investigation of the Debye plots along the elution volume axis may bring information about the separation process, as illustrated in Figure 4.17 for a sample of highly branched polystyrene. As explained in Section 6.2.1, the slices at high elution volumes consist of major parts of smaller molecules separated by steric exclusion that are contaminated by small amounts of large branched molecules.

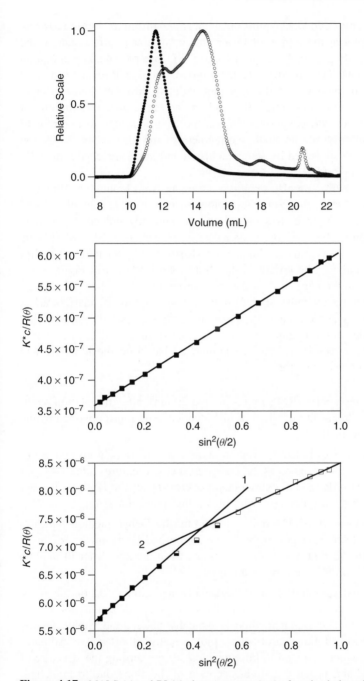

Figure 4.17 MALS (•) and RI (○) chromatograms (top) of randomly branched PS and Debye plots (Zimm formalism) taken at elution volumes of 11.7 mL (center) corresponding to $M \approx 2.8 \times 10^6$ g/mol and 14.4 mL (bottom) corresponding to $M \approx 1.8 \times 10^5$ g/mol. Lines 1 and 2 are linear fits for points (■) and (□), respectively.

Figure 4.17 shows two Debye plots obtained at two different elution volumes. A linear Debye plot at the region of lower elution volumes corresponds to the narrow fractions eluting from the column. Since the Debye plot at a higher elution volume gives a molar mass of about 180,000 g/mol, one would expect a straight plot. However, a more complicated Debye plot with two different slopes strongly resembles the plot shown in Figure 2.9, which indicates presence of a mixture of molecules of markedly different sizes. In fact the data presented in Figure 4.17 provide additional evidence about the correctness of the idea of the co-elution of smaller molecules separated by steric exclusion and large branched molecules delayed by anchoring effect.

The plot of K^*c/R_θ can also exhibit a curvature from branching. Burchard showed[5] that branching causes in Zimm plots an upturn of the angular distribution. Polydispersity reduces this upturn and in randomly branched polymers the upturn is balanced by the large polydispersity. However, in SEC-MALS the angular variation is measured for almost monodisperse polymer molecules eluting from SEC columns. The upturn in the Zimm plot is shown in Figure 4.18, where the slice from the beginning of the chromatogram of branched polystyrene (Figure 4.17) is contrasted with a slice from narrow polystyrene standard of about the same molar mass. A strong deviation from linearity suggests elution of highly branched molecules at the beginning of the chromatogram.

The following conclusions can be summarized based on the literature and experimental results shown here:

1. For small molecules (RMS radius around 20 nm), all extrapolation methods require linear fit and yield results identical with those obtained by the random coil method.

2. For medium-size molecules (up to about 70 nm), the Zimm and Berry methods are linear, whereas the Debye fit method requires a higher-order fit. The first-order Zimm extrapolation overestimates the RMS radius and the overestimation increases with increasing molecular size.

3. For molecules with RMS radius >100 nm, the Debye method typically requires a fifth-order polynomial. The Zimm and Berry methods can be fitted by the second-order polynomial and even the first-order fit may be sufficient if only lower angles are used.

4. Choice of extrapolation method has usually negligible effect on the molar mass, but it is more significant in the case of the determination of the RMS radius. The Zimm method is typically not a good choice for the determination of the conformation plot, because it often overestimates the slope, which can lead to false conclusions concerning the structure of polymer chain, because a slight decrease of the slope due to branching can be balanced by the overestimation of the slope by Zimm formalism.

5. The random coil method works well also for branched molecules with a high degree of branching.

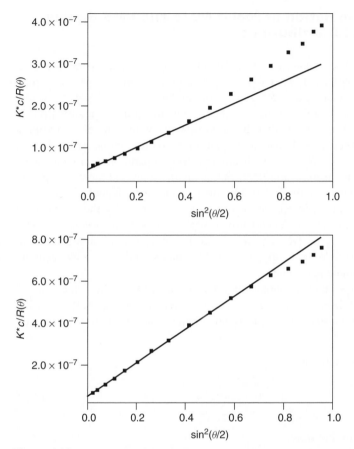

Figure 4.18 Debye plots (Zimm formalism) for a slice from the beginning of the chromatogram shown in Figure 4.17 (elution volume of 10.4 mL) (top) and for a slice of narrow polystyrene standard (bottom). $M \approx 2 \times 10^7$ g/mol for both plots. An upturn on the Zimm plot indicates a high degree of branching of molecules eluting at the beginning of the chromatogram.

6. In the case of large molecules of RMS radius around 100 nm and more, processing the light scattering data should be always based on carefully and critically evaluated Debye plots.

7. Zero-order fit, that is, line parallel with $\sin^2(\theta/2)$ axis, can be used for small polymers with negligible angular variation of scattered light intensity. The zero-order fit calculates an average intensity measured at particular angles.

8. The Berry fit method appears to be the most universal extrapolation method for polymers of various conformation over a broad molar mass range.

4.2.2 Determination of Molar Mass and RMS Radius Averages and Distributions

Regardless of the light scattering formalisms used for data processing, the SEC-MALS measurement yields for each slice of elution volume V_i the molar mass M_i and the mean square radius R_i^2. In fact, the molar mass is the weight-average and the mean square radius is the z-average, but the primary assumption for all SEC-MALS calculations is that the polydispersity within the elution volume slice can be neglected, and that both quantities are just molar mass M_i and the mean square radius R_i^2. Although efficient SEC separation is crucial for the determination of molar mass and RMS radius distributions and moments, it must be emphasized that even if SEC resolution is poor (for whatever reason), the M_w and R_z are always correct, because they are measured by the fundamental principle of light scattering independently of separation performance. On the other hand, the averages M_n, M_z, R_n, and R_w are correct only if the volume slices are truly monodisperse. Consequently, SEC-MALS has a tendency to overestimate M_n and underestimate M_z values.

Using the slice values of molar mass and mean square radius together with the concentrations determined by a concentration-sensitive detector, it is possible to calculate the molar mass and RMS radius moments according to the following equations:

Number-average molar mass:

$$M_n = \frac{\sum c_i}{\sum \frac{c_i}{M_i}} \tag{4.14}$$

Weight-average molar mass:

$$M_w = \frac{\sum c_i M_i}{\sum c_i} \tag{4.15}$$

Z-average molar mass:

$$M_z = \frac{\sum c_i M_i^2}{\sum c_i M_i} \tag{4.16}$$

Number-average mean square radius:

$$R_n^2 = \frac{\sum \frac{c_i}{M_i} R_i^2}{\sum \frac{c_i}{M_i}} \tag{4.17}$$

Weight-average mean square radius:

$$R_w^2 = \frac{\sum c_i R_i^2}{\sum c_i} \tag{4.18}$$

Z-average mean square radius:

$$R_z^2 = \frac{\sum c_i M_i R_i^2}{\sum c_i M_i} \tag{4.19}$$

The quantities M_i, R_i^2, and c_i in the above equations are the molar mass, the mean square radius, and the concentration, respectively, and the subscript i indicates the ith elution volume slice. All sums are typically taken over one chromatographic peak. The RMS radius moments are obtained as the square roots of the appropriate mean square radii. The SEC-MALS measurement yields the weight-average molar mass and the z-average RMS radius, which are obtainable by a batch light scattering experiment. That means the values of M_w and R_z from the SEC-MALS experiment should equal those determined in batch mode. This is typically well fulfilled as demonstrated in Table 4.3. A discrepancy in SEC-MALS and batch MALS results indicates problems with chromatography, such as shearing degradation of molecules with ultra-high molar mass or adsorption of part of the analyzed sample in the columns.

Note that the molar mass averages in Equations 4.14–4.16 are true molar mass moments (i.e., higher powers of M_i are used in the successive moments). On the other hand, the RMS radius moments in Equations 4.17–4.19 are different, because they use the same power of R_i^2. That means the radius averages are not true moments of the radius distribution. For example, the true z-average mean square radius is defined as:

$$R_z^2 = \frac{\sum \frac{c_i}{M_i}(R_i^2)^3}{\sum \frac{c_i}{M_i}(R_i^2)^2} \tag{4.20}$$

Table 4.3 Weight-Average Molar Masses and z-Average RMS Radii Determined by Batch and SEC-MALS

Sample	$M_w (10^3$ g/mol)		R_z (nm)	
	Batch MALS	SEC-MALS	Batch MALS	SEC-MALS
PS (A)	344 ± 8	326 ± 4	34.4 ± 0.6	36.2 ± 0.8
PS (AN)	303 ± 1	294 ± 3	29.5 ± 0.1	30.1 ± 0.1
PS (NIST 706)	288 ± 2	285 ± 3	27.9 ± 0.2	28.4 ± 0.2
PS (1)	386 ± 1	370 ± 2	40.3 ± 0.4	42.7 ± 0.4
PS (2)	398 ± 6	375 ± 1	39.5 ± 0.4	43.3 ± 0.4
PS (K)	239 ± 3	238 ± 1	27.2 ± 1.0	27.9 ± 0.6
PS (S)	236 ± 1	225 ± 1	27.3 ± 0.4	27.2 ± 0.8
PMMA (J)	96 ± 1	89 ± 1	13 ± 1	13 ± 2
PMMA (Y)	604 ± 10	558 ± 7	41.3 ± 0.5	44.3 ± 0.1
EP (1)	3.5 ± 0.1	3.4 ± 0.1	—	—
EP (2)	8.3 ± 0.1	8.1 ± 0.1	—	—

Note: Results uncertainty is based on repeated experiments and/or uncertainty calculated by ASTRA software.

where c_i/M_i is the molar amount n_i of molecules having the mean square radius R_i^2. A similar equation can be written for the average R_w that would be the true moment of radius distribution. However, the average RMS radius that is obtained by the batch light scattering measurement of a polydisperse polymer is that defined by Equation 4.19 and not by Equation 4.20 (compare Equation 2.20). Although the SEC-MALS experiment permits calculation of true R_z, the calculation according to Equation 4.19 provides a result that is consistent with the batch measurement.

The physical meaning of the mean square radius defined by Equation 4.19 is that it is the mean square radius of molecules having the z-average molar mass as evident from the following: For a random coil polymer in theta conditions, the mean square radius is proportional to the molar mass ($R^2 = A \times M$). For a polydisperse random coil polymer, the mean square radius of each species is proportional to the molar mass of that species. That means that using Equation 4.19 one gets:

$$R_z^2 = \frac{\sum c_i A M_i^2}{\sum c_i M_i} = A \times M_z \tag{4.21}$$

where A is the proportionality constant. Equations 4.17 and 4.18 are equivalent and yield the RMS radii corresponding to the number- and weight-average molar masses, respectively. The values of R_n and R_w permit comparison of RMS radius averages with the same types of molar mass averages, but their applicability for polymer characterization appears rather limited. In addition, the accuracy of R_n is often affected by the presence of small molecules for which the RMS radius cannot be directly determined and can be only estimated by the extrapolation procedure, as will be shown later.

Besides molar mass and RMS radius moments, the SEC-MALS data allow for the determination of differential and cumulative distributions using a similar procedure as in the case of conventional SEC. In contrast to conventional SEC, the combination of SEC-MALS also allows the determination of the distribution of RMS radius and relation between the RMS radius and molar mass. The latter relation is often called a *conformation plot*,[i] because it can provide information about the conformation and architecture of polymer chains. The values of the slopes of the conformation plots for linear random coils in theta solvents, rods, and compact spheres are 0.5, 1.0, and 0.33, respectively. Due to the expansion of polymer chains in thermodynamically good solvents, the slopes of the con-formation plots of linear random coils are typically 0.58–0.60, and larger values indicate expansion of polymer chain due to chain rigidity or repulsive electrostatic forces. On the other hand, the decrease of the slope to lower values indicates a more compact molecular structure, usually due to the presence of branched

---

[i]The term *conformation* is related to the spatial arrangement of a polymer chain that results from the rotation of its segments about single bonds. However, the relation RMS radius versus molar mass (conformation plot) is typically used with respect to linearity or branching. That means the term *conformation* is often incorrectly associated with polymer branching.

macromolecules. The assumption of monodisperse slices is even more serious in the case of a conformation plot than in the case of molar mass distribution, because the z-average RMS radius is more sensitive to polydispersity than the weight-average molar mass. That means that if the conformation plot of a linear polymer in a thermodynamically good solvent is perfectly linear with the slope around the value of 0.58, the separation performance of SEC columns can be considered appropriate.

If the polydispersity of the slices is increased due to secondary separation mechanisms or low column performance, the conformation plots become distorted. The non–size exclusion separation usually increases the polydispersity at higher elution volumes and the conformation plots show a typical upswing at the region of lower molar masses. The upswing is given by the higher sensitivity of the z-average RMS radius to polydispersity. The presence of molecules with molar masses above the exclusion limit increases the polydispersity at the beginning of the chromatogram and the upswing may appear at the region of high molar masses, but the upswing at the region of lower molar masses is usually more evident.

4.2.3 Chromatogram Processing

Data processing in SEC-MALS is generally identical with that in conventional SEC and involves setting the baseline and peak limits. In addition, one has to decide which formalism and fit degree are appropriate for the acquired light scattering data. To process the SEC-MALS data, the signals from both MALS and RI detectors must be considered and their different responses to molar mass and concentration taken into account. In contrast to conventional SEC, the baseline has to be set for signals corresponding to all scattering angles. This procedure can be performed using the autobaseline feature of the light scattering software. The baseline for other detectors (RI, UV, viscometer) should be set independently because of their different shapes and sensitivities to chromatographic conditions. For example, the RI detector often shows positive and negative solvent and impurity peaks at the end of the chromatogram that are completely unseen by the light scattering detector. A typical feature of the MALS detector is almost no drift of the light scattering signals, which is due to very low sensitivity to temperature and pump fluctuations and slight changes of the mobile phase composition during column stabilization. As in the case of conventional SEC, moving the starting and ending points of the baseline backward or forward should have negligible impact on the obtained results. To select the peak limits it is necessary to take into account the signals from at least two detectors of different sensitivity to molar mass and concentration. Besides the most frequent combination MALS-RI, this can often be UV-MALS-RI or MALS-VIS-RI, and other detector combinations are possible as well. Detection of the peak beginning is usually easier than detection of the peak end. Issues concerning the determination of the end of the RI chromatogram are discussed in Section 3.5.2. If a polymer sample contains a low level of oligomers causing the tailing of the RI chromatogram, the light scattering signal at the peak end can drop below the detection limit.

A typical example of RI and MALS chromatograms for a polymer containing an oligomeric tail is shown in Figure 4.19. The determination of the peak beginning in this particular example is straightforward while the determination of the peak end is complicated by a long oligomeric tail showing weak RI and almost no MALS response. No appreciable signal is obtained by the light scattering detector beyond about 15 mL, although the RI detector signal remains evident. The low signal from MALS results in inaccurate molar masses and consequently typical scattering of molar mass data points.

The accuracy of the determination of molar mass can be improved by the extrapolation of the molar mass–versus–elution volume relation from the high-concentration region toward the end of the chromatogram, where the data are scattered due to low concentration, as demonstrated in Figure 4.19. In this particular example, the peak end is established at the elution volume corresponding to about 1,000 g/mol, which yields reasonable M_n in good correlation with the value obtained by conventional SEC (compare Table 3.4). The extrapolation uses the molar masses determined in the region where the signals from both detectors are high to establish the relation between the molar mass and elution volume (i.e., SEC calibration for a given polymer under given SEC conditions) to calculate the molar masses in the region where the detector signal is weak and the data are uncertain. The extrapolation procedure can be also used to improve the determination of lower moments of RMS radius. The values determined for NIST SRM 706 polystyrene are shown in the caption of Figure 4.19. It is worth mentioning that the obtained R_n and R_w agree well with the experimental values of M_n and M_w. Using the relation $R = 0.014 \times M^{0.585}$ for linear polystyrene in THF, the M_n and M_w calculated from the values of R_n and R_w are 72,000 g/mol and 277,000 g/mol, respectively.

The region of reliable M is usually limited by the sensitivity of the RI detector at the high-molar-mass part of the chromatogram and by the sensitivity of the MALS detector at the low-molar-mass part. The extrapolation of molar mass and RMS radius can improve accuracy of measurements at the peripheral regions of the chromatogram where the signal of the MALS or RI detector is low. However, the peak limits should not be extended to the regions where signals of both detectors drop to baseline. In such a case the software calculates solely randomly scattered data points associated with zero concentration of eluting molecules. It is also important to keep in mind that any extrapolation behind the region of experimental values is always an uncertain procedure and the extrapolated data should not go too far behind the experimental ones. The *data extrapolation* should not be confused with the *results fitting*, when all data points, including those determined from weak detector signals, are taken into account. When too many scattered points are included, the fit may become incorrect. The user should get familiar with the features and possibilities of a particular software package.

Although results extrapolation or results fitting may improve the accuracy of the M_n and the low-molar-mass part of the molar mass distribution to a great

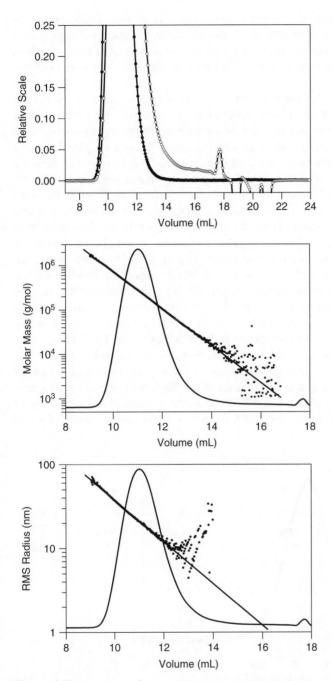

Figure 4.19   MALS at $90°$ (●) and RI (○) chromatograms of polystyrene NIST SRM 706 (top) and the extrapolation of molar mass (center) and RMS radius (bottom) toward high elution volumes. $M_n = 66,000$ g/mol, $R_n = 9.7$ nm, $R_w = 21.4$ nm.

extent, the more reliable way to get correct results is to improve the signal-to-noise ratio. It must be emphasized at this point that the electronic noise generated by a light scattering photometer is far below the noise generated from the chromatography system by particles from the mobile phase or bleeding from the column packing.

Although samples that show an imperceptible MALS signal at the end of the chromatogram, but still an evident signal of the RI detector, are more frequent, the reverse situation can occur at the beginning of chromatograms of samples containing trace amounts of high-molar-mass species. For such samples the strong difference in sensitivities between the RI detector and the light scattering detector results in an intensive signal of MALS and a barely detectable signal of the RI detector, as shown in Figure 4.20. Note that even if the weak RI signal does not allow for the determination of reliable molar mass, the RMS radius can still be determined accurately from the angular variation of the light scattering intensity. If the concentration signal is very close to zero, noise can make the concentrations randomly positive and negative, and the Debye plots may look confusing. If this happens, it is possible to manually adjust the baseline for the concentration detector to be slightly below the baseline, thus insuring that all concentration values are positive regardless of being incorrect. This procedure will drastically affect the molar masses, but they are incorrect anyway. Manually adjusting the RI baseline has no effect on the accuracy of the RMS radii.

Since the molar mass is calculated from the responses of both the light scattering and RI detectors, a random fluctuation of the RI signal can lead to upward or downward curvature of the molar mass at the beginning of a chromatogram, as shown in Figure 4.21. Since the RMS radius is calculated independently of the signal of the RI detector, the RMS radius–versus–elution volume plot can

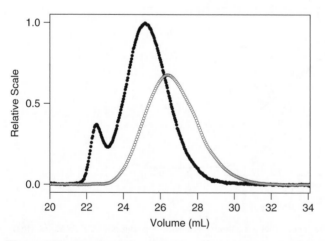

Figure 4.20   MALS at 90° (●) and RI (○) chromatograms of NIST SRM 1476 polyethylene showing intensive light scattering signal at the beginning of the chromatogram.

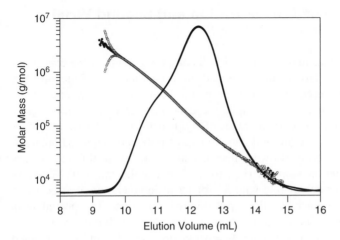

Figure 4.21 Molar mass–versus–elution volume plots and RI chromatograms from three consecutive SEC-MALS runs of broad polystyrene. The discrepancy of the plots at the beginning of chromatograms is caused by random fluctuation of RI signal.

reveal that the curvature is an artifact and does not give a true picture of the polymer separation.

The effect of random noise level on the accuracy of molar masses and RMS radii determined by SEC-MALS was studied by Tackx and Bosscher,[6] who theoretically simulated MALS and RI chromatograms and showed systematic deviations of molar mass in the region of low and high elution volumes. They showed that in the low-molar-mass region of the chromatogram the log molar masses are too high, while in the high-molar-mass region they are systematically too low. This was explained by the random noise in the MALS and RI signals. When the noise level becomes too high, positive and negative molar masses are obtained, but only the logarithm of positive molar masses has mathematical meaning. Therefore, all the negative points are omitted in the log M –versus–elution volume plot. According to my own experience, the majority of polymer samples show no (or negligible) systematic deviation at the region of lower molar masses.

The calculation at reference 6 starts way before the real beginning of MALS and RI chromatograms. Also in the region of high elution volumes the integration limit seems to go too far after the real end of the RI chromatogram. Although the explanation given in reference 6 appears reasonable, it remains unclear why this systematic deviation is not observed for all samples. For instance, the molar masses in Figures 4.9 or 4.19 evidently do not show any strong systematic deviation. Experience also suggests, but does not directly prove, that the systematic deviations on molar mass plots appear more often when older SEC columns are used, which might be caused by release of previously retained impurities due to a slight pressure pulse created by sample injection.

4.2.4 Influence of Concentration and Second Virial Coefficient

The calculation of molar mass in SEC-MALS commonly neglects the concentration effect; that is, the term with the second virial coefficient in Equations 4.4, 4.8, and 4.10 is considered to be small enough to be neglected without a significant effect on molar mass. In principle, the concentration effect can be easily corrected for by using appropriate values of A_2. However, the relation between A_2 and molar mass is generally unknown for most polymers requiring analysis. In addition, light scattering software may not allow for using different values of A_2 at different elution volumes (i.e., for different molar masses). Nevertheless, since A_2 varies only weakly with M for most polymers, an average value corresponding to M_w can be used for polydisperse polymers. The procedure is that the sample under analysis is first processed using $A_2 = 0$ mol mL/g^2, then the obtained M_w is used to calculate $A_2 =$ from Equation 2.4 and the obtained value is used to recalculate the data. Nevertheless, the differences in the obtained results are mostly negligible.

According to Equation 4.2, the product $2A_2cM \ll 1$; then $\frac{R_0}{K^*c} = M$. That means the effect of A_2 must be considered with respect to the concentration of molecules eluting at a given elution volume and their molar mass. Table 4.4 shows the effect of A_2 on the molar masses determined across the peak of typical polydisperse polymer. The molar mass at each volume slice was at first determined using $A_2 = 0$ mol mL/g^2, and then for each molar mass the A_2 was calculated according to Equation 2.5 and used to calculate the true molar mass.

The data in Table 4.4 show that the concentration of injected solution is diluted by a factor of at least 25 and that neglecting the concentration effect results in an error of about 2% at maximum. The molar mass error is equal to $100 \times 2A_2cM\,(\%)$, which means the only parameter directly influencing the molar mass correctness is the concentration, which should be therefore kept as low as possible, though without sacrificing the signal-to-noise ratios of the applied detectors. It is obvious that the concentration-related errors strongly decrease with decreasing molar mass and thus that the concentration is typically not an important parameter in the case oligomers, when solutions of about 10 times higher concentrations than those used for the analysis of higher-molar-mass polymers can be used in order to assure an intensive light scattering signal.

4.2.5 Repeatability and Reproducibility

It has been shown that flow rate is an absolutely key parameter for the accurate determination of molar mass by conventional SEC with column calibration and that the obtained results are also strongly affected if the temperature used for the analysis deviates from that used for the column calibration. Very low sensitivity of MALS detection to both flow rate and temperature deviations results in significantly improved repeatability and reproducibility of the measurements

Table 4.4 Influence of the Second Virial Coefficient on Molar Masses of Polydisperse Polymer Determined by SEC-MALS (Injection 100 μL 0.2% w/v)

Volume (mL)	Concentration (g/mL)	$M(A_2 = 0)$ (g/mol)	A_2 (mol mL/g^2)	$M_{corrected}$ (g/mol)	Error (%)
9.0	1.24e−6	2.316e+6	2.56e−4	2.320e+6	0.17
9.1	2.25e−6	2.028e+6	2.65e−4	2.033e+6	0.25
9.3	6.20e−6	1.631e+6	2.79e−4	1.640e+6	0.55
9.4	9.59e−6	1.466e+6	2.87e−4	1.478e+6	0.81
9.5	1.41e−5	1.329e+6	2.95e−4	1.344e+6	1.12
9.6	1.95e−5	1.204e+6	3.02e−4	1.221e+6	1.39
9.7	2.55e−5	1.083e+6	3.10e−4	1.102e+6	1.72
9.8	3.17e−5	9.753e+5	3.18e−4	9.948e+5	1.96
9.9	3.75e−5	8.739e+5	3.27e−4	8.930e+5	2.14
10.0	4.28e−5	7.806e+5	3.36e−4	7.985e+5	2.24
10.1	4.76e−5	6.972e+5	3.46e−4	7.135e+5	2.28
10.2	5.18e−5	6.247e+5	3.56e−4	6.394e+5	2.30
10.3	5.56e−5	5.619e+5	3.65e−4	5.751e+5	2.30
10.4	5.90e−5	5.053e+5	3.75e−4	5.169e+5	2.24
10.5	6.22e−5	4.562e+5	3.85e−4	4.664e+5	2.19
10.6	6.51e−5	4.136e+5	3.94e−4	4.226e+5	2.13
10.8	7.04e−5	3.404e+5	4.14e−4	3.473e+5	1.99
11.0	7.48e−5	2.801e+5	4.35e−4	2.853e+5	1.82
11.2	7.79e−5	2.315e+5	4.56e−4	2.354e+5	1.66
11.4	7.89e−5	1.904e+5	4.79e−4	1.932e+5	1.45
11.6	7.70e−5	1.565e+5	5.03e−4	1.584e+5	1.20
11.8	7.15e−5	1.281e+5	5.29e−4	1.294e+5	1.00
12.0	6.30e−5	1.061e+5	5.54e−4	1.068e+5	0.66
12.2	5.30e−5	8.630e+4	5.83e−4	8.680e+4	0.58
12.5	3.77e−5	6.420e+4	6.28e−4	6.440e+4	0.31
12.8	2.42e−5	4.760e+4	6.77e−4	4.770e+4	0.21

compared with conventional SEC. The negligible impact of the temperature and the flow rate on the molar mass determined by the MALS detector is demonstrated in Table 4.5. Repeatability and reproducibility are illustrated in Table 4.6. Considerably better repeatability and reproducibility of the SEC-MALS method should be of interest even for those who demand only mutual comparison of samples of equal chemical composition and molecular structure and does not require absolute molar mass values, and such may consider the light scattering detector for their applications needless. Note that very good reproducibility is found especially for M_w values that are determined by the MALS detector alone and are not affected by SEC resolution.

Table 4.5   Effect of Flow Rate and Column Temperature on M_w of Broad Polystyrene Determined by SEC-MALS

Temperature (°C)	M_w (10^3 g/mol)	Flow Rate (mL/min)	M_w (10^3 g/mol)
20	332	1.00	337
24	333	1.01	338
28	333	1.03	337
32	333	1.05	336
35	333	1.07	337
—	—	1.10	336

Note: Compare Tables 3.1 and 3.6.

Table 4.6   Repeatability and Reproducibility of M_w of Broad Polystyrene Determined by SEC-MALS

Measurements	Number	M_w (10^3 g/mol)	RSD (%)
One day, single SEC-MALS setup	12	329	0.2
22 months, single SEC-MALS setup	34	332	1.7
17 months, 10 MALS and 4 RI detectors in various combinations	28	331	1.6

Note: Wyatt Technology Corporation miniDAWN™ and DAWN® EOS photometers were used for the study.

4.2.6 Accuracy of Results

Random errors are caused by accidental noise in detector signals (in the MALS signal mostly due to randomly eluting dust particles; in RI due to flow rate and temperature fluctuations), run-to-run variations of the SEC separation process, and drift of the baseline (the MALS drift is usually negligible while the RI detectors are more apt to drift). Systematic errors include calibration constants of MALS, RI, and UV detectors; physical constants such as refractive index and especially *dn/dc* values; errors from improper normalization; errors of interdetector delay volume; errors caused by interdetector band broadening; errors due to improper type of light scattering formalisms and/or polynomial fit degree; and baseline selection and peak integration limits. MALS detector calibration and normalization and *dn/dc* accuracy are discussed in Sections 2.2.1.3 and 2.4. Errors due to the interdetector volume determination are specific for SEC with combined detectors. Although the determination of the interdetector volume appears straightforward, the obtained values can be affected by polydispersity of standard used for alignment, and flow rate deviations, as well as by errors due to the subjective approach of the operator. The effect of errors of interdetector volume on the exponents of the Mark-Howink and conformation plots can be

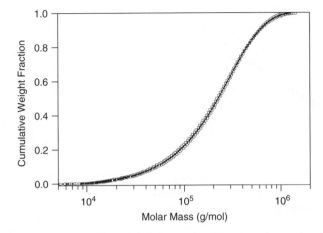

Figure 4.22 Cumulative distribution curves of broad polystyrene obtained using correct (■) interdetector volume, and values incorrect by +10% (○) and −10% (□). Molar mass moments: $M_n = 112{,}000$ g/mol, $116{,}000$ g/mol, $108{,}000$ g/mol; $M_w = 289{,}000$ g/mol, $289{,}000$ g/mol, $289{,}000$ g/mol; $M_z = 485{,}000$ g/mol, $466{,}000$ g/mol, $507{,}000$ g/mol for correct volume delay, and values incorrect by + 10% and −10%, respectively.

quite significant, whereas the impact on the molar mass distribution curves is usually negligible.

Figure 4.22 shows the influence of interdetector volume errors on the molar mass distribution and the molar mass moments. Of note is the fact that errors in the interdetector volume do not appreciable affect the determination of M_w and that errors within about 10% will have only a moderate effect on the molar mass distribution and the moments M_n and M_z.

Figure 4.23 shows a typical effect of interdetector volume on the RMS radius–versus–molar mass plot. It is evident that a relatively small error in the interdetector volume results in a relatively significant change of the slope of the conformation plot and for some samples may even lead to false estimation of polymer conformation and branching, especially if, for instance, the increase of the slope caused by using Zimm formalism is combined with that given by incorrect volume delay.

4.3 APPLICATIONS OF SEC-MALS

4.3.1 Determination of Molar Mass Distribution

Let us at first ignore the problems associated with polymer blends and heterogeneous copolymers, whose *dn/dc* values may vary significantly from slice to slice, and whose slice concentrations cannot be obtained from the signal of RI detector. The primary application of SEC-MALS is the determination of molar mass averages and molar mass distribution. This kind of information is needed in all

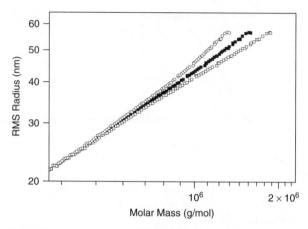

Figure 4.23   Effect of interdetector volume on shape of conformation plot. Data for broad polystyrene. Conformation plots obtained using correct (■) interdetector volume, and values incorrect by +10% (○) and −10% (□). Slopes of the plots = 0.58; 0.64 and 0.53 for correct volume delay, and values incorrect by +10% and −10%, respectively.

areas of polymer science covering fundamental research and theoretical consider-ations, manufacturing quality control, control of properties, and final applications. In contrast to conventional SEC, the use of a MALS detector yields true molar mass averages and molar mass distribution and eliminates errors resulting from improper calibration. Although the results obtained by the SEC-MALS tech-nique typically differ by several tens or even hundreds percent from the values generated by conventional SEC with improper calibration, incidental agreement between conventional SEC and SEC-MALS can happen even in the case of obscure calibration procedure.

For curiosity's sake, let me relate an experience that happened many years ago in my laboratory. The goal was to determine the molar mass averages of poly-electrolyte based on sodium salts of acrylic and maleic acids. The only available columns allowing the analysis of aqueous polymers were those based on bare silica gel, which could be also used in THF. Since the laboratory had no suit-able water-soluble standards, the columns were at first calibrated by polystyrene standards in THF and then flushed with water and finally with aqueous sodium sulphate. The polyelectrolyte samples were analyzed and processed by means of the polystyrene calibration determined in THF. The obtained molar masses were considered to be far from the true ones, but they were useful for sample comparison and allowed us to optimize the polymerization process. Surprisingly, almost-identical results were determined after several years when the same poly-mers were analyzed by SEC-MALS.

Accuracy of M_n can be questioned due to limited resolution of SEC columns. The values of M_n of several polydisperse polymers determined by SEC-MALS and membrane osmometry are compared in Table 4.7. The differences are not

Table 4.7 Number-Average Molar Masses of Polydisperse Polymers Determined by Membrane Osmometry and SEC-MALS

Sample	$M_n (10^3$ g/mol)	
	SEC-MALS	MO
PS (AN)	118 ± 2	120
PS (1)	121 ± 2	139
PS (2)	108 ± 4	153
PS (S)	49 ± 2	66
PS (K)	92 ± 1	98
PMMA (Y)	184 ± 4	185
PMMA (J)	43 ± 2	33

unusually large considering the results from substantially different methods are compared. The results listed in Table 4.7 may be to some extent overestimated due to the polydispersity of volume slices in SEC-MALS and permeation of small amounts of oligomeric fractions through the membrane in membrane osmometry,[i] but it is obvious that SEC-MALS can yield M_n values with the same accuracy as traditionally used, but more laborious and currently less common, membrane osmometry. Results published in reference 4 for a series of 17 oligomeric compounds with molar mass covering 200–5,800 g/mol showed very good agreement of M_n values determined by SEC-MALS with those obtained by VPO, NMR, or HPLC. Favorable results of experiments comparing M_n and M_w of oligomers determined by SEC-MALS with those obtained by NMR, VPO, MALDI, and supercritical fluid chromatography were also reported in references 7 and 8. Taking into account the limitations of absolute methods of the determination of M_n, that is, errors due to permeation of oligomeric fractions in MO and very high sensitivity to even trace amounts of low-molar-mass compounds in the case of VPO, and laboriousness of the measurements (both osmometric methods require measurements at multiple exactly known concentrations), SEC-MALS appears to be a fast and reliable technique for determination of M_n.

The low-molar-mass limit for which SEC-MALS can be used is a frequent question that users and potential users of SEC-MALS ask. As already discussed, the response of the MALS detector is proportional to the molar mass and concentration and thus the MALS detector yields several orders of magnitude lower response for oligomers than for polymers. A typical concentration for the SEC-MALS analysis of polymers is around 0.2% w/v. In the case of oligomers the concentration can be safely increased by a factor of 10 to 25 without significant impact on the error caused by neglecting the term with the second virial coefficient, because for very low molar masses the term $2A_2cM \ll 1$ even for high concentrations.

---

[i]As a matter of fact, both SEC-MALS and membrane osmometry have a tendency to overestimate M_n, though for different reasons.

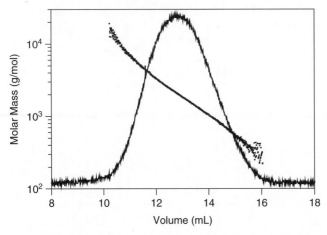

Figure 4.24 Molar mass–versus–elution volume plot and 90° MALS signal for polybutandiol. Injection 200 μL 5% w/v, dn/dc = 0.069 mL/g, M_n (SEC-MALS) = 950 g/mol, M_n (HPLC) = 940 g/mol.[9] Columns: 2 × PLgel Mixed-E 300 × 7.5 mm, THF at 1 mL/min.

An example in Figure 4.24 shows that an acceptable signal-to-noise ratio of a light scattering detector and consequently reliable molar mass can be measured down to a few hundreds g/mol in spite of relatively low dn/dc of the analyzed sample. Despite high injected mass, the data show good separation without noticeable signs of column overloading. In a given example, the highest concentration of the eluting molecules was about 3 mg/mL at molar mass of about 1,200 g/mol. That means the error caused by neglecting the concentration term is most likely well below 3% unless the second virial coefficient is unusually high. It has been explained that the lower the angle of measurement, the more likely the signal will show light scattered by impurities or dust. This is of particular importance for oligomers, which generally yield low light scattering signal. However, since no information can be obtained from the measurement of the angular variation of light scattering intensity for small molecules, the lower angles can be safely dropped and only those angles showing the minimum noise can be used for the molar mass calculation. A zero-fit degree (i.e., line parallel with $\sin^2(\theta/2)$ axis) can be used for processing the data of oligomers since they scatter light equally in all angles. In the case of poor signal-to-noise ratio, only a 90° detector, which usually shows the lowest noise level, can be used to process the data of oligomers.

The only assumptions for the determination of true molar mass distribution by SEC-MALS of homopolymers are negligible polydispersity of the elution volume slices, that is, negligible band broadening in the columns and other parts of chromatographic setup, and negligible concentration effects. Both assumptions are usually fulfilled to a great extent and thus SEC-MALS can be considered a reliable and fast method for characterization of molar mass distribution with precision limited only by the correctness of the MALS calibration constant, normalization coefficients, calibration constant of the concentration detector, and dn/dc. More complicated polymers for the SEC-MALS measurements are heterogeneous copolymers, polymer blends, and branched polymers, where the obtained results may be negatively affected by sample heterogeneity and limited separation efficiency of SEC.

A challenge may be estimation of molar mass distribution of narrow polymers, because possible inaccuracies arising from interdetector peak broadening are significantly more serious than for the polydisperse polymers. Since the calculation of the RMS radius is independent of concentration, the RMS radius can be used instead of molar mass to assess the sample polydispersity. Once the RMS radius distribution is established, it can be converted to molar mass distribution using the RMS radius–versus–molar mass relation. However, the determination of polydispersity of narrow polymers of higher molar mass is limited by SEC resolution.

4.3.2 Fast Determination of Molar Mass

One possible application of SEC-MALS is combination with so-called high-throughput, fast or rapid SEC columns of reduced dimension (e.g., 150×7.5 mm or 100×10 mm). A serious weakness of the fast columns, which recently achieved a certain popularity, is their lower selectivity and efficiency. This fact is of particular importance for conventional SEC, where the calculation of molar mass distribution is based on the assumption of negligible peak broadening. The application of a MALS detector at least partly eliminates the reduced separation efficiency of the fast columns. It has been already explained that determination of M_n, M_z, and the entire molar mass distribution by means of SEC-MALS is based on the assumption of almost monodisperse elution volume slices. However, determination of M_w and R_z is based on the fundamental principle of light scattering and as a matter of fact does not require sample fractionation at all. Therefore, the M_w and R_z values determined by SEC-MALS do not depend on the SEC resolution and are always correct even if other molar mass averages and molar mass distribution are affected by poorer SEC separation efficiency.

Molar mass averages determined by fast SEC-MALS measurements are compared with those obtained by regular SEC column sets in Table 4.8. According to the expectation, differences in M_w are negligible for all column systems, but the use of a single high-throughput column results in a slight overestimation of M_n and underestimation of M_z, which corresponds to decreased resolution and increased polydispersity within the elution volume slices. The graphical comparison of regular and high-throughput analysis of two polymers covering different molar mass range is presented in Figures 4.25 and 4.26. The obtained data indicate slight differences in the molar mass distribution especially at the margins of molar mass distribution. However, it is apparent that high-throughput SEC-MALS analysis can evidently provide not only correct M_w and R_z, but also an acceptable estimation of molar mass distribution, and can be used especially in applications that require short analysis times.

Table 4.8   Molar Mass Averages Obtained by SEC-MALS Analysis Using Four and Two Regular Columns (4R, 2R) and a Single High-Throughput Column (1H)

Sample	$M_n (10^3$ g/mol)			$M_w (10^3$ g/mol)			$M_z (10^3$ g/mol)		
	4R	2R	1H	4R	2R	1H	4R	2R	1H
PES	1.56	1.57	1.81	2.49	2.56	2.54	3.66	3.83	3.44
EP (1)	1.58	1.56	1.80	3.40	3.42	3.45	5.83	5.99	5.49
EP (2)	3.20	3.24	3.58	8.20	8.22	8.15	19.0	18.9	16.2
PMMA (J)	48.7	47.8	50.3	89.2	88.9	88.4	137.8	140.3	127.5
Alkyd	5.90	5.77	6.38	151.3	151.7	151.0	1003	1023	904
PS (AN)	117.0	111.9	120.5	292.6	289.5	287.6	484.5	487.2	441.7

Regular columns: 300 × 7.5 mm PLgel Mixed-C; *high-throughput column:* 150 × 7.5 mm PLgel HTS-C; THF at 1 mL/min; *injection volume:* 200, 100, and 25 μL for four, two, and single columns, respectively. *Samples:* PES = polyester, EP = epoxy resin, PMMA = poly(methyl methacrylate), PS = polystyrene.

The fastest SEC measurement happens when the separation is minimized to the separation of polymer from low-molar-mass compounds, such as solvents, additives, and impurities. This type of measurement, sometimes called *FIPA* (flow injection polymer analysis), provides M_w and R_z. In addition, the analysis can be completed by the measurement of the average intrinsic viscosity by an online viscometer. The SEC-MALS measurement with only a guard column yields the same type of information that can be obtained by a classical batch light scattering experiment. However, a classical light scattering experiment involves measurements of a series of solutions of different concentrations, which is rather a laborious procedure that requires precise concentrations of all solutions. In the case of fast SEC-MALS with a guard column, the polymer is separated from the low-molar-mass compounds and the concentration of polymer molecules eluting from the guard column is measured online by a concentration detector, which means the accurate concentration of the measured solution need not be known. The results can be obtained by a single injection in an analysis time of less than three minutes, which makes the method uniquely suitable for industrial applications for the control of polymerization processes.

Other advantages are very low consumption of solvent and very low system backpressure. The RI signal yields the total injected mass, which allows determination of polymer concentration in the analyzed sample as additional information that is important, for example, in the case of samples taken from the reactor during the manufacturing process. For linear polymers the M_z value can be estimated from R_z provided the relation between the RMS radius and molar mass is known for the polymer under investigation. Knowledge of M_z is important from a practical viewpoint, because comparison of M_z and M_w permits estimation of sample polydispersity. Comparison of results obtained by guard column measurements compared with those determined by means of regular SEC columns or other techniques is shown in Table 4.9.

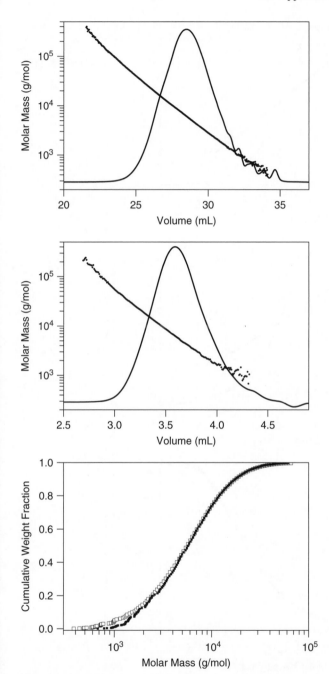

Figure 4.25   Molar mass–versus–elution volume plots and RI chromatograms obtained by 4 ×
PLgel Mixed-C 300 × 7.5-mm columns (top) and a single PLgel HTS-C 150 × 7.5-mm
high-throughput column (center). Overlay of cumulative distribution plots (bottom) obtained with
four columns (□) and a single high-throughput column (■). Sample: epoxy resin.

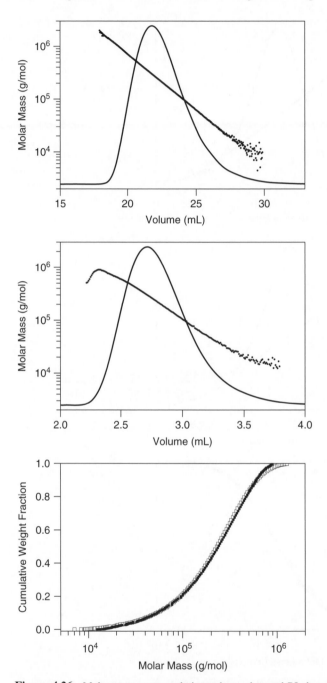

Figure 4.26   Molar mass–versus–elution volume plots and RI chromatograms obtained by 4 × PLgel Mixed-C 300 × 7.5-mm columns (top) and a single PLgel HTS-C 150 × 7.5-mm high-throughput column (center). Overlay of cumulative distribution plots (bottom) obtained with four columns (□) and a single high-throughput column (■). Sample broad polystyrene.

Table 4.9 Comparison of Results Obtained by Fast SEC-MALS Analysis with a Single Guard Column with Nominal (N) Values and Results Obtained by Means of Batch MALS and SEC-MALS Using Regular Columns

	Guard column		M_z from R_z	Two 300 × 8 mm columns		M_w
Sample	$M_w (10^3$ g/mol)	R_z (nm)	$(10^3$ g/mol)	M_w (g/mol)	M_z (nm)	$(10^3$ g/mol)
EP (1)	3.46	—	—	3.42	—	3.50^{MALS}
Narrow PS	208	18.2	210	204	—	200^N
PS (NIST 706)	281	27.5	426	283	421	288^{MALS} 285^N
PS (A)	326	33.9	609	326	644	344^{MALS}
PBZMA (38)	418	27.5	679	399	660	—
PMMA (Y)	544	41.7	1185	558	1051	612^{MALS}

4.3.3 Characterization of Complex Polymers

Although the experimental results indicate that the slices of linear homopolymers are almost monodisperse, the situation may be significantly less favorable in the case of branched polymers and heterogeneous copolymers, where molecules of identical hydrodynamic volume, but different molar mass, can co-elute within the elution volume slices. Increasing SEC performance does not help since the slices may be perfectly monodisperse with respect to the hydrodynamic volume, yet polydisperse from the viewpoint of molar mass. To assess polydispersity within the elution volume slice one can, at least theoretically, compare the number-average molar mass determined by SEC-MALS with that measured by membrane osmometry. Significantly higher M_n obtained by SEC-MALS compared to that from MO would indicate increased slice polydispersity. However, such a comparison would be relatively laborious and unfortunately also uncertain due to the drawbacks of membrane osmometry, that is, penetration of oligomeric fractions through the membrane and low sensitivity to high-molar-mass fractions. It may be worth noting that due to the polydispersity within elution volume slices, the slice molar masses obtained by MALS detection are the weight-averages, while the combination of an online viscometer with the universal calibration yields the number-averages.

4.3.3.1 Branched Polymers

Branching of any kind results in the reduction of molecular dimensions, which are typically expressed in terms of RMS radius, hydrodynamic radius, or intrinsic viscosity. With the exception of regular branched structures, such as perfect stars composed of the same number and length of arms, most natural and synthetic branched polymers have random structure, which means they consist of molecules of different molar masses, different numbers of branch units, and different branch lengths and topologies. Since the separation of polymers by SEC is controlled by hydrodynamic volume and not by molar mass or molecular architecture, molecules eluting at a given elution volume from the columns are polydisperse with respect to their molar mass and branching characteristics. However, if the band-broadening effect is negligible, they are monodisperse with respect to the hydrodynamic volume.

Note that this kind of polydispersity is different from that resulting from delayed elution of large-branched molecules due to their anchoring in the pores of SEC column packing. In the case of polydisperse slices the MALS detection yields M_w and R_z that are valid for a given elution volume (hydrodynamic volume). The weight-average calculated from Equation 4.15 by entering $M_{w,i}$ instead of M_i provides the weight-average, and the same is true for R_z. However, using $M_{w,i}$ in Equations 4.14 and 4.16 results in overestimation of M_n and underestimation of M_z and the polydispersity M_w/M_n. The entire molar mass distribution is shifted to higher (in the lower-molar-mass region) and lower (in the higher-molar-mass region) molar masses.

The authors of reference 10 defined a distribution function $N(M, V_h)$, the number of chains with molar mass M and hydrodynamic volume V_h, which was suggested to be used to reveal mechanistic information about polymer synthesis. They showed how true M_w and M_n can be obtained from processing the hydrodynamic volume distribution. Although reference 10 can serve as an important source of the theory of characterization of complex branched polymers using multiple detection, practical application is limited by the fact that commercially available SEC-MALS-VIS software assumes that the polydispersity within the elution volume slices is negligible. This assumption, although rarely completely fulfilled, significantly simplifies the processing and interpretation of the experimental data. My experience indicates that the SEC-MALS and SEC-MALS-VIS results obtained under the simplifying assumption of negligible slice polydispersity mostly provide a valuable characterization of various branched polymers, and that the characterization of branching is often affected by other limitations of the same significance.

4.3.3.2 Copolymers and Polymer Blends

In the most favorable, and also unusual, case of homogeneous copolymers (i.e., when all molecules present in the sample have identical chemical composition), one can use the weighted average of dn/dc of parent homopolymers to determine the molar mass at each elution volume. In the case of heterogeneous copolymers the chain dimensions depend not only on the molar mass, but also on the chemical composition. Molecules of different composition and molar mass may have the same hydrodynamic volume. Consequently, in SEC separation of chemically heterogeneous copolymers a given elution volume slice can consist of molecules of different molar mass and chemical composition (different dn/dc). Changing the dn/dc along the elution volume axis is theoretically possible, but mostly would be of little use, because the composition as a function of molar mass is unknown for most polymers being analyzed. Therefore, an average dn/dc calculated on the basis of bulk copolymer composition or determined experimentally must be used for processing the entire peak.

In contrast to branched homopolymers, where branching affects the polydispersity within the elution volume slice but the dn/dc remains constant, the chemical heterogeneity also affects dn/dc at a given elution volume and consequently the concentration obtained by the RI detector and the molar mass calculated from the light scattering intensity. That means, in contrast to branched polymers, even M_w cannot be considered to be correct. The influence of chemical heterogeneity of copolymers on the separation process in SEC was studied theoretically by Netopilik et al.[11] with these two main conclusions:

1. The molar mass of a chemically heterogeneous copolymer at a given elution volume may vary with the chemical composition to a considerable extent (e.g., by about 30%).

2. Significant deviations of experimental molar mass averages obtained by SEC with light scattering detection from the true values due to the effect of composition heterogeneity may be expected only for significantly different dn/dc of the parent homopolymers.

The results published in reference 4 for blends of polystyrene, poly(methyl methacrylate), and poly(benzyl methacrylate) showed acceptable accuracy of M_n and M_w determined by SEC-MALS. The blends were processed using an average dn/dc calculated on the basis of composition. However, the investigated polymers were of almost overlapping molar mass distribution. In such a case, where the molar mass of one polymer is overestimated due to the incorrect dn/dc, the molar mass of the other polymer is underestimated and thus the total error is at least partly compensated. If the molar mass distributions of particular polymers are not completely overlapping, there is at least one region of the chromatogram where molecules of only one chemical composition elute. Using an average dn/dc results in over- or underestimation of molar mass that is not compensated by molecules of another chemical composition.

In the case of copolymers, the actual heterogeneity is usually significantly lower compared to the polymer blends, which represents the ultimate example of chemically heterogeneous systems. Consequently, the deviations of the average dn/dc from the true values for molecules of particular chemical composition will be lower. Another parameter to consider is the difference in dn/dc of parent homopolymers, which is in many practical cases small. For example, for various acrylic copolymers the differences in dn/dc of parent homopolymers are relatively low.

It can be concluded that chemically heterogeneous copolymers cannot be accurately characterized by SEC-MALS and the possible errors may be considerable. Errors of molar mass are less serious in the case of moderately heterogeneous samples and/or where the differences between the dn/dc values of parent homopolymers are low. The combination of RI and UV detection can be applied for copolymers if one of the monomers is UV absorbing.

4.3.4 Conformation Plots

The conformation plot relates RMS radii and molar masses obtained at particular elution volumes. The ability to relate the RMS radius to the molar mass is important for polymer characterization and also permits testing of polymer theories. The conformation plots may be affected by the light scattering formalism utilized for data processing. The conformation plots obtained for a linear polydisperse polymer using various light scattering formalisms are depicted in Figure 4.27. The plots show typical behavior, and similar results have been obtained for other polymers. It can be concluded that the slopes obtained by the Zimm method are usually overestimated, especially if the sample contains large molecules more than about 100 nm. The Berry and random coil methods yield almost identical conformation plots. The slopes of the conformation plots obtained by the Debye methodology with the first-order fit are underestimated, and the underestimation increases strongly toward high molar masses.

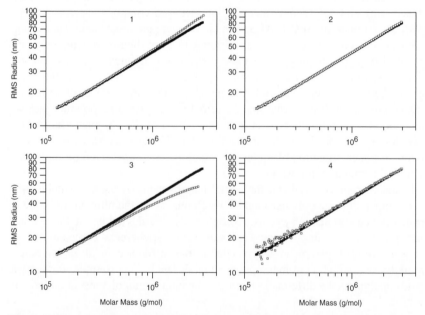

Figure 4.27 Conformation plots of broad linear polystyrene obtained by different processing methods. (1) Berry (■) versus Zimm (□), slopes = 0.57 and 0.60, respectively; (2) Berry (■) versus random coil (□), slope = 0.57; (3) Berry (■) versus Debye first-order fit (□), slope = 0.49; (4) Berry (■) versus Debye third-order fit (□), slope = 0.55.

The application of the first-order Debye formalism to large molecules can easily lead to false conclusions about polymer conformation, because even for linear random coils in thermodynamically good solvent the first-order Debye fit can yield a slope below 0.5 (i.e., characteristic value for branched macromolecules). By fitting a higher-degree polynomial to the Debye formalism, more accurate data providing a reliable slope of the conformation plot are obtained. However, a higher-order fit increases the scattering of the data points in the region of lower molar masses, because the extrapolation becomes very sensitive to data containing random errors. Although a difference in the slope of several hundredths may appear negligible, it may play an important role in the identification of branching, since, for example, a slope equal to 0.58 is typical for linear molecules in thermodynamically good solvents, while a slope of 0.54 may indicate the presence of a certain amount of branched molecules. Both slopes can be easily obtained for a polymer under investigation by just changing from the Zimm processing method to the Berry or random coil method. Conclusions based on the slope of the conformation plot should therefore be made with caution, always considering the accuracy of the M and R from which the conformation plot is constructed and the limitations of particular processing methods. This is of particular importance for very large scattering species where the differences between the particular processing methods are greater.

Although in principle the slope of the conformation plot itself is sufficient for making conclusions about conformation and branching, the risk of misinterpretation is significantly reduced if a model linear counterpart is available. Since the slope of the conformation plot is very sensitive to the processing method used to determine the molar mass and RMS radius, for polymer samples covering very broad molar mass range it may be beneficial to divide the chromatogram into several sections and evaluate each section separately using the most appropriate processing method. Alternatively, the particular data points of the conformation plot can be obtained by means of individually evaluated Debye plots, where the most appropriate angular intervals and fit orders are used for each volume slice. The molar mass and RMS radius pairs obtained for a sufficient number of slices can be processed and plotted by Excel or Origin software.

Besides sensitivity to the light scattering method used for data processing, the conformation plots can be affected by delayed elution of branched polymers (see Section 6.2.1) and resolution of SEC columns. And as already mentioned, errors in the interdetector volume delay can affect the accuracy of the slope (see Figure 4.23).

4.3.5 Mark-Houwink Plots

Intrinsic viscosity is one of the most fundamental quantities in polymer science. It has had considerable historical importance in establishing the very existence of polymer molecules, and can still provide considerable insight into polymer structure and behavior. A modern way to generate Mark-Houwink plots is to

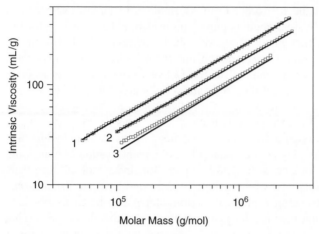

Figure 4.28   Mark-Houwink plots of (1) polystyrene, (2) poly(methyl methacrylate), and (3) poly(benzyl methacrylate) determined by SEC-MALS-VIS (symbols) compared with the literature Mark-Houwink equations (solid lines): $[\eta]_{PS} = 0.0117 \times M^{0.717}$, $[\eta]_{PMMA} = 0.0107 \times M^{0.70}$, $[\eta]_{PBZMA} = 0.00438 \times M^{0.738}$ (references 12-14, respectively), THF at room temperature.

couple an SEC instrument to an online viscometer and a light scattering detector. The hyphenated SEC-MALS-VIS technique yields molar masses and intrinsic viscosities for narrow fractions eluting from SEC columns with the limitations discussed in relation to complex polymers.

Figure 4.28 compares Mark-Houwink plots obtained by SEC-MALS-VIS with those generated by classical measurements using a capillary viscometer. The results confirm that at least for homopolymers the SEC-MALS-VIS hyphenated technique represents a reliable and fast way to obtain the Mark-Howink plots.

Figure 4.28 shows excellent agreement for polystyrene and poly(methyl methacrylate), where the classical determination of Mark-Houwink parameters was performed using narrow PS and PMMA standards, whereas a slight discrepancy in the plots for poly(benzyl methacrylate) may be due to using fractions isolated by precipitation fractionation (i.e., samples of larger polydispersity than that of narrow SEC standards). The polydispersity within the elution volume slice due to co-elution of molecules of the same hydrodynamic volume, but different structure and molar mass, or due to band broadening in the SEC columns, is in most cases lower than the polydispersity of the fractions prepared by classical methods of polymer fractionation. Consequently, the Mark-Houwink parameters determined by means of SEC-MALS-VIS may often be more accurate than the values listed in the older scientific references.

Similarly as in the case of determination of the conformation plot, polydisperse samples covering at least one order of magnitude of molar mass are needed. The Mark-Houwink and conformation plots are affected by low signal intensities, and thus the beginning and end regions of the chromatogram with randomly scattered data points should be eliminated from the calculation.

The effect of the erroneous interdetector volume delay is similar to that in the case of the conformation plot; this issue is even more serious because of generally larger interdetector volumes between the viscometer and other detectors.

4.4 KEYNOTES

- MALS detection in SEC provides absolute molar mass distribution and molar mass moments, and completely eliminates the need for constructing the calibration curves with standards. A MALS photometer also yields the RMS radius distribution.
- The dependence of the RMS radius on molar mass (conformation plot) yields information about the molecular conformation of polymers in solution. The conformation plot is the most direct method for detection and characterization of branching.
- SEC-MALS data acquired at particular elution volume slices are processed by means of Debye plots. The processing can be performed using various light scattering formalisms. The selection of the processing method is typically not critical for smaller polymers, but it becomes more and more important with increasing molecular size.
- Combination of MALS detection with an online viscometer is a fast and accurate method for determination of a Mark-Houwink plot.
- The determination of molar mass distribution, conformation plot, and Mark-Houwink plot assumes efficient SEC separation yielding almost monodisperse fractions within the elution volume slice. However, the M_w and R_z values determined by SEC-MALS are based on the fundamental principle of light scattering and are completely independent of the efficiency of SEC separation.
- One of the important advantages of SEC-MALS compared to conventional SEC is significantly reduced sensitivity to operation variables such as flow rate, temperature, or column performance, which assures very good repeatability and reproducibility.

4.5 REFERENCES

1. Wyatt, P. J., *Analytica Chimica Acta*, **272**, 1 (1993).
2. Wyatt, P. J., in *Handbook of Size Exclusion Chromatography and Related Techniques*, 2nd Edition, Wu, C.-S. (editor), Marcel Dekker, New York (2004), p. 623.
3. Andersson, M., Wittgren, B., and Wahlund, K.-G., *Anal. Chem.*, **75**, 4279 (2003).
4. Podzimek, S., in *Multiple Detection in Size-Exclusion Chromatography*, A. M. Striegel (editor), ACS Symposium Series **893**, Washington, D.C. (2004), p. 109.
5. Burchard, W, *Macromolecules*, **10**, 919 (1977).
6. Tackx, P. and Bosscher, F., *Analytical Communications*, **34**, 295 (1997).
7. Xie, T., Penelle, J., and Verraver, M., *Polymer*, **43**, 3973 (2002).

8. Saito, T., Lusenkova, M. A., Matsuyama, S., Shimada, K., Itakura, M., Kishine, K., Sato, K., and Kinugasa, S., *Polymer*, **45**, 8355 (2004).
9. Rissler, K., Socher, G., and Glockner, G., *Chromatographia*, **54**, 141 (2001).
10. Gaborieau, M, Gilbert, R. G., Gray-Weale, A., Hernandez, J. M., and Castignolles, P., *Macromol. Theory Simulations*, **16**, 13 (2007).
11. Netopilik, M., Bohdanecky, M., and Kratochvil, P., *Macromolecules*, **29**, 6023 (1996).
12. Kolinsky, M. and Janca, J., *J. Polym. Sci.: Polym. Chem. Ed.*, **12**, 1181 (1974).
13. Podzimek, S., *J. Appl. Polym. Sci.*, **54**, 91 (1994).
14. Podzimek, S. and Vlcek, T., *J. Appl. Polym. Sci.*, **82**, 454 (2001).

Chapter 5

Asymmetric Flow Field Flow Fractionation

5.1 INTRODUCTION

The history of *field flow fractionation* (FFF) started with the work of Giddings[1] of the University of Utah in Salt Lake City. He and his team invented the principle of FFF, derived fundamental theory, developed various types of FFF, and constructed the first instruments. Although invented in the same decade when the first *size-exclusion chromatography* (SEC) instruments became commercially available, FFF suffered for a long time from the fact that reliable instruments were not commercially available and the work concentrated in several research groups working with their own instrumentation. For several decades, academic works concerning theoretical aspects of separation by various kinds of FFF prevailed over real applications; FFF did not become a widely applied and routine analytical tool. SEC thus became the dominant method used for the characterization of molar mass distribution of synthetic and natural polymers. Further progress in SEC occurred with the development of advanced detector systems such as online viscometers, which allowed full usage of the principles of universal calibration, and especially light scattering detectors that completely eliminated the need for column calibration. However, despite wide application range and versatility, SEC sometimes suffers from serious drawbacks that can be significantly reduced or even completely eliminated by the application of FFF.

This chapter focuses on *asymmetric flow field flow fractionation* (A4F), which currently represents the most instrumentally developed type of FFF with readily available instrumentation and numerous applications covering synthetic polymers, natural polymers, colloidal particles, proteins, vaccines, various biological materials, and environmental samples. Due to the recent development of a new generation of A4F instruments, the method has finally achieved the mature state

Light Scattering, Size Exclusion Chromatography and Asymmetric Flow Field Flow Fractionation: Powerful Tools for the Characterization of Polymers, Proteins and Nanoparticles, by Stepan Podzimek
Copyright © 2011 John Wiley & Sons, Inc.

where it can be used as routinely as SEC; especially in combination with a multi-angle light scattering (MALS) detector A4F represents a powerful method of polymer analysis and characterization with several advantages over traditionally used SEC.

One of the most serious limitations of SEC is the possibility of shearing degradation of polymer molecules with molar mass above about 10^6 g/mol.[2] Since most synthetic and natural polymers are polydisperse, many samples contain fractions running into the molar mass range of several millions g/mol, and for such samples the experimenter often does not know whether the high-molar-mass region of the molar mass distribution curve provides a true picture of the real molar mass distribution, or whether it was affected by shearing degradation. Finding out whether shearing degradation has occurred requires analysis under different conditions (flow rate, particle size of column packing) or crosschecking with batch light scattering. The critical molar mass at which shearing degradation occurs can be to a certain extent controlled by SEC separation conditions. The use of columns packed with 20-μm packing at a flow rate of 0.2–0.5 mL/min can especially extend the shearing degradation limit far beyond 10^6 g/mol. However, the possibility of shearing degradation in SEC columns can never be completely eliminated. The likelihood of shearing degradation is significantly reduced in A4F, where the system backpressure is several times lower than in SEC columns. For example, two 300×8-mm columns packed with 5-μm packing at room temperature with THF at 1 mL/min yield pressure of about 60 bar, whereas the pressure in the A4F channel is mostly below 10 bar; thus the method can be supposed to be relatively nondestructive.

Another very important advantage of A4F is that the lack of stationary phase strongly eliminates the enthalpic interactions of macromolecules with column packing. Various polymers containing polar functional groups have a strong tendency to interact with SEC columns and thus to suppress the size-based separation. Proteins and various polyelectrolytes are other examples of important polymers that often show non–size exclusion separation mechanisms. Although the macromolecules analyzed by A4F are in contact with the semipermeable membrane of the accumulation wall, the total contact surface is much lower compared to the packed porous SEC columns and thus enthalpic interactions are significantly less likely. The SEC separation according to hydrodynamic size is also affected in the case of branched polymers due to their specific elution behavior (see Section 6.2.1). This problem does not exist in A4F separation.

The characterizations of (1) ultra-high-molar-mass polymers, supermolecular structures, and particles, (2) polymers having a tendency to enthalpic interactions with SEC columns, and (3) branched polymers represent the key application areas of A4F. However, the method offers other advantages that are outlined in the following.

The injection volume in SEC is limited to a maximum of about 200 μL per 300×8-mm column, and actually lower injection volumes in the range of 50–100 μL should be preferred. Injection of, for example, 200-μL solution at 1-mL/min flow rate takes 12 seconds. That means the molecules injected into the column at the end of injection are 12 s delayed after the molecules injected at the beginning. Using two regular 300×7.5-mm columns, the polymer molecules elute in the range of about 8–17 minutes. That means the 12 s injection time is obviously acceptable because the molecules have enough time to "catch" each other and sort out according to their hydrodynamic size. However, this simple example shows that it is impossible to significantly increase the injected volume without excessive contribution to the peak broadening. On the other hand, in A4F the injected molecules do not start their passage through the channel immediately after the injection; they stay close to injection inlet and are focused after the injection to a common starting line. This allows for increasing the injection volume to several milliliters without affecting the resolution. The possibility of large-volume injections may be useful for the analysis of samples that can be prepared only at very low concentrations (e.g., environmental samples) or that are available as a small drop. The drop can be diluted to a larger volume that can be easily handled and injected.

In SEC, resolution can be controlled mainly by using different numbers of columns or columns of different particle size and/or porosity. In A4F, resolution can be controlled by separation conditions and thus polymer samples of extremely broad molar mass range of $\approx 10^4 – 10^9$ g/mol can be analyzed using the same separation device.

Changing the solvent in SEC requires many hours to replace the solvent completely and to stabilize the baseline of the RI detector because typical SEC packing is swollen gel and thus it takes a long time for the old solvent to be completely replaced by the new solvent. The solvent change in A4F is fast because of low total system volume and no stationary phase. In addition, absence of stationary phase also eliminates possible column bleeding, which disturbs the signal of a light scattering detector.

SEC columns are relatively delicate devices that can be damaged by improper use (e.g., solvent evaporation during storage, use of incompatible solvent, injection of samples containing insoluble matters or sample precipitation in the columns, excessive pressure, or mechanical shocks). In addition, supermolecular structures and molecules interacting with column packing can be irreversibly retained in SEC columns, which not only causes loss of information about the sample under analysis, but also shortens column lifetime. The A4F channel is practically indestructible, and channel membrane, when contaminated, can be easily replaced at a cost that is a small fraction of that of an SEC column.

5.2 THEORY AND BASIC PRINCIPLES

Fundamental FFF theory can be found in the articles of Giddings and Wahlund and their coworkers,[3–6] books such as references 7 and 8, or the *FFF Handbook*,[9] which may be also used as sources of further references. In the context of A4F, one must point out the article in reference 6, which represents the fundamental description of theory related to the asymmetric channel.

Although by the applications and instrumentation A4F resembles liquid chromatography, there is a principal difference from any kind of liquid chromatography that is due to the absence of stationary phase. The separation in A4F is achieved with no stationary phase, solely by a flow in an empty channel where a perpendicular flow force is applied. The channel consists of two plates jointed together that are separated by a spacer. The bottom plate is permeable, made of porous frit covered by a semipermeable membrane with a typical cutoff of 5 or 10 kDa (kg/mol). The membrane is permeable for the molecules of eluent (the term *carrier* is often used in FFF literature), but impermeable for the polymer molecules and colloidal particles and therefore keeps the sample in the channel so that it is directed by flow to the channel outlet.

In the originally developed symmetrical design, the channel with both walls permeable to cross flow was used and the field was generated by the solvent being pumped into the channel through the upper wall and leaving the channel through the accumulation wall. An obvious disadvantage of this configuration is that it requires two HPLC pumps, which increases the cost of the apparatus as well as maintenance requirements and probability of system breakdown. In the asymmetric A4F equipment, the upper plate is impermeable and only the bottom plate is permeable. This design permits making all flows with a single pump.

A typical asymmetric channel and its main parts are depicted in Figure 5.1. In contrast to SEC, the sample is not injected directly into the main stream, but the channel has a separate inlet for sample injection. The channel spacer is cut out of a thin plastic (e.g., Teflon) sheet (Figure 5.2) of usual thickness of 350 μm, which is sandwiched between two blocks to create a ribbon-like channel. The cross section of the channel is rectangular and usually decreases from the channel inlet toward the channel outlet. The laminar flow of the carrier creates a parabolic flow profile within the channel (Figure 5.3); that is, the carrier moves more slowly closer to the channel walls compared to the channel center. The analyzed molecules and particles are driven by the cross flow toward the bottom wall of the channel. Diffusion creates a counteracting motion due to which smaller particles, which have higher diffusion coefficients, move closer to the channel center, where the axial flow is faster. The velocity gradient inside the channel separates the molecules and particles according to their hydrodynamic size in such a way that smaller molecules elute before the larger ones. This means that A4F separation is the opposite of SEC separation, in which the large molecules elute first.

The schematic view of the separation channel used in A4F is in Figure 5.3. The axial (longitudinal, channel) flow velocity profile is parabolic, with the axis of symmetry located in the center of the channel:[10]

$$v(x) = 6\langle v \rangle \left(\frac{x}{w} - \frac{x^2}{w^2} \right) \tag{5.1}$$

where x is the distance from the bottom (accumulation) wall, w is the channel thickness (i.e., distance between the accumulation wall and upper wall), and $\langle v \rangle$ is the average cross-sectional carrier velocity (cm/s) along the axis of the channel.

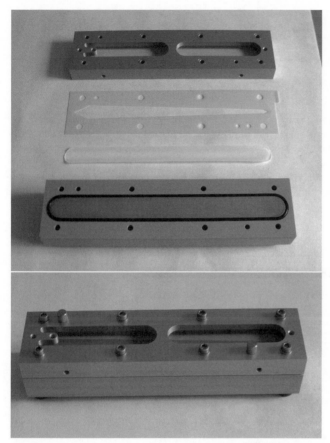

Figure 5.1 Components of A4F channel (top) and assembled channel (bottom). Components (from top to bottom): upper block consisting of metal part and transparent polycarbonate inlay, spacer, semipermeable membrane; bottom block with a frit and sealing o-ring. Dimensions: $290 \times 70 \times 50$ mm.
Source: Courtesy of Wyatt Technology Europe.

The equations given in the following text describe the main parameters of the separation process.[6] They can be used as guidelines for understanding the experimental results and optimizing the experimental variables such as channel flow rate, cross flow rate, channel length and thickness, and temperature.

The cross flow velocity (cm/s) as a function of the distance from the accumulation wall is:

$$u(x) = u_0 \left(1 - \frac{3x^2}{w^2} + \frac{2x^3}{w^3}\right) \tag{5.2}$$

where u_0 is the cross flow velocity at the accumulation wall. The cross flow velocity is zero at the upper wall and increases toward the bottom wall, as

Figure 5.2   Top view of A4F spacer (trapezoidal shape). Tip-to-tip length = 265 mm, length of the triangular tapered inlet = 20 mm, length of the triangular tapered outlet = 5 mm, breadth at the widest point of the tapered inlet = 22 mm, breadth at the widest point of the tapered outlet = 6 mm, area ≈ 36 cm².
Note: This channel shape and 5-kDa regenerated cellulose membrane were used to collect the experimental results presented in this chapter.
Source: Courtesy of Wyatt Technology Europe.

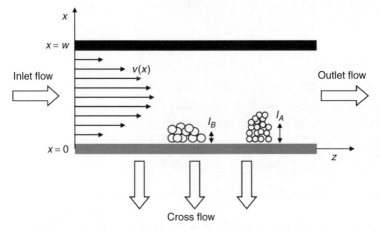

Cross flow

Figure 5.3   Schematic illustration of side view of A4F channel with zones of compounds A and B flowing in the direction of longitudinal axis z; w is the channel thickness; $v(x)$ is the flow velocity as a function of distance from the channel bottom (accumulation wall); l_A and l_B are the centers of gravity of zones of compounds A and B.

shown in Figure 5.4. The change of the cross flow velocity with the distance from the accumulation wall is a principal difference from the symmetrical flow FFF, where the cross flow velocity is constant across the channel.

It is worth noting that close to the accumulation wall the cross flow velocity is approximately constant, which means that for highly retained samples the separation can be described in the same way as in symmetrical flow FFF. As shown in reference 11, the cross flow velocity does not change along the longitudinal axis of the channel.

The longitudinal flow velocity as a function of axis z decreases linearly according to relation 5.3:[6]

$$\langle v \rangle = \langle v \rangle_0 - \frac{u_0}{w} z \tag{5.3}$$

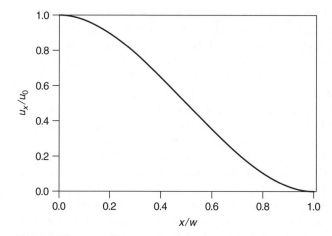

Figure 5.4 Cross flow velocity related to the cross flow velocity at the accumulation wall (u_0) as a function of distance x from the accumulation wall related to total channel thickness w (Equation 5.2).

where $\langle v \rangle_0$ is the average longitudinal flow velocity at the channel inlet. For a given channel thickness the steepness of the decrease of the longitudinal flow velocity is controlled by the cross flow velocity at the accumulation wall. Note that using the trapezoidal channel (see Figure 5.2) the decrease of the longitudinal flow velocity is compensated by the decreasing channel cross section. Comparison of velocity gradients in the rectangular and trapezoidal channels was published in paper[12]. The obtained results favor the trapezoidal channel over the rectangular one, and also the asymmetrical versus the symmetrical channel.

The concentration profile of the analyzed sample with respect to the distance from the accumulation wall can be approximated by:[6]

$$c = c_0 e^{-\frac{x}{l}} \tag{5.4}$$

where c_0 is the concentration at the channel accumulation wall and l is the distance of the center of the sample layer from the accumulation wall, equal to:

$$l = \frac{D}{u_0} \tag{5.5}$$

and D is the sample diffusion coefficient (cm^2/s). To account for different channel dimensions, the distance l is related to the channel thickness:

$$\lambda = \frac{l}{w} \tag{5.6}$$

Retention parameter λ defines the distance between the accumulation wall and the center of gravity of the sample zone relative to the channel thickness. As seen from Equations 5.5 and 5.6, the retention parameter is related to the properties of the analyzed molecules, the strength of the cross flow, and the channel thickness.

Assuming the usual cross flow of 3 mL/min and membrane area of about 36 cm^2, the corresponding $u_0 \approx 0.0014$ cm/s. The diffusion coefficients of polystyrene molecules of molar masses of 10^4 g/mol, 2×10^5 g/mol, and 2×10^6 g/mol are approximately 2×10^{-6} cm^2/s, 4×10^{-7} cm^2/s, and 1.2×10^{-7} cm^2/s, respectively. The distances l corresponding to these molar masses are 14.3 μm, 2.9 μm, and 0.9 μm, respectively; that is, assuming channel thickness of 300 μm the corresponding retention parameters λ are about 0.048, 0.010, and 0.003. It is evident that even for relatively small molecules with molar mass close to the cutoff of membrane the center of the sample layer is in close proximity to the channel bottom.

The concentration profiles for polystyrene molecules of different molar mass are shown in Figure 5.5. The curves illustrate how the sample concentration is distributed across the channel and how it depends on the diffusion coefficient. The injected sample forms a layer whose thickness is determined by the diffusion coefficient and the cross flow velocity.

The effect of the cross flow on the concentration profile is demonstrated in Figure 5.6. The concentration of monodisperse molecules in the layer decreases exponentially with the distance from the accumulation wall. For a polydisperse sample, the total concentration profile is a superimposition of the profiles of individual species in the sample and the molecules are distributed within the layer according to their diffusion coefficients. It has been shown in reference 6 that for identical λ the concentration profile in the asymmetrical channel is less compressed against the accumulation wall than that in the symmetrical channel, but the difference is negligible for λ less than about 0.1.

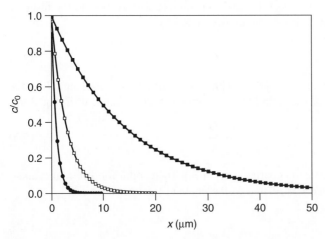

Figure 5.5 Concentration profiles for three polystyrene molecules of molar mass 10^4 g/mol (■), 200×10^3 g/mol (□), and 2×10^6 g/mol (•). Centers of sample layers $l = 14.3$ μm (■), 2.9 μm (□), and 0.9 μm (•). Cross flow = 3 mL/min, accumulation wall area = 36 cm^2.

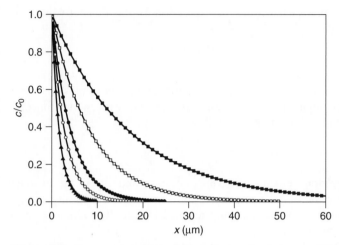

Figure 5.6 Concentration profiles for polystyrene of molar mass of 200,000 g/mol at various cross flow rates: 0.5 mL/min (■), 1 mL/min (□), 2 mL/min (•), 3 mL/min (○), and 5 mL/min (▲). Accumulation wall area $= 36$ cm^2.

In real experiments, the volumetric flow rates in mL/min are measured instead of velocities in cm/s:

$$u_0 = \frac{\dot{V_c}}{bL} \tag{5.7}$$

$$\langle v \rangle_0 = \frac{\dot{V}_{in}}{bw} \tag{5.8}$$

$$V_0 = bLw \tag{5.9}$$

where V_0 is the channel volume; $\dot{V_c}$ and $\dot{V}_{in}$ are the volumetric cross flow and the volumetric inlet flow (at $z = 0$), respectively; and L, b, and w are the channel length, the channel breadth, and the channel thickness, respectively. The product Lb equals the membrane surface, but for a channel with uneven channel breadth the membrane surface must be calculated according to a given spacer shape. All the flow rates can be easily measured directly and their mutual relation is:

$$\dot{V}_{in} = \dot{V_c} + \dot{V} \tag{5.10}$$

where $\dot{V}$ is the volumetric channel outlet flow (detector flow) at $z = L$. Substituting Equations 5.7 and 5.9 into the expressions for the retention parameter (Equations 5.5 and 5.6) we get:

$$\lambda = \frac{DV_0}{w^2 \dot{V_c}} \tag{5.11}$$

Equation 5.11 can be rearranged using the expression for the channel volume to the form:

$$\lambda = \frac{DbL}{V_c w} \tag{5.12}$$

Retention ratio, which describes the retention of the molecules of given diffusion coefficient compared to the molecules that are under given experimental conditions unretained by the applied cross flow, is defined as:

$$R = \frac{v}{\langle v \rangle} \tag{5.13}$$

where v is the migration velocity of given molecules and $\langle v \rangle$ is the average longitudinal carrier velocity. The retention ratio can range from unity for an unretained sample to zero for a completely retained sample. The retention ratio can be expressed as follows:[13]

$$R = 6\lambda \left(coth\frac{1}{2\lambda} - 2\lambda \right) \tag{5.14}$$

The bracketed function of hyperbolic cotangent approaches unity with decreasing λ and thus for many practical cases the retention can be described by the simple relationship:

$$R = 6\lambda \tag{5.15}$$

or

$$R = 6\lambda - 12\lambda^2 \tag{5.16}$$

Equation 5.15 is accurate within 5% when $\lambda < 0.02$; Equation 5.16 is accurate within 2% when $\lambda < 0.2$. The above equations show that retention is exclusively controlled by the retention parameter, that is, by the diffusion coefficient related to the cross flow velocity and channel thickness. The term *retention* refers to the fact that the particles are retained by forcing them close to the channel bottom into streamlines that move with slower velocity than the average. Alternatively, the retention ratio can be expressed as:

$$R = \frac{t_0}{t_R} \tag{5.17}$$

where t_0 is the void time of unretained component travelling with the average carrier velocity, t_R is the retention time of the investigated component. The retention time in A4F can be approximated by the following expressions:[6]

$$t_R = \frac{w^2}{6D} \ln\left(1 + \frac{\dot{V}_c}{\dot{V}}\right) = \frac{w^2 \pi \eta R_h}{kT} \ln\left(1 + \frac{\dot{V}_c}{\dot{V}}\right) \tag{5.18}$$

$$t_R = \frac{w^2}{6D} \ln\left(\frac{z_f/L - \dot{V}_{in}/\dot{V}_c}{1 - \dot{V}_{in}/\dot{V}_c}\right) = \frac{w^2 \pi \eta R_h}{kT} \ln\left(\frac{z_f/L - \dot{V}_{in}/\dot{V}_c}{1 - \dot{V}_{in}/\dot{V}_c}\right) \tag{5.19}$$

where w is the channel thickness, D is the diffusion coefficient related with the hydrodynamic radius R_h by Stokes-Einstein relation, k is the Boltzmann's constant, T is the absolute temperature, η is the viscosity of the carrier, z_f is the position of the focusing point, and $\dot{V}_{in}$, $\dot{V}_c$ and $\dot{V}$ are the inlet flow rate, the cross flow rate and the detector flow rate, respectively. Equation 5.18 applies for the case when sample elution starts directly at the channel inlet, i.e., $z_f = 0$. Note that for given experimental conditions, the sample components are separated solely according to their hydrodynamic radius, since R_h is the only quantity in Equations 5.18 and 5.19 that is characteristic of the analyzed molecules. Molar mass, which is a far more desirable quantity for the characterization of polymer samples, is related to the diffusion coefficient by Equation 1.20.

The A4F analysis consists of three steps: (1) *injection*, (2) *focusing* and *relaxation*, and (3) *elution*. During the first two steps, injection and focusing, the flow delivered by the HPLC pump is split, enters the channel from the inlet and outlet, and is balanced to meet near the injection port. At this step the eluent moves only in cross direction, permeates the membrane, and exits from the cross outlet of the channel. The sample moves and becomes focused at a certain distance from the channel inlet as a narrow line, from which the elution begins after the flow is switched into the elution mode. The focusing step can be visualized by injection of a colored polymer (blue dextran can be used for this purpose; see Figure 5.7). After complete injection of the sample solution, the injection flow is stopped and the sample is focused for several additional minutes before the third step (elution), when the eluent flows from the channel inlet to the channel outlet that is connected to detectors. Sample focusing is the transport of the injected sample axially to the focusing line. The time needed for this step (focusing time t_f) was derived in reference 6:

$$t_f \approx \frac{w^2 \ln(1 - f)}{-6D} \tag{5.20}$$

where f is the factor <1 (e.g., 0.99). Equation 5.20 shows that the focusing time is independent of the flow rate, and it is directly proportional to the square of channel thickness and indirectly proportional to the diffusion coefficient. It is evident that for high-molar-mass components with low diffusion coefficients the focusing time might range from minutes to hours. The position of the focusing line can be controlled by the ratio of the flow rates entered into the channel inlet and outlet:

$$z_f = L \frac{\dot{V}_{in,f}}{\dot{V}_{in,f} + \dot{V}_{out,f}} \tag{5.21}$$

where z_f is the position of the focusing line, L is the channel length, $\dot{V}_{in,f}$ is the flow rate that enters the channel inlet and $\dot{V}_{out,f}$ is the flow rate that enters the channel from the outlet, and the index f refers to the focusing and

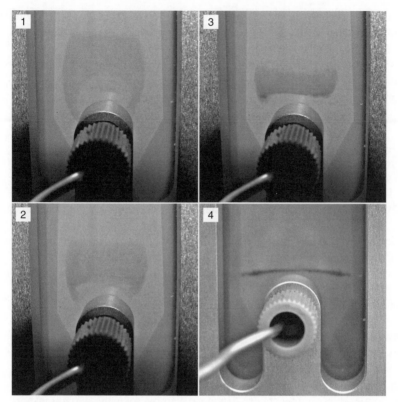

Figure 5.7 Images taken during the focusing of blue dextrane at approximately (from 1 to 4) 3 min, 5 min, 7 min, and 18 min after the injection. Channel thickness ≈ 300 μm.

relaxation process. To maximize the length of the channel for elution and separation, the focusing point should be near the channel inlet (i.e., a few millimeters from the injection inlet toward the channel outlet; see (Figure 5.7). In order to achieve the position of the focusing point near the channel inlet, $\dot{V}_{out,f}$ must be $\gg \dot{V}_{in,f}$.

Simultaneously with the focusing, the sample is concentrated near the accumulation wall. At the time of injection into the channel the sample is distributed across the entire channel thickness. After the injection, the sample moves toward the accumulation wall with a velocity equal to that of the cross flow; the focus flow during the sample injection and subsequent focusing period causes the sample relaxation. Relaxation means establishing the concentration equilibrium against the accumulation wall. The relaxation time (τ) needed for complete sample relaxation in symmetric flow FFF is given by the ratio $V_0/\dot{V}_c$. Since in asymmetric flow FFF the cross flow velocity decreases with increasing distance from the accumulation wall and approaches zero at the upper wall, the species close to the upper wall would require very long times to be transported across the channel; that is, a part of the injected sample would create an almost stagnant layer at the proximity of the upper wall. However, even the molecules that are situated very close to the upper wall do not stay there indefinitely, because they are driven from this stagnant layer by diffusion. The combination of both cross flow and diffusion can be described by the following equation:[6]

$$\tau \approx \left(\frac{V_0}{3\dot{V}_c} \right)^{\frac{2}{3}} \left(\frac{w^2}{mD} \right)^{\frac{1}{3}} \tag{5.22}$$

where m is a constant of order unity. For example, for polystyrene molecules of $M = 2 \times 10^6$ g/mol, using a typical channel of $V_0 \approx 1$ mL, $w \approx 300$ μm, and $\dot{V}_c = 3$ mL/min, the relaxation time is roughly 70 sec.

5.2.1 Separation Mechanisms

The major A4F separation mechanism is called *normal* or *Brownian separation* and governs submicrometer colloidal particles and macromolecules. The normal mode is based on the transport of relatively small particles driven by the cross flow toward the accumulation wall and their concentration in the proximity of the wall. The increase of concentration near the wall creates a concentration gradient that causes the diffusion away from the wall. As the transport of molecules continues and the concentration gradient increases, the counteracting diffusion increases proportionally until it balances the cross flow–driven transport toward the wall. The balance of the two counteracting transport processes forms a cloud of particles having an equilibrium concentration distribution. The concentration of particles decreases exponentially with distance from the accumulation wall and different components have different concentration profiles (Figure 5.5). The separation in normal mode is based on the fact that particles of different diffusion coefficients form layers of different thickness and different mean elevation above the accumulation wall. In parabolic flow profile the thickest layers flow fastest, followed by layers that are more and more compressed to the accumulation wall. Thus the normal elution order is from the smallest to the largest particles.

In the case of much larger particles with diameter approximately above 1 μm, the diffusion coefficients are very low, the Brownian motion becomes negligible, and the diffusion does not create a sufficient counteracting force against the flow force. The large particles are driven by the cross flow directly to the accumulation wall and stay there in direct contact with the wall. The layer thickness in this case is controlled solely by the geometrical dimensions of the particles; the larger particles extend further into the faster flowing streamlines of the channel. This is called *steric separation*, and the separation order is inverted compared to the normal separation (i.e., larger particles elute before the smaller ones). This idealized model of steric separation is affected by hydrodynamic lift forces that disturb direct contact of particles with the wall. The hydrodynamic forces increase with increasing channel flow rate as well as with increasing particle size. The hydrodynamic lift forces are oriented from the wall and thus they move the particles away from the accumulation wall. The movement of the particles from the wall is counteracted by the cross flow until equilibrium is reached. As a consequence of the balance of hydrodynamic lift forces and cross flow, particles of different sizes form narrow bands at different elevations above the wall. This mechanism is called *hyperlayer separation*. Retention in hyperlayer A4F is given by the ratio of the cross flow strength and the counteracting lift forces. The elution order is identical as in the case of steric separation.

Both separation modes are illustrated in Figure 5.8. In real experiments, it is practically impossible to distinguish between steric and hyperlayer separation, because they are closely related and produce similar separation results. Also the inversion point between the normal and steric-hyperlayer separation cannot be expected to be sharp. The steric-hyperlayer separation mode can occur simultaneously with the normal separation. This can happen in the case of polydisperse samples with a broad size distribution that spans across the steric inversion diameter. In the case of such samples the smaller components elute in the normal mode and the larger in the steric-hyperlayer mode. The co-elution of normally and sterically migrating components of a polydisperse sample increases the polydispersity of fractions eluting at a given time from the channel.

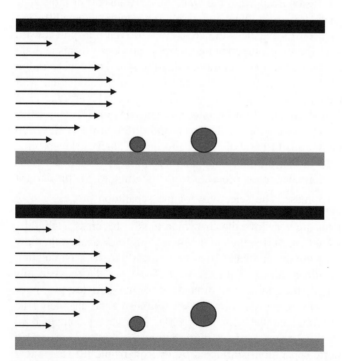

Figure 5.8   Schematic illustration of steric separation (top) and hyperlayer separation (bottom).

The steric inversion diameter, at which the transition between the normal and steric-hyperlayer modes occurs, has been reported between 0.3–3 μm. [14,15] It can be shifted by changing experimental conditions.[16] For polymers, the separation conditions are mostly chosen so as to work completely in the normal separation mode before the inversion point. When steric-hyperlayer separation becomes active for high-molar-mass fractions, the normal mode can be enhanced by raising the temperature, that is, intensifying Brownian movement, using channels of larger thickness (i.e., making the molecules relatively smaller), or decreasing $\dot{V}$ and $\dot{V}_c$. Although the steric-hyperlayer mode is usually unwanted in the case of separation of polydisperse polymers, it can be effectively used for the separation of larger particles.

For midsized particles, where both separation mechanisms play role in retention, the retention ratio can be written as:[14]

$$R = 6\lambda + 3\frac{\gamma d}{w} \tag{5.23}$$

where γ is the dimensionless steric correction parameter of the order of unity and d is the diameter of the particles. For small particles, the second term of the above equation becomes negligible and the retention can be approximated by Equation 5.15. For very large particles, the retention parameter becomes insignificant because of very slow diffusion, while the ratio d/w becomes significant. The factor γ can be set as unity to simplify the above expression. The values of $\gamma > 1$ apply to hyperlayer separation when hydrodynamic lift forces lift the particles away from the wall.

5.2.2 Resolution and Band Broadening

As with SEC, the A4F technique is often used to separate and characterize the molar mass distribution of polydisperse polymers. The *selectivity* of the separation process can be defined in the same way as in SEC by the slope of molar mass–versus–retention (elution) time dependence. This slope reflects how well a separation device (A4F channel, SEC column) can separate monodisperse polymers of different molar mass. The selectivity of the separation process can be defined as:

$$S = \left| \frac{d(\log t_R)}{d(\log M)} \right| \tag{5.24}$$

where the absolute value reflects the fact that retention time increases (in A4F) or decreases (in SEC) with molar mass. Selectivity according to Equation 5.24 is based on molar mass. Alternatively, selectivity can be defined in terms of size:

$$S = \left| \frac{d(\log t_R)}{d(\log d)} \right| \tag{5.25}$$

where d is the particle diameter. Indexes M or d can be used to distinguish the selectivity based on molar mass or size, respectively.

The relation between selectivity and sample retention is given by the following equation:[17]

$$S = \left| 3 \left(\frac{R}{36\lambda^2} + 1 - \frac{1}{R} \right) \right| \left| \frac{d(\ln \lambda)}{d(\ln M)} \right|$$ (5.26)

The above relation shows that the value of S increases as R decreases and approaches its maximum value S_{max} at low R (high retention) when Equation 5.15 becomes a true approximation of the relation between R and λ. The maximum selectivity:

$$S_{max} = \left| \frac{d(\ln \lambda)}{d(\ln M)} \right|$$ (5.27)

equals the exponent β of the molar mass dependence of diffusion coefficient as evident from Equation 5.11 and molar mass dependence of diffusion coefficient (Equation 1.20). That means the selectivity depends on the conformation of the sample being analyzed and thermodynamic quality of the solvent and typically varies in the range of about 0.3–0.7. The dependence of selectivity on molecular conformation indicates that selectivity will decrease with increasing degree of branching. Generally, the typical selectivity of A4F is markedly more than that of SEC (generally around 0.1).

A graphical illustration of Equation 5.26 is shown in Figure 5.9. The comparison of selectivity of A4F and of SEC obtained for a typical polydisperse polymer measured under typical A4F and SEC experimental conditions is shown in Figure 5.10. The data were obtained by a MALS detector and a commercially

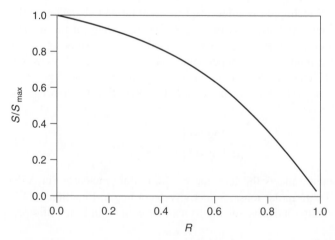

Figure 5.9   A4F selectivity as a function of retention. S_{max} is the selectivity in the limit of very high retention that equals the exponent β in the molar mass dependence of diffusion coefficient.

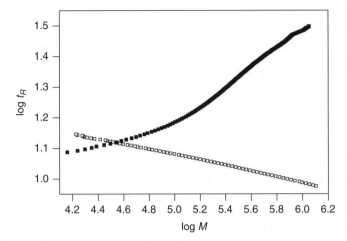

Figure 5.10   Selectivity plot log t_R versus log M for broad linear polystyrene analyzed by SEC-MALS (□) and A4F-MALS (■) using typical separation conditions. SEC: $2 \times 300 \times 7.5$ mm PLgel Mixed-C 5-μm columns, injection 100 μL 0.25% w/v, THF at 1 mL/min, 35°C; A4F: detector flow 1 mL/min, cross flow 3 mL/min to 0.15 mL/min within 30 min plus 5 min at 0.15 mL/min, injection 100 μL 0.25% w/v, 350 μm spacer, temperature 60°C. SEC/A4F slopes = 0.08/0.10, 0.08/0.21, 0.09/0.35, and 0.1/0.30 for log M range of 4.3–4.8, 4.8–5.2, 5.2–5.6, 5.6–6.0, respectively.

available A4F channel and high-performance SEC columns. Selectivity for a certain molar mass range is determined as the slope of the plot log t_R versus log M. Selectivity can also be estimated from the plot of log M versus log t_R (i.e., SEC calibration curve). Then a decrease of the slope indicates increased selectivity.

High selectivity means that there is a significant change of retention time with a small change of molar mass or particle size. However, the quantity that describes the quality of separation of two components is the *resolution* that takes into account both selectivity and *efficiency* (i.e., *band-broadening* effects).

The resolution of the two components can be expressed as:[18]

$$R_S = \frac{\Delta t_R}{4\bar{\sigma}} = \frac{1}{4} \frac{\Delta R}{\bar{R}} \left(\frac{L}{\bar{H}} \right)^{\frac{1}{2}} \tag{5.28}$$

where Δt_R is the difference in retention time of the two components, $\bar{\sigma}$ is the average standard deviation of the two peaks, ΔR is the difference of retention ratio, $\bar{R}$ and $\bar{H}$ are the average values of R and H for the two components, and L is the channel length. This equation shows that resolution increases with $N^{1/2}$. Band broadening (also called *zone broadening*, *zone dispersion*, *zone spreading*, *axial dispersion*) is expressed in terms of *plate height* (H) or number of theoretical plates $N = L/H$. High selectivity means high resolution only if zone broadening is small, because high selectivity can be distorted by excessive band broadening.

There are several contributions to H, for example, axial diffusion, nonequilibrium, polydispersity, instrumental effects (channel irregularities, reversible adsorption of components on the accumulation wall, overloading), and contributions outside the channel such as spreading in connecting tubings and detector cells. Note that unlike SEC the contribution caused by injection of finite sample volume is eliminated by sample focusing.

The contribution of the axial diffusion can be expressed as:

$$H_d = \frac{2D}{R\langle v \rangle} \tag{5.29}$$

where R is the retention ratio, D is the diffusion coefficient, and $\langle v \rangle$ is the average carrier velocity. The effect of axial diffusion increases with increasing sample retention. However, the contribution of H_d is usually negligible because polymers have generally low diffusion coefficients.

For highly retained components (small λ) the nonequilibrium contribution to plate height can be approximated as:[19]

$$H_n = 24\lambda^3(1 - 8\lambda + 12\lambda^2)\frac{w^2\langle v \rangle}{D} \tag{5.30}$$

In the case of lower retention levels, the above equation includes more complicated functions of λ, while for $\lambda \to 0$ the bracketed term in Equation 5.30 can be neglected. Nonequilibrium is the most important source of band broadening in A4F, which originates from the fact that particular molecules of the sample layer flow in the channel at different velocities. The molecules of identical hydrodynamic radius do not form a thin layer at a given streamline, but they protrude from the wall to streamlines of different velocity (see Figure 5.5). As a consequence of that, molecules of a given kind do not move toward the channel outlet with the same velocity, but they move with the velocity given by the distance x of a particular streamline from the accumulation wall. Different velocities of different streamlines spread the sample axially, because the molecules located closer to the channel center forerun those located closer to the bottom wall. The concentration gradients created by spreading are counteracted by tangential diffusion so that the rapidly moving sample components diffuse toward the accumulation wall while slowly moving components diffuse away from the wall.

The way to reduce nonequilibrium dispersion is to reduce λ, that is, to form more compact sample layers by increased cross flow. The nonequilibrium dispersion is also reduced by using small channel thickness w and low carrier velocity. Note that in A4F $\langle v \rangle$ changes along the longitudinal axis and the time-average carrier velocity $\langle \bar{v} \rangle$ must be used in the relations for plate height.

A significant contribution to plate height is caused by sample polydispersity. In contrast to other plate height contributions, this broadening actually represents the separation process and is not destructive. In fact, sample polydispersity is not a true broadening factor; H_p represents band broadening as a result of true sample separation. Typically, the polydispersity term of plate height is nonnegligible even for samples of low polydispersity, which indicates the high resolving power of A4F. For samples of low-to-moderate polydispersity, the contribution of polydispersity to plate height is:[20]

$$H_p = LS^2 \left(1 - \frac{1}{\mu} \right) \tag{5.31}$$

where μ is polydispersity M_w/M_n, and selectivity S can be calculated from the average retention ratio via Equation 5.26, or determined experimentally from the slope of log t_R–versus–log M relation.

The general expression for the plate height is given as the sum of individual contributions:

$$H = H_d + H_n + H_p + \Sigma H_i \tag{5.32}$$

where subscripts d, n, p, and i indicate contributions of axial diffusion, nonequilibrium, polydispersity, and the sum of other contributions, respectively. The nonequilibrium term shows a significant dependence on carrier velocity (Equation 5.30). That means extrapolation of H to zero velocity eliminates the nonequilibrium contribution to the plate height. Then, assuming H_d and ΣH_i to be negligible, one can determine H_p from the intercept at zero velocity and thus calculate the sample polydispersity from Equation 5.31. The determination of plate height can be performed by the usual procedure based on peak retention time and baseline width or half-height width.

Another parameter describing the separation in A4F is *fractionating power*, which can be defined on the basis of diameter (d) or molar mass (M):[21]

$$F_d = \frac{R_S}{dd/d} = \frac{d}{4\sigma} \frac{dt_R}{dd} = \frac{t_R}{4\sigma} S_d \tag{5.33}$$

$$F_M = \frac{R_S}{dM/M} = \frac{M}{4\sigma} \frac{dt_R}{dM} = \frac{t_R}{4\sigma} S_M \tag{5.34}$$

where d is the particle diameter, M is the molar mass, dt_R/dd and dt_R/dM are the changes in retention time with d or M, respectively, and σ is the standard deviation of the peak. According to the above equations the fractionating power can be understood as the resolution between particles whose diameters differ by the relative increment dd/d, or whose molar masses differ by the relative increment dM/M.

5.3 INSTRUMENTATION

In asymmetric flow FFF instruments, splitting the main flow delivered by a chromatographic pump into the channel generates the cross flow. The A4F instrumentation is currently manufactured by Wyatt Technology Europe and Postnova. The Postnova AF2000 system utilizes (along with the main pump) a second syringe pump, which withdraws liquid from the channel cross flow outlet, and a third, separate pump that is used for sample injection. The Wyatt Technology Europe Eclipse™ A4F system is based on a single HPLC pump to generate three separate flow streams (detector flow, cross flow, and injection flow). The Eclipse™ system is based on an Agilent or a Shimadzu HPLC pump. Using a single pump from a renowned HPLC manufacturer benefits the reliability and maintenance requirements.

The Eclipse™ A4F separation system consists of a channel and a control chassis that includes a CoriFlow device, a LiquiFlow device, two software-controlled motor-driven needle valves, two Rheodyne switching valves, and two pressure sensors. The cross flow generates at the channel inlet; that is, the flow entering the channel flows in two directions: channel (detector) direction and cross direction. The motor-driven switching valves are used to direct the flow during the three phases of the separation process (focusing and relaxation, injection, and elution). To achieve sufficient and stable cross flow the backpressure in the detector direction must be several times higher than the backpressure across the membrane. The pressure in the channel is also useful for extending the boiling temperature of volatile solvents such as THF, which can be safely used up to a temperature of about 90°C.

A simplified scheme of the single-pump Eclipse™ A4F instrument is shown in Figure 5.11. In the focusing mode, the flow delivered by the HPLC pump flows through tubing 1 and splits into lines 2, 3, and 4. The inject needle valve is closed and the channel is entered from the inlet (line 2) and from the outlet (line 3). The split of the focusing flow into the channel inlet and outlet is adjusted with a software-controlled motor-driven focus needle valve. During the injection mode, the inject system (consisting of a needle valve and a LiquiFlow measuring device) is opened and flow is controlled at specified flow rate (usually 0.2 mL/min), which delivers the sample from the loop of autosampler or manual injector into the channel inject port (line 5). When the injection is completed, the inject system is closed while the flow remains flowing in lines 2, 3, and 4 for sample focusing and relaxation.

Note that in the focusing mode the detector flow actually bypasses the channel. During the elution mode the inject line is closed and the solvent flows from the pump via line 1 into the channel inlet (line 2) and the channel outlet flows through lines 3 and 4 into detectors. The cross flow rate (line 6) is controlled and measured by a CoriFlow device. If a cross flow gradient is applied, the flow from the pump is gradually decreased along with the equivalent decrease of the flow through the CoriFlow controller to keep the detector flow constant.

In general, an A4F instrumental setup is similar to that used in SEC, only SEC columns are replaced by a channel and a chassis. Similarity between SEC and A4F allows shearing the pump, injector, detectors, and measuring and processing software. In other words, an A4F instrumental setup can be easily converted to an SEC setup by disconnecting the channel and chassis and connecting columns instead. This is a simple procedure than can be done manually in a few minutes. In addition, the Eclipse™ A4F system offers a GPC option, which allows easy

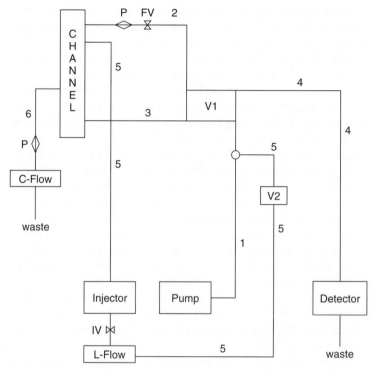

Figure 5.11 Simplified scheme of A4F instrument Eclipse™ utilizing a single HPLC pump to generate all flows. C-Flow = CoriFlow Coriolis style mass flow meter/mass flow controller, L-Flow = LiquiFlow liquid flow meter operating on a thermal through-flow measuring principle, P = pressure sensor, FV = focus valve, IV = injection valve, V1, V2 = Rheodyne valves.
Source: Courtesy of Wyatt Technology Europe.

switching between the channel and columns using an additional switching valve controlled by the software. When SEC columns are not being used, they can be continuously flushed by an additional pump connected to the chassis.

The profile of the channel is cut in the spacer. The channel shape is usually a trapezoid with two tapered ends. Due to the trapezoidal shape the channel breadth decreases toward the outlet, which maintains the longitudinal flow velocity despite the loss of the carrier through the membrane. In addition, peak dilution is reduced compared to a rectangular channel. The channel parameters are tip-to-tip length, thickness, length of the triangular tapered inlet, length of the triangular tapered outlet, breadth at the widest point of the tapered inlet, and breadth at the widest point of the tapered outlet.

The top and bottom channel walls must be parallel to assure a uniform parabolic flow profile across the channel breadth. It is also necessary that the walls are smooth, even though the membrane surface is always sort of irregular concerning porosity and roughness. An ultrafiltration membrane made of regenerated cellulose can be used for both aqueous and organic solvent applications. Other membrane materials include, for example, polyethersulfone or cellulose acetate.

The membrane must be compatible with the carrier and must not swell or dissolve. The membrane must not interact with the sample since the interactions would affect retention. The interactions depend on the membrane surface chemical composition, composition of the sample, and eluent properties. Although sample–membrane interactions represent an issue for some samples, interactions-related problems are significantly less frequent compared to SEC. The most important parameter of the membrane is cutoff, which must be low enough to keep the sample in the channel. The typical membrane cutoffs are 5 kDa, 10 kDa, or 30 kDa. The nominal cutoffs provide only a rough estimation of real molar masses that will be retained by the membrane, because the critical parameter is sample size. Thus various polymers will have different molar mass cutoffs according to their actual chain dimensions. The semipermeable membrane consists of a thin skin with very fine, regular pores that is supported by a significantly thicker layer with large pores. It is important to make sure that the thin, polished side of the membrane faces the channel. The membrane is supported by a frit; when the channel is mounted together the membrane is compressed by the spacer, whereas in the area where the membrane is not in contact with the spacer it maintains its original thickness. As a result of compression by the spacer, the membrane protrudes into the channel and the effective channel thickness is less than the nominal spacer thickness.

Membrane compression can be easily measured by a micrometer after channel disassembly. The experience shows that the regenerated cellulose membrane is typically compressed by 60–90 μm. When the actual channel thickness is of interest, as in the case of the determination of retention ratios or λ parameters (and consequently diffusion coefficients and hydrodynamic radii), it can be determined from the retention time of a compound of well-known diffusion coefficient and Equation 5.19. The proteins bovine serum albumin or ferritin in aqueous solvents and narrow polystyrene standards in organic solvents can be used for the determination of the actual channel thickness. However, when the A4F system is coupled with a MALS detector the exact channel thickness is actually not needed.

As already mentioned, an A4F instrument is similar to an SEC (HPLC) instrument with the only difference that of using channels instead of columns. Although the HPLC pump, injector, connecting tubings, and detectors are identical with those used in SEC, the channel pressure is substantially lower than in SEC columns. Most HPLC pumps require a certain backpressure for proper check-valve operation. Nevertheless, this requirement is usually fulfilled since pump pressure for the entire system from injector to detector is usually sufficient (for THF and detector flow of 1 mL/min the pressure is around 20 bar and more when the cross flow is applied). When needed for the sake of proper cross flow, a short piece of 0.13-mm ID tubing connected before the detector can be used to increase the pressure in the detector direction (1 m of 0.13-mm ID tubing at 1-mL/min THF yields pressure of about 12 bar).

Submicrometer particles in the carrier can completely disturb any information obtainable from the MALS detector. SEC columns are a potential source of particles bleeding from the packing material, but when the column system gets flushed and stabilized, the SEC columns actually serve as an efficient eluent filter. Since the packing material is not used in A4F, both effects are missing. The channel is usually not a serious source of particles, but on the other hand, the filtration effect is missing. Therefore, the carrier must be filtered with an online filter placed between the pump and injector. When aqueous buffers are used, pre-filtration with 0.1 μm should be performed. However, the pre-filtration is not needed in the case of organic solvents such as THF. As a result of using two flows, the cross flow being usually several times higher than the detector flow, the solvent consumption is significantly higher than in SEC. That means the solvent reservoirs and waste bottles should be of larger volume than those regularly used in SEC.

Similarly as in SEC, refractive index (RI) and UV detectors are the most frequently used concentration-sensitive detectors in A4F (the principles of particular detectors are described in Section 3.3.3). Other detectors for polymer analysis include online viscometer or evaporative light scattering detectors, and in principle other HPLC detectors can be used. An interesting application is the characterization of aquatic colloids in natural water using an ICP-MS (inductively coupled plasma-mass spectrometry) detector.[22,23]

5.4 MEASUREMENTS AND DATA PROCESSING

The first step in an A4F experiment is sample preparation, which is similar to that used for SEC (Section 3.5.1). As in SEC, the major requirement of successful determination of molar mass distribution is sample solubility. Although A4F can measure particles up to several tens of micrometers, filtration of polymer solutions before injecting into the A4F system is strongly recommended. The filtration can remove any large particles or swollen gels that might block connecting or inner-detector tubings. Some polymer samples can contain trace amounts of fractions with ultra-high molar mass and size beyond the steric inversion diameter, which can co-elute with the main polymer peak and disturb the light scattering signal by intensive spikes. The co-elution of very large particles with the main polymer peak occurs due to inversion of the separation mechanism from normal to steric when particles become too large. If the species disturbing the light scattering signal cannot be completely separated from the polymer peak, they can be removed prior to analysis by filtration using a filter size that is large enough for polymer molecules and small enough for the components that are intended to be removed.

Figure 5.12 shows the smoothing effect of filtration on the light scattering signal without noticeable effect on the molar mass distribution. However, the filtration step can strongly affect the molar mass distribution, as demonstrated in Figure 5.13. Selection of an appropriate filter size may not always be straightforward and samples containing high-molar-mass fractions should be filtered with

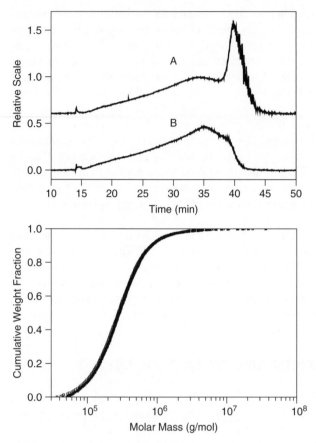

Figure 5.12 Effect of sample filtration: light scattering fractograms (top) and cumulative distribution curves (bottom) for a sample of polybutadiene rubber. Sample solution filtered with 0.45-μm ($\square$, B) or 1-μm ($\blacksquare$, A) filter.

different filters in order to select a suitable filter porosity that will not corrupt the molar mass distribution, but will remove the species disturbing the light scattering signal. Since one of the strong features of A4F lies in the analysis of polymers containing ultra-high-molar-mass species, the filtration step of sample preparation represents a generally more important issue than in the case of SEC, where filtration with a 0.45-μm filter is usually appropriate.

In contrast to SEC, the A4F separation technique allows measurement of insoluble particles. In this case the sample must be completely dispersible in the carrier with no aggregation and sedimentation when left in a vial for several hours.

In the simplest case, the measurements are carried out solely with the concentration-sensitive detector and the direct comparison of fractograms from similar samples measured under identical conditions is used to determine qualitative

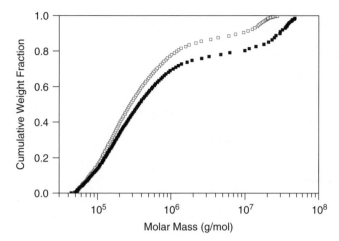

Figure 5.13   Effect of sample filtration: molar mass distribution plots of styrene-butadiene rubber determined by A4F-MALS with sample filtration using 0.45-μm (□) or 5-μm (■) filter.

differences among samples. For example, one can identify the presence of high-molar-mass fractions, polymer blends, or nanoparticles used as polymer modifiers. For more detailed analysis, the retention time axis must be converted to molar mass or particle size. The particle size can be theoretically determined directly from the retention data and molar mass can be calculated from the relation between the molar mass and diffusion coefficient. Alternatively, calibration curves can be established on the basis of standard materials (i.e., polymer narrow standards or particle size standards). However, direct measurement of molar mass by a MALS detector is undoubtedly the most accurate and reliable way to obtain molar mass, molecular size, and conformation information. Determination of molar mass and RMS radius distribution with a MALS detector is identical with that described in Chapter 4.

Similarly as in SEC-MALS, the M_w results are independent of the separation efficiency, while other molar mass averages, molar mass distribution curves, and conformation plots are calculated on the basis of the assumption that each data slice is monodisperse. Thus it is important to choose separation conditions that provide sufficient resolution for the sample to be analyzed. Data processing, that is, setting the baseline and peak integration limits, follows the same rules as those applied in SEC and SEC-MALS. Since oligomeric portions of polymer samples are lost through the accumulation membrane, the RI peaks do not show oligomeric tail or the peaks of low-molar-mass impurities, and the determination of the peak beginning is typically easier than the determination of the peak end in SEC. In contrast to SEC, the RI signal may be affected by switching from focusing to elution and by cross flow gradient. This RI baseline instability can be eliminated by subtraction of a blank signal from the run signal, as demonstrated in Figure 5.14. The blank signal is obtained by injection of pure solvent under identical conditions as those used for the sample analysis.

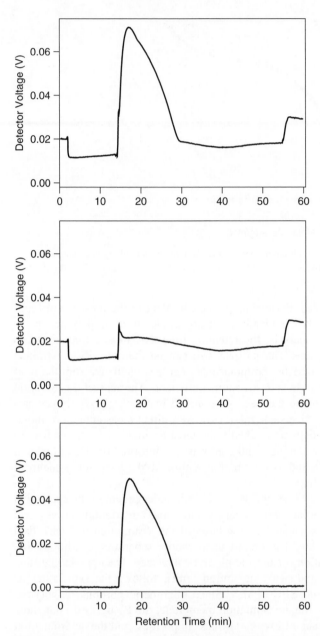

Figure 5.14   RI fractograms from analysis of polydisperse polystyrene. From top to bottom: raw signal, blank signal, and difference obtained by subtraction of the blank from the raw. Cross flow: 3 mL/min for 5 min and then linear gradient to 0.1 mL/min within 20 min + 10 min at 0.1 mL/min, 6 min at 0 mL/min, and 5 min at 3 mL/min. Elution starts at 14 min.

The extent of RI signal instability may depend on solvent, gradient type, and membrane cutoff. For successful baseline subtraction it is recommended to perform the sample sets in sequence, sample–blank injection–sample–blank injection, and so forth, instead of using just a single blank for subtraction from several sample runs. A certain disadvantage of this approach is that it actually doubles the run time per sample. For THF it is beneficial to use helium sparging to prevent chemical changes brought about by the presence of oxygen. The use of both vacuum degassing together with helium sparging was found to yield the most stable baseline and accurate RI signal subtraction.

As in SEC, the results should be reported together with instrumental details necessary for their interpretation and/or remeasurement. In addition to the parameters stated in Section 3.5.3, the membrane material, channel dimensions, detector flow, cross flow, and cross flow gradient (when applied) should be reported with the experimental results. Since part of a sample can be lost through the membrane, mass recovery should be stated with the reported results. Mass recovery is calculated from the injected mass and the mass calculated from the RI detector response and sample dn/dc. For polydisperse polymers, mass recovery less than 100% usually indicates presence of oligomeric fractions that were lost through the membrane. In such a case, additional SEC analysis can be performed in order to characterize the lower-molar-mass portion of the sample. Although a 100% mass recovery is ideal, partial recovery may be acceptable for analytical purposes assuming that sample loss does not occur selectively to specific size or molar mass range.

5.4.1 Influence of Separation Conditions

The experimental variables that control the separation process are the cross flow rate, the axial flow (detector flow) rate, the temperature, and the channel dimensions. Equations 5.28, 5.30, 5.33 and 5.34 for resolution, plate height, and fractionating power can be expressed in terms of variables that can be used to control the A4F experiments:

$$R_S = 0.051 \left(\frac{\Delta R}{\bar{R}} \right) \left(\frac{w^2}{\bar{D} V_0} \right) \left(\frac{\dot{V}_c^3}{\dot{V}} \right)^{1/2} \tag{5.35}$$

$$H_n = 24 \frac{L D^2 V_0^2}{w^4} \frac{\dot{V}}{\dot{V}_c^3} \tag{5.36}$$

$$F_{d,M} = S_{d,M} \frac{0.051 \, w \, \dot{V}_c^{3/2}}{b L D} \frac{}{\dot{V}^{1/2}} \tag{5.37}$$

Subscripts d and M in Equation 5.37 refer to the diameter or molar mass–based parameters. The above equations were derived for the symmetrical channel where $\langle v \rangle$ can be expressed as $\dot{V} L/\dot{V}_0$ In A4F the detector flow rate does not equal to the longitudinal flow rate in the channel, which depends on the position z and on the cross flow rate. However, the two flow rates are directly proportional and such the equations can be used to demonstrate the effect of experimental variables on the separation process in A4F as well. A general goal is to keep the resolution and fractionating power high, whereas plate height should be low. An increase of cross flow rate has a strong enhancing effect on resolution because of 1.5-power dependence. On the other hand, an increase of the detector flow decreases resolution, but at a lower rate of 0.5-power. Consequently, simultaneous increase of $\dot{V}_c$ and $\dot{V}$ will increase the resolution without affecting retention time. Low detector flow rate will reduce peak broadening and may also be beneficial in high-molar-mass polymer analysis because of decreased probability of shearing degradation. Fractionating power increases with the rate of cross flow and channel thickness and decreases with the rate of detector flow. However, simultaneous increase of both flow rates yields an increase in fractionating power.

Fractionating power is also inversely proportional to the channel bottom area bL. This is because at a given $\dot{V}_c$ the linear velocity cross flow (i.e., actual strength of the flow field) increases. Smaller channels may therefore provide better separation, but there is also a limit of channel pressure and small channels are also more prone to overloading effects. Also the total number of plates will decrease with decreasing channel length. Small channels may be efficient for the separation of some compounds, such as proteins, when samples are available in very low quantities. An additional advantage of small channels is lower consumption of eluent, because of generally lower cross flow rates.

The strength of the cross flow field is probably the most important separation variable; it has a strong effect on resolution and can be varied widely and rapidly without need to change the channel geometry. Note that cross flow and detector flow rates are limited by the HPLC pump and pressure in the channel (for Eclipse™ the maximum pressure in the channel is 30 bar). Although the cross flow has a positive effect on the resolution, excessive cross flow may push the sample through the membrane and thus increase the sample loss. Too-high cross flow may also promote interactions of sample with the membrane, sample aggregation, or mutual entanglement of macromolecules, because the molecules are pushed to a thin, concentrated layer near the accumulation wall.

Temperature of the channel is an often-overlooked experimental variable. A4F experiments are traditionally carried out at ambient temperatures. However, the entire channel can be placed into a sufficiently large HPLC oven or another suitable thermostating device. The retention time decreases with increasing temperature. Although temperature has generally negative impact on A4F separation, because it decreases resolution and fractionation power and increases plate height via increase of diffusion coefficient, it may be beneficial for polymer samples containing fractions with very high molar mass. For such a sample, the elevated temperature increases the steric inversion diameter and thus extends the range of normal separation. The elevated temperature can also hasten the elution of large, slowly eluting sample components.

Another experimental variable that can be used to control the separation is channel thickness. A certain disadvantage of this parameter is that its change requires manual disassembly of the channel. Thick channels decrease the probability of overloading and occurrence of steric separation and thus are suitable for polymer samples containing ultra-high-molar-mass fractions. On the other hand, thin channels can be used when steric separation of larger particles is of interest.

5.4.1.1 Isocratic and Gradient Experiments

The detector flow rate is usually constant throughout the injection, focusing, and elution steps. The cross flow can be either kept constant during the entire experiment or programmed to decay over the course of analysis to speed up the elution of strongly retained components. Programmed cross flow may also improve detectability, because isocratic runs of highly polydisperse polymers may result in very broad peaks and consequently weak detector response, especially at the end of fractograms.

A possible advantage of the isocratic run is reduction in baseline drift of the RI detector, which is sensitive to minor pressure changes due to decay of the cross flow. Also the prediction of retention time or estimation of the diffusion coefficient from retention data may be easier from the isocratic measurements. Isocratic experiments are usually sufficient for the separation of simple mixtures such as proteins containing dimer and trimer without the presence of large aggregates. However, for most polydisperse polymers, measurements using the cross flow gradient should be preferred. The cross flow can be decreased in a linear or exponential way. The speed and the pattern of the decay depend on the sample type and also on requested information.

A rule of thumb is to start the gradient at a cross flow that assures sufficient retention of early-eluting components; then the cross flow is decreased, either linearly or in an exponential pattern. Linear cross flow decay is obtained by setting the initial cross flow and the ramp time to reach the final cross flow rate. During the gradient run the channel inlet flow rate and the flow of one flow stream are accurately controlled, which assures control of the second stream as well.

Let us give a concrete example: When the starting cross flow of 3 mL/min and detector flow of 1 mL/min are required, the HPLC pump must generate a flow rate of 4 mL/min. Since the backpressure in the detector direction is significantly more than that in the cross direction, the liquid has a tendency to flow in the cross direction, which is regulated by the CoriFlow device to 3 mL/min, and thus the remaining flow of 1 mL/min must flow in the detector direction. To achieve the cross flow gradient, the flow stream generated by the pump is decreased at the same rate as the cross flow is decayed such that the outlet flow is maintained at 1 mL/min. When the A4F software can generate only the linear cross flow gradients, the exponential cross flow decay can be approximated by a series of consecutive linear segments. However, linear cross flow decay with possibly one or two isocratic sections is usually sufficient for the separation of even extremely broad polymers.

When the cross flow reaches zero value while part of the sample is still in the channel, the remaining sample is swept out without further separation, which may result in underestimation of molar mass and size distribution in the high-molar-mass region. To avoid this effect, the cross flow gradient is programmed in such a way that the cross flow reaches a small non-zero value that is maintained for a certain period of time in order to reduce the possibility of some large species being eluted unfractionated. On the other hand, the "swept effect" can help to detect small amounts of high-molar-mass species. The presence of high-molar-mass fractions in the sample is evident from the MALS signal shortly after the cross flow is switched to zero.

5.4.1.2   Overloading

Overloading in thermal and flow FFF analysis was theoretically described and experimentally studied by Caldwell et al.[24] A rule of thumb is to take caution in A4F measurements when injecting larger masses of samples. For most samples the separation takes place in a relatively thin layer near the accumulation wall. During the sample relaxation the local concentrations of molecules can become very high and sample components may start to affect each other. At very high concentrations, the mutual interference between the particles may result in a steric effect; that is, there is not enough room for the particles to reach equilibrium. Consequently, the relaxation equilibrium in the channel may not be fully reached, which can result in the shift of the retention time and perturbation of the shape of the fractogram. Other possible effects resulting from too-high injected mass are aggregation of particles, entanglement of macromolecular chains, or repulsion forces between macromolecules or particles.

The major factors that affect the amount of sample that can be injected into the channel are channel volume, sample molar mass, polydispersity, cross flow, and ionic strength in the case of aqueous solvents. The cross flow determines the extent at which the sample is concentrated near the wall. The sample amount that can be injected into the channel without overloading decreases as the molar mass increases, and narrow distribution samples should be injected at lower amounts compared to the broad samples. The injected amount can increase with the channel volume (channel breadth, length and thickness). For aqueous solvents the injected mass can usually increase as the ionic strength increases. In routine experimental practice, it is necessary to balance the injected amount with the response from the light scattering and concentration detectors. Especially the latter detector may show weak response in the case of polymers with high molar mass that require analysis under low sample load.

Unknown samples should be analyzed at different injected amounts. Overloading is missing if there is no significant change of retention time and peak shape; in particular, the absence of overloading is proved by the consistency of distribution curves obtained with different sample loads. Although keeping sample concentration as low as possible is generally advantageous, one has to keep in mind that in the A4F channel, after initial sample concentration during the relaxation process, elution is accompanied by a significant sample dilution. As a general rule, a sample mass of 0.2–0.3 mg can be safely injected without overloading for polydisperse polymers of M_w up to about 5×10^5 g/mol, with possibly higher amounts for polymers with $M_w < 1 \times 10^5$ g/mol and lower amounts if $M_w > 5 \times 10^5$ g/mol (for the channel outlined in Figure 5.2 and 350-μm spacer).

5.4.2 Practical Measurements

In contrast to SEC, in A4F there are several experimental parameters that can be readily adjusted to control the separation. This feature makes the method more flexible and powerful, but it requires a certain amount of development to optimize the conditions in order to get appropriate separation within an acceptable time. The experimental variables to consider are $\dot{V}$, $\dot{V}_c$, w, channel dimensions, temperature, and to some extent also the eluent. In order to get a first impression of an unknown sample, one can start with a 350-μm spacer and the generic conditions outlined in Figure 5.15. In this particular example the analysis starts with two-minute elution and focusing steps that stabilize the system; then the sample is injected for three minutes into the channel, and after the injection is completed the focusing is kept for an additional five minutes in order to allow the sample to focus and relax.

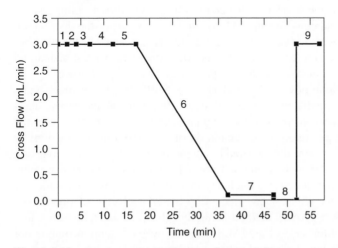

Figure 5.15 Generic conditions for separation of polydisperse polymers by A4F. Step: 1 = elution, 2 = focusing, 3 = focusing + inject, 4 = focusing, 5 to 9 = elution.
Note: Unless otherwise stated, the focus flow identical with the cross flow at the beginning of elution was used to obtain results presented in this chapter.

Note that length of the injection step depends on the injection volume and inject flow. The injection must be long enough not only to flush the sample loop, but also to transfer the sample from the injector through the A4F chassis into the channel. It is possible to use a few minutes' longer inject time than the minimum value in order to flush the inject line properly. The elution starts with a five-minute isocratic step to improve the separation of the lower-molar-mass components. After the isocratic elution, the cross flow decays linearly within 20 minutes to 0.1 mL/min, which is kept for an additional ten minutes to separate possibly occuring fractions with very high molar mass. Then the cross flow is completely stopped and the channel is flushed for five minutes. During this flush period there is typically no sample elution, but some samples may contain aggregates or other large species that are retained in the channel and released when no cross flow is applied. The final step is returning back to the initial cross flow that will be used for the next measurement. After collecting the first results, the conditions can be adjusted to increase retention of the lower-molar-mass fractions, to hasten elution of the high-molar-mass fractions, or to diminish steric separation.

The effect of particular experimental variables on the separation of a typical polydisperse polymer is illustrated in Figures 5.16–5.24. With the exception of Figure 5.22, the data were obtained by the measurements of the same broad polystyrene. The results depicted in these figures are mostly self-explanatory and can be used for planning your own experiments. The general message from this comparison is that once an appropriate separation is achieved, further increase of resolution does not lead to a significant difference in the obtained molar mass distribution plots. Almost identical molar mass distribution curves can be determined using different separation conditions. However, when too-low cross flow is used (Figure 5.17), the local polydispersity within the elution volume slices becomes significant and consequently the low-molar-mass part of the distribution is shifted to higher molar masses while the high-molar-mass part of the distribution is shifted to lower molar masses; that is, M_n values are overestimated and M_z values are underestimated and the sample appears virtually narrower.

If too-low focusing flow is applied (Figure 5.19), the lower-molar-mass fractions are swept immediately after switching to elution mode and appear unresolved as the intensive void peak at the beginning of fractogram. Note also the positive effect of temperature on the separation of high-molar-mass fractions as shown in Figure 5.22, where the elevated temperature shifted the steric inversion point to larger molecules and maintained normal separation over an extremely broad molar mass range. It can be concluded that the molar mass distribution of a polydisperse polymer will not be affected by a minor change of operational conditions (i.e., the A4F-MALS method is highly robust). This is, however, true only in the case of online use of a MALS detector, where retention time is not a principal parameter used for sample characterization.

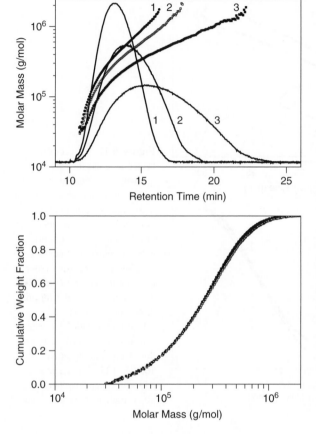

Figure 5.16 Effect of cross flow for isocratic separation: light scattering fractograms and molar mass–versus–retention time plots (top) and molar mass distribution curves (bottom) obtained by A4F-MALS measurements of broad polystyrene using isocratic cross flow of 1.5 mL/min (1,●), 2 mL/min (2,□), and 3 mL/min (3,■). A4F conditions: THF, detector flow 1 mL/min, injection 25 μL 0.25% w/v, spacer 250 μm, channel temperature 80°C. Elution starts at 10 min.

5.5 A4F APPLICATIONS

In contrast to SEC, which is strongly limited to soluble polymers, A4F is applicable to samples that can be not only dissolved, but also suspended in the carrier. Originally, FFF was developed to determine the diffusion coefficient and consequently hydrodynamic radius calculated on the basis of the retention data. That means it served as a separation and characterization method simultaneously. Although A4F theory allows the calculation of hydrodynamic radius from the retention time, significantly more detailed characterization can be achieved with a MALS detector, which allows the calculation of molar mass and RMS radius directly from the intensity of scattered light (see Chapter 2). The MALS detector can be equipped with a QELS option for direct measurement of hydrodynamic radius and thus RMS and hydrodynamic radii can be measured simultaneously.

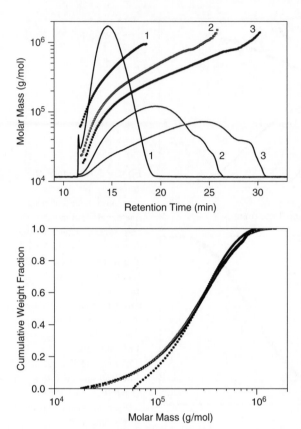

Figure 5.17 Effect of beginning cross flow for gradient separation: light scattering fractograms and molar mass–versus–retention time plots (top) and molar mass distribution curves (bottom) obtained by measurements of broad polystyrene using cross flow starting at 1 mL/min (1,•), 2 mL/min (2,□), and 3 mL/min (3,■), linear gradient to 0.1 mL/min within 20 minutes. A4F conditions: THF, detector flow 1 mL/min, injection 100 μL 0.25% w/v, spacer 350 μm, channel temperature 80°C. Elution starts at 11 min.

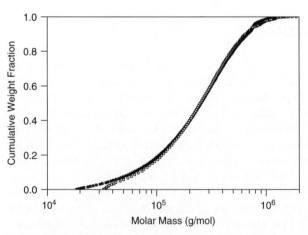

Figure 5.18 Distribution curves of polydisperse polystyrene determined by isocratic (□) and gradient (■) experiments using cross flow 3 mL/min (see Figures 5.16 and 5.17).

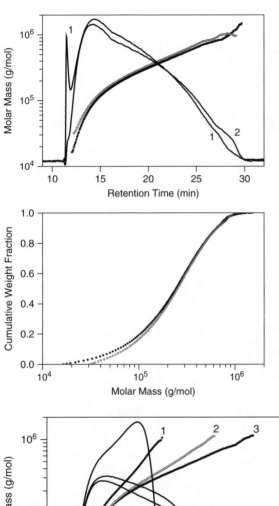

Figure 5.19 Effect of focus
flow: RI fractograms and molar
mass versus retention time plots
(top) and molar mass distribution curves (bottom) obtained by
measurements of broad polystyrene using focus flow 1 mL/min
(1,□) and 3 mL/min (2,■). A4F
conditions: THF, detector flow
1 mL/min, cross flow 3 mL/min
for 5 min and then linear decay to
0.1 mL/min within 20 min plus
10 min at 0.1 mL/min, injection
100 μL 0.25% w/v, spacer
350 μm, channel temperature
60°C. Elution starts at 11 min.

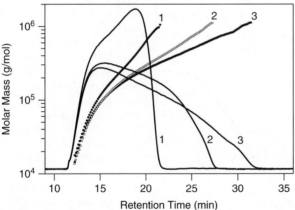

Figure 5.20 Effect of gradient decay time: RI fractograms and molar mass–versus–retention
time plots (top) and molar mass distribution curves (bottom) obtained by A4F-MALS
measurements of broad polystyrene using linear cross flow gradient with different decay time. A4F
conditions: THF, detector flow 1 mL/min, cross flow 3 mL/min to 0.15 mL/min within 10 min
(1,•), 20 min (2,□), and 30 min (3,■) plus 5 min at 0.15 mL/min, injection 100 μL 0.25% w/v,
350 μm spacer, channel temperature 60°C. Elution starts at 11 min.

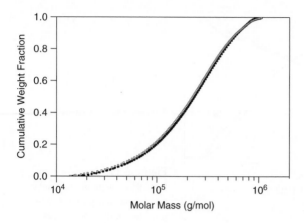

Figure 5.20   (*Continued*)

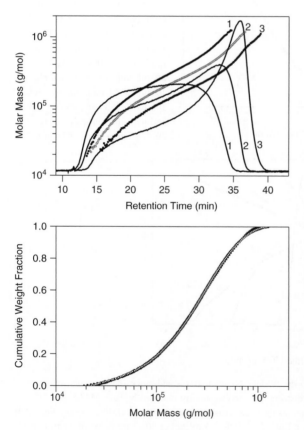

Figure 5.21   Effect of temperature: RI fractograms and molar mass–versus–retention time plots (top) and molar mass distribution curves (bottom) obtained by A4F-MALS measurements of broad polystyrene at channel temperature 25°C (3,■), 60°C (2,□), and 90°C (1,•). A4F conditions: THF, detector flow 1 mL/min, cross flow 3 mL/min for 5 min and then linear gradient to 0.1 mL/min within 20 min plus 10 min at 0.1 mL/min, injection 100 μL 0.25% w/v, spacer 490 μm. Elution starts at 12 min.

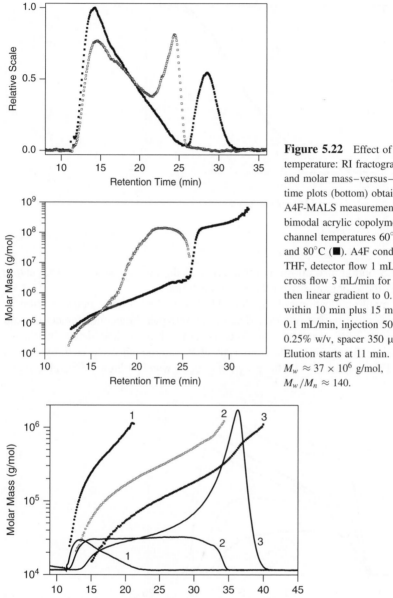

Figure 5.22 Effect of temperature: RI fractograms (top) and molar mass–versus–retention time plots (bottom) obtained by A4F-MALS measurements of bimodal acrylic copolymer using channel temperatures 60°C (□) and 80°C (■). A4F conditions: THF, detector flow 1 mL/min, cross flow 3 mL/min for 5 min and then linear gradient to 0.1 mL/min within 10 min plus 15 min at 0.1 mL/min, injection 50 μL 0.25% w/v, spacer 350 μm. Elution starts at 11 min. $M_w \approx 37 \times 10^6$ g/mol, $M_w/M_n \approx 140$.

Figure 5.23 Effect of channel thickness: RI fractograms and molar mass–versus–retention time plots (top) and molar mass distribution curves (bottom) obtained by A4F-MALS measurements of broad polystyrene using spacer thickness of 250 μm (1,●), 350 μm (2,□), and 490 μm (3,■). A4F conditions: THF, detector flow 1 mL/min, cross flow 3 mL/min for 5 min and then linear gradient to 0.1 mL/min within 20 min plus 10 min at 0.1 mL/min, injection 25 μL, 100 μL, and 200 μL 0.25% w/v for increasing spacer thickness. Elution starts at 11 min.

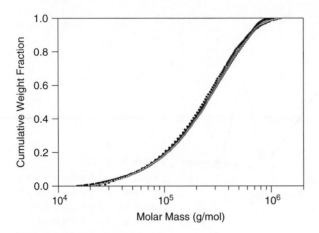

Figure 5.23   (*Continued*)

The QELS option also allows confrontation of directly measured hydrodynamic radii with those obtained from the retention time.

The main applications of the A4F-MALS method are the determination of molar mass distribution, characterization of polymer branching, detection and characterization of aggregates in proteins and polymers, and determination of particle size distribution. The materials that can be characterized by means of flow FFF include synthetic polymers, both water soluble[25,26] and organic

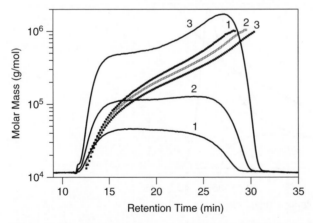

Figure 5.24   Effect of injected mass: RI fractograms and molar mass–versus–retention time plots (top) and molar mass distribution curves (bottom) obtained by A4F-MALS measurements of broad polystyrene using different injected mass: 125 μg (1,●), 250 μg (2,□), and 500 μg (3,■). A4F conditions: THF, detector flow 1 mL/min, cross flow 3 mL/min to 0.15 mL/min within 20 min plus 10 min at 0.15 mL/min, injection 50 μL, 100 μL, and 200 μL 0.25% w/v, spacer 350 μm, ambient temperature. Elution starts at 11 min.

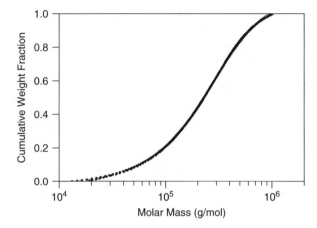

Figure 5.24   (*Continued*)

soluble,[27,28] natural polymers,[29-32] proteins[33-35] and protein conjugates, DNA,[36] antibodies,[37,38] viruses,[39-41] ribosomes,[42] liposomes,[43-45] micelles,[46] polymer latexes,[47] nanoparticles,[48-51] and natural colloids.[52]

Determination of molar mass distribution is the major application of A4F-MALS for the characterization of synthetic and natural polymers. Not only does successful application of A4F-MALS benefit from the A4F separation advantages when analysis of SEC-difficult polymers is requested; the method should provide comparable results with SEC-MALS even for samples that can be characterized by SEC-MALS without considerable problems. A typical separation of polydisperse polymer with A4F-MALS is demonstrated on NIST SRM 706 polystyrene in Figure 5.25, which also compares the corresponding molar mass distribution plot with that determined by SEC-MALS (see SEC separation in Figure 4.19). The molar mass distribution determined by A4F-MALS misses the oligomeric part below roughly 10^4 g/mol, which was lost by permeation through the semipermeable membrane. In addition, the molar mass distribution in the region of low molar masses may be affected by low retention of low-molar-mass fractions. Nevertheless, the agreement of the two distribution plots is quite acceptable. It is worth noting that the M_n value determined by A4F-MALS is in excellent agreement with that determined initially by membrane osmometry, that is, when the result was similarly affected by the permeation of oligomeric fractions through the membrane of a membrane osmometer.

Acceptable agreement between M_n values determined by A4F-MALS and MO was found also for other samples, as shown in Table 5.1. Although for the determination of molar mass distribution of regular polymers A4F does not provide significant improvement over SEC, the power of the method appears when the SEC separation is affected by non-SEC mechanisms. This is presented in Figure 5.26, which compares SEC-MALS and A4F-MALS data for randomly branched polystyrene for which the lower-molar-mass part of the distribution is affected by abnormal SEC elution (see Section 6.2.1). In this case, A4F provides

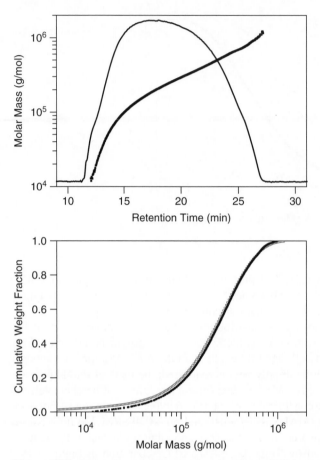

Figure 5.25 Molar mass–versus–retention time plot and RI fractogram for NIST SRM 706 PS determined by A4F-MALS (top) and comparison of cumulative molar mass distribution plots determined by SEC-MALS (□) and A4F-MALS (■) (bottom). SEC-MALS: $M_n = 66,000$ g/mol; $M_w = 279,000$ g/mol; $M_z = 442,000$ g/mol. A4F-MALS: $M_n = 137,000$ g/mol; $M_w = 284,000$ g/mol; $M_z = 423,000$ g/mol. SEC conditions: $2 \times$ PLgel Mixed-C 300×7.5-mm columns, THF at 1 mL/min, injection 100 μL 0.25% w/v. A4F conditions: THF, detector flow 1 mL/min, cross flow from 3 to 0.15 mL/min within 20 min, injection 100 μL 0.25% w/v, spacer 350 μm, channel temperature 60°C. Elution starts at 11 min. For SEC separation, see Figure 4.19.

a more accurate description of molar mass distribution; the higher M_n from SEC can be attributed to abnormal SEC elution while the lower M_z indicates either shearing degradation of fractions with very high molar mass or poor SEC resolution of high-molar-mass fractions exceeding the total exclusion limit.

Figure 5.27 presents another example where A4F provides better separation compared to SEC, namely analysis of polymer containing amine functionality resulting in strong polymer interaction with SEC column packing. No indication of interactions is found for the data obtained by A4F-MALS.

Table 5.1 Number-Average Molar Masses of Polydisperse Polymers Determined by A4F-MALS and Membrane Osmometry

	M_n (g/mol)	
Sample	A4F-MALS	MO
PS (AN)	135 ± 9	120
PS (1)	138 ± 2	139
PS (2)	145 ± 3	153
PS (S)	82 ± 2	66
PS (K)	117 ± 3	98
PMMA (Y)	225 ± 5	185
PMMA (J)	62 ± 1	33

A4F is particularly suited for the separation and characterization of ultra-high-molar-mass polymers and polymers containing supermolecular structures. Shear degradation of ultra-high-molar mass fractions in the A4F channel is minimized due to mild conditions. In contrast to SEC, where the separation range is limited by total exclusion limit, there is no exclusion limit in A4F, and as for the molar mass the method is practically unlimited.

An example of the analysis of a polymer with molar mass range spanning over four orders of magnitude and polydispersity of about 25 is presented in Figure 5.28. The molar mass–versus–retention time plot shows separation up to molar masses over 10^8 g/mol. Such separation is typically impossible with SEC, because these molar masses are over the separation range of regularly available SEC columns and they are also highly prone to shearing degradation. Figure 5.28 also demonstrates the ability of A4F to yield conformation plots with no artifacts caused by abnormal SEC elution.

Another example of polymer containing fractions with ultra-high molar mass is presented in Figure 5.29. In this example, the ultra-high-molar-mass fractions are present at very low concentrations, yielding almost no RI detector signal; nevertheless, they are baseline separated from the main polymer peak due to the high separation power of A4F and no filtration and shearing effect, and are sensitively detected by the MALS detector.

Proteins and protein-related materials represent another important application area of the A4F-MALS technique. A typical property of proteins is their ability to form aggregates by association of protein molecules. Detection and characterization of protein aggregates is a common goal in protein science. In the pharmaceutical industry, the protein aggregation is strongly related to drug activity and undesirable side effects. The protein aggregation can be a major limitation in the development of protein-based drugs and thus understanding these phenomena is important for development of new products. The protein samples can range from single molecules to insoluble precipitates, which makes AF4 an ideal separation tool. SEC is conventionally used for the characterization of various proteins, but the effects of shear and adsorption can affect the obtained

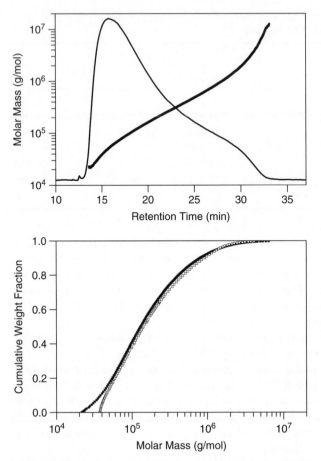

Figure 5.26 Plot of molar mass–versus–retention time with RI fractogram for randomly branched polystyrene obtained by A4F-MALS (top) and comparison of cumulative molar mass distribution plots determined by SEC-MALS (□) and A4F-MALS (■) (bottom). SEC-MALS: M_n = 108,000 g/mol; M_w = 353,000 g/mol; M_z = 1.2 × 10⁶ g/mol. A4F-MALS: M_n = 92,000 g/mol; M_w = 350,000 g/mol; M_z = 1.9 × 10⁶ g/mol. SEC conditions: 2 × PLgel Mixed-C 300 × 7.5-mm columns, THF at 1 mL/min, injection 100 μL 0.2% w/v. A4F conditions: THF, detector flow 1 mL/min, cross flow from 3 to 0.2 mL/min within 20 min + 10 min at 0.2 mL/min, injection 100 μL 0.35% w/v, spacer 350 μm, ambient channel temperature. Elution starts at 12 min. For SEC separation, see Figure 6.2.

results. In addition, regulatory bodies may request verification of the absence of aggregates by an additional analytical technique.

Most protein-based samples are water soluble and the most common A4F carriers include phosphate buffer or sodium nitrate solution. Because pH and ionic strength affect protein association and denaturation, they can be adjusted to match the pH and ionic strength of the sample solution. Figure 5.30 shows a

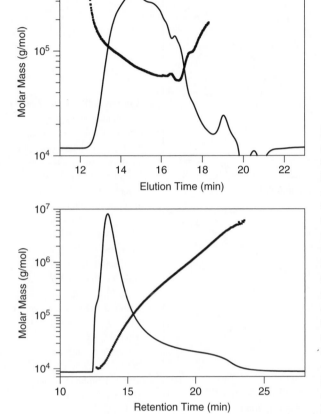

Figure 5.27 RI signals and molar mass−versus−retention time plots from SEC-MALS (top) and A4F-MALS (bottom) analysis of polymer containing amine functionality. SEC conditions: 2 × PLgel Mixed-C 300 × 7.5-mm columns, THF at 1 mL/min, injection 100 μL 1% w/v. A4F conditions: THF, detector flow 1 mL/min, cross flow from 3 to 0.15 mL/min within 10 min + 10 min at 0.15 mL/min, injection 100 μL 1% w/v, ambient channel temperature. Elution starts at 12 min.

typical protein fractogram. The concentrations of protein monomer, dimer, and trimer can be determined from the relative peak areas of the fractogram recorded by the RI or UV detector and the molar mass of protein molecules is determined directly by the MALS detector.

5.6 KEYNOTES

- A4F is an alternative to SEC with its main advantages including ability to analyze polymers up to ultra-high molar masses and significantly reduced possibility of shearing degradation and enthalpic interactions. The method has been proven to be very efficient for the separation of branched polymers.

- Besides analysis of soluble polymers the A4F method is capable of separating particles up to the micrometer scale.

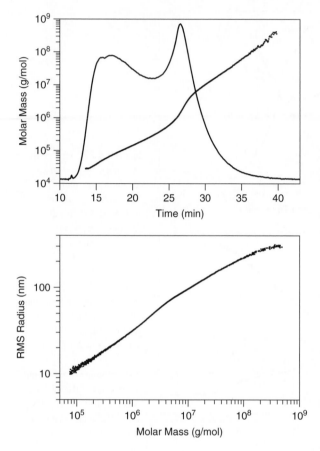

Figure 5.28 Molar mass distribution along the retention time axis overlaid on the RI response (top) and conformation plot (bottom) for a branched polystyrene sample of very broad molar mass distribution. $M_n = 134,000$ g/mol; $M_w = 3.3 \times 10^6$ g/mol; $M_z = 64.7 \times 10^6$ g/mol. A4F conditions: THF, detector flow 1 mL/min, cross flow 3 mL/min for 5 min and then linear decay to 0.1 mL/min within 10 min + 40 min at 0.1 mL/min, injection 200 μL 0.2% w/v, spacer 490 μm, ambient channel temperature. Elution starts at 11 min.

- The principal difference of A4F compared to SEC is its lack of stationary phase. Separation is achieved by flow in an empty channel where perpendicular cross flow is applied.
- Separation power of A4F is adjustable by operational variables.
- Separation of soluble macromolecules and particles in the nanoscale range is based on the equilibrium of the cross flow field and diffusion, and is called *normal* or *Brownian*. The separation mechanisms for micrometer-sized particles is based on either the steric effect of the particles pressed closely to the bottom wall or the equilibrium of the cross flow force and hydrodynamic lift forces.
- The main contributions to plate height in A4F are nonequilibrium and sample polydispersity. However, the polydispersity contribution is actually given by the real sample separation.
- Selectivity of A4F is usually significantly higher than that of SEC.

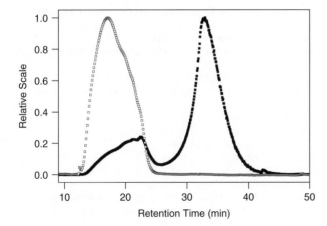

Figure 5.29 RI (□) and MALS (■) fractograms of polymer containing trace amounts of ultra-high-molar-mass fractions. A4F conditions: THF, detector flow 1 mL/min, cross flow from 3 mL/min to 0.2 mL/min within 20 min + 10 min at 0.2 mL/min, injection 200 μL 0.2% w/v, spacer 490 μm, ambient channel temperature. Elution starts at 12 min.

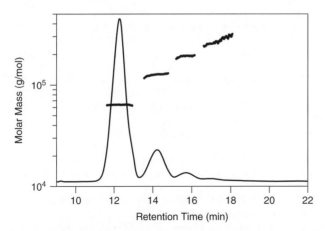

Figure 5.30 RI fractogram and molar mass for particular peaks of BSA. A4F conditions: 50 mM sodium nitrate aqueous solution, detector flow 1 mL/min, isocratic cross flow 5 mL/min, focus flow 3 mL/min, injection 15 μL 0.2% w/v, spacer 350 μm, ambient channel temperature, beginning of elution at 6 min. Relative peak area = 76.1% (monomer), 17.1% (dimer), 4.6% (trimer), and 2.2% (tetramer + higher).

5.7 REFERENCES

1. Giddings, J. C., *Separation Sci.*, **1**, 123 (1966).
2. Barth, H. G. and Carlin, F. J., *J. Liq. Chromatogr.*, **7**, 1717 (1984).
3. Giddings, J. C., Yang, F. J., and Myers, M. N., *Science*, **193**, 1244 (1976).
4. Giddings, J. C., Yang, F. J., and Myers, M. N., *Anal. Chem.*, **48**, 1126 (1976).

5. Wahlund, K.-G., Winegarner, H. S., Caldwell, K. D., and Giddings, J. C., *Anal. Chem.*, **58**, 573 (1986).
6. Wahlund, K.-G. and Giddings, J. C., *Anal. Chem.*, **59**, 1332 (1987).
7. Janca, J., *Field-Flow Fractionation: Analysis of Macromolecules and Particles*, Marcel Dekker, New York (1988).
8. Janca, J., *Microthermal Field-Flow Fractionation*, HNB P, New York (2008).
9. Schimpf, M. S., Caldwell, K., and Giddings, J. C. (editors), *Field-Flow Fractionation Handbook*, Wiley-Interscience, New York (2000).
10. Giddings, J. C., *J. Chem. Ed.*, **50**, 667 (1973).
11. Giddings, J. C., *Sep. Sci. Technol.*, **21**, 831 (1986).
12. Litzen, A. and Wahlund, K.-G., *Anal. Chem.*, **63**, 1001 (1991).
13. Giddings, J. C., Myers, M. N., Caldwell, K. D., and Fisher, S. R., *Methods of Biochemical Analysis*, Glick, D. (editor), John Wiley & Sons, New York (1980), Vol. **26**, p. 79.
14. Lee, S. and Giddings, J. C., *Anal. Chem.*, **60**, 2328 (1988).
15. Ratanathanawongs, S. K. and Giddings, J. C., *Chromatography of Polymers: Characterization by SEC and FFF*, ACS Symposium Series **521**, Provder, T. (editor), American Chemical Society, Washington D.C. (1993), Chapter 2.
16. Jensen, K. D., Williams, S. K. R., and Giddings, J. C., *J. Chromatogr. A*, **746**, 137 (1996).
17. Caldwell, K. D., *Modern Methods of Polymer Characterization*, Barth, H. G. and Mays, J. W. (editors), John Wiley & Sons, New York (1991), Chapter 4.
18. Giddings, J. C., Martin, M., and Myers M. N., *J. Chromatogr.*, **158**, 419 (1978).
19. Giddings, J. C., Yoon, U. H., Caldwell, K. D., Myers. M. N., and Hovingh, M. E., *Sep. Sci.*, **10**, 447 (1975).
20. Smith, L. K., Myers, M. N., and Giddings, J. C., *Anal. Chem.*, **49**, 1750 (1977).
21. Giddings, J. C., Williams, P. S., and Beckett, R., *Anal. Chem.*, **59**, 28 (1987).
22. Dubascoux, S., Hecho, I. Le, Gautier, M. P., and Lespes, G., *Talanta*, **77**, 60 (2008).
23. Siripinyanond, A., Worapanyanond, S., and Shiowatana, J., *Environ. Sci. Technol.*, **39**, 3295 (2005).
24. Caldwell, K. C., Steven, L. B., Gao, Y., and Giddings, J. C., *J. Appl. Polym. Sci.*, **36**, 703 (1988).
25. Leeman M., Islam, M. T. Haseltine. W. G., *J. Chromatogr. A*, **1172**, 194 (2007).
26. Benincasa, M.-A. and Caldwell, K. D., *J. Chromatogr. A*, **925**, 159 (2001).
27. Mes, E. P. C., de Jonge, H., Klein, T., Welz, R. R., and Gillespie, D. T., *J. Chromatogr. A*, **1154**, 319 (2007).
28. Bang D. Y., Shin D. Y., Lee, S., and Moon, M. H., *J. Chromatogr. A*, **1147**, 200 (2007).
29. Kwon, J. H., Hwang, E., Cho, I.-H., and Moon, M. H., *Anal. and Bioanal. Chem.*, **395**, 519 (2009).
30. Benincasa, M.-A., Cartoni, G., and Delle Fratte, C., *J. Chromatogr. A*, **967**, 219 (2002).
31. Wittgren, B. and Wahlund, K.-G., *J. Chromatogr. A*, **760**, 205 (1997).
32. Leeman, M., Wahlund, K.-G., and Wittgren, B., *J. Chromatogr. A*, **1134**, 236 (2006).
33. Giddings, J. C., Yang, F. J., and Myers, M. N., *Anal. Biochem.*, **81**, 395 (1977).
34. Min-Kuang Liu, Ping Li, and Giddings J. C., *Protein Science*, **2**, 1520 (2008).
35. Silveira, J. R., Hughson, A. G., and Caughey, B., *Methods Enzymol.*, **412**, 21 (2006).
36. Liu, M.-K. and Giddings, J. C., *Macromolecules*, **26**, 3576 (1993).
37. Hawe, A., Friess, W., Sutter, M., and Jiskoot, W., *Anal. Biochem.*, **378**, 115 (2008).
38. Gabrielson, J. P., Brader, M. L., Pekar, A. H., Mathis, K. B., Winter, G., Carpenter, J. F., and Randolph, T. W., *J. Pharmaceutical Sci.*, **96**, 268 (2007).
39. Giddings, J. C., Yang, F. J., and Myers, M. N., *J. Virology*, **21**, 131 (1977).
40. Citkowicz, A., Petry, H., Harkins, R. N., Ast, O., Cashion, L., Goldmann, C., Bringmann, P., Plummer, K., and Larsen, B. R., *Anal. Biochem.*, **376**, 163 (2008).
41. Pease, L. F., Lipin, D. I., Tsai, D.-H., Zachariah, M. R., Lua, L. H. L., Tarlov, M. J., and Middelberg, A. P. J., *Biotechnology and Bioengineering*, **102**, 845 (2009).
42. Nilsson, M., Birnbaum S., and Wahlund, K.-G., *J. Biochem. Biophys. Methods*, **33**, 9 (1996).
43. Arifin, D. R. and Palmer, A. F., *Biotechnology Progress*, **19**, 1798 (2003).

44. Hupfeld, S., Ausbacher, D., and Brandl, M., *J. Sep. Sci.*, **32**, 1465 (2009).
45. Hupfeld, S., Ausbacher, D., and Brandl M., *J. Sep. Sci.*, **32**, 3555 (2009).
46. Wittgren, B., Wahlund, K.-G., Derand, H., and Wesslen, B., *Macromolecules*, **29**, 268 (1996).
47. Frankema, W., van Bruijnsvoort, M., Tijssen, R., and Kok, W. T., *J. Chromatogr. A*, **943**, 251 (2002).
48. Lesher, E. K., Ranville, J. F., and Honeyman, B. D., *Environmental Science and Technology*, **43**, 5403 (2009).
49. Zattoni, A., Rambaldi, D. C., Reschiglian, P., Melucci, M., Krol, S., Garcia, A. M. C., Sanz-Medel, A., Roessner, D., and Johann C., *J. Chromatogr. A*, **1216**, 9106 (2009).
50. Kang, D. Y., Kim, M. J., Kim, S. T., Oh, K. S., Yuk, S. H., and Lee, S., *Anal. Bioanal. Chem.*, **390**, 2183 (2008).
51. Dubascoux, S., Kammer, F. V. D., Le Hecho. I., Gautier, M. P., and Lespes, G., *J. Chromatogr. A*, **1206**, 160 (2008).
52. Bouby, M., Geckeis, H., and Geyer, F. W., *Anal. Bioanal. Chem.*, **392**, 1447 (2008).

Chapter 6

Characterization of Branched Polymers

6.1 INTRODUCTION

Branching of polymeric chains is one type of polymer nonuniformity. It is one of the most important molecular parameters that determines various physical, mechanical, and end-use properties of synthetic and natural polymers. Polymer properties affected by branching include glass transition temperature, mechanical properties (e.g., strength, tack, peel), viscoelastic properties, ability to crystallize, thermodynamic behavior (A_2 or θ-temperature), solubility, chemical resistance, phase separation of polymer mixtures, solution viscosity, melt viscosity, melt elasticity, rheological behavior of solutions and melts, workability, curing behavior of synthetic resins, and ability to swell. This last property can be of primary importance for drug delivery systems since the drug release rate can be controlled by the degree of branching of a polymer used as a drug carrier. Indeed, various branched polymers, such as branched polylactides, are of great interest for medical, biomedical, and pharmaceutical applications.

The effect of branching on polymer properties can be either negative or positive, depending on particular circumstances and polymer applications. For some properties the effect of branching may not be straightforward. For example, branching increases glass transition temperature (T_g) as a consequence of decreased mobility of polymer segments; on the other hand, increased concentration of chain ends increases free volume and thereby contributes to the decrease of the T_g. Similarly as for molar mass it may be necessary to find optimum values of branching, because positive effects on one property may result in negative effects on another property. The relation between type and degree of branching and polymer properties can be studied experimentally by means of model branched polymers or derived theoretically. Although it

Light Scattering, Size Exclusion Chromatography and Asymmetric Flow Field Flow Fractionation: Powerful Tools for the Characterization of Polymers, Proteins and Nanoparticles, by Stepan Podzimek
Copyright © 2011 John Wiley & Sons, Inc.

may be difficult to predict properties of branched polymer unambiguously due to the variety of possible branching structures and the fact that many branching reactions occur in a random fashion, branching in polymers is undoubtedly an important structural variable that can be used to modify the processing and application properties of polymers. Some properties of specific branched polymers (e.g., dendrimers) are unachievable by means of linear polymers.

Some properties of branched polymers are very attractive. For example, branching is an efficient means of inhibition of polymer crystallization. That is because branching discourages the chains from fitting closely together so that the structure is amorphous with relatively large amounts of empty space, whereas structures with little or no branching allow the polymer chains to fit closely together, forming a crystalline structure. Another result of branching is that it reduces the molecular size. Consequently, branching is an effective way of reduction of solution viscosity. Some polymers may be easily melt-processable, but favorable processability may be associated with bad mechanical properties. Improvement in mechanical properties may cause processsability to deteriorate, and branching of polymer chains may help to balance the two counteracting properties. The impact of branching on shear thinning (i.e., decrease of viscosity with increasing rate of shear stress) may play an important role in extrusion and injection molding. The impact on polymer dielectric properties may improve breakdown voltage and energy loss of capacitors. Various types of polyethylene play an important role as engineering polymers and thus the influence of long-chain branching on flow properties is of primary interest in the polyolefin industry. For instance, long-chain branches in polyethylene lead to significantly improved processing behavior due to increased shear thinning and strain hardening in elongation flow. An important fact is that even a very low level of branching can improve processing behavior in a sufficient manner.

One of the most important properties in rheology of polymer melts is the temperature dependence of the viscosity (Equation 1.89). The higher the activation energy E, the more temperature sensitive is the polymer melt. The value of E varies from polymer to polymer depending on chain composition. From a certain minimum value of molar mass (equal to the entanglement molar mass), E is independent on molar mass, but increases with the amount of long-chain branches. Branching is not only important from the viewpoint of structure–property and structure–property–processing relationships, but it brings valuable information about the polymerization mechanisms and extent and type of side reactions.

The term *degree of branching* refers to number of branch units in a macromolecule or number of arms in a starlike macromolecule. A *branched polymer* is a polymer containing at least one *branch unit* (branch point, junction point), that is, a small group of atoms from which more than two long chains emanate. A branch point from which f linear chains emanate is called an f-functional branch point. A distinguishing feature of branched polymers compared to linear polymers is presence of more than two chain ends. A branch is an oligomeric or polymeric offshoot from a macromolecular chain. An oligomeric branch is called a *short-chain branch* while a polymeric branch is termed a *long-chain branch*. If the addition or removal of one or a few structural units has a negligible effect on the molecular properties, then the material can be considered to be a polymer and a branch can be considered to be a long-chain branch. Alternatively, if the branch length is comparable to the length of the backbone, the branched polymer can be termed as long-chain. However, the line between short-chain and long-chain is never explicit.

Branched molecules can be of varied structures, namely randomly branched molecules, combs, stars, dendrimers, and hyperbranched molecules. Even unusual branched structures such as centipedes or barbwire molecules have been reported in the literature. Branched molecules can arise as a result of various side reactions or can be prepared purposefully by simple or sophisticated synthetic pathways. Chain transfer to polymer, addition of water and glycols on double bonds during the synthesis of unsaturated polyesters, reaction of epoxy groups with aliphatic hydroxyl groups in epoxy resins, or polymerization of the second double bond in polyisoprene or polybutadiene are examples of side reactions leading to branching.

Free radical polymerization of a monovinyl monomer (e.g., styrene, acrylates, methacrylates) with the addition of a small amount of a divinyl monomer[i] (e.g., divinyl benzene, ethyleneglycol dimethacrylate) represents the simplest example of the synthesis of branched polymers. The reaction technique is very straightforward and typically involves addition of free radical initiator to solution of monomers in a suitable solvent and polymerization at a temperature appropriate to the initiator. The resulting polymer has randomly (statistically) branched structure. At early stages of polymerization, predominantly only one of the vinyl groups of the divinyl monomer reacts and thus branch units appear at higher conversions. The length and number of branches can be partly controlled by the reaction temperature; varying the concentrations of divinyl monomer, initiator, and solvent; and using transfer agent.

Depending on the conversion and concentration of a polyfunctional monomer, the branching reaction can reach the gel point, that is, the stage when an infinitesimal fraction of indefinite structure appears in the system. At the gel point the weight-average molar mass reaches infinite value while the number-average molar mass is finite. With increasing reaction conversion after the gel point, the fraction of infinite insoluble structure (gel) increases while the amount of soluble fraction (sol) decreases and the M_w of sol decreases as well. A typical feature of a polymer prepared by free radical polymerization of a mixture of mono and divinyl monomers is a large fraction of linear molecules even at the proximity of gel point.

Another characteristic of randomly branched polymers is a presence of high-molar-mass tail and broadening of the molar mass distribution in comparison to the distribution that would arise under identical reaction conditions without

---

[i]In free radical polymerization, each vinyl group is difunctional (i.e., divinyl monomer is tetrafunctional).

presence of polyfunctional monomer. The probability that a randomly selected molecule will contain a branch unit increases with increasing molar mass and thus the degree of branching increases with increasing molar mass.

Randomly branched macromolecules can also arise by the radical polymerization of vinyl monomers due to the chain transfer to polymer. In contrast to the polymerization of monovinyl and divinyl monomer mixture, the resulting branch units are trifunctional. Another synthetic way to prepare randomly branched polymers is polycondensation of difunctional (e.g., hydroxy acid, mixture of dicarboxylic acids and glycols) and polyfunctional monomers (e.g., glycerol, pentaerythritol), which leads to polymers of equivalent structure to those prepared by radical polymerization with the exception of the presence of relatively high levels of oligomeric fractions even at high conversions. Other reactions leading to randomly branched polymers are heat treatment or irradiation of linear polymers. Branched macromolecules may be also formed during polymer aging.

Special synthetic methods can lead to branched polymers with well-defined structure compared to randomly branched polymers. Anionic polymerization is particularly suited for the preparation of branched polymers with controlled architecture. The technique yields polymers that retain their active chain ends even after all monomer has been consumed. These living polymers can react with multifunctional linking agents under formation of starlike polymers with the number of arms equivalent to the functionality of the linking agent. Possible heterogeneity is given by formation of stars containing different numbers of arms, as in the case of the reaction of poly(styryl)lithium with silicon tetrachloride, which usually leads to a mixture of three-arm and four-arm stars. Another synthetic way to form starlike polymers is living polymerization using multifunctional initiators.

Group transfer polymerization (GTP) is a technique applicable to the polymerization of methacrylic monomers, which in contrast to anionic polymerization yields living polymers at room temperature. The technique applies special initiators that enable synthesis of various polymethacrylates with controlled molecular structure. GTP synthesis of branched polymers involves two steps. Living arms from methacrylate are prepared in the first step, and then in the second step they react with a dimethacrylate monomer (e.g., ethyleneglycol dimethacrylate, EGDMA), which can react with up to four arms and partly also with itself. The resulting polymer consists of a dense EGDMA center from which numerous arms protrude. The method can provide starlike polymers with up to hundreds of arms and possibly functional groups located at the arm ends or in the arm chains. The properties can be further modified by the preparation of arms consisting of block copolymer.

Dendrimers are highly branched well-defined macromolecules with a branch point at each monomer unit, resembling the structure of trees. Their synthesis consists of numerous protection/deprotection and purification steps leading to products of increasing generation and molar mass. Dendrimers of higher generations of eight to ten reach the molar mass of several hundreds of thousands g/mol. Many interesting potential applications of dendrimers are based on their molecular uniformity, highly dense structure, and multifunctional surface. On the other hand, the extreme laboriousness of their preparation, which requires multiple steps, is a serious limitation on their broader application.

Hyperbranched polymers represent a simpler alternative to dendrimers. They are prepared by a single-step polycondensation of AB_x ($x \geq 2$) monomers with different types of functional groups A and B capable of reacting with each other (e.g., dimethylol propionic acid, dimethylol butanoic acid). Hyperbranched polymers are expected to be highly branched like dendrimers; however, experimental results often indicate the presence of significant amounts of linear segments.

Macromonomers, that is, polymeric or oligomeric monomers with polymerizable functional groups at one end, can be used for the preparation of comb-shaped branched polymers with the distance between the branch units controllable with a regular low-molar-mass comonomer. In principle, macromonomers enable synthesis of polymers with a very high branch density.

6.2 DETECTION AND CHARACTERIZATION OF BRANCHING

In most branched polymers the distribution of degree of branching coexists with the distribution of molar mass and possibly with the distribution of other characteristics such as chemical composition. In addition, branched polymers are typical in variety of branching topology, as can be seen from the following simple example. Let us imagine a polymer molecule with two trifunctional branch units. The two branches can protrude from the backbone or one branch can protrude from the other, and they can have various positions with respect to the chain ends. If the branches are located close to the backbone ends, then the entire molecule virtually behaves as a linear molecule. This simple example shows that a branched molecule of a given molar mass and branching degree can occur in numerous positional isomers.

Another reaction complicating the architecture of branched polymers is the possible formation of intramolecular rings due to the reaction that occurs between two segments of the same macromolecule. Branched polymers mostly consist of molecules of different molar mass and degree of branching, which is particularly true in the case of randomly branched polymers. Molecules of identical molar mass can differ in the number of branch units, and macromolecules of identical molar mass and number of branch units can differ in the position of branch units in a polymer chain.

Thorough characterization of branched polymers involves not only the determination of the distribution of molar mass, but also the distribution of branching degree. However, there is no experimental technique that would enable separation of polymer molecules purely according to the degree of branching and regardless of their molar mass. The temperature rising elution fractionation (TREF) and crystallization analysis fractionation (CRYSTAF) techniques utilize the different crystallizability brought in by short-chain and long-chain branches to fractionate polymer on the basis of the branching degree. In principle, this mechanism permits characterization of the branching distribution, but the applicability of both methods is limited mainly for the analysis of polyethylene and polyethylene-based copolymers. In addition, although the separation is expected to be according to the degree of branching, there is also simultaneous separation on the basis of molar mass. In a conventional arrangement with a concentration detector only (infrared detector) there is no information about the molar mass of fractions eluting from the TREF/CRYSTAF system, and the obtained fractions must be further characterized.

Another method attempting the separation of molecules according to their branching degree has been reported by Meunier et al.[1] The method is based on the idea of anchoring of branched molecules in porous matrix, described in reference 2. It uses monolithic columns to create a tortuous path where the flow-through channels have size of the order of polymer molecular dimensions. Because the long-chain branched molecules are retained more than linear molecules of the same size, the method can be used to separate branched chains from linear molecules. The method may succeed in separating simple branched polymers, such as mixtures consisting of four-arm and three-arm stars and residual linear molecules. However, it is unlikely that the method would allow efficient separation of complex branched polymers solely according to the branching degree.

Two-dimensional chromatography may offer another possible approach in the more detailed characterization of various branched polymers, where fractions of identical hydrodynamic volume obtained by SEC in the first step can be further separated according to the branching degree by a suitable type of interaction chromatography in the second step.

Currently, the only really measurable branching distribution is the degree of branching as a function of molar mass. In fact, the branching ratio determined at a given molar mass is an average for all molecules having this molar mass, because the molecules may have different degrees of branching at the same molar mass. That means molecules of a given molar mass consist of molecules with a certain distribution of branching degree. Another complicating fact is that the available analytical separation techniques separate on the basis of hydrodynamic volume and not molar mass. Consequently, the molecules eluting from the separation system at a given elution time have the same hydrodynamic volume, but different degrees and topologies of branching and also different molar masses. In a rigorous manner the obtained result is the relation between the average branching degree (average for a given average molar mass) and the average molar mass (average for a given hydrodynamic volume).

Various methods providing relationships between branching degree and molar mass will be described in this chapter. Taking into account the complexity of branched polymers with respect to their molar mass distribution, number of branch units per molecule, and branching topology, a detailed description of branching is virtually impossible. In reality, detection of the presence of branched molecules and estimation of the degree of branching as a function of molar mass and the determination of an average branching characteristic for the entire sample is the only obtainable information for most branched polymers.

The characterization of branching is often complicated by the impossibility of preparing well-defined branched model polymers. Although sophisticated synthetic routes can yield well-defined stars or combs, the preparation of well-defined models for randomly branched polymers is impossible. For some polymers even the preparation of a linear counterpart of identical chemical composition is impossible, as in the case of various synthetic resins (e.g., alkyds, phenol-formaldehyde resins, epoxies). Also the tendency of various acrylates (not methacrylates) to chain transfer to polymer is so strong that it may be difficult to prepare purely linear homopolymers by free radical polymerization.

It may be noted that structure and number of branches in a polymer sample can be revealed from the rheological measurements.[3,4] Rheological experiments conducted on irradiated polypropylene were shown[4] to be more sensitive with respect to long-chain branching than characterization by means of SEC-MALS. However, rheological measurements are not applicable to all kinds of branched polymers and the obtained results are merely average characteristics with no information about branching distribution. Although the presence of branch units can be in principle detected by spectral techniques, the level of long-chain branching is often that low that it cannot be detected by standard analytical methods. Branched molecules can be detected and characterized indirectly by the measurement of dilute solution properties. A review of solution properties of branched macromolecules can be found in reference 5.

The fundamental principle of detection and characterization of branching is based on the fact that at a given molar mass the molecular size decreases with increasing degree of branching. In other words, branching reduces the molecular size and increases the compactness of the macromolecules. This means that to get branching information it is necessary to determine both molar mass and molecular size. The analytical methods commonly applicable for the investigation of branching can be divided into two categories: (1) characterization methods, mainly light scattering, dynamic light scattering, and viscometry, and (2) separation methods, mainly SEC and field flow fractionation techniques. The characterization methods can be used in batch mode or in combination with a separation technique.

A numerical description of the degree of branching can be achieved by the parameter defined by Zimm and Stockmayer:[6]

$$g = \left(\frac{R_{br}^2}{R_{lin}^2} \right)_M \tag{6.1}$$

where R^2 is the mean square radius of a branched (br) and a linear (lin) macromolecule of the same molar mass (M). The parameter g is called *branching ratio* (*branching index*, *branching parameter*, or *contraction factor*) and is equal to unity for linear polymers and decreases with increasing extent of branching. For polymers with a very high degree of branching, the ratio g approaches values around 0.1, but never zero. It is also evident from the definition of g that it is never larger than 1.

Branching frequency (λ), defined as the number of branch units per 1,000 repeat units, is another parameter that can be used for the description of branched polymers and is frequently used in the analysis of polyolefins. Branching ratio itself can serve for the characterization of branching degree, and it can be used for further calculations. As shown by Zimm and Stockmayer, the branching ratio g allows calculation of number of branch units per molecule or number of arms in starlike molecules. Various theoretical equations describing the structure of branched polymers can be found in the literature.[6–11] The most frequently stated relations are those derived by Zimm and Stockmayer for randomly branched and star-branched polymers.[6]

For randomly branched polymers with tri- or tertafunctional branch units, the ratio g is given by the equations:

$$g_3 = \left[\left(1 + \frac{m}{7} \right)^{\frac{1}{2}} + \frac{4m}{9\pi} \right]^{-\frac{1}{2}} \tag{6.2}$$

$$g_4 = \left[\left(1 + \frac{m}{6} \right)^{\frac{1}{2}} + \frac{4m}{3\pi} \right]^{-\frac{1}{2}} \tag{6.3}$$

where m is the average number of branch units per molecule and subscripts 3 and 4 refer to trifunctional and terafunctional branch units. For instance, the polymerization of vinyl monomers with an addition of a small amount of a divinyl monomer results in tetrafunctional branch units while chain transfer to polymer creates trifunctional branch units. In a given example, both functionalities can occur simultaneously, the former being relatively easily controllable by the amount of divinyl monomer and conversion, the latter being indirectly controllable by reaction conditions. Simultaneous existence of branch units of different functionality may be also expected in the case of polyfunctional monomers, for instance in polycondensation of a mixture of difunctional monomers with addition of a tetrafunctional monomer, which can undergo reaction completely, giving tetrafunctional branch units, or incompletely, giving trifunctional branch units. Trifunctional randomly branched topology can be assumed in branched polymers formed by irradiation, because formation of branch points with higher functionality is statistically improbable.

Equations 6.2 and 6.3 were derived for the case where a material of heterogeneous molar mass containing randomly distributed branch units is fractionated into a series of samples, each of them being monodisperse but of different molar mass, and the branch units in each fraction are still randomly distributed. As a matter of fact, these conditions approximately correspond to the situation occurring in SEC or FFF. Since m is the average number of branch units (actually the number-average), the ratio g given by Equations 6.2 and 6.3 is the average value valid for molecules of the same molar mass but different numbers of branch units.

The expressions for polymers heterogeneous with respect to molar mass are:

$$g_{3,w} = \frac{6}{n_w} \left[\frac{1}{2} \left(\frac{2+n_w}{n_w} \right)^{\frac{1}{2}} \ln \left(\frac{(2+n_w)^{\frac{1}{2}} + n_w^{\frac{1}{2}}}{(2+n_w)^{\frac{1}{2}} - n_w^{\frac{1}{2}}} \right) - 1 \right] \tag{6.4}$$

and

$$g_{4,w} = \frac{\ln(1+n_w)}{n_w} \tag{6.5}$$

where n_w is the weight-average number of branch units per molecule and the subscripts indicate functionality of the branch unit and the fact that the ratios are the weight-averages.

For the z-average values of g, the formulas derived in reference 6 are:

$$g_{3,z} = \frac{1}{1 + \frac{n_w}{3}} \tag{6.6}$$

and

$$g_{4,z} = \frac{1}{1 + n_w} \tag{6.7}$$

For stars with f arms of random length, the branching ratio is given by the equation:[6]

$$g = \frac{6f}{(f+1)(f+2)} \tag{6.8}$$

And for regular stars with arms of equal length:

$$g = \frac{3f - 2}{f^2} \tag{6.9}$$

Note that for the linear molecules, $f = 2$ or $m = 0$, the above formulas yield the value unity. The graphical representation of Equations 6.2, 6.3, and 6.8 is shown in Figure 6.1.

It must be emphasized that the above equations were derived assuming unperturbed chain statistics for both linear and branched macromolecules, that is, assuming no excluded volume effect. However, the real measurements are mostly carried out in thermodynamically good solvents where the molecules expand due to intensive interactions between polymer molecules and solvent. The expansion of a polymer chain is more pronounced in the case of linear molecules, and thus the ratio g determined in thermodynamically good solvent is smaller than it would be in theta solvent and the branching degree is overestimated. The correction for the excluded volume effect yields equations that are complicated, with several parameters generally unknown or at least uncertain. In reality, the effect of the excluded volume is probably lower than experimental and other uncertainties and thus can be neglected. The selection of an appropriate equation relating g with a parameter characteristic for a given branched structure may not

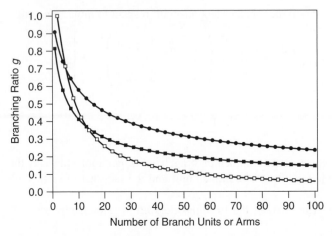

Figure 6.1   Branching ratio as a function of number of branch units per molecule for three (•) and four (■) functional branch units in randomly branched polymer and number of arms (□) in star-branched polymer (Equations 6.2, 6.3, and 6.8).

be always straightforward, because the functionality of branch unit or the type of branched structure may not be known. In addition, the theoretical equations were typically not verified by experiments due to lack of well-defined branched standards.

Although the application of theoretical equations may not provide an absolutely accurate description of a branched polymer, especially when the measurement is carried out far from θ-conditions, the equations can yield an estimate of m or f, mutual comparison of various samples with respect to their branching degree, and indicate the effectiveness of synthetic techniques with respect to their ability to yield branched polymers of desired structure. The equations relating g with the number of branch units or arms are only one part of branching characterization; before they can be applied, a reliable value of g must be determined by a suitable experimental technique. The main objective of this chapter is to show the most important methods available for the characterization of branching and discuss their advantages and limitations. The focus is on methods that have general applicability and that are relatively easy to apply.

As described in Chapter 2, the molar mass and the root mean square radius, which are necessary to characterize branched molecules by parameter g, can be obtained by multi-angle light scattering. One serious limitation of MALS is that when applied on unfractionated polymer sample it yields M and R of different moments, namely the weight-average molar mass and the z-average RMS radius. The two averages are of different sensitivies to high-molar-mass species; consequently, batch MALS usually can provide branching information only for monodisperse or very narrow polymers. Combination of MALS with a separation method overcomes this limitation to a great extent since M and R are measured for narrow fractions eluting from the separation device.

Traditional methods of polymer fractionation, such as precipitation fractionation, are of limited applicability due to their laboriousness and low efficiency with respect to the polydispersity of obtained fractions. SEC and A4F, or other FFF techniques, are separation methods that can be used in combination with MALS for the characterization of branched polymers. However, in comparison to linear polymers, where the local slice polydispersity is given solely by the band broadening in the SEC columns or A4F channel and can be reduced by optimization of separation conditions, branching represents another contribution to the polydispersity of fractions eluting from an SEC or AF4 system. Since both methods separate molecules according to their hydrodynamic volume, in a given elution volume all molecules eluting from the separation device have identical hydrodynamic volume (assuming absence of band broadening), but different molar masses, numbers of branch units, and chain architectures. Consequently, a MALS detector does not measure the molar mass and RMS radius, but their averages. Nevertheless, as there is no simple method providing information about the polydispersity of fractions eluting from SEC columns or A4F channel, the polydispersity effect must be neglected and the molar masses and RMS radii must be treated as if they were determined for monodisperse fractions. This is a basic assumption that one has to accept at least for routine characterization.

Characterization of branching on the basis of RMS radius is also limited by the fact that RMS radius cannot be determined for smaller polymer molecules with RMS radii below about 10 nm, which roughly corresponds to molar mass of 10^5 g/mol. That means the size of many branched polymers cannot be determined by elastic light scattering. There are two alternative techniques that can be used for the determination of size below 10 nm: dynamic light scattering and viscometry of dilute solutions. The latter one is especially efficient for the determination of size of polymer molecules covering broad molar mass range. The intrinsic viscosity, $[\eta]$, describes the size of polymer molecules in a dilute solution. An alternative branching ratio was defined as the ratio of the intrinsic viscosities of a branched (*br*) and a linear (*lin*) macromolecule of identical molar mass:[12]

$$g' = \left(\frac{[\eta]_{br}}{[\eta]_{lin}} \right)_M \tag{6.10}$$

The mutual relationship of the two branching ratios can be expressed by a simple relation:

$$g' = g^e \tag{6.11}$$

where e is a parameter related to drainability of a polymer chain, the values of which are usually assumed to vary from 0.5 to 1.5. The value of e depends on solvent, molar mass, temperature, and branching. An important advantage of intrinsic viscosity is that it can be accurately determined even for small polymer molecules with molar masses down to about a thousand g/mol with practically no upper limit. In contrast to RMS radius, the weight-average intrinsic viscosity, $[\eta]_w$, is measured for an unfractionated sample and therefore the combination of MALS with a capillary viscometer can be used for the determination of average

g' by batch measurements. However, considerably more information can be obtained by means of an online viscometer combined with an SEC-MALS setup.

6.2.1 SEC Elution Behavior of Branched Polymers

As already mentioned, the polydispersity of elution volume slices in SEC separation of branched polymers is generally larger compared to linear polymers separated under the same separation conditions because of co-elution of molecules of the same hydrodynamic volume, but different molar masses and branching. However, for many branched polymers there is an additional, substantially more serious contribution that increases the local polydispersity within the elution volume slices. Many branched polymers show the SEC elution that is demonstrated in Figure 6.2, which depicts plots of molar mass and RMS radius versus elution volume for a randomly branched polymer. At lower elution volumes both molar mass and RMS radius decrease evenly with increasing elution volume, as is typical of SEC separation. At a certain point in the chromatogram both quantities

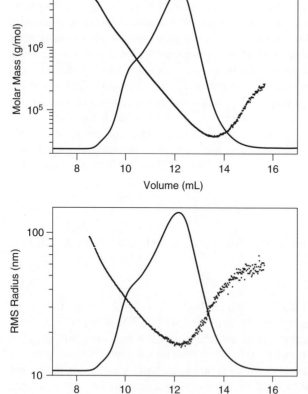

Figure 6.2 Abnormal SEC elution of branched polymer: Molar mass–versus–elution volume (top) and RMS radius–versus–elution volume (bottom) plots of randomly branched polymer. RI chromatogram is also shown here.

start to increase with increasing volume. For the RMS radius this upturn appears at a lower elution volume than for the molar mass. Although this SEC elution behavior is not observed for all branched polymers, it is frequent and becomes more pronounced with increasing degree of branching.

This behavior was reported by several authors and explained in different ways. It was systematically studied in reference 2, where the influence of various experimental parameters (e.g., flow rate, column type, temperature) on the pattern of molar mass and RMS radius plots was investigated. The most important finding was that the abnormal elution behavior was not found in the separation by thermal FFF, which was afterward repeatedly verified by separation using A4F, as shown later in this chapter.

On the basis of the obtained results, the abnormal SEC elution was explained by the entanglement of large, highly branched molecules in the pores of column packing. Since the previously used term *entanglement* is also used with respect to the concentration at which the polymer chains in solution start to overlap, the term *anchoring* will be used in this book to avoid confusion of the two phenomena. Anchoring is where particular parts of large branched molecules can behave as separate molecules, penetrate into the pores, and anchor the entire molecules.

This anchoring idea is sketched in Figure 6.3. Due to the anchoring effect some large molecules are delayed and elute at higher elution volumes than would correspond to their hydrodynamic volumes. The fractions eluting from SEC columns at the region of high elution volumes consist of smaller molecules separated by a purely SEC mechanism and very large branched molecules that were delayed by the anchoring effect. In such a case, the assumption of monodisperse fractions eluting from the SEC columns is not valid, the local polydispersity cannot be neglected, and the MALS detector yields values of M_w and R_z instead of values of M and R. The R_z is more sensitive to the presence of small amounts of fractions with very high molar mass than the M_w, and thus the upswing on the RMS plot appears at a lower elution volume compared to the molar mass. Simple estimation leads to the conclusion that the concentration of contaminating large molecules is rather low. For example, taking data from Figure 6.2 it is possible to estimate that the fractions of smaller macromolecules eluting at about

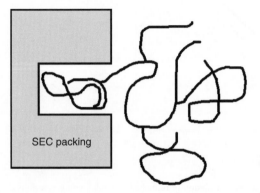

SEC packing

Figure 6.3   Illustration of the anchoring of a large branched molecule in a pore of column packing by a part of the polymer chain.

15 mL have molar mass around 10,000 g/mol. This number can be obtained by extrapolation of a molar mass–versus–elution volume plot from the region of low elution volumes to the high elution volumes. The molar mass measured by the MALS detector at this elution volume is about 140,000 g/mol. Assuming that the delayed molecules are of molar mass of the order of magnitude of 5×10^6 g/mol, that is, approximately the molar mass measured at the beginning of the chromatogram, one can estimate the fraction of the delayed molecules as follows:

$$140{,}000 = w_1 \times 10{,}000 + (1 - w_1) \times 5 \times 10^6$$

where w_1 is the weight fraction of molecules with molar mass 10,000 g/mol and $(1 - w_1) = w_2$ is the weight fraction of molecules with molar mass 5×10^6 g/mol. For this particular example, we get $w_2 = 0.026$; that is, delayed macromolecules represent approximately 2.6% of eluting molecules. The reality is certainly more complicated; co-eluting macromolecules are of different molar mass and the upturn on the molar mass and RMS radius plots can be caused by a trace level of ultra-high-molar-mass, highly branched fractions. For some highly branched polymers the anchoring effect is so strong that the light scattering signal tails to the elution volumes beyond the limit of total permeation of SEC columns. That means the molecules are delayed by anchoring for a relatively long time and they elute at very low concentrations detectable only by a light scattering detector.

An unwelcome consequence of the anchoring effect is the upswing on the conformation plot, as shown in Figure 6.4. It must be emphasized that the upswing on the conformation plot from SEC-MALS is totally virtual; that is, there are no molecules in the analyzed sample that would have two different radii at the same molar mass as one can see at the lower-molar-mass part of the conformation plot in Figure 6.4. The upswing is caused by the nonnegligible polydispersity of the fractions eluting at the end of the chromatogram and the higher sensitivity of R_z to the polydispersity compared to M_w.

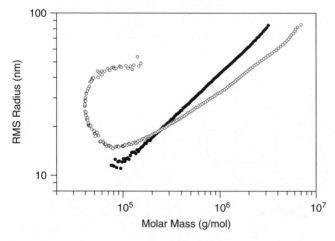

Figure 6.4   Conformation plots for linear (•) and randomly branched (○) polystyrene determined by SEC-MALS. The same polymer as in Figure 6.2.

6.2.2 Distribution of Branching

Branching distribution is usually characterized by the relation between the branching ratio and molar mass and consequently the relation between the number of branch units and molar mass. As already explained, this is usually the only measurable branching distribution, because the available separation techniques separate according to hydrodynamic volume and there is no generally working separation technique separating primarily on the basis of branching degree. The acquisition of requested experimental data requires hyphenation of an analytical separation technique and characterization methods capable of providing molar mass and molecular size. The characterization of branching distribution is based on the assumption of almost monodisperse fractions eluting from a separation system. That means an efficient separation analytical technique is crucial for accurate branching characterization.

Although the separation of branched polymers is generally less efficient than the separation of corresponding linear polymers, the experimental results for many branched polymers suggest that SEC and especially A4F are usually sufficient to provide useful branching information. For routine characterization of branched polymers, the slice polydispersity is neglected and the effort is focused on optimization of separation conditions, especially in the sense of minimizing the undesirable anchoring effect. Generally, there are three methods that can be used for determination of ratio g as a function of molar mass: (1) *radius method* (calculation of g from R at the same M), (2) *mass method* (calculation of g from M at the same elution volume), and (3) *viscosity method* (calculation of g' from $[\eta]$ at the same molar mass).

Figures 6.4–6.6 show typical results obtained by SEC-MALS or SEC-MALS-VIS analysis of a branched polymer and a corresponding linear counterpart. They are: (1) RMS radius–versus–molar mass plot (radius method

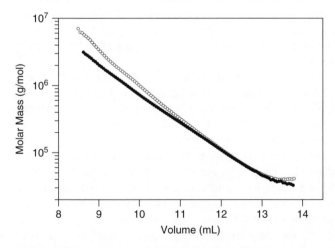

Figure 6.5 Molar mass–versus–elution volume plots for linear (•) and randomly branched (○) polystyrene (conformation plots are in Figure 6.4).

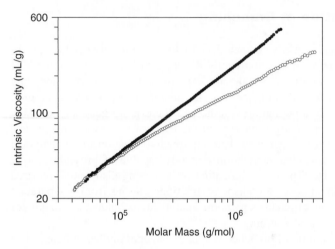

Figure 6.6   Mark-Houwink plots for linear (•) and randomly branched (○) polystyrene (conformation plots are in Figure 6.4).

of calculation of g), (2) molar mass–versus–elution volume plot (mass method), and (3) Mark-Houwink plot (viscosity method). All three plots relate the size information with the molar mass. The first two plots can be obtained using only SEC-MALS; the determination of the Mark-Houwink plot requires coupling the SEC-MALS instrument with an online viscometer.

The relation between the RMS radius and molar mass (*conformation plot*) is one of the basic tools for detection and characterization of branching. The conformation plots of linear polymers in thermodynamically good solvents are linear with the slopes around 0.58. In the case of branched polymers the slope of the conformation plot decreases with increasing degree of branching. Some conformation plots show leveling off toward high molar masses. The conformation plot can prove the presence or absence of branching even if the linear counterpart is unavailable. However, the decrease of the slope at low branching degrees is small (for example, from 0.58 to 0.55), which makes the detection of branching without a linear counterpart uncertain.

The slope of the conformation plot can be also influenced by the light scattering formalisms used for data processing (see Section 4.3.4) and possibly also by the separation range of the SEC columns. If the sample contains molecules with molar mass over the exclusion limit of the SEC columns, these molecules elute nonseparated at the beginning of the chromatogram and the MALS detector measures for that region M_w and R_z, which may result in the increase of the slope of the conformation plot. In such a case, the data from the beginning of the chromatogram should not be used for the determination of the slope; the same is true for the data points at the end of the chromatogram, which are often scattered due to the low concentration and molar mass of eluting molecules.

As already discussed, the conformation plots from SEC-MALS of many branched polymers show a noticeable upswing at the region of lower molar

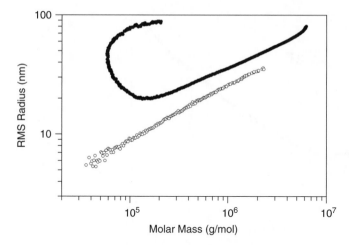

Figure 6.7 Conformation plots of two different branched samples with and without upswing. M_w, g_{M_z} and g'_{M_w} are listed in Table 6.1 for PS B2 ($\bullet$) and PS 20 ($\circ$).

masses. In fact, the upswing is so typical that it may serve as evidence of the presence of branching. It has been explained that the upswing is a consequence of the anchoring of branched macromolecules in the pores of column packing during their flow through the SEC columns.

Although the upswing on the conformation plot is quite frequent, it may be missing on the conformation plots of some branched polymers as shown in Figure 6.7. The upswing is negligible especially in the case of polymers with lower degree of branching, branched samples that do not contain molecules with very high molar masses or polymers with relatively short branches. This is because smaller branched molecules or molecules with relatively short branches cannot be anchored in the way shown in Figure 6.3. The upswing also becomes less pronounced with decreasing sample polydispersity.

The intensity of the upswing is also related to the type of SEC columns used for the separation, as demonstrated in Figure 6.8, which shows that the upswing is more pronounced when SEC columns with smaller particle size and smaller pore size are used for the SEC separation. The upswing can be also partly eliminated using very low flow rates such as 0.1 mL/min. However, working at such low flow rates increases run time and the change in the flow rate typically results in increased noise level of a MALS detector due to release of particles from SEC columns. Stabilization of the signal at a very low flow rate takes a long time, and thus working at extremely low flow rates appears impractical. The only effective solution to the anchoring problem is change of the separation technique.

The upswing is completely eliminated when A4F is applied instead of SEC, as shown in Figure 6.9. The comparison of SEC-MALS and A4F-MALS results for various branched polymers proved unambiguously A4F to be a more efficient separation technique for branched polymers. This is certainly true mainly for

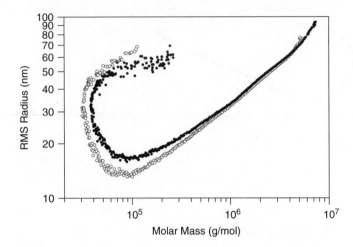

Figure 6.8   Conformation plots of randomly branched PS obtained by two different columns: 2 × Plgel Mixed-C (●) and 2 × Plgel Mixed-B (○). Mixed-C is 5-μm particle size column packing with operating range 200–2 × 10^6 g/mol; Mixed-B is 10-μm packing with operating range 500–10^7 g/mol.

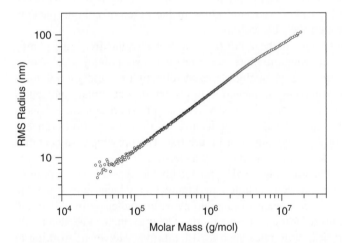

Figure 6.9   Conformation plot of branched polystyrene determined by A4F-MALS. (For comparison with SEC-MALS, see Figure 6.4).

large polymers with a tendency to anchor in the SEC column packing, whereas for smaller molecules SEC usually provides satisfactory results.

SEC elution volume is related to molecular size and therefore the relation between the molar mass and elution volume can be used for branching studies. Using the approach of Yu and Rollings,[13] the molar mass–versus–elution volume plots of a branched polymer and a linear polymer of the same chemical composition measured under identical separation conditions can yield the branching ratio according to the equation:

$$g = \left(\frac{M_{lin}}{M_{br}}\right)_V^{\frac{1+a}{e}} \tag{6.12}$$

where M is the molar mass of linear (*lin*) and branched (*br*) molecules eluting at the same elution volume V, a is the exponent of the Mark-Houwink equation of a linear polymer in a given solvent and at a given temperature, and e is the draining parameter.

Calculation of g from the molar mass−versus−elution volume plots appears to be the least appropriate method of branching characterization, because it strongly depends on separation conditions and has no general validity. It requires knowledge of two parameters, a and e, which may not be always known. On the other hand, the molar mass−versus−elution volume plot is applicable over a broad molar mass range including oligomers. The molar mass−versus−elution volume plots can be valuable for mutual graphical comparison of samples measured under identical separation conditions. The mass method does not require the measurement of RMS radius, which makes it less sensitive to anchoring phenomena. Since the molecular size is not measured directly, but merely estimated from the elution volume, the method is highly sensitive to any kind of non-SEC separation.

In contrast to the molar mass−versus−elution volume plot, the conformation plot and the Mark-Houwink plot relate absolute quantities obtained by MALS and VIS detectors. They do not require concurrent measurement of a linear counterpart, but previously established relationships R versus M or $[\eta]$ versus M for a corresponding linear polymer can be used for branching detection and characterization.

An important finding is the significantly lower sensitivity of the Mark-Houwink plot to the anchoring effect. The explanation is that in the case of poor SEC separation the conformation plot relates R_z with M_w, whereas the Mark-Houwink plot relates $[\eta]_w$ with M_w, and thus the effect of poor separation is at least partly eliminated because both quantities are equally affected by increased polydispersity.

Figure 6.10 compares branching ratios g and g' obtained from the data depicted in Figure 6.4 and Figure 6.6. Branching ratio g' starts at about unity and decreases with increasing molar mass. This is typical behavior for randomly branched polymers, because the probability that a randomly selected macromolecule contains a branch unit increases with increasing polymerization degree. Consequently, in randomly branched polymers the fractions with lower molar mass are mostly linear while the fractions with higher molar mass are more branched. The branching ratio g calculated from the RMS radius shows different behavior. At the region of high molar masses it goes approximately parallel with g', while at the region of lower molar masses it goes for above unity, which is a maximum value of this parameter. As already explained, this is caused by increased polydispersity of fractions eluting at the region of higher elution volumes.

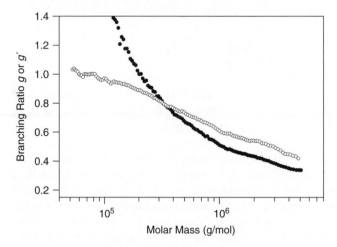

Figure 6.10   Molar mass dependence of branching ratios g (•) and g'(○) calculated using the data from Figures 6.4 and 6.6.

To avoid misunderstanding, it must be stressed that g of branched polymers is always smaller than unity if the branched and linear molecules are compared at the same molar mass or at the same molar mass moment. It has been shown in reference14 that g of a branched polymer can be larger than unity in the case of an unfractionated sample when the RMS radii are compared at the same M_w, which is exactly the case in SEC-MALS when some elution slices are polydisperse due to the anchoring effect.

Figure 6.11 contrasts the parameter g–versus–molar mass plots determined by A4F-MALS and by SEC-MALS. The two plots overlap at the region of

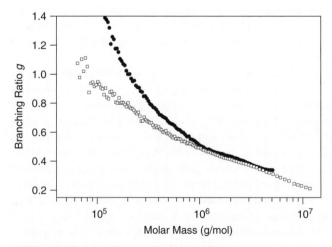

Figure 6.11   Branching ratio g–versus–molar mass plots of randomly branched polystyrene determined by SEC-MALS (•) and A4F-MALS (□).

high molar masses, where the anchoring effect in SEC is imperceptible and elution slices are narrow, and deviate toward lower molar masses due to the rising polydispersity of the SEC slices. The contamination with large branched molecules decreases toward lower elution volumes and obviously there is no contamination at the very beginning of the chromatogram. It may be noted that the anchoring affects not only the ability of SEC-MALS to characterize branching, but also the ability to determine the molar mass distribution and molar mass averages. Since light scattering measures M_w and R_z by its first principle, these two quantities are correct even in the case of poor SEC separation. On the other hand, the value of M_n is affected strongly, because the lower-molar-mass fractions are the most contaminated. As the contamination effect is small at the beginning of the chromatogram, the value of M_z that counts strongly the high-molar-mass fractions can be considered to be almost correct (assuming no shearing degradation).

The distribution of branching along the molar mass axis (i.e., plots of g versus M or m versus M) can be overlaid with the molar mass distribution curve as shown in Figure 6.12. Such a plot permits the determination of branching distribution with respect to the molar mass. Using the data presented in Figure 6.12 one can see that molecules with molar mass below about 10^5 g/mol are linear. This represents almost 40% of molecules in the given sample. The fraction of molecules with molar mass over 10^6 g/mol represents about 8%, which means about 8% of molecules in the sample have more than seven branch units on average.

Figure 6.13 compares branching results obtained by means of radius and viscosity methods. It shows the relation between g obtained directly from the

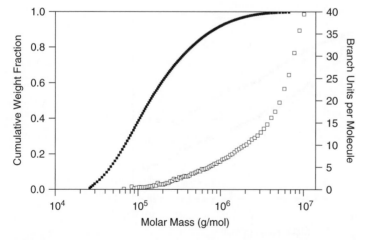

Figure 6.12 Cumulative molar mass distribution plot of randomly branched polystyrene (■) and number of branch units per molecule (□). Branch units calculated from g determined by A4F-MALS (see Figure 6.11).

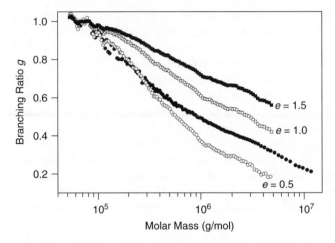

Figure 6.13 Molar mass dependence of branching ratio g determined by radius method (●) and calculated from g' using different values of draining parameter e.

conformation plot (Figure 6.9), and calculated from the Mark-Houwink plot (Figure 6.6) using Equations 6.10 and 6.11 with various values of draining parameter. None of the boundary values of e provide good agreement of results generated from the RMS radii and intrinsic viscosities. The best agreement of the two plots is obtained using e of about 0.7. However, since the parameter e depends on the properties of polymer chain and experimental conditions, no general validity of this value can be assumed.

The comparison of results obtained by the radius and mass methods is depicted in Figure 6.14. Similarly as in the viscosity method, the

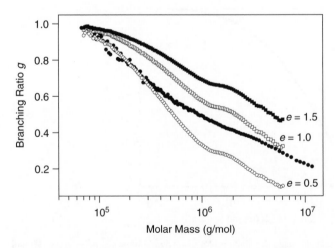

Figure 6.14 Molar mass dependence of branching ratio g determined by radius method (●) and mass method using different draining parameters e.

obtained results strongly depend on the draining parameter and the best agreement of the mass method with the radius method is achieved for e around 0.7.

Although g calculated by means of viscosity and mass methods depends on the value of parameter e and generally does not concur with g obtained by the radius method, with the same e the viscosity and mass methods yield consistent results, as demonstrated in Figure 6.15.

Despite uncertainty of the calculation of g from the intrinsic viscosity due to the commonly unknown parameter e, the intrinsic viscosity has the great advantage of the possibility of measurement of relatively small branched molecules for which the RMS radius cannot be determined. Characterization of lower-molar-mass polymers is demonstrated in Figure 6.16 for star poly(DL-lactic acid) with tripentaerythritol as a branching agent. The RMS radius of such polycondensates is below the detection limit of elastic light scattering and thus the intrinsic viscosity appears as a suitable alternative. The Mark-Houwink plot allows clear detection of branching in the analyzed sample. As molar mass increases, the difference of the intrinsic viscosity between the branched molecules and the linear reference increases, which indicates that molecules of higher molar mass consist of more arms.

Although more detailed analysis of the obtained experimental data regarding the number of arms is limited by the uncertainty of draining parameter e, the data undoubtedly prove the presence of branched macromolecules, and allow mutual comparison of different samples, as well as give a certain feeling of the number of arms. Figure 6.16 also presents the molar mass–versus–elution volume plot that can serve as a simple alternative for branched oligomers when a viscometer is not available.

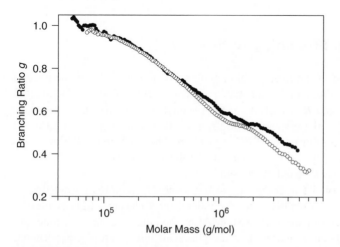

Figure 6.15 Molar mass dependence of branching ratio g determined by viscosity method (•) and mass method (○) using $e = 1$.

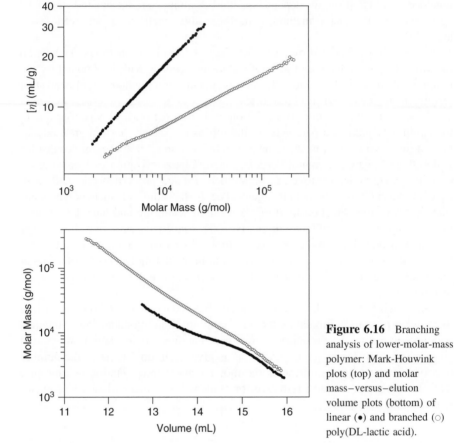

Figure 6.16 Branching analysis of lower-molar-mass polymer: Mark-Houwink plots (top) and molar mass–versus–elution volume plots (bottom) of linear (•) and branched (○) poly(DL-lactic acid).

6.2.3 Average Branching Ratios

A possible advantage of the average branching ratios is that some of the needed information can be obtained by the batch measurements of an unfractionated sample, namely M_w and R_z by classical light scattering and $[\eta]_w$ by a capillary viscometer. Moreover, these averages are not at all or only negligibly affected by the anchoring effect during the SEC separation. The average branching ratios are (1) average branching ratio g_{M_z} calculated for the same M_z, and (2) average branching ratio g'_{M_w} calculated at the same M_w.

The calculation of the average g_{M_z} at the same M_z requires knowledge of R_z and M_z. The former quantity can be obtained by a classical light scattering experiment in the batch mode or by SEC-MALS; the latter can be determined by SEC-MALS. It has been explained that the classical light scattering experiment provides R_z by its fundamental principle and consequently the value of R_z is not influenced by the anchoring effect in SEC column packing. The anchoring effect on M_z is relatively small, because this quantity counts mainly the high-molar-mass fractions that are not contaminated by molecules delayed by anchoring. In addition to the experimentally determined R_z and M_z the relation between RMS radius and molar mass for the linear polymer of the identical chemical composition must be known. Let us illustrate the calculation on the branched polystyrene (see conformation and Mark-Houwink plots in Figures 6.9 and 6.6, respectively). The experimental results obtained by SEC-MALS are as follows:

$R_z = 37.0$ nm

$M_z = 1.24 \times 10^6$ g/mol

The relation RMS radius versus molar mass $R = 0.014 \times M^{0.585}$ can be found in the literature.[15] Using this relation and experimental M_z one can calculate R_z of a linear polymer of the same M_z as that of branched polymer: $R_z = 0.014 \times (1.24 \times 10^6)^{0.585} = 51.4$ nm, which yields average branching ratio $g_{M_z} = 0.52$.

A similar calculation leads to the average branching ratio g'_{M_w} at the same M_w:

$[\eta]_w = 68.4$ mL/g

$M_w = 370,000$ g/mol

Using the literature[16] Mark-Houwink equation $[\eta] = 0.0117 \times M^{0.717}$, the intrinsic viscosity of a linear polymer having the same M_w as the branched polymer is 115.0 mL/g and consequently average branching ratio g'_{M_w} for the entire sample is 0.59. Note that the average g'_{M_w} can be determined solely by batch measurements with no influence of chromatography separation.

A possible drawback of g_{M_z} is that a part of fractions with very high molar mass can be retained in SEC columns and/or undergo shearing degradation. The R_z and M_z determined by A4F for our branched polystyrene can be used for the sake of demonstration:

$R_z = 43.7$ nm

$M_z = 2.27 \times 10^6$ g/mol

Significantly larger R_z and M_z determined by A4F and smaller average g_{M_z} of 0.36 indicate that some of the high-molar-mass fractions were degraded by shear or retained in SEC columns.

Comparison of g_{M_z} and g'_{M_w} for several branched polymers is shown in Table 6.1. It is worth mentioning that the two average ratios are not supposed to be identical, because they are based on different molar mass averages with different sensitivities to the fractions with very high molar mass, and they are mutually related with the draining parameter e, which is generally not equal to unity. It must be also pointed out that g_{M_z} and g'_{M_w} are not true moments of their distribution, and especially g_{M_z} reflects mainly the branching degree of the most-branched fractions (i.e., those with the highest molar mass).

The calculation of the average branching ratios as well as the determination of the relation between g and M require log-log relations R versus M and $[\eta]$ versus

Table 6.1 Comparison of g_{M_z} and g'_{M_w} for Various Branched Polymers Obtained by SEC-MALS-VIS

Polymer	M_w (10^3 g/mol)	g_{M_z}	g'_{M_w}
Randomly branched PS (8)	180	0.71	0.81
Randomly branched PS (9)	220	0.63	0.77
Randomly branched PS (42)	270	0.53	0.71
Randomly branched PS (B1)	250	0.53	0.80
Randomly branched PS (10)	370	0.52	0.59
Randomly branched PS (19)	55	0.49	0.80
Randomly branched PS (40)	330	0.47	0.64
Randomly branched PS (B2)	540	0.45	0.60
Randomly branched PS (20)	80	0.42	0.67
Star PBZMA (197/1)	2,900	0.09	0.09
Star PBZMA (199/1)	2,100	0.09	0.08
Star PS 8-arm	350	0.32	0.51
Brush PS	7,200	0.07	0.04

M for linear polymer of the same chemical composition as the branched polymer under investigation. The latter can be found in the literature for many polymer solvent systems. The published data are not always consistent and careful selection of the literature values and if possible their experimental verification should be carried out. Published relations R versus M are significantly less frequent.

Table 6.2 lists parameters of conformation plots for several common polymers. The determination of the conformation and Mark-Houwink plots requires measurements of a series of narrow standards or a broad polymer sample with polydispersity $M_w/M_n \approx 2$ and more. Narrow standards can be measured either in batch or online mode, whereas broad polymers require online measurement. The necessary condition is that the data cover a sufficiently broad range of molar mass of at least one order of magnitude. It must be noted that measurement of a single narrow polymer can lead to false results and conclusions about branching.

Table 6.2 Constants of Relation $R = k \times M^b$

Polymer	k	b
Polystyrene	0.014	0.585
Poly(methyl methacrylate)	0.012	0.583
Poly(benzyl methacrylate)	0.0114	0.580
Polybutadiene	0.016	0.597
Polyisobutylene	0.0145	0.581
Polyisoprene	0.012	0.611
Polyethylene	0.0186	0.596

Note: THF and room temperature with exception of PE (TCB at 135°C).

6.2.4   Other Methods for the Identification and Characterization of Branching

A certain limitation of the conformation plot based on the RMS radius is the impossibility of characterizing smaller polymers with a majority of molecules with RMS radii below 10 nm. Since the hydrodynamic radius can be accurately determined down to about 1 nm, the relation between the hydrodynamic radius and molar mass (i.e., R_h conformation plot) may become a suitable alternative.

Figure 6.17 shows the conformation plot for a relatively small polymer. Due to the scattering of RMS radii the slope of the conformation plot is determined with high uncertainty. In contrast, the R_h conformation plot appears significantly more reliable and allows accurate determination of the slope. In fact, the same information about the molecular structure can be obtained from the Mark-Houwink plot. However, the R_h conformation plot is a closer equivalent of the RMS radius conformation plot with the slope approximately equal to that based on the RMS radius.

A possible advantage of the R_h conformation plot over that based on the RMS radius is low sensitivity to abnormal SEC elution, as demonstrated in Figure 6.18. In addition, the R_h conformation plot is less sensitive to the light scattering formalisms used for data processing than the RMS radius conformation plot. A conformation plot based on R_h can be determined by SEC-MALS-VIS or SEC-MALS-DLS.

Comparison of the slopes of the conformation plots based on the RMS radius and hydrodynamic radius is shown in Table 6.3, which also includes the slopes of the Mark-Houwink plots. It may be noted that larger differences of the slopes of Mark-Houwink plots of linear and branched polymers compared

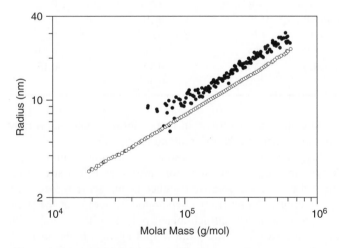

Figure 6.17   RMS radius (•) and hydrodynamic radius (○)−versus−molar mass plots of linear poly(isobutyl methacrylate) determined by SEC-MALS-VIS.

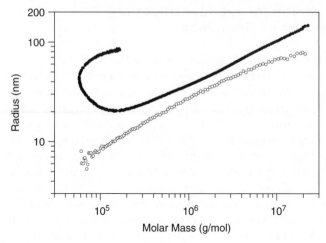

Figure 6.18   RMS radius (•) and R_h (○) conformation plots of randomly branched polystyrene determined by SEC-MALS-DLS.

Table 6.3   Slopes of Conformation Plots Based on RMS Radius and Hydrodynamic Radius and Exponents of Mark-Houwink Relation for Linear and Branched Polymers

| | Slope of Conformation Plot Based on | | Mark-Houwink |
Polymer	RMS Radius	Hydrodynamic Radius	Exponent
Linear PS	0.57	0.57	0.70
Linear PS (NIST 706)	0.57	0.57	0.70
Linear PMMA (Y)	0.57	0.57	0.71
Linear PBZMA (38)	0.57	0.57	0.70
Linear PIBMA	—	0.57	0.73
Randomly branched PS (10)	0.46	0.50	0.51
Randomly branched PS (40)	0.46	0.50	0.50
Randomly branched PS (42)	0.45	0.51	0.51
Star PBZMA (197/1), high f	0.49	0.39	0.15
Star PIBMA, high f	—	0.36	≈ 0

to the differences of the slopes of conformation plots confirm higher sensitivity of the Mark-Houwink plot to detect and differentiate branching.

Burchard et al.[14] suggested the characterization of branching on unfractionated samples and derived theoretical equations for various branched structures. In contrast to g, where the radii are compared either at the same molar mass (in the case of narrow fractions eluting from a separation device) or at the same z-average molar mass (in the case of unfractionated polymer), they came with the definition of g on the basis of R_z at the same weight-average molar mass. A potential advantage of this approach is given by the fact that the quantities R_z and M_w are directly measured by a batch light scattering experiment, while the

z-average molar mass is not directly accessible by light scattering and requires an online fractionation. In contrast to g defined at the same molar mass or M_z, where g decreases with the branching degree and is always smaller than unity, the g defined at the same M_w can be larger than unity and increase with the increasing number of branch units. The explanation of this behavior is that polydispersity causes a larger increase of R_z than the corresponding increase of M_w. The g value is then a result of two counteracting effects: polydispersity, which causes an increase, and branching, which causes a decrease. A typical feature of randomly branched polymers is very high polydispersity, which overwhelms the decrease due to the branching.

Another quantity for the characterization of branching suggested by Burchard et al. is the dimensionless ratio ρ:[14]

$$\rho = \frac{R_z}{R_{h,z}} \tag{6.13}$$

where $R_{h,z}$ is the z-average hydrodynamic radius obtained by dynamic light scattering. A potential advantage of the ratio ρ is that it does not require data for linear chains as reference.

The ratio ρ is readily measurable by combination of static and dynamic light scattering. In addition, with the combination of SEC-MALS-DLS(VIS) or A4F-MALS-DLS the ratio ρ can be measured across the molar mass distribution of polymers and thereby indicate the change of polymer architecture along the molar mass axis. The values of the dimensionless ratio ρ were derived for various polymer architectures in theta and thermodynamically good solvents.[5,14]

The ability of the dimensionless ratio ρ to detect and characterize branching is based on the fact that the two radii are of different definitions and reflect different properties, namely distribution of mass around the center of gravity and the hydrodynamic dimensions. Comparison of RMS radii and hydrodynamic radii of linear polystyrene is presented in Figure 6.19. It may be noted that the difference of R_h based on diffusion coefficient and intrinsic viscosity is small and that the molar mass dependencies of R and R_h are parallel.

Graphical representation of ratio ρ as a function of molar mass for a linear polymer, a randomly branched polymer, and a star polymer with a large number of arms is shown in Figure 6.20. Almost constant values of ρ across the molar mass distribution for the linear chains indicate identical molecular architecture across the molar mass distribution. Decreasing ρ for randomly branched molecules toward high molar masses indicate change of the molecular structure with molar mass as is typical of randomly branched polymers. However, it is not clear why the parameter ρ for the fractions with lower molar mass is larger than in the case of linear molecules. The lowest ρ for the stars consisting of large numbers of arms can be interpreted as a consequence of highly compact molecular structure.

Branching information can be also revealed from the pattern of particle scattering function. Functions $P(\theta)$ for various branched structures can be found in

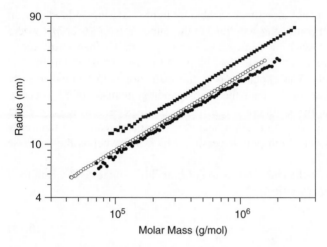

Figure 6.19 Molar mass dependence of RMS radius (■) and hydrodynamic radius determined by DLS (●) and viscometry (○) for linear polystyrene.
Data obtained by SEC-MALS-DLS and SEC-MALS-VIS, THF at 25°C.

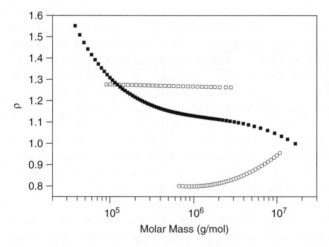

Figure 6.20 Ratio $\rho = R/R_h$ as a function of molar mass for linear polystyrene (□), randomly branched polystyrene (■), and star poly(benzyl methacrylate) of large f (○).

the literature.[14,17] In order to infer branching information the molecules must have large dimensions, which represents a significant limitation of applicability.

Another possible parameter characterizing the branching degree is the ratio of the weight-average molar mass determined by conventional SEC, M_w(SEC), to the true M_w determined by light scattering. Both M_w(SEC) and M_w can be obtained in a single SEC-MALS experiment when the obtained RI signal is processed also using the column calibration. The ratio M_w(SEC)/M_w is certainly the least fundamental parameter describing the degree of branching that has no direct relation to the number of branch units or arms in branched macromolecules. However, it may appear useful especially in the case of smaller branched polymers that are characterized by SEC-MALS without a viscometer. Since SEC separates molecules according to their hydrodynamic volume, the elution volume itself bears information about the molecular size, and consequently the M_w(SEC) calculated from a calibration curve is related to the average hydrodynamic size. If polystyrene calibration is used for data processing, the M_w(SEC) can be interpreted as the weight-average molar mass of a hypothetical sample of polystyrene having the same distribution of hydrodynamic volume as a polymer under investigation. Similarly to g and g', the ratio M_w(SEC)/M_w decreases with increasing degree of branching. However, depending on the relation between the calibration applied for data processing and the true calibration for a polymer under investigation, the ratio M_w(SEC)/M_w for a linear polymer may be larger or smaller than unity.

6.3 EXAMPLES OF CHARACTERIZATION OF BRANCHING

It has been shown that there are several possible methods useful for the characterization of polymer branching. Some of them are of rather theoretical applicability and their real ability to provide information about the branching seems to be limited. That is the case with the particle scattering function, which can provide structural information only for very large macromolecules. Additional experimental data may be needed to further verify the usability of parameter ρ, because not all results appear as straightforward as those shown in Figure 6.20. Some methods are purely empirical and have no direct relation to branched structure, such as the ratio M_w(SEC)/M_w. However, simplicity and applicability over broad molar mass range is the great advantage of this ratio, which can be readily determined by combination of conventional SEC and SEC-MALS.

The conformation plot appears to be the soundest way to obtain branching information, but it is significantly restricted by the impossibility of studying smaller molecules and moreover by frequently poor SEC separation of large, highly branched macromolecules. Consequently, the combination of A4F with MALS appears to be the most efficient method for branching studies of larger macromolecules, while for smaller branched molecules SEC-MALS-VIS will usually be the best choice. However, there is no general method for branching characterization. The results obtained by the particular methods need not be identical, but they should provide the same trends. The combination of results obtained by different methods may confirm the presence or absence of branching and indicate the structure of branched molecules. For successful branching investigation it is important to understand the limits and realize that really detailed characterization is usually impossible, and that any obtainable information may be valuable. The purpose of this chapter is to show several typical examples of data for various branched polymers, which should facilitate proper interpretation of the reader's own experimental results and help avoid making false conclusions.

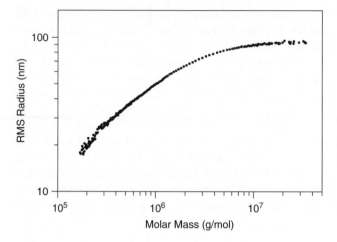

Figure 6.21 Conformation plot of randomly branched polymer with slope leveling off toward high molar masses.

Figure 6.21 is an example of a conformation plot with the slope decreasing toward high molar masses. At the high-molar-mass end, the conformation plot reaches a plateau, which means that a further increase of molar mass is absorbed inside the polymer coil without increase of molecular size. This pattern of conformation plot is typical for randomly branched polymers containing high-molar-mass fractions with high degree of branching.

Figure 6.22 shows an example of Mark-Houwink plot for a linear oligomer. The slope of about 0.5 could lead to a false conclusion of the presence of branching. However, the slope of the Mark-Houwink plots of linear oligomers is close to 0.5 even in thermodynamically good solvents, since the excluded volume

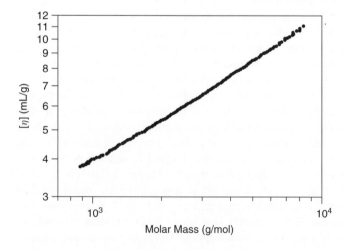

Figure 6.22 Mark-Houwink plot of linear oligoester: Mark-Houwink exponent $a = 0.48$.

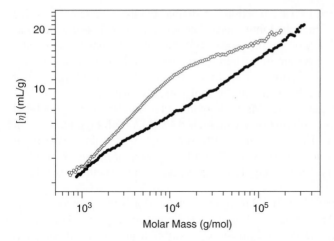

Figure 6.23 Mark-Houwink plots of unsaturated polyester resins of different distributions of branching.

effect is absent in the case of oligomeric chains that are too short to obey the intramolecular self-avoiding behavior.

Figure 6.23 depicts Mark-Houwink plots for two unsaturated polyester (UP) resins. One of the UP resins prepared from difunctional monomers (○) shows different slopes of the Mark-Houwink plot. The slope up to the molar mass of about 10,000 g/mol is roughly 0.5, which is in good agreement with the Mark-Houwink exponents for model UP resins that were prepared with maleic anhydride replaced with succinic acid that has no double bond capable of side reactions. Above a molar mass of about 10,000 g/mol the slope is significantly lower (≈ 0.16),

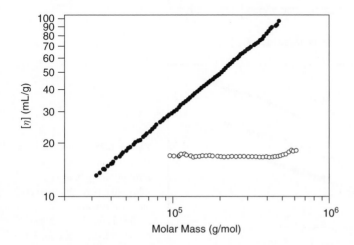

Figure 6.24 Mark-Houwink plots of linear (●) and star high-f (○) poly(isobutyl methacrylate). Zero slope indicates compact sphere-like structure of star macromolecules.

which can be explained by side reactions of double bonds of maleic anhydride leading to formation of branched molecules. The second UP resin (•), which was prepared with an addition of six-functional dipentaerythritol, shows a markedly lower Mark-Houwink exponent of ≈ 0.26 from the region of low molar masses. In this resin the addition of polyfunctional monomer resulted in the formation of branched molecules from the beginning of molar mass distribution.

Figure 6.24 compares the Mark-Houwink plot of linear poly(isobutyl methacrylate), PIBMA, with that of star-branched PIBMA with high numbers of arms prepared by GTP. The plot of linear PIBMA is linear as is characteristic of linear polymer with typical slope of ≈ 0.7, while the slope of the star PIBMA is close to zero, which indicates a highly compact molecular structure resembling compact spheres.

Another example of Mark-Houwink plots with the slopes in proximity to zero is shown in Figure 6.25 for a series of high-f star-branched polymers. The mutual shift of the plots along the axis of intrinsic viscosity indicates different compactness of the sphere-like macromolecules. It is worth noting that the Mark-Houwink

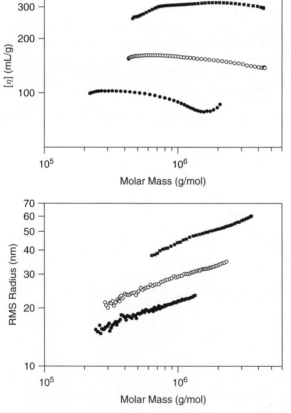

Figure 6.25 Mark-Houwink plots with the slopes close to zero for highly branched sphere-like polymers (top) and the corresponding conformation plots obtained by A4F-MALS (bottom). Slopes of the conformation plots ≈ 0.23.

plots show regions of decrease of the intrinsic viscosity with increasing molar mass. Such behavior was described for dendrimers[18] and explained as due to the polymer growing faster in density than in radial growth, when the intrinsic viscosity initially increases with molar mass, reaches a maximum, and then steadily decreases with increasing molar mass. The Mark-Houwink plots are completed by the corresponding conformation plots with the slopes confirming compact sphere-like structure of the analyzed molecules.

Figure 6.26 shows conformation and branching ratio g –versus–molar mass plots for starlike poly(benzyl methacrylate) prepared by GTP. Unlike the g-versus-M plot of randomly branched polymers (e.g., Figure 6.11), the plot in this figure starts at very low g and further decreases with increasing molar mass. This indicates that the sample does not contain, with the exception of residual unreacted arms that are not included in the plot, linear molecules. That is a significant difference compared to randomly branched polymers that typically contain a large part of linear molecules of lower molar mass. According to

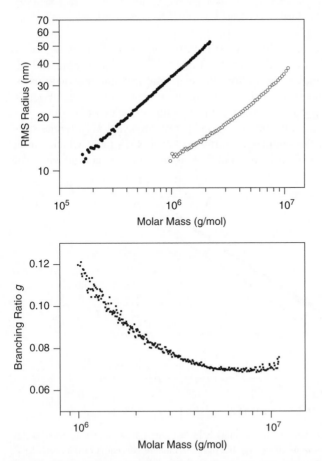

Figure 6.26 Conformation plots of linear (•) and star high-f (○) poly(benzyl methacrylate) (top) and branching ratio–versus–molar mass plot (bottom).

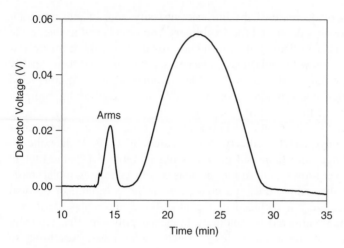

Figure 6.27 A4F RI fractogram showing separation of unreacted arms from stars for poly(benzyl methacrylate) prepared by GTP. Conformation and g-versus-M plots are shown in Figure 6.26.

Equation 6.8, the branching ratio g of 0.1–0.07 corresponds to about 60–80 arms. The presence of unreacted arms (see Figure 6.27) in the sample and the determination of their molar mass allow the estimation of number of arms by dividing M_n of stars and arms. This results in an average number of about 70 arms, which is in very good agreement with the value estimated from g.

Figure 6.28 compares conformation plots of two branched polystyrene samples. The plot of one of the samples is shifted to lower RMS radii, which can be explained by higher branching degree and thus higher polymer coil density. The

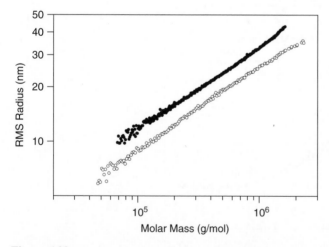

Figure 6.28 Conformation plots of two randomly branched polystyrene samples prepared using identical amounts of divinylbenzene (0.5% wt), but different initiator concentrations, namely 0.1% (•) and 1% (○). Conversion: 26% and 63% for initiator concentration of 0.1% and 1%, respectively.

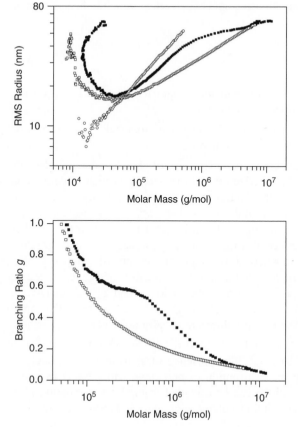

Figure 6.29 Conformation plots (top) and g-versus-M plots (bottom) of linear polyethylene NIST SRM 1475 (o) and two branched polyethylenes: NIST SRM 1476 (■) and a commercial sample (□). Columns 3× PLgel Mixed-B, TCB at 1 mL/min, 160°C.

samples were prepared using identical levels of branching co-monomer (divinyl-benzene), but different concentrations of initiator of radical polymerization. The higher radical concentration resulted in higher conversion of divinylbenzene (i.e., higher number of branch units), and also shorter chains due to termination by recombination of growing radicals.

Figure 6.29 contrasts conformation plots of three polyethylenes and the branching ratios of the two branched samples. The conformation plots of both branched samples cross that of linear polyethylene and show typical upturn in the region of lower molar masses. Except for the upturn, the conformation plot of standard NIST SRM 1476 polyethylene (■) can be separated into two regions with different slopes suggesting different branching topology, while the slope of the commercial sample (□) is approximately constant. The obtained data indicate that up to molar mass of about 5×10^6 g/mol, NIST SRM 1476 is less branched than the commercial sample since its conformation plot is shifted to higher RMS radii and the branching ratio is larger. At the region of high molar masses the plots of the two branched polyethylenes meet each other, which indicates similar degree of branching.

6.4 KEYNOTES

- The fundamental principle of branching characterization is based on the fact that branching reduces molecular size.
- The molar mass and a parameter characterizing molecular size are necessary to determine the degree of branching.
- The effect of long-chain branching on the size of branched macromolecules in dilute solution is characterized with parameter g, which is defined as the ratio of the mean square radius of a branched molecule to that of a linear molecule at the same molar mass. The parameter g decreases with increasing degree of branching and can be used to calculate the number of branch units or number of arms in branched molecules. An alternative branching ratio g' is defined as the ratio of the intrinsic viscosity of a branched molecule divided by the intrinsic viscosity of a liner molecule at the same molar mass. The branching ratios g and g' are related via the draining parameter e, whose values are supposed to be in the range $0.5-1.5$. However, the exact e values are usually unknown.
- SEC-MALS, A4F-MALS, and SEC-MALS-VIS data can be processed in three different ways to detect the presence of branched molecules and get information about the distribution of branching with respect to molar mass. The branching ratio g can be obtained from (1) conformation plot, (2) molar mass–versus–elution volume plot, and (3) Mark-Houwink plot.
- Two average branching characteristics g_{M_z} and g'_{M_w} can be calculated, the latter being obtainable even without chromatographic separation. In addition, the ratio $M_w(SEC)/M_w$ can be used as a simple parameter capable of identifying branching and differentiating branching degree of polymers of identical chemical composition.
- The conformation plot, which is the most direct way to obtain parameter g and further branching characteristics, is often affected by the anchoring of branched molecules in the pores of SEC column packing. Anchoring increases the polydispersity of slices at the region of higher elution volumes. Since polydispersity causes a larger increase of the z-average RMS radius compared to the weight-average molar mass, the conformation plots become upward at the region of lower molar masses.
- A crucial requirement for the determination of a true conformation plot is that the molecules eluting from the separation device are almost monodisperse. For many large branched polymers this requirement is not fulfilled using SEC due to the anchoring effect in the pores of column packing. The anchoring effect is totally eliminated in the case of A4F, where the separation occurs in an empty channel and such highly efficient separation

of branched molecules is achieved resulting in perfectly linear conformation plots with no artifacts. Another requirement for the determination of accurate conformation plot is avoiding the marginal scattered data points.

- Lower sensitivity to the anchoring effect and the possibility of characterizing small branched polymers are advantages of the SEC-MALS-VIS measurements.

6.5 REFERENCES

1. Meunier, D. M., Smith, P. B., and Baker, S. A., *Macromolecules*, **38**, 5313 (2005).
2. Podzimek, S., Vlcek, T., and Johann, C., *J. Appl. Polym. Sci.*, **81**, 1588 (2001).
3. Trinkle, S., Walter, P., and Fredrich, C., *Rheol. Acta*, **41**, 103 (2002).
4. Auhl, D., Stange, J., Munstedt, H., Krause, B., Voigt, D., Lederer, A., Lappan, U., and Lunkwitz, K., *Macromolecules*, **37**, 9465 (2004).
5. Burchard, W., *Adv. Polym. Sci.*, **143**, 113 (1999).
6. Zimm, B. H. and Stockmayer, W. H., *J. Chem. Phys.*, **17**, 1301 (1949).
7. Orofino, T. A., *Polymer*, **2**, 305 (1961).
8. Kurata, M. and M. Fukatsu, *J. Chem. Phys.*, **41**, 2934 (1964).
9. Burchard, W., *Macromolecules*, **10**, 919 (1977).
10. Burchard, W., *Adv. Polym. Sci.*, **48**, 1 (1983).
11. Stockmayer, W. H. and Fixman, M., *Ann. N.Y. Acad. Sci.*, **57**, 334 (1953).
12. Zimm, B. H. and Kilb, R. W., *J. Polym. Sci.*, **37**, 19 (1959).
13. Yu, L. P. and Rollings, J. E., *J. Appl. Polym. Sci.*, **33**, 1909 (1987).
14. Burchard, W., Schmidt, M., and Stockmayer, W. H., *Macromolecules*, **13**, 1265 (1980).
15. Podzimek, S., *J. Appl. Polym. Sci.*, **54**, 91 (1994).
16. Kolinsky, M. and Janca, J., *J. Polym. Sci.: Polym. Chem. Ed.*, **12**, 1181 (1974).
17. Burchard, W., *Macromolecules*, **5**, 604 (1972); **7**, 835, 842 (1974); **10**, 919 (1977).
18. Mourey, T. H., Turner, S. R., Rubinstein, M., Frechet, J. M. J., and Wooley, K. L., *Macromolecules*, **25**, 2401 (1992).

Symbols

A	Absorbance
A	Peak area, slice area
A_2	Second virial coefficient
A_3	Third virial coefficient
a	Exponent of Mark-Houwink equation
a	Extinction coefficient
a	Parameter of Schulz-Zimm distribution
a	Radius of sphere
b	Channel breadth
b	Exponent of the molar mass dependence of the root mean square radius
b	Parameter of Schulz-Zimm distribution
C_∞	Asymptotic value of the characteristic ratio C_n
c	Concentration
c_0	Concentration at accumulation wall
c^*	Critical concentration
D	Diffusion coefficient
D	Sphere diameter
d	Density
d	Particle diameter
dn/dc	Specific refractive index increment
E	Flow activation energy
e	Draining parameter
F_d	Fractionating power based on particle diameter
F_M	Fractionating power based on molar mass
$F_w(\log M)$	Weigh-differential molar mass distribution in logarithmic scale
f	Functionality of branch unit
f	Instrumental constant of light scattering photometer
f	Monomer functionality

Light Scattering, Size Exclusion Chromatography and Asymmetric Flow Field Flow Fractionation: Powerful Tools for the Characterization of Polymers, Proteins and Nanoparticles, by Stepan Podzimek
Copyright © 2011 John Wiley & Sons, Inc.

f	Number of arms
$f_n(M)$	Number-differential molar mass distribution
$f_w(M)$	Weigh-differential molar mass distribution
$G(V)$	Detector signal (chromatogram)
g	Branching ratio (branching index) based on the root mean square radius
g	Gravitational acceleration (9.81 m/s^2)
g'	Branching ratio based on the intrinsic viscosity
$g(V)$	Normalized detector signal (normalized chromatogram)
$g(\tau)$	Autocorrelation function
H	Plate height
H	Peak or slice height
H_d	Contribution of axial diffusion to plate height
H_n	Contribution of nonequilibrium to plate height
H_p	Contribution of polydispersity to plate height
IP	Inlet pressure
$I_n(M)$	Number-cumulative molar mass distribution
$I_w(M)$	Weight-cumulative molar mass distribution
I_0	Intensity of the incident radiation
I_θ	Intensity of scattered radiation at angle θ
K	Constant of Mark-Houwink equation
K_d	SEC distribution coefficient
K^*	Optical constant
k	Boltzmann's constant (1.38065 $\times$ 10^{-23} JK^{-1})
k	Capacity factor in liquid chromatography
k	Proportionality constant of molar mass dependence of root mean square radius
k_H	Huggins constant
k_K	Kraemer constant
L	Channel length
L	Column length
L	Flow cell length
L	Length of tube
L	Rod length
l	Segment length
l	Distance of sample layer center from the accumulation wall
M	Molar mass
M_n	Number-average molar mass
M_{PEAK}	Molar mass at peak
M_v	Viscosity-average molar mass
M_w	Weight-average molar mass
M_z	z-average molar mass
M_{z+1}	$z+1$-average molar mass
M_w/M_n	Polydispersity
M_0	Mass of monomeric unit
M_0	Parameter of log-normal distribution

m	Mass
m	Average number of branch units per molecule
m	Relative refractive index of scattering molecules
m_0	Slope of the angular variation of scattered light intensity at zero angle
N	Number of measurements
N	Number of theoretical plates
N_A	Avogadro's number (6.022×10^{23} mol^{-1})
n	Number of moles
n	Number of segments
n	Refractive index of polymer solution
n_w	Weight-average number of branch units per molecule
n_0	Solvent refractive index
P	Polymerization degree
P_n	Number-average polymerization degree
P_w	Weight-average polymerization degree
ΔP	Pressure difference
$P(\theta)$	Particle scattering function
p	Hydrostatic pressure
Q	Volumetric flow rate
q	Reaction conversion
R	Correction factor based on nominal and experimental polydispersity
R	Gas constant (8.314 JK^{-1} mol^{-1})
R	Inner radius of capillary
R	Retention ratio
R	Root mean square radius (radius of gyration)
RF_{UV}	Response factor of UV detector
R_h	Hydrodynamic radius
R_n	Number-average root mean square radius
R_S	Resolution
R_{SP}	Specific resolution
R_w	Weight-average root mean square radius
R_z	z-average root mean square radius
R_θ	Excess Rayleigh ratio at angle θ
R^2	Mean square radius
$R(\theta)$	Excess Rayleigh ratio at angle θ
$\langle R^2 \rangle$	Mean square radius
$\langle R^2 \rangle^{\frac{1}{2}}$	Root mean square radius (radius of gyration)
r	Inner radius of column
r	Molar ratio of monomers A and B
$r_{1,2}$	Relative retention in liquid chromatography
$\langle r \rangle^2$	Mean square end-to-end distance
$\langle r^2 \rangle^{1/2}$	Root mean square end-to-end distance
$\langle r^2 \rangle_0^{1/2}$	Root mean square end-to-end distance in theta conditions

S	Selectivity
s	Peak symmetry
s	Sedimentation coefficient
T	Temperature
T_g	Glass transition temperature
t	Time
t_R	Retention time
t_0	Retention time of unretained component
u_0	Cross flow velocity at the accumulation wall
$u(x)$	Cross flow velocity at distance x from the accumulation wall
V	Volume
V	Elution volume
V_c	Volume of SEC column, $V_c = V_0 + V_i + V_g$
V_e	Elution volume
V_g	Volume of SEC packing matrix
V_h	Hydrodynamic volume
V_i	Volume of the solvent inside the pores of SEC packing
V_i	Voltage at the ith elution volume slice
V_R	Retention volume
V_t	Total solvent volume in SEC column, $V_t = V_0 + V_i$
V_0	Channel volume
V_0	Volume of the mobile phase outside the SEC packing particles
$\dot{V}$	Volumetric flow rate
$\dot{V}$	Volumetric detector flow rate
$\dot{V}_c$	Volumetric cross flow rate
$\dot{V}_{in}$	Volumetric channel inlet flow rate
V_R'	Net retention volume
ΔV	Slice elution volume (volume between two data points)
v	Migration velocity of sample molecules
$v(x)$	Carrier velocity at distance x from the accumulation wall
$\langle v \rangle$	Average cross-sectional carrier velocity
W	Peak width at the baseline
$W_{1/2}$	Peak width at half height
w	Channel thickness
w	Weight fraction
x	Distance from the accumulation wall
x	Molar (number) fraction
$\bar{x}$	Arithmetic average
z	Axial dimension of channel
z_f	Position of focusing line
α	Expansion factor based on the root mean square end-to-end distance
α	Calibration constant of RI detector
α_R	Expansion factor based on the root mean square radius
α_η	Expansion factor based on the intrinsic viscosity
β	Exponent of the molar mass dependence of diffusion coefficient

β	Parameter of log-normal distribution
β	Signal amplitude of the autocorrelation function
γ	Shear rate
δ	Solubility parameter
ε	Parameter in Ptitsyn-Eizner equation
η	Viscosity (dynamic viscosity)
η_{inh}	Inherent viscosity
η_{red}	Reduced viscosity (viscosity number)
η_{rel}	Relative viscosity
η_{sp}	Specific viscosity
$[\eta]$	Intrinsic viscosity
ϑ	Exponent of the molar mass dependence of the second virial coefficient
λ	Branching frequency
λ	Retention parameter
λ	Wavelength of light in medium
λ_0	Vacuum wavelength of light
μ	Scattering vector
μ	Polydispersity M_w/M_n
$\bar{v}$	Partial specific volume
π	Osmotic pressure
ρ	Ratio of R to R_h
σ	Standard deviation
σ	Steric factor
σ^2	Variance of the chromatographic peak
τ	Delay time
τ	Relaxation time
Φ	Flory constant after correction for non-theta conditions
Φ_0	Flory constant
φ	Volume fraction of polymer in solution

Abbreviations

A4F	Asymmetric flow field flow fractionation
BHT	2,6-Ditert-butyl-4-hydroxy toluene
BSA	Bovine serum albumin
CRYSTAF	Crystallization analysis fractionation
DGEBA	Diglycidyl ether of bisphenol A
DLS	Dynamic light scattering
DMF	N,N-dimethyl formamide
DMSO	Dimethyl sulfoxide
DNA	Deoxyribonucleic acid
DRI	Differential refractive index
DVB	Divinylbenzene
EGDMA	Ethyleneglycol dimethacrylate
ELSD	Evaporative light scattering detector
EP	Epoxy resin
ESI	Electrospray ionization
FFF	Field flow fractionation
GPC	Gel permeation chromatography
GTP	Group transfer polymerization
HA	Hyaluronic acid
HEMA	Hydroxyethyl methacrylate
HFIP	Hexafluoroisopropanol (1,1,1,3,3,3-Hexafluoro-2-propanol)
HPLC	High-performance liquid chromatography
ICP	Inductively coupled plasma
IR	Infrared
LALLS	Low-angle laser light scattering
LC	Liquid chromatography
MALDI	Matrix-assisted laser desorption ionization
MALS	Multi-angle light scattering

Light Scattering, Size Exclusion Chromatography and Asymmetric Flow Field Flow Fractionation: Powerful Tools for the Characterization of Polymers, Proteins and Nanoparticles, by Stepan Podzimek
Copyright © 2011 John Wiley & Sons, Inc.

MO	Membrane osmometry
MS	Mass spectrometry
NMP	N-methylpyrrolidone
NMR	Nuclear magnetic resonance
NIST	National Institute of Standards and Technology
PBZMA	Poly(benzyl methacrylate)
PDA	Photodiode array
PE	Polyethylene
PEG	Poly(ethylene glycol)
PES	Polyester
PET	Poly(ethylene terephthalate)
PIBMA	Poly(isobutyl methacrylate)
PMMA	Poly(methyl methacrylate)
PTFE	Polyetetrafluoro ethylene (Teflon)
PS	Polystyrene
QELS	Quasielastic light scattering
RALS	Right-angle light scattering
REACH	Registration, Evaluation, and Authorization of Chemicals
RGD	Rayleigh-Gans-Debye
RI	Refractive index
RMS	Root mean square
RSD	Relative standard deviation
SD	Standard deviation
SEC	Size-exclusion chromatography
SP	Separation performance
SRM	Standard reference material
TCB	1,2,4-Trichlorobenzene
THF	Tetrahydrofuran
TOF	Time-of-flight mass analyzer
TREF	Temperature-rising elution fractionation
UC	Ultracentrifugation
UP	Unsaturated polyester
UV	Ultraviolet
VIS	Viscometry (viscosity detection)
VPO	Vapor pressure osmometry (vapor-phase osmometry)

Index

A

Absorbance, 130
Absorption, correction for in light
    scattering, 84
Accuracy, 193
Additives in mobile phase, 114, 121
Adsorption, 113
Anchoring of branched polymers in SEC,
    319
Asymmetric Flow Field Flow
    Fractionation (A4F), 259, 323
Autocorrelation, 59
    function, 60
Autosampler, 117

B

Band broadening in SEC-MALS, 210
Baseline subtraction, 283
Berry formalism, *see* Light scattering,
    formalism
Bond angle, 7
Bovine serum albumin (BSA), 57, 203,
    303
Branch unit(s), 308
    number of per molecule, 314, 327
Branching
    degree, 308
    frequency, 314
    long-chain, 309
    random, 309
    ratio, 313–315, 317
        averages, 330
        mass method, 321, 325

radius method, 321
    viscosity method, 321
    short-chain, 309
Burchard-Stockmayer-Fixman method, 9

C

Calibration
    curve in SEC, 110, 111, 144, 145
    of column in SEC, 143
    of detector response, 180
    of IR detector, 134
    of light scattering photometer, 55
    of RI detector, 79
    of UV detector, 131
    standards
        narrow, 144
        polydisperse, 147
    universal, 149
Capacity factor in HPLC, 109
Channel
    components, 263
    thickness, 262
    volume, 267
Characteristic ratio, 8
Chromatogram, 105, 159, 162
Column, 122
    effect of temperature, 184–186
    guard, 123
    narrow-bore, 122
    packing, 123
    preparative, 122
    rapid, 122
    volume, 103

*Light Scattering, Size Exclusion Chromatography and Asymmetric Flow Field Flow Fractionation:
Powerful Tools for the Characterization of Polymers, Proteins and Nanoparticles*, by Stepan Podzimek
Copyright © 2011 John Wiley & Sons, Inc.

355

Concentration
  critical, 29
  effect of in SEC, 186–191
  profile in A4F channel, 265–267
Configuration, 2
Conformation, 2
  plot, 254, 302, 320, 322–324, 334,
    338, 340–343
  plot based on R_h, 333
Contour length, 10
Conversion, 18
Copolymer, 2
  alternating, 3
  block, 2
  conversion heterogeneity of, 4
  graft, 3
  random, 2
  statistical heterogeneity of, 4
Correlator, 59
Cross flow, 267
  gradient, 287, 289
  isocratic, 287
  velocity, 263, 265
Crystallization analysis fractionation
  (CRYSTAF), 311

D

Debye formalism, *see* Light scattering,
  formalism
Debye plot, 54, 86, 87, 216, 223–225,
  227–229, 231
Dendrimer, 310
Derivatization, 159
Detection limit, 127
  of MALS detector, 54
Detector, 127
  drift, 128
  evaporative light scattering, 134
  infrared, 133
  light scattering, 140
  photodiode array, 130
  refractive index (RI), 131
  signal-to-noise ratio, 128
  time constant, 129
  UV, 130
  viscosity, 135
Dialysis, 68
Diameter of sphere, 45, 48
Diffusion coefficient, 11, 265, 276, 285

Distribution coefficient in SEC, 102

E

Elution volume, 102, 105
End group method, 23
Enthalpic interactions, *see* Interactions,
  enthalpic
Excluded volume, 5
Expansion factor, 6
Exponent
  a, *see* Mark-Houwink, exponent
  b, 33, 332
  β, 11, 33, 274
Extinction coefficient, 130
  of epoxy resins, 181
  of polystyrene, 82
Extrapolation of data in SEC-MALS, 236

F

Fast SEC-MALS, 247
Flory constant, 9
Flory distribution, *see* Molar mass,
  distribution, most probable
Flory-Fox and Ptitsyn-Eizner
  equation, 32
Flory-Fox equation, 8
Flow activation energy, 34, 308
Flow cell of MALS photometer, 64
Flow marker, 152
Flow rate, in SEC, 152, 154, 242
Fluorescence, correction for in light
  scattering, 84
Focusing, 269
  position of focusing line, 269
  time, 269
Fractionating power, 277, 285

G

Gel permeation chromatography (GPC),
  see Size-exclusion chromatography
Glass transition temperature, molar mass
  dependence of, 11
Goniometer, 63
Group transfer polymerization, 310

H

Height equivalent to theoretical plate
  (HETP), *see* Plate height
Homopolymer, 2

Huggins
  constant, 27
  equation, 27
Hydrodynamic radius, 32, 61
Hydrodynamic volume, 31

I

Interactions
  enthalpic, 113
  hydrophobic, 114
  polymer-polymer, 4, 41
  polymer-solvent, 4, 41
Interdetector volume, *see* Volume delay
Interference of scattered light
  intermolecular, 57, 59
  intramolecular, 57

K

Kraemer
  constant, 28
  equation, 27
Kuhn segment, *see* Statistical segment

L

Lambert-Beer law, 130
Light scattering
  composition gradient multi-angle, 95
  dynamic, 39, 59
  excess, 38
  fit method, *see* Light scattering,
    formalism
  formalism
    Berry, 54, 221
    Debye, 54, 208, 220
    Zimm, 54, 220
  in batch mode, 72, 84
  in chromatography mode, 72
  low-angle (LALLS), 64
  multi-angle (MALS), 63
  quasielastic, *see* Light scattering,
    dynamic
  Rayleigh, *see* Light scattering, static
  static, 39, 40
Limit
  of total exclusion, 102, 104
  of total permeation, 102
Limiting viscosity number, *see* Viscosity,
    intrinsic
Liquid chromatography, 99

M

Mark-Houwink
  constants, 29, 149
  equation, 29
  exponent, 29, 30, 33, 149, 325
  plot, 31, 255, 322, 330, 338–340
Matrix-assisted laser desorption mass
    spectrometry, 34
Mean square end-to-end distance, 5, 45, 48
Model of polymer chain
  equivalent chain, 10
  freely jointed chain, 6
  freely rotating chain, 7
Molar mass
  accuracy of determination by light
    scattering, 78
  distribution, 13
    cumulative, 13, 163
    differential, 13, 162
    log-normal, 18
    most probable, 19
    number, 13
    Schulz-Zimm, 17
    Tung, 17
    weight, 13
  number-average, 21, 138, 161, 232
  PEAK average, 144, 147, 162
  viscosity-average, 22, 162
  weight-average, 21, 42, 161, 232
  z-average, 22, 142, 162, 232, 234
  z+1 average, 22, 142, 162
Molecular weight, *see* Molar mass
Monomer unit, 2

N

NIST SRM 706 polystyrene standard, 82,
    83
  A4F-MALS results, 298
  effect of peak end on molar mass
    averages, 169
  molar mass distribution plots, 14
  SEC-MALS results, 237
  Zimm plot, 85
Nonequilibrium contribution to plate
    height, 276
Normalization, 55
Number fraction, 15
Number of arms in star polymers, 315
Number of theoretical plates, 107

O

Oligomer, 2, 175, 202
   chromatogram, 176–178, 181, 183
Optical constant K^*, 41
Osmometry
   membrane, 25
   vapor pressure, 24
Overloading in A4F, 288

P

Parabolic flow profile in A4F, 262
Particle scattering function, 42, 47
   of homogeneous spheres, 48
   of linear random coils, 47
Peak broadening, 105, 275
Peak symmetry, 108
Plate height, 107, 276, 285
Polarizability, 38
Polydispersity, 2, 22
   contribution to plate height, 277
Polymer, 1
   branched, 3, 308
   hyperbranched, 311
Polymerization degree, 12
   of polycondensation products, 19

R

Radius of gyration, *see* Root mean square
   radius
Radius of sphere, 45, 48
Random coil, 4
Ratio
   M_w(SEC)/M_w, 336
   ρ, 335
Rayleigh-Gans-Debye approximation, 53
Rayleigh ratio, 40
REACH (Registration, Evaluation, and
   Authorization of Chemicals), 182
Repeatability, 192, 240
Reproducibility, 192, 240
Resolution, 108, 275, 285
Results fitting in SEC-MALS, 236
Retention parameter, 265, 267
Retention ratio, 268
Retention time, 268
Root mean square end-to-end distance, 5,
   150

Root mean square radius, 6, 44
   number-average, 232
   versus molar mass relation, 33, 332
   weight-average, 232
   z-average, 46, 233
Round robin test, 193

S

Scattering vector, 43
Second virial coefficient, 25, 40
   in theta state, 5
   molar mass dependence of, 42
Segment length, 6, 8, 10
Selectivity, 109, 110, 273
Semipermeable membrane, 25, 263, 279
Separation mechanism
   in A4F
      hyperlayer, 272
      normal (Brownian), 271
      steric, 271
   in SEC
      restricted diffusion, 103
      separation by flow, 103
      steric exclusion, 102
Separation performance, 112
Size-exclusion chromatography (SEC), 23,
   99, 207, 259, 313
Solubility parameter, 155
Solvent(s)
   for SEC, 118
   thermodynamically good, 5, 27, 41
   thermodynamically poor, 5, 27
Spacer, 262–264
Specific refractive index increment, 41, 65
   in multicomponent solvents, 68
   of aqueous sodium chloride, 80
   of polymer blends and copolymers, 70,
      252
   of various polymers, 71
Specific resolution, 109
Standard deviation
   of measurement, 192
   of peak, 105, 275, 277
Statistical segment, 10
Steric factor, 7
Stokes-Einstein relation, 61
Structural repeating unit, 2
Synthetic resins, 2

T

Tacticity, 3
Temperature rising elution fractionation
    (TREF), 311
Tetrahydrofuran, 118
Theta conditions, 5
Theta solvent, 41
Tung's equation, 105

U

Ultracentrifugation, 35
Unperturbed dimensions, 5

V

Viscometry of dilute solution, 26
Viscosity
    inherent, 27

intrinsic, 27, 136
molar mass dependence of polymer
    melts, 11
reduced, 27
relative, 26
specific, 26, 135
temperature dependence of, 34
Viscous fingering, 187
Volume delay, 209

W

Weight fraction, 15

Z

Zimm formalism, *see* Light scattering,
    formalism
Zimm plot, 54, 85, 88, 90–93

Philadelphia Preserved

Published with the
Philadelphia Historical Commission

The Historic American Buildings Survey
is a program of
the National Park Service,
United States Department
of the Interior.

Philadelphia Preserved

Catalog of
the Historic American
Buildings Survey

Richard J. Webster

With an Introduction by
Charles E. Peterson

Temple University Press
Philadelphia

Temple University Press, Philadelphia 19122
© 1976 by Temple University. All rights reserved
Published 1976
Printed in the United States of America

International Standard Book Number: 0-87722-089-1
Library of Congress Catalog Card Number: 76-18669

Contents

Illustrations   vii

Foreword by Richard Tyler   xiii

Preface by Richard J. Webster   xv

HABS—In and Out of Philadelphia
by Charles E. Peterson   xxi

                                         Introduction/Catalog

Penn's City: Society Hill   1/9

Penn's City: Old City   37/59

Penn's City: Center Square   103/121

South Philadelphia: Moyamensing and Passyunk
Townships   151/165

West Philadelphia: Blockley and Kingsessing
Townships   191/203

Fairmount Park   221/227

Northwest Philadelphia: Germantown and
Roxborough Townships   243/259

North Philadelphia   281/293

Delaware River Corridor: Northern Liberties,
Kensington, Richmond, Frankford, and
Northeast Philadelphia   303/321

Notes   341

Index   383

Illustrations

HABS—In and Out of Philadelphia

Following page xlvi:
Independence Hall
Bingham House
Second Bank of the United States, iron gates
Anthony Benezet House
Wyck
Central High School cupola
Second Bank of the United States
Philadelphia Exchange Company (Merchants Exchange)
Whitby Hall
Vernon
Leicester B. Holland
E. Perot Bissell
Joseph P. Sims
James C. Massey

Philadelphia Preserved

Facing page 1:
Map of Philadelphia

Penn's City: Society Hill

Page 2:
Map of Society Hill
Following page 36:
Bethel African Methodist Episcopal Church
Man Full of Trouble Tavern
Pennsylvania Hospital
Samuel Powel House
St. Peter's (Protestant Episcopal) Church

Shippen-Wistar and Norris-Cadwalader Houses
Water Trough and Fountain
Joseph Wharton House
William H. Winder Houses

Penn's City: Old City

Page 38:
Map of Old City

Following page 102:
Arch Street Friends Meeting
Athenaeum of Philadelphia
John Brock Sons & Co. Warehouse
Christ Church
Elfreth's Alley Area Study
Farmers' and Mechanics' Bank
First Bank of the United States
Independence Hall, Assembly Room, tabernacle frame
Independence Hall, Central Hallway
Insurance Patrol
Leland Building
Pennsylvania Fire Insurance Company Building
Philadelphia Bourse
Philadelphia Exchange Company (Merchants Exchange)
Philadelphia Fire Department, Engine Co. No. 8
Philosophical Hall
Provident Life and Trust Company Bank
Reynolds-Morris House
Second Bank of the United States

Penn's City: Center Square

Page 104:
Map of Center Square

Following page 150:
Academy of Music
Bulletin Building
Girard Trust Corn Exchange Bank
Harrison Building
Market Street National Bank Building
Pennsylvania Academy of the Fine Arts
Pennsylvania Institution for Deaf and Dumb
Pennsylvania Railroad, Broad Street Station
Philadelphia and Reading Railroad, Terminal Station, Train Shed

Philadelphia City Hall
Philadelphia City Hall, toilet stalls
St. Clement's (Protestant Episcopal) Church
Scott-Wanamaker House
Sims-Bilsland House (Franklin Row)
Spruce Street Area Study
West Arch Street Presbyterian Church

South Philadelphia

Page 152:
Map of South Philadelphia

Following page 190:
Gloria Dei
Jesse Godley Warehouse
Nathaniel Irish House
Nathaniel Irish House, fireplace detail
George Pearson Houses
Philadelphia County Prison
Ridgway Branch of Library Company of Philadelphia
United States Naval Asylum, Surgeon's Residence

West Philadelphia

Page 192:
Map of West Philadelphia

Following page 220:
George Burnham House
Church of St. Francis de Sales
Fort Mifflin, Sally Port
Professor Henry Charles Lea House
Mount Moriah Cemetery Gate House
Pennsylvania Hospital for Mental and Nervous Diseases
Pennsylvania Institution for Instruction of the Blind
Tabernacle Presbyterian Church
University of Pennsylvania, Men's Dormitories
University of Pennsylvania, University Museum
Woodland Terrace
Woodlands

Fairmount Park

Page 222:
Map of Fairmount Park

Following page 242:
Belmont Mansion
Cedar Grove
Fairmount Waterworks
International Exhibition of 1876, Memorial Hall
International Exhibition of 1876, Ohio State Building
Laurel Hill Cemetery Gatehouse
Philadelphia Zoological Gardens, Bear Pits
Strawberry Mansion Bridge

Northwest Philadelphia

Page 244:
Map of Northwest Phildelphia

Following page 280:
Belfield
Cliveden
Concord School
Germantown Cricket Club
Grumblethorpe
Grumblethorpe, Water Pump
Hood Cemetery Entrance
Houston-Sauveur House
Ebenezer Maxwell House
Mennonite Meeting House
Stenton Barn
Trinity Lutheran Church
Upsala

North Philadelphia

Page 282:
Map of North Philadelphia

Following page 302:
Cast-Iron Sidewalk
Church of St. James the Less
Church of the Gesu
Eastern State Penitentiary
Girard College, Founder's Hall
Green Street Area Study (Kemble-Bergdol House)
Northern Saving Fund, Safe Deposit and Trust Company
Preston Retreat
Ridge Avenue Farmers' Market Company
Shibe Park
Peter A. B. Widener House

Delaware River Corridor

Page 304:
Map of Delaware River Corridor

Following page 340:
John Bromley & Sons, Lehigh Mills
Burholme
Henry Disston & Sons Keystone Saw, Tool, and Steel Works
Thomas Holloway Houses
Hospital of the Protestant Episcopal Church in Philadelphia
Carl Mackley Apartments
Trinity Church, Oxford

Foreword

Since the creation of the Philadelphia Historical Commission in 1955, the Commission and the Historic American Buildings Survey have maintained a continuously evolving relationship. Together they have striven to identify and protect the historically and architecturally significant structures of Philadelphia. They have also sought to preserve at least an archival record of the tremendous variety of buildings which make this city one of the nation's richest sources of architectural, social, and cultural history. Finally, the Commission and the Survey have collaborated in compiling the historical data essential to the scholarly utility of a comprehensive collection of measured drawings and photographs. After twenty-one years, their collaboration has reached fruition in this excellent publication, *Philadelphia Preserved: Catalog of the Historic American Buildings Survey*.

In the beginning, the already long established Survey provided a documented base for the Commission's first efforts, and lent stature to the recommendations and decisions of the municipal agency. The Commission also adopted several of the Survey's proven procedures and methods for the preparation of an inventory of buildings. As the Commission and its reputation grew, and as its staff (under the supervision of Dr. Margaret B. Tinkcom, the Commission's first historian) undertook fundamental research, the Survey and this Catalog came to rely increasingly on the resources of the Commission. Without the scores of eighteenth-, nineteenth-, and twentieth-century insurance surveys, the hundreds of chains-of-title, and the countless notes on builders and architects in the Commission's files, the Catalog as it now appears simply would have been an impossibility.

The participation of the Philadelphia Historical Commission in the publication of this work extends far beyond the sharing of information. It made office space and services available to Richard Webster, the editor. It alerted the Survey to the approaching demolition of important buildings, such as Moyamensing Prison and the entire neighborhood of early houses now replaced by the Delaware Expressway, so that these structures could be recorded. Through its own local activities, it contributed to the expansion of the Survey's listings in Philadelphia from some 60 primarily eighteenth-century structures in 1959 to the present 600 buildings of every period and genre, including a castiron sidewalk. And, perhaps most importantly, it agreed in 1969 to publish the Catalog and to raise the necessary subvention.

Much of the credit for the Commission's constant involvement with the Catalog and for the quality of the Catalog itself belongs to Margaret B. Tinkcom. She persuaded the Commission to underwrite its publication and directed the staff research which complements Mr. Webster's work. Her vision of preservation and significance brought professional standing to a much larger appreciation of the City's architecture. She played an instrumental part in securing the service of Mr. Webster as editor and acted as his unfailing advisor. Finally, she handled much of the requisite administrative detail which resulted in *Philadelphia Preserved: Catalog of the Historic American Buildings Survey.*

Richard Tyler
Philadelphia Historical Commission
Philadelphia, Pennsylvania

Preface

Since 1933, when the Historic American Buildings Survey was established as a Works Progress Administration program, HABS has grown into a national archive of significant American architecture. Under a tripartite agreement dating from 1934, HABS administers the program and is responsible for qualitative standards, selection of subjects for recording, and organization of the recording projects. The Library of Congress is the depository of the records, which are serviced by the Prints and Photographs Division, and the American Institute of Architects provides professional advice.

The Survey's goal is to develop the broadest possible coverage for all areas, periods, styles, and types of American architecture. Structures represented in the Survey span the period from both prehistoric and colonial times to the early twentieth century and include examples from all fifty states, the District of Columbia, Puerto Rico, and the Virgin Islands. The collection now includes approximately 16,000 buildings represented by more than 31,000 drawings, 47,000 photographs, and 23,500 pages of written data.

Those who are interested in consulting the HABS records may either visit the Division of Prints and Photographs of the Library of Congress, or consult the published HABS catalogs. A comprehensive, geographically arranged *Catalog* was published in 1941, and was updated by a *Supplement* in 1959. More recently, because of the extensiveness of HABS holdings, new catalogs are being published for states and metropolitan areas. To date, state catalogs have appeared for New Hampshire (1963), Massachusetts (1965), Wisconsin (1965), Michigan (1967), Utah (1969), Maine (1974), and Texas (1974). The catalogs for Chicago and nearby Illinois areas and for Washington,

D.C., were published in 1966 and 1976 respectively. Most of these publications can be found in major university or public libraries. Further questions regarding the consultation of records and ordering of reproductions may be addressed to:

> Division of Prints and Photographs
> Library of Congress
> Washington, D.C. 20540.

Questions regarding HABS recording and publishing programs or materials temporarily held in HABS offices may be addressed to:

> Historic American Buildings Survey
> National Park Service
> Washington, D.C. 20240.

In 1970 the National Park Service published *Recording Historic Buildings,* compiled by Professor Harley J. McKee. Based on the Survey's long experience and widely regarded as the definitive work in the field, this volume establishes standards and procedures for gathering both historic and architectural documentation. Copies may be purchased from:

> Superintendent of Documents
> U.S. Government Printing Office
> Washington, D.C. 20402.

More than 600 examples of Philadelphia's structures and street furniture are listed in this catalog. They represent the total number of items in the city that have been recorded by the Historic American Buildings Survey through the summer of 1976. The Philadelphia survey includes all of the entries from the national 1941 *Catalog* and the 1959 *Supplement,* which have been updated and expanded to provide more useful information. During the 1960's and 1970's, approximately 550 structures were added to the Survey to make it more comprehensive and illustrative of the city's multiplicity of building types. Most of the additions of the 1960's resulted from two major cooperative projects that recorded buildings in the path of Interstate Route 95 (Delaware Expressway). The Southwark Project, funded by the American Philosophical Society and executed by HABS and the Philadelphia Historical Commission, led to the publication of Margaret B. Tinkcom's excellent architectural history of the area in the August 1970 *Proceedings of the American Philosophical Society.* The

Pennsylvania Historical Salvage Council, a joint project of HABS, the Philadelphia Historical Commission, and the University of Pennsylvania, centered its attention on Northern Liberties and Kensington, north of the original city limits.

The catalog is arranged much as a museum script. The central focus is on the material objects, or, more accurately, the catalog entries, which are organized alphabetically within nine historical-geographical districts. Preceding each group of entries is an architectural history of that district which relates its buildings to the district's social and economic development. Following each section is a gallery of illustrations, also arranged alphabetically for the convenience of the reader.

As far as possible, both the introductions and entries reflect the latest historical research. In many cases it has been possible to correct dates for buildings or to note them more precisely. More historically accurate names have been assigned to some buildings, and abundant cross-references under formerly or popularly used names have been employed. The identification and history of the buildings are based on a variety of sources, but the most frequently used sources were briefs of title, fire insurance surveys, building permits, newspapers, and old maps and atlases. While briefs of title (compiled for the most part by the Philadelphia Historical Commission), newspapers, and maps are familiar sources, the fire insurance survey remains a frequently overlooked but invaluable research tool. There is an abundance of surveys for Philadelphia buildings in particular, because of the early and prosperous history of fire insurance companies in the city, dating from the 1752 founding of the Philadelphia Contributionship for Insuring Houses against Loss by Fire. Surveys and re-surveys often possess detailed information on buildings' functions, plans, materials, and technology, which is useful for both the historian and the restorer. Building permits, on the other hand, are useful for only the twentieth century. Although surviving permits date from the late 1880's, an index system was not begun until 1902.

The format of the entries consists of: the historic name; HABS number; address; brief physical description, including construction materials, dimensions, number of stories, roof type, and important interior and exterior details; statement of architectural significance; date of erection; architect and builder, if known; alterations and additions; statement of historical significance; citations by public agencies; and a listing of the measured drawings, photographs, photocopies, and data pages available in the HABS archives.

The following abbreviations, terms, and symbols have been used in the catalog entries:

PA-100   All structures recorded by the Survey are assigned a Historic American Buildings Survey number. These numbers have no historical significance but serve only to facilitate processing. They should be used when inquiring about a structure or ordering reproductions.

London   Adopted from William Murtagh's article on the
house plan   Philadelphia row house in the December 1957 issue of the *Journal of the Society of Architectural Historians*, this term applies to a house plan two rooms deep with a side hall leading to stairs between the two rooms.

city   Also from Murtagh's article, this plan consists of a
house plan   one-room front building and a rear ell which includes a piazza, or stair tower.

town   The largest and most sophisticated of Philadelphia
house plan   house plans, the town house plan contains a wide front section with two rooms and a stair hall and narrower backbuildings.

certified,   The building has been judged historic by the
PHC   Philadelphia Historical Commission, affording the building an element of protection from wanton demolition and inaccurate or unsympathetic alteration.

Pennsylvania   The building has been entered on the Pennsyl-
Register   vania Register of Historic Sites and Landmarks.

NR   The building has been entered on the National Register of Historic Places.

sheets   The number of sheets of measured drawings available for study and reproduction is cited near the end of the catalog entry. Sheets are a standard size, 15¹/₂ inches by 20 inches inside border lines. The number of sheets in the set and kinds of drawings (plans, elevations, sections, details) are listed.

Prints of measured drawings are made at actual size.

photos

The number of exterior and interior photos available for reproduction is also cited near the end of the entry. HABS negatives normally are 5 by 7 inches, but are sometimes in other sizes, especially 4 by 5 inches or 8 by 10 inches. Duplicate prints of HABS photographs are the same size as the originals.

data pages

Data pages include information on the structure, ideally its physical history and a technical architectural description. Original data pages are typewritten and may be duplicated.

HABSI form

The Historic American Buildings Survey Inventory form is a single-page record that provides for identification of the structure, concise written historical and/or architectural data, a small photograph, a location diagram, and source references. HABSI forms are duplicated at actual size (8 by $10^{1}/_{2}$ inches).

n.d.

The date is not ascertainable.

*

An asterisk after a date indicates that the measured drawings, photographs, or data pages made on the indicated date are temporarily being held for editing and processing at the Washington, D.C., office of the Historic American Buildings Survey, prior to transmittal to the Library of Congress.

I am indebted to many people and institutions who have helped to make this book possible. First tribute must go to Dr. Margaret B. Tinkcom, former historian of the Philadelphia Historical Commission, for her valuable counsel and scholarly direction. I am also indebted to the Philadelphia Historical Commission staff for its many favors, and especially to Mrs. Beatrice Kirkbride for sharing with me her expert knowledge of Philadelphia insurance surveys. I appreciate the generosity of the many contributors to the Harry A. Batten Memorial Fund of the Philadelphia Historical Commission, which made the book's publication economically possible. Some of the city's leading business institutions that made major contributions are First Pennsylvania Banking and Trust Company, Germantown Savings Bank, Girard Bank, Mutual Assurance Company for Insuring Houses against Loss by Fire, Old Original Bookbinder's, Penn Mutual Life Insurance Company, Philadelphia Contributionship for Insuring Houses against Loss by Fire, Philadelphia Saving Fund Society, Daniel J. Walsh's Sons, and Western Saving Fund Society. Major individual contributors include Dorothy Collins, C. Jared Ingersoll, Olan Lowry, Esq., Lawrence M. C. Smith, Charles R. Tyson, members of the Washington Square Association, Henry M. Watts, Jr., and members of the Philadelphia Historical Commission staff.

I also appreciate the cooperation of the many present and former members of the Historic American Buildings Survey staff, particularly James C. Massey, John C. Poppeliers, S. Allen Chambers, Jr., and Christine St. Lawrence; Allen Weinberg, John Daly, and Ward Childs of the Philadelphia City Archives; Joseph McDevitt of Philadelphia Records Storage; Lee Nelson and Penelope Hartshorne Batcheler, architects with the National Park Service; Carolyn Pitts of the Historic Sites Survey, National Park Service; Isadore Lichstein, formerly of the *Evening and Sunday Bulletin*; Reeves Wetherill, Public Relations Director of John Wanamaker; John F. Harbeson, F.A.I.A.; Charles E. Peterson, F.A.I.A.; George E. Thomas; the staffs of the Philadelphia City Planning Commission, Free Library of Philadelphia, Historical Society of Pennsylvania, Library Company of Philadelphia, and Philadelphia Museum of Art; and three fine men and excellent photographers, Jack E. Boucher, George A. Eisenman of James Dillon and Company, and Cortlandt V. D. Hubbard.

I am also grateful for the nearly three centuries of Philadelphia architects, builders, craftsmen, and patrons who have produced a city rich in material culture to be enjoyed by all who care to look about them.

And for my cheerful and loving wife, Yvonne Sanwald Webster, I reserve the last and most grateful credit line.

Richard J. Webster
West Chester State College
West Chester, Pennsylvania

HABS—In and Out
of Philadelphia

For more than forty years the Historic American Buildings Survey has been an active institution and a source of ready aid to a large community scattered from coast to coast, and beyond. But published accounts of the Survey are scarce—perhaps because few architects write for publication and the founders of the Survey were all architects.[1] In addition to architects in different parts of the country, an energetic and dedicated Park Service staff in Washington was essential to the success of the new enterprise. The tiny headquarters staff worked up directives to the men in the field, collected the harvest of drawings, and pushed it on to the Library of Congress, where it was and still is available to all. A distinguished Advisory Commission, early appointed and convened, lent prestige and support to HABS, the bureau's first national venture into historiography.

Those pioneers of the 1930's and their successors with antlike industry built, and have never stopped building, the great collection of architectural drawings, photographs, and historical data in the Library of Congress. But many of the official files that dealt with administration and personnel have been dispersed and for the time being at least are out of reach. To assemble a definitive chronicle of the HABS achievement now would entail major archival campaigns.[2] But the *architectural* records, far from lying forgotten, have been actively used from the first days of their arrival at the Library. It has been shown that HABS is one of the most actively consulted archives in that vast collection of collections. As works of graphic analysis, they are more than just mindless mechanical exercises; they provide a basic understanding and appreciation of the American builders' art.

The original proposal still works as a comprehensive, loose-fit charter that in later years has allowed development in several directions. Notably, since 1957, the interpretive and artistic standards of HABS photography have been built up. The policy was then developed that photographs should be of publishable quality, with great care being taken as to camera position, lighting, and other technical details. And the historical research that supplies the textual component of the Survey has come to incorporate some of the best-informed work being done today.

Much of the Survey as now conducted follows procedures and techniques developed at Philadelphia in the period 1950–1962. The program was deliberately used at that time for the recruitment and training of young architects considering the field of historic conservation as a career. Some of them stayed with the Survey and others moved into the field of restoration; a number are today in important positions nationally.

From Philadelphia the new wave of HABS spread to San Francisco and then from the Virgin Islands and Puerto Rico to Hawaii and Alaska. The Survey is now well rooted and promises to flourish indefinitely. Certainly there is plenty to do as public moneys flow into historic conservation programs. Today hard data are more needed than ever before.

Early Portraits of Philadelphia

The reader consulting this catalog will soon note that the buildings of Philadelphia have had an extraordinary amount of attention from HABS. With more than six hundred items, the collection for this city is more extensive than for any other; more buildings were recorded here than in many states. The Philadelphia collection was not made too soon, for there have been mighty changes in the architectural face of the city in the last twenty-five years. At best, our fine old buildings have been carefully restored, HABS drawings often serving as the base upon which the architects' plans were developed. At worst, a tragic number have been removed, among several reasons, for highway rights of way, by the lethal hand of urban renewal, and through the negligence—if not animosity—of city planners. In more than one case HABS was on hand during the last agonies of buildings under demolition.

Appreciation of Philadelphia's landmarks as such took time to develop. We do have one exceptionally early example reported by

Peter Kalm, the agricultural explorer from Upsala, who spent much time here. In the year 1749 he wrote that Philadelphians were already preserving an old structure—the Swanson house—as a reminder of pioneer days![3] Unfortunately, we have no sure record of it, and afterward it disappeared.

We have very few portraits of individual buildings during Philadelphia's first century. But beginning in 1750 the situation improved. In that year Scull and Heap's map of the city carried an engraved view of the State House. Then came the "East Prospect" of the city drawn by George Heap in 1752 from a point across the Delaware River.[4] In the Heap view (as engraved, over seven feet long) the steeples are identifiably portrayed and the warehouses on the river bank have a characteristic gambrel-roofed individuality. A splendid start.[5]

First Records by Architects

Scale drawings—floor plans, elevations, sections and details—are the architect's peculiar idiom. Such precise documents are required to fix building projects that can then be turned over to mechanics for execution at the site. They have other uses, too. When an architect wants to analyze an existing building closely, he makes a measured drawing of it. This reveals to him the peculiar relationship between what the eye sees in perspective and how the facts appear when formally projected on a sheet of paper. Plans and sections speak to the professional of things not seen by the eye.

European architects had long been making measured drawings of the buildings that survived antiquity. First Roman structures, and then Greek, provided a treasury of details that could be copied or adapted in new buildings across Britain and Europe. Their books, published in London, could be found in American libraries, and details from engraved plates might be seen on new American buildings.[6] But while the measuring of antiquities did not start early in America, we do have the example of a Boston architect picking up ideas in Philadelphia. In 1789, Charles Bulfinch carried home an elevation of William Bingham's new and fashionable house on South Third Street,[7] which he used soon afterward as the design for the Harrison Gray Otis House on Cambridge Street, Boston, for many years now the headquarters of the Society for the Preservation of New England Antiquities.

Picture books of Philadelphia began in earnest at the end of the eighteenth century, when the English-trained William Russell Birch engraved his attractive and well-known set of *Views* intended as "a memorial of progress for the first century." Today highly prized by collectors and often reproduced, these charming perspectives delineate buildings now regarded as historic which in their time were just a part of the contemporary scene.

The first outright *antiquarian* record by an *architect* may well have been William Strickland's drawing of Anthony Benezet's house on Chestnut Street commissioned by Roberts Vaux just before it was pulled down in March, 1818. The street elevation appeared as a frontispiece to *The Port Folio* of Philadelphia for October, 1818. The thought was expressed that it was gratifying to look upon the buildings that our ancestors occupied. "There remains," according to the anonymous staff writer, "scarcely a vestige of the style of architecture adopted by our goodly and adventurous predecessors. It is but a few months since one of the oldest, *if not the first brick house* erected in Philadelphia, was torn down, to give space to a more spacious structure, and we believe that edifice to have been the last *specimen* in this city, toward which the curious inquirer in these matters, might have been directed."[8]

In 1824, the triumphant return of General Lafayette to America on its fiftieth anniversary triggered waves of patriotic nostalgia that have continued to inspire writers and artists. At that point there fortunately appeared in Philadelphia an Englishman—William L. Breton—who had, he said, learned to draw on the way across the Atlantic. Soon after his arrival he sketched the Stritzel house in Church Alley and the Carpenter house on Chestnut, both soon to be torn down.[9] Breton's views first appear in *The Casket* (another local magazine) for 1827 and they quickly became familiar to the reading public in the form of engravings, woodcuts, and lithographs.

Local historians, beginning with Robert Proud, did not seem to fancy architectural landmarks. It remained for John Fanning Watson, shopkeeper, banker, and railroad executive immigrating here from Burlington County, New Jersey, to become the tireless antiquarian investigator of "Olden Time Philadelphia," as he liked to call it. Watson's avocation was interviewing oldtimers, salvaging documents, and souveniring all kinds of relics. In 1824, the historian proposed a scheme for purchasing and rehabilitating the old Rising Sun Tavern (which he called the Letitia Penn House) to "preserve its exterior walls with inviolate faithfulness," at the same time adapting its interior as a historical museum. Realizing the value and interest of pictures for promoting his project, Watson engaged Breton to make a sketch of the tavern in 1828.[10] Watson's grand ambition was heralded in *The Casket* for June, 1828, in a little article on the Old Court House then still standing in the middle of Market Street at Second but pulled down shortly afterward.[11] Watson's answer to such threats was his book, the now famous *Annals* of 1830, one of the first illustrated American city histories.[12] In it old buildings became heroes and they have remained so to many people.

An interesting architectural record of the year 1824 was the floor plan of Wyck in Germantown drawn by the office of William Strickland just before he enlarged the house, completely denaturing its original seventeenth-century character (illus.).[13]

More Measured Drawings: Photography Arrives

A few years later, Mrs. Louisa C. Tuthill, a popular Connecticut writer, took on the whole world of architecture from its beginning. In a work published here and inscribed "to the Ladies of the United States of America, the acknowledged Arbiters of Taste," she presented a chapter on the current "State of Architecture," including a blue-ribbon roll call of Philadelphia buildings. But beyond citing the layout of Babylon as prototype for William Penn's city plan, she says little about Philadelphia history. Her lions were the new buildings admired at the time.[14] Another variety of historian, Benson J. Lossing, came from New York, systematically seeking out Revolutionary War landmarks and personally sketching "physical vestiges of that struggle." He tells us that he traveled eight thousand miles to build up his "pictorial field book." He did well by Philadelphia and his indignant description of Carpenters' Hall in 1848—then in use as an auction house—may well have led directly to its restoration by the Carpenters' Company nine years later.[15]

On September 25, 1839, Joseph Saxton, following directions from France and using a homemade box camera, took what is believed to be the first American outdoor photograph. A scruffy specimen, though historic, it is a small daguerreotype view from a window in the first U.S. Mint looking toward the high school in Center City. The next year an Englishman, Fox Talbot, invented the calotype or talbotype. This process was brought to Philadelphia in 1849 by William and Frederick Langenheim who introduced it commercially—though not successfully.[16] The "Langenheim Talbotypes" give us some remarkable views of local landmarks, usually circular in shape.[17] Other daguerreotypes followed, and then glass plates. The collections of the 1850's and later are so vast—so far—as to defy all attempts at comprehensive listing. As a record of what meets the eye the camera cannot be beat. But the eye cannot see everything, and the architect and his drawings are still needed.

As the Victorian rage for complicated effects was topping out—and probably in reaction to it—architects became interested for the first time in early American buildings. The Colonial Revival began formally when Richard Upjohn of New York, president of the American Institute of Architects, addressing its third annual convention in 1869, recommended the collection of information on our early buildings and a close study of remaining examples. By 1872, when Charles Follen McKim of New York had restored a colonial room in Newport,[18] the need by draftsmen for details in the old/new style—such as sections of moldings and other ornament—must have been evident.[19]

The centennial celebration at Philadelphia quickened popular interest. *The Historic Mansions and Buildings of Philadelphia* by Thompson Westcott (Philadelphia, 1877) was a product of the times and a prototype for many later works. But it didn't do much for the architectural offices expected to imitate the best features of the eighteenth-century mansions still standing in numbers. *The American Architect and Building News* of Boston took up the challenge. In 1881, it published a piece on "colonial architecture" illustrated with sketches and a few measured floor plans and elevations. Two issues carried the report by George Champlin Mason, Jr., of Newport, Rhode Island, chairman of a committee of the American Institute of Architects, titled "Practice of American Architects and Builders during the Colonial Period and the first Fifty Years of our Independence." The first installment (August 13) referred to origins of style in England, laid out the lines for an American architectural history, reproduced twelve sketches of old Newport houses and some fireplace details. The second part (August 20) carried eight more drawings, a detailed specification from an early contract to build a house. It concluded with a wistful hope that "the little we have done may serve as an incentive for fuller investigation and research, begun and carried on in a systematic and business-like manner."[20]

Books by and for Architects

Two Philadelphia architects produced a book of measured drawings exclusively aimed at drafting rooms: *Old Colonial Architectural*

Details in and around Philadelphia. 50 Plates of Scaled and Measured Drawings, offered to William Helburn's architectural publishing house in New York on October 31, 1890.[21] In it William Davenport Goforth and William J. McAuley presented a collection of "as good specimens . . . of their respective classes as could be found in and around Philadelphia . . . arranged as nearly as possible in Historical order." The plates reproduced competently executed drawings of about twenty buildings—mostly houses—in Philadelphia and Germantown. The shaky identification by names and dates and the scant textual content reflect the undeveloped historiography of the times.[22]

Such studies did not come too soon. The buildings on Independence Square, an early focus of historic sentiment, were about to be restored *again.* Unfortunately, the earlier work there, designed by capable *modern* architects such as William Strickland and John Haviland, was not considered acceptable by later standards of authenticity and most of it was fated to be removed.

Today, early restoration drawings—as elsewhere—are scarce. We have little more than artists' views and photography to explain those pioneer restorations. Perhaps the oldest working drawing surviving is on a single sheet of paper of the period 1895–97 done by George C. Mason. The Newport architect had by that time opened an office in Philadelphia where City Councils invited the various "patriotic hereditary societies" to restore some of the historic rooms on the Square. Mason's drawing proposed certain details including that special Yankee feature—the open, twisted stair newel.[23] Although executed, such work was not at home in Philadelphia and much of it was removed a few years later. Mason may have used the wrong book; perhaps he did not realize that New England details, however attractive they may be, are not always exportable.[24]

The architects in many American cities did not have old Georgian buildings to study first hand, and the importance of scale drawings to the revival of colonial architecture sweeping the country can hardly be exaggerated. The demand for such architecture continues even now, when there are fewer and fewer professionals who can meet it with confidence. It takes extreme skill to produce a good Georgian building—though our leading campus theoreticians (among others) hardly admit that their former students now facing apprenticeship often do not even know what a molding is. The truly competent restorationist today, practicing what might be called the geriatrics of the building business, has become a rare breed indeed.

More books came forth to meet the need. William Rotch Ware's great three-volume work, *The Georgian Period, A Collection of Papers*

Dealing with "Colonial" or XVIII-Century Architecture in the United States, together with References to Earlier Provincial and True Colonial Work, was a collection of photographs, sketches, and measured drawings published in 1901, 1902, and 1908. It illustrates English precedents and carries thoughtful essays on many subjects. Philadelphia is represented by some of its well-known eighteenth-century structures, such as Christ Church, Old Swedes', Carpenters' Hall, Mount Pleasant, and Woodford Mansion, many of which were recorded by that skilled draftsman Charles L. Hillman.[25]

The well-known "White Pine Series of Monographs," begun in 1915 under the direction of Russell F. Whitehead and continued for many volumes, provided measured drawings and photographs distributed from New York to as many as ten thousand addresses. The "Monographs" did not include any Philadelphia buildings,[26] but our city is well represented in twentieth-century publications—as twelve pages in the Roos bibliography attest.[27] *Old Philadelphia Colonial Details* (New York, 1914), by Joseph Patterson Sims and Charles Willing, is a notable example of a book useful to practitioners of the Georgian style. And the photographic picture books of Philip B. Wallace include fine measured drawings by William Allen Dunn and M. Luther Miller.[28]

The reports of the historical committee of the Philadelphia AIA Chapter have not all been located, but the collection they built up included Independence Hall (drawings of 1898, 1918, 1921, 1923), Congress Hall (1901, 1912), the Bartram House on the west bank of the Schuylkill (undated), the Blue Bell Tavern (1916) and Washington's Headquarters at Valley Forge (1931). At least some of this material is in The Historical Society of Pennsylvania.

Local architects had become deeply involved with our early buildings during the early twentieth century. One of the high points was another restoration of Independence Hall, this time under the direction of Horace Wells Sellers in the 1920's. Notably, it was backed up by the architect's extensive historical scholarship. But the great stock market crash of 1929 signaled the end of an era in many fields, including architecture.

The Old Philadelphia Survey

The dank winds of the Great Depression did blow some good. The architectural profession, badly hit all over the world, found measures to relieve its distress by setting up various campaigns for recording historic architecture. Those operating independently in London, New York, Pittsburgh, and Philadelphia were notably successful.[29]

The movement here had begun in the 1920's with Frances A. Wister, President of the Civic Club of Philadelphia and kingpin of "The Old Philadelphia Survey." Miss Wister had firsthand knowledge of the old parts of the city, which she felt could be rehabilitated "in harmony with the traditional architecture of the City." Housing and zoning agencies got interested, but when Miss Wister read in the newspapers of September 17, 1929, that Mayor Mackey was having his *engineers* work on plans for the reconstruction of the district, she realized that the rehabilitation program was in serious trouble. She and others like her began a campaign that led to the harmonious restoration she advocated.

As a part of that early proposal for "urban renewal" it was felt that a systematic collection of historical and architectural data was needed for reference. But an attempt to involve the students at the School of Architecture at the University of Pennsylvania failed and the program was temporarily shelved. Conditions in 1930 nevertheless cried for action. The Philadelphia AIA Chapter set up a temporary committee with D. Knickerbacher Boyd as chairman. A comprehensive survey of what they called "Old City"—the Delaware to Ninth Street and Spring Garden Street to Washington Avenue—was proposed. Mrs. Cyrus H. K. Curtis, widow of the publisher, gave $5,000 and a committee of architects was appointed.[30] The committee met in December, studied old maps and assigned a schedule of field work to twenty-two members of the chapter. Sydney E. Martin succeeded to the chairmanship and by October, 1931, the work as originally projected was substantially completed.[31] Fifty-seven draftsmen produced 407 excellent measured drawings of outstanding structures in the Old City and along the banks of the Schuylkill. Reports covering the whole area were made: 125 photographs had been taken, and architect William M. Campbell, a mapmaker of note and later an important HABS team supervisor, was engaged to compile the results on an attractive map.[32]

A book with a selection of the new drawings, such as New York City had, was planned.[33] Mr. Martin prepared a foreword and Horace Wells Sellers an extended essay, "Philadelphia's Early Architecture and Architects."[34] Although the book never did appear, the original drawings survived and are carefully preserved in the Art Department of the Free Library.[35] And through the Old Philadelphia

Survey the architectural community of Philadelphia was readied for the Historic American Buildings Survey, which was soon to follow.

The Beginnings of HABS

Washington, D.C., was agog with excitement in 1933. President Roosevelt, after his inauguration on March 4, immediately began his dramatic war against the Depression. New Deal agencies, administrators, and idea men rose, spoke, and faded month by month. New schemes were regularly announced in the press and on the radio. Some of them were to change the American way of government for good. Just how much was pure idealism will be debated long into the future. But there is no question that a real spirit of innovation was in the air and some profitable experiments were made. On November 9, the President by executive order established the Civil Works Administration with Harry L. Hopkins as Administrator. His mandate was to create four million jobs to help the vast army of unemployed over the winter. The executive departments were invited to bring forth programs for those needing work, including professional people. Speed was essential, and four days later the National Park Service proposed a comprehensive operation calling for work by architects and architectural draftsmen, titled *The Historic American Buildings Survey*. Dated November 13, 1933, it proposed the allocation of $448,000 to employ 1,200 persons under the National Recovery Act.[36]

The proposal moved along rapidly. On November 17, the program was approved by Secretary of the Interior Harold Ickes and on December 1, by Hopkins, with the full backing of the Executive Committee of the American Institute of Architects.[37] The Advisory Committee of the Williamsburg Restoration endorsed it, too.[38] A picture story in the *Washington Post* for November 29 predicted that work would start immediately—and it did, with HABS under the broad authority of the National Recovery Act.

The original work outline proposed to employ the thousand persons for a period of six months, the time limit then specified for all such programs. One of the very first steps was to contact Dr. Leicester B. Holland, Philadelphia scholar and architect then in Washington. Fortuitously, he was both head of the Fine Arts Department of the Library of Congress and Chairman of the AIA's national Committee on Historic Buildings. Dr. Holland had been active in Philadelphia's own campaign and he had already initiated a collection at the Library called "The Pictorial Archives of American Architecture."[39] He was ready to undertake a national project,[40] and at the request of Holland and myself, Edward C. Kemper, the alert and able AIA Executive Secretary, sent out urgent telegrams to the presidents of the chapters across the country asking each to nominate a district officer to direct HABS work in his respective area.

By December 27, E. Perot Bissell of Philadelphia had been appointed District Officer for Eastern Pennsylvania (east of the West Fork of the Susquehanna and the Genessee) and was assigned forty-two men. Philadelphia work began at once in the Germantown area, with care not to repeat the drawings made by the Old Philadelphia Survey. Among the first buildings tackled were "Wister's Big House"—now called Grumblethorpe—Vernon, Wyck, Germantown Academy, and the Blair House, all along Germantown Avenue or north of the Old City.[41]

The Survey Made Permanent

On January 8 and 9, 1934, the first meeting of the National Advisory Committee was held in Washington. It was a distinguished group, the majority architects, and the first committee of historical advisors ever assembled by the Park Service. Many problems of policy and technique encountered by the new Survey were discussed. The report reveals many issues of the moment, some of bureaucratic nature, rather than professional.[42] Drafting-room techniques were reviewed. The standard paper selected for final record drawings was of a 40-pound weight and 100% rag content—considered the most permanent drawing base available—and a uniform sheet size was established.[43]

Each district was encouraged to assemble its own advisory committee. The budget for photography was small. While the drawings would be government property, they were in the public domain and therefore available at the Library of Congress for consultation and copying. Some doubt was expressed by the committee that architects could make measured records of prehistoric remains without help from archaeologists.

Apprehension that the program might have to terminate as early as February 15 motivated a recommendation that field men make up lists of important buildings remaining to be documented. The sessions closed with a recommendation to the Secretary of the Interior for a "reasonable extension and enlargement of the present

program."[44] The life expectancy of the Survey was precarious indeed.
On February 13, the staff was advised of the imminent end of the
project, but working with inspired desperation, the men in the field
matched wits with local administrators, and HABS somehow came
through.[45]

At the peak of employment 772 men were at work but no more
than $196,267.63 had been spent. Telegrams from concerned ar-
chitects and preservationists urged that the Survey be continued
under the authority of the states using federal funds.[46] An exhibition
of the new records from the entire nation was promptly assembled
and hung at the National Museum in Washington from April 5 to 26.
Panels displayed drawings supplemented by photographs, texts, and
maps. Instructions to the field had advised that "this will be a public
exhibition. Although it is naturally assumed that its greatest appeal
will be to architects and historians, a distinct effort will be made to
give also a strong popular appeal."[47] It was a success.

A nationwide total of 880 buildings represented by over five
thousand ink-on-paper drawings of good quality and three thousand
photographs had been made—all in four months. Index cards for two
thousand subjects showed that much more work was in prospect.
The pay for the field work was only $39.60 per 32-hour week, but the
program had made a hit—men were working at something they could
like and value.[48]

By July 23, a consensus national in scope between the National
Park Service,[49] the American Institute of Architects, and the Library
of Congress was reached. Known as "The Tripartite Agreement," it
survives today with only modest changes, as the cooperative base on
which HABS rests.[50]

It has not been possible to trace in detail Philadelphia operations
under Joseph P. Sims, who had been active in the Old Philadelphia
Survey and succeeded Mr. Bissell in 1935 as District Officer. The 1938
Catalog of HABS measured drawings in the Library of Congress lists
substantial contributions such as those for Trinity Church, Oxford (6
sheets), Old Swedes' Church (10 sheets) and the Cove Cornice House
at the foot of Spruce Street. Other structures, such as those in
Workman Place (Southwark), in Fort Mifflin, and nearby Cannonball
Farm, were represented by only a few photographs.

A comprehensive report by staff architect Waterman dated May
19, 1937, showed that in the whole state of Pennsylvania 156 buildings had already been measured (total, 1034 sheets of drawings) and
151 photographed (616 negatives).[51] Waterman was particularly

pleased that the eighteenth-century mansions Port Royal and Chalkley Hall in the Frankford area of Philadelphia could be recorded at
that time. He pointed out that, although they had been abandoned
and their fine woodwork stripped, it was possible to locate the missing parts and reassemble them on paper for HABS.[52]

The National Parks Come East

The presence of National Park Service architects in the east was a
new development. Until 1930, the bureau had been almost exclusively a western outfit with its professional branches centered in field
headquarters in San Francisco. But in that year, as the result of much
promotion, two new projects emerged in Virginia: the George Washington Birthplace and the Colonial National Monument (the latter at
Jamestown-Williamsburg-Yorktown).[53] A minimal staff of landscape
architects, architects, and engineers was set up to plan and supervise.
But that staff was vastly augmented from 1933 on by temporary ap-
pointees engaged to direct New Deal ventures, such as the great Civilian Conservation Corps program. And under the Roosevelt reorganization the War Department gave up its historic battlefield parks,
adding new responsibilities from Maine to Louisiana. All this came
about while the three scenic national parks—the Great Smoky Moun-
tains, the Shenandoah, and the Mammoth Cave—were coming into
official existence.

Many historic sites and buildings were involved. Where architectural study was required, HABS records were made early in the
process. In each case structures were measured and drawn prepara-
tory to making new plans: for example, at Morristown, New Jersey,
in 1934–35, the Tempe Wick House in Jockey Hollow (13 sheets of
drawings), the Guerin House (9 sheets), and the Ford Mansion
(called "Washington's Headquarters," 26 sheets). In quite another
field, the extensive seventeenth-century remains being excavated at
Jamestown Island, Virginia, were recorded in detail. Another archaeological project undertaken by the CCC was digging the foundations of Mannsfield, a great plantation house on the Rappahannock
River below Fredericksburg that had burned in the Civil War.

One of the most intensive architectural studies of all was undertaken at St. Louis, beginning in 1936. Incidental to it, the local
HABS collection was extended by a great addition of records made in
the old riverfront area on the Levee.[54] Many of the interesting old
structures on those eighty acres were soon to be removed as de-

manded by the local sponsors who wanted to create a "City Beautiful" development, vintage 1900, on the site.[55]

The Independence National Park

The first Philadelphia restoration project of the federal government, in what was to become Independence Park, was the rehabilitation of the old Custom House (now called the Second Bank of the United States) on Chestnut Street. Abandoned by the Treasury Department, it was transferred in order to save it. The Carl Schurz Foundation agreed to occupy the building and provided most of the cash needed. Park Service architects working out of the Washington office drew the plans and, as a matter of course, first prepared a fine set of nineteen drawings of existing conditions for—and deposited in 1939 with—HABS, along with thirty-five record photographs.

A dozen years (and one war) later, the Park Service set up in Philadelphia a project office to acquire the land necessary to create the great Independence Park.[56] An intensive study of the existing buildings in the area was undertaken and measured drawings were begun in order to analyze the condition and needs of those that could be proven "historic." The postwar prosperity of 1951 kept architects generally busy, so the office expediently employed gifted young draftsmen, whose first efforts produced conspicuously valuable results.[57] When it was discovered that the U.S. Army Engineers had been authorized to hire undergraduate professional students during their summer academic recess, the HABS Philadelphia office decided to invoke the same privilege.[58] Accordingly, in the summer of 1952 teams of architectural students were organized to make detailed scale drawings, systematically, building by building, on the new lands being acquired. Supervisors were selected on the basis of their personal interest in American buildings as well as their ability to lead. Students were individually recruited for their drafting ability and/or their tentative interest in history, and they were deliberately chosen from among different schools. Gradually, HABS won the confidence of architectural faculties, who recommended their best students to compete for its jobs.

While both quality and quantity of draftsmanship was exacted, the careful study of each fabric was emphasized. The difference between surviving original features and later modifications was to be observed and reflected in the drawn record. Lecturers on Philadelphia architectural history were brought in to give depth to the summer's experience and, when actual restoration work was started, the students had the chance to become acquainted with building mechanics and their work.

Through the summer of 1962, and later, valuable files of analytical drawings were built up for the use of the architectural planners at a minimum cost. Best of all, young architects had developed a fund of experience.[59]

In 1954, the Park Service opened a new professional office to handle planning and construction work in Philadelphia for the eastern half of the United States. This consortium of architects, landscape architects, and engineers was called the Eastern Office of Design and Construction (EODC). The architects from the Independence Park staff were assimilated and their geographic responsibilities greatly increased. HABS was then exported to other National Park areas where restoration work was programmed. In the summer of 1955, for instance, a team was placed at the Adams Mansion, Quincy, Massachusetts, a complex more than two centuries old that had been recently given to the nation. Another was set up at Harpers Ferry, West Virginia, a picturesque industrial village on the Potomac. The historical project sent architectural teams to make the basic drawings on which plans for restoration would be based. A team was placed at Greenville, Tennessee (1957), where the home of President Andrew Johnson was to be restored, another at Appomattox Courthouse, where General Robert E. Lee surrendered, and one at Gettysburg Battlefield (1957), which still had farmhouses surviving the great Civil War battle. Their work in accumulating the necessary records was financed by the Interior Department's annual construction appropriation as a preliminary step toward plans for preservation. Measured drawings of government-owned buildings continued to be made with an eye, whenever suitable, for deposit in HABS.[60]

In 1957, HABS was suddenly catapulted to a new level of responsibility by the on-coming program known as "Mission 66," which required that the architectural stock of the country as a whole be reviewed.

Mission 66

"Mission 66" was a promotional package to obtain funds for bringing the National Parks, neglected during the years of World War II, up to physical standards by 1966, the fiftieth anniversary of the Park Service. Congress appropriated annual funds for a vast ten-year program.

The Historic Sites and Buildings Act of 1935 had granted

permanent authority to HABS to make records of historic architecture—either in or out of federal ownership—but no financing had been provided. In the Mission 66 appropriation HABS was now allocated $200,000, a staggering sum considering that no personnel were on hand for such an enterprise. All the new money was to be spent in one fiscal year by the Philadelphia office, the only one with relevant experience. While there had always been a certain amount of loose talk (not shared by this writer) about "finishing" HABS, that is, completing drawings, photographs, and write-ups for *all* historic buildings in the whole country, nobody seemed to know how many such buildings there were.

In the emergency, Agnes Addison Gilchrist, an experienced architectural historian, was brought from New York to help the writer analyze the situation. On a large blank county map of the United States the statistics for HABS work already done were compiled, state by state.[61] Large areas across the country still had no records at all, but none of them was entirely without "historic" buildings. No fixed and accepted standards of judgment, probably wisely, had ever been established. And a quarter-century had increased the published knowledge of American places and the spectrum of architectural appreciation.[62] The Survey in 1957 could not be reactivated on its original basis of unemployment relief. So the successful annual summer student program at Philadelphia was emulated on a large scale.

Because it seemed then (as it does now) that to prepare detailed, elaborate drawings for *all* historic buildings in the United States worth attention would be impossible, it was decided to develop a wide coverage by photography and to emphasize quality so that the pictures could be used by writers and editors for publication. Each picture published can do as much good as hundreds that merely exist in archives. The problem was to get good negatives at a reasonable cost. Well-known lensmen like Fritz Henle and Cervin Robinson of New York and Cortlandt Van Dyke Hubbard of Philadelphia were induced to undertake sample projects at modest rates. Jack E. Boucher, then of Pleasantville, New Jersey, became the in-house architectural photographer. A large photographic project under outside contract was arranged for Providence, Rhode Island, where the redevelopment of College Hill was in prospect. Another New England project was inspired by the threatened abandonment of a branch of the Boston & Albany Railroad with a fine series of H. H. Richardson stations. The so-called "photo-data book" concept was developed, in which a selected community having a group of interesting buildings would provide the historical data and the government would arrange for photography.

One way of estimating the work needed to "complete" HABS was to consider one county at a time and multiply the number of counties by the estimated number of historic buildings in each. For a sample, we chose nearby Chester County, Pennsylvania, where the county's historical society had built up a splendid file on local building history. The society and HABS selected one hundred buildings of varied type and period, for which the former wrote up the historic data, and HABS engaged a photographer, Ned Goode of Frazer. Cooperation was excellent, and the project concluded with an exhibition of salon prints and a celebratory dinner at West Chester.

Of course, Chester County had more than a hundred buildings worth recording, whereas a county in the west would probably have many less. But the arbitrary figure of one hundred each for all three thousand American counties produced a tremendous total of three million buildings. Because no one could recommend recording on such a scale the idea of "finishing HABS" was quietly dropped. Instead the term "open-end archive" was adopted.[63]

Society Hill and Other Places

During the 1950's, urban renewal was getting under way on Society Hill, where a large number of buildings was scheduled to come down in line with city planning. Many good ones did. HABS was able to make last-minute records before some disappeared; other buildings probably survived as a result of the attentions of the Survey. HABS under James O'Gorman made drawings of the great Neave and Abercrombie houses (both from 1759) on Second Street after they had been emptied through urban renewal and while the original woodwork was nightly being stolen.[64] In another case HABS junior architect John Milner was dispatched to record a small but choice eighteenth-century house on Delancey Street that had retained much good woodwork. The new owner, his architect, and his builder were inexperienced but willing to learn, so that the active presence of HABS saved a great deal of the original fabric that might otherwise not have been kept in place.

As the idea got around that an HABS record in the Library of Congress was a real distinction, we were called to many places besides Philadelphia to confer that form of immortality. Many owners were surprised to learn that the Secretary of the Interior and the Library of Congress found their houses to be of significance, but the little certificate conferred generated a new pride by both owners and occupants.

We found that cooperative projects could be set up with the local people footing part of the bill, especially for historic research. In the summer of 1959 we mounted three projects in the middle Connecticut Valley. In a sort of barter process Dartmouth College agreed to provide a drafting room, inexpensive summer housing for the team in a faculty house, and other considerations. The team measured a selection of buildings on campus and in the nearby countryside on both sides of the river. Professor Frank J. Roos of the University of Illinois prepared data for buildings around Bellows Falls later recorded by our photographer. And lower down in the Valley a team was operating in Old Deerfield under the direction of Professor Harley J. McKee of Syracuse University.

Two notable New England projects were those on Cape Cod in 1959 and 1962 where studies and records were made under Ernest A. Connally's direction while the acquisition of lands for the National Seashore project was just beginning. The anxiety of the local people (who feared that a federal agency from out of the Wild West would thoughtlessly destroy all the old houses they had treasured) was much alleviated. Another attractive scene of operations was the American Virgin Islands, where eighteenth- and nineteenth-century Danish architecture had made a picturesque compromise with tropical needs and practices. The Royal Academy of Fine Arts of Copenhagen was sending an architectural expedition to survey the scene (with Rockefeller assistance) and we were able to field an American team of students to work alongside them.

New Techniques

During this activity, the original field handbook, "Specifications for . . . Measurement and Recording"—hurriedly mimeographed in December, 1933, and added to and revised in bits up until July, 1958—was re-examined and found wanting. The original criteria for selecting buildings had become somewhat dated and field techniques had been more highly developed. Professor McKee was brought in to Philadelphia for the summers of 1960–62 to work on a new manual.

Various aspects of the recording process were written up individually and issued in parts for testing in the field. For instance, very few people, even architects, can describe a building systematically, succinctly, and thoroughly. For that purpose a form called the "Architect's Work Sheet" was devised as well as "Historian's Work Sheet," handy checklists of the many parts that make up even the simplest building. The *Manual*—generously illustrated with samples of drawings and photographs—was completed by 1962.[65]

At the same time, experiments were made in photographing existing architectural drawings onto plastic sheets which could thus be built into HABS like new drawings. The Philadelphia Blueprint Company showed that duPont's "Cronar"—believed to be of archival permanence—was suited to this purpose and that technique of adding valuable material drawn by others became standard procedure.

We were proud to have a part in introducing architectural photogrammetry to the recording of historic American buildings in 1957. Photogrammetry makes possible quite accurate scale drawings of details impossible to reach without architectural steeplejacks or extensive scaffolding. The highly specialized technique was originally brought here by Perry E. Borchers of the School of Architecture at Ohio State University, who had studied it in Europe.[66]

Our first order to his department was for sixteen drawings of a tavern in Lancaster, Ohio, the Congregational Church in Oberlin, and the facade of Isaac M. Wise (Plum Street) Temple in Cincinnati. Professor Borchers's recording of the lofty oriental minarets of the temple was a spectacular contribution to HABS that would have been impossible to produce by any other technique.[67] Also included were sixty-four "stereopairs," or photographic records on glass to include other buildings.[68] The expensive equipment of Ohio State's Engineering Experiment Station was rented and used by Mr. Borchers's architectural students.[69] This pilot project was a decided success and led to another contract the following year. In 1959, the Ohio team made stereopairs of the great row of banks at 407–431 Chestnut Street, Philadelphia, opposite the Independence Park.[70] It was done by good fortune shortly before Frank Furness's great Provident National Bank (No. 407) was demolished.

In the summer of 1962, when this chronicle ends, faculty and students from twenty-three architectural schools across the country were working on HABS teams under the direction of the Philadelphia central office. A small but growing body of young architects had earned their stripes in this unique opportunity to acquire a healthy respect for native design and construction. And the value of the Survey's work—both in the treatment of buildings nationally owned and to the preservation community as a whole—had been demonstrated. At the end of that summer, an all-day "Carpenters' Carnival" was staged when the Building Restoration Specialists of the Independence Park gave an outdoor exhibition with the men at work. By invitation people came from all over the east to see it.[71]

Architect James C. Massey, who first worked at the Independence Park as a student, took over HABS in the east, assisted by historian John Poppeliers. In 1966, when EODC was closed down, Mr. Massey moved to Washington to direct the whole national program, continuing HABS with notable energy and imagination until 1972.

Charles E. Peterson, F.A.I.A.

Philadelphia, Pennsylvania

Notes

1. The leading figures from the beginning, most of whom I have known, have been interesting people. Among the first generation was Leicester B. Holland of Philadelphia, classical archaeologist, teacher, and inventor as well as architect. William Graves Perry and Frank Chouteau Brown were from Boston. Perry, the architectural genius of the Williamsburg Restoration, was the first chairman of the HABS Advisory Committee. Brown, a well-known master draftsman, though dying of cancer, produced sheet after sheet of fine drawings—as long as he could hang onto his drafting board. Charles Morse Stotz of Pittsburgh, after completing his outstanding Western Pennsylvania Architectural Survey, was invited to join us. Albert Simons of Charleston, well established as an authority in South Carolina, put his shoulder to the wheel. Richard Koch of New Orleans—practically indestructable, it seemed—built up his post of operations on the lower Mississippi. Earl H. Reed of Chicago was to become a national figure in the Preservation movement. And then there was that squad of Denver architects who went into the desert to join John Gaw Meem of Santa Fe and make pioneer records of the Indian pueblos. Through all the years from 1933 until he left Washington in 1961, Thomas C. Vint, though basically a landscape architect, was a tireless and effective proponent of HABS.

2. It is hoped that the notes in this brief history will encourage others to develop the story more fully. This sketch relates chiefly to the period before October, 1962, when I left the Park Service. I am indebted to Dr. Margaret B. Tinkcom for suggestions.

3. It was probably a cabin of horizontal logs, a type introduced to the Delaware Valley by the Swedes. Kalm called it "a wretched old wooden building . . . preserved on purpose as a memorial to the poor condition of the place before the town was built on it. . . . Its antiquity gives it a kind of superiority over all the other buildings. . . . But . . . [it] is ready to fall down" (Adolph B. Benson, ed., *Peter Kalm's Travels in North America*, New York, 1966, pp. 33–34).

4. Published in 1752, see Nicholas B. Wainwright, "Scull and Heap's East Prospect of Philadelphia," *PMHB*, 73 (Jan., 1949), 16–25. Engraved by Gerard Vandergucht of London. The crude delineations in the much-reproduced Peter Cooper painting "ca 1720" at the Library Company of Philadelphia can hardly be regarded as reliable architectural portraits.

5. Five valuable books of old Philadelphia views available today are Theodore B. White, ed., *Philadelphia in the Nineteenth Century*, Philadelphia, 1953 and 1973; Nicholas B. Wainwright, *Philadelphia in the Romantic Age of Lithography*, Philadelphia, 1958; George B. Tatum, *Penn's Great Town*, Philadelphia, 1961; Martin P. Snyder, *City of Independence*, New York, 1975; Edwin Wolf 2nd, *Philadelphia, Portrait of an American City*, Harrisburg, Pa., 1975.

6. The first American building to bring Europeans across the Atlantic was the Walnut Street Jail by Robert Smith. But the eighteenth-century interest in that remarkable pre-Revolutionary fireproof building was for its experiments in penology rather than its design or construction.

7. Designed by an unidentified London architect (Robert C. Alberts, *The Golden Voyage, the Life and Times of William Bingham*, Boston, 1969, pp. 157–63).

8. Benezet was a well-known Quaker philanthropist who had died years before. The house, according to the article, was built by David Breitnall and was later rented to the Governor of Bermuda. It stood on the right bank of Dock Creek and directly between Benjamin Franklin's mansion and Carpenters' Hall.

9. Martin P. Snyder, "William L. Breton, Nineteenth-Century Philadelphia Artist", *PMHB*, LXXXV, 2. At the same time C. G. Childs, Philadelphia engraver, published a collection called *Views in Philadelphia from Original Drawings Taken in 1827–30* by several artists. But in their own time they were not historical subjects.

10. See Deborah Dependahl Waters, "Philadelphia's Boswell: John Fanning Watson," *PMHB*, XCVIII (Jan., 1974), 38–41. The project proved to be fifty years ahead of its time, but the building was suffered to remain and was eventually moved to Fairmount Park in honor of the City's 200th anniversary.

11. Watson's project was described in *The Casket*: "This once venerable building, long divested of its original honours . . . had long been regarded by us and others, as a rude and *undistinguished* edifice: But our feelings and sympathies in its favour, have been strongly awakened by seeing its former renown and even greatness depicted in a lively manner in the MS. "Annals of Philadelphia", by J. F. Watson, Esq. at some future day to be published with pictorial embellishments, when we venture to pronounce, that our cotemporaries [sic] will be surprised, amused and charmed with the hidden facts reserved for his amusement and gratification. . . ."

12. John F. Watson, *Annals of Philadelphia*, Philadelphia, 1830, p. 147, first of several well-thumbed editions. Somewhere in this period B. R. Evans began his colorful series of watercolors of Philadelphia street views of which some three hundred survive for the period 1830–1879. And from David Kennedy, even more prolific, we have about nine hundred watercolors from the years 1838–1873.

13. Published here, courtesy Thomas Wistar, Jr., The Wyck Charitable Trust, and the Athenaeum of Philadelphia.

14. *The History of Architecture from the Earliest Times*, Philadelphia, 1848, said to be the first American architectural history. Illustrated with competent perspectives were Girard College, the mansions of James Dundas and Mathew Newkirk (all by Thomas U. Walter), the Athenaeum (John Notman), Monument Cemetery Chapel, and the Bank of North America. Mrs. Tuthill had the use of Ithiel Town's library at New Haven (Agnes Addison Gilchrist in *Notable American Women 1607–1950*, Cambridge, Mass., 1971, pp. 487–88.

15. Benson J. Lossing, *The Pictorial Field Book of the Revolution*, New York, 1851, Vol. II, pp. 521–23. Lossing reproduced two views of the Hall. His interior is the oldest known. On its restoration, see Charles E. Peterson, "Carpenters' Hall," *Transactions*

American Philosophical Society, Vol. 43, Part I, pp. 113, 114. No drawings have been found for that operation.

16. Beaumont Newhall, *The History of Photography,* New York, 1964, pp. 20, 31–44.

17. Charles van Ravenswaay, then Director of the Missouri Historical Society, St. Louis, kindly loaned us in 1951 an album of Langenheim paper prints and permitted us to photocopy them. They turned out to be the oldest known sets of photographs of Philadelphia, Washington, and Mount Vernon.

18. For a chronology see Charles E. Peterson, *Antiques Magazine,* February, 1966, pp. 229–32.

19. That such work could be well done up that way is illustrated by the excellent record drawings by John H. Sturges of the Hancock House on Beacon Hill, Boston, just before it was torn down in 1863. These were not published until November, 1926, in *Architecture,* pp. 333–36.

20. In the *AABN* issue for August 27 (p. 102), A. J. Bloor of New York claimed credit for proposing the committee at the Thirteenth AIA convention. Bloor was a national secretary of the AIA and a founding member of the Metropolitan Museum of Art.

21. The book is now scarce and its compilers little known. Goforth appears in Gopsill's Philadelphia directories for 1891–97 as an architect-partner of Albert E. Yarnell at 14 South Broad. He is listed as living in Jenkintown but disappears by 1900. McAuley appears in 1889 as a draughtsman at 3711 Powelton, in 1895 as an architect but is not listed after 1896.

22. The "Thomas House at Angora, Pa." turned out (after some detective work) to be Whitby Hall on Cobb's Creek near Kingsessing. In this century the building was taken apart and two paneled rooms and the main stairway sent to the Detroit Institute of Arts, some parts to Haverford, Pennsylvania, and the rest to the dump.

23. Mason's report on this work appeared in *The American Historical Register* (issue not located).

24. For an evaluation of Mason's work as an architect and restorationist in Newport, see Antoinette F. Downing and Vincent J. Scully, *The Architectural Heritage of Newport, Rhode Island,* 2d ed., New York, 1967, p. 5. Not to be outdone, T. Mellon Rogers, for the Daughters of the American Revolution, went ahead and installed some woodwork in Independence Hall reminiscent of early McKim, Mead and White "Colonial," but it, too, was removed soon afterward. For the story of recent architects' work at Independence Hall, see Lee H. Nelson and Penelope Hartshorne Batcheler in *Building Early America,* Radnor, Pa., 1976, pp. 277–318.

25. Issued serially, those dates according with the title pages in this writer's set. The Hillman drawings exist, at least in part, at the Free Library of Philadelphia.

26. This seems surprising now—in view of the fact that Whitehead was a native of Trenton and studied at Drexel Institute. See Charles Magruder, "The White Pine Monograph Series," *Journal of the Society of Architectural Historians,* XXII (March, 1963), 39–41.

27. Frank J. Roos, *Bibliography of Early American Architecture,* 2d ed., Urbana, 1968, pp. 179–90.

28. Dunn, in *Colonial Ironwork in Old Philadelphia,* New York, 1930; Miller, in *Colonial Houses, Pre-Revolutionary Period,* New York, 1931.

29. "The Work of the R.I.B.A. Unemployment Committee," *Journal of the Royal Institute of British Architects,* June 23, 1934. In the background were the Society for Photographing Relics of Old London (1875) and the London Survey Committee (1893).

Hermione Hobhouse, *Lost London*, New York, 1971, p. 7. For New York, see note 33. For Pittsburgh, see Charles Morse Stotz, *The Early Architecture of Western Pennsylvania. A Record of Buildings before 1860 Based upon the Western Pennsylvania Architectural Survey a Project of The Pittsburgh Chapter of the American Institute of Architects . . .*, New York, 1936, pp. 280–83.

30. Half of the Curtis fund was allocated for the purpose, and $13,500 was raised by the Chapter's Unemployment Committee. Members were Walter H. Thomas, H. Louis Duhring, Leicester B. Holland, Horace Wells Sellers, Franklin D. Edmunds, and Charles S. Hillman.

31. "Mr. Martin organized and conducted all of the work of the Chapter that has been done along these lines in the last two years" (C. C. Zantzinger to L. B. Holland, November 28, 1933, Library of Congress [MS], Prints & Photographs file).

32. Copies were sold to benefit the program; the map is still in print.

33. The New York "Architects' Emergency Committee" published two handsome volumes of measured drawings and photographs of buildings scattered from Maine to Louisiana in 1933 and 1937. Though they were titled *Great Georgian Houses of America*, the examples extended into the Greek Revival period. The books have recently been re-published by Dover.

34. He also wrote an unpublished history of the Survey which is the basis of this account. A copy is in my possession.

35. They have just been photocopied under the direction of Miriam L. Lesley, head of the Art Department, where prints may be ordered.

36. The original longhand draft is still in my possession. It had the old-fashioned title "The Relief Employment Under the Civil Works Administration of a Substantial Number of the Architectural Profession in a Program Recording Interesting and Significant Specimens of American Architecture." It is published in full in "American Notes," *Journal of the Society of Architectural Historians*, XVI (Oct., 1957), 29–31. Although the word "Survey" was loosely used for promotional reasons, as surveys were popular at the time, the index cards used from the beginning did perform a true survey function. But they can be consulted only at the Library of Congress and few people know about them.

37. AIA Document No. 259, p. 7. Special meeting, Nov. 18–21, 1933.

38. The telegram, signed by perhaps the most distinguished group of American restorationists that ever met, seems well worth reproducing in full:

WESTERN UNION
STRAIGHT MESSAGE
CHG.PSH, WMSBG.

WILLIAMSBURG, VIRGINIA
NOVEMBER 16, 1933

CHARLES E. PETERSON
3065 NAVY BUILDING
WASHINGTON, D.C.

RESOLVED THAT THE UNDERSIGNED COMMA MEMBERS OF THE ADVISORY COMMITTEE OF ARCHITECTS FOR AND OTHERS ASSOCIATED WITH THE RESTORATION OF WILLIAMSBURG VIRGINIA COMMA HAVING RECEIVED THE PLAN OF MAKING A NATIONAL SURVEY AND RECORD OF EXISTING BUILDINGS AND OTHER STRUCTURES WORTHY OF PRESERVATION COMMA EITHER PHYSICALLY OR BY SUITABLE RECORD COMMA HEREBY EXPRESS APPROVAL OF THE

PROJECT STOP THE PLAN AS DETAILED IMPRESSES US AS AN ADMIRABLE METHOD OF AC-
COMPLISHING A WORK OF HISTORIC IMPORTANCE STOP

                           SIGNED: [ROBERT P.] BELLOWS, [EDMUND S.] CAMP-
                                     BELL, DR. [W.A.R.] GOODWIN, [ANDREW H.]
                                     HEPBURN, [A. LAWRENCE] KOCHER, [W.
                                     DUNCAN] LEE, MRS. [SUSAN HIGGINSON]
                                     NASH, [WILLIAM G.] PERRY, [THOMAS MOTT]
                                     SHAW, [ARTHUR A.] SHURCLIFF, [HAROLD R.]
                                     SHURTLEFF, [PHILIP N.] STERN, [THOMAS E.]
                                     TALLMADGE, [R.E. LEE] TAYLOR, [D. EVERETT]
                                     WAID, [MARCELLUS E.] WRIGHT.

39. *American Institute of Architects Journal of Proceedings*, 1930, pp. 130–31. *The Octagon*, June, 1930, pp. 7–8. That collection, though of considerable size, remains obscure and it's still hard to find out what's in it.

40. In *The Octagon, A Journal of the AIA*, for February, 1933 (Vol. 5, No. 2), pp. 15, 16, Dr. Holland called for nominations for a national list of historic buildings—with what success has not been determined.

41. For a complete list, see John P. O'Neill, comp. and ed., *Historic American Buildings Survey Catalog* . . ., January 1, 1938, Washington, GPO, n.d., pp. 186–87.

42. For instance, rulings handed down from the Civil Works Administration first declared HABS personnel as "skilled workers instead of technicians." However, when such persons could not be found locally they could be imported from other places, in which case they would be classified as "technical or supervisory employees."

43. Dr. Holland explained the format adopted: "There are three reasons for using these small size drawing sheets: (1) It is possible to take it on the ground for field work. (2) It can be bound up in albums for consultations at the Library of Congress. (3) When it is reduced to one half it makes a page which is about standard for architectural books at the present time." The idea of publishing the drawings was to keep coming up again and again for years. At one point there was word that Dr. Nicholas Murray Butler of Columbia had asked Secretary Ickes if the Survey could be published in toto and for how much. The answer came, as I remember it, to a couple of million dollars, which sum ended the discussion.

44. A copy of the report of the proceedings was distributed as "HABS Bulletin No. 20" dated February 8, 1934, and is found in a collection of circular letters for the years 1933–1938 kept by the late Earl H. Reed (xerox in HABS office, Washington).

45. An "informal communication" by Thomas C. Vint, Chief Architect, to the district officers on February 19 gives a good idea of the hectic atmosphere within the federal establishment. Herewith an excerpt: "The status of the Survey was completely changed almost hourly. Instructions wired to District Officers on February 15, were the result of a ruling by the Federal Civil Works Administrator's office that the HABS would be classed as a Statistical Project and was therefore to operate after February 15, without having personnel reduced. On the night of February 15, that ruling was rescinded by the Federal Civil Works Administration, and the HABS was immediately placed under the classification of a Federal Project operating upon Public Property. According to this classification the HABS was to accept an immediate reduction in personnel of approximately 50% of present work force and to continue operations for ten weeks following February 15, reducing the total personnel 10% each week. As a result of further negotiations with the Supervisors of other National Parks Projects the National Headquarters of the HABS was enabled to alleviate this drastic reduction of 50%

by the addition of more than two hundred persons to the reduced rolls of the Survey."

46. The same district officers were continued, and still under the direction of Washington.

47. HABS Bulletin No. 22, March 5, 1934. *Journal of the American Institute of Architects*, April, 1934, pp. 15, 16.

48. The ingenuity and skill of the staff in Washington was provided by three young architects: Thomas T. Waterman of Virginia, a veteran of the early days of Williamsburg; John P. O'Neill, a graduate of Notre Dame with some experience in archaeology; and Dudley C. Bayliss of Minnesota.

49. On March 2, 1934, Congress changed the name of the bureau back to the original "National Park Service," much to everyone's gratitude.

50. The "Memorandum of Agreement" was published in a 15-page "Circular of Information" by the U.S. Department of the Interior (U.S. Government Printing Office, Washington, 1936).

51. Washington National Archives, Record Group No. 64, WPA, Division of Information, Primary File A 36–42—785, HABS (typescript).

52. Both houses were demolished but Chalkley Hall not before 1950. Mr. Waterman was then working for Henry Francis du Pont at Winterthur where some of the Port Royal woodwork ended up. Waterman was able to use a lot of his HABS studies in *The Dwellings of Colonial America*, (Chapel Hill, 1950).

In 1941, early in World War II when we were both in the Navy Department, Waterman would bring out of hiding in his drafting table the illustrations for *Mansions of Virginia* (Chapel Hill, 1946) then in preparation.

53. For some of the earliest ventures of the National Park Service in the field of history, see Horace M. Albright, "Origins of National Park Service Administration of Historic Sites" (a pamphlet), Eastern National Park & Monument Association, Philadelphia, 1971.

54. For instance, before the Old Courthouse was restored we extended the original set of 17 HABS drawings to over one hundred. The area was then being developed as the Jefferson National Expansion Memorial. HABS documented many interesting structures now gone, including the Old Rock House or Manuel Lisa Warehouse, the oldest building in St. Louis that was demolished *after* it was completely restored.

55. A progress report of that period is Charles E. Peterson, "Our National Archives of Historic Architecture," *The Octagon*, July, 1936.

56. A land purchasing office was opened late in 1949; I reported as Resident Architect soon afterward.

57. David Krumbhaar of the University of Pennsylvania and Richard E. Pryor, graduate of the University of Miami (Ohio), recommended by Dr. Holland, were the first.

58. I recently noted that the fine drawings in Part VII (1900) of Ware's *Georgian Period* were made by "Sundry Pupils of the Architectural Department of the M.I.T. during the Summer Schools of 1894–95."

59. A report on this subject is Ernest A. Connally, "Preserving the American Tradition," *Journal of the American Institute of Architects*, May, 1961.

60. At the same time, the work of summer teams was being brought to National Park areas such as Fort McHenry, Baltimore (1958), Christiansted, St. Croix (1959), and the Minute Man park, Lexington to Concord (1961), among others.

61. Four special field offices complete with station wagons and travel money had been active in 1940–41.

62. Even the Greek Revival was hardly well established as a subject. When HABS began, the only volume on the subject was Howard Major, *Domestic Architecture in the Early American Republic,* Philadelphia, 1926.

63. In a somewhat parallel project Robert L. Raley of Wilmington, Delaware worked on a complete sweep of buildings in neighboring Mill Creek Hundred. He used the new Historic American Buildings Survey Inventory forms but the project was never completed and HABSI (familiarly called "Habsy") eventually fizzled out as a national program.

HABSI was formed in answer to the challenge of Frederick L. Rath of the National Trust who was then proposing to start a new national survey of his own. He needed a wider coverage than could be afforded by HABS. For quite a period large numbers of people used to meet in Washington and haggle over the mechanics of the proposed operation. It finally got started and Earl Reed of Chicago became its chief salesman. But although extensive work was done in a few areas, the records were hard to handle, were seldom looked at in the Library and the program eventually expired.

64. At the time I lived in the area and was serving on the Philadelphia Historical Commission, on the board of The Landmarks Society, and on the boards of the two successive neighborhood associations during this traumatic period. Fifty-nine certified historic houses came down within two blocks of my house.

65. It was subsequently published in letterpress under the title *Recording Historic Buildings,* Harley J. McKee, comp., by the Washington office in 1970. Available from the Superintendent of Documents, U.S. Government Printing Office, Washington, D.C. 20402—I 29.2H62/12, $6.50.

66. The procedure is very complicated though Prof. Borchers won't admit it. The more he explains, the more obscure it becomes. See *JSAH,* XVI (Dec., 1957), 29.

67. Illustrated in "American Notes", *JSAH,* XVII (May, 1958), 27 as "Photogrammetry, the Magic Scaffold."

68. From such pairs the architectural drawings are made. It is thus possible to make emergency records on glass which can be transposed to drawings at any time later.

69. The first contract amounted to $6,824.00 and included a report on the cost and practicability of such work.

70. The value of the new technique was by then well demonstrated, and we helped the group obtain the commission to record Trinity Church and St. Paul's Chapel in Manhattan. The latest (1976) is a series of record drawings of Indian pueblos in the Southwest—huge complexes on many levels and with few, if any, right angles, a draftsman's nightmare by conventional methods of measurement.

71. A report issued in that period is Charles E. Peterson, "Thirty Years of HABS," *Journal of the American Institute of Architects,* November, 1963.

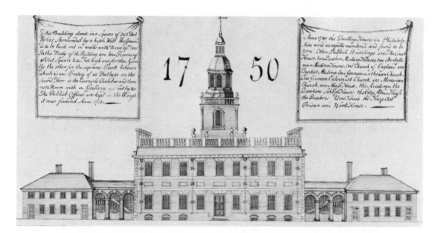

Independence Hall, Chestnut Street front. *Engraving, 1750, the first of a Philadelphia building.*

Bingham House. *Drawing by Charles Bullfinch, 1789, courtesy Library of Congress.*

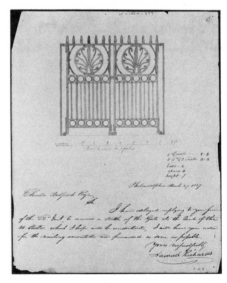

Second Bank of the United States, iron gates. *Drawing by Samuel Richards, Philadelphia ironmonger, for Charles Bullfinch, 1827, courtesy Library of Congress.*

Anthony Benezet House. *Engraving published in* The Port Folio *for October, 1818, from a drawing by William Strickland, courtesy Historical Society of Pennsylvania.*

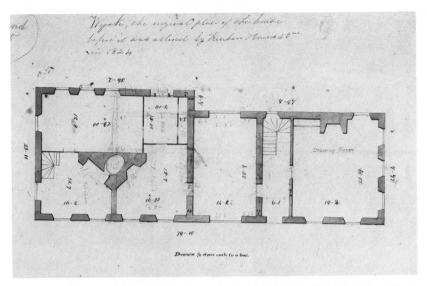

Wyck. *Drawing by William Strickland before remodeling, 1824, courtesy Wyck Charitable Trust.*

Oldest surviving American exterior daguerreotype view, of Central High School cupola from United States Mint. *Daguerreotype by Joseph Saxton, September 16, 1839, courtesy Historical Society of Pennsylvania.*

Second Bank of the United States. *Langenheim "talbotype," 1850, courtesy Missouri Historical Society.*

Philadelphia Exchange Company (Merchants Exchange). *Langenheim "talbotype," 1850, courtesy Historical Society of Pennsylvania.*

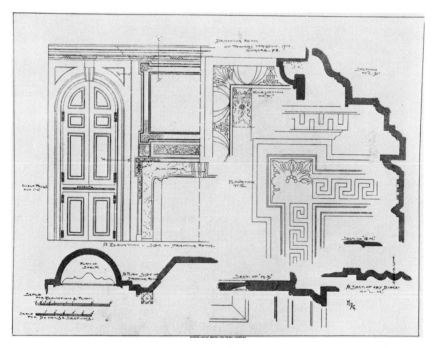

Whitby Hall, chimney wall paneling details. *Plate 7 in W. Davenport Goforth and William J. McAuley,* Old Colonial Architectural Details, *1890, a pioneer work.*

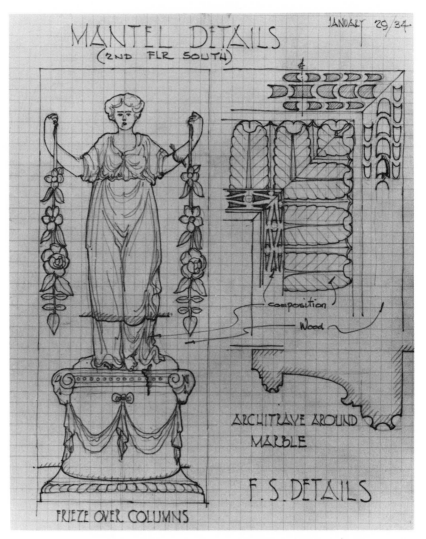

Vernon, mantelpiece woodwork and composition ornament details. *Pencil drawing from one of the first HABS field notebooks used in Philadelphia, January 29, 1934.*

Four Philadelphia HABS architects: *Upper left:* Leicester B. Holland, FAIA (1882–1952), former chairman of the HABS Advisory Committee. *Upper right:* E. Perot Bissell, FAIA (1873–1944), first District Officer for HABS in eastern Pennsylvania. *Lower left:* Joseph P. Sims, FAIA (1890–1953), second Philadelphia District Officer. *Lower right:* James C. Massey (1932–), director of the HABS national program from Washington between 1966 and 1972 and currently Vice President of the National Trust for Historic Preservation in charge of Historic Properties.

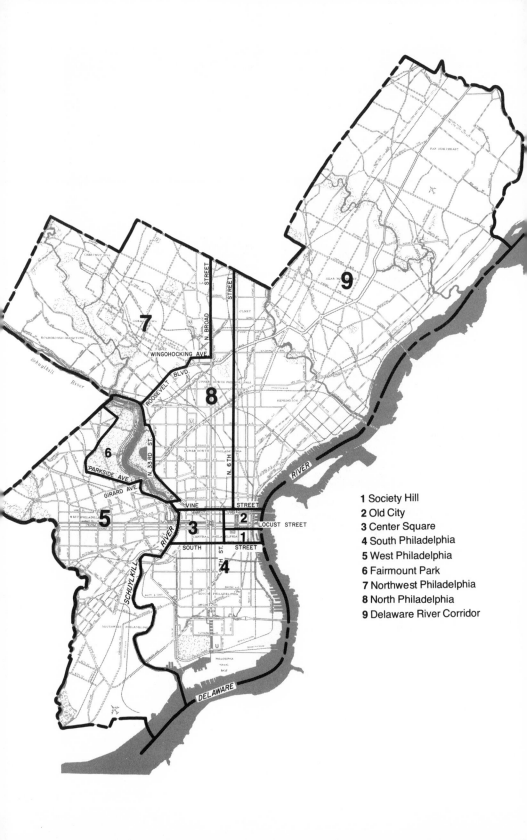

1 Society Hill
2 Old City
3 Center Square
4 South Philadelphia
5 West Philadelphia
6 Fairmount Park
7 Northwest Philadelphia
8 North Philadelphia
9 Delaware River Corridor

Penn's City

Society Hill

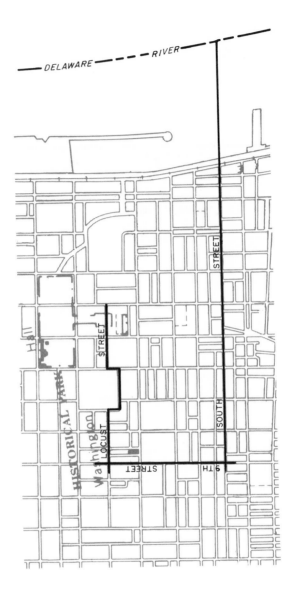

1

Introduction

"The brick walls, as well as the framework of the houses, are painted yearly. The doors are usually white, and kept delicately clean, which, together with the broad slabs of white marble spread before them, and the trees, now gay with their first leaves, which, with some intervals, line the pavements, give an air of cheerfulness and elegance to the principal streets quite unknown to the black and crowded streets of Europe."[1]

$\mathcal{F}$RANCES WRIGHT'S description of residential Philadelphia in the spring of 1819 conforms closely to the present state of Society Hill, an area whose origins go back to the founding of the Quaker Commonwealth. Shortly after William Penn launched his Holy Experiment, he deeded to the Free Society of Traders a raised strip of land between the present Spruce and Pine streets and extending west from the Delaware River to the Schuylkill River. The Free Society was a joint-stock company organized in London to stimulate settlement in Pennsylvania. It overextended itself and was dissolved in 1723,[2] but the neighborhood of its property is still called Society Hill.

Society Hill in conjunction with neighboring Independence National Historical Park is popularly known today as historic Philadelphia. Such terminology betrays an eighteenth-century bias, of course, but, as a living museum of early American architecture, Society Hill has few peers. The boundaries of Society Hill depend largely on who is describing them. The National Register of Historic Places defines Society Hill as the area between Walnut and Lombard streets and Front and Eighth streets. In this catalog, however, the area is defined as that between Locust and South streets and the Delaware River and Ninth Street. This includes buildings east of Front Street that were surveyed by HABS but are now gone, and it runs the boundary to the southern limits of Penn's original city to encompass an area that is in general socially and architecturally homogeneous.

Much of Philadelphia's early development was concentrated along the Delaware River and to the north in Old City. Among the

3

earliest houses built in Society Hill are the two matching dwellings at 117–19 Lombard Street (illus.), which were erected about 1743.[3] Their verticality, intensified by abutting the third-story windows against the plain box cornice, combine with the segmental arches above the 9 over 9 windows and the plain, transomed doorways to illustrate the relative awkwardness of early Georgian proportions. Slightly later in date, between 1743 and 1745, are the two-and-a-half-story bandboxes (that is, one room to a floor) at 220–22 Spruce Street. Since they were built for William Arbour, laborer and porter, they are more modest than the Lombard Street examples, which were built for skilled craftsmen, John Palmer, bricklayer, and Joseph Wharton, cooper. Originally separate but similar houses, the Spruce Street dwellings were made into one during their 1958 rehabilitation when one door was remodeled into a window.[4]

About a decade later, sometime between 1750 and 1758, the master builder and politician Samuel Rhoads built a comfortable two-and-a-half-story dwelling on an empty lot at 217 Delancey Street, and in 1758 sold it to Alexander Barclay, who had just become comptroller of the Port of Philadelphia.[5] Although the Rhoads-Barclay House retains the shed dormer, pent eave, and simple, transomed doorway of the Arbour House, it reflects a growing sophistication in the use of Georgian architectural elements. Its greater width and pent eave combine with the water-table at the street level and a line of bricks between the second-story windows and the modillion cornice to give it a greater horizontality and more classical proportions than the earlier houses exhibited.

The random distribution of these early houses—all built on vacant lots—suggests that the development of Society Hill was not a systematic march from either the north or the east. For many years during the mid-eighteenth century dwellings stood by themselves, surrounded by empty land. For example, a 1762 map of the city shows only two dwellings on Third Street between Spruce and South.[6] Beginning in the late 1750's and lasting until the outbreak of the American Revolution, a burst of construction filled many of the empty spaces east of Fourth Street.

Two of the finest town houses of this period are the Abercrombie House at 268 South Second Street and the Powel House (illus.) at 244 South Third Street. The Abercrombie House is the older and the larger of the two. It was built about 1759 for Captain James Abercrombie, a prosperous seagoing merchant, and was one of the tallest dwellings in the city, four-and-a-half stories high with a decked gable roof capped by a balustrade. It was a self-conscious expression of Georgian architecture complete with a Doric frontispiece and mutule cornice.

Although the Powel House was not as large, the English traveler Nicholas Pickford in 1786 found that it was "of admirable Design both without and within, and might do credit to London. It is evident that no expense was spared that might contribute to either Elegance or Comfort."[7] Its owner, "typical and best of Philadelphia's cosmopolitan gentry," was Samuel Powel, a successful merchant and evidently an equally successful politician who served as the city's last mayor under the monarchy and its first mayor under the new Republic.[8] Unlike the Abercrombie House, which later became a warehouse and was gutted, the Powel House, although used commercially, preserved its elegant interiors. When the house was threatened with demolition in the 1920's, the paneling and woodwork from several rooms were removed and installed in the Philadelphia Museum of Art and New York's Metropolitan Museum of Art.[9] After the Powel House was saved, its missing interior elements were carefully reconstructed beginning in 1931 and its exterior restored, and today it is open to the public as a historic house museum; even the two-hundred-year-old pear tree in the garden of the Powel House remains and blooms each spring probably much as it did when it was enjoyed by Washington and Adams and other notables who visited the Powels.

Contrasting with merchants' town houses were workingmen's bandboxes, often set back from the street in courts.[10] A good example is Drinker's Court at 236–38 Delancey Street. In about 1765 John Drinker erected his dwelling on Pine Street and five income-producing bandboxes at the rear of his lot facing Delancey Street; other apartments were added during the early nineteenth century.[11] This practice was an obvious deviation from Penn's original plan of single dwellings and large garden lots, but it provided cheap housing for workingmen close to their places of employment and gave the area an economic mix.

Post-Revolutionary Society Hill saw a resumption of construction, most of it farther west, that is, on Fourth Street and westward. Some of the new residences were grand and were freestanding within large lots. The Hill-Physick House at 321 South Fourth Street was built in 1786 for the prosperous wine merchant Henry Hill.[12] It ranks among the city's finest examples of the Federal style that became so popular after the Revolution. During this period the robust sculptural quality of the colonial houses was replaced by delicate, flat surfaces, attained in this house by the use of a flat, finely carved frontispiece and windows without lintels. Its large leaded fanlight with delicately wrought muntins serves as the focal point of the street facade and coordinates other large-scale elements to indicate the house's great size: thirty-two rooms within three-and-a-half stories.

Closer to Philadelphia's row house tradition is the town house of the merchant-banker John Clement Stocker, which was built at 402 South Front Street during the early 1790's.[13] Although it was constructed on a much larger scale than its colonial neighbors and had an abundance of delicately carved ornamentation on its fireplaces and ceilings, it adjoined houses on both sides and continued the earlier town house plan of side hall and rear stairtower, or piazza, as Philadelphians called this small ell which enclosed two or three flights of stairs. The house still stands, starkly alone, awaiting restoration.

By the end of the eighteenth century, Society Hill's population had grown to the point that the market stalls at Second and Lombard streets had to be extended along the middle of Second Street, first to South Street in 1797 and then to Pine Street in 1804. There the present Head House was erected to house a volunteer hose company and a volunteer engine company.[14]

Another response to the city's population growth was the extension of the Pennsylvania Hospital (illus.) at Eighth and Pine streets, then on the fringes of the city. Outskirts were favorite sites for hospitals since they afforded the open spaces and tranquil settings that were believed to be the most healthful for body and mind and reduced the threat of fire, an ever present menace to early cities. Pennsylvania Hospital is another product of the indefatigable Benjamin Franklin, this time working in conjunction with Dr. Thomas Bond, and is another Philadelphia first—the country's earliest public hospital. Its eastern wing had been standing since 1757, and in 1794 work was begun on the center and western sections. In 1805 the splendid Adamesque-Federal center section with its circular surgical amphitheater (the nation's oldest) was completed.[15] The result was not only a physical plan that was to be emulated in American hospitals for a century but also an outstanding medical institution that helped to make Philadelphia a center of medicine.

Part of Society Hill's post-Revolutionary growth was created by the migration of free Negroes into the area. The nucleus of this early black neighborhood, which still exists, was Bethel African Methodist Episcopal Church (illus.).[16] In 1787 the Reverend Richard Allen, a former slave who would later become the first bishop of the African Methodist Episcopal Church, led his fellow blacks out of St. George's Methodist Episcopal Church in protest against its segregation practices and bought a lot at 419 South Sixth Street, north of Lombard.[17] The present late-nineteenth-century Romanesque Revival church, the third built on this site by the congregation of Mother Bethel, stands on what is thought to be the oldest piece of ground continuously owned and occupied by blacks in the United States.[18]

Society Hill remained relatively stable throughout most of the nineteenth century. Physical changes were minimal. Shops were put in the ground stories of some dwellings, particularly those at corners. Some new churches went up to accommodate growing congregations, and some new warehouses were built along the waterfront. Because of Philadelphians' continuing affinity for brick and the buildings' relatively low height, the new structures blended into the older cityscape. The Norris-Cadwalader House (Mutual Assurance Company) (illus.) at 240 South Fourth Street, for example, was built about 1828, yet it fits well with its colonial neighbor, the Shippen-Wistar House (illus.).[19] It was such homogeneity and regularity, with its "air of cheerfulness, cleanliness, and comfort," that gave Frances Wright "so much satisfaction" during her visit to the city in 1819.[20]

After the Civil War and particularly after electric trolleys were introduced in the 1890's, Society Hill residents started moving to Rittenhouse Square and farther out to the new residential "suburbs." Their homes were taken by the immigrants who began flooding into Philadelphia from southern and eastern Europe after 1880. The shift can be traced in the changing ownership of old churches and the erection of new houses of worship. In 1900, for example, Chevia B'nai Aviohome Mi Russe began its synagogue at 523 Lombard Street; in 1912 a Rumanian congregation converted Spruce Street Baptist Church into a synagogue; in 1922 St. Andrew's Protestant Episcopal Church became St. George's Greek Orthodox Church.[21]

Dwellings throughout the area were altered to accommodate stores, warehouses, light manufacturing establishments, or tenements, but because Society Hill became a low-rent district quickly there was little demolition or rebuilding. That left the material

culture of an earlier era under galvanized iron shopfronts, layers of
wallpaper, and bricked-up fireplaces to be uncovered when Society
Hill's rebirth would come a half-century later.

That renaissance occurred during the late 1950's and 1960's.
Through the cooperation of three government agencies and one
nonprofit corporation, Society Hill became known as one of the
nation's most successful urban renewal projects. The Philadelphia
City Planning Commission prepared the comprehensive plan; the
Redevelopment Authority acquired the properties for development;
the Philadelphia Historical Commission certified historically
significant buildings and collected the materials needed for accurate
restorations; and the Old Philadelphia Development Corporation
made many of these houses available for purchase and restoration.[22]
Through this system nearly eight hundred houses have been restored
and approximately two hundred new houses have been built.
Visually anchoring the revitalized neighborhood are the Society Hill
Towers. These three high-rise apartment buildings designed by Ieoh
Ming Pei serve not only as focal points for Society Hill in relation to its
primary historic sites and open spaces but also in relation to the rest
of downtown Philadelphia and its traffic flow.[23] In spite of their
radically different scale and style they relate to the older buildings
about them through their articulated fenestration and adjacent town
houses.

Society Hill today stands as a fine example of how a historically
significant square mile can be restored from slum status, its old
buildings integrated with modern structures, and converted into one
of the most desirable living areas in the region.

Catalog

Abercrombie, Captain James, House (now Perelman Antique Toy Museum)

(PA-1315), 268–70 S. Second St. Museum. Brick with marble trim, 30'-5" (four-bay front) × 45'-2½" originally with two-story rear ell, four-and-a-half stories, decked gable roof with balustrade, Doric frontispiece, mutule cornice, arched alleyway to rear yard, side-hall town house plan.

Notable Georgian town house unusual for its height. Built c. 1759; ground-story storefront built before 1857; ground story rebuilt later. Rear ell rebuilt 1913, enlarged 1922, 1929. Rear ell demolished, front section restored 1967–68; John Lloyd and Associates, architects. Certified, PHC 1957. 5 sheets (1959–60, including plot plan, plans, elevations, section);* 9 ext. photos (1957, 1959, 1972),* 4 int. photos (1960);* 4 data pages (1957, including 1770 insurance survey).*

Apartment Building

(PA-1316), 113–17½ South St. Five brick dwelling units with ground-story shops, each approx. 16' (three-bay front) × 60', three stories, flat roof, projecting upper-story molded metal bay with pilasters and pediment on each unit, every ninth row of bricks recessed to simulate rustication, originally one-family apartment on each upper floor, L-plan for each unit.

Example of modest early-20th-century apartment building in a degenerate expression of the classical tradition. Built 1913–14; Solomon Brothers, contractors. New storefront for No. 113 built 1941; permastone facade added to No. 115, 1950. 1 ext. photo (1967);* 3 data pages (1973, including 1913 building permit).*

9

Arbour, William, House

(PA-1051), 220 Spruce St., at S.E. corner American St. Brick, 14'-10" (two-bay front) × 18'-1", two-and-a-half stories on raised basement, gable roof, one room to a floor. Built 1743–45; two-story rear ell added c. 1746, demolished 1958; front section rehabilitated 1958. Probably the oldest surviving house in Society Hill. Certified, PHC 1957. 1 sheet (1959, including plot plan, plans, details); 2 ext. photos (1957), 1 int. photo (1957); 4 data pages (1959).

Bake House and Oven

(PA-1317), Rear of 423 S. Second St. House: brick, approx. 35' (four-bay front) × 17', two stories, shed roof, one room to a floor. Oven in basement: brick, recessed approx. 3' within low arch approx. 9' wide × 6' high. Built before 1796. Certified, PHC 1957. 1 ext. photo (1963),* 1 int. photo (1963);* 5 data pages (1961, including 1802 insurance survey).*

Bethel African Methodist Episcopal Church (now Mother Bethel African Methodist Church). Illustrated

(PA-1318), 419 S. Sixth St., at S.E. corner Addison St. Brick with random rough-hewn granite front and rusticated base, approx. 64' (three-bay front) × 82' with three-story rear Sunday-School building, two stories, gable roof with two cross-gables, four-story square corner tower and belfry with gargoyles and clustered-colonette pinnacles, large semicircular entrance arch on clustered columns, large semicircular window on three sides, three-aisle plan with balconies, hexagonal chancel.

Church with architectural emphasis on its Romanesque Revival facade. Built 1889–90; Hazlehurst and Huckel, architects. The fourth African Methodist Episcopal church building on this lot, purchased 1787 and the oldest piece of real estate owned by blacks in the United States. Richard Allen, first ordained black man in America and founder of the African Methodist Episcopal Church, is buried in basement crypt. Certified, PHC 1964; Pennsylvania Register 1970; designated National Historic Landmark 1974. 2 ext. photos (1973);* 3 data pages (including 1890 insurance survey);* HABSI form (1971).*

Blackwell, Reverend Robert, House (now St. Peter's Church House)

(PA-1319), 313 Pine St. Brick with granite trim, approx. 24' (three-bay front) × 38' with three-story rear ell, three-and-a-half stories on raised basement, gable roof, arch-head entrance with fanlight, first- and second-story windows set in blind arches with recessed solid tympanums, side-hall plan, large columnated parlor.

Fine Federal-style urban dwelling. Built 1809; possibly by Robert Mills, architect. Ground-story interior altered, kitchen ell enlarged c. 1840; west wall altered 1965. St. Peter's Church House since 1922. Certified, PHC 1956; Pennsylvania Register 1970. 1 ext. photo (1972);* 6 data pages (1958, including 1809, 1880 insurance surveys);* HABSI form (1971).*

Bridges, Robert, House

(PA-1320), 507 S. Front St. Brick, approx. 22' (three-bay front) × 40', three-and-a-half stories on raised basement, gable roof, brick string-courses, London house plan. Built c. 1791; altered to tavern 1865; store and dwelling after 1918; demolished 1966. Certified, PHC 1957. 1 ext. photo (1958);* 7 data pages (1961, including 1791–1918 insurance surveys);* HABSI form (1957).*

Bridges-LaTour House

(PA-1321), 509 S. Front St. Brick, approx. 18' (two-bay front) × 40', three-and-a-half stories on raised basement, gable roof, brick string-courses, round-arch entrance with fanlight and Adamesque frontispiece, probably London house plan. Built 1791; demolished 1966. Between 1813 and 1850 home of John LaTour, merchant, who owned dwelling and rear adjoining lot with warehouse (PA-1056). Certified, PHC 1957. 2 ext. photos (1958, 1961),* 8 int. photos (1966);* 7 data pages (1957, including 1791 insurance survey);* HABSI form (1957).*

Burgin, Dr. George H., House

(PA-1322), 319 S. Fifth St. Brick with stone trim, approx. 16' (two-bay front) × 36' with three-story rear ell, three-and-a-half stories, gable roof, originally side-hall town house plan.

Built between 1821 and 1823; rear kitchen ell raised to two stories 1827; altered to store and rear ell raised to three stories after 1912; demolished 1964. Built for George H. Burgin, physician, druggist, glass manufacturer, and public servant who served as director of public schools 1838–53, manager of Wills Eye Hospital 1848–56, trustee of Philadelphia Gas Works 1842–60. Burgin lived here c. 1825–31, rented house 1831–70. Certified, PHC 1957. 1 ext. photo (1960);* 6 data pages (1959, including 1823–1912 insurance surveys).*

Bussey-Poulson House

(PA-1323), 320 S. Fourth St. Brick, 18'-0" (three-bay front) × 26'-8" with three-story rear ell, three-and-a-half stories, gable roof, side-hall plan. Built 1783; rehabilitated 1967–70. Certified, PHC 1957. 4 sheets (1959, including plot plan, plans, elevations, details)* courtesy of Philadephia Historical Commission; 3 ext. photos (1959, 1972);* 5 data pages (1957, including 1796–1908 insurance surveys);* HABSI form (1958).*

Cove Cornice House. See Man Full of Trouble Tavern

Currie, Dr. William, House

(PA-191), 271 S. Fifth St. Brick, approx. 23' (three-bay front) × 40', three-and-a-half stories, gable roof, side-hall plan. Built 1811; entire ground story removed later; demolished 1957. 2 ext. photos (1957); 2 data pages (1957, including 1811 insurance survey).

Davis-Lenox House

(PA-1324), 217 Spruce St. Brick, approx. 22' (three-bay front) × 30' with two-story rear ell, three-and-a-half stories on raised basement (originally two), gable roof, originally side-hall London house plan.

Built c. 1758; James Davis, house carpenter. Raised one-and-a-half stories, kitchen altered to stair tower, new two-story kitchen built 1784; kitchen altered to dining room, new one-story kitchen built 1848; kitchen raised one story later. Restored to 1784 state 1960; George B. Roberts, architect. Major David Lenox, Revolutionary hero and later president of Bank of the United States and American minister to the Court of St. James, lived here 1779–1810. Certified, PHC 1957. 2 ext. photos (1959, 1961);* 6 data pages (1958, including 1781–1848 insurance surveys).*

Drinker, John, House

(PA-1325), 241 Pine St. Brick, approx. 18' (two-bay front) × 30' with three-story rear ell (originally two stories), three-and-a-half stories, gable roof, pent eave between first and second stories, side-hall plan.

Fine example of modest Philadelphia town house whose lot originally included rear tenement court (Drinker's Court) behind 236–38 Delancey St. Built c. 1765; John Drinker (bricklayer), builder. Rear ell enlarged before 1870. Restored 1956–57; Edward Parnum, architect. Certified, PHC 1957; Pennsylvania Register 1970. 1 ext. photo (1973),* 2 int. photos (1959);* 6 data pages (1957, including 1806, 1870 insurance surveys).*

Drinker's Court

(PA-1326), 236–38 Delancey St. A complex of two front buildings and six rear apartments and a rear building flanking a courtyard.

No. 236: brick, 16'-4" (two-bay front) × 14'-9", two-and-a-half stories, half-gambrel roof, pent eave, one room to a floor; three brick rear apartments (originally four), each approx. 14' (two-bay front) × 12', two stories, shed roof, one room to a floor.

No. 238: brick, 15'-2" (two-bay front) × 16'-10", two-and-a-half stories, half-gambrel roof, one room to a floor; three brick rear apartments, each approx. 16' (two-bay front) × 11', two stories, shed roof, one room to a floor; rear brick building, 13'-3" (two-bay front) × 15'-7", three stories, shed roof, one room to a floor. Originally balanced by three-story rear building at rear of apartments of No. 236.

The construction of small row houses on a court with one row facing the other was a common building practice in 18th-century and early 19th-century Philadelphia, as the large city blocks were subdivided by secondary streets. These buildings were on the northern extremity of a lot owned by John Drinker and on which he built his dwelling at 241 Pine St. See Drinker, John, House (PA-1325). Court later known as Deimling Place, later as Bodine St.

No. 236 and rear kitchen (later northernmost apartment), No. 238 and three rear apartments built c. 1765; probably by John Drinker, bricklayer. Rear apartments of No. 236 built after 1806. Three-story rear building of No. 238 built between 1827 and 1829. No. 236 altered to store c. 1870; reverted to dwelling between 1920 and 1931. Southernmost three-story apartment of No. 236 demolished after 1939. No. 238 and apartments rehabilitated 1958–59; No. 236 and apartments rehabilitated 1965–66. Shutters, window and door elements at Architectural Study Collection, Museum Division, Independence National Historical Park. Certified, PHC 1956; Pennsylvania Register 1970; NR.

4 sheets (1958, 1959, including plot plan, plans, elevations, section, details)* courtesy of Philadelphia Historical Commission; 5 ext. photos (1959, 1972),* 1 int. photo (1959);* 13 data pages (1958, including 1806–70 insurance surveys);* HABSI form (1957).*

Eastburn Mariners' Bethel (Presbyterian) Church (also known as Mariners' Bethel)

(PA-1327), 328–30 S. Front St., at S.W. corner Delancey St. Brick, approx. 54' (three-bay front) × 76', two-and-a-half stories, gable roof originally with square clock tower near front peak, name stone in gable, chapel and reading rooms on ground story, three-aisle sanctuary on second story.

Built 1868–69; possibly by Stephen D. Button, architect. Clock tower removed after 1912; building vacated 1959; interior destroyed by fire 1963. Organized by Joseph Eastburn 1819, Mariners' Bethel was one of the oldest seamen's churches in the country; merged with Third Presbyterian Church 1959. Certified, PHC 1957. 2 ext. photos (1961),* 1 photocopy of ext. photo (1959),* 9 photocopies of int. photos (1959);* 19 data pages (1959).*

Estlack, Thomas, House

(PA-1328), 413 Lombard St. Brick, approx. 16' (two-bay front) × 35' with three-story rear stair tower, three-and-a-half stories on raised basement, gable roof, arch-head entrance, side-hall plan. Built 1821; probably by Thomas Estlack, bricklayer. Demolished 1964. Certified, PHC 1957. 1 ext. photo (1960);* 3 data pages (1959, including 1822 insurance survey).*

Fitzgerald, Thomas, House

(PA-1329), 437 Lombard St. Brick with marble trim, approx. 19' (three-bay front) × 30' with three-story rear ell, three-and-a-half stories, gable roof, side-hall town house plan. Built 1799; rear ell built 1844; demolished 1964. Certified, PHC 1957. 1 ext. photo (1960);* 5 data pages (1959, including 1811, 1845 insurance surveys).*

Girard Row

(PA-1330), 326–34 Spruce St. Five brick row houses with marble ground story and trim, each approx. 21' (two-bay front) × 40' with two- and three-story rear ell, three-and-a-half stories on raised basements, gable roofs. Row houses built as investment for Stephen Girard, early millionaire merchant and financier. Built 1831–33; William Struthers, marble mason. Nos. 332–34 restored 1954–   ; Charles E. Peterson, architect. Certified, PHC 1957. 1 sheet (1958, downspout detail);* 2 ext. photos (1959),* 4 int. photos (1959, including 4 fireplaces);* 20 data pages (1964, including 1845 insurance survey).*

Hall, John, House

(PA-1331), 327 S. Third St., at N.E. corner Delancey St. Brick, 16'-0" (two-bay front) × 35'-4", three-and-a-half stories, gable roof, brick stringcourses. Built between 1783 and 1799; John Hall, house carpenter. Ground story altered to store c. 1854; rehabilitated 1965–

66. Certified, PHC 1957. 5 sheets (1957, 1959, including plot plan, plans, elevations, section, details)* courtesy of Philadelphia Historical Commission; 3 ext. photos (1959, 1972),* 2 int. photos (1959);* 2 data pages (1958);* HABSI form (1959).*

Hall-Wister House

(PA-1332), 330 S. Third St. Brick, approx. 16' (two-bay front) × 20' with two-story rear ell, two-and-a-half stories, gable roof, brick stringcourse, well-preserved interior woodwork, one-room ground story. One of a pair of row houses. Built between 1771 and 1774; John Hall, house carpenter. Altered c. 1890. Restored, rear ell replaced 1959–61; Robert T. Trump, restorer. In possession of Wister family 1778–1853. Certified, PHC 1957. 2 ext. photos (1959, 1972),* 1 int. photo (1959);* 3 data pages (1958, including 1774 insurance survey).*

Hill, David, House

(PA-1333), 309 S. Third St. Brick with marble trim, 25'-5$^{1}/_{2}$" (three-bay front) × 49'-6$^{1}/_{2}$" with three-story rear ell, three-and-a-half stories on raised basement, gable roof, round-arch entrance with fanlight, side-hall town house plan. Built 1831–39; ground story altered to store later. Restored 1970–71; Hugh M. Zimmers, architect. Certified, PHC 1957. 6 sheets (1958, including plans, elevations, section, details)* courtesy of Philadelphia Historical Commission; 4 ext. photos (1959, 1973),* 3 int. photos (1959);* 2 data pages (1958);* HABSI form (1958).*

Hill-Physick House

(PA-1334), 321 S. Fourth St., at S.E. corner Cypress St. Historic house museum. Brick with marble trim, approx. 48' (three-bay front) × 51', three-and-a-half stories on raised basement, hipped roof, round-arch entrance with sidelights and notable fanlight, center-hall plan, springhouse extends from cellar under Cypress St. sidewalk.

Notable example of a large Federal-style residence. Built 1786 incorporating earlier house into its northeast end; one-story wing built at rear of north side, interior remodeled 1815; one-story wing extended to front c. 1838; interior extensively remodeled 1850's; three-story rear ell demolished before 1868; interior remodeled probably c. 1920; three-story elevator shaft built on south side 1921; elevator shaft, two-story rear stable demolished 1966. Restored by Philadelphia Society for Preservation of Landmarks 1965–66; George B. Roberts, architect.

Home of Col. Henry Hill, wine merchant active in Revolutionary affairs, 1786–98, and of Dr. Philip Syng Physick, father of American surgery, 1815–37. Headquarters of the Society of the Cincinnatus of Pennsylvania since 1966. Certified, PHC 1957; Pennsylvania Register 1970. 3 ext. photos (1972, 1976),* 1 int. photo (1959);* 17 data pages (1958, 1964, including 1815–1921 insurance surveys).*

Hilyard, Eber, House

(PA-1335), 427 Lombard St. Brick with marble trim, approx. 16' (two-bay front) × 40', three-and-a-half stories on raised basement, gable roof, side-hall plan. One of a pair of houses. Built 1807; probably by Eber Hilyard, bricklayer. Demolished 1964. Certified, PHC 1957. 1 ext. photo (1960);* 4 data pages (1959, including 1807, 1919 insurance surveys).*

Holy Trinity (German) Roman Catholic Church

(PA-1336), 601–9 Spruce St., at N.W. corner Sixth St. Brick, approx. 100' (five-bay front) × 60', one story, hipped roof, round-arch windows, three-aisle plan, polygonal chancel. Built 1789; damaged by fire, altered 1860. Altered 1902; George I. Lovatt, architect. 3 ext. photos (1973),* 1 photocopy of int. lithograph (c. 1880);* 1 data page (1964).*

House

(PA-1809), 532 S. Water St., at N.W. corner South St. Brick with marble trim, approx. 18' (two-bay front) × 38', three-and-a-half stories, rusticated marble flat-arch lintels with keystones, brick stringcourse. Built c. 1790; altered to store before 1860; demolished 1959. Certified, PHC 1958. 1 ext. photo (1958).*

Houses

(PA-1337), 319–21 S. Front St. Brick covered with plaster, approx. 18' (three-bay front) × 40', three-and-a-half stories, gable roof. Built late 18th century; altered to store before 1861; demolished 1967. 2 ext. photos (1961).*

House

(PA-1338), 532 S. Front St., at N.W. corner South St. Brick with stone trim, approx. 18' (two-bay front) × 36', three-and-a-half stories, gable roof. Built late 18th century; altered to store later; demolished 1959. Certified, PHC 1957. 1 ext. photo (1958);* HABSI form (1957).*

Houses

(PA-1339), 722–30 Spruce St., at S.E. corner Eighth St. Eight brick houses with marble trim, each approx. 25' (three-bay front) × 50' with two-, three-, four-story rear ells, three-and-a-half stories (No. 730 four stories), mansard, gable, flat roofs, side-hall plans (No. 730 center-hall plan). Nos. 722–26 built before 1845, mansard added later; No. 728 built before 1860; No. 730 built 1837. Demolished 1958. Certified, PHC 1958. 5 ext. photos (1958, including two of iron stair railing);* 26 data pages (1963, including 1837–1937 insurance surveys).*

Jefferson, Joseph, House

(PA-1340), 600 Spruce St., at S.W. corner Sixth St. Brick with marble trim, approx. 20' (two-bay front) × 42', three-and-a-half stories, gable roof. Built 1804–5; probably by Edward Bonsall, house carpenter. Altered to store 1835; renovated c. 1905; two-story rear ell demolished 1969; rehabilitated 1969–70. Joseph Jefferson, perhaps America's most celebrated 19th-century actor and comedian, was born here 1829. Certified, PHC 1957. 1 ext. photo (1958);* 4 data pages (1958, including 1805–1914 insurance surveys).*

Jordan-Stoddart House

(PA-1341), 409 S. Fifth St., at N.E. corner Addison St. Brick with stone trim, approx. 16' (two-bay front) × 38' with two-story rear wing, three-and-a-half stories, gable roof, originally London house plan. Built between 1804 and 1806; rear two-story wing built 1885; altered to store 1909; demolished 1964. Certified, PHC 1957. 1 ext. photo (1960);* 5 data pages (1959, including 1811–1909 insurance surveys).*

Kosciuszko, General Thaddeus, House

(PA-1342), 301 Pine St., at N.W. corner Third St. (originally 342 S. Third St.), Independence National Historical Park. Museum. Brick, approx. 31' (three-bay front) × 15', three-and-a-half stories, gable roof, London house plan. Built 1774–75; probably by Joseph Few, carpenter. Altered to tavern 1865; altered to store and dwelling later; altered to dwelling 1942. Restored by National Park Service 1974–76. Gen. Thaddeus Kosciuszko, Polish patriot and chief engineer of the Continental Army 1780–83, lived here 1797–98 while in exile. Certified, PHC 1957; Pennsylvania Register 1971; NR. 2 ext. photos

(1972);* 6 data pages (1957, including 1775–1922 insurance surveys);*
HABSI form (1971).*

LaTour Warehouse
(PA-1056), 508 S. Water St. Brick, 21' (three-bay front) × 50'-6", two-and-a-half stories, gambrel roof, double doors on first and second stories front and rear, hooded dormer covering hoist tackle, one room to a floor.

Notable early-19th-century warehouse. Built 1817–18; demolished 1967. Certified, PHC 1957. 8 sheets (1958, including plot plan, plans, elevations, sections, details); 6 ext. photos (1958), 3 int. photos (1958); 5 data pages (1959, 1960); HABSI form (1957).*

Lombard Street Area Study
(PA-1678), 323–33 Lombard St. Six adjoining brick houses with marble trim (No. 331 with coral stone rear ell and cellar), each approx. 18' (two-bay front) × 38' with two- or three-story rear ell (except No. 333), three-and-a-half stories on raised basement (No. 331 originally two-and-a-half stories), gable roofs (No. 331 originally gambrel), side-hall plans.

No. 331 built 1763–66, enlarged and interior altered 1818, raised to three stories and new front built 1927; No. 333 built 1813, altered to store later, reverted to dwelling 1928; No. 325 built 1817; Nos. 327–29 built c. 1835; No. 323 built 1839, renovated as part of St. Peter's Choir School after 1888. All but No. 323 demolished 1964. Certified, PHC 1958. 4 ext. photos (1960);* 18 data pages (1959, 1963, including 1763–1849 insurance surveys).*

McDonald, Malcolm, Houses
(PA-1343), 315–17 S. Fifth St. Brick with stone trim, each approx. 20' (three-bay front) × 45' with two- and three-story rear ell, three-and-a-half stories, gable roof, originally side-hall town house plan. Built between 1795 and 1797; altered to stores before 1912; demolished 1964. Certified, PHC 1957. 1 ext. photo (1960);* 5 data pages (1959, including 1799 insurance survey).*

McKean, Thomas, Jr., House
(PA-190), 269 S. Fifth St. Brick, approx. 23' (three-bay front) × 30', three stories on raised basement, flat roof, side-hall plan. Built c. 1816; demolished 1957. 1 sheet (1957, including details); 2 ext. photos

(1957), 3 int. photos (1957); 4 data pages (1957, including 1837 insurance survey).

McMullin, Robert, House

(PA-1344), 411 Pine St. Brick with marble trim, 19' (two-bay front) × 35'-6" with three-story rear ell, three-and-a-half stories on raised basement, gable roof, arch-headed entrance with fanlight, side-hall plan. Built c. 1831; exterior restored 1965. Certified, PHC 1957. 2 ext. photos (1965),* 6 int. photos (1965);* 3 data pages (1963).*

Man Full of Trouble Tavern (also known as Cove Cornice House and Stafford's Tavern). Illustrated

(PA-128), 127 Spruce St., at N.W. corner Mattis St. Historic house museum. Brick, approx. 22' (two-bay front) × 20', two-and-a-half stories, gambrel roof with original cove cornice, front pent eave with cove cornice, balcony on east end, two rooms to a floor.

Fine example of 18th-century city tavern. Built 1760; projecting one-story frame shed built at northeast corner before 1835; interior altered 1879, 1900. Frame shed demolished c. 1963. Restored by Knauer Foundation for Historic Preservation 1963–65; W. Nelson Anderson, architect. Wallpaper sample (c. 1785) in Architectural Study Collection, Museum Division, Independence National Historical Park. Certified, PHC 1957.

4 sheets (1936, 1958, including plans, elevations, details); 10 ext. photos (1958, 1961, 1973),* 3 int. photos (1958, 1959),* 1 photocopy of old wash drawing (1835),* 1 photocopy of old ext. photo (1850);* 26 data pages (1958, 1966, including 1769–1900 insurance surveys).*

Manning Street Area Study. See Marshall's Court Area Study

Mariners' Bethel. See Eastburn Mariners' Bethel (Presbyterian) Church

Marshall's Court Area Study (also known as Manning Street)

(PA-1345), 403–11 Marshall's Court, between 256–58 S. Fourth St. and 275–77 S. Fifth St. N. of Spruce St. Independence National Historical Park. Five contiguous brick houses, each approx. 15' (two-bay front) × 32', three-and-a-half stories, gable roofs, London house plans. Built c. 1812; demolished 1959. Mantels, stairway elements, shingles in Architectural Study Collection, Museum Division, Independence National Historical Park. Certified, PHC 1957. 5 ext. photos

(1958),* 9 int. photos (1958);* 3 data pages (including 1816, 1837 insurance surveys).*

Mears-Heaton House

(PA-1070), 240 Delancey St. Brick, 14'-11¹/₂" (two-bay front) × 30'-6" with two-story rear brick stair tower and two-story brick-pane (half-timber frame with brick nogging) rear ell, two stories, mansard roof (originally gable), originally pent eave, two rooms to a floor. The rear ell was the only known extant brick-pane building in Philadelphia. Rear ell probably built before 1765. Front building built 1765–66; John Mears (joiner), builder. Mansard added later. Demolished 1954. 4 sheets (1954, 1962, including plot plan, plan, section, details); 5 data pages (1962, including 1782 insurance survey).*

Mellon, Thomas, House

(PA-1346), 716 Spruce St. Brick with marble trim, approx. 25' (three-bay front) × 54' with three- and four-story rear ell, four stories, flat roof, side-hall town house plan, notable Greek Revival interior woodwork. Built 1836; demolished 1959. Certified, PHC 1957. 2 sheets (1959, including plan, details)* courtesy of Philadelphia Historical Commission; 1 int. photo (1959),* 7 photocopies of int. photos (1959);* 3 data pages (1958, including 1836 insurance survey).*

Mifflin, John, Houses

(PA-1347), 521–23 S. Front St. Brick, each approx. 14' (two-bay front) × 35', three-and-a-half stories, gable roof, round-arch entrances with fanlights, London house plans. Built before 1779; demolished 1966. Certified, PHC 1957. 13 data pages (1961, including 1784, 1785 insurance surveys);* 3 HABSI forms (1957).*

Mitchell, Thomas, House

(PA-1348), 276 S. Third St., at N.W. corner Spruce St. Brick, approx. 18' (two-bay front) × 30' with rear addition, four stories on raised basement, half-hipped roof, blind arch with recessed tympanum above second-story window, elliptical windows at fourth story. Built c. 1815; probably by Robert Mills, architect. Flat-roof addition along Spruce St. built c. 1850. Demolished 1964. Certified, PHC 1957. 3 ext. photos (1961, 1962),* 4 int. photos (1961),* 1 aerial photo (1962);* 1 data page (1957).*

Mother Bethel African Methodist Episcopal Church. See Bethel African
Methodist Episcopal Church

Mutual Assurance Company. See Norris-Cadwalader House

Neave, Samuel, House and Store

(PA-1349), 272–74 S. Second St., at N.W. corner Spruce St. Brick with
marble trim, 29'-9" (three-bay front) × 44'-2¼" originally with one-
and two-story rear ell, three-and-a-half stories on raised basement,
gable roof, pent eave on south facade (not original), pedimented frontispiece, originally side-hall town house plan with store in front.

Fine Georgian house of a prominent merchant. Built c. 1759; rear
ell raised to three stories 1829; ground-story interior extensively altered, one-story lateral rear addition built 1841; ground-story
storefront built before 1857; rear ell, storefront rebuilt 1905; three-
story rear ell demolished 1959. Front section restored 1967–70; Carl
Massara and Associates, architects. Certified, PHC 1957.

10 sheets (1959, 1960, including plot plan, plans, elevations, section, details);* 9 ext. photos (1957, 1959, 1960, 1972),* 11 int. photos
(1959, 1960);* 9 data pages (1957, including 1760–1841 insurance surveys).*

Nevel, Thomas, House

(PA-1350), 338 S. Fourth St. Brick, approx. 22' (three-bay front) × 37'
with two-story rear ell, two-and-a-half stories, gambrel roof with
mutule cornice, round-arch entrance with fanlight and Tuscan frontispiece, side-hall plan.

Fine Georgian house originally with shop in front. Built 1770;
Thomas Nevel, house carpenter. Rear ell enlarged and altered 1844;
interior altered 1865; altered to tavern 1888. Exterior restored 1961;
John A. Bower, architect. Certified, PHC 1957. 3 ext. photos (1964,
1972),* 3 int. photos (1964);* 13 data pages (1960, including 1770–1880
insurance surveys).*

"New Market"

(PA-1351), in middle of Second St. between S. side Pine St. and N.
side Lombard St. Market House (originally quarters of Hope Hose
Co. No. 6 and Fellowship Engine Co. No. 29, later known as Head
House): brick with marble trim, approx. 30' (three-bay front) × 28',
two stories, gable roof with octagonal cupola, arched passage

through center of ground story to shambles (open market sheds). Shambles: brick piers, approx. 30' × 245', one story, gable roof, vaulted plaster ceiling.

Probably the oldest extant firehouse and market structure in the United States. First eight stalls built at Lombard St. 1745; extended north to Stamper's Alley 1795–97; extended north to Pine St. and Head House built 1804. Shambles widened 1809, enclosed 1921. Head House restored 1959–61; shambles restored 1962–63. Certified, PHC 1956; Pennsylvania Register 1970; designated National Historic Landmark 1966. 2 ext. photos (1963).*

Norris-Cadwalader House (now Mutual Assurance Company). Illustrated (PA-1352), 240 S. Fourth St. Brick with marble trim, approx. 30' (three-bay front) × 50' with three- and four-story rear ell, four-and-a-half stories, gable roof; notable interior paneling and plasterwork, side-hall town house plan with ground-story offices.

Fine early-19th-century residence-office. Built c. 1828 for attorney Joseph P. Norris, Jr., and after 1837 home of prominent attorney John Cadwalader. Interior altered 1912; rear ell enlarged 1941, 1947. In possession of Cadwalader family until 1912 when it became offices of Mutual Assurance Company for Insuring Houses from Loss by Fire. Certified, PHC 1957. 1 ext. photo (1972),* 8 int. photos (1959, of fireplaces),* 2 photocopies of old ext. photos (c. 1900),* 8 photocopies of int. photos (c. 1900);* 3 data pages (1959, including 1837 insurance survey).*

Old Pine Street Church. See Third Presbyterian Church

Old St. Mary's. See St. Mary's Church

Palmer, John, House
(PA-1353), 117 Lombard St. Brick, approx. 18' (three-bay front) × 32' with two- and three-story rear ell, three-and-a-half stories, gable roof, segmental-arch heads over windows, brick stringcourse with glazed headers, alleyway through ground story to rear yard, originally London house plan.

One of a pair of row houses. Built 1743; probably by John Palmer, bricklayer. Rear kitchen and stair ell built c. 1763; altered and enlarged c. 1809. Exterior restored 1966–68; interior rehabilitated 1973. Certified, PHC 1957. 1 ext. photo (1973);* 5 data pages (1961, including 1764, 1810 insurance surveys).*

Pancake, Philip, House

(PA-1354), 333 S. Fifth St. Brick, approx. 20' (two-bay front) × 34' with three-story rear ell, three-and-a-half stories, gable roof, side-hall plan. Built c. 1775; two-story rear kitchen built 1773; rear kitchen raised to three stories, ground story altered to store, new dormer and galvanized metal cornice added 1911. Demolished 1964. Certified, PHC 1957. 1 ext. photo (1960);* 3 data pages (1964).*

Pancoast-Lewis-Wharton House

(PA-1083), 336 Spruce St. Brick with marble trim, approx. 20' (three-bay front) × 36' with three-story rear ell, three-and-a-half stories, gable roof, round-arch entrance with fanlight and Tuscan frontispiece, side-hall town house plan.

Fine Federal-style town house. Built 1790; probably by Samuel Pancoast, house carpenter. Rear ell enlarged 1834; house enlarged and altered 1916–17. Restored 1962; George A. Robbins, architect. Birthplace and home of Joseph Wharton, founder of the Wharton School of Finance and Commerce, University of Pennsylvania. Christ Church Rectory since 1962. Certified, PHC 1957.

7 sheets (1932, including elevations, sections, details) from Survey, Philadelphia Chapter, A.I.A., 4 sheets (1961, including plot plan, plans); 2 ext. photos (1961, 1962), 4 int. photos (1962), 1 photocopy of old ext. photo (c. 1890); 10 data pages (1961, 1964).

Paschall, Benjamin, House. Illustrated

(PA-1355), 129A Spruce St. Brick, approx. 16' (two-bay front) × 20', two-and-a-half stories, half-gambrel roof with cove cornice, front pent eave with cove cornice, one room to a floor. Built 1760; altered to store c. 1827; altered to warehouse before 1931. Restored 1963–65; W. Nelson Anderson, architect. Certified, PHC 1957. 2 sheets (1957, including plans, elevation, details);* 2 ext. photos (1958, 1961),* see Man Full of Trouble Tavern (PA-128) for 1 ext. photo (1973),* 1 int. photo (1968);* 15 data pages (1958, 1966).*

Pennsylvania Hospital. Illustrated

(PA-1123), N.W. corner Eighth and Pine sts. Museum. Brick with granite and marble trim, approx. 285' (twenty-three-bay front) × 110' including wings, two-and-a-half and three stories on raised basement, hipped roof with circular balustraded skylight on octagonal base on center section, gable and hipped roofs with octagonal cupolas on wings, marble ground story and six giant pilasters at second and third stories of center section, round-arch entrance with fanlight and modified Corinthian frontispiece on center section, central-hall plan with offices and library in center section and wards in wings.

Notable architectural complex whose center building is an outstanding example of American Adamesque architecture and whose plan influenced later hospital design. East wing built 1755–57; Samuel Rhoads, architect. West wing built 1794–96, center section built 1794–1805; David Evans, Jr., architect. West wing altered 1846–47, center section altered 1848, east wing altered 1853; John McArthur, contractor; John McArthur, Jr., superintendent. East and west wings altered 1896; Addison Hutton, architect. Founded 1751 by Benjamin Franklin and Dr. Thomas Bond, Pennsylvania Hospital is the earliest established hospital in the United States with the country's oldest surgical amphitheater. It pioneered in psychotherapy from its beginning and established one of the country's first mental health clinics 1885. Nursing museum in east wing since 1976. Certified, PHC 1956; designated National Historic Landmark 1965.

19 sheets (1940, including plans, elevations, section, details) from Survey, Philadelphia Chapter, A.I.A. (1931–32); 4 ext. photos (1974),* 3 int. photos (1976),* 3 photocopies of measured drawings (1897),* 19 photocopies of architectural competition drawings for building never erected (1834, including four by John Haviland, four by Samuel H. Kneass, four by William Strickland, three by John C. Trautwine, four by Thomas U. Walter),* 1 photocopy of heating plans by Morris, Tasker & Morris (1846);* 23 data pages (including 1761–1896 insurance surveys).*

Perelman Antique Toy Museum. See Abercrombie, Captain James, House

Physick-Conner House
(PA-1356), 240 Pine St. or 401 S. Third St., at S.E. corner Third and Pine sts. Brick with marble trim, approx. 20' (two-bay front) × 43' with two-story rear ell, three-and-a-half stories, gable roof, elliptical window in gable end, side entry hall. Built c. 1830; altered to store c. 1900; rehabilitated 1967–68. Certified, PHC 1957. 1 photocopy of old ext. photo (c. 1890),* 5 photocopies of old int. photos (c. 1890);* 2 data pages (1958).*

Pickands, Thomas, House
(PA-1357), 307 S. Third St. Brick with marble trim, 26'-3" (three-bay front) × 42'-9" with three-story rear ell, three-and-a-half stories on

raised basement, gable roof, round-arch marble entrance with fanlight, marble lintels with keystones, originally side-hall town house
plan.

Fine early-19th-century town house. Built 1808–9; Thomas
Pickands, house carpenter. Altered to store 1868; reverted to dwelling
1895. Restored 1966–68; Hugh D. Zimmers, architect. Hallway arches
at Smithsonian Institution, Washington, D.C. Certified, PHC 1957.

6 sheets (1958, including plans, elevations, section, details)*
courtesy of Philadelphia Historical Commission; 3 ext. photos (1959,
1972),* 1 int. photo (1959);* 8 data pages (1958, including 1813–96 insurance surveys);* HABSI form (1958).*

Piles, John, House

(PA-1358), 328 S. Third St. Brick, approx. 16' (two-bay front) × 20',
two-and-a-half stories, gable roof, brick stringcourse, one-room
ground story. One of a pair of row houses. Built 1772; John Piles,
house carpenter. Restored 1959–61. Certified, PHC 1957. 2 ext.
photos (1959, 1973);* 3 data pages (1958, including 1772 insurance
survey).*

Powel, Samuel, House. Illustrated

(PA-1359), 244 S. Third St. Historic house museum. Brick with stone
trim, approx. 30' (three-bay front) × 50' with two- and three-story
rear ell, three-and-a-half stories, gable roof, Doric frontispiece, stone
stringcourses, side-hall town house plan, notable interior.

One of the finest Georgian town houses in Philadelphia. Built
1765 for Charles Stedman, merchant. Interior altered 1770; Robert
Smith, builder. Ground-story interior altered mid-19th century; in-
terior elements removed 1921, 1925. Restored 1931–33; Louis Duhr-
ing, architect. Ballroom woodwork and ornamental plaster ceiling in
Philadelphia Museum of Art; rear parlor woodwork in Metropolitan
Museum of Art, New York; reproductions in house. Home of Samuel
Powel, city's last mayor under the Crown and first mayor in the New
Republic, 1769–93. Owned by Philadelphia Society for the Preserva-
tion of Landmarks since 1931. Certified, PHC 1956; Pennsylvania
Register 1970.

5 ext. photos (1967, 1969, 1972),* 14 int. photos (1958, 1962,
1967),* 2 photocopies of int. photos (1925, 1929);* 5 data pages
(including 1769–1859 insurance surveys).*

Reed Houses

(PA-1615), 518–20 S. Front St. Two brick row houses with stone trim, each approx. 24' (three-bay front) × 42' with original two-story rear ell and later one-story additions, three-and-a-half stories, gambrel roofs, originally frontispiece with columns, stone stringcourses, lintels with keystones, side-hall town house plans.

Fine merchants' town houses with later alterations. Built 1796–98. No. 518 altered to store, one-story rear side brick addition built 1913–14; Herman H. Kline, architect. No. 520 interior altered to warehouse later. Certified, PHC 1957; Pennsylvania Register 1970. 3 ext. photos (1958, 1961, 1966);* 8 data pages (1960, including 1798 newspaper advertisement and 1803, 1851 insurance surveys);* 2 HABSI forms (1957).*

Rhoads-Barclay House

(PA-1057), 217 Delancey St., at N.E. corner Philip St. Brick, 22'-6" (three-bay front) × 37'-0" with two-story rear ell, two-and-a-half stories, gambrel roof, pent eave, notable interior woodwork and staircase, side-hall London house plan.

Fine example of mid-18th-century Philadelphia dwelling originally with detached kitchen. Built between 1750 and 1758; Samuel Rhoads, builder. Rear ell demolished 1959. One-story rear kitchen wing built, exterior restored, interior altered 1960; W. Nelson Anderson, architect; Robert T. Trump, restorer. Built by Samuel Rhoads, master builder and Mayor of Philadelphia 1774, for Alexander Barclay, Comptroller of Port of Philadelphia 1758–71. Certified, PHC 1957; Pennsylvania Register 1970.

7 sheets (1961, including plot plan, plans, elevations, details); 4 ext. photos (1959, 1973*), 3 int. photos (1959); 6 data pages (1959, 1963).

Rich-Truman House

(PA-1074), 320 Delancey St. Brick, 18'-11" (two-bay front) × 18' with two-story rear kitchen ell, two-and-a-half stories, gable roof, pent eave, one-room ground story. Fine example of modest craftsman's house built as one of a pair of row houses. Rear ell built before 1771. Front section built 1771; John Rich (plasterer), builder. Restored 1962; Joseph A. Praissman, architect. Certified, PHC 1957; Pennsylvania Register 1970. 12 sheets (1962, including plot plan, plans, elevations, sections, details); 2 ext. photos (1963), 1 int. photo (1963), 1 photocopy of old ext. photo (n.d.); 8 data pages (1963).

Richardson, Nathaniel, House
(PA-1360), 524 S. Front St. Brick, approx. 20' (three-bay front) × 37' with two- and three-story rear ell, three-and-a-half stories, gable roof, originally side-hall town house plan. Built 1792–93; altered to store before 1860; two-story rear brick addition built 1919; interior altered 1920. Certified, PHC 1957; Pennsylvania Register 1970. 1 ext. photo (1967);* 5 data pages (1961, including 1893 insurance survey);* HABSI form (1958).*

Robinson, William, House
(PA-1361), 23 Clymer St. Wood frame with clapboard sides and rear, approx. 16' (two-bay front) × 18' with one-story rear wing, two-and-a-half stories, half-gambrel roof. Built c. 1795; demolished 1967. 1 ext. photo (1966);* 9 data pages (1966).*

St. Andrew's (Protestant Episcopal) Church (now St. George's Greek Orthodox Cathedral)
(PA-1362), 250–54 S. Eighth St. Brick covered with rough cast to simulate ashlar blocks, approx. 65' × 89', two stories on raised basement, gable roof, Ionic hexastyle portico, notable iron fence, three-aisle plan with balconies.

Fine Greek Revival church with historical antecedents. Built 1822–23; John Haviland, architect. Three-story rear Sunday School and vestry building built 1840. Haviland belonged to this congregation and was buried under the church; body reinterred at Philadelphia Divinity School 1922. Acquired by St. George's Greek Orthodox congregation 1922; became Cathedral 1971. Certified, PHC 1956; Pennsylvania Register 1970. 1 ext. photo (1958);* 13 data pages (1965, including 1824–54 insurance surveys).*

St. George's Greek Orthodox Cathedral. See St. Andrew's (Protestant Episcopal) Church

St. Mary's (Roman Catholic) Church (also known as Old St. Mary's)
(PA-1363), 244 S. Fourth St. Brick with marble trim, 71'-7" (three-bay front) × 110'-10", one story, gable roof, rectangular front facade with Gothic-, Tudor-, ogee-arch windows and entrance, round-arch side and rear windows; three-aisle plan, galleries, semicircular chancel.

Built 1763; interior renovated 1782. Enlarged 1810–11; Charles Johnson (master carpenter), architect. Crucifix carved by William Rush c. 1810; new organ front designed by Thomas U. Walter (architect) 1839. Altar moved to western end, rounded bay added 1886; church remodeled 1963. The chief Roman Catholic house of worship during the Revolutionary period, St. Mary's in 1779 was the site of the first public religious service in commemoration of the Declaration of Independence. Commodore John Barry, father of the American Navy, is buried in the churchyard. Cathedral Church of the See of Philadelphia 1808–38. Certified, PHC 1957. 1 ext. photo (1972),* 1 photocopy of old int. lithograph (c. 1880),* 1 photo (1958)* of crypt; 1 data page (1959).*

St. Peter's Church House. See Blackwell, Reverend Robert, House

St. Peter's (Protestant Episcopal) Church. Illustrated
(PA-1118), S.W. corner Third and Pine sts. Brick with marble trim, approx. 61' (three-bay front) × 95', two stories with six-story square tower and octagonal spire (originally only octagonal cupola), gable roof, round-arch windows with large Palladian window in eastern end, quoins; three-aisle plan with box pews, galleries, rectangular chancel.

Fine Georgian church design. Built 1758–61; Robert Smith, architect. Organ loft built over chancel 1789. Tower and spire built 1842; William Strickland, architect. Iron spiral staircases installed c. 1846; Thomas U. Walter, architect. Interior altered 1960. Many members of this congregation were prominent in Revolutionary politics and military affairs. Certified, PHC 1957.

17 sheets (1940, including plot plan, plans, elevations, sections, details) from Survey, Philadelphia Chapter, A.I.A. (1932); 6 ext. photos (1976),* 3 photocopies of old ext. photos (1898, c. 1910),* 4 photocopies of old int. photos (1898, c. 1910);* 3 data pages (1965, including 1843 insurance survey).*

Schively, Henry, House
(PA-1364), 329 S. Third St., at S.E. corner Delancey St. Brick, approx. 17' (two-bay front) × 30', three stories on raised basement, gable roof. Built c. 1826; altered to store 1877; rehabilitated 1969. Certified, PHC 1957. 3 ext. photos (1961);* 3 data pages (1958, including 1877 insurance survey).*

Shippen-Wistar House. Illustrated
(PA-1365), 238 S. Fourth St., at S.W. corner Locust St. Brick, approx. 33' (three-bay front) × 42' originally with two-story back buildings on

Locust St. attached by stair tower, three-and-a-half stories, gable roof, Tuscan pilaster frontispiece with fanlight, side entrance foyer.

Fine example of corner-house plan with its more efficient use of interior space. Built c. 1765; back buildings demolished 1912. Between c. 1765 and 1797 the home of William Shippen, a leading colonial physician and medical educator who headed the medical department of the Continental Army 1777–81. Caspar Wistar, a physician and chemistry professor who served as third president of the American Philosophical Society, lived here 1798–1809. Offices of Mutual Assurance Company for Insuring Houses from Loss by Fire since 1912. Certified, PHC 1956; Pennsylvania Register 1970. 1 ext. photo (1963),* 2 int. photos (1959);* 12 data pages (1959, including 1785–1913 insurance surveys).*

Sink-Burgin House

(PA-1366), 331 S. Fifth St. Brick with marble trim, approx. 20' (three-bay front) × 38' with two-story rear wing, three-and-a-half stories, gable roof. Built between 1810 and 1815; altered to store, new cornice and dormers added after 1890; shopfront altered 1941; demolished 1964. Built for Lawrence Sink, cabinetmaker. Between 1831 and 1870 home of George H. Burgin, physician-druggist known for his glass manufacturing and civic activities, which included roles in the founding of Philadelphia College of Pharmacy 1821 and Women's Medical College 1853. Certified, PHC 1957. 1 ext. photo (1960);* 3 data pages (1964).*

Smith, Daniel, Jr., House

(PA-1367), 505 S. Front St. Brick with marble trim, approx. 20' (two-bay front) × 40', three-and-a-half stories on raised basement, gable roof, arch-head entrance with fanlight, two rooms and entry with rear stair ell. Built 1828; demolished 1966. Certified, PHC 1957. 1 ext. photo (1961);* 3 data pages (1961, including 1828 insurance survey);* HABSI form (1957).*

Society Hill Synagogue. See Spruce Street Baptist Church

South Water Street Area Study

(PA-1619), 516–26 S. Water St., W. side Water St. in block between Lombard and South sts. Six contiguous brick houses, each approx. 14' (two-bay front) × 30', two-and-a-half and three-and-a-half stories, gable roofs, probably London house plans except No. 526

(side-hall plan with rear stair tower). Nos. 516–18 built 1819–20; Nos. 520–22 built between 1779 and 1783; No. 524 built 1784; No. 526 built 1832–34. Demolished 1967. Certified, PHC 1957. 3 photocopies of ext. photos (1957, c. 1960);* 13 data pages (1961);* 5 HABSI forms (1957).*

Southern Loan Company of Philadelphia (later known as Tradesmen's National Bank of Philadelphia)

(PA-1368), 300 S. Second St., at S.W. corner Spruce St. Brick with marble ground story (now covered with scored stucco) and stucco top story, approx. 16' (two-story front) × 60', four stories, flat roof. Built c. 1837; ground story altered 1922; rehabilitated 1967–68. First home of Tradesman National Bank. 2 ext. photos (1959, 1973);* 3 data pages (1960).*

Southwark Hose Company No. 9

(PA-1369), 512 S. Third St. Brick with stone and iron trim, approx. 20' (three-bay front) × 75', three stories, flat roof, round-arch doors and windows, vertical emphasis to facade from projecting piers and recessed spandrels. Built 1858; possibly by Stephen D. Button, architect; Charles T. Brown, superintendent of construction. Altered to store later; shopfront remodeled 1906, 1914, 1934; renovated 1972. 1 ext. photo (1963).*

Spruce Street Baptist Church (now Society Hill Synagogue)

(PA-1370), 426 Spruce St., at S.W. corner Lawrence St. Brick with smooth-rusticated granite basement and scored-plastered granite front, approx. 60' (three-bay front) × 82' with rear projection, one story on raised basement with square corner towers (originally with cupolas), gable roof, front windows with granite surrounds and lintels on corbels; three-aisle plan, rear balcony (originally balconies on three sides), rectangular chancel with elliptical arch.

An early work of a major early-19th-century architect. Built 1829–30; Thomas U. Walter, architect. Enlarged, new front built 1851; Thomas U. Walter, architect; R. Morris Smith, supervisory architect. Interior altered, two-story rear projection built 1871; rear projection replaced by larger one 1877; interior altered 1901; cupolas removed after 1911. Exterior restored (except for cupolas) 1968; Henry J. Magaziner, architect. Interior restored 1971– ; Francis, Cauffman, Wilkinson and Pepper, architects, in association with John D. Milner, architect. Synagogue since 1912. Certified, PHC 1957; Pennsylvania Register 1970. 1 ext. photo (1976),* 1 photocopy of old ext. print (c. 1860);* 6 data pages (1963, including 1852–77 insurance surveys).*

Stafford's Tavern. See Man Full of Trouble Tavern

Stewart, Thomas, House

(PA-189), 410 Locust St., Independence National Historical Park. Brick, 17'-8" (two-bay front) × 18'-10" with one- and three-story rear ell, three-and-a-half stories, gable roof, round-arch entrance with fanlight, city house plan. Built c. 1824; rehabilitated 1954. Hardware samples, cornice and molding elements in Architectural Study Collection, Museum Division, Independence National Historical Park. Certified, PHC 1957. 6 sheets (1957, including plot plan, plans, elevations, section, details); 2 ext. photos (1957); 7 data pages (1957).

Stiles, William, House

(PA-1371), 310 Cypress St. Brick with marble trim, approx. 20' (three-bay front) × 20' with two- and three-story rear ell, three-and-a-half stories, gable roof, side-hall plan. Built before 1792; probably by William Stiles, stonecutter. Certified, PHC 1957. 2 ext. photos (1962);* 5 data pages (1963, including 1810 insurance survey).*

Stocker, John Clement, House

(PA-1068), 402 S. Front St. Brick with marble trim, 27'-2" (three-bay front) × 45'-0" with two- and three-story rear wing, three-and-a-half stories on raised basement, gable roof, originally round-arch entrance with fanlight and frontispiece with columns and pediment, side-hall town house plan.

Fine example of a wealthy merchant's Federal town house. Built between 1791 and 1795; rear addition built c. 1803; altered to store before 1910; altered to warehouse c. 1936. Certified, PHC 1957. 6 sheets (1960, 1961, including plot plan, plan, elevation, section, details); 2 ext. photos (1961), 7 int. photos (1961), 2 photocopies of old ext. photos (c. 1910), 4 photocopies of measured drawings of second-floor fireplace walls and details (1914); 11 data pages (1962).

Stride-Madison House

(PA-1073), 429 Spruce St. Brick, 17'-7" (three-bay front) × 32'-3½" with three-story rear ell (originally two- and three-story ell), three-and-a-half stories, gable roof, round-arch entrance with fanlight and pedimented frontispiece, side-hall plan.

Built c. 1791; altered to warehouse 1916. Exterior restored, interior rehabilitated 1962–63; W. Nelson Anderson, architect. James Madison, fourth President of the United States, and his wife Dolley rented this house 1795–96 when he was a member of Congress.

Certified, PHC 1957. 6 sheets (1961, including plot plan, plans, sections, details); 4 ext. photos (1961, 1965), 5 int. photos (1962, 1965), 1 photocopy of old ext. photo (c. 1914); 8 data pages (1964).

Sully, Thomas, House

(PA-1372), 530 Spruce St. Brick with marble trim, approx. 22' (three-bay front) × 44' with one- and three-story rear ell, three-and-a-half stories, gable roof, rear bow, floor-to-ceiling second-story windows with iron screens, side-hall town house plan. Built 1797; Griffith Coombe, house carpenter. Remodeled, three-story rear ell built 1842; one-story rear kitchen built 1896. Rehabilitated 1965. Thomas Sully, famous and prolific 19th-century portraitist, lived here 1828–29. Certified, PHC 1957; designated National Historic Landmark 1965. 1 ext. photo (1967),* 5 int. photos (1967);* 5 data pages (1957, including 1841–96 insurance surveys).*

Summers-Worrell House

(PA-1373), 505 Delancey St., at N.E. corner Reese St. Brick, approx. 14' (two-bay front) × 19' with two-story rear ell, three stories, gable roof, one room to a floor. Built 1796; George Summers and Joseph Worrell, house carpenters. Demolished 1964. Certified, PHC 1957. 1 ext. photo (1960);* 5 data pages (1964, including 1810, 1893 insurance surveys).*

Third Presbyterian Church (also known as Old Pine Street Church)

(PA-1374), 422 Pine St. Stuccoed brick, approx. 63' (three-bay front) × 93', two stories with Corinthian octastyle portico (originally with pedimented frontispiece and Palladian window above), gable roof, pilasters on side walls, three-aisle plan with galleries.

Colonial church with fine 19th-century classical revival remodeling. Built 1766–68; interior remodeled, exterior stuccoed 1837. Exterior remodeled 1857; John Fraser, architect; A. Catanach, carpenter. Interior renovated 1867, 1952. Many members of this congregation served in the American Revolution. With the merger of Holland-Scots Church 1953 and Eastburn Mariners' Bethel Church 1958, the church's corporate name is now Third, Scots, and Mariner's Church. Certified, PHC 1957. 1 ext. photo (1974),* 4 photocopies of int. photos (1956);* 2 data pages (1965).*

Tradesman National Bank of Philadelphia. See Southern Loan Company of Philadelphia

U.S. Bonded Warehouse

(PA-1375), 415–19 S. Front St. and 416–18 S. Water St., at S.E. corner Front and Lombard sts. Brick with stone trim, approx. 150' (sixteen-bay front) × 130', three stories on Front St., five stories on Water St., flat roof, segmental-headed windows and doorways, chamfered stone piers on ground-story front, rectangular plan. Built 1853; John McArthur, Jr., architect. Demolished 1967. 4 ext. photos (1960, including one of iron sign hook).*

Waln, Isaac, House

(PA-1376), 259 S. Third St. Brick with marble trim, approx. 27' (three-bay front) × 44' with three-story rear ell, three stories, low gable roof, Ionic frontispiece with pediment, side-hall plan. Built 1791; altered to store later; demolished 1960. Certified, PHC 1957. 2 ext. photos (1960),* 4 int. photos (1960);* 5 data pages (1960, including 1791–1849 insurance surveys).*

Warehouse

(PA-1377), 105 Delancey St. Brick with stone trim, approx. 18' (three-bay front) × 40', three stories, half-gambrel roof, rectangular plan. Built early 19th century; demolished 1961. 1 ext. photo (1961).*

Warehouse

(PA-1378), 329 S. Water St. and 330 S. Delaware Ave., at N.E. corner Water and Panama sts. Brick with stone trim, 29'-11" (three-bay front) × 44'-7" with four-story rear addition, three stories, gable roof, rectangular plan with interior wood columns. Built c. 1810; rear addition built c. 1830; demolished 1967. 5 sheets (1966, including plot plan, plans, elevations, section);* 3 ext. photos (1966).*

Water Trough and Fountain. Illustrated

(PA-1379), E. side Ninth St. between Spruce and Pine sts., opposite Clinton St. Granite, carved monolithic fountain pedestal (approx. 5'-6" height) with attached trough on west (street) side, pedestal approx. 4'-9" × 1', trough approx. 4'-9" × 2', gadrooned fountain basin on east (sidewalk) side; "Edward Wetherill 1821–1908" inscribed above fountain, "A Merciful Man/Is Merciful To His Beast" inscribed above trough. Fine example of functional street furniture before the age of the auto. Built c. 1910. 2 photos (1975).*

Wetherill, Joseph, House

(PA-1380), 233 Delancey St. Brick with marble trim, approx. 17' (two-bay front) × 40', three-and-a-half stories on raised basement, gable roof, arch-head entrance, London house plan. Built 1811. Certified, PHC 1957. 1 ext. photo (1972),* 3 int. photos (1959, of fireplaces);* 3 data pages (1959, including 1824 insurance survey).*

Wharton, Isaac, House

(PA-1381), 510 S. Front St., at S.W. corner Naudain St. Brick, approx. 23' (three-bay front) × 40' with two- and three-story rear ell, three-and-a-half stories on raised basement, round-headed entrance, brick stringcourse, side-hall town house plan.

A merchant's town house built for rental income in a commercial-residential neighborhood. Built 1793; interior altered 1871. Certified, PHC 1957; Pennsylvania Register 1970. 2 ext. photos (1958, 1967);* 5 data pages (1961, including 1793–1879 insurance surveys);* HABSI form (1957).*

Wharton, Joseph, House. Illustrated

(PA-1382), 119 Lombard St. Brick, approx. 18' (three-bay front) × 32' with two-story rear ell, three-and-a-half stories, gable roof, segmental-arch heads over windows, brick stringcourse with glazed headers, alleyway through ground story to rear yard, London house plan.

One of a pair of row houses. Built c. 1743; possibly by John Palmer, bricklayer. Rear ell built c. 1767. Restored 1964–66; Duncan W. Buell, architect. Certified, PHC 1957. 2 ext. photos (1962, 1972),* 4 int. photos (1962);* 3 data pages (1961, including 1767 insurance survey).*

Williams-Hopkinson House

(PA-1084), 338 Spruce St. Brick, approx. 18' (three-bay front) × 36' with two- and three-story rear ell, three-and-a-half stories, gable roof, pedimented frontispiece, brick stringcourses, side-hall town house plan.

Modest Federal town house. Built between 1785 and 1791; rear addition built 1907; altered to store 1909; restored 1959–64. Old wallpaper samples in Architectural Study Collection, Museum Division, Independence National Historical Park. Joseph Hopkinson, writer of "Hail, Columbia!" lived here 1794–1800. Certified, PHC 1957.

5 sheets (1932, including elevations, details) from Survey,

Philadelphia Chapter, A.I.A., 3 sheets (1961, including plot plan, plans); 2 ext. photos (1961, 1962), 5 int. photos (1959, 1961, 1962), 3 photocopies of measured drawings of second-floor fireplace walls and details (1914); 10 data pages (1961, 1964).

Williams-Mathurin House

(PA-1383), 427 Spruce St. Brick with marble trim, 22'-9" (three-bay front) × 34'-5" with three-story rear ell, three-and-a-half stories, gable roof, round-arch entrance with fanlight and Adamesque frontispiece, marble stringcourses and lintels with keystones, mutule cornice, side-hall town house plan.

Fine Federal-style town house. Built 1791; William Williams, house carpenter. Altered to warehouse c. 1920. Exterior restored 1969–72; Carl Massara, architect. Antoine René Charles Mathurin, French Consul General, lived here 1791–93. Certified, PHC 1957. 3 ext. photos (1961, 1972),* 6 int. photos (1961);* 6 data pages (1959, including 1793, 1812 insurance surveys).*

Winder, William H., Houses. Illustrated

(PA-1384), 232–34 S. Third St. Two brick row houses with marble trim, each approx. 20' (four-bay front) × 49' with three- (originally two-) and four-story rear ell, four-and-a-half stories with marble ground story, gable roofs, Ionic frontispieces, notable ornamental ironwork balcony at second story, side-hall plans.

Fine Greek Revival row; originally a unit of three. Built 1843–44; possibly by Thomas U. Walter, architect. Rear ell raised to three stories 1855; exterior restored, interior rehabilitated 1962–64. No. 232 houses Executive Offices of Episcopal Community Services. Certified, PHC 1957. 4 ext. photos (1962, 1972);* 10 data pages (1958, including 1844, 1855 insurance surveys).*

Winemore, Phillip, House

(PA-1050), 222 Spruce St. Brick, 15'-5" (two-bay front) × 21'-10" with two-story rear ell, two-and-a-half stories, gable roof, one room to a floor. Built c. 1746; rear ell built later; rehabilitated 1958. Certified, PHC 1957. 1 sheet (1959, including plot plan, plans, section); 2 ext. photos (1957), 1 int. photo (1957); 5 data pages (1959).

Wood, George, Houses

(PA-1385), 335–37 S. Fifth St. Brick with stone trim, each approx. 20' (two-bay front) × 35' with three-story rear ell (originally two stories), three-and-a-half stories, gable roof, side-hall plan. Built after 1767; probably by George Wood, house carpenter. Rear ell raised to three stories c. 1853; altered to stores, cornices and dormers changed later; demolished 1964. Certified, PHC 1957. 1 ext. photo (1960);* 8 data pages (1959, including 1851–53 insurance surveys).*

Woods, Captain John, House
(PA-1111), 500 S. Front St., at S.W. corner Lombard St. Brick, approx. 21' (two-bay front) × 44', originally with two- and three-story rear wing, three-and-a-half stories on raised stone basement, gable roof, round-arch side entrance with fanlight, center-hall plan. Built 1783; originally a store in east room. Communicated with adjacent house on Front St. 1904; destructively altered and reduced to two stories 1955. 9 sheets (1940, including elevations, details) from Survey, Philadelphia Chapter, A.I.A. (1932); 3 data pages (including 1783, 1785, 1918 insurance surveys).*

Bethel African Methodist Episcopal Church. *Jack E. Boucher photo, 1973.*

Man Full of Trouble Tavern *(right)* and Benjamin Paschall House *(left)*. Jack E. Boucher photo, 1972.

Pennsylvania Hospital. *Cortlandt V. D. Hubbard photo, 1976.*

Samuel Powel House. *Jack E. Boucher photo, 1972.*

St. Peter's (Protestant Episcopal) Church. *James L. Dillon & Co. photo, 1976.*

Shippen-Wistar *(right)* and Norris-Cadwalader *(left)* Houses. *Cortlandt V. D. Hubbard photo, 1963.*

Water Trough and Fountain. *Jack E. Boucher photo, 1974.*

Joseph Wharton House. *Jack E. Boucher photo, 1972.*

William H. Winder Houses. *Jack E. Boucher photo, 1972.*

Penn's City

Old City

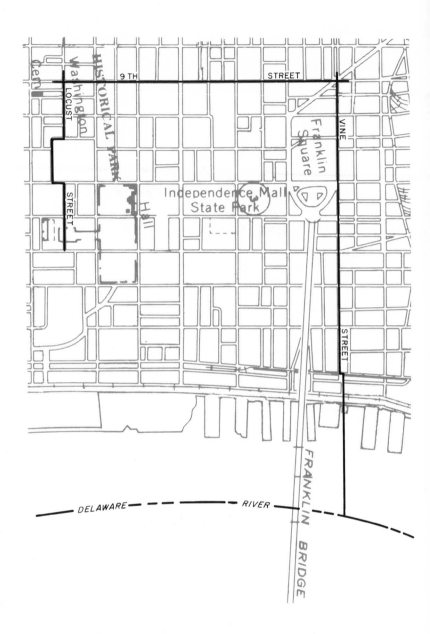

2

Introduction

O LD CITY IS ONE of Philadelphia's richest neighborhoods in terms of history, architectural diversity, and potential development. Its history incorporates the history of the city from its beginning; its architecture includes examples of nearly every type of building erected since the mid-eighteenth century; its future could be that of a combined New York Greenwich Village and Washington Georgetown. Old City's georgaphical limits are arbitrarily defined to include the area bounded by the Delaware River on the east, Vine Street on the north, Ninth Street on the west, and Locust Street on the south. It encompasses Independence National Historical Park, two historic districts listed by the National Register of Historic Places, and eight National Historic Landmarks, among them the U.S.S. *Olympia,* Admiral Dewey's flagship at the Battle of Manila in 1898.[1]

Part of the original (1682) plan of the city of Philadelphia as laid out by Thomas Holme, William Penn's surveyor-general, this neighborhood was the center of much of the early political history of Pennsylvania. The first site of city and provincial government was the Old Court House, or Town Hall, which in the medieval English fashion stood in the middle of High Street, at Second Street, with market stalls extending westward behind it. Built in 1701–10, the Old Town Hall was demolished in 1837 during renovation of the market sheds, and the sheds, in turn, were removed in 1859 in favor of a street railway.[2]

Always functionally separate, city and colonial governments were physically separated in 1748, when the new State House (Independence Hall) (illus.) opened on what was then the western edge of the city at Sixth and Chestnut streets. This latter building, an out-

standing example of Georgian public architecture, was to play an
even greater role in the founding of the American nation when it
served as the meeting place for the Second Continental Congress and
saw the adoption of the Declaration of Independence and in 1787 the
drafting of the United States Constitution. It has been a national
shrine ever since.[3] In 1791 the city government moved to the same
site, when City Hall was built as an eastern wing to the State House,
balancing the County Court Building, built two years earlier on the
western end. Both buildings were also used for federal purposes
when Philadelphia served as the nation's capital during the 1790's.
Local government offices began moving out of the two buildings and
into the present City Hall at Center Square after 1877.[4]

The city's early commercial district concentrated in this area and
has done so ever since the first wharf was built along the Delaware
River at Walnut Street in 1685. Contrary to Penn's intentions (he had
planned a promenade there), the waterfront by the mid-eighteenth
century was congested with buildings catering to maritime needs.[5]
All these eighteenth-century structures between the river and the
east side of Front Street and their later replacements fell to Interstate
95 (Delaware Expressway) in the late 1960's. HABS, however,
recorded some examples, such as the warehouse of Paul Beck, Jr., on
the waterfront at Delaware Avenue and two eighteenth-century
houses on North Water Street.

Beck's warehouse was generally thought to be a rare eighteenth-
century structure, and at one point its removal to another site was
under consideration.[6] Fire insurance surveys uncovered by Mrs.
Earle Kirkbride of the Philadelphia Historical Commission, however,
show the building to have been newly built in December, 1805.[7] The
gambrel roof, thought to have been its eighteenth-century hallmark,
was actually added about 1870, presumably for the same reasons it
was used in the eighteenth century—to save building materials and
provide more space in the garret.[8] The warehouse had a simple rectangular plan with open stairs to the upper floor and was divided by a
solid brick wall that extended the length of the building on both
stories. The dividing walls were evidently standard features at the
time. For example, three warehouses in Philadelphia's Southwark
section were similarly built: Thomas Penrose's three-story warehouse
on Water Street below South (c. 1797), the 1810 warehouse at 329
South Water Street, and Jesse Godley's extensive structure at 19–27
Queen Street, built as late as 1868.[9] Such interior walls offered two
advantages. First, the wall enabled the owner to rent the parts

separately, as is suggested in Beck's warehouse by the presence of four plastered counting rooms, presumably one on each floor of each section, and, second, it made the warehouse more fire-resistant. Evidently the latter was also its intent, since iron window shutters were put up along the sides and the flat dormers of the sailmaker's loft were cased with iron, an early use of this material.[10]

The colonial houses at 113–15 North Water Street, on the other hand, were initially dwellings. It was not unusual for eighteenth-century merchants to live along the waterfront, either next to or above their stores; in fact, in the 1790's, Stephen Girard, the wealthiest of these entrepreneurs, had his dwelling built at 23 North Water Street with a private counting room on the ground floor and a public counting room in the adjacent building. These houses were not temporary quarters to be endured until the merchant could afford to move to a more fashionable address. They were first-class dwellings, well built and handsomely finished, as seen in the case of the Dowers-Okill House at 115 North Water Street, which had carved mantles and cornices, stucco and gilt decoration, and a balcony overlooking the river. These appointments had been removed long before the house's demolition in the early 1960's, but its rare scalloped lintels suggested its earlier elegance.[11] As the only known surviving example of scalloped lintels in Philadelphia, they were removed to the Museum Division of Independence National Historical Park when the building gave way to the Delaware Expressway.[12]

With the rise of business activity along the waterfront came the need for a common meeting place where merchants could exchange information on cargoes, prices, and sales. The first of these was the London Coffee-House, which was built in 1702 as a residence at the southwest corner of Front and Market streets. In 1754 it began twenty years' service as a merchants' exchange and center of commercial and political activity, until it was overshadowed by the newer, more modish City Tavern, which opened in 1773, on Second Street above Walnut.[13] While merchants gathered at City Tavern, the old London Coffee-House spent a brief period as a printing and bookseller's shop before it was put to mercantile use in 1789 by the retail hardware merchant, James Stokes; it remained a store, albeit one with a declining status, until it was demolished in 1884.[14]

The eastern end of Old City Philadelphia continued to develop and to increase its commercial diversity during the first half of the nineteenth century. In addition to the maritime commerce and retail trade of the eighteenth century, light manufacturing, wholesale houses, and the financial institutions needed to underpin these entrepreneurial activities, banks and insurance companies, began to fill the area.[15]

This increasingly complex commercial and financial activity quickly exceeded the limited space and facilities the old coffee houses had afforded Philadelphia's merchants. In response to the new needs, the Philadelphia Exchange Company (illus.) was organized in 1831 and the next year it commissioned William Strickland to design a building on the raised triangular lot bounded by Third, Walnut, and Dock streets.[16] Its location a block west of City Tavern reflected the westward shift of commerce and the concern to be as accessible to the city's three major financial institutions—Girard's Bank (illus.) on Third Street and the Bank of North America and the Second Bank of the United States (illus.) on Chestnut Street—as to the waterfront warehouses. As the exchange was opening in 1834 the formerly splendid City Tavern had degenerated into a rooming house, its elegant interior extensively altered.[17] The Exchange Building, on the other hand, with its curved Corinthian colonnade and lantern copied after the Choragic Monument of Lysicrates, was to dominate the neighborhood architecturally as well as financially for the next two generations. It housed the Philadelphia Merchants' Exchange, the Maritime Exchange, and the Stock Exchange intermittently until the end of the century, but its fortunes declined after the Civil War. An attempt to revitalize the building's commercial prospects in 1900 by gutting and rebuilding the interior proved an eventual failure, and in 1922 the former exchange became the center of the wholesale food market district.[18]

Part of the difficulties of the Philadelphia Exchange after the Civil War was the consolidation of a group of grain and flour dealers who inaugurated the Commercial Exchange, a reorganization of the earlier Corn Exchange, and challenged the primacy of the older Philadelphia Exchange. For their new building the Commercial Exchange acquired and cleared a lot at the southeast corner of Second and Sansom streets, then the site of the Slate Roof House, the city's oldest building and once the residence of William Penn. Cosmic retribution struck in December, 1869, when the new exchange was severely damaged by fire, less than a year after it had opened. It was rebuilt in 1870 from the designs of James H. Windrim, who salvaged most of the original walls but gave the building a stylish mansard and a more impressive tower. Thereafter, the Commercial Exchange emerged as one of the city's largest and most important commercial organizations until its building was enlarged and altered to house a telephone exchange and offices in 1901–2, accounting for its later ungainly proportions.[19]

By this time mercantile hopes and interests had shifted to a new organization and building, the Philadelphia Bourse (illus.) on Fifth Street below Market. Literally as big as a city block, this steel-frame pile could boast of a handsome interior court, industrial exhibition halls, and a grain and stock exchange. Its great size and special accommodations bore ample testimony to the scale of the city's commerce and industry.[20]

Another reflection of the growing specialization of industrialism was the practice of concentrating one kind of business activity within a one- or two-block area. Front Street north of Market evolved into a wholesale warehouse district and almost the whole block was rebuilt during the 1830's, giving it remarkable visual homogeneity. The style, a utilitarian classicism, was popular for nearly a quarter-century. Its architectonic features, granite or marble piers with molded caps and a plain granite or marble architrave, were limited to the ground story; above it rose a simple brick superstructure, often with stone sills and lintels and a molded brick cornice. Although the North Front Street stores were probably designed by builders, some later examples with clustered stone columns on the second story were architect-designed.[21] Built as warehouses, these buildings have ample dimensions, sometimes as wide as twenty-six feet and as deep as eighty feet, and except for an area partitioned off on the first or second floors for the counting-room and fire-proof, there is but one large room to a floor. Straight stairs rise along one wall for four or five stories, and a hoist machine is on the top floor with hatchways cut through each floor at the front of the building or, if it parallels an alley, at the side.[22] Probably the finest of these warehouses today is that of Nathan Trotter and Company. Built in 1833, it has been altered only slightly and still has its cast-iron columns extending along the middle of the large rooms of the first two floors.[23] While this is an early example of structural cast iron, it is not the earliest. Iron piers were used on the ground-story front of two warehouses around the corner on Church Alley in 1831, and as early as 1829 William Strickland had incorporated iron columns into the balconies of the United States Naval Asylum.[24]

As America entered the modern world of industrial technology in the second quarter of the nineteenth century, architects groped toward an appropriate historical style that would gratify the aesthetic and social needs of merchants as well as the economic and functional demands of their businesses. In Philadelphia John Haviland employed the immutable Egyptian in 1838 to suggest the permanence of the Pennsylvania Fire Insurance Company (illus.); Samuel Sloan used the castellated Gothic to make Bennett and Company's Tower Hall one of the most distinctive stores of its day; John Fraser applied the exotic Moorish in 1854 to one of Caleb Cope's dry goods bazaars.[25] Yet they all failed to provide the adaptable formula that would gratify the ego as well as the eye. Instead, that formula was found in Europe's more recent past: "There had never before been any considerable amount of architecturally ambitious commercial building; and the Renaissance cities of Italy seemed to provide the most impressive architectural results of the patronage of a rich merchant class."[26] Designers turned to the Italian style for its flexibility in attaining the proper impression of conservatism and stability, yet with a spirit of progressive enterprise and liberality. In fine, the objective was to make every merchant a miniature Medici.

During the 1850's Chestnut Street between Second and Third streets became a showcase for the Italianate, as the many forms of the Italian became known. The south side of the street has been demolished by the federal government, the east end in the 1930's for the Customs House and the west end in the 1950's for Independence National Historical Park. The north side remains, however, as a fine example of a mid-nineteenth-century center of dry goods commission merchants. Because of the wholesale nature of the block's business many stores were rendered in a utilitarian Italianate. This mode was handled either as an updating of the utilitarian Greek Revival or a variant of the orthodox Italian Renaissance Revival. The former, best illustrated by the store at 233 Chestnut, employed a more elaborate ground story, either in iron or stone, and a modest console cornice instead of the molded brick.

More innovative is the other handling of the utilitarian Italianate, sometimes called Philadelphia Functionalism.[27] The Jayne Building, which stood across the street at 242–44 Chestnut until early 1958, was in the Venetian Gothic style rather than the Italianate, but it probably was the source of inspiration for the utilitarian Italianate examples that followed its erection in 1849–50.[28] Two extant examples of this latter style can be seen side by side at 235–37 Chestnut Street, the Elliott and Leland buildings respectively, and at 239–41 Chestnut Street, the Lewis Building.[29] Designed by Stephen D. Button in 1852, the Lewis Building is the more radical departure from the norm. At a time when M. Field, a published New York spokesman on good taste in urban architecture, was pointing out "the unnecessary insertion of perpendicular lines as the prevalent fault of street architecture,"[30] Button went to extremes to stress perpendicular lines. Field suggested that "to produce real as well as apparent solidity, the solid parts should preponderate" in a building, allowing a rich variety of architectural elements to ornament the street front,[31] as in the case of the brownstone store at 229 Chestnut Street at the west corner of Bank Street. Button, on the other hand, stripped the wall of excessive ornament and designed thin piers and wide voids to open the wall to light. To indicate a skeletal framework, the spandrels were recessed to emphasize the continuous upward flow of the plain piers.

Less extreme examples of Philadelphia Functionalism are the Elliott and Leland buildings, built in 1854 and 1856.[32] On these identical buildings the architect, Joseph C. Hoxie, employed many elements of the Italianate but, by recessing the spandrels and restricting the stone carving, he made the design simpler and more coherent than was common for commercial buildings at the time.

Like the Front Street warehouses a generation earlier, these stores were built with one room to a floor with one or two counting rooms partitioned off, usually at the rear. The stores of the 1850's, however, were larger and incorporated many new appointments made available by the Industrial Revolution. The depth of the Leland Building, for example, was twice that of Nathan Trotter's warehouse, which gave it more than double the storage space since it was a story higher. Equally significant was the difference in fire-proofs, or safes. Fire-proofs in the Front Street stores were not described by insurance surveys of the 1830's except to mention that they were in the counting rooms, hence no bigger than closets. By the 1850's, however, fire-proofs had become large and separate parts of the buildings. Although most fire-proofs of this period were only one story high, the one in the Leland Building was two stories high and measured 7 feet 6 inches by 11 feet 8 inches. Like most commercial buildings of the time, the Leland store had a water closet, washstand, and urinal on the ground story, a skylight whose illumination extended into the basement, and plate-glass front windows protected at night by iron sliders rather than iron shutters.[33]

Around the corner on Third Street was Brokers' Row. Within the two-block square area of Third Street between Market and Walnut streets were 94.7 per cent of the city's brokers and investment bankers during the 1850's. The greatest concentration was on Third

Street between Market and Chestnut where 33.6 per cent of the city's
113 brokerage houses were located.[34] The architecture of mid-
nineteenth-century brokers' offices was not significantly different
from that of wholesale dry goods stores. The offices of Hutchinson
and Company at 20 South Third Street were in a structure that had
once served as a dwelling and shop but that was remodeled in late
1847 into a four-story, flat-roofed commercial building whose
brownstone storefront featured battered Egyptian Revival doors, a
not so subtle suggestion of the firm's sobriety.[35]

Drexel and Company, on the other hand, under the leadership of
Francis M. Drexel, commissioned G. Runge in 1853 to design a new
building at 34 South Third. The architect produced "one of the most
imposing business edifices" in that part of the city.[36] Its chaste marble
front was executed in the Tuscan mode of the Italianate, in which a
balance of horizontal and vertical lines was attained through the use
of molded pilasters and cornices at each floor.

Most brokers, however, rented office space. For example, both
Edwin Booth and B. W. Tingley and Company found accommodations in the new building at 37–39 South Third, (illus.) which had
been built for Charles Leland in 1855. As the finest example of
Stephen D. Button's utilitarian Italianate,[37] it stands in vivid contrast
to the Drexel Building across the street. Evidently in appreciation of
the building's office functions, the elements of the Leland Building
are more subtly synthesized than in the Lewis Building around the
corner. The upward thrust of the Leland Building's granite piers
gracefully terminate at the fourth story in round arches with plain
moldings that are duplicated by the arcade on the ground story.

The interiors of these office buildings were similarly finished
with higher quality materials and craftsmanship, reflecting the different nature and status of their functions. The Tuscan mode of the
Drexel Building's front was continued in the large banking room,
which was impressively furnished with a fifty-foot-long counter with
a built-in ventilator near the encaustic tile floor. The rear of the building was illuminated by a large skylight that extended into the basement restaurant and the sub-basement's furnace room. The two
wainscoted offices on the second story were finished with stucco
cornices, ceiling centerpieces for gas lamps, and figured and painted
glass in the paneled partitions.[38] As a rental property, the Leland
Building across the street was done more plainly. Yet its two ground-
story rooms were finished with pilasters, brackets, and enriched
cornices, and deeply set square panels graced the foyer and stair-

way.[39] In both buildings water closets and iron wash basins were placed on nearly every floor, and sometimes there were two to a floor; and throughout both the windows were cased and finished with architraves rather than simply plastered to the sashes as in the Chestnut Street wholesale stores.[40]

Bank Row had eighteenth-century beginnings at 307 Chestnut Street with the Bank of North America, chartered in 1781 as America's first commercial bank, and followed in 1797 with the opening of Samuel Blodgett's architectural and Alexander Hamilton's financial masterpiece, the First Bank of the United States (illus.) on Third Street below Chestnut.[41] The institution that determined the nineteenth-century development of the row, however, was the Second Bank of the United States (illus.).[42] Chartered by Congress in 1816 and headquartered in Philadelphia, the Second Bank was to play a prominent role in both national politics and architecture, as a chartered monopoly with a unique and lucrative relationship to the federal government and as one of the country's first public buildings in the Greek Revival style. The former role aroused the presidential wrath of Andrew Jackson and led to the bank's financial demise, and the latter role established the popularity and viability of this variant of Romantic Classicism, catapulting the building's young architect, William Strickland, into the profession's front ranks.[43] In Philadelphia it was to cement banking functions on Chestnut Street, where they stayed even after the Second Bank became the United States Customs House in 1844. This was evidenced by the decision of the Bank of North America in 1847 to erect a new building at its original site there.[44]

Afterward, a series of granite and marble banks designed by a host of notable Philadelphia architects rose on the street.[45] John M. Gries was responsible for the two fine examples of the 1850's, the Philadelphia Bank and the Farmers and Mechanics Bank (illus.), adjacent to each other at 421 and 427 Chestnut Street respectively. The Farmers and Mechanics Bank, built in 1854–55, is the earlier of the two structures, and is significant not only for its "highly enriched white marble front" but also for its initial interior plan.[46] Instead of placing the banking room in the front, as was customary at the time, Gries followed a plan that was reminiscent of the nearby Second Bank of the United States. The front section had three doorways, two at the ends (since converted to windows) opening onto stairways and one in the middle with Corinthian pilasters that led down a passage to the rear, skylighted banking room. Richly decorated to echo key exterior elements, this room rose the full three stories to the arched ceiling.[47] Iron was used throughout the building, as part of the structure in the arched girders of the roof and straight girders of floor joists, as part of the furnishings in the iron counter with its built-in vents, and as decorative elements in the form of lions' heads on the front door that matched the marble examples on the keystones of the flanking windows.[48]

Although built of granite rather than marble and less elaborately carved, the Philadelphia Bank is the more imposing building.[49] This is partly because it is greater in size and scale, fifteen feet wider across the front and an additional story taller. Its Venetian Italianate style, "boldly modelled and open in feeling," gives the building a vertical articulation and less ornate appearance that complements its bulk.[50] Its plan was very similar to that of the Farmers and Mechanics Bank. The rear banking room rose the building's full height to an iron dome, whose top was sixty-five feet above the floor, giving an impressive interior space.[51]

Undaunted by the Panic of 1873, Philadelphia bankers added the greatest number of new banks to the row during the 1870's. Frank Furness, who was then emerging as the city's leading architect, contributed two banks, the Guarantee Trust and Safe Deposit Company at 318 Chestnut (1874–75) and the Provident Life and Trust Company (illus.) at 409 Chestnut (1876–79).[52] James H. Windrim designed two in 1874, the Philadelphia Trust, Safe Deposit and Insurance Company, 415 Chestnut, and the Peoples Bank, 435 Chestnut;[53] and Addison Hutton, the talented Quaker, was responsible for the finely modeled Pennsylvania Company for Insuring Lives and Granting Annuities, 431 Chestnut, in 1872–73.[54] All except Hutton's granite building are gone now, but for nearly eighty years these banks made this one of the most imposing stretches of streetscape in the city. Furness's banks, eyesores to some, were the visual anchors for the blocks.[55] Their robust detailing and monumental scale evoked the free-wheeling dynamism of a sanguine era. The last nineteenth-century addition to the row was the third Bank of North America, another Windrim work, this time designed in conjunction with his son, John T. Windrim.[56] Until it fell to make space for a parking lot in 1972, its Roman facade related well to John McArthur's chaste Italianate design for the First National Bank (1865–67) at 315 Chestnut.[57]

Insurance Row, the area on Walnut Street and adjacent streets between Third and Fifth, evolved around the venerable Philadelphia

Contributionship for the Insuring of Houses from Loss by Fire, organized by Benjamin Franklin in 1752, the nation's oldest mutual fire
insurance company. Its present offices at 212 South Fourth Street
were erected in 1835–36 from the Greek Revival designs of Thomas
U. Walter.[58] Its domestic scale fits its original function of both dwelling and office. The insistence of the company's directors on the dual
role was initially a point of contention between them and the architect, but Walter appears to have given his clients what they
wanted, that is, a house with ground-story offices, a cellar kitchen,
and upper chambers.[59] After the Civil War the company hired Collins
and Autenreith to replace the porch, whose low-quality marble was
deteriorating rapidly, and to add a mansard in order to gain an additional five rooms.[60] To take advantage of new technology, more
alterations were made in the 1890's. Furness, Evans and Company
drew up the designs for a new cellar kitchen and toilet room and
other alterations to accommodate telephones in the offices and toilet
rooms on the upper floors, with the top story remodeled into a well-
appointed apartment.[61]

Meanwhile, other insurance companies moved into the neighborhood. Buildings to house their operations were erected along
Walnut Street well into the twentieth century. Not much larger than
the Philadelphia Contributionship, usually only four stories high,
they were nevertheless unmistakably commercial buildings with
ciphers or plaques or distinctively grotesque facades to make them
stand out among their competitors. Yet the practice of maintaining an
apartment on the top floor was continued, at least in the Reliance Insurance Company, 429 Walnut Street, and apparently in the adjacent
Spring Garden Insurance Company.[62]

The availability of structural steel allowed early-twentieth-
century office buildings to rise higher and be distinctive for that
reason alone, but the inertia of nineteenth-century taste meant that
they would be finished in an historical style. The Irvin Building (formerly the Fire Association of Philadelphia) at 401 Walnut Street, for
example, was built in 1911–12 from the restrained classical designs of
Edgar V. Seeler.[63] The Doric distyle portico in antis and other classical
elements were proportionately scaled and low enough to be easily
viewed by pedestrians. When an addition was required seventeen
years later, however, another architect carried the new section to
seventeen stories, piled another ten stories on the old part, and introduced a series of juxtaposed elements to produce what might be
called a modernistic classicism.[64] A Greek fret frieze above the

seventh story and freestanding spread eagles at the corners of the tenth story were suggestive of an imperial Greek style, while recessed bays with rounded corners and the building's name carved in shallow relief made the building very up-to-date.

The Irvin Building illustrates not only shifting taste but also changes in building technology and scale of business. Whereas the Philadelphia Contributionship employed traditional building materials and techniques, brick-bearing walls with oak joists and heart pine floors, the Irvin Building had a steel frame with brick and limestone curtain walls and concrete floors, construction materials that did not exist in 1835.[65] Building costs embody similar contrasts. Philadelphia Contributionship cost $20,946, while the Irvin Building was estimated to cost $300,000, or more than fourteen times the former.[66] These figures testify to the tremendous expansion of the national economy in general and the insurance business in particular during the last three-quarters of the nineteenth century, as well as to a company's willingness to spend its earnings on notable architecture.

But like Bank Row, Insurance Row has known architectural change in recent years. When the National Park Service tore down the Irvin Building in 1974, it completed the decimation of the north side of the 400 block of Walnut Street, while the former American Fire Insurance Company Building, with one of Furness, Evans and Company's more restrained business facades, is one of the last two insurance company buildings to survive in the 300 block. One member of the row, however, has put its best face forward and remains a part of the streetscape. As a result of the imagination and determination of the Penn Mutual Life Insurance Company and its architects, Mitchell and Giurgola Associates, the Egyptian Revival facade of the Pennsylvania Fire Insurance Company's old building (illus.) at 508–10 Walnut Street has been preserved as a courtyard screen in front of a new high-rise office building.[67]

Despite the onslaught of automobiles and bulldozers, the commercial row pattern persists. Clothing jobbers, for example, have been on Third Street above Market since the middle of the last century, Jewelers' Row still occupies the 700 block of Sansom Street and much of the north side of Walnut Street in that block as well, and the Washington Square neighborhood has been known as Publishers' Square for nearly eighty years. Among the major publishing companies still on the Square are J. B. Lippincott and Company, W. B. Saunders Company, the Farm Journal, and Lea and Febiger, the nation's oldest continuous publishing house.[68]

By the time of the Civil War, commercial Philadelphia was beginning to encroach upon what had been residential Old City. Starting in the 1850's, the brick dwellings along Arch and Cherry streets were replaced by factories and commercial buildings as far west as Fifth and Sixth streets. Most of these buildings stood for nearly a century. Joseph C. Hoxie's stark and imposing Horstmann Factory (1852–53) at the northeast corner of Fifth and Cherry streets, for example, was not replaced until the 1930's.[69] Others, such as the Schenck Building at the southeast corner of Sixth and Arch, survived until they were demolished in order to create a vista in front of Independence Hall.[70]

Farther south, the shopping district moved westward along Market and Chestnut streets. Three retail stores devoted to the carriage trade, for example, opened on Chestnut Street west of Eighth in 1857 alone, including the elegant L. J. Levy & Co. and Sharpless Brothers, whose interior arrangement anticipated the modern department store.[71] About the same time the first of the former dwellings in the 700 block of Market Street was replaced by a five-story commercial building for James M. Maris and Company at 711 Market, followed two years later by J. B. Lippincott's adjacent marble-fronted bookstore with its rear printing shop and bindery.[72] Although a retail district by the time of the Civil War, this block's appearance changed more slowly than did some of the wholesale blocks during the 1850's. Change became more rapid after 1893 when the Lit brothers began to expand their dry goods business, which within a remarkably short time grew from a single store at the northeast corner of Eighth and Market to a department store occupying the entire block by 1907, consolidating the hitherto disparate elements composing that block into one of the city's great architectural monuments and department stores.[73] Evidently pleased with Collins and Autenreith's design for the new corner store in 1893, they retained the architects to continue this design on 731–33 Market in 1895, 723–29 Market in 1896, and the other end of the block, 701–7 Market, in 1907.[74] Lit Brothers' reputation as a great example of cast-iron architecture, however, is exaggerated. (It received a plaque from the Philadelphia Chapter of the Victorian Society of America in 1973, commemorating the store as a great surviving example of the nineteenth-century building material.) The only two cast-iron buildings in the group are the former Joel Bailey store at 719 Market and its neighbor to the west.

Old City also retains some of the nation's most famous eighteenth-century residences. The twenty-nine brick houses of Elfreth's Alley (illus.), a six-foot-wide alley between Front and Second streets above Arch, illustrate the typical side-street dwellings

of eighteenth-century artisans on the oldest continuously inhabited
street in Philadelphia. It is one of the best known and, through the
joint efforts of the Elfreth's Alley Association and the Philadelphia
Historical Commission, one of the best researched streets in the city.
Since late 1957 Hannah Benner Roach, the historian and researcher of
the Association, has been poring over a wide variety of documentary
materials to develop an accurate history of this byway. Opened
between 1702 and 1704, it was initially called Gilbert's Alley, after
John Gilbert, who owned the land on the north side of the alley. By
the time that Jeremiah Elfreth acquired title to properties on both
sides of the street and it accordingly became known as Elfreth's Alley
in the mid-eighteenth-century, most of the houses on the south side
and six on the north side had been built. [75] During this time the
residents' livelihoods were tied to the water, as they were mariners,
river pilots, shipwrights, and the like. After the Revolution, when
houses at each end and in the middle of the north side were built,
most of the occupants were employed in domestic trades, such as
carpentry, furniture-making, and pewtering. [76] In spite of Elfreth's
Alley's reputation for its eighteenth-century charm, a few houses
were erected in the nineteenth century, and one, Number 125, not
until the late 1830's, as suggested by its fanlight transom, symmetry
of fenestration, and flat gable roof. [77]

Less famous than Elfreth's Alley but equally significant is the
German Reformed Church district on North Fourth Street below
Race. Many of these properties, including a school at 327 Cherry
Street, were owned by German Reformed or Lutheran parishes that
played a prominent role in this predominantly German neighborhood for more than a century after the 1730's. The finest of these
dwellings is the three-and-a-half-story house at the corner of Fourth
and Cherry, next to the German School and opposite the site of the
Zion Lutheran Church, which was demolished in 1869. Built in 1787
for the German Lutheran congregation, its ample proportions,
modillion returning cornice, and circular window in the gable cause it
to stand out from its older neighbors, [78] which were built during the
colonial period, as suggested by the pent eaves and lower ceilings. [79]
The German Lutheran Church acquired four of these properties, 129–
35 North Fourth, during the 1760's and 1770's and for more than
seventy years used one as a parsonage and rented the others to
people of modest means: tailors, barbers, a laborer, and a huckster. [80]
The Fourth Street houses and the German Reformed Church at
Fourth and Race are currently being restored, and it is expected that

the area will become the kind of mixed neighborhood of dwellings, workshops, and stores that it was two hundred years ago.

Another kind of neighborhood development, one that would become increasingly familiar to Philadelphians during the nineteenth century, was begun at the very end of the eighteenth century west of Seventh Street along Walnut and Sansom streets. This was Sansom's Row, the city's first speculative row development in which the entire block was planned and built as a row rather than as individual houses built at different times adjacent to one another. In about 1799 William Sansom had Benjamin Latrobe design a row of brick houses with traditional London house plans for Walnut Street and a year or two later Thomas Carstairs did similar drawings for Sansom Street.[81] Although eventually an apparent financial success, the row was an aesthetic disappointment. Instead of treating the row as a unit, the designers planned the dwellings with individual but identical Georgian facades. The only concession to comprehensive planning was the pairing of doorways and placing of battlements at extreme ends of every pair to create the impression of ten pairs of dwellings. Whereas Elfreth's Alley has the charm of houses with varying widths and heights, and different street lines, doorways, and brickwork, Sansom's Row was a flat, repetitive expanse, pointing out one of the major design problems that has plagued speculative developments ever since.

Robert Mills, America's first professionally trained native-born architect,[82] tried to solve this problem in Franklin Row (illus.), a stretch of ten row houses on Ninth Street above Locust being promoted by John Savage and John Meany. In the spring of 1809 Mills prepared two sets of plans, one with doorways alternating and the other with doorways paired under a large arch, but both with the traditional London house plan and cellar kitchen. The houses were built according to the first plan, Mill's second choice, but he "gave it added grandeur by placing a tripartite window and its recessed arch on the second floor as well as on the first floor."[83] For the first time a conscious effort had been made to lend some distinction to a practical row development. Later examples in other parts of the city, Thomas U. Walter's Portico Square and John Haviland's Colonnade Row, would employ freestanding porticoes to give rows even greater plasticity.

Old City also had a number of comfortable town houses, most of which have been removed as the real estate has been put to other uses. Two of the finest late-eighteenth-century town houses to be found anywhere, however, survive in this area and have been recorded by HABS. They are the Reynolds-Morris House (illus.), 225 South Eighth Street, and Bishop William White's House, 309 Walnut Street, both built in 1787. Bishop White's three-and-a-half-story house follows the pattern of most Philadelphia town houses of its time, with a side entrance hall, a sitting room and dining room to the side, and a rear ell consisting of a piazza, or stairwell, two-story kitchen with scullery and laundry below, and a two-story nursery. It was a large house with a comfortable twenty-six-foot front and unusually large windowpanes, and it was an elegantly designed house with a pedimented Doric frontispiece, marble lintels with keys, and decorative scrolls flanking its dormers. In fine, it was a proper residence appropriately placed for one of the nation's leading Episcopal clergymen, who served both Christ Church to the north and St. Peter's to the south.[84]

The facade of the Reynolds-Morris House incorporates the same elements found on the bishop's house, but the former's great size is unusual, made possible by the double lot on which it was built. The house's interior lived up to its exterior dimensions, having a central hall with stairs, rare for a town house.[85] Each floor was finished with chimney breasts, wainscoting, and double architraves on the doors and single architraves on the windows, luxurious appointments for its day.[86] The rear has been changed over the years and today only the facade remains of the original, expressing the elegance of an early merchant-manufacturing class.[87]

In response to Penn's call for industrious family men and his assurance of religious liberty, a wide range of religious sects arrived during the early years of the province, a fact that is materially represented by a series of religious structures in Old City that were built through the first half of the nineteenth century. Later, the area's commercial-industrial development militated against the growth of residential neighborhoods that support churches. Reflecting both the city's Quaker origins and later crises of state and soul are the Arch Street Friends Meeting (illus.) at the southeast corner of Fourth and Arch and the Free Quakers Meeting House at the southwest corner of Fifth and Arch. The former, although begun as late as 1803 (late for the faith that founded the colony in 1682), is the city's oldest extant Quaker meeting house and a notable example of the Quaker plain style. The Free Quakers Meeting House vividly illustrates the impact of the American Revolution on the pacifist Society of Friends and of the subsequent spirit of nationalism on the Quaker plain style. It is

the older of the two, erected in 1783–85, but as its name implies, it
was never an orthodox Quaker meeting house, since it was built for
Friends who had been disowned by their meetings for participating in
the Revolutionary War. Its interior plan followed the general plan of
Friends meeting houses of the time except for the introduction of
wine storage vaults beneath the street, an income-producing feature
for the enterprising congregation. The building's significance,
however, rests in its exterior design, which in addition to its up-to-
date architectural details includes a stone tablet in the north gable that
patriotically proclaims:[88]

> By General Subscription
> for the
> FREE QUAKERS
> Erected in the YEAR
> of OUR LORD 1783
> of the EMPIRE 8.

Many of the pietist sects did not stay in the city but moved into
the fertile countryside where they could remain farmers. The more
worldly Anglicans, however, settled in and about the city, where as
early as 1695 they built their first church on Second Street north of
Market. Beginning in 1727 this structure was replaced by the present
Christ Church (illus.), a *tour de force* of Palladian church architecture
strongly influenced by the London city churches of Sir Christopher
Wren.[89] Within a decade after the Anglicans' beginnings in the
Quaker city, Philadelphia's first Presbyterian congregation organized
and in 1704 moved into a frame building at Market and Bank streets
that served as its church throughout the colonial period. It was re-
placed in 1793 by a handsome structure with a Corinthian portico, but
when the columns rotted, the congregation in 1822 left the noise and
stench of the market sheds extending along the middle of Market
Street for the open air and fashionable address of Washington
Square, where John Haviland had designed a Greek Revival church.
Out of the evangelical turbulence of the Great Awakening emerged
Philadelphia's second Presbyterian congregation, which in 1750 built
its church at Third and Arch streets. Its spire, which rivaled that of
nearby Christ Church for half a century until it had to be dismantled
in 1803, suggested that the congregation was as devoted to public display as to the new Calvinist preaching. This trait was repeated as the
congregation moved westward with the city's expansion, first in 1837

to a comfortable classical church on Seventh below Arch and again in 1872 to its present site at Twenty-first and Walnut streets.[90]

Other houses of worship were erected in the area by a host of faiths, many of them, like the German Lutherans, reflecting the city's ethnic pluralism. Lutherans outgrew St. Michael's within a generation after they built it at Fifth and Appletree streets in the 1740's, and in 1766 they began the Zion Church, one of the colonies' largest churches, at Fourth and Cherry.[91] Meanwhile the German Reformed Church, which had been renting space in the neighborhood since 1727, began its own structure, a hexagonal church with a tall cupola, in 1747 at the southeast corner of Fourth and Race streets, where its third structure still stands.[92] Philadelphia's Methodists also located nearby, at Fourth and New streets. Architecturally plain but historically significant as the world's oldest Methodist church in continuous use, St. George's dates from 1769, when it was acquired from a Presbyterian congregation that lacked the funds to finish construction.[93] Across Fourth Street from St. George's stands St. Augustine's Roman Catholic Church, which was rebuilt in 1848 as a living monument to the triumph over religious persecution. The discrimination suffered by Philadelphia's Jews (a traditionally persecuted people) in the erection of the first Mikveh Israel Synagogue (1782) on Cherry near Third was mild compared to that endured by the city's Catholics during the eighteenth and early nineteenth centuries. St. Joseph's, located on Willing's Alley where Old City merges with Society Hill, was threatened by anti-Catholic mobs twice during the colonial era and again during the most violent outburst of intolerance, the "Native American" riots of 1844, which led to the burning of St. Michael's in Kensington and the late-eighteenth-century St. Augustine's.[94]

Despite its religious origins and multiplicity of faiths, Philadelphia easily assumed the secular ways of the Enlightenment and became a center for the exchange of ideas as well as goods. Benjamin Franklin, the Boston-trained printer who retired to become one of the foremost theoretical physicists of his day, played a key role in this stage of the city's history. Aware that the scientific society had offered a successful means of promoting the arts and sciences ever since the founding of England's Royal Society in 1662, he helped to establish the American Philosophical Society in 1743, making it now the oldest learned society in the country. Because of local apathy the society languished until it was reorganized in the 1760's. After the Revolutionary War, the society built its hall (illus.) on South Fifth

Street, in what is now Independence National Historical Park,[95] while directly across the street another pioneer intellectual institution begun by Franklin and his friends was going up. Founded in 1731, the Library Company of Philadelphia built a handsome new library in 1789–90 from the designs of Dr. William Thornton, later the architect of the United States Capitol. The present structure is not affiliated with the Library Company and instead houses the library of the Philosophical Society. It is a reconstruction, since the original Library Company building was razed in 1887 and replaced by the Drexel Building.[96]

Similar institutions continued to be established in the neighborhood during the early nineteenth century. Two of the most notable examples, quite different in goals and membership, were the Franklin Institute and the Athenaeum of Philadelphia (illus.). The former was founded in 1823 as part of the early-nineteenth-century Lyceum Movement, an effort to provide a practical scientific education for workingmen. The institute's first building still stands at 15 South Seventh Street as a major example of the Greek Revival. By basing the small building on the Choragic Monument of Thrasyllus, architect John Haviland produced a compact monumental facade whose romantic allusions to democratic learning remain appropriate for its present use as the Atwater Kent Museum.[97] The Athenaeum was organized in 1814 by a group of literate young gentlemen in order to provide a "place of common resort" with a reading room and library. In 1847 it moved from its Philosophical Hall quarters to Washington Square, where its new brownstone Renaissance Revival building helped to popularize locally both the material and the style. Since then the basic design and function of the Athenaeum has not changed, making it a notable example of the Italian style and the city's oldest library building in continuous use.[98]

The theater was one expression of secularism that was opposed throughout most of the eighteenth century, but after Philadelphia was designated the temporary capital of the new nation in 1789, the last ordinance that had prohibited—but had not prevented—theatrical performances was repealed. As the governmental and economic center of the nation at the time, Old City also became the nucleus of the city's theater district when the Chestnut Street Theatre was begun west of Independence Hall in 1791. It was followed four years later by Ricketts' Circus, which, until it burned down in 1799, presented drama, pantomime, and equestrian acts, diagonally across Chestnut Street.[99]

   The 1820's were an even busier period of theater construction. In
that decade William Strickland designed three theaters: a second
Chestnut Street Theatre to replace the burned-out original, the Arch
Street Theatre two blocks to the north, and the Musical Fund Society's Hall at 808 Locust Street. At about the same time John
Haviland remodeled into the Walnut Street Theatre the blank brick
Olympic Theatre, which had been built at Ninth and Walnut streets
in 1809. The latter two buildings still stand, both with long and varied
theatrical and architectural histories. The Walnut Street Theatre has
recently had its Haviland facade restored and carries on its active life
as the oldest theater in continuous use in the English-speaking world.
The Musical Fund Hall, however, has not been as fortunate. Although it was saved from humiliating service as a tobacco warehouse
in 1964, it has remained empty and unused, its famous acoustics
wasted and its fabric deteriorating. [100]
   As the city's commercial district moved westward during the late
nineteenth and early twentieth centuries, the eastern end of Old City
declined. Although this was a blow to the neighborhood's economy,
it was a blessing for its historic preservation. Surviving examples of
urban Philadelphia include relics of its early commerce, such as the
1806 Singer Warehouse at the rear of 319 Market Street, the city's
oldest extant warehouse; significant but obscure examples of architectural history, such as the Brock Warehouse (illus.) at 242–44 North
Delaware Avenue, the country's oldest known surviving cast-iron
building; and modest dwellings of common folk, such as Henry Harrison's little known mid-eighteenth-century row on Cuthbert near
Front. [101]
   The creation of Independence National Historical Park in 1948
and urban renewal and highway projects beginning in the 1960's
have destroyed some of this Old City, but they have also revitalized
other segments, particularly in the area of Franklin Court and along
Independence Mall, a three-block expanse north of Independence
Hall now lined with government buildings and commercial structures. [102] The restoration of the nineteenth-century commercial-
residential Old City Historic District is going forward; a major
contribution to the section's renaissance will be Penn's Landing, a
thirty-acre, $30 million development of offices, shops, hotels,
museum, marina, and promenades along the Delaware River
waterfront. [103]

Catalog

American Fire Insurance Company Building
(PA-1386), 308–10 Walnut St. Brick with stone and terra cotta trim, approx. 46' (five-bay front) × 60' with three-story rear ell, five stories (originally four), mansard roof (originally gable), rectangular plan. Built 1840. Altered 1881; Furness and Evans, architects. Remodeled, enlarged 1886; Furness, Evans & Co., architects; Jacob Myers, builder. Certified, PHC 1973. 1 ext. photo (1958);* 6 data pages (including 1840 insurance survey, 1886 news article).*

American Life Insurance Company Building (also known as the Manhattan Building)
(PA-1064), 330–36 Walnut St. Gray stone ashlar with interior iron columns and beams, 46'-9" (four-bay front) × 101'-3", eight stories with three-story tower, mansard roof with semicircular projecting hoods, offices disposed around central stair, elevator, and fire-escape shaft.

Early Philadelphia skyscraper. Built 1888; Thomas P. Lonsdale, architect. Exterior partially remodeled c. 1891; probably by Will Decker, architect. Demolished 1961. 7 ext. photos (1960, 1961), 1 int. photo (1960), 1 photocopy of old ext. print (n.d.); 10 data pages (1960).

American Philosophical Society. See Philosophical Hall

Arch Street Area Study
(PA-1387), 501–27 Arch St., on N. side Arch St. between Fifth and Sixth sts. Twelve contiguous commercial buildings ranging from three-and-a-half-story late-18th-century house with gable roof to mid- and late-19th-century brick and cast-iron buildings of two, four, and six stories. Includes Insurance Patrol (PA-1433) and Robert Ralston House (PA-1016). Representative of the architectural variety of a Philadelphia commercial block. Built at various times between late 18th and late 19th centuries; demolished 1959. 4 ext. photos (1959).*

Arch Street Friends Meeting. Illustrated
(PA-1388), 330 Arch St. Brick, approx. 180' (thirteen-bay front) × 85', two stories, hipped roof with cross-gable, projecting central pavilion, three porches with Tuscan columns across north side and two each on east and west sides, 1804 date stone in gable end.

Notable example of Quaker plain style. East wing and center building built 1803–5; Owen Biddle, architect. West wing built 1810–11. One-story addition behind west wing built 1896; demolished 1968. Two-story addition behind center building built 1902, 1906; demolished 1968. Interior east wing renovated and two-story addition behind center building built 1968–69; Cope and Lippincott, architects. Structural, exterior, interior elements in Architectural Study Collection, Museum Division, Independence National Historical Park. Oldest Friends Meeting in the city and headquarters of the Philadelphia Yearly Meeting of the Society of Friends. On ground owned by Friends since the city's beginning. Certified, PHC 1956; Pennsylvania Register 1970; NR. 4 ext. photos (1958, 1974).*

Armory of the National Guards. See National Guard's Hall

Athenaeum of Philadelphia. Illustrated
(PA-1389), 219 S. Sixth St., at S.E. corner St. James St. on E. side of Washington Sq. Stuccoed brick and coursed brownstone ashlar, approx. 50' (three-bay front) × 128', three stories, flat roof, projecting lintels, balustrade between first and second stories, quoins, center-hall plan.

An early and influential example of the Italian Renaissance Revival style. Built 1845–47; John Notman, architect. Interior altered (A.I.A. Meeting Rooms) 1870; Frank Furness of Frazer, Furness and Hewitt, architect. Exterior restored 1975–76; John Dickey, architect. Interior restored, mechanical systems renovated 1975–76; George D. Batcheler, Jr., architect. Founded 1814 as a private literary association. Certified, PHC 1956; Pennsylvania Register 1970; NR. 3 ext. photos (1972),* 2 photocopies of architectural drawings (1845, one by Notman, one competition drawing by John Haviland);* 9 data pages (including architect's agreement and specifications).*

Athenaeum of Philadelphia, Privy
(PA-1092), rear of 219 S. Sixth St. Brick, approx. 8' × 10', one story, low gable roof, six private compartments arranged in two back-to-back rows of three. Built c. 1847; John Notman, architect. Demolished by falling tree 1962. 1 sheet (1962, including plot plan, plan, detail);* 2 ext. photos (1962);* 2 data pages (1962, including architect's specifications).*

Atwater Kent Museum. See Franklin Institute

Ayer, N. W., & Son, Inc., Building (later N. W. Ayer & Company)
(PA-1390), 204–12 S. Seventh St., at S.W. corner St. James St., on W. side Washington Sq. Steel skeleton, brick curtain walls faced with dressed limestone ashlar, approx. 105' (ten-bay front) × 115', fifteen stories with setbacks, flat roof, two carved three-story limestone figures on each facade above first setback, notable brass sculpture on entrance and in lobby.
Fine example of Art Deco mode integrating sculpture with design of the tall office building. Built 1927–29; Ralph B. Bencker, architect; J. Wallace Kelly and Raphael Sabatini, sculptors; Ketcham and McQuade, contractors. Founded 1869, N. W. Ayer & Company is the oldest advertising company in the United States. 2 ext. photos (1976).*

Bank of North America
(PA-1391), 305–7 Chestnut St. Brick with steel joists, Jonesboro granite front with polished Scotch granite trim, 56'-6" (three-bay front) × 118'-4" with four-story rear wing, one story (two-story facade), flat roof with front pediment, smooth rusticated ground story, central balcony with balustrade supported by coupled Tuscan columns.

Built 1893–95; James H. and John T. Windrim, architects; George Watson & Son, contractors. Demolished 1972. Bank of North America was chartered 1781 as the nation's first commercial bank, and occupied this site until it merged with Pennsylvania Company for Banking and Trusts 1929. Certified, PHC 1970. 4 ext. photos during demolition (1972),* 2 photocopies of old ext. photos (c. 1900, 1959);* 9 data pages (including 1894–1945 insurance surveys);* HABSI form (1971).*

Bank of Pennsylvania (also known as Philadelphia Bank)
(PA-1392), 421 Chestnut St. Brick with Quincy granite ashlar front, approx. 70' (five-bay front) × 45' with four- and six-story rear additions, four stories originally with cast-iron dome, flat roof, monolithic pilasters and round arches on ground story, tall arches with carved keystones rise through second and third stories, rusticated quoins, originally center corridor to rear banking room in one story.

Fine example of mid-19th-century Renaissance Revival style in Venetian mode. Built 1857–59; John M. Gries, architect; H. C. Oram & Co., iron work. Four-story addition built between front and rear buildings 1892, rear banking room demolished and replaced by five-story office building 1892–93, raised one story 1903; Theophilus Parsons Chandler, Jr., architect. Bank of Pennsylvania failed during Panic of 1857 and building was completed by Philadelphia Bank. Certified, PHC 1969; Pennsylvania Register 1970. 4 ext. photos (1972),* 1 photogrammetric stereopair (1959),* 1 photocopy of old ext. photo (c. 1910).*

Baugh Warehouse. See Beck Warehouse

Beck Warehouse (also known as Baugh Warehouse)
(PA-1188), 18–20 S. Delaware Ave. Brick, 40' (four-bay front) × 75'-8", three-and-a-half stories (originally two-and-a-half stories), gambrel roof (originally gable roof), rectangular plan with longitudinal brick partition.

Fine example of early-19th-century waterfront warehouse. Built 1805; raised one story, gambrel roof built before 1870; No. 18 altered to storefront 1915. Demolished 1967. Paul Beck, Jr., a prominent, civic-minded merchant, owned building until his death 1844. Headquarters of Baugh & Sons Co., fertilizer manufacturers, 1860–1954. 5 sheets (1965, 1968, including plot plan, plans, elevation, hoist machinery);* 2 ext. photos (1958, 1965), 3 int. photos (1965); 3 data pages (including 1805–1915 insurance surveys).*

Berry-Coxe House

(PA-1062), 413 Locust St., Independence National Historical Park. Brick with marble trim, approx. 24′ (three-bay front) × 44′ with two- and three-story rear ell, three-and-a-half stories, gable roof, Tuscan frontispiece with fanlight, side-hall town house plan.

Fine early-19th-century town house. Built c. 1802–4; probably by Peter L. Berry, house carpenter. Interior extensively remodeled 1959– 60. Tenanted 1804–7 by Tench Coxe, political economist and father of American cotton industry. Certified, PHC 1957; Pennsylvania Register 1970; NR. 10 sheets (1954, 1962, including plot plan, plans, elevations, sections, details); 3 ext. photos (1958, 1973*), 12 int. photos (1958); 8 data pages (1958).

Blair, Reverend Samuel, House

(PA-1063), 415 Locust St., Independence National Historical Park. Brick with marble trim, approx. 23′ (three-bay front) × 45′ originally with one-story rear ell, three-and-a-half stories, gable roof, Adamesque frontispiece with fanlight, side-hall town house plan. Built c. 1807; possibly by Peter L. Berry, house carpenter. Rear ell raised to three stories, two-story addition built before 1860; ground story altered to store before 1890. Rear ell demolished, exterior re- stored, interior extensively rebuilt 1959–60. Hardware samples in Ar- chitectural Study Collection, Museum Division, Independence Na- tional Historical Park. Certified, PHC 1957; Pennsylvania Register 1970; NR. 2 ext. photos (1958), see also ext. photo (1973)* of Berry- Coxe House (PA-1062; 4 data pages (1959).*

Blight Warehouse

(PA-1393), 101–3 S. Front St., at S.E. corner Chestnut St. Brick with marble trim, approx. 40′ (five-bay front) × 40′, five stories, flat roof, marble-pier storefront, 1840 date stone in end parapet. Built 1840; de- molished 1967. 1 ext. photo (1966);* 9 data pages (1966).*

Bonsall, John, House

(PA-1394), 706 Locust St., on S. side of Washington Sq. Brick with marble trim, approx. 24′ (three-bay front) × 42′ with four-story rear ell, three-and-a-half stories on raised basement, gable roof. One of five row houses. Built c. 1818–19. Certified, PHC 1957. 1 ext. photo (1975),* 5 int. photos (1959);* 2 data pages (1959).*

Brock, John, Sons & Co. Warehouse. Illustrated
(PA-1395), 242–44 N. Delaware Ave. and 243–45 N. Water St. Brick with cast-iron front on Delaware Ave. and cast-iron storefront on Water St. (rear), approx. 42' (six-bay front) × 148' (originally approx. 136' × 148'), five stories on Delaware Ave., four stories on Water St., flat roof, originally balcony with iron railing at second story, rectangular plan.

Fine early example of cast-iron design with iron plates simulating smooth rusticated masonry; perhaps nation's oldest iron-front building. Built 1850; Hoxie and Button, architects; William and N. W. Ellis, contractors; iron work by Reeves, Buck & Co., Cresson & Co., J. & S. Field; Robert Wood, iron railing. Iron railing and balcony removed 1865; first- and second-story doors and windows bricked up 1951. Southern section (Nos. 238–40) demolished 1939; northern section (Nos. 246–48) demolished 1959. 5 ext. photos (1972, 1973);* HABSI form (1970).*

Butcher & Brother Warehouse
(PA-1396), 142–44 N. Front St., at N.W. corner Quarry St. Brick with granite trim, approx. 27' (four-bay front) × 33', five stories, flat roof, storefront of granite piers and lintel, rectangular plan. Built between 1853 and 1861; demolished c. 1970. 1 ext. photo (1967);* 12 data pages (1968).*

Carpenter, Joshua, House (also known as Carpenter's Mansion)
(PA-1397), 615–19 Chestnut St., originally grounds occupied N. side Chestnut St., between Sixth and Seventh sts. Brick, approx. 36' × 24' with rear stair tower, two-and-a-half stories, jerkin-head roof, pent eaves at first and second stories, originally three rooms to a floor.

Built between 1701 and 1722; Joshua Carpenter, builder. Remodeled and gable-roof front addition built for John Dickinson 1774. Demolished 1826. Occupants included Provincial Governor George Thomas 1738–47, Pennsylvania Chief Justice William Tilghman 1798–1826, and the French legation 1779–85. 1 photocopy of old watercolor drawing (c. 1826); 3 data pages (including 1766–85 insurance surveys).*

Carpenters' Hall
(PA-1398), 320 Chestnut St., Carpenters' Court, Independence Na-

tional Historical Park. Museum. Brick, approx. 50' (five-bay front) × 50', two stories on raised basement, gable roof with cross-gables on north and south sides, projecting central pavilion on north and south sides, pedimented frontispiece, round-arch entrance with fanlight, round-arch windows with false balustrades on second story of north pavilion, octagonal cupola, cruciform plan.

Fine colonial example of Georgian guild hall whose cruciform plan gives it four distinct facades. Built 1770–74; Robert Smith, architect. Interior renovated 1857, 1904; south doorway restored 1964; interior restored 1967–68. Interior and exterior architectural elements in Architectural Study Collection, Museum Division, Independence National Historical Park. Founded 1724, Carpenters' Company is the oldest builders' organization in the United States and still owns and maintains the Hall. First Continental Congress met here 1774. Offices of Secretary of War 1790–91; offices and vaults of First Bank of the United States 1791–97; used by Musical Fund Society 1821–24, by Franklin Institute 1824–26; site of country's first exhibit of American manufactures 1824; auction hall 1828–56; opened to public as museum 1857. Certified, PHC 1956; Pennsylvania Register 1970; designated National Historic Landmark 1970. 3 sheets (1932, including elevation, section, details);* 8 ext. photos (1939, 1975,* including 1 of reconstructed Pemberton House on east side of court), 1 int. photo (1975),* 1 photocopy of ext. stereo view (n.d.),* 1 photocopy of int. stereo view (n.d.),* 4 photocopies of measured drawings (1897),* 1 photocopy of proposed plan for Carpenters' Court (c. 1768);* 1 data page (including 1773 insurance survey).*

Front Store. (PA-1399), 322 Chestnut St. Stuccoed brick (originally brick), 26'-6" (three-bay front) × 71' (originally 26'-6" × 44' with three-story back building), four stories, hipped roof, monolithic stone piers and Doric cornice at ground story, projecting corbeled lintels, corbeled cornice. Built 1810–11. New storefront, mastic added to upper stories, interior altered 1846; William Johnston, architect. New front, interior alterations, fourth story added to rear 1861; Collins and Autenreith, architects. Demolished 1957. 3 sheets (1958, including plot plan, plan, elevations, details);* 4 ext. photos (1957),* 1 aerial photo (1957);* 7 data pages (1954, including 1811 insurance survey).*

New Hall. (PA-1400), rear of 322 Chestnut St., on W. side of Carpenters' Court, Independence National Historical Park. Museum. Brick, approx. 61' (eight-bay front) × 20', four stories (originally two stories), flat roof (originally hipped roof). Built 1791; third story added

1833; fourth story added 1861; demolished 1957–58; reconstructed 1958–60 to 1791 appearance for U.S. Marine Corps Museum. 2 ext. photos (1957);* 6 data pages (1960, including 1791–1857 insurance surveys).*

Carriage House
(PA-1401), rear of 212 S. Fourth St. and 422 Walnut St. on E. side of Lawrence St. (not cut through), Independence National Historical Park. Brick, 46'-11¹/₂" (three-bay front) × 30'-2³/₈", two stories, shed roof, rectangular plan. Example of colonial vernacular architecture. Built mid-18th century; demolished 1960. Shutter, door in Architectural Study Collection, Museum Division, Independence National Historical Park. 2 sheets (1958, including plans, elevations, sections);* 2 ext. photos (1959),* 9 int. photos (1959).*

Chamber of Commerce. See Commercial Exchange

Chestnut Street Area Study
(PA-1402), 213–43 Chestnut St., on N. side Chestnut St. between Strawberry and Third sts. Thirteen contiguous commercial buildings, brick with brick, brownstone, granite, and cast-iron fronts, mostly three- and four-bay fronts, four and five stories, flat roofs, rectangular plans.

One of the few surviving examples of a formerly stylish mid-19th-century wholesale-retail block. Includes: Keen Building (No. 217), built 1851, Samuel Sloan, architect, cast-iron storefront by Architectural Iron Works (Daniel D. Badger), New York; Gendell and Landreth Buildings (Nos. 219–21), built c. 1853 with cast-iron front by Architectural Iron Works; Elliott Building (No. 235), built 1856, Joseph C. Hoxie, architect; Leland Building (No. 237), built 1854, Joseph C. Hoxie, architect; Lewis Building (Nos. 239–41), built 1852, Stephen D. Button, architect, cast-iron storefront by Architectural Iron Works; Borie Brothers Bank (No. 243), built 1897, Wilson Eyre, Jr., architect. 2 ext. photos (1972),* 1 photocopy of old ext. photo (1930);* HABSI form (1970).*

Christ Church (Protestant Episcopal). Illustrated
(PA-1071), 22–26 N. Second St., at N.W. corner Church St. Brick, 87'-0¹/₂" (eight-bay side) × 61'-6", two stories with square tower and three-stage spire, gable roof, round-arch windows, notable Palladian window at east end, Doric frieze, mutule cornice, balustrade with

urns; three-aisle plan, clerestory, galleries, rectangular chancel framed by Doric pilasters and pediment.

An outstanding example of Palladian church architecture in the United States. Built 1727–44; Dr. John Kearsley (physician), supervisor and probable architect. Steeple built 1751–54; John or Daniel Harrison, designer; John Palmer, mason; Robert Smith and John Armstrong, carpenters. West gallery enlarged 1732, 1782. Interior altered 1834; Thomas U. Walter, architect. Interior altered 1852. Interior altered 1882; G. W. Hewitt, architect. Steeple rebuilt after fire 1908. Certified, PHC 1956; Pennsylvania Register 1970, designated National Historic Landmark 1970.

21 sheets (1933, c. 1959,* 1971, including plot plan, plans, elevations, spire section, spire plan, sections, details, conductor head*); 21 ext. photos (1939, 1965, including 6 of tower renovations*), 9 int. photos (1965, 1969, including one of tower),* 1 photocopy of old ext. photo (c. 1910),* 2 photocopies of measured drawings (1897).*

Commercial Buildings
(PA-1404), 105–11 N. Water St. and 106–12 N. Delaware Ave. Four adjoining brick stores with granite trim, each approx. 25' (three-bay front) × 123', four stories, flat roofs, Tuscan granite piers and architraves at ground story, rectangular plans. Built c. 1835–40; demolished 1968. 1 ext. photo (1967).*

Commercial Buildings
(PA-1405), 125–27 S. Second St., at S.E. corner Gatzmer St. Two brick stores, northern one (No. 125) with cast-iron front, southern one with granite-pier storefront, each approx. 20' (three-bay front) × 125' (No. 125) and 85' (No. 127), five stories, flat roof. No. 127 built c. 1840. No. 125 built 1852 for J. Howell and Co.; John Riddell, architect. Storefront elements altered later; both demolished 1974. 1 sheet (1974, including plot plan, elevation).*

Commercial Exchange (also known as Chamber of Commerce)
(PA-1406), 135 S. Second St., at S.E. corner Sansom St. Brick with brownstone and granite trim, approx. 105' (seven-bay front) × 150', six stories (originally with central tower), flat roof (originally mansard as sixth story), central four-story entrance pavilion.

A post–Civil War office building that has retained its character in spite of later alterations. Built 1867–68; John Crump, architect.

Partially destroyed by fire 1869. Rebuilt 1870; James H. Windrim, architect; Benjamin Ketchum, builder. Cupola added before 1875. Tower removed, interior extensively altered 1901–2; Seymour Davis, architect; George Kessler, contractor. Mansard removed 1910. Interior altered, reinforced 1927; Clarence E. Winder, architect. Demolished 1976. Largest and most important commercial organization in late-19th-century Philadelphia; site of world's first automatic telephone exchange during occupancy of Keystone Telephone Co. 1902–44. Vacated 1961. Certified, PHC 1970. 2 ext. photos (1972);* HABSI form (1970).*

Commercial Union Assurance Company Building

(PA-1076), 416–20 Walnut St. Brick with limestone trim, 54'-5" (three-bay front) × 80', eight stories, gable roof with flat roof over rear wings, central service core with iron and marble stairs rising height of building. Built 1888–89; Theophilus Parsons Chandler, Jr., architect; J. E. and A. L. Pennock, contractors. Demolished 1963. 4 ext. photos (1962, 1963), 3 int. photos (1962); 6 data pages (1963).

Cooper, Jacob, House

(PA-1407), 118 Cuthbert St. Brick, approx. 12' (two-bay front) × 20', two-and-a-half stories, gable roof, cove cornice, originally pent eave. One dwelling in an original unit of three. Built late 18th century. Certified, PHC 1957. 2 ext. photos (1958);* 2 data pages (1958).*

Cope, Caleb, & Co. Store (late Goldberg's Army-Navy Store)

(PA-1408), 429 Market St. Brick with granite front, approx. 28' (four-bay front) × 205', five stories, flat roof, arcaded windows at second and fifth stories, corbeled cornice, parapet, rectangular plan. Fine example of mid-19th-century retail store built for a leading dry goods merchant and later banker. Built 1853; Joseph C. Hoxie, architect; Charles Rubicam, builder; Solomon K. Hoxsie, granite-cutter. Demolished 1970. 2 ext. photos (1970);* 7 data pages (including 1854–92 insurance surveys).*

Dilworth-Todd-Moylan House (also known as Dolley Madison House)

(PA-1409), 343 Walnut St., at N.E. corner Fourth St., Independence National Historical Park. Historic house museum. Brick, approx. 16' (two-bay front) × 35' with two-story rear kitchen ell, three stories, gable roof, brick string-course, London house plan.

Fine Georgian row house. Built c. 1775; Jonathan Dilworth,

house carpenter. Altered to store before 1859; new storefronts built 1909, 1923, 1931. Restored by National Park Service and two adjacent 1775 houses rebuilt 1958–70. John Todd, Jr., lawyer, lived here 1791–93; his widow, Dolley, married congressman James Madison 1794, who leased the house to Stephen Moylan, merchant and Continental Army officer. Certified, PHC 1956; NR. 4 sheets (1961, including floor and framing plans);* 1 ext. photo (1974).*

Dock Street Sewer
(PA-1072), beneath Dock St. between Third St. and Delaware River. Elliptical brick vault, span approx. 50'. Built for Girard Estate c. 1900; James H. Windrim, architect. Demolished 1962. 1 sheet (1962, including plot plan, sections).*

Dowers-Okill House
(PA-1410), 115 N. Water St. Brick, approx. 23' (three-bay front) × 45', three stories, gable roof, flat-arch brick lintels with scalloped edges at second story, London house plan. Built between 1747 and 1750; demolished c. 1963. Lintel in Architectural Study Collection, Museum Division, Independence National Historical Park. 4 ext. photos (1957).*

Drexel & Co. Building
(PA-1411), 34 S. Third St. Brick with coursed marble ashlar front, approx. 20' (three-bay front) × 83', four stories with one-story rear extension, flat roof, carved Tuscan pilasters and projecting cornice at each story, acroteria on pediment, "Drexel & Co." carved on ground-story cornice, rectangular plan.
Fine example of Tuscan mode of Italianate style office building. Built 1853–54; G. Runge, architect. Altered later. Founded 1837 by Francis M. Drexel, Drexel & Co. became one of the nation's foremost 19th-century investment banks; at this site 1854–85. Certified, PHC 1973; Pennsylvania Register 1975; NR. 2 ext. photos (1975),* 1 phtocopy of old ext. photo (c. 1900).*

Drinker, John, House (later known as Krider Gun Shop)
(PA-1055), 133–35 Walnut St., at N.E. corner Second St. Brick, 27'-1¹/₂" (four-bay front) × 47'-1¹/₂", three-and-a-half stories, gable roof, pedimented west end. Built 1751 as two dwellings in one unit; demolished 1955. Served as gun shop 1856–97. 16 sheets (1953, including plans, elevations, sections, details), 2 sheets (1953, including ele-

vation, sections, isometric assembly of casement window);* 4 ext. photos (1953, 1958, including details of original casement window), 1 int. photo (1953), 1 photocopy of old ext. photo (n.d.); 9 data pages (1963).

Duncannon Iron Company Warehouse and Office

(PA-1412), 122–24 Race St. Brick, approx. 36' (four-bay front) × 76'with two-story rear extension, two stories, gable roof, large vehicle door on ground story, rectangular plan with counting room in northeastern quarter. Built c. 1867; cornice, vehicle door altered after 1893; wooden trusses replaced by steel beams 1914; rear extension enlarged 1920–21; demolished c. 1969. Duncannon Iron Company here 1868–1909. 1 ext. photo (1967);* 4 data pages (including 1893 insurance survey, 1914–20 building permits).*

Elfreth, Jeremiah, House

(PA-1413), 126 Elfreth's Alley. Historic house museum. Brick, approx. 16' (two-bay front) × 22' with two-story rear ell, two-and-a-half stories, gambrel roof, pent eave, originally one room to a floor. Fine example of a mid-18th-century artisan's house. Built between 1741 and 1762; rear ell added later. Partially restored by Elfreth's Alley Association 1958–66. One of many houses on the alley owned by Elfreth for rental income. Elfreth's Alley Association Headquarters since 1962. Certified, PHC 1956. 2 ext. photos (1959),* 3 int. photos (1959),* 1 photocopy of old ext. photo (n.d.);* 1 data page (including 1762 insurance survey).*

Elfreth's Alley Area Study. Illustrated

(PA-1103), both sides of Elfreth's Alley between Front and Second sts., including Bladen's Court. Twenty-nine adjoining brick houses, mostly two-bay fronts, two, two-and-a-half, and three-and-a-half stories, gable and gambrel roofs. Includes Jeremiah Elfreth House (PA-1413).

Built at various times between mid-18th and early 19th centuries. The six-foot-wide street is the oldest unchanged and continuously inhabited street in Philadelphia and illustrates the typical side-street urban dwellings of 18th-century artisans. Residents included cabinetmaker Daniel Trotter, silversmith Philip Syng, pewterer William Will. Alley opened between 1702 and 1704; Bladen's Court opened between 1749 and 1753. Certified, PHC 1956; Pennsylvania Register 1970; designated National Historic Landmark 1963.

19 sheets (1940, including plans, elevations, sections, details) from Survey, Philadelphia Chapter, A.I.A. (1932); 9 ext. photos (1972, 1976),* 1 aerial photo (1965);* 40 data pages (1960, 1961, including insurance surveys).*

Ellison, John B., & Sons, Building

(PA-1414), 22–26 S. Sixth St. Brick with brownstone trim and cast-iron storefront, approx. 49' (five-bay front) × 173', five stories, mansard roof, floral patterns between third and fourth stories. Built 1881; demolished 1961. One of the largest cloth-importing houses in late-19th-century America. 1 ext. photo (1961),* 1 photocopy of old ext. photo (c. 1890).*

Farmers' and Mechanics' Bank. Illustrated

(PA-1415), 427 Chestnut St. Brick with marble ashlar front, approx. 55' (five-bay front) × 45' with two-story passage and three-story rear building, three stories, flat roof, round-arch windows with corbeled lintels, bracketed cornice with parapet, notable iron front door, originally stairs in front building and center corridor to rear three-story banking room in one story with skylight.

Fine Italianate bank design for one of three surviving banks on what was known as Bank Row. Built 1854–55; John M. Gries, architect; John Rice, builder; William Struthers, marble mason; Robert Wood, iron work. Interior altered 1873. Interior altered, basement enlarged 1903; W. Woodward Potter, architect. Rear one-story addition built 1917; C. Henry Wilson, architect. Interior altered 1919; William Steele and Sons Co., architects and builders. Altered to museum c. 1965. Owned by Philadelphia National Bank since 1918. Site of Philadelphia Maritime Museum 1965–74. Certified, PHC 1969; Pennsylvania Register 1970.

4 ext. photos (1972),* 1 photogrammetric stereopair (1959),* 2 photocopies of old ext. photos (c. 1910, n.d.),* see also photogrammetric stereopair of Pennsylvania Company for Insurances on Lives and Granting Annuities (PA-1452);* 8 data pages (including 1855–1904 insurance surveys, 1915–19 building permits).*

Fife, Matthew, House

(PA-1416), 136 Race St. Brick with marble trim, approx. 16' (two-bay front) × 25' with three-story rear ell, three stories on raised basement, flat roof, round-arch entrance with fanlight and arched marble hood on carved marble corbels, carved lintels, heavy bracketed cornice, side-hall plan with rear stairwell. Fine example of mid-19th-century city dwelling. Built 1854–55 for Matthew Fife, box manufacturer. 1 ext. photo (1967);* 2 data pages (including 1855 insurance survey).*

Fire Association Building. See Irvin Building

Fireman's Hall. See Philadelphia Fire Department, Engine Company No. 8

First Bank of the United States (also known as Stephen Girard's Bank).
Illustrated
(PA-1417), 120 S. Third St., Independence National Historical Park. Brick with marble front and trim, 90'-11" (seven-bay front) × 81'-9", three stories with Corinthian hexastyle portico, hipped roof with balustrade and gable skylight over dome (not original), notable mahogany pediment sculpture.

Notable early example of Classical monumental building. Built 1795–97; Samuel Blodgett, Jr., architect; Claudius LeGrand, sculptor. Extensive interior renovations and dome added 1901; James H. Windrim, architect; Doyle and Doak, contractors. Restored by National Park Service 1974–76. Origional iron entrance gates at Henry Francis duPont Winterthur Museum, Delaware. Construction elements, pieces of tympanum carving, parts of galvanized iron cornice (1901) in Architectural Study Collection, Museum Division, Independence National Historical Park. One of the nation's oldest bank buildings, it served as the Bank of the United States until 1812 and then as Stephen Girard's Bank and its successor, Girard National Bank, 1812–1926. Certified, PHC 1956; Pennsylvania Register 1970; designated National Historic Site 1948.

17 sheets (1958, 1960, including plot plan, plans, elevations, truss section, portico framing plan, details);* 22 ext. photos (1939, 1959,* 1961,* 1965,* 1975* including 1 photo of lantern, 11 photos of entrance gates), 2 int. photos (c. 1975, 1956),* 1 aerial photo (c. 1965),* 1 photocopy of old ext. photo (n.d.).*

First German Reformed Church Area Study
(PA-1418), 129–51 N. Fourth St., between Cherry and Race sts., and 322–30 Race St. Seven contiguous brick houses (Nos. 129–39, 151), each approx. 18' (two-bay front) × 18'with two- or three-story rear ells, three-and-a-half stories (No. 131 originally two-and-a-half stories), gable roofs, brick stringcourses or second-story pent eaves (Nos. 133–35).

First (German) Reformed Church (322–30 Race St.): brick with stone trim, approx. 64' (three-bay front) × 82', two stories, gable roof, round-arch windows, three-aisle plan, galleries. This church with the nearby school (327 Cherry St.) and dwellings served as a center of the city's German community for a century and a half.

Houses built at various times between 1745 and 1804; Nos. 129, 131 raised one story c. 1865, c. 1800 respectively; all altered to shops later. Restored 1969–; P. Richard Frantz, architect. Dwellings (Nos. 129–35) owned by German Lutheran Church 1771–1848. Church built 1836–37; altered to factory 1882. Restored 1967–; P. Richard Frantz, architect. Plaster ceiling ornament, wallpaper sample in Architectural Study Collection, Museum Division, Independence National Historical Park. Church founded 1727; congregation's third house of worship on this site; now called Old First Reformed Church, United Church of Christ. Certified, PHC 1965. 5 ext. photos (1974);* 15 data pages (1965, including 1769–1819 insurance surveys).*

First National Bank
(PA-1011), 315 Chestnut St. Brick with Quincy granite ashlar front, approx. 60' (three-bay front) × 82', two stories, flat roof with balustrade, central round-arch entrance with engaged columns and piers, round-headed windows, originally central banking room with balustraded galleries.

Restrained Italian Revival design. Built 1865–67; John McArthur, Jr., architect; John Rice, builder. Rear addition built 1912; interior remodeled 1953. First national bank chartered in the United States; now First Pennsylvania Banking and Trust Company. 1 ext. photo (1957), 1 photocopy of old int. photo (n.d.); 4 data pages (1958).

First Presbyterian Church (also known as Washington Square Presbyterian Church)
(PA-1117), S.E. corner Seventh and Locust sts., on S. side of Washington Sq. Stuccoed brick with stone trim, approx. 71' (five-bay front) × 88', two stories with Ionic hexastyle portico, gable roof, octagonal cupola; four-aisle plan, galleries, apsidal chancel.

Early use of radiating seating on inclined floor in a significant Greek Revival church. Built 1820–22; John Haviland, architect; John Clark, builder; John Struthers, stonecutter. Vacated 1928; demolished 1939. 15 sheets (1940, including plans, elevations, sections, details) from Survey, Philadelphia Chapter, A.I.A. (1932); 1 photocopy of old photolithograph (c. 1900).*

Franklin Institute (now Atwater Kent Museum)

(PA-121), 15 S. Seventh St. Museum. Brick with marble ashlar front and interior cast-iron columns, 61'-1$^{1}/_{2}$" (three-bay front) × 98'-6$^{1}/_{4}$", three stories on raised basement, flat roof, four Tuscan piers, wreathes along frieze, originally front offices flanked center hall leading to auditorium.

Greek Revival building with ancient design precedents. Built 1825–27; John Haviland, architect. Interior renovated, two-story rear wing built 1897; John R. Wiggins, contractor. Rear wing removed later, interior renovated 1937–38. Between 1827 and 1933 headquarters of Franklin Institute, scientific research and study center named in memory of Benjamin Franklin, which also offered draftsmanship and architecture instruction during its early decades. Since 1939 it has served as Atwater Kent Museum, devoted to the history of Philadelphia. Certified, PHC 1956; Pennsylvania Register 1970; NR.

7 sheets (1936, 1937, including plans, elevation, sections, details); 2 ext. photos (1972),* 1 photocopy of architect's drawing (c. 1825);* 1 data page (including 1867 insurance survey).*

Free Quakers Meeting House

(PA-1120), 500 Arch St., at S.W. corner Fifth St., Independence National Historical Park. Historic museum. Brick, 48'-4" (three-bay front) × 36'-3", two stories, gable roof, pedimented doorway, rusticated stucco lintels, dentil cornice, inscribed stone plaque in north gable; auditorium plan with galleries along three interior walls.

Fine example of nationalist influences on traditional Quaker plain style. Built 1783; Samuel Wetherill and Timothy Matlack, architects. Renovated, second floor installed 1788. Interior altered 1858, 1868. Two-story western addition built 1868; Stephen D. Button, architect. Masonry vaults and basement demolished when building was moved 33' west and 8' south 1961. Restored 1961–69; F. Spencer Roach, architect. Free Quakers were Friends who were expelled from the Religious Society of Friends for participation in the American Revolution; they proposed no new doctrines and dissolved 1836. Building served as Apprentices' Library 1841–97. Maintained by Junior League of Philadelphia, Inc., whose headquarters are in basement. Certified, PHC 1956; Pennsylvania Register 1970; NR.

2 sheets (1940, including elevation, details) from Survey, Philadelphia Chapter, A.I.A. (1932), 21 sheets (1963–64, including plot plan, plans, elevations, sections, details);* 9 ext. photos (1961,

1976*), 3 int. photos (1961), 1 aerial photo (1961), 3 photocopies of architectural drawings (1783); 4 data pages (including 1795–1882 insurance surveys).*

Garden, C. H., & Co. Building
(PA-1419), 606 Market St. Brick with cast-iron front, approx. 40' (five-bay front) × 118', six stories (originally five), mansard roof (originally flat with ornamental parapet), arcaded windows above ground story, originally columned ground story. Built c. 1866. Mansard added, interior altered 1899; Collins and Autenreith, architects. Ground story altered 1940; Israel Demchick, architect. Demolished c. 1965. 1 ext. photo (1962);* 1 photocopy of old ext. print (n.d.).*

Godley's Stores. See Granite Street Vaults

Goldberg's Army-Navy Store. See Cope, Caleb, & Co., Store

Gordon, George, Building
(PA-1065), 300 Arch St., at S.W. corner Third St. Cast iron with brick party walls, 15'-9" (two-bay front) × 37'-4", five stories, flat roof, arcaded windows above ground story. Fine early example of mid-19th-century cast-iron commercial building. Built 1856; demolished 1963. 4 sheets (1963, including plan, elevations, cast-iron structural details); 4 ext. photos (1959, 1962, 1963), 1 int. photo (1963); 5 data pages (1961, 1963).

Granite Street Vaults (also known as Godley's Stores)
(PA-1420), 100–12 and 101–27 Granite St., beneath sidewalks and street, between Front and Dock sts. Rubble-and-brick walls, longitudinal brick-arch arcade, granite-slab ceiling (street floor above), approx. 25' × 300', approx. 9' height.
An unusual example of granite-covered storage vaults under the length and width of a street. East section built 1849–50; John Rice, builder and superintendent of construction. West section built 1850–53; Jessee Godley, superintendent of construction. Demolished 1960. 2 ext. photocopies (1959),* 2 int. photocopies (1959).*

Hansell, John, House
(PA-1012), 153 N. Sixth St. Brick, approx. 18' (two-bay front) × 45' with three-story rear ell, three-and-a-half stories, gable roof. Built between 1788 and 1802; altered to store later; demolished 1959. 1 ext. photo (1959); 4 data pages (1959).

Harrison, Henry, Houses

(PA-1421), 112–16 Cuthbert St. (originally Coombe's Alley). Three brick row houses, each approx. 14' (two-bay front) × 37', two-and-a-half stories, gambrel roofs, cove pent eaves, two rooms to a floor with rear stair tower.

Fine example of 18th-century Philadelphia side-street rental properties. Built c. 1760. Fireplace and paneling of No. 116 in Queen Anne Bed Room, Henry Francis duPont Winterthur Museum, Delaware. Certified, PHC 1957; Pennsylvania Register 1970. 8 sheets (1957, 1959, including plot plan, plans, elevations, sections, details)* courtesy of Philadelphia Historical Commission; 1 ext. photo (1951),* 3 int. photos (1951);* 9 data pages (1957, including 1781–1850 insurance surveys);* 3 HABSI forms (1957).*

Hensel, Colladay & Co. Factory

(PA-1422), 45–51 N. Seventh St. Brick with cast-iron storefronts, approx. 90' (five-bay front) × 100', six stories, steep mansard roof, large windows with thin piers and spandrels, dark brick bands and carved stone panels on piers at each story, rectangular plan. Built 1881; Allen B. Rorke, contractor. Demolished 1965. Hensel, Colladay & Co. were manufacturers of dress trimmings. 2 ext. photos (1965);* 1 data page (1965).*

House

(PA-1423), 620, Arch St. Brick with cast-iron storefront, approx. 25 (three-bay front) × 65', three-and-a-half-stories, gable roof. Built late 18th century; renovated to store later; demolished 1965. 1 ext. photo (1965);* 1 data page (1965).*

House

(PA-1424), 628–30 Arch St. Brick with wooden and pressed metal storefronts, approx. 35' (four-bay front) × 45', three-and-a-half stories, gable roof. Built late 18th century; renovated to store later; demolished 1965. 1 ext. photo (1965);* 1 data page (1965).*

House

(PA-1425), 113 N. Water St. Brick, approx. 25' (three-bay front) × 46', three stories, gable roof. Built mid-18th century; demolished c. 1963. 2 ext. photos (1957).*

House

(PA-1426), 523–25 Quarry St., at N.E. corner Fairhill St. Brick, approx. 38' (four-bay front) × 18' with two-story rear ells, two-and-a-half stories, gable roof, two dwellings in one unit. Built mid-18th century; demolished c. 1960. 1 ext. photo (1959).*

Houses

(PA-1427), 113–15 Summer St. (originally Swarthmore Pl.). Two contiguous brick houses, each approx. 15' (two-bay front) × 30' with rear two-story ell, three stories, gable roof. Originally part of a row of six tenement houses on court extending east from between 229 and 231 N. Second St. Built early 19th century; demolished 1968. Swarthmore Pl. became Summer St. 1911. 1 ext. photo (1967).*

Howell & Brothers Building

(PA-1428), 12–14 S. Sixth St. Brick with iron front, approx. 41' (six-bay front) × 173', five stories, flat roof, straight-arch windows. Built c. 1870; demolished 1961. Howell & Brothers were one of the country's largest manufacturers of paper hangings. 1 ext. photo (1961).*

Independence Hall (originally known as the State House of Pennsylvania).
Illustrated

(PA-1430), S. side Chestnut St. between Fifth and Sixth sts., Independence National Historical Park. Brick with marble and soapstone trim, approx. 107' (nine-bay front) × 45', two stories, decked gable roof with balustrade; on north facade are marble panels between marble stringcourses at second story, marble keystones, soapstone quoins, wooden Corinthian cornice with carved modillions; south facade dominated by brick-veneered stone tower with wooden steeple and octagonal cupola, Doric frontispiece, Palladian window, Doric, Ionic, Corinthian pilasters on successive stories, urns and balustrade mark setbacks of steeple; open triple-arch arcades connect main building to flanking two-story hipped-roof wings. Central Hallway with engaged Doric columns and entablature flanked by Assembly Room (east room) fully paneled on east wall with fluted Ionic pilasters and cove ceiling and Supreme Court Room (west room) fully paneled on all walls with fluted Doric pilasters; Long Room extends length of second floor.

An outstanding colonial example of Georgian public architecture. Built 1732–48; Edmund Wooley (master carpenter) and Andrew

Hamilton (lawyer), architects. Tower and steeple built 1750–53; steeple removed 1781. Clock on soapstone case structure built on west end 1752. Wings and arcades removed, replaced with office wings 1813–15; Robert Mills, architect. Soapstone clock case structure removed 1815. Steeple reconstructed, clock on west end of Hall removed 1828; William Strickland, architect. Assembly Room and second story renovated 1802; Assembly Room remodeled c. 1818. Assembly Room "restored" 1831; John Haviland, architect. Second story renovated 1828 for federal courts 1828–54; John Haviland, architect. Second story renovated 1854 for City of Philadelphia Councils' Chambers 1854–95. Interior "restored," wings and arcades reconstructed by Philadelphia Chapter, Daughters of American Revolution 1896–98; T. Mellon Rogers, architect. Second story restored by City of Philadelphia 1922, plans and supervision by committee of Philadelphia Chapter, A.I.A.; Horace Wells Sellers, chairman. Restored by National Park Service 1951–73.

Pennsylvania State House 1735–99, housing Assembly, Council, and Supreme Court of the Province and later Commonwealth. Second Continental Congress and Congress of Confederation held most of their sessions here. Site of the adoption of Declaration of Independence 1776, enactment of Articles of Confederation 1781, drafting of U.S. Constitution 1787. Charles Willson Peale's museum (Philadelphia Museum), America's first modern museum, occupied second story 1802–28. Owned by City of Philadelphia since 1818, administered by National Park Service. Certified, PHC 1956; Pennsylvania Register 1970; designated National Historic Site 1943.

18 ext. photos (1939, 1959,* 1975*), 4 ext. photos of tower (1959),* 14 int. photos of tower (1959),* 47 photos of Central Hallway (1959, 1975),* 27 photos of Assembly Room (1959, 1969, 1975),* 11 photos of Supreme Court Room (1959, 1975),* 7 photos of second floor (1959, 1975),* 6 photos of garret (1959),* 33 photos of stairhall (1959),* 1 photocopy of old drawing (c. 1732),* 58 photocopies of measured drawings (1898, 1899, 1918, 1923, n.d.),* 1 photocopy of old ext. photo (c. 1865–69);* 4 data pages (including 1818–29 insurance surveys).*

Congress Hall (originally known as Philadelphia County Court House). (PA-1431), S.E. corner Sixth and Chestnut sts., Independence National Historical Park. Brick with marble trim, approx. 50' (five-bay front) × 98', two-and-a-half stories with slightly projecting three-bay pedimented pavilion and round-arch entrance with fanlight, hipped roof with open octagonal cupola, rear two-story semiocta-gonal bay, round-arch windows on ground story; House Chamber on ground floor, four semicircular tiers, gallery on north wall, Senate Chamber on second floor.

Fine example of Federal-style public architecture complementary to notable adjacent Georgian building. Built 1787–89; enlarged to ac-commodate U.S. Congress 1793–95. Senate Chamber restored by So-ciety of Colonial Dames in Pennsylvania 1895–96; George C. Mason, architect. House of Representatives Chamber restored by City of Philadelphia 1912–13, plans and supervision by committee of Philadelphia Chapter, A.I.A.; Frank Miles Day, chairman. Restored by National Park Service 1959–62. Cornice and interior elements in Architectural Study Collection, Museum Division, Independence Na-tional Historical Park.

U.S. Congress met here 1790–1800, making it the oldest extant building in which U.S. Congress has met. Site of the ratification of the Bill of Rights 1791, of Washington's second inauguration 1793 and his Farewell Address 1797, and of John Adams's inauguration 1797. Philadelphia County Court House 1800–1890; later public uses. Certified, PHC 1956; Pennsylvania Register 1970; designated National Historic Site 1948. 3 ext. photos (1975),* 14 int. photos (1963, 1964, 1975),* 13 photocopies of measured drawings (1901, 1912);* 4 data pages (including 1823, 1851 insurance surveys).*

Philadelphia City Hall (also known as U.S. Supreme Court Building). (PA-1432), S.W. corner Fifth and Chestnut sts., Independence National Historical Park. Brick with marble trim, approx. 50' (five-bay front) × 70', two-and-a-half stories with slightly projecting three-bay pedimented pavilion and round-arch entrance with fanlight, hipped roof with open octagonal cupola, exterior appearance same as Congress Hall.

Built 1790–91; David Evans, master carpenter. Restored by City of Philadelphia 1921–22; plans and supervision by committee of Philadelphia Chapter, A.I.A.; Horace Wells Sellers, chairman. Re-stored by National Park Service 1951–74. U.S. Supreme Court, U.S. Circuit Court, U.S. District Court met here 1791–1800. Used as Philadelphia City Hall 1791–c. 1895. Certified, PHC 1956; Pennsyl-vania Register 1970; designated National Historic Site 1948. 2 sheets (1961, including elevations);* 1 ext. photo (1975),* 4 int. photos (1975),* 18 photocopies of measured drawings (1918);* 1 data page (including 1821 insurance survey).*

Independence National Historical Park. See Carpenters' Hall, Dilworth-Todd-Moylan House, First Bank of the United States, Independence Hall, Philadelphia Exchange Company, Philosophical Hall, Second Bank of the United States, Bishop William White House

Insurance Patrol. Illustrated
(PA-1433), 509 Arch St. Brick with cast-iron front, approx. 27' (three-bay front) × 205', two stories, flat roof. Built c. 1879; demolished 1965. Sponsored by insurance companies, the Insurance Patrol protected property during and after fires. 1 ext. photo (1959),* 1 photocopy of old ext. photo (n.d.);* 3 data pages (1964).*

Irvin Building (originally known as Fire Association Building)
(PA-1434), 401–13 Walnut St., at N.W. corner Fourth St. Steel frame, brick curtain walls faced with coursed limestone ashlar, approx. 128' (ten-bay front) × 100' originally 53'-3½" × 80'-3¾"), ten, fourteen, seventeen stories (originally seven), flat roof, two ground-story flat Doric distyle porticoes in antis, second-story windows with bold lintels and Ionic pilasters, recessed center bay above sixth story, carved spread eagle on fasciae at ninth-story corners, fluted panels at seventh and eighth stories, rectangular plan with elevators at rear wall.
    Fine example of evolution of an early-20th-century office building in a late Victorian eclectic classicism complementary to neighboring structures. Built 1911–12; Edgar V. Seeler, architect; Cramp and Company, contractors. Raised four stories, lateral fourteen- and seventeen-story and rear ten-story additions built 1928–30; Ernest J. Matthewson, architect; Doyle and Company, contractors. Mezzanine floor filled in to create another floor 1955; C. H. Woolmington, architect. Demolished 1974. 6 ext. photos (1974),* 1 photocopy of old ext. photo (1915);* 2 data pages (including building permits 1911–55).*

Iungerich Warehouse
(PA-1403), 147 S. Front St. Brick with sandstone and granite trim, approx. 28' (four-bay front) × 25'; four stories, mansard roof, squat central tower with pointed roof, rectangular plan. Built after 1867; demolished 1967. 1 ext. photo (1966);* 7 data pages (1966).*

Jayne Building
(PA-188), 242–44 Chestnut St. Brick with granite front and interior iron columns and wooden girders supporting floors, 42'-2" (seven-

bay front) × 133'-6", eight stories with two-story castellated tower (destroyed 1872), flat roof, clustered-column piers terminate with pointed arches at seventh story and quatrefoil windows at eighth story, rectangular plan.

Prototype skyscraper design. Built 1849–50; William Johnston, architect; Thomas U. Walter, architect, finished building at Johnston's death 1849; ironwork by J. K. and E. K. Smith (Tamaqua, Pa.). Six-story side wings (238–40 and 246 Chestnut St.) added 1851; Thomas U. Walter, architect. Interior rebuilt, rear granite wall replaced by brick, steam elevator installed after fire 1872. Altered 1907; Furness, Evans & Co., architects. Demolished 1957–58. Dr. David Jayne was a prominent patent-medicine producer and real estate developer.

16 sheets (1957, including plot plan, plans, elevations, sections, details); 15 ext. photos (1951, 1957, 1958), 2 photocopies of insurance survey plans (c. 1862, 1873);* 3 data pages (1957).

Jefferson Fire Insurance Company Building

(PA-1435), 423 Walnut St. Brick with stone and tile trim, approx. 22' (three-bay front) × 76' with three-story rear addition, four stories, flat roof, parapet with acroterion. Built c. 1885. Interior altered, three-story addition built 1907; Arnold H. Moses, architect. Demolished 1959. 2 ext. photos (1959),* 1 photocopy of old ext. photo (1915).*

Keystone Telephone Company Building. See Commercial Exchange

Kid-Chandler House

(PA-1436), 323 Walnut St., Independence National Historical Park. Brick with marble trim, approx. 21' (three-bay front) × 42', three-and-a-half stories, gable roof, notable interior details. Originally apothecary shop on ground story with separate entrance at eastern (right) end and entrance hall to rear stair tower at western end.

Early Philadelphia example of use of second story as main living floor. Built c. 1810–15; altered c. 1840, c. 1870, c. 1880, c. 1909; rear kitchen wing demolished 1959; interior extensively remodeled and exterior restored 1963–64. Shutter and many elements of top-floor woodwork in Architectural Study Collection, Museum Division, Independence National Historical Park. Present quarters of Pennsylvania Horticultural Society and Philadelphia Society for Promoting Agriculture. Certified, PHC 1957; Pennsylvania Register 1970; NR. 9 ext. photos (1959, 1960, 1961, 1972, 1975),* 36 int. photos (1959, 1961);* 35 data pages (1959, 1961).*

Kid-Physick House
(PA-1437), 325 Walnut St., Independence National Historical Park.
Brick with marble trim, approx. 18' (two-bay front) × 42', three-and-
a-half stories, gable roof; originally side-hall plan with two counting
rooms on ground story, living quarters on second story.

Built c. 1811–12; altered c. 1865, c. 1909; rear wing demolished
1959; interior extensively remodeled and exterior restored with 18th-
century-style garden 1963–64. With adjacent dwelling this house has
been headquarters of Pennsylvania Horticultural Society and
Philadelphia Society for Promoting Agriculture since 1964. Certified,
PHC 1957; NR. 2 ext. photos (1959),* see also ext. photo (1973)* of
Kid-Chandler House (PA-1436), 18 int. photos (1959);* 46 data pages
(1958, 1959).*

Krider Gun Shop. See Drinker, John, House

Lee, Robert M., House and Law Office
(PA-1052), 109–11 N. Sixth St. Brick, approx. 34' (four-bay
front) × 28' with three-story rear ell, three-and-a-half stories, gable
roof, handsome Greek Revival interior details. Built between 1769
and 1774 as two dwellings; remodeled into one building with Greek
Revival detailing c. 1840; demolished 1959. 3 sheets (1959, including
plot plan, second-floor plan, details); 1 ext. photo (1959), 6 int. photos
(1959), 1 photocopy of old ext. photo (c. 1870); 11 data pages (1959,
1960).

Leland Building. Illustrated
(PA-1086), 37–39 S. Third St. Brick with granite ashlar front, approx.
34' (five-bay front) × 80', five stories, flat roof, rectangular plan with
rear ell, facade has vertical emphasis from continuous granite piers
and recessed spandrels. One of a complex of buildings around a
central courtyard. Early architectural solution for tall, narrow commercial building. Built 1855; Stephen D. Button, architect; Solomon
K. Hoxsie, granite-cutter; John Kilgore and John R. Hudders,
carpenters. Certified, PHC 1971. 5 ext. photos (1963, 1965), 1 int.
photo (1965); 7 data pages (1964).

Lit Brothers
(PA-1438), 701–39 Market St., composing block bounded by Seventh,
Market, Eighth, Filbert sts. Brick end pavilions with terra cotta and

galvanized iron trim (701–7 and 723–39 Market St.), brick center buildings with painted granite and marble ashlar fronts, reinforced concrete and buff brick building on northeast end (10–34 N. Seventh St.), approx. 400' (fifty-bay front) × 350', five, six, and seven stories, flat roof, semihexagonal bays at three corners, at third-story rear arched concrete bridge with brick walls and carved limestone trim over Filbert St.

Outstanding example of the architectural and business evolution of a major Philadelphia department store. No. 711 Market St. built for J. M. Maris & Co. 1859; Nos. 715–17 Market St. built for J. B. Lippincott & Co. 1861; Nos. 723–25 Market St. built c. 1865; No. 709 Market St. built c. 1870; No. 713 Market St. built 1872. Nos. 719–21 Market St. built for Joel J. Bailey & Co. 1873; William D. Price, architect. Nos. 737–39 Market St. built 1893, Nos. 727–35 Market St. and Nos. 7–13 N. Eighth St. built 1895–96; Collins and Autenreith, architects. Interior of Nos. 715–21 Market St. altered 1898–99, Nos. 15–21 N. Eighth St. built 1900; Collins and Autenreith, architects. Nos. 701–7 Market St. built 1907; Charles M. Autenreith, architect. Bridge at 720 Filbert St. built 1912; Stearns and Castor, architects. Nos. 10–34 N. Seventh St. built 1917–18; Simon and Bassett, architects. Many later interior alterations.

Enterprise expanded from store at N.E. corner Eighth and Market sts. (37' × 96') 1893 to include three store lots 1895, another three 1897, and three stores 1898 to measure 236' × 231'. By 1907 twenty-three more stores were incorporated at Seventh and Market sts. and along Seventh St. to give present dimensions. 3 ext. photos (1972, 1973, including 1 of bridge).*

McClare-Hutchinson Building

(PA-1439), 20 S. Third St. Stuccoed brick with brownstone storefront (originally brick), approx. 20' (three-bay front) × 36' with one-story rear extension (originally three-story kitchen ell), four stories (originally three-and-a-half stories), flat roof (originally gable), lotus-leaf piers, battered doors, rectangular plan (originally two shops and parlor on ground story).

Fine example of 18th-century shop-dwelling remodeled to mid-19th-century office building with notable Egyptian Revival doors. Built for Samuel McClare 1789; raised one story, new front built 1847–48 for Hutchinson & Co., stockbrokers; rear ell replaced later. Certified, PHC 1973. 2 ext. photos (1958, 1972).*

McCrea, James, House
(PA-1440), 108–10 Sansom St. Brick, approx. 47' (six-bay front) × 28' with three-story rear ell, four stories (originally three), shed roof, pedimented round-arch entrances. Built between 1794 and 1812 as two dwellings in one unit. Certified, PHC 1959; Pennsylvania Register 1970. 1 ext. photo (1959);* 4 data pages (1965).*

Manhattan Building. See American Life Insurance Company Building

Market Street Area Study
(PA-1441), 621–35 Market St., on N. side of Market St. between Sixth and Seventh sts. Nine contiguous brick and cast-iron commercial buildings, three- and four-bay fronts, four, five, and six stories, flat and mansard roofs. Typical late-19th-century retail block. Built at various times during the last half 19th century; demolished 1965. 5 ext. photos (1965);* 4 data pages (1965).*

Mason, James S., & Co., Store
(PA-1442), 138–40 N. Front St., at S.W. corner Quarry St. Brick with granite ashlar front and interior iron columns and wooden girders supporting floors, approx. 28' (five-bay front) × 95', five stories, flat roof, large Tudor-arch windows with engaged clustered columns on second-story front, flat-arch lintels on upper-story windows, rectangular plan.
Notable example of mid-19th-century wholesale store. Built 1850–51; demolished 1973. James S. Mason & Co., a prominent manufacturer of blacking and inks, was here until 1919. 5 ext. photos (1973),* 1 int. photo (1973),* 1 photocopy of old ext. print (1855).*

Mechanics Bank (now Norwegian Seamen's Church)
(PA-1443), 22 S. Third St. Coursed granite ashlar front and stuccoed brick sides and rear, approx. 33' (three-bay front) × 58', two stories (originally one) with Corinthian distyle portico in antis, gable roof, originally one-story banking room with square skylight.
Fine Greek Revival bank, one of Strickland's last designs in Philadelphia. Built 1837; William Strickland, architect; John Struthers, stonemason. Interior renovated 1874; James H. Windrim, architect. Interior altered, two-story rear brick addition built 1930; Edwards and Green, architects. Norwegian Seamen's Church since 1930. Certified, PHC 1961; Pennsylvania Register 1970. 2 ext. photos (1958);* 5 data pages (1958, 1965, including 1837, 1855 insurance surveys).*

Megargee Brothers Building
(PA-1444), 18–20 S. Sixth St. Brick with iron storefront, approx. 41′ (four-bay front) × 173′, five stories, flat roof. Built c. 1868; upper stories replaced c. 1880; demolished 1961. 1 ext. photo (1961).*

Merchants' Exchange. See Philadelphia Exchange Company

Merchants' Hotel
(PA-1445), 40–50 N. Fourth St. Brick with marble trim, 88′ (nine-bay front) × 190′, five stories, flat roof, colonnade at ground story, recessed five-bay balcony at second story, central-hall plan. Built 1837; William Strickland, architect. Destroyed by fire 1966. Columns, part of iron balcony railing in Architectural Study Collection, Museum Division, Independence National Historical Park. At the time of construction it was the largest, most distinguished hotel in the city, and served as James Buchanan's presidential campaign headquarters 1856. Certified, PHC 1956. 6 ext. photos (1962),* 5 int. photos (1962),* 1 photocopy of old print (1857);* 19 data pages (1962, 1963, including 1837 news article and 1837–51 insurance surveys).*

Meredith House. See Stone-Penrose House

Morris Brewery Vaults
(PA-1446), 210 Chancellor St., at S.E. corner American St. Rubble foundation and piers with galleting and brick-vaulted ceiling, approx. 35′ × 85′, approx. 11′ height. Built 1789; demolished 1961. 6 photocopies of int. photos (1959);* 3 data pages (1962, including 1790–1831 insurance surveys).*

Musical Fund Society Hall
(PA-1447), 808 Locust St. Brick with buff brick front and copper and terra cotta trim, approx. 76′ (three-bay front) × 94′ (originally approx. 60′ × 110′), three stories (originally two), flat roof with pediment, giant Corinthian pilasters on upper stories, large arched entrance.
    Fine 19th-century music hall whose appearance and function reflect changes in architectural styles and the city's cultural life. Built 1824; William Strickland, architect; John O'Neill, builder; Daniel Groves, master mason. Enlarged, new front built, stage moved to rear 1847; Napoleon LeBrun, architect. New front built, interior remodeled into three saloons and one banquet room 1891; Addison Hutton, architect. Entrance altered, interior converted to warehouse 1946; Howard Carter Hill, architect.

Oldest surviving music hall in the United States, noted for its acoustics. Site of first Republican Party National Convention 1856. Musical Fund Society was formed 1820 to cultivate musical taste and help needy musicians and used hall for concerts and lectures until 1924 when it became Philadelphia Labor Institute; used for various social and athletic events 1934–45; warehouse 1946–64; vacated 1964. Certified, PHC 1957; Pennsylvania Register 1970; designated National Historic Landmark 1974.

1 ext. photo (1976);* 15 data pages (1959, including architects' estimates and specifications 1825, 1847).*

National Guard's Hall (also known as Armory of the National Guards)
(PA-1015), 518–20 Race St. Brick, approx. 60′ (five-bay front) × 130′, three stories, flat roof, round-arch windows. Built 1856–57; demolished 1959. 3 ext. photos (1959), 4 int. photos (1959, including chandelier), 1 photocopy of old ext. print (n.d.); 6 data pages (1959).

North Front Street Area Study
(PA-1448), 2–66 N. Front St., on W. side Front St. between Market and Arch sts. Twenty-six contiguous commercial buildings, brick with granite storefronts (one with cast-iron facade), mostly two- and three-bay fronts, four and five stories, flat, gable, and mansard roofs, rectangular plans.

Well-preserved block of early-19th-century wholesale stores and warehouses. Three-and-a-half story Nos. 16, 46 built c. 1780, 1785; Nos. 2–6, 14, 18–44, 48–66 built between 1828 and 1836; five-story iron-front No. 8 built c. 1855. Nos. 10–12 demolished c. 1915; No. 54 demolished between 1960 and 1969. Nos. 18–30 part of Stephen Girard estate since 1830. Nos. 46, 50 certified, PHC 1960, 1970. 9 ext. photos (c. 1958, 1974);* 40 data pages (1960, including 1829–1933 insurance surveys).*

North Seventh Street Area Study
(PA-1449), 21–33 N. Seventh St., on E. side Seventh St. between Commerce and Filbert sts. Six contiguous brick and cast-iron commercial buildings, three- and four-bay fronts, two, three, four, and five stories, flat and gable roofs. Includes Philadelphia Hose Company (PA-1460). Built at various times during last half of 19th century; demolished 1965. 1 ext. photo (1965).*

North Third Street Area Study
(PA-1450), 17–63 N. Third St., E. side Third St. between Church and Arch sts. Twenty-two contiguous brick commercial buildings with brick, granite, brownstone, and cast-iron fronts, mostly three-bay fronts, three-and-a-half, four, and five stories, gable and flat roofs.

A treasury of mid-19th-century stores. Built between c. 1840 and 1875. Includes Nos. 47–49 built 1853, Joseph C. Hoxie, architect; Nos. 61–63 built 1852; John Riddell, architect. Nos. 53–55 enlarged, new fronts added 1913–14, Sauer and Hahn, architects; new front added to No. 59 1926. No. 49 demolished 1964–65. Certified, PHC 1974. 7 ext. photos (1974),* 1 photocopy of old lithograph (c. 1855).*

Norwegian Seamen's Church. See Mechanics Bank

Old Customs House. See Second Bank of the United States

Old St. Paul's. See St. Paul's (Protestant Episcopal) Church

Penn Mutual Life Insurance Company Building (known as Penn Mutual Building)
(PA-1451), 129 S. Third St. and 241 Dock St., at N.E. corner Third and Dock sts. Cast-iron with brick party walls, 19'-11½" (three-bay front) × 32'-10", five stories, flat roof, originally cast-iron statue of William Penn in second-story niche, round-arch windows, rectangular plan.

One of the earliest cast-iron buildings in the country, introducing a new method of jointing and bolting cast-iron plates. Built 1850–51; G. P. Cummings, architect; Joseph Singerly, builder. Demolished 1956. William Penn statue removed to lobby of Penn Mutual Building, 530 Walnut St., Philadelphia 1945. 1 ext. photo (1951).*

Pennsylvania Company for Insurances on Lives and Granting Annuities
(PA-1452), 431 Chestnut St. Brick with Quincy granite ashlar front, approx. 30' (three-bay front) × 164', three stories, flat roof, three-arch arcades with foliated carvings on keystones on first and second stories, third-story six-arch arcade with squat columns, front two-story banking room with skylights in vaulted ceiling.

Only extant post–Civil War bank on Bank Row. Built 1871–73; Addison Hutton, architect; William A. Armstrong, builder. Interior remodeled 1880; Addison Hutton, architect. Chartered 1812, Pennsylvania Company is the oldest institution of its kind in the United States. Pennsylvania Company here 1873–90, Tradesmen's National Bank 1910–30, Philadelphia National Bank since 1940. Certified, PHC 1958.

2 ext. photos (1959),* 2 photogrammetric stereopairs (1959),* 1 photocopy of old ext. photo (1929),* 1 photocopy of old int. photo (c. 1910).*

Pennsylvania Company for Insurances on Lives and Granting Annuities
(PA-1453), 304 Walnut St. Brick with Pictou stone front, approx. 25′ (three-bay front) × 97′, two stories, flat roof, incised pilasters and spandrels on ground story, smooth rusticated upper story, round-arch windows and entrance with carved keystones, large one-room banking room with galleries and rectangular skylight.

Fine Italianate bank design. Built c. 1859. Restored 1966–67; Stewart, Noble, Class & Partners, architects. Now a branch bank of First Pennsylvania Banking and Trust Company. Certified, PHC 1961; Pennsylvania Register 1970. 3 ext. photos (1958, 1961),* 3 int. photos (1961);* 13 data pages (1961, including 1859–75 insurance surveys).*

Pennsylvania Fire Insurance Company Building. Illustrated
(PA-1454), 508–10 Walnut St. Brick with marble front, approx. 45′ (six-bay front) × 138′ (originally approx. 22′ × 50′ with three-and four-story rear ell), four stories, flat roof, fluted papyrus-leaf columns at ground story, windows with battered surrounds and winged orb on cavetto lintels at second story, fluted cavetto parapet with winged orb and company's name (not original).

Fine example of Egyptian Revival architecture with later matching addition. No. 510 built c. 1839; probably by John Haviland, architect. No. 508 built, No. 510 enlarged 1902; Theophilus Parsons Chandler, Jr., architect; Doyle and Doak, contractors. Building demolished, facade preserved as courtyard screen before twenty-one-story office building 1971–74; Mitchell/Giurgola Associates, architects. Certified, PHC 1962. 8 ext. photos (1959, 1974, 1976).*

Perseverance Hose Company No. 5
(PA-1455), 316 Race St. Brick with Pictou stone ashlar front, approx. 16′ (one-bay front) × 75′, two stories, flat roof, large straight-arch equipment door, triple-arch window above, rectangular plan. Built 1867; upper story removed 1948; demolished 1959. 1 photocopy of old ext. photo (c. 1935).*

Philadelphia Bank. See Bank of Pennsylvania

Philadelphia Bourse. Illustrated
(PA-1456), 11–21 S. Fifth St., composing block bounded by Fifth, Ludlow, Fourth, Ranstead sts. Steel frame, rusticated sandstone lower stories and buff brick curtain walls with terra cotta trim, approx. 132' (eleven-bay front) × 362', nine stories, flat roof, giant columns and piers mark two-story entrance, rusticated quoins on lower three stories; long central interior court with gallery and skylights, originally exhibition halls in basement and eighth story.

Built 1893–95; G. W. and W. D. Hewitt, architects; Allen B. Rorke, contractor. Interior court altered later. Ground-story interior and exterior west (Fifth St.) and south sides altered 1973–74; Thalheimer and Weitz, architects. Modeled after European bourses, it was the only American institution of its kind when built. Certified, PHC 1971. 6 ext. photos (1973, 1974, 1976),* 1 photocopy of old ext. photo (c. 1925);* 4 data pages (including 1933 insurance survey).*

Philadelphia Contributionship for the Insuring of Houses from Loss by Fire
(PA-1457), 212 S. Fourth St. Brick with marble trim, approx. 50' (three-bay front) × 54', four stories (originally three) with Corinthian tetrastyle porch, mansard roof (originally flat), double-entry stairway, double windows with marble lintels and pilasters, center-hall plan.

Fine example of Greek Revival office-dwelling. Built 1835–36; Thomas U. Walter, architect; Daniel Knight, Robert Knight, and Joseph S. Walter, builders. Mansard added and porch replaced 1866–67; Collins and Autenreith, architects; J. Struthers and Son, marble masons; Michael Errickson, carpenter. Interior altered 1898; Furness, Evans & Co., architects; John Duncan, contractor. Renovated and partially restored 1930, 1970. Formed by Benjamin Franklin in 1752, the Philadelphia Contributionship is the nation's oldest mutual fire insurance company. Certified, PHC 1956; Pennsylvania Register 1970; NR.

2 ext. photos (1958),* 3 int. photos (1958),* 1 photocopy of architect's drawing (1835);* 34 data pages (1959, including 1835–37, 1866–67 correspondence with architect and marble mason).*

Philadelphia Exchange Company (later known as Merchants' Exchange). Illustrated
(PA-1028), 143 S. Third St., on triangular lot bounded by Third, Walnut, Dock Sts., Independence National Historical Park. Brick faced with marble ashlar, 95'-4³/₄" (five-bay front) × 152'-8¹/₂", three

stories, hipped roof, notable semicircular Corinthian prostyle portico on east facade flanked by stairs, circular lantern with engaged Corinthian columns, plain sides with tripartite windows, Corinthian tetrastyle portico in antis with pediment on west facade.

A masterpiece of Greek Revival architecture that exploits the break in the city's grid afforded by its irregular lot. Built 1832–33; William Strickland, architect; John Struthers, marble mason. Recumbent marble lions imported from Italy 1838. Interior altered 1891; Frank Miles Day, architect. Interior gutted and rebuilt 1901–3; Louis C. Hickman, architect. Exterior and lantern tower restored 1964–65. For over half a century Philadelphia's commercial center, replacing earlier taverns. Certified, PHC 1956; Pennsylvania Register 1970; designated National Historic Site 1948. 14 sheets (1954, including plot plan, plans, portico plan, section, elevation, details);* 19 ext. photos (1939, 1964,* 1975*), 1 photocopy of old ext. print (c. 1835);* 17 data pages (including 1834–1903 insurance surveys).*

Philadelphia Fire Department, Engine Company No. 8 (now known as Fireman's Hall). Illustrated

(PA-1459), 149 N. Second St., at N.E. corner Quarry St. Museum. Brick with stone trim, approx. 30' (three-bay front) × 105', two-and-a-half stories, stepped gable roof with triple dormer, diapering above second story, rectangular plan. Built 1898; Thomas M. Seeds, Jr., contractor. Altered later. Fire Department Museum since 1967, known as Fireman's Hall since 1975. 1 sheet (1974, including elevation);* 1 ext. photo (1976).*

Philadelphia Hose Company No. 1

(PA-1460), 33 N. Seventh St., at S.E. corner Filbert St. Brick with brownstone ashlar front, approx. 15' (two-bay front) × 58', three stories, flat roof, paired windows framed by Tuscan pilasters and architrave, quoins. Built 1849–50; demolished 1965. Established 1803 as the first hose company in the United States and in 1857 became the country's first regular fire company to employ a steam fire engine. 1 ext. photo (1965),* 3 photocopies of old ext. photos (c. 1880, c. 1935);* 1 data page (1965).*

Philadelphia Saving Fund Society

(PA-1461), 306 Walnut St. Brick with marble ashlar front, approx. 28' (three-bay front) × 94', two stories (originally one) with Ionic distyle portico in antis with pediment (originally balustrade), flat roof.

Fine example of small Greek Revival bank design. Built 1839–40; Thomas U. Walter, architect; J. Struthers & Sons, masons. Pediment added 1881; rear raised to two stories 1896. Interior of front altered to two stories, remodeled 1950–51; Leopold Hauf, Jr., architect. Founded 1816 as the first savings bank in the United States, Philadelphia Saving Fund Society was here until 1881. Certified, PHC 1957; Pennsylvania Register 1970.

2 ext. photos (1958),* 1 photocopy of old ext photo (n.d.),* 1 photocopy of old engraving (n.d.);* 6 data pages (1959, including 1882–96 insurance surveys, 1907–50 building permits).*

Philadelphia Saving Fund Society
(PA-1462), 700–710 Walnut St., at S.W. corner Seventh St. Brick faced with granite ashlar, approx. 152' (eight-bay front) × 170' (originally approx. 55' × 125' with three-bay front), two stories, flat roof with balustrade, symmetrical facade with two round-arch rusticated entrances, two-story banking room with gallery.

Fine Italianate bank building with complementary additions. Built 1868–69; Addison Hutton of Sloan and Hutton, architect. Rear addition built 1885–86; Addison Hutton, architect. Western addition built 1888; Addison Hutton, architect. Another western addition built 1897–98; Furness, Evans & Co., architects. Interior altered 1919; Furness, Evans & Co., architects. Gallery erected 1925, altered 1927; Mellor, Meigs and Howe, architects. Interior altered 1929–31; Howe and Lescaze, architects. Second headquarters of the bank. Certified, PHC 1957; Pennsylvania Register 1970. 2 ext. photos (1975);* 17 data pages (1959, including 1869 news article and 1888 insurance survey);* HABSI form (1970).*

Philadelphia Trust, Safe Deposit and Insurance Company
(PA-1181), 415 Chestnut St. Brick with marble ashlar front, approx. 50' (three-bay front) × 164', two stories in front, seven stories in rear, flat roof, projecting central bay with grouped pilasters, statuary at cornice, round-arch entrance and windows, rusticated quoins. Built 1873–74; James H. Windrim, architect; William A. Armstrong, builder. Rear raised one story 1898, interior remodeled 1908; John T. Windrim, architect. Rear raised four stories 1922. Demolished 1959. 1 sheet (1965, including elevation);* 2 ext. photos (1959),* 1 photogram-metric stereopair (1959),* 1 photocopy of old ext. photo (c. 1910).*

Philosophical Hall (also known as American Philosophical Society).
Illustrated
(PA-1464), 104 S. Fifth St., Independence National Historical Park. Brick with marble trim, approx. 70' (five-bay front) × 50', two-and-a-half stories, hipped roof, flat lintels with keystones, round-arch entrance with keystone and fanlight, oculus above entrance, central-hall plan.

Restrained Federal style that complements adjacent public buildings. Built 1785–89; Samuel Vaughan, architect. Third floor added 1890; Wilson Brothers & Co., architects; Charles McCaul, contractor. Building restored and third floor removed 1948; Sydney Martin, architect. Founded in 1743 by Benjamin Franklin, the oldest learned society in the United States; reorganized 1769 and officially entitled American Philosophical Society, held at Philadelphia, for Promoting Useful Knowledge. Certified, PHC 1956; Pennsylvania Register 1970; designated National Historic Landmark 1965. 3 ext. photos (1967, 1975),* 6 int. photos (1967);* 14 data pages (including 1828–1908 insurance surveys).*

Provident Life and Trust Company Bank. Illustrated
(PA-1058), 409 Chestnut St. Granite ashlar with marble and tile trim, approx. 40' (three-bay front) × 160', one story, flat roof, gigantic ornamental membering and projecting center bay; one monumental skylighted story inside with decorated iron trusses, tiled walls and floors, notable and unusual ornament, L-plan with a front at 44 S. Fourth St.

One of the most significant works of Frank Furness. Built 1876–79; Frank Furness, architect. Ten-story eastern addition (401–7 Chestnut St.) built 1888–90; Furness, Evans & Co., architects. Extended north on Fourth St. to Ranstead St. 1892, 1895, 1902; Furness, Evans & Co., architects. Buildings on Fourth St., 401–7 Chestnut St. demolished 1945; bank demolished 1959–60.

2 sheets (1962, including elevations, details); 7 ext. photos (1959), 3 int. photos (1959), 8 photogrammetric stereopairs (1959), 1 photocopy of old view (c. 1880); 11 data pages (1960); HABSI form (1962).*

Ralston, Robert, House
(PA-1016), 521 Arch St. Brick, approx. 27' (three-bay front) × 50' with one-story rear addition, three-and-a-half stories on raised basement,

gambrel roof, Tuscan pilaster frontispiece, Greek Revival interior details. Built between 1804 and 1838; altered to office, rear one-story warehouse built 1920; demolished 1959. Ralston was a prominent merchant; George R. Graham, publisher of *Graham's Magazine,* lived here 1843–48. 3 ext. photos (1959, including photo of iron work), 5 int. photos (1959); 3 data pages (1959).

Reliance Insurance Company of Philadelphia Building

(PA-1465), 429 Walnut St. Brick with brownstone front, approx. 22' (three-bay front) × 50' with two- and three-story rear ell, four-and-a-half stories, gable roof with cross-gable and gargoyles at roofline, apartment on top floor. Built 1881–82; Furness, Evans & Co., architects. Interior altered, two-story rear ell addition built 1907; Brockie and Hastings, architects. Demolished 1960. 1 ext. photo (1959),* 2 int. photos (1959),* 2 photocopies of old ext. photos (c. 1895, 1915).*

Reynolds-Morris House. Illustrated

(PA-1107), 225 S. Eighth St. Brick with marble trim, approx. 42' (five-bay front) × 24' with three-story rear ell, three-and-a-half stories, gable roof with modillion cornice, pedimented round-arch entrance with fanlight, marble lintels with keystones, central-hall plan.

Notable Georgian town house. Built 1786–87; restored 1959. Certified, PHC 1956; Pennsylvania Register 1970; designated National Historic Landmark 1967. 9 sheets (1940, including plans, elevations, details) from Survey, Philadelphia Chapter, A.I.A. (1932); 2 ext. photos (1972),* 2 photocopies of measured drawings (1897);* 9 data pages (1960, 1964, including 1830 insurance survey).*

Riggs & Brother, Navigator Statue

(PA-1466), 310 Market St. Painted wooden statue of navigator sighting sextant, approx. 5' height, originally anchored on exterior at second story. A notable example of 19th-century advertising sculpture. Carved c. 1875; Samuel Sailor, woodcarver. Probably moved to Market St. from S.E. corner of Dock and Walnut sts. 1896; dismantled and deposited at Philadelphia Maritime Museum 1973. Riggs & Brother, chronometer and nautical instrument-makers, were in business 1818–1973 and at this site since 1896. Certified, PHC 1962. 1

photocopy of photo (c. 1970);* 6 data pages (1962);* HABSI form (1970).*

Roney, John, House
(PA-1017), 117 N. Sixth St. Brick, approx. 17' (two-bay front) × 30' with three-story rear ell, three-and-a-half stories, gable roof. Built early 19th century; altered to store later; demolished 1959. 1 ext. photo (1959); 3 data pages (1959).

Rowley-Pullman House
(PA-1467), 238 S. Third St. Brick, approx. 20' (two-bay front) × 44' with two- and three-story rear ell, three-and-a-half stories, gable roof, round-arch entrance with fanlight, side-hall town house plan. Built c. 1823–25. Remodeled and notable new facade added c. 1888; Wilson Eyre, Jr., architect. Restored to 1825 facade 1963; Joseph A. Praissman, architect. Certified, PHC 1963. 5 ext. photos (1962, 1963, 1972);* 2 data pages (1961).*

Royal Insurance Company Building
(PA-1468), 212 S. Third St. Brick with stone trim, approx. 26' (four-bay front) × 97', four-and-a-half stories, gable roof, facade dominated by piers ending in pointed arches. Built 1882; demolished 1960. 3 ext. photos (1960).*

Rumpp, C. F., & Sons, Inc., Factory
(PA-1469), 114–30 N. Fifth St., at S.W. corner Cherry St. Buff brick with blue Indiana limestone and terra cotta trim, approx. 83' (five-bay front) × 103' (originally approx. 55' × 103'), seven stories, flat roof, rusticated first and second stories, giant pilasters at third and fourth stories, arches at fifth story, rectangular plan. Built 1893; Hazlehurst and Huckel, architects; John R. Wiggins, contractor. Two seven-story southern additions (122–24 and 114–20 N. Fifth St.) built 1902, 1912; Carl P. Berger, architect. Rear three-story boiler–engine house built 1910; Carl P. Berger, architect. Demolished 1965. Manufacturers of pocketbooks and fancy leather goods since 1850, C. F. Rumpp & Sons were here 1893–1958. 1 ext. photo (1959),* 2 photocopies of old ext. photos (1894, c. 1905).*

Russell Building
(PA-1797), 152–54 N. Front St., at S.W. corner Race St. Brick with iron shopfront and stone trim, approx. 52' (six-bay front) × 48', five

stories (originally four stories), flat roof, rectangular plan. Built c. 1868; ground story altered, raised to five stories between 1935 and 1945; demolished c. 1969. 1 ext. photo (1967),* 1 photocopy of old ext. photo (1919).*

Russell, Captain James, House
(PA-1470), 113 Spring St. (originally 6 Brooks Court). Brick, approx. 16' (two-bay front) × 18' with two-story rear ell (originally tenements), three stories, gable roof, brick stringcourses at second and third stories, probably originally London house plan. Built c. 1752 as house in court extending west between 206 and 208 N. Front St.; excavated around basement to make four-story house c. 1922; demolished 1973. Brooks Court became Augusta Place 1858, Spring St. 1911. 1 ext. photo (1967);* 3 data pages (1960).*

St. Augustine's Roman Catholic Church
(PA-1471), N.W. corner Fourth and New sts. Brick with granite trim, approx. 61' × 152', one story, gable roof, center square three-stage tower with stone quoins and small dome steeple on octagonal lantern, carved limestone volutes between tower and parapet, blind arch on each side of tower, large Doric pedimented frontispiece, three-aisle plan, galleries, rectangular chancel with dome (not original).
    Example of Italian Revival architecture in Palladian mode. Built 1847–48; Napoleon LeBrun, architect; William Moroney, supervisor of construction. Steeple added 1867; Edwin F. Durang, architect. Frescoes by Philip Costaggini 1882, 1884; shrine of Our Lady of Good Counsel added to north side of sanctuary 1891, extensively renovated 1911–12. Interior renovated including dome over sanctuary c. 1895, c. 1923; entrance altered when street was lowered 1924. Three-story brick school built on north side 1870; Edwin F. Durang, architect. Three-and-a-half-story brick and limestone rectory built adjoining southwest corner 1924; Folsom, Stanton and Graham, architects. Certified, PHC 1963. 1 ext. photo (1974).*

St. Charles Hotel
(PA-1472), 60–66 N. Third St. Brick with cast-iron front and interior iron columns, approx. 66' (nine-bay front) × 31', five stories, flat roof, ornamental lintels, balcony at second story originally with ornamental iron railing, ground story originally in one room with three vestibules. Rare surviving early hotel. Built 1851; Charles A.

Rubicam, builder. Ground story altered to store 1920; second floor rehabilitated as offices 1975. 3 ext. photos (1962);* 5 data pages (including 1851–1905 insurance surveys).*

St. George's Methodist Episcopal Church
(PA-1473), S.E. corner Fourth and New sts. Brick, approx. 50' (three-bay front) × 80', three stories on raised basement, gable roof; three-aisle plan, shallow rectangular apse, balconies. Built c. 1769; balconies built 1792; basement added 1804, enlarged 1836; street lowered 1924. St. George's is the world's oldest Methodist church in continuous service. Francis Asbury, first Bishop of American Methodism, preached his first sermon in America here 1771; Richard Allen, founder of the African Methodist Episcopal Church, licensed here as first Negro Methodist to preach in America. Historical center and museum adjoins church. Certified, PHC 1956; Pennsylvania Register 1970; NR. 2 ext. photos (1974),* 3 int. photos (1972);* HABSI form (1970).*

St. James Street Area Study
(PA-1474), 202–14 St. James St., on S. side St. James St. between Second and American Sts. Seven contiguous brick houses, each approx. 16' (two-bay front) × 20', three-and-a-half stories, gable roofs (No. 202 shed roof). Built middle and late 18th century; demolished 1959. 4 ext. photos (1959),* 3 int. photos (1959).*

St. Paul's (Protestant Episcopal) Church (also known as Old St. Paul's)
(PA-1475), 225 S. Third St. Brick with stone trim, approx. 65' (three-bay front) × 90', two stories, gable roof, stone Tuscan pilaster frontispiece, round-arch windows, originally three-aisle plan with galleries.
        Built 1760–61; John Palmer (bricklayer), architect; Robert Smith, builder. Organ installed 1806, its wooden figures carved by William Rush 1812; Palladian window in east end walled up 1822. Altered 1830; William Strickland, architect; interior details by Thomas U. Walter, architect. Altered to three stories of offices, organ figures removed to St. Peter's Protestant Episcopal Church 1904; Duhring, Okie & Ziegler, architects. Since 1904 headquarters of Episcopal Community Services of the Diocese of Pennsylvania. Certified, PHC 1960. 1 ext. photo (1959);*3 data pages (including 1823, 1831 insurance surveys).*

Sansom, William, House

(PA-1476), 707 Walnut St., on N. side Washington Sq. Brick with marble trim, approx. 20' (three-bay front) × 40' with three-story rear ell (not original), three-and-a-half stories, gable roof, side-hall town house plan (originally London house plan).

Built c. 1799; Benjamin Henry Latrobe, architect. Renovated and enlarged in mid-19th century. One of an original row of twenty houses on Walnut St. and twenty-two on Sansom St. known as Sansom's Row; first example of large-scale speculative development in the city. Certified, PHC 1957. 1 ext. photo (1958),* 1 photocopy of architectural plan and elevation (c. 1800);* 6 data pages (1958).*

Schenck Building

(PA-1078), 535–37 Arch St., at N.E. corner Sixth St. Cast iron on south and west facades, brick party walls on north and east facades, approx. 45' (nine-bay front) × 81', five-and-a-half stories with central mansard tower, mansard roof, arcades on all stories, four truncated towers. Built c. 1869–72; Stephen D. Button, architect. Upper stories removed 1938; surviving ground story demolished 1959. 2 ext. photos (1958), 1 photocopy of old ext. print (c. 1875); 3 data pages (1963).

Second Bank of the United States (also known as the Old Customs House). Illustrated

(PA-137), 420 Chestnut St., Independence National Historical Park. Museum. Brick faced with Pennsylvania marble ashlar, approx. 87' (three-bay front) × 161', two stories with Doric octastyle porticoes at each end, gable roof, barrel-vaulted ceiling supported by double range of six Ionic columns in central one-story banking room.

As the first public building based on the design of the Parthenon, the bank was widely acclaimed and established the popularity and viability of the Greek Revival in the United States. The original interior was an outstanding example of classical progression of space. Built 1818–24; William Strickland, architect; John Struthers, marble mason. Rehabilitated 1844; William Strickland, architect. Interior altered 1864–72; John McArthur, Jr., architect. Partially restored by Works Progress Administration 1939–41; partially restored by National Park Service 1964–73. Served as Second Bank of the United States until 1836 and Philadelphia Customs House 1844–1935; since 1975 portrait gallery of eminent early Americans. Certified, PHC 1956; Pennsylvania Register 1970; designated National Historical Site 1939.

19 sheets (1939, including plans, elevations, sections, details, iron gates); 27 ext. photos (1939, 1975,* including 2 of iron gates), 11 int. photos (1939, 1975*), 1 aerial photo (1965),* 2 photocopies of competition drawings (plan, section) by Benjamin Henry Latrobe (1818),* 1 photocopy of elevation drawing by William Strickland (c. 1821).*

Singer, John, Warehouse

(PA-1478), 319½ Market St., at rear of 319 Market St. Brick, approx. 16' (two-bay front) × 58', three-and-a-half stories, gable roof. Built 1806; altered to dwelling c. 1848; reverted to warehouse between 1911 and 1928. Oldest known extant warehouse in the city. Certified, PHC 1970. 1 ext. photo (1972);* 4 data pages (including 1806–81 insurance surveys);* HABSI form (1970).*

Smythe Building

(PA-1479), 101–5 Arch St., at N.W. corner Front St. Brick with cast-iron front, approx. 75' (nine-bay front) × 73' (originally approx. 150' × 73' and 97'), five stories, flat roof, arcade on each story (originally with Corinthian caps), rectangular plan.

Fine example of cast-iron commercial architecture in Northern Italian Renaissance style. One half of original building which extended through 111 Arch St. Built 1855–57; Brown and Allison, carpenters; George Creely, bricklayer; Tiffany and Bottom (Trenton, N.J.), iron work. No. 107 demolished for trolley loop 1913; cornice removed after 1920; No. 101 demolished 1975, its front dismantled and stored by Fairmount Park Commission. Certified, PHC 1976. 1 ext. photo (1966);* 6 data pages (including 1873–95 insurance surveys);* HABSI form (1970).*

Sons of Temperance Fountain

(PA-1480), octagonal cast-iron water fountain with Gothic motifs, approx. 6' high. Made 1876, errected by Grand Division of Sons of Temperance on Centennial grounds at Belmont and Fountain aves.; removed to Independence Sq. 1877–1969; now in storage. 1 photo (1961).*

Souder, Charles F., House

(PA-1018), 514 Race St. Brick, approx. 18' (two-bay front) × 58', three-and-a-half stories, gable roof. Built 1834–35; altered to store later; demolished 1959. 1 ext. photo (1959); 3 data pages (1959).

South Water Street Area Study
(PA-1810), 100-50 S. Water St., W. side Water St. between Chestnut and Walnut sts. Twenty-four contiguous brick warehouses, each approx. 25' (two- or three-bay front) × 27' or 35', four, five, and six stories, gable and flat roofs, rectangular plans. Includes Blight Warehouse (PA-1393). Built early and mid-19th century. 1 ext. photo (c. 1967).*

Southeast Square. See Washington Square Area Study

Spring Garden Insurance Company Building
(PA-1481), 431 Walnut St. Brick with marble facade, approx. 25' (four-bay front) × 42' with three-story rear ell, three-and-a-half stories, mansard roof. Built 1880–81; Theophilus Parsons Chandler, Jr., architect. Extensive interior alterations 1897; Addison Hutton, architect. Demolished 1960. 1 ext. photo (1959),* 1 photocopy of old ext. photo (1915).*

State House of Pennsylvania. See Independence Hall

Stephen Girard's Bank. See First Bank of the United States

Stiles, Edward, House
(PA-1482), 128 N. Front St., at N.W. corner Elfreth's Alley. Brick with marble trim, approx. 23' (three-bay front) × 85' (originally approx. 23' × 42' with three- and two-story rear ell), four stories (originally three-and-a-half), flat roof (originally gable), cast-iron storefront, marble stringcourse and flat-arch lintels, rectangular plan (originally side-hall plan with rear stair tower).
Built 1787–88; altered to store before 1860; enlarged and converted to manufacturing use before 1875; demolished c. 1969. Built for rental income for Edward Stiles, prominent merchant. 1 ext. photo (1967);* 4 data pages (1960, including 1787–88 insurance surveys).*

Stone-Penrose House (also known as Meredith House)
(PA-1483), 700 Locust St., at S.W. corner Seventh St., on S. side of Washington Sq. Brick with marble trim, approx. 27' (three-bay front) × 70', three-and-a-half stories on raised basement, gable roof, side-hall plan, rear addition with veranda. Built c. 1818–23; altered

later. Restored 1967–68; Henry J. Magaziner, architect. Residence of British Consul General since 1968. 3 ext. photos (1975),* 1 photocopy of old photolithograph (c. 1900);* 5 data pages (1959, including 1877 insurance survey);* HABSI form (1970).*

Stortz, John, & Son, Store
(PA-1484), 210 Vine St. Brick with molded galvanized metal trim, approx. 17' (two-bay front) × 48' with three-story rear wing, four stories, flat roof, molded metal cornice and pedimented parapet capped by spread eagle. Built c. 1885; ground story altered later. Owned by family cutlery firm since it was built. 1 ext. photo (1967).*

Tutleman Brothers & Faggen Building
(PA-1485), 56–60 N. Second St., at N.W. corner Cuthbert St. Brick with cast-iron front, approx. 44' (six-bay front) × 50', five stories (originally four), flat roof, articulated front with continuous piers and wide spandrels, scroll-shaped posts on upper stories, center parapet with small swag-decorated pediment, rectangular plan. One of the last cast-iron fronts erected in the country. Built as three brick stores before 1836. Altered to factory, raised one story, cast-iron front built 1900–1901; Thomas Stephen (Camden, N.J.), architect. Ground story altered later. 1 ext. photo (1976).*

U.S. Supreme Court Building. See Independence Hall, Philadelphia City Hall

Vaults
(PA-1486), beneath sidewalks at S.W. corner Front and Walnut sts. Rubble walls and brick-vaulted ceiling, approx. 12' × 40' along Front St., approx. 12' × 80' along Walnut St., approx. 8' height. Storage vaults under sidewalks at basement or sub-basement levels were common in the 19th century but extant examples are rare. Built c. 1840; demolished 1959. 7 photocopies of int. photos (1959).*

Walnut Street Theatre
(PA-1487), 829–33 Walnut St., at N.E. corner Ninth St. Brick faced with marble ashlar, approx. 90' (six-bay front) × 140', two stories, flat roof with balustrade (originally dome), smooth rusticated ground story with Doric colonnade flanked by round-arch niches, second-story windows set in blind arches.

Fine example of the Greek Revival style applied to theater design. Built 1808–9 as circus; altered to theater 1816. Extensively renovated, new front built 1827–28; John Haviland, architect; John Randall (master carpenter), builder. Interior renovated 1852; Joseph C. Hoxie, architect. Interior renovated c. 1903; Willis G. Hale, architect. Exterior, interior renovated 1920–21; William Lee, architect. Exterior restored to 1828 appearance 1970–72; Dickey, Weissman, Chandler and Holt, architects. Interior remodeled 1970–72; F. Bryan Loving, architect. Oldest surviving theater in the United States; began presenting legitimate drama 1811. Certified, PHC 1957; Pennsylvania Register 1970; designated National Historic Landmark 1968.

26 ext. photos (1958, 1969, 1972),* 51 int. photos (1969),* 1 photocopy of 1830 ext. drawing, 1 photocopy of 1831 ext. engraving, 10 photocopies of old ext. photos (n.d., 1865, c. 1890, c. 1900. c. 1905, 1913).*

Washington Hose Company No. 10
(PA-1488), 35 N. Ninth St. Brick with cast-iron front, approx. 15' (three-bay front) × 55', three stories, flat roof, originally arched pediment with bust of George Washington, rectangular plan. Built c. 1852; altered to store after 1872; storefront altered 1918. New storefront installed, upper stories altered to office and dwelling 1939–40; Sidney Laub, architect. New storefront installed, interior altered to store 1946; Walter Russell, architect. 1 photocopy of old ext. photo (c. 1935).*

Washington Square Area Study (originally known as Southeast Square)
(PA-1489), Block bounded by Sixth, Seventh, Walnut, Locust sts. Twenty-one commercial and domestic buildings ranging from the three-and-a-half-story brick William Sansom House (1799) to the thirty-two-story concrete Hopkinson House (1962–63). Includes Athenaeum of Philadelphia (PA-1389), N.W. Ayer & Son, Inc., Building (PA-1390), John Bonsall House (PA-1394), Philadelphia Saving Fund Society (PA-1462), William Sansom House (PA-1476), Stone-Penrose House (PA-1483), and between 1822 and 1939 First Presbyterian Church (PA-1117).

One of four original squares established as parks by William Penn, it served as potter's field and pasturage 1706–c. 1813. Landscaped as park 1816–18; George Vaux (merchant), director; George Bridport (decorative painter), designer. Officially named Washington Square 1825, and residential in nature until after Civil War. Noted for law offices in last quarter of 19th century and for publishing houses after c. 1900. Landscaped 1913–15; Olmsted Brothers and Co.

(Boston), landscape architects. Revived as residential square since 1957. 2 aerial photos (1965),* 2 photocopies of ext. photos (c. 1957, c. 1962),* 4 photocopies of old photolithographs (c. 1900, 1914),* 1 photocopy of 1842 plot map.*

Washington Square Presbyterian Church. See First Presbyterian Church

White, Bishop William, House
(PA-1490), 309 Walnut St., Independence National Historical Park. Historic house museum. Brick with marble trim, approx. 26' (three-bay front) × 44' with two- and three-story ell, three-and-a-half stories, gable roof, pedimented Doric frontispiece, round-arch entrance with fanlight, marble lintels with keystones, large glass panes, side-hall town house plan.

Fine example of a post-Revolution town house. Built 1786–87; ground story altered to commercial use c. 1858, c. 1880, 1946. Restored by National Park Service 1951–63. Bishop White, rector of Christ Church and St. Peter's, was instrumental in restructuring Anglicanism in the form of the Episcopal Church after the American Revolution. Certified, PHC 1957; NR. 1 ext. photo (1974);* 1 data page (including 1795 insurance survey).*

Arch Street Friends Meeting. *Jack E. Boucher photo, 1974.*

Athenaeum of Philadelphia. *Jack E. Boucher photo, 1972.*

John Brock Sons & Co. Warehouse. *Jack E. Boucher photo, 1972.*

Christ Church. *James L. Dillon & Co. photo, 1969.*

Elfreth's Alley Area Study. *Jack E. Boucher photo, 1972.*

Farmers' and Mechanics' Bank. *Jack E. Boucher photo, 1972.*

First Bank of the United States. *Jack E. Boucher photo, 1965.*

Independence Hall, Central Hallway. *James L. Dillon & Co. photo, 1969.*

Insurance Patrol. *Photo c. 1880, courtesy Historical Society of Pennsylvania.*

Independence Hall, Assembly Room tabernacle frame. *Jack E. Boucher photo, 1959.*

Leland Building. *Jack E. Boucher photo, 1965.*

Philadelphia Exchange Company (Merchants Exchange). *Jack E. Boucher photo, 1973.*

Pennsylvania Fire Insurance Company Building. *Cortlandt V. D. Hubbard photo, 1976.*

Philadelphia Bourse. *James L. Dillon & Co. photo, 1976.*

Philadelphia Fire Department, Engine Co. No. 8. *Hugh McCawley, del., 1974.*

Philosophical Hall. *Cortlandt V. D. Hubbard photo, 1967.*

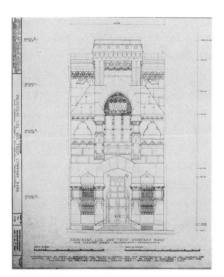

Provident Life and Trust Company Bank. *Photogrammetric drawing, 1959, plotted by Perry E. Borchers 1962, Dellas H. Harder, del., 1962.*

Reynolds-Morris House. *Jack E. Boucher photo, 1972.*

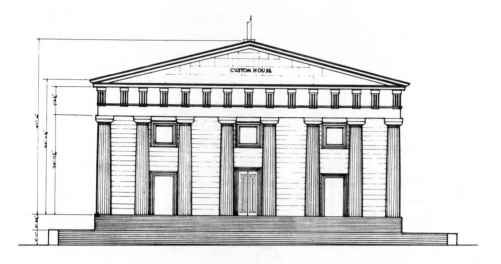

NORTH (CHESTNUT) ELEVATION

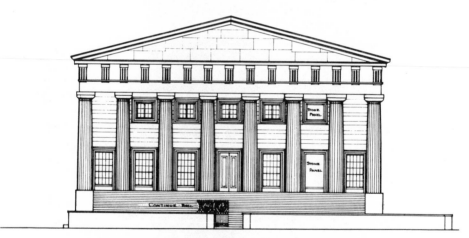

SOUTH (SANSOM ST) ELEVATION

Second Bank of the United States. *A. G. Guttersen, del., 1939.*

Penn's City

Center Square

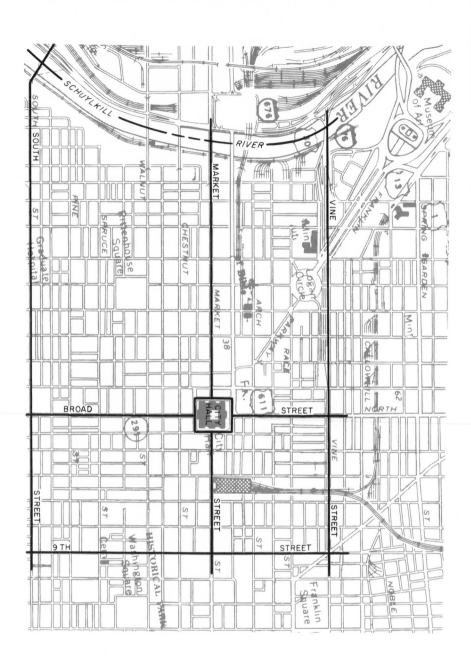

3

Introduction

*P*HILADELPHIA'S CITY HALL (illus.), a marble masterpiece of High Victorian Picturesque Electicism, fills Center Square and visually challenges every edifice in the vicinity. Once the largest structure in the United States, it consumes all of the Square's five acres, and with its tower and statue of William Penn it rises 584 feet, still the tallest building in the city. By erecting City Hall in Center Square, one of the city's original five public places, Philadelphians strategically placed it in the middle of the intersection of their two main thoroughfares, Broad and Market streets. Its location also put it in the geographical center of the original city, a point reinforced by the massive brass compass inlaid in the center of City Hall courtyard. Because of its central location and massive size, City Hall fixes the city's central design scheme between the Delaware and Schuylkill rivers and serves as a beacon for those approaching the city from any direction. It also has contributed to a rare American use of the term "center city," which for Philadelphia makes more sense than the more common American designation, "downtown."

Three decades in the making, City Hall is the paragon of the French Renaissance style that became the rage for public buildings after the Civil War. It was begun in 1871 with John McArthur, Jr., as its chief architect, in association with the highly respected Thomas U. Walter. Both men had died, however, before the building was officially completed in 1901 under the supervision of W. Bleddyn Powell.[1] Its plan is quite simple: a squarish rectangle whose interior court is one of the finest urban public spaces in the country. It is entered through four vaulted portals that are large enough to accommodate buses that pass through them each weekday to deliver prisoners for trial in the building's county and state courts, which occupy more than half of City Hall's fourteen and a half acres of floor space. Its exterior walls are bristling with symbolic sculpture: caryatids, bas-reliefs, statues, all from the designs of Alexander Milne Calder.[2] As Nature abhors a vacuum, City Hall abhors a blank surface; even the windows are filled—with air conditioners. Its spacious interior is equal to its exterior. Many of its finest rooms are the work of George Herzog, the best local interior designer of the day. Gilded ornament and opulent materials, such as polished marbles, encaustic tiles, and mother-of-pearl inlays, become all the more impressive within such great volumes. In fine, City Hall is the central structure of the Center Square area, a part of the original city with arbitrary boundaries at Ninth Street on the east, South Street on the south, Schuylkill River on the west, and Vine Street on the north.

City Hall was not the first structure to occupy Center Square.[3] In 1800, when most of the present Center Square area was undeveloped, the engine house for the city's water system was erected here. Benjamin Henry Latrobe's plans for the waterworks established his national reputation as an engineer and his bold design for the Pump House promoted both the Greek Revival style and his reputation as an architect.[4] After the Fairmount Waterworks on the Schuylkill superseded Latrobe's system, the Pump House was torn down and the square appears to have remained vacant for the next fifty years. At the time of its demolition in 1827 the Pump House was no longer isolated; the city's westward development had extended beyond Ninth Street. The area's earliest construction of note was Franklin Row (illus.), a range of ten dwellings designed by Robert Mills for John Savage and John Meany in 1809. Mills's plan for the development that extended along the west side of Ninth Street north of Locust expanded upon the practice begun a decade earlier in Sansom's Row in the Old City area. Instead of designing each house separately, as was the case of all Philadelphia rows up to that time, Mills treated the entire row as a design project. Through the use of large recessed blind arches over the tripartite windows on the first and second stories, Mills was able to impart an element of plasticity to each facade and a visual unity to the row.[5] The only survivor, the Sims-Bilsland House at 228 South Ninth Street, is being restored for Thomas Jefferson University.

It was Thomas U. Walter and John Haviland, however, who made the Greek Revival an overt part of the row house design and carried the Philadelphia row development to new heights of grandeur. Walter was just emerging as a young architect when he received the commission from John Savage in 1830 to erect another of his rows, sixteen dwellings along the south side of Spruce Street between Ninth and Tenth.[6] The source of its name, Portico Row, originally Portico Square, is obvious. A flat portico supported by three Ionic marble columns graced each of the paired doorways. This scheme gives the development both symmetry and monumentality, no mean feat for speculative housing. Haviland, on the other hand, was well established in his profession when he designed Colonnade Row on the south side of Chestnut Street between Fifteenth and Sixteenth in the same year that Walter was working on Portico Square. By running an Ionic colonnade the length of the ten-house row he was able to achieve even greater grandeur and continuity.[7] While Colonnade Row was replaced by commerical buildings long ago, Portico Square still stands as an example of how regal the row house can look.

When fashionable Philadelphians began moving west of Broad Street into the Rittenhouse Square neighborhood in the mid-nineteenth century, the row development followed. Some of these later examples, such as Samuel Sloan's 1856 Harrison Row on Locust Street east of Eighteenth, were as visually impressive as the Walter and Haviland examples.[8] Harrison Row is gone now, however, and more representative of the mid-century housing construction is the brick row on the south side of Delancey Street between Eighteenth and Nineteenth. The ten easternmost dwellings, 1800–1818 Delancey Street, were built with a common plan and design between 1854 and 1856 by different carpenters for three different clients, who promptly sold the completed houses to gentlemen, attorneys, and merchants.[9] The western end of the block was similarly developed from somewhat different plans in 1857 for John McCrea, gentleman.[10] After a hundred and twenty years the block retains much of its old character and has been enlivened with some later "modernizations," particularly the delicate detailing and carved caryatids by Eyre and McIlvaine on Number 1810 and the robust Georgian Revival front by Wilson Eyre, Jr., on Number 1824.[11]

After the Civil War, real estate developers in the neighborhood became more style-conscious, probably in response to the tastes of their prospective clients. Row developments of the 1870's and 1880's appear to be the work of architects rather than builders or house carpenters. The twelve dwellings on the north side of Delancey Street between Twenty-third and Twenty-fourth, for example, are stylistically quite sophisticated, in the up-to-date High Victorian Gothic. They were built c. 1875 as a carefully conceived unit with contrasting colors, textures, and silhouettes.[12] A four-story brick and brownstone house stands at each end, and four of them anchor the center of the block, flanked by three two-and-a-half-story dwellings of random Trenton sandstone. This composition is an improvement over the repetitive brick rows to be found in less affluent parts of the city.

In addition to the attractive rows in the Rittenhouse Square neighborhood were many fine individual dwellings. Each decade of the last half of the nineteenth century seems to have spawned a new generation of domestic architecture. Some of the finest examples of the 1850's and 1860's have fallen to later development. Joseph Harrison's magnificent Italian mansion, for example, designed by Samuel Sloan in 1855, on the east side of Rittenhouse Square gave way to an apartment building in 1925.[13] Yet some brownstones can still be found along Locust Street. One of the finest is the four-story town house at 1618 Locust, built c. 1850 for Charles C. Harrison. It is particularly interesting because it has a bay window added in 1888 from the arabesque designs of Wilson Eyre, Jr.[14] One of Philadelphia's finest town houses of the 1870's is the 1875 Thomas Hockley House at 235 South Twenty-first Street. Frank Furness, who was just emerging as the leading and most innovative local architect of his generation, applied to this High Victorian Gothic house expressive brick patterns and bold details to make a heavy and uncommon design all the more massive and eccentric.[15]

The fashion *au courant* in the 1880's is exhibited in the Scott-Wanamaker House (illus.) at 2032 Walnut Street. The popular and prolific Theophilus Parsons Chandler, Jr., designed this Jacobean Revival mansion in 1883 for James P. Scott, head of a sugar-refining firm. In 1894 the great department-store pioneer John Wanamaker purchased it and made it his city home until his death in 1922.[16] The original exterior has remained intact and its richly textured granite front with bays, minarets, and intricately carved balustrades makes it an architectural gem of its era. Although the upper floors were altered to make apartments in the 1930's and elements of the ground story were vandalized in 1972, its present owners are restoring the main floor and the massive open stairway to their former grandeur. Beaux-Arts Classicism was the vogue of the 1890's, and one of its most patrician examples is the Fell–Van Rensselaer House at the northwest corner of Eighteenth and Walnut streets. Peabody and Stearns of Boston, arbiters of architectural good taste of the time, were com-

missioned by Mrs. Sarah Drexel Fell in 1896 to design this Renaissance palace. [17] Mrs. Fell was the widow of the president of A.
Pardee & Company, a leading Pennsylvania coal operator, but shortly
after the house was completed in 1898 she married Alexander Van
Rensselaer of the baronial New York family. [18] Although the Scott-
Wanamaker House and the Fell–Van Rensselaer House are stylistic
opposites—the stippled granite walls and medieval elements of the
former are diametrically opposed to the smooth marble surfaces and
classical details of the latter—they have shared similar fates. Both
houses have changed very little on the outside and very much on the
inside. Until it was gutted in October, 1974, the Fell–Van Rensselaer
House had one of the richest residential interiors in the city. All that is
left is the stained-glass dome in the entrance hall and the spectacular
Doge Room, a dining room whose ceiling is filled with ninety-four
portrait medallions of Venetian doges. [19] Nearly all of its original interior may be gone, but the building still stands on the corner of
Rittenhouse Square, a marble monument to the luxurious life of
inherited wealth and high fashion in the days before income taxes
and regulatory commissions.

Churches were always highly visible parts of Victorian communities, and the status of the community usually determined the sect, if
not the style of the churches. The upper strata were generally Epis-
copalians and Presbyterians, and below them were the Baptists and
Methodists. [20] Consequently, it is not surprising to find a plethora of
Episcopal and Presbyterian churches in the Center Square area. One
of the earliest examples is the 1822 St. Stephen's Protestant Episcopal
Church at Tenth and Ludlow streets. To worshipers St. Stephen's is
best known for its healing services and work with the sick. To architectural historians it is recognized as an early and significant Gothic
Revival design by an architect best remembered for his fine Greek Revival works, William Strickland. [21] To most Philadelphians, however,
it is an unchanging center city landmark. Its twin octagonal towers
with their crenelated fringes and connecting granite screen have been
part of the streetscape for more than 150 years. The same cannot be
said about the interior, which has a history of many redecorations,
additions, and alterations, including a number of notable pieces of
sculpture and stained glass.

St. Mark's Protestant Episcopal Church at 1625 Locust Street
serves as another chapter in the architectural and religious history of
Center Square. St. Mark's was formed in June, 1847, as an early
American response to the religious revival that began in the Anglican

faith during the 1830's. This movement attempted to restore to Anglican practice elements of Roman Catholicism that had been purged during the Reformation. Directed by the Camden Society, later called the Ecclesiological Society, the movement identified Gothic architecture with "High Church" rituals. Accordingly, the architect of St. Mark's, John Notman, was directed to design an archaeologically accurate church for the congregation. The result, completed in 1851, is one of America's finest Gothic Revival churches in the English decorated and perpendicular styles.[22] The architectural history of the church did not end with Notman's masterpiece. Of the later additions the most notable is the sensitive rendering by Cope and Stewardson of the Lady Chapel at the east end of the church.[23] As in the case of medieval abbeys there was no attempt to hide the fact that it was an addition; its version of the Gothic, executed in red sandstone, is distinctive but not discordant. The magnificence of the interior of both the church and the chapel is enhanced by one of the country's finest collections of ecclesiastical treasures. Of these the most splendid is the 1908 sterling silver altar in the Lady Chapel.

Notman designed two other Episcopal churches in the area, St. Clement's (illus.) at Twentieth and Cherry streets and Church of the Holy Trinity at Nineteenth and Walnut streets, both completed in 1859.[24] St. Clement's spire was dismantled ten years later,[25] and the church is not as imposing as Notman had intended; yet its bowed chancel and Romanesque arcade remain to impress the careful observer. Holy Trinity still has its tower and although it no longer looms over its neighborhood, since high-rise apartment buildings are all about it, it overshadows them architecturally. Holy Trinity has long been considered a permanent fixture on Rittenhouse Square. Other buildings have disappeared with barely a whimper, but when it was rumored in 1968 that Holy Trinity was to be replaced by a thirty-three-story apartment-office tower, parishioners, preservationists, neighbors, and strangers rallied to its cause.[26] As a result, this seminal work of the Romanesque Revival still graces Rittenhouse Square.

In spite of the preponderance of Episcopal churches, the Center Square area was not an Episcopal ghetto in the nineteenth century. In 1853 the First Baptist congregation began its brownstone church from the designs of Stephen D. Button at the northwest corner of Broad and Arch streets, and in 1869 the Arch Street Methodist Episcopal Church was erected diagonally across the intersection.[27] The Baptist Church was demolished in 1898 to make room for the UGI Building,[28]

and the congregation now worships in Edgar V. Seeler's Byzantine
gem at Seventeenth and Sansom, but Arch Street Methodist still
gleams in the shadow of City Hall. Presbyterians built earlier and
farther west. Three fine examples of their churches are still standing:
Joseph C. Hoxie's 1853 West Arch Street Presbyterian (illus.) at
Eighteenth and Arch, John McArthur's 1854 West Spruce Street
Presbyterian at Seventeenth and Spruce (now Tenth Presbyterian),
and Henry Sims's 1869 Second Presbyterian Church (now First
Presbyterian) at Twenty-first and Walnut.[29] The prefix "West" for the
former church differentiated it from the Arch Street Presbyterian,
which once stood near Tenth and Arch, and where, according to
romantic theory, the young Emily Dickinson swooned over its
minister, the Reverend Charles Wadsworth.[30] Romantic tales of unre-
quited love and secret poets, however, are not part of the lore of
Joseph C. Hoxie's West Arch Street church. It does not need them to
be noteworthy; its architecture alone does that. It was built at a cost of
$100,000,[31] a great sum for a church in its day, but it brought results.
Praised for its architectural beauty and convenience in 1855,[32] and, although its bell towers and central cupola are now gone, it still has
probably the richest Italianate interior in the city. Part of the reason
this well-proportioned church is often overlooked by Philadelphians
is that it is dwarfed by its neighbor at Eighteenth and Race streets, the
Cathedral of SS. Peter and Paul. Three architects, John Notman, John
T. Mahoney, and Napoleon LeBrun, and two priests, Mariano Maller
and John B. Tornatore, worked on the cathedral's plans, and all can
take some credit for this great brownstone pile.[33] No one, however,
may care to take credit for the interior renovations, lateral protrusions, and northern chapel addition of the 1950's, since they have
destroyed the cathedral's former symmetry.

The fashionable mid-nineteenth-century neighborhood that built
so many splendid churches also supported a host of social and
cultural institutions. Many of them can still be found extending along
Broad Street north and south from Center Square. One of the earliest,
at the corner of Pine Street, is the former Pennsylvania Institution for
the Deaf and Dumb (illus.), which in spite of many later additions still
suggests the spirit of the ancient Greek style. It was built in the 1820's
from the designs of John Haviland[34] when the area was sufficiently
underdeveloped to possess qualities of a pastoral countryside, which
was considered the ideal environment for such well-intentioned
projects. Probably the best known of Philadelphia's cultural institutions is the Academy of Music (illus.) at Broad and Locust streets.

World-famous for its acoustics since its opening in January, 1857, the Academy was the work of LeBrun and Runge, who won the architectural competition in 1854.[35] Marble facing was intended for the exterior walls but a shortage of funds caused the directors to postpone this final step,[36] and like so many temporary solutions, it has become permanent. Consequently, the "Old Lady of Locust Street," as the Academy is affectionately known, remains clothed in modest brick. Yet its lavish interior and unsurpassed acoustics make it a suitable home for the Philadelphia Orchestra, one of the world's greatest. Perhaps not as well known but equally significant is the Pennsylvania Academy of the Fine Arts (illus.) at Broad and Cherry streets, the nation's oldest art school and museum. The Academy is a notable expression of America's High Victorian Gothic and the most important and best-known work of Furness and Hewitt, a Philadelphia architectural partnership that ended in 1875, a year before the Academy's completion.[37] Both men, Frank Furness and George W. Hewitt, continued their energetic work throughout the last quarter of the century in other partnerships. The exterior of the Academy was cleaned in 1966, exposing to even the most casual observer a fascinating variety of forms, materials, and decorations. A restoration project completed in 1976 brought back the same richness to the interior and replaced some missing elements on the front. No nineteenth-century theaters have survived on Broad Street, but an extant example in the area is the former Arch Street Opera-House at 1005 Arch Street. It was rather casually named, since it was originally built in 1870 for minstrelsy, which made its conversion to a burlesque house in 1896 barely noticeable.[38] It now holds the dubious distinction as the oldest, if not the only, burlesque theater in Philadelphia, but its long tenure in this barely remunerative enterprise has preserved it from ruinous remodelings.

Clubs and fraternal societies were also part of nineteenth-century social life. The oldest and most prestigious of the city social clubs is the Philadelphia Club at Thirteenth and Walnut streets.[39] Built in 1836 as a city mansion and the club's headquarters since 1850,[40] the building's scale is so great and its condition so pristine that many casual observers mistake it for a twentieth-century example of the Georgian Revival. Architecturally more conspicuous is the Union League of Philadelphia at Broad and Sansom streets. The Union League was founded in 1862 as a semipolitical organization of businessmen and civic leaders who pledged "unqualified loyalty to the government of the United States and unwavering support of its measures for the suppression of the Rebellion." In this effort they raised a half-million dollars and 10,000 volunteers for the Civil War's battlefields. Originally a third of its members were Democrats, but in 1898 it became explicitly a Republican club, allowing the Union League to call itself the oldest Republican Party club in continuous existence in the United States.[41] John Fraser designed the house in 1864 and it was completed the next year as one of the city's early examples of the new "Mansard Mode."[42] Its striking but lugubrious brick and brownstone eminence inspired the late Philadelphia wit Alfred Bendiner to suggest that club members hang black crepe to cheer it up.[43] A large granite and steel-frame annex was built at the rear between 1908 and 1911 from the designs of Horace Trumbauer,[44] but the Broad Street front has changed little. The ornamental iron roof fringe and the tower's mansard were removed during the 1920's, but everything else is intact, including one of the handsomest sweeping stairs to be found anywhere.

Equally prominent for its architecture and history is the Philadelphia Masonic Temple at Broad and Filbert streets. James H. Windrim won the architectural competition in 1868, but it was five years before the construction ended, primarily because all the stone had to be cut, squared, and marked at the quarries according to Masonic tradition. The design is a fine academic rendering of the Norman style and the asymmetrical towers and splendidly carved entrance porch are especially noteworthy. Yet its reputation as one of the world's greatest Masonic temples rests more on its rich interior, which took more than fifteen years to complete. Great rooms are decorated in different historical styles intended as symbolic instruction in the principles of Freemasonry.[45]

Other notable clubs and fraternal headquarters also stood along Broad Street, such as the Odd Fellows Hall at Race Street, the Manufacturers Club at Walnut Street, and the Philadelphia Art Club at Chancellor Street. Since it was built as a high-rise office building, the Manufacturers Club building has a bright future, but the others are gone. The last to fall was the Art Club. This notable Italian design was the first major work of Frank Miles Day, who won the architectural competition in 1887 shortly after returning from Europe.[46] The club fell victim to the Great Depression, and many present-day Philadelphians knew the building only as the Keystone Automobile Club, which had its headquarters there between 1946 and 1968.[47] After that, it endured long periods of vacancy broken by occasional occupancies, often as political campaign headquarters. Finally in late 1975 came the final blow, one suffered by so many historic buildings: the Art Club was demolished to make space for a parking lot.[48]

As churches and clubhouses went up along Broad Street and to the west, commercial Philadelphia started to edge into the Center Square area, primarily along Market and Chestnut streets. Partly because a railroad did not run down the middle of Chestnut Street as it did down Market Street, Chestnut Street attracted greater investment in commercial structures, and its stores became known for their carriage trade. Except for the chaste Italianate stores at 920–22 Chestnut Street, designed by Stephen D. Button in 1853,[49] examples of the 1850's have been replaced by later buildings. Many fine stores of the 1860's and 1870's could be found in the 1000 block of Chestnut Street until the greater part of its south side was torn down in 1975.[50] Still standing at the northwest corner of Tenth and Chestnut, however, is the imposing New York Mutual Life Insurance Company Building, or Victory Building, as it is almost universally known about the city. One of the country's largest insurance companies after the Civil War, New York Life retained New York architects to design its many office buildings across the land. In 1873 Henry Fernbach of New York was that architect, and it was he who was responsible for the French Renaissance design of the initial four-story office building.[51] Philip W. Roos of New York was the company's choice for architect in 1890, when three stories were sandwiched between the massive French roof and the lower three stories, accounting for its present appearance.[52]

Many Philadelphians were surprised and perplexed when John Wanamaker converted the old Pennsylvania Railroad Freight Depot at Thirteenth and Market into the Grand Depot Store in time for the centennial celebration. It seemed then to be on the western outskirts of the city's shopping area. Indeed it was. Wanamaker, however, had observed the westward movement of the commercial district since he had opened his Oak Hall Store at Sixth and Market less than fifteen years earlier, and he suspected that the erection of the new City Hall in Center Square would stimulate this shift.[53] Also, he evidently expected that the removal of the rails from Market Street (which made the Freight Depot available, of course) would enhance its value as a retail street. So Wanamaker both anticipated and contributed to the continuation of the westward direction of the retail district.

Two other responses and stimuli to the area's commercial development were the Broad Street Station (illus.) at the northwest corner of Broad and Market and the Reading Terminal at the

northeast corner of Twelfth and Market.[54] The Pennsylvania Railroad's Broad Street Station was completed in 1882, but the greatest
commercial development west of Broad Street coincided with the station's enlargement in 1892–94.[55] Nearly all these early skyscrapers,
such as Cope's and Stewardson's Harrison Building (illus.) and Furness, Evans and Company's Arcade Building at opposite corners of
Fifteenth and Market,[56] have been replaced by the extensive Penn
Center renewal project. The opening of the Terminal Station of the
Philadelphia and Reading Railroad in 1893 set off a similar flurry of
building activity on both sides of Market Street between Ninth and
Thirteenth streets. One of the finest was the six-story Monarch Building at Tenth and Filbert. It was designed by Cope and Stewardson as
twelve separate stores,[57] unified by a single monumental facade
whose classical ornament was sensitively scaled for the passing
pedestrian, while the deeply recessed windows punched out of the
plain brick upper stories emphasized its sense of mass. The greatest
of all, however, is the terminal itself. The Head House, whose florid
Renaissance style is executed in rich cream-colored terra cotta on
buff brick, is a peerless example of railroad grandeur, executed for a
company that went bankrupt within months after the station
opened. The cavernous train shed (illus.) extending from the rear of
the Head House to Arch Street is even more important because of its
engineering significance. Fundamental to its construction was the
three-hinged arch, which allowed engineers to bridge great distances
and bring under a single span unprecedented expanses of clear space.
Such single-span train sheds were not uncommon at the turn of the
twentieth century, and its architects and engineers, Wilson Brothers
and Company, built an even larger one for the Broad Street Station.[58]
Since the Broad Street shed burned in 1923, the rarity of the Reading
Terminal's shed has made it all the more valuable.

These building ventures continued into the early years of the
twentieth century, virtually remaking the center of the city. Some of
the offices and hotels are now gone, but many of the best still stand.
Among them is the Land Title Building by D. H. Burnham and Company of Chicago, one of America's great architectural firms and a
leader of commercial design at the time. The first stage, built in 1897–
98 at the southwest corner of Broad and Chestnut streets, is a fine
example of what became known in the 1890's as the Commercial
style.[59] This style had its roots in mid-nineteenth-century Philadelphia and New York but had come into its own in Chicago after 1890.
In fact, the first stage of the Land Title Building bears a close resem-

blance to the Fisher Building, which Burnham had built in Chicago a year earlier.[60] To maximize light and space Burnham designed bay windows into the steel frame of both buildings, but while in the Chicago building a balance of horizontal and vertical lines is maintained, in the Philadelphia example the frame construction was accentuated by recessing the spandrels to suggest a continuous upward thrust of the piers that end in arches at the next-to-top floor. Burnham also designed the twenty-two-story addition at the northwest corner of Broad and Sansom.[61] Built in 1902, it departs from the earlier Commercial style and incorporates classical elements into the facade, placing a range of giant pilasters along the ground story and giant engaged Corinthian columns at the top of the building. Nevertheless, Chicago influenced the later design as much as the early ones, for Chicago's Columbian Exposition in 1893 had turned the attention of many architects to an academic classicism. For want of a better term, the handsome 1902 section might be called a Neoclassical Revival skyscraper.

Burnham's other great work in Philadelphia is the John Wanamaker Department Store. Built in three stages between 1902 and 1910, this enormous building fills an entire city block, between Thirteenth and Juniper and Market and Chestnut streets.[62] Here Burnham turned for inspiration to the Renaissance Revival with great success. The traditional Renaissance division of a building into base, body, and top resolved the problem of organizing architectural elements over twelve stories, allowing the firm to bring to bear its experience and skill in handling familiar architectural motifs. At the lower level giant pilasters cover steel piers and form the necessary sense of support for the large show windows, and giant Tuscan columns mark the recessed entrances. The body of the four facades is composed of rusticated limestone ashlar with well-spaced fenestration, and to prevent top-heaviness the last three stories are finished in dressed ashlar and relieved by a two-story arcade and small square windows below the cornice. The large-scale grace of the exterior is continued inside. The five-story court is one of the city's grandest interior spaces. Its organ, the world's largest when acquired from the 1904 Louisiana Purchase Exposition in St. Louis, has been enjoyed by generations of Philadelphia shoppers, and its 2,500-pound bronze eagle, from the same exposition, has marked a favorite rendezvous since its installation in 1911.[63] The richness of the Egyptian Hall and Greek Hall has been diminished by alterations, but the Crystal Tea Room remains virtually unchanged as one of the city's largest and most elegant public dining places. Philadelphia can be grateful to John Wanamaker for his good taste. His store remains the handsome offspring of a long relationship in which for sixty years citizens and merchant patronized each other. The local prince of merchandising got his Renaissance palace and the people of Philadelphia got an architectural monument that can be enjoyed with no purchase necessary.

Of all Center Square's new-century buildings probably the most eye-catching is the Girard Bank (illus.) at the northwest corner of Broad and Chestnut streets, not because it is so large but because it is so small and chaste, making it stand out among its tall neighbors. It is Philadelphia's best example of the Neoclassical Revival traditionally claimed by the noted New York firm of McKim, Mead and White. Construction was begun in 1905 under the supervision of Allen Evans of Furness, Evans and Company. When it was completed in 1908, a glistening white marble version of Rome's second-century Pantheon graced the center of the banking district.[64] To base the design of a twentieth-century bank on an ancient temple to pagan gods suggests satirical allusions that the architects probably never intended. Instead they almost certainly were following the nineteenth-century architectural tradition of employing an historical style that would suggest a bank's fiscal stability, an architectural venture into the subliminal that seems to have done little in preventing bank failures. It was a practice continued well into the twentieth century, even after the infamous Crash of 1929.[65]

Probably the best of the later classically derived banks is the Federal Reserve Bank of Philadelphia at the northeast corner of Tenth and Chestnut streets, built in 1932–35.[66] It is a fine example of the stripped classicism of Paul P. Cret, who in 1903 was induced to leave his studies at the Paris Ecole des Beaux-Arts, then the fountainhead of academic classicism, to come to Philadelphia, a city that at the time seemed addicted to this architectural mode. The decision was never regretted by Cret or by Philadelphia, or by hundreds of his devoted architectural students at the University of Pennsylvania.[67] As in much of Cret's work, the exterior design of the Bank is remarkably simple. It is composed of two functional units: engaged piers and a simplified classic cornice on the lower part suggest a classical temple in which the large, carefully crafted banking room is located; and the repetitive fenestration of the four-story upper section marks it as office space. Further, Cret's use of Boittau's sculpture makes the building's symbolism as clear as does its design. The bas-relief to the left of the main entrance depicts the Federal Reserve Bank as advisor of the country's banking system, a figure representing the institution rests upon the shield of Athena, the goddess of wisdom, and looks to her idol for guidance while a rugged oak, the symbol of strength, fills the background. On the right, emblems of agriculture, industry, and trade suggest the foundations of national wealth, and overhead hovers a spread eagle, indicating the national scope of the Federal Reserve System.[68] The symbolism became incongruous, however, when the Federal Reserve Bank in 1976 moved into the new Federal Building at Seventh and Arch, a collaborative high-rise that for better or worse forsakes Philadelphia's tradition of bank architecture.

At the same time that Cret was updating Philadelphia's classical tradition at Tenth and Chestnut, two other architects were striking out in a new direction at Twelfth and Market. In 1929 George Howe, a Philadelphia architect with important social connections, and William Lescaze, a Swiss-born architect with European training and a New York practice, formed a partnership with the express intention of designing a new office building for the Philadelphia Saving Fund Society to be modeled on "the new system" then being practiced in Europe by such men as Walter Gropius, Mies van der Rohe, and Le Corbusier.[69] Even before its completion in 1932 it was being hailed as the country's only skyscraper that was thoroughly dedicated to the principles of what was then becoming known as the International style. By way of contrast, one of the best of its contemporaries, the nearby Market Street National Bank Building (illus.), with its modish Art Deco elements, was modernistic rather than modern. Howe's and Lescaze's building has worn well, and although it is now over forty years old, it still looks up-to-date, a tribute to the architects and to the bank's directors, whose architectural judgment over the institution's 160-odd years has compared favorably with their financial wisdom. Philadelphians have been calling it the PSFS Building ever since 1933, when its twenty-seven-foot red neon sign was perched atop its thirty-two stories.[70] A precursor of supergraphics, the sign remains one of the best examples of architectural advertising to be found anywhere; it is often imitated but rarely equaled.

While the PSFS Building is universally recognized as a trailblazer in American architecure—and accordingly was proclaimed the "Building of the Century" by the American Institute of Architects at their centennial in 1969—the Benjamin Franklin Parkway on the other side of Center Square is similarly acknowledged as a successful pioneer venture in urban planning. The first plans for the parkway

were submitted in 1907 by Paul P. Cret. This should be expected since
the nation was then swept up in the City Beautiful movement that
had spun out of the Columbian Exposition, and Cret's credentials
as a Beaux-Arts architect were impeccable. Cret was not the final
designer, however. That distinction went to another Frenchman, city
planner Jacques Gréber, who was indebted to a considerable degree
to a host of local architects and art-conscious organizations who had
contributed plans and opinions for nearly twenty years before Gréber
got the nod. [71] The end result was a bold diagonal cutting across the
city's restrictive grid plan, extending northwest from the appropriately French Renaissance City Hall through Logan Square, which
was converted to a circle for ease of traffic flow, to the Neoclassical
Revival Museum of Art on Fairmount, Philadelphia's own Acropolis.
The parkway's earliest buildings, with low, classically derived
facades parallel to the street, are compatible with its baroque design.
Its post–World War II architecture, on the other hand, reflects the
newer idea of buildings as large, isolated volumes surrounded by
their own green space and oriented to one another rather than to the
street. [72] Although these recent high-rise buildings violated the
designers' original intent, the parkway's great width, combined with
its abundant plantings and strategically placed fountains, enables it to
hold its own as one of the nation's most spectacular and successful
examples of baroque city planning.

The most recent urban renewal project in the Center Square area
is Penn Center, a glass and steel menagerie for button-down businessmen and their minions. It was begun in 1956 on the fourteen-acre
site of the Broad Street Station and its Chinese Wall, the elevated
tracks that had carried trains and its own peculiar form of blight into
the heart of the city. Lined with parking lots and decaying stores, the
grim brick wall was less a fortress than a noose, strangling the city's
westward commercial development. [73] When it was first proposed
about 1952, the city hoped to stimulate private investment in the
project by tearing down City Hall, then considered "dismal and
depressing with an unfortunate psychological effect on all people
who use it," and replace it with a "new and inspiring municipal
building" on Reyburn Plaza. [74] The Municipal Services Building was
indeed erected on Reyburn Plaza during the early 1960's and, while it
may not be inspiring, it is well designed. [75] City Hall, however, was
spared. It was found that cleaning the exterior and painting the interior was infinitely cheaper than demolition. [76] What psychological
effects this has had on its occupants is still open to question. Because

the Penn Center project was considered a high risk during its early years, the initial buildings are of the glass-slab genre, but later structures, particularly Vincent G. Kling's IBM Building at Seventeenth and Market and his Transportation Building at Eighteenth and Market, exhibit a more sanguine spirit and consequent investment. Most successful are the recently built office towers that came in the wake of Penn Center's financial achievements. One of the best is the Centre Square Towers at Fifteenth and Market, a Kling venture into the New Brutalism mode. It opened in 1974 on the site of the former Harrison Building (illus.) at the southwest corner of Fifteenth and Market, and shows that urban renewal does not necessarily mean uglification. Under the geodesic dome of its lobby bright brass and unfinished concrete contrast with great effectiveness. The handling of Penn Center's open space has followed a similar historical sequence. Although no one equates Penn Center with New York's Rockefeller Center, improvements have been made in recent years and it is now considered by some one of the more successful examples of extensive urban renewal. In the final analysis, the plan's one great virtue was the decision to work around City Hall, the city's focal point, which is where we came in.

Catalog

Academy of Music (originally known as American Academy of Music).
Illustrated
(PA-1491), 232–46 S. Broad St., at S.W. corner Locust St. Brick with sandstone trim, approx. 132' (seven-bay front) × 220', three stories, hipped roof, round-arch entrances and windows, cast-iron bracketed cornice, parapet with 1857 date stone, octagonal ventilation cupola, highly ornate interior.

Fine example of the Italianate mode with a neobaroque emphasis on interior. Built 1855–57; Napoleon LeBrun and G. Runge, architects; Charles Conrad, superintendent of construction; John D. Jones, contractor; Robert Wood, iron work. Planned exterior marble sheathing never erected. Interior altered 1940–41; Silverman and Levy, architects. Restored 1957–66; Robert A. Class of Martin, Stewart, Noble and Class, architect. Home of the Philadelphia Orcestra since 1900, the Academy is the oldest musical auditorium in the United States still in use and remains famous for its acoustics. Certified, PHC 1956; Pennsylvania Register 1970; designated National Historic Landmark 1963. 3 ext. photos (1957, 1963),* 15 int. photos (1967),* 2 photocopies of architects' drawings (1856),* 1 photocopy of architectural competition elevation by Stephen D. Button (1854);* 2 data pages (1965).*

Academy of Notre Dame
(PA-1492), 208 S. Nineteenth St. on W. side of Rittenhouse Sq. Brick with brownstone trim, approx. 53' (three-bay front) × 70' with four-story rear addition, three stories, flat roof. Built c. 1867. Four-story rear addition built 1896; Edwin F. Durang, architect. Demolished 1969. 4 ext. photos (1969);* 1 data page (1969).*

American Academy of Music. See Academy of Music

Arcade Building (also known as Commercial Trust Building)
(PA-1493), 1428–34 Market St., composing block bounded by Market, Broad, Fifteenth sts., S. Penn Sq. Steel frame with coursed dark ashlar and terra cotta curtain walls, aprox. 120' (twelve-bay front) × 230' (originally 50' on Market St. × 230' on Fifteenth St. × 120' on S. Penn Sq.), thirteen stories with twenty-one-story tower, flat roof, originally L-plan, later developed into U-plan, pedestrian arcade along two sides, pedestrian bridge at second story over Market St. to Pennsylvania (Broad St.) Railroad Station.

Rare example of a highrise building designed to accommodate pedestrian traffic and to incorporate later expansion as adjoining properties were acquired. Built 1900–1902; Furness, Evans & Co., architects; Samuel Hart & Sons, contractors. Enlarged 1904–1906, tower added 1913, further additions until c. 1930; Furness, Evans & Co., architects. Demolished 1969. 7 ext. photos (1962–63, 1964),* 3 int. photos (1962–63),* 1 photocopy of architectural rendering (c. 1901);* 9 data pages (1962).*

Arch Street Methodist Episcopal Church
(PA-1494), S.E. corner Broad and Arch sts. Brick faced with coursed Massachusetts white marble ashlar, approx. 75' (three-bay front) × 96' with two-story rear random marble ashlar chapel (approx. 38' × 76'), one story, gable roof with cross-gable of chapel, three-stage square corner entrance and bell tower with marble pinriacles and spire at northwest corner, pointed-arch windows with flanking buttresses, large pointed-arch window with tracery on front of church (north end) and chapel, aedicule at each end; four-aisle plan with galleries suspended on carved brackets, Gothic vaulted ceiling.

Fine "decorated" English Gothic Revival church whose landmark qualities have been enhanced by creation of open space about City Hall during the 1950's and 1960's. Chapel built 1864–65; Addison Hutton, architect. Church built 1869–70; Addison Hutton, architect. Chapel remodeled, new entrance door installed 1956; Charles L. Bolton, architect. 1 ext. photo (1974);* 2 data pages (1964).*

Arch Street Opera-House (now known as The Troc)
(PA-1495), 1003–5 Arch St. Brick, approx. 48' (three-bay front) × 170', two stories, flat roof, large round-arch windows with

ornamental lintels, originally lyre-sculpture atop cornice, notable interior details. Built 1870; burned, rebuilt 1872; interior burned, rebuilt 1883. Interior altered 1891, 1894, 1896; George W. Plowman, architect. Built for minstrelsy; burlesque house since 1896. Name changed to Continental Theatre 1888, The Trocadero 1896, The Troc 1949. Certified, PHC 1972. 1 ext. photo (1973),* 1 photocopy of old ext. photo (1916).*

Arch Street Presbyterian Church. See West Arch Street Presbyterian Church

Baltimore and Ohio Railroad Station
(PA-1220), S.W. corner Twenty-fourth and Chestnut sts. Brick with iron girder and brownstone and copper trim, approx. 173' (twenty-bay front) × 132', three-and-a-half stories, hipped and gable roofs, porte-cochere and picturesque clock tower mark entrance, iron-beam arcade at lower level; irregular plan on two levels, ornamental iron and brick interior.

Notable urban railroad station whose eclectic design made it a landmark. Built 1886–88; Frank Furness of Furness, Evans & Co., architect. One-story steel-frame addition built 1912; M. A. Long (Baltimore), architect. Part of interior remodeled 1943; L. P. Kimball (Baltimore), architect. Demolished 1963. 4 sheets (1966, 1968, including plans, elevation); 3 ext. photos (1959), 3 int. photos (1959), 7 photocopies of old ext. photos (1940, n.d.), 4 photocopies of old int. photos (n.d.).

Bartram, John, Hotel. See Hotel Walton

Broad Street Station. See Pennsylvania Railroad, Broad Street Station

Bulletin Building (now Penn Square Building). Illustrated
(PA-1496), 1315–25 Filbert St., at N.E. corner Juniper St. Steel frame, brick curtain walls faced with enameled terra cotta on south and west upper stories and Indiana limestone ground story, 105'-4$^5/8$" (nine-bay front) × 147'-4$^3/4$" (originally 63'-1$^1/2$" × 147'-4$^3/4$"), nine and ten stories, flat roof originally with balustrade, twelve-story round corner tower with green and white terra cotta dome, rounded corner, rectangular plan.

Fine example of an early-20th-century commercial building that complements neighboring City Hall and with later alteration making significant use of glass in design. Built 1906–8; Edgar V. Seeler, architect; Frank C. Roberts & Co., engineers; Doyle and Doak, contractors. Ten-story eastern addition (1315–17 Filbert St.) built 1915–16; Edgar V. Seeler, architect; Doyle and Co., contractors. Cornice removed 1936. Interior, ground-story exterior remodeled in glass blocks 1938–43; George Howe, architect. Offices and printing facilities of *The Evening and Sunday Bulletin* until 1955. 1 ext. photo (1973),* 1 photocopy of old ext. photo (1911);* 7 data pages (1973, including 1906–56 building permits).*

Butler, Thomas, House. See Philadelphia Club

Cathedral of SS. Peter and Paul (Roman Catholic)
(PA-1497), N.E. corner Eighteenth and Race sts., at intersection of Benjamin Franklin Parkway, on E. side Logan Sq. Coursed brownstone ashlar, approx. 130' (five-bay front) × 216', one story with large central copper dome, gable roof on nave, hipped roofs on transept, shed roofs on aisles, Corinthian tetrastyle portico on nave pavilion, statuary in pedimented niches, round-arch windows; cruciform plan with extended nave and semicircular chancel, three aisles with altars along sides, richly ornamented interior.

Outstanding Palladian Revival church that dominates Logan Sq. and serves as a major focal point along Benjamin Franklin Parkway. Built 1846–64; initial designs by Rev. Mariano Maller, C.M., and Rev. John B. Tornatore, C.M.; ext. plans by John Notman, architect, in association with John T. Mahoney, architect; int. plans by Napoleon LeBrun, architect; int. decorations by Constantino Brumidi. Interior renovated 1914–15; Henry D. Dagit, architect. Enlarged, interior extensively renovated 1957; Eggers and Higgins (New York), architects. Three-story steel-frame Bishop's House with brownstone ashlar curtain walls built at southeast corner (1723 Race St.) 1912; Henry D. Dagit, architect. One-story brownstone ashlar Chapel of Our Lady of the Blessed Sacrament built on north side 1954; Eggers and Higgins (New York), architects. Certified, PHC 1957; Pennsylvania Register 1971; NR. 2 ext. photos (1973),* 1 photocopy of old ext. photo (1928).*

Chestnut Street Bridge
(PA-1054), Chestnut St. across Schuylkill River. Cast iron with stone abutments and pier, approx. 390' × 42', two arches of six parallel tubular ribs, vertical supporting members composed of cast-iron Gothic Revival arches. As the third iron arch bridge in the United

States, it marked a notable advance in the evolution of tublar-arch construction; it was also considered one of the handsomest bridges in the country. Built 1861–66; Strickland Kneass, architect and engineer. Demolished 1958. 4 photos (1957), 1 photocopy of old photo (n.d.); 3 data pages (1957).

Chinatown YMCA (also known as Chinese Cultural and Community Center)

(PA-1498), 125 N. Tenth St. Museum. Brick, smooth stuccoed front with tile and enameled wood trim imported from Taiwan (Republic of China), approx. 25' × 95' (originally 20' × 36' with three-story rear ell), three-and-a-half stories with front tiled gable-roof penthouse, gable roof, balcony at second and third stories, projecting tile pent eaves with "clud" brackets at first and third stories, imported carved stone bas-reliefs and ornamentation flanking recessed entrance, entry foyer and large front room with imported wood-and-gilt tiles.

Rare and outstanding example of the Peking Mandarin Palace style in the United States. Built 1831–32. One-story side addition and three-story rear addition built 1906, new front with ground-story storefront and upper bay windows built and second story put onto side addition 1910; Clyde S. Adams, architect. New front built and interior extensively altered 1967–71; C. C. Yang (Taipei, Taiwan), architect; Stephen Burczynski, Jonathan Bugbee, and Joseph Nowicki, consulting architects. Houses Chinese Cultural and Community Center. 5 ext. photos (1974);* 3 data pages (1974, including 1906–10 building permits.)*

Church of the Holy Trinity (Protestant Episcopal)

(PA-1085), 200 S. Nineteenth St., at S.W. corner Walnut St., on W. side of Rittenhouse Sq. Brownstone ashlar, approx. 74' (three-bay front) × 126', two stories, gable roof, three-stage corner tower with pinnacles, round-arch triple entrances and windows; three-aisle plan, clerestory, balconies, apsidal chancel.

Pioneer example of the Romanesque Revival by a leading mid-19th-century church architect. Built 1857–59; John Notman, architect. Tower completed 1868; George W. Hewitt, architect. Interior renovated 1880, 1914, 1970. Three-and-a-half-story brick and stone parish house at west end built 1890–91; G. W. and W. D. Hewitt, architects. Small two-story stone choir building at southwest corner built 1898; Cope and Stewardson, architects. Certified, PHC 1958;

Pennsylvania Register 1970; NR. 3 ext. photos (1959), 3 int. photos (1959), 1 photocopy of architect's drawing (c. 1857); 7 data pages (1964).*

Church of St. Luke and the Epiphany (Protestant Episcopal) (originally Church of St. Luke)

(PA-1499), 330 S. Thirteenth St. Stuccoed stone and brick, approx. 65' (six-bay front) × 130', two stories, gable roof, Corinthian octastyle portico with cast-iron capitals, twin front doors with corbeled lintels; second-story auditorium, three-aisle plan, galleries, semicircular chancel.

Fine example of late Greek Revival church design with notable iron fence and gates. Built 1839–40; Thomas S. Stewart, architect. Parish house adjoining south side built 1874–75; Furness and Hewitt, architects. Addition to south rear corner 1891. Vestibule enlarged, new front doors and windows added, new furnishings for chancel 1906; probably by Charles M. Burns, architect; George M. Newhall Engineering Co. Ltd., engineer. Second-story organ loft added to north corner 1925. Church of the Epiphany (founded 1834) and Church of St. Luke (founded 1839) merged 1898. Certified, PHC 1961. 3 ext. photos (1974, including 1 of fence);* 7 data pages (1961, including 1885–1906 insurance surveys).*

Commercial Trust Building. See Arcade Building

Cope, Edward Drinker, Houses

(PA-1500), 2100–2102 Pine St., at S.W. corner Twenty-first St. Brick with green serpentine ashlar front, approx. 39' (six-bay front) × 37' with three-story rear wing, three-and-a-half stories on raised basement, mansard roof, semi-hexagonal bays extending full height, segmental-arch windows with carved lintels, round-arch entrances, side-hall plans.

Built c. 1876; interior altered to apartments 20th century. Edward Drinker Cope, one of America's greatest paleontologists and geologists, lived here 1878–97, leasing the corner house after 1886; properties owned by his descendants until 1950. No. 2102 on Pennsylvania Register 1975; designated National Historic Landmark 1975. 1 ext. photo (1976).*

Delancey Street Area Study

(PA-1501), 1800–36 Delancey St., on S. side Delancey St. between

Eighteenth and Nineteenth sts. Nineteen brick row houses with marble trim, each approx. 20' (two- and three-bay fronts) × 75', four stories on raised basements, flat roofs, side-hall plans.

Mid-19-century middle-class residential street that has retained its visual character in spite of subsequent remodelings. Nos. 1816–18 built 1854; Thomas Dugan, builder. Nos. 1800–1806 built 1855; Joseph Gilbert, builder. Nos. 1808–14 built 1856; Joseph M. Hancock, builder. Nos. 1820–36 built c. 1857; John McCrea, developer. Interior No. 1820 remodeled 1891; Furness, Evans and Co., architects. No. 1824 remodeled 1895; Wilson Eyre, Jr., architect. No. 1836 remodeled c. 1895; Cope and Stewardson, architects. Interior No. 1814 altered 1898; Charles M. Burns, architect. Interior No. 1826 altered 1900–1901; Mantle Fielding, architect. No. 1832 altered, two-story rear addition built 1906–7; Edgar V. Seeler, architect. No. 1818 remodeled, three-story rear addition built 1908; Lloyd Titus, architect. Interior No. 1810 remodeled, four-story rear addition built 1908; Baker and Dallett, architects; exterior remodeled 1923; Eyre and McIlvaine, architects. No. 1804 altered, four-story rear addition built 1910; Thomas, Churchman and Molitor, architects. Interiors Nos. 1816, 1818 altered 1940; Erling Pederson, architect. Certified, PHC 1971. 8 ext. photos (1974);* HABSI form (1970).*

Delancey Street Area Study
(PA-1502), 2301–23 Delancey St., on N. side Delancey St. between Twenty-third and Twenty-fourth sts. Six brick and six brownstone row houses with sandstone trim, each approx. 20' (two-bay front) × 75', three and four stories on raised basements, mansard roofs. A row-house development in the High Victorian Gothic style. Built c. 1874; Robert Kaighn, developer. 3 ext. photos (1973);* HABSI form (1970).*

Drexel & Company
(PA-1503), 135–43 S. Fifteenth St., at N.E. corner Walnut St. Steel frame, smooth-rusticated granite curtain walls with dressed granite trim, 96'-4" (five-bay front) × 73'-0¾", six stories with penthouse, flat roof, balustraded balcony at second story, bas-relief panels of signs of zodiac above ground-story windows, carved escutcheons at corners and above large round-arch entrance, heavily worked wood and brass door, notable marble banking floor. Fine example of Italian Renaissance palazzo manner adapted to 20th-century office use. Built 1925–27; Day and Klauder, architects; Doyle and Co., contractors.

Banking floor altered 1943; Charles Z. Klauder, architect. Drexel & Company here 1927–43; vacated 1957. 2 ext. photos (1976).*

Dunlap-Eyre House

(PA-1504), 1003 Spruce St. Brick with marble trim, approx. 20' (two-bay front) × 38' with three-story rear ell, three-and-a-half stories on raised basement, gable roof, round-arch entrance with fanlight, side-hall town house plan. Built between 1830 and 1835. Interior remodeled, three-story rear addition built 1909; Wilson Eyre, Jr., architect. Home and office of Wilson Eyre, Jr., prominent Philadelphia architect, 1909–40. Certified, PHC 1960. 1 ext. photo (1974);* 8 data pages (1962, including 1852 insurance survey);* HABSI form (1971).*

Far East Chinese Restaurant

(PA-1505), 907–9 Race St. Two adjoining brick houses, each approx. 18' (two-bay front) × 40' with one-, two- and three-story rear ells, three and four stories (originally two), flat and mansard roofs (originally probably gable), ornamented iron balcony across second stories, recessed balcony at third story of No. 907.

Earliest known work and originally dwellings of a prominent early-19th-century architect and his brother; later remodeled to fit the Chinatown image. Built as two-story houses c. 1827; Thomas U. Walter (bricklayer), builder. No. 909 raised two stories later; No. 907 raised one story after 1889. Second story of both altered to restaurant, iron balcony erected, rear ells enlarged 1906–7; storefront of No. 909 replaced 1922, altered to grocery 1963. Far East Chinese Restaurant 1908–52. Certified, PHC 1973. 2 ext. photos (1973),* 1 photocopy of wash drawing (1923);* 4 data pages (1973, including 1836–49 insurance surveys, 1906–22 building permits).*

Federal Reserve Bank of Philadelphia

(PA-1506), 921–39 Chestunt St., at N.E. corner Tenth St. Steel frame with Vermont blue marble ashlar walls, approx. 170' (eleven-bay front) × 113' with earlier rear wing and later rear addition, seven stories (originally six), flat roof with simplified classical cornice, lower unit of engaged piers and narrow windows with ornamental bronze grilles, plain four-story upper unit, bas-relief sculpture flank entrance, center entrance lobby with main banking room running depth of west side.

One of the finest examples of Paul P. Cret's stripped classicism, continuing the bank design tradition of the Second Bank of the

United States. Seven-story steel-frame northern wing with marble ashlar walls built c. 1925; William L. Baily of Baily and Bassett, architect. Main building built 1931–35; Paul P. Cret, architect; Irwin and Leighton, contractors; Alfred Boittau, sculptor. Seven-story steel-frame rear eastern addition with marble ashlar walls and front formal garden built 1940–42; Paul P. Cret, architect; Irwin and Leighton, contractors. Recessed seventh story added 1952–53; Harbeson, Hough, Livingston and Larson, architects. 3 ext. photos (1951, 1975),* 4 int. photos (1951).*

Fell-Van Rensselaer House (later known as Penn Athletic Club)
(PA-1507), 1801–3 Walnut St., at N.W. corner Eighteenth St., on N. side Rittenhouse Sq. Coursed Indiana limestone ashlar with smooth-rusticated ground story and pecked-rusticated raised basement, approx. 124' (ten-bay front) × 44' including two-story northern extension, three-and-a-half stories, low half-hipped roof behind balustrade, curved bays on east front, balustraded entrance porch with coupled Ionic columns, large recessed round-arch entrance, round-headed ground-story windows with carved side panels, carved swags in panels between second- and third-story windows, quoins; central-hall plan open to notable stained-glass dome, notable paneling, marble elements, and ceilings.

Notable example of Beaux-Arts domestic architecture and one of the last extant 19th-century mansions on Rittenhouse Square. Built 1896–98; Peabody and Stearns (Boston), architects; George Watson & Son, contractors. Door cut into south exterior wall, most of interior gutted and remodeled 1974–75. Penn Athletic Club established here 1922, remained until 1964. John Fell was president of A. Pardee & Co., a major anthracite coal operator; his widow married Alexander van Rensselaer 1898. Certified, PHC 1970. 1 ext. photo (1976),* 9 int. photos (1974).*

First Unitarian Church
(PA-1508), 2125 Chestnut St., at N.W. corner Van Pelt St. Coursed granite ashlar (originally rock-faced), approx. 73' (seven-bay front) × 150', one story with open stone porch, hipped roof with three cross-gables, rose window and aedicule on south (front) gable, originally massive square tower with porte-cochere at southeast corner; cruciform plan with short transepts, two aisles, exposed wood-and-iron trusses; two-story Chapel-Parish House at northeast corner.

Important High Victorian church whose interior has survived 20th-century renovations. Built 1885–86; Frank Furness of Furness, Evans & Co., architect. Church and Parish House altered 1902; Savery and Sheetz, architects. Tower removed before 1939; squat entrance arch demolished, rock-facing reduced 1955. Certified, PHC 1957; Pennsylvania Register 1971; NR. 2 ext. photos (1973);* HABSI form (1970).*

Franklin Row. See Sims-Bilsland House

George, Henry, House
(PA-1509), 413 S. Tenth St. Museum. Brick with marble trim, approx. 18' (two-bay front) × 16' with three-story rear ell (originally two-and-a-half stories), three stories on raised basement (originally two-and-a-half stories), mansard roof (originally gable), galvanized iron cornice and semihexagonal second-story bay window, one room to a floor with box entry.

Built c. 1820; Edward Harlan, carpenter. Bay window and mansard added, rear ell raised to three stories later; interior rehabilitated, partially restored 1957. Birthplace of Henry George, political economist, social philosopher, and author of single-tax theory. Acquired by Henry George Foundation 1926, Philadelphia Extension of Henry George School of Social Science since 1957. Certified, PHC 1958. 1 ext. photo (1974);* 5 data pages (1961, including 1831 insurance survey.)*

Girard Trust Corn Exchange Bank. Illustrated
(PA-1510), 34–46 S. Broad St., at N.W. corner Chestnut St. Steel frame with marble ashlar curtain walls, approx. 138' (five-bay front) × 144', two stories with Ionic hexastyle portico on east facade and Ionic tetrastyle portico in antis on south facade, flat roof with Roman dome, large two-story domed banking room with side galleries.

A fine example of a Classical Revival design reflecting the architectural influence of Chicago's Columbian Exposition 1893. Built 1905–8; design by Frank Furness of Furness, Evans & Co., architects; detailing by McKim, Mead and White (New York), architects; Allen Evans of Furness, Evans & Co., superintendent of construction. Name changed to Girard Trust Bank 1964; known as Girard Bank since 1969. Certified, PHC 1963; Pennsylvania Register 1970. 2 ext. photos (1973);* HABSI form (1970).*

Gladstone Hotel (later known as Greystone Apartments)
(PA-1511), 328–38 S. Eleventh St. and 1101–11 Pine St., at N.W.
corner Eleventh and Pine sts. Brick with granite trim and steel floor
beams, approx. 120' (eighteen-bay front) × 52', ten stories, flat roof,
octagonal bays through sixth story, balconies at fourth and sixth
stories, large arched entry to circular lobby, reversed L-plan with rear
courtyard.

Philadelphia's first multistory apartment building. Built 1889–90;
Theophilus Parsons Chandler, Jr., architect; J. E. and A. L. Pennock,
contractors; Levering and Garrigues, steel contractors. Demolished
1971. Fashionable apartment hotel through 1920's; became Greystone
Apartments 1958. 2 photocopies of ext. photos (1971).*

Greystone Apartments. See Gladstone Hotel

Harrison Building I. Illustrated
(PA-1088), 4 S. Fifteenth St., at S.W. corner Market St. Steel frame
with dark ashlar and terra cotta curtain walls, approx. 115' (nine-bay
front) × 40', eleven-and-a-half stories, hipped roof, central-hall plan,
elaborate French Renaissance details. Built 1894–95; Cope and
Stewardson, architects; Louis D. Havens, contractor. First floor altered 1902; Cope and Stewardson, architects. Demolished 1969. 7 ext.
photos (1962), 3 int. photos (1964), 1 photocopy of old ext. photo (c.
1900), 4 photocopies of published prints of architects' drawings
(1895); 7 data pages (1962, 1964).

Harrison Building. See Monarch Building

Hockley, Thomas, House
(PA-1512), 235 S. Twenty-first St., at N.E. corner St. James St. Brick
with brownstone trim, approx. 35' (five-bay front) × 78', three stories
on raised rusticated brownstone basement, mansard roof with brick
dormers, shed roof on two-story rear ell, second-story bay on
brackets with pointed hipped roof, notable carved stone floral ornament on corner entrance tympanums, bands of canted and black
brick.

An early Furness town house that established a pattern of texture
and silhouette copied by Philadelphia developers for the next two
decades. Built c. 1875; Frank Furness of Furness and Hewitt, architect. Altered, enlarged 1894; Furness, Evans & Co., architects. Interior altered to apartments later. Certified, PHC 1962. 1 ext. photo
(1973);* HABSI form (1962).*

Holy Redeemer Chinese Catholic Church and School
(PA-1513), 915 Vine St., at N.W. corner Ridge Ave. Cinder block
faced with buff brick and limestone trim, approx. 42' × 72' with two-
story rear wing (school), one story on raised basement, gable roof,
entrance and oculus set within deeply recessed arch, straight double-
entry stairs to entrance, three-aisle plan, auditorium in basement and
communicating with school.

Built 1941; Henry D. Dagit and Sons, architects; McCloskey and
Co., contractors. Founded 1941 as the first Chinese Roman Catholic
church and school in the United States; remains the social and religious center of Philadelphia's Chinatown. Not a parish but a
mission of St. John's Roman Catholic Church. 2 ext. photos (1973).*

Hotel Metropole. See Hotel Walton

Hotel Walton (later John Bartram Hotel)
(PA-1091), 233–47 S. Broad St., at S.E. corner Locust St. Brick with
brownstone trim, approx. 170' (seventeen-bay front) × 130', main
(north) section ten-and-a-half stories, older (south) section eight-
and-a-half stories, mansard and gable roofs, turrets at corners of
main section, stylized Flemish gables, "standardized" plan with
rooms arranged around rectangular light well.

Older (south) section, former Hotel Metropole, built c. 1892;
Angus S. Wade, architect. Main (north) section built 1893–95; Angus
S. Wade, architect. Interior renovated 1948; demolished 1966. 4 ext.
photos (1963), 1 photocopy of old ext. photo (1896), 1 photocopy of
architectural rendering of Hotel Metropole (1894); 11 data pages
(1964).

Keystone Automobile Club. See Philadelphia Art Club

Land Title Building
(PA-1514), 100–118 S. Broad St., between S.W. corner Chestnut St.
and N.W. corner Sansom St. Steel frame with buff brick (northern
building) and pale granite-faced brick (southern building) curtain
walls, approx. 230' (sixteen-bay front) × 120' (originally 105' × 100'),
fifteen and twenty-two stories, flat roof, rusticated Ionic ground-
story arcade and semihexagonal bays above on northern building,
range of giant modified Tuscan pilasters at lower stories and giant engaged Corinthian columns and pilasters at upper stories of southern
building.

Excellent study in the development of the "Commercial style" at the turn of the 19th century by a leading designer of tall office buildings. Northern fifteen-story building and one-story connection built 1897–98; D. H. Burnham & Co. (Chicago), architects; Charles Mc-Caul, contractor. Southern twenty-two-story building built 1902; D. H. Burnham & Co., architects; Doyle and Doak, contractors. Banking room altered 1907; Horace Trumbauer, architect. Banking room altered 1946; James S. Hatfield, architect. Banking room and entrance altered 1955–56; Thalheimer and Weitz, architects. Banking room altered 1964; Vincent G. Kling and Associates, architects. Chartered as The Real Estate Title Insurance Co. 1876, Land Title Bank and Trust Co. is the world's oldest title insurance company. 1 ext. photo (1973),* 1 photocopy of architectural rendering (c. 1901);* 2 data pages (1973).*

Leidy, Dr. Joseph, Jr., House (now Poor Richard Club)
(PA-1515), 1319 Locust St. Brick with sandstone trim, approx. 46' (three-bay front) × 80', three-and-a-half stories on raised basement, gable roof with cross-gables, two semihexagonal bays, round-arch entrance with projecting voussoirs, central-hall plan.

Fine example of the Georgian Revival. Built 1893–94; Wilson Eyre, Jr., architect. Interior renovated, tiered stair tower added to east wall 1925; interior renovated 1937. Since 1925 the quarters of the Poor Richard Club, world's oldest club of advertising men. Certified, PHC 1963. 2 ext. photos (1972),* 1 photocopy of architect's plan and elevation (1894);* 2 data pages (1964);* HABSI form (1971).*

Lippincott, Joshua B., House
(PA-1516), 204 S. Nineteenth St., on W. side of Rittenhouse Sq. Brick with marble front, approx. 25' (three-bay front) × 85', four stories, mansard roof, side-hall plan. Example of comfortable town houses of prominent Philadelphians that lined Rittenhouse Square in last half of 19th century. Built 1869–70; Addison Hutton, architect. Demolished 1972. Joshua B. Lippincott, founder of J. B. Lippincott & Co. publishing house, lived here 1870–86. 2 ext. photos (1969),* 6 int. photos (1969);* 3 data pages (1969).*

Market Street National Bank Building (now One East Penn Square Building). Illustrated
(PA-1517) 1–21 N. Juniper St., at N.E. corner Market St. Steel frame, yellow brick curtain walls with yellow terra cotta lower stories and polychromatic terra cotta trim, approx. 180' (twenty-five-bay front) × 58', seventeen, nineteen, twenty-one, twenty-three, and twenty-four stories produced by setbacks, flat roofs, projecting piers,

simulated buttresses above nineteenth story, rectangular plan with second-story banking room.

Notable example of modernistic, or Art Deco, architecture with Mayan-inspired decoration. Built 1930; Ritter and Shay, architects; William Steele & Sons Co., contractors. Extensive interior alterations 1962; fenestration of lower stories altered 1971. Name changed to One East Penn Square Building 1962. 4 ext. photos (1973);* 3 data pages (1973).*

Mask and Wig Club

(PA-1518), 310 S. Quince St. Stuccoed brick with brick trim, approx. 28' (five-bay front) × 85', three stories, gable roof, shingled gable peak, large hood over entrance, hood over second-story windows, center-hall plan. Built as coach house and stable between 1843 and 1853. Remodeled into clubhouse 1894; Wilson Eyre, Jr., architect. Enlarged, altered 1901–3; Wilson Eyre, Jr., architect. Since 1894 the headquarters of the Mask and Wig Club, a University of Pennsylvania dramatic group. Certified, PHC 1961. 1 ext. photo (1975);* 11 data pages (1962, including 1853–1914 insurance surveys).*

Meeting House of Central Philadelphia Meeting of Friends. See Twelfth Street Meeting House

Mellor and Meigs Architectural Offices

(PA-1519), 205–7 S. Juniper St., at S.E. corner Chancellor St. Brick with stone trim, 50' (three-bay front) × 24', two stories, gable roof with cross-gables, large windows on north and west facades, center entry room with "Big Room" in one story in southern half.

Offices and studio of and by a noted early-20th-century architectural firm. Built as carriage house and stable mid-19th century. Extensively remodeled to architectural offices 1912; Mellor and Meigs, architects; A. H. Williams' Sons, contractors. Altered to restaurant 1946. Mellor and Meigs (Mellor, Meigs and Howe 1917–28) were known for their distinguished suburban house designs and had offices here 1912–40. Mitchell's Restaurant since 1947. Certified, PHC 1973. 2 ext. photos (1973).*

Monarch Building (also known as Harrison Building)

(PA-1520), 1001–23 Filbert St., at N.W. corner Tenth St. Pompeiian brick with terra cotta trim and interior iron columns, approx. 308' (thirty-six-bay front) × 135', six stories, flat roof, ground-story brick piers with terra cotta bands and ornamented terra cotta architrave, deeply recessed windows, round-arch windows at top story, rectangular plan divided into twelve sections or stores. Monumental commercial building with well-scaled ornament; part of area's development stimulated by construction of Reading Terminal. Built 1893–94; Cope and Stewardson, architects; William C. McPherson and Son, contractors. Later minor interior alterations. 2 ext. photos (1976).*

Moore, Clarence B., House

(PA-1521), 1321 Locust St., at N.E. corner Juniper St. Buff brick with limestone trim, approx. 26' (two-bay front) × 80', three-and-a-half stories (originally three-and-a-half), mansard roof (originally tiled roofs on corner towers, recessed pointed-arch windows, medallions and gargoyles at entrance, great-hall plan.

Notable eclectic Gothic design by an imaginative architect of city and country houses. Built 1890; Wilson Eyre, Jr., architect. Interior altered later. Certified PHC 1963; Pennsylvania Register 1970; NR. 1 ext. photo (1972),* see Dr. Joseph Leidy, Jr., House (PA-1515) for photocopy of architect's plan and elevation (1894);* 1 data page (1962);* HABSI form (1971).*

New Century Club

(PA-1522), 124 S. Twelfth St. Buff brick with Indiana limestone trim, approx. 40' (five-bay front) × 35' with three-story rear wing, four stories (originally three-and-a-half), mansard room (originally tiled hipped roof) and rear shed roof, asymmetrical facade with curved and semihexagonal bays, center cross-gable with crockets and large fanlight (originally open), center-hall plan with front stairs.

An early architectural work by a woman architect. Built 1891; Minerva Parker (later Minerva Parker Nichols), architect; J. E. and A. L. Pennock, contractors. Interior altered, dormers removed, recessed balcony and fanlight enclosed 1907; Horace Wells Sellers, architect. Demolished 1973. Founded 1877, the New Century Club of Philadelphia is one of the nation's oldest women's clubs devoted to protection and promotion of women's interests and general social questions. Certified, PHC 1965. 2 ext. photos (1973),* 5 int. photos (1973),* 1 photocopy of old ext. photo (c. 1892).*

New York Mutual Life Insurance Company Building (also known as Victory Building)

(PA-1523), 1001–5 Chestnut St., at N.W. corner Tenth St. Brick faced with Rhode Island granite rusticated on ground story, interior iron posts and girders, approx. 58' (five-bay front) × 176', seven stories on raised basement (originally four), high mansard roof, rounded corner, engaged Ionic and Corinthian columns, balustrade at each original story, mansard windows with heavy arched lintels and pilasters; reverse E-plan with court on each side of stair tower, ornate iron and marble longitudinal corridors.

An early office building in the Second Empire mode with well-integrated later enlargement and florid Beaux-Arts addition. Built 1873–75; Henry Fernbach (New York), architect. Raised three stories, old mansard preserved 1890–91; Philip W. Roos (New York), architect. Ten-story cream brick and terra cotta addition (1007–13 Chestnut St.) with interior steel posts built 1901–2; Roos and Brosam (New York), architects; J. E. and A. L. Pennock, contractors. Rear steel-frame addition with brick curtain walls built, corner building's interior altered 1940–41; Max A. Bernhart, architect. Regional headquarters of one of the nation's largest insurance companies 1875–1920. Certified, PHC 1972. 2 ext. photos (1975),* 1 photocopy of architectural rendering (1901).*

One East Penn Square. See Market Street National Bank Building

Parry, Charles T., House
(PA-1524), 1921 Arch St. Brick with brownstone ashlar front, approx. 45' (five-bay front) × 52' with three-story rear ell, four stories, flat roof, asymmetrical facade, two projecting bays of paired windows, elaborately carved interior woodwork and marble.

Fine example of post–Civil War town house. Built 1870–71; demolished 1968. Parts of interior architectural elements preserved in Philadelphia Museum of Art. Home 1871–87 of Charles T. Parry, president 1874–87 of Burnham, Parry, Williams & Co., proprietors of Baldwin Locomotive Works. 2 ext. photos (1968),* 15 int. photos (1968),* 1 ext. photo of carriage house (1968),* 2 photocopies of old ext. photos (c. 1900),* 4 photocopies of old int. photos (c. 1890).*

Penn Athletic Club. See Fell-Van Rensselaer House

Penn Square Building. See Bulletin Building

Pennsylvania Academy of the Fine Arts. Illustrated
(PA-1525), 118–24 N. Broad St., at S.W. corner Cherry St. Museum. Brick with sandstone, granite, marble, ceramic tile trim, approx. 100' (three-bay front) × 265', two stories, complex roof; textured,

polychromatic picturesque facade with pointed-arch windows and bas-relief panels; notable interior grand staircase, functional skylight system, notable interior ornament.

An outstanding work and an early example of expressed iron construction on the interior by an important late-19th-century architect. Built 1872–76; Frank Furness of Furness and Hewitt, architect; William A. Armstrong, builder. Galleries altered early 20th century; street lamps removed before 1930; statue in middle of center pointed-arch window removed c. 1935. Monitors over front galleries removed, lobby made smaller and simpler 1949; Harbeson, Hough, Livingston and Larson, architects. Roof cresting removed, door embrasures and grilles removed c. 1950. Restored 1973–76; Day and Zimmerman, Inc., architects.

Founded 1805, Pennsylvania Academy, a combined art museum and art school, is the oldest art institution in the United States. Certified, PHC 1956; Pennsylvania Register 1970; designated National Historic Landmark 1975. 3 ext. photos (1967),* 8 int. photos (1967),* 2 photocopies of architect's drawings (c. 1872).*

Pennsylvania Institution for the Deaf and Dumb (now Philadelphia College of Art). Illustrated

(PA-1526), 320 S. Broad St., composing block bounded by Broad, Pine, Fifteenth, Delancey sts. Coursed granite ashlar with plastered stone rear wings and brick back buildings, approx. 200' (fifteen-bay front) × 160' (originally approx. 96' × 92'), two and three stories, gable and low hipped roofs, two bays with two-story blind arches (niches) flank central Doric tetrastyle portico in antis, elongated O-plan with wings (originally U-plan).

Fine application of Greek Revival elements to an institutional building. Built 1824–26; John Haviland, architect. Rear three-story stone addition to center building built, rear wings extended 1838. Plastered stone lateral wings added to north and south ends of front 1852; possibly by William Strickland, architect. Rear wings raised to three stories 1863. Two three-and-a-half-story brick buildings with hipped roofs built on Fifteenth St. and connected to rear wings 1875; Furness and Hewitt, architects. Interior front building and lateral wings altered 1876; one-story hipped-roof brick building built between Fifteenth St. buildings 1879. Detached one-story brick and concrete-block buildings built in yard 1914, c. 1950, 1963, 1967. Pennsylvania Museum and School of Industrial Art (later known as Philadelphia Museum College of Art) 1893–1964; Philadelphia

College of Art since 1964. Certified, PHC 1956; Pennsylvania Register 1971; NR. 5 ext. photos (1973, 1974),* 1 photocopy of architectural drawing (n.d.);* 25 data pages (1965, including 1826–89 insurance surveys).*

Pennsylvania Railroad Station, Broad Street Station. Illustrated
(PA-1527), N.W. corner Broad and Market sts. Granite ashlar and brick with terra cotta trim, approx. 306' (thirteen-bay front) × 212' with three-hinged wrought-iron arch train shed approx. 600' × 300', north section five-and-a-half stories with eight-story tower, south section eleven stories with thirteen-story tower, gable roof with cross-gables, rusticated ground story, pointed arches, pedestrian arcade, enclosed pedestrian bridge over Market St. to Arcade Building, main entrance at base of thirteen-story tower at corner Broad and Market sts., waiting rooms on second story.

One of the world's great railroad terminals. North section built 1880–82; Wilson Brothers & Co., architects. Station expanded and south section built for executive offices 1892–94; Furness, Evans & Co., architects. Notable architectural sculpture in terra cotta and interior bas-relief in plaster by Karl Bitter, sculptor. Train shed built 1892–93; Wilson Brothers & Co., engineers. Pedestrian bridge built c. 1902; Furness, Evans & Co., architects. Train shed destroyed by fire 1923; Karl Bitter's bas-relief "Spirit of Transportation" removed to Thirtieth St. Station 1933; station demolished 1952–53. When built, the station was the world's largest railroad passenger terminal, and the train shed had the world's largest permanent roof with a span of 300'-8". 5 photocopies of old ext. photos (1893, c. 1894, 1903, c. 1925),* 1 photocopy of architectural rendering (1893).*

Philadelphia and Reading Railroad, Terminal Station (also known as Reading Terminal). Illustrated
(PA-1528), 1115–41 Market St., at N.E. corner Twelfth St. Pink brick with steel beams and posts and cream terra cotta trim, originally rusticated granite ground story, approx. 267' (thirteen-bay front) × 107' with two-story concourse approx. 267' × 50' to three-hinged wrought-iron arch train shed approx. 267' × 559', eight stories originally on raised basement, flat roof, front corners and central section above second story recessed, elaborately ornamented oriel window in northwest corner recess, originally roof balustrade and second-story open arcade in central section, "stub end" plan of head house (passenger facilities, shops, offices) and rear train shed, waiting room on second story and level with tracks.

Significant engineering work and rare use of a composite Renaissance style on a railroad terminal. Built 1891–93; train shed designed by Wilson Brothers & Co., architects and engineers; head house designed by Francis H. Kimball of Kimball and Thompson (New York), architect; Charles McCaul, general contractor; J. J. Ryan & Co., excavation contractor; Pencoyd Iron Works, steel contractor. Sidewalk canopy added at main entrance before 1906, replaced by cantilevered stainless steel canopy 1948. Skylights of train shed roofed over, head house interior and exterior altered 1948. Train shed is world's oldest long-span roof structure and America's only surviving single-span arched train shed. Reading Terminal Market, in shed's ground story, remains on market space dating from 1860. Pennsylvania Register 1972; NR.

5 ext. photos (1974),* 2 photocopies of old ext. photos (c. 1892 during construction, c. 1937),* 3 photocopies of old int. photos (c. 1893 of train shed, c. 1948 before alterations).*

Philadelphia Art Club (later Keystone Automobile Club)
(PA-1529), 220 S. Broad St., at S.W. corner Chancellor St. Buff brick with limestone trim, approx. 64' (four-bay front) × 116', four stories on raised basement with seven-story rear addition, low hipped tile roofs, veranda on three-story northeast corner, quoins, center-hall plan.

An early work of a noted late-19th-century architect making fine use of architectural ornament. Built 1887; Frank Miles Day, architect. Ground story altered 1891; Frank Miles Day, architect. Interior altered 1902; Watson, Huckel and Co., architects. Rear addition built 1907; Newman and Harris, architects. Interior altered 1926; Day and Klauder, architects. Interior altered 1946; George I. Lovatt, architect. Demolished 1975–76. Occupied by Keystone Automobile Club 1946–68. Certified, PHC 1966; Pennsylvania Register 1970. 4 ext. photos (1972);* HABSI form (1970).*

Philadelphia City Hall. Illustrated
(PA-1530), Penn Square, intersection of Broad and Market sts. First story of granite blocks, upper stories of brick faced with white marble, upper stages of tower of iron, approx. 486' (nineteen bays) × 470', seven stories with 548-foot tower, mansard roofs, square plan enclosing central courtyard, towered pavilions mark arched pedestrian passages to courtyard, truncated towers at corners, sculpture on all facades and tower, lavish interior of marble, mosaics, carved woodwork.

One of the finest American examples of picturesque eclectic architecture in the Second Empire style, City Hall remains the nation's largest municipal governmental building and its tower makes it the world's tallest bearing-wall building. Built 1871–1901; John McArthur, Jr., chief architect; Thomas U. Walter, consulting architect; John Ord, assistant architect; Alexander Milne Calder, sculptor; William C. McPherson, superintendent of construction; Struthers & Sons, marble contractor; George Herzog, decorator of Supreme Court Room and Judges' Consultation Room. John Ord, chief architect 1890–94; W. Bleddyn Powell, chief architect 1894–1901. Occupied in stages after 1877; William Penn statue placed on tower 1894. Rehabilitated 1963–. 15 ext. photos (1958, 1959, 1962, 1963, 1965),* 5 int. photos (1959),* 2 aerial photos (1965),* 1 photocopy of old ext. photo (c. 1925),* 1 photocopy of old photo of William Penn statue (1894),* 2 photocopies of architect's drawings (1872, 1897).*

Philadelphia Club (originally known as Thomas Butler House)

(PA-1531), 1301–3 Walnut St., at N.W. corner Thirteenth St. Brick with Quincy granite trim, approx. 70' (seven-bay front) × 58' with one- and three-story rear additions (originally with two-story rear ell at northwest corner), three-and-a-half stories, gable roof, large round-arch entrance with notable fanlight, side entrance with transverse corridor and two intersecting stairhalls.

Notable example of 19th-century urban mansion with traditional, large-scale Georgian facade. Built 1837; probably by William Strickland, architect. Interior altered, kitchen moved to cellar 1850; one-story billiard room attached to east side of ell and connected to house by passageway 1858. Rear ell, billiard room, and passageway raised to three stories with mansard roof and second-story elliptical bay window 1888–89; Furness, Evans & Co., architects. One-story rear addition built at northeast corner of house 1892. Since 1850 the quarters of the Philadelphia Club, America's oldest city social club; founded 1834. Certified, PHC 1956. 1 ext. photo (1976);* 10 data pages (1964, including 1837–92 insurance surveys).*

Philadelphia College of Art. See Pennsylvania Institution for the Deaf and Dumb

Philadelphia Masonic Temple

(PA-1532), 1 N. Broad St., at N.E. corner Filbert St. Granite ashlar, approx. 138' (five-bay front) × 250', three stories with five- and seven- story towers, hipped roof, elaborate one-story Norman entrance portico, rounded-headed windows, turrets, lavish interior with separate apartments in different historical architectural styles.

One of the world's outstanding Masonic structures. Built 1868–73; James H. Windrim, architect. Much of the interior decorated in 1890's; George Herzog, decorator of Egyptian, Ionic, and Norman halls. Serves as Grand Lodge for Pennsylvania Masonry. Certified, PHC 1962; Pennsylvania Register 1970; NR. 9 ext. photos (1963),* 4 int. photos (1963),* 11 photocopies of architectural competition drawings (1867),* 1 photocopy of architect's plan (n.d.),* 14 photocopies of old photos of construction and interior (c. 1870, c. 1900);* 12 data pages (1967, 1971).*

Philadelphia Saving Fund Society Building (also known as PSFS Building)
(PA-1533), 12 S. Twelfth St., at S.W. corner Market St. Steel frame with granite and glass curtain walls, 164'-4¹/₄" (nine-bay front) × 134'-4¹/₄", thirty-two stories and three mezzanines with penthouse, flat roof; twenty-seven-story office tower with horizontal ribs rises from five-story curved glass and polished granite banking rooms, black brick service spine rises thirty-six stories, great neon PSFS sign (27' height) incorporated into design at top, rectangular plan with second-story banking room.

In its reconciliation of architectural form with modern technology the building stands as a landmark in architectural evolution. Often considered the first modern skyscraper in 20th-century United States, it marked the early use on a large scale of stainless steel and central air-conditioning. Built 1930–32; Howe and Lescaze, architects; Walter Behrman, interior designer; George A. Fuller Co., contractors. Ground-story shops altered 1949; Alexander T. C. Hazell, architect. Judged "Building of the Century" by Philadelphia Chapter, A.I.A. 1969. Certified, PHC 1968; Pennsylvania Register 1970. 2 ext. photos (1976),* 2 photocopies of architectural renderings (1932).*

Poor Richard Club. See Leidy, Dr. Joseph, Jr., House

Portico Square (later known as Portico Row)
(PA-1534), 900–930 Spruce St., composing block bounded by Ninth, Spruce, Tenth, Cypress sts. Sixteen brick row houses with marble trim, each approx. 24' (three-bay front) × 50' with three-story rear ell, three-and-a-half stories on raised basements, gable roofs, each pair of houses with common Ionic tristyle porch, side-hall plans.

Philadelphia's last surviving stylish residential development of the Jacksonian era and an early plastic solution to row house design. Built 1831–33; Thomas U. Walter, architect. Exterior No. 922 remodeled after 1870; No. 928 altered to store 1924. Commodore Issac Hull, commander of U.S.S. *Constitution* during War of 1812, lived at No. 908 1842–43; Mrs. Sarah Josepha Hale, editor of *Godey's Lady's Book* and crusader for Thanksgiving as a national holiday, lived at No. 922 1859–61; No. 910 Philadelphia High School for Young Ladies 1836–41. Certified, PHC 1962. 5 ext. photos (1975);* 58 data pages (1962, including 1832–1920 insurance surveys).*

PSFS Building. See Philadelphia Saving Fund Society Building

Reading Terminal. See Philadelphia and Reading Railroad, Terminal Station

Reed's, Jacob, Sons
(PA-1535), 1424–26 Chestnut St. Reinforced concrete, diapered brick curtain walls with marble and polychromatic terra cotta trim, 50'-1" (seven-bay front) × 230'-5", four stories in front, five stories in rear, front low gable roof, rear flat roof, giant recessed arched entrance on monolithic Byzantine columns with terra cotta designs on extrados, open arcade at top story, marble cavetto cornice with terra cotta medallions, date stones (of founding and erection) at front corners, quoins, rectangular plan.

Notable example of early reinforced concrete building in an eclectic style. Built 1903–4; William L. Price of Price and McLanahan, architect; murals by C. T. Monaghan. Rear fifth story added 1909; Spencer Roberts, architect. Jacob Reed's Sons have been retail clothiers since 1824. Certified, PHC 1966. 2 ext. photos (1976).*

Restaurant and Apartment Building
(PA-1536), 110 N. Ninth St. Brick with granite ashlar front, approx. 18' (three-bay front) × 46', six stories including mansard roof, arcades of horseshoe-arch windows with "Moorish" details, rectangular plan originally with living quarters on upper floors. Built c. 1870; ground story altered later. 2 ext. photos (1963);* 3 data pages (including 1891 insurance survey).*

Rogers-Cassatt House
(PA-1537), 202 S. Nineteenth St., on W. side of Rittenhouse Sq. Brick with brownstone and copper trim, approx. 50' (five-bay front) × 140',

four stories, mansard roof (probably originally flat), bay windows, ornate interior woodwork.

Example of a remodeled mid-19th-century city mansion on a fashionable square. Built c. 1856. Remodeled, renovated c. 1888; probably by Wilson Brothers and Co., architects. Demolished 1972. Alexander J. Cassatt, president of the Pennsylvania Railroad 1899–1906 and brother of Mary Cassatt, expatriot painter and etcher, lived here 1888–1906. Home of Fairman Rogers, prominent scientist, civil engineering professor and horseman, c. 1856–87. Church House for Episcopal Diocese of Pennsylvania 1921–72. Certified, PHC 1963. 3 ext. photos (1969),* 6 int. photos (1969);* 2 data pages (1969).*

St. Clement's (Protestant Episcopal) Church. Illustrated

(PA-1538), 128 N. Twentieth St., at S.W. corner Cherry St. Brownstone ashlar, approx. 68' × 128', one story with two-story square tower at northeast corner (originally three-story tower with octagonal spire), gable roof, pronounced apse with smooth rusticated base on east end, round-arch windows; three-aisle plan, apsidal chancel.

Fine Romanesque Revival church by a noted mid-19th-century church architect. Built 1856–59; John Notman, architect. Crypt Chapel built beneath apse 1898; Horace Wells Sellers, architect; William Halsey Wood (architect), designer of altar. Clergy house built onto southwest corner 1901–2, parish house built onto northwest corner 1906–7, church's sanctuary altered, enlarged 1908; Horace Wells Sellers, architect. Interior renovated 1915. Church moved 40' west 1929. Certified, PHC 1963; Pennsylvania Register 1971; NR. 7 ext. photos (1974);* 8 data pages (including 1902–15 insurance surveys).*

St. Mark's (Protestant Episcopal) Church

(PA-1093), 1625 Locust St. Random brown freestone and red sandstone ashlar, approx. 147' (ten-bay front) × 60', one story with four-stage main entry side tower and spire, gable and side-aisle shed roofs; notable interior design, three-aisle plan with rectangular chancel and appended Lady Chapel.

One of the earliest American applications of the concepts of the Anglican "High Church" religious revival to an urban site, producing a notable Gothic Revival church. Built 1848–51; John Notman, architect. Parish house at west end built 1892–93; Hazlehurst and Huckel, architects. Lady Chapel at southeast end built 1899–1902;

Cope and Stewardson, architects. Certified, PHC 1956; Pennsylvania Register 1970. 4 ext. photos (1958), 3 int. photos (1958), 2 photocopies of architect's drawings (c. 1848, including plan, sections); 10 data pages (1964).

St. Stephen's Protestant Episcopal Church

(PA-1539), 19 S. Tenth St., at N.E. corner Ludlow St. Coursed granite ashlar front and towers, brick sides and rear walls faced with coursed granite ashlar, approx. 61' (five-bay front) × 102', one story, gable roof, twin five-stage octagonal towers, pointed-arch windows; three-aisle plan, galleries, rectangular chancel.

Rare surviving example of a Gothic Revival design by William Strickland, known for his Greek Revival architecture. Built 1822–23; William Strickland, architect; Daniel Groves, master mason; John Struthers, marble mason. Stenciled interior decorations added, transept and vestry room built on north side 1878–79; Frank Furness, architect. Three-story rusticated granite ashlar parish house built on west side (front) of vestry room 1888; George C. Mason & Son, architects. Interior stenciled decorations painted over 1940's. Interior highlighted by notable memorials, which include "Angel of Purity" sculpture by Augustus Saint-Gaudens (c. 1900) and three Tiffany stained-glass windows. Certified, PHC 1957. 3 ext. photos (1972);* 6 data pages (including 1823–88 insurance surveys).*

Schuylkill Hose, Hook & Ladder Company No. 24

(PA-1540), 1227 Locust St. Stuccoed brick, approx. 20' (three-bay front) × 58', four stories, flat roof, projecting lintels with corbels, rectangular plan. Built c. 1857. Ground-story storefront added, second-story iron balcony removed 1934–35; R. R. Neely, architect. Storefront closed up, interior altered to restaurant 1955. Vacated 1967, condemned 1969. 1 photocopy of old ext. photo (c. 1934);* 2 data pages (1972, including 1934–55 building permits).*

Scott-Wanamaker House. Illustrated

(PA-1531), 2032 Walnut St. Brick faced with rusticated granite ashlar, approx. 50' (five-bay front) × 100', three-and-a-half stories on raised deeply vermiculated basement, hipped and gable roofs, asymmetrical facade with two-story semihexagonal bays capped by elaborately carved stone balustrade, picturesque roofline; center-hall plan with large open staircase to third story, elaborately carved and paneled interior.

Notable example of Jacobean Revival style. Built 1883; Theophilus Parsons Chandler, Jr., architect. Altered to apartment house, exterior stair tower added to west side 1938; Gabriel B. Roth, architect. Vacated 1972; rehabilitated, ground story partially restored 1973–. John Wanamaker lived here 1894–1922. As a major contributor to the concept of department store merchandising, Wanamaker founded the John Wanamaker Department Store, and as an active participant in Republican politics he served as Postmaster General in President Benjamin Harrison's Cabinet 1889–93. James Scott, sugarrefiner, lived here 1884–94. Certified, PHC 1962; Pennsylvania Register 1971; NR.

3 ext. photos (1973),* 8 int. photos (1973);* 17 data pages (including 1883, 1894 insurance surveys);* HABSI form (1970).*

Sims, Joseph, House. See Sims-Bilsland House

Sims-Bilsland House (originally known as 2 Franklin Row). Illustrated
(PA-1186), 228 S. Ninth St. Brick with stone trim, 23'-6" (two-bay front) × 40'-3", three-and-a-half stories, gable roof, triple window on first and second stories set in blind arch with recessed solid tympanum, side-hall plan with two rooms to a floor.

Last survivor of an early ten-house speculative development known as Franklin Row, significant for its use of monumental forms within a unified architectural composition. Built 1809–10; Robert Mills, architect; John Savage and John Meany, developers. Doorway lowered, third-story window replaced by double window 1935. Vacated 1966. Restored 1975–. Fanlight, woodwork, and plaster elements in Architectural Study Collection, Museum Division, Independence National Historical Park. Investment property of merchant Joseph Sims; Ann Bilsland, widow, was first occupant 1811–29. Facilities of Booth, Garrett and Blair, America's oldest commercial chemical laboratory, here 1935–66. Certified, PHC 1961.

5 sheets (1965, including plot plan, plans, elevation, section, details); 3 ext. photos (1962, 1965), 4 int. photos (1962), 2 photocopies of old ext. photos (1870, 1917), 1 photocopy of architect's elevation drawing (1809); 14 data pages (1961, 1962, including 1811 insurance survey*).

Spruce Street Area Study. Illustrated
(PA-1532), 2009–47 Spruce St., on N. side Spruce St. between Twentieth and Twenty-first sts. Twenty brick row houses faced with

brownstone ashlar, seven easternmost houses approx. 22' (three-bay fronts) × 105', thirteen westernmost houses approx. 19' (two-bay fronts) × 72', three-and-a-half stories on raised basements, mansard roofs, side-hall plans.

Fine study of a decade's change in brownstone row-house design. Nos. 2009–21 built c. 1868; Nos. 2023–47 built c. 1880. No. 2037 altered 1894; Hazlehurst and Huckel, architects. No. 2039 altered 1910; Spencer Roberts, architect. Interiors altered to apartments and professional offices between 1925 and 1937. 11 ext. photos (1974);* 4 data pages (including 1905 insurance survey);* HABSI form (1970).*

Troc. See Arch Street Opera-House

Twelfth Street (Friends) Meeting House (also known as Meeting House of Central Philadelphia Meeting of Friends)

(PA-1533), 20 S. Twelfth St., at N.W. corner Clover St. Brick, 90'-8" (five-bay front) × 57'-4", two stories, gable roof, entry porch on east end, two entry porches on north and south sides; meeting room with three longitudinal aisles and two transverse aisles, balconies, rear room.

Fine example of traditional Philadelphia architecture in early 19th century. Built 1812–13; John D. Smith, master carpenter and builder. Six trusses, attic roof, and floor framing, and some first-floor joists salvaged from Market Street Meeting House, S.W. corner Second and Market sts. (built 1755, demolished 1809). Friends Institute built at S.W. corner 1892, second story added 1909, Walter Smedley, architect. Second story of Meeting House remodeled 1936. Dismantled 1972, re-erected on grounds of George School, Newtown, Bucks Co., Pa. Meeting House served as a center for dissenting reform groups ever since the abolition movement in 1840's. Certified, PHC 1956; Pennsylvania Register 1970.

9 sheets (1971, including plot plan, plans, elevations, sections, section and details of 1755 truss, profile details);* 9 ext. photos (1971),* 32 int. photos (1971, including 8 of trusses),* 2 photocopies of old ext. photos (c. 1885, 1902);* 8 data pages (1963, 1972, including 1876–1908 insurance surveys).*

Union League of Philadelphia

(PA-1534), 140 S. Broad St., at S.W. corner Sansom St. Brick with brownstone trim, approx. 100' (seven-bay front) × 130', two-and-a-half stories on raised basement, mansard roof, central Corinthian porch with curved double-entry stairway, square tower, central-hall plan, notable interior.

Notable and early Philadelphia example of the Second Empire style. Built 1864–65; John Fraser, architect; John Crump, builder. Interior burned, remodeled 1889; James H. Windrim, Theophilus Parsons Chandler, Jr., Charles M. Burns, architects; George Herzog, decorator. Two-and-a-half-story brick rear annex built c. 1881; Theophilus Parsons Chandler, Jr., architect. Interior altered, rear addition built 1893; Robert G. Kennedy, architect. Annex enlarged 1896; Keen and Mead, architects. Annex demolished c. 1910. Five-story rear steel-frame annex with granite ashlar curtain walls built in two phases: west section on Fifteenth St. built 1909–10; Horace Trumbauer, architect; James G. Doak & Co., contractors; middle section built 1910–11; Horace Trumbauer, architect; Cramp & Co., contractors. Mansard roof on tower removed after 1927. Interior altered 1945, 1956, 1953; Ralph B. Bencker, architect. Interior altered 1963, 1972.

Organized 1862, the Union League is the oldest Republican Party club in continuous existence. Certified, PHC 1957; Pennsylvania Register 1970. 1 ext. photo (1972),* 1 photocopy of architectural rendering (1864);* 11 data pages (1965, including 1865–1911 insurance agreements and surveys).*

Victory Building. See New York Mutual Life Insurance Company Building

Wanamaker, John, Store
(PA-1535), 1300–1324 Market St., composing block bounded by Market, Thirteenth, Chestnut, Juniper sts. Steel frame, brick curtain walls faced with coursed granite ashlar on upper stories and limestone ashlar on two lower stories, approx. 250' (eleven-bay front) × 480', twelve stories, flat roof, modified Tuscan distyle entrance in antis on each facade, great arches at ninth and eleventh stories; notable interior volumes include spacious five-story Grand Court on ground floor, Egyptian Hall on third floor, Crystal Dining Room on ninth floor; fluted columns, piers, pilasters, and ornate friezes throughout store.

Fine application of Second Renaissance Revival style to a large department store by one the nation's leading architects at the turn of the century. Built in three stages 1902–10; Daniel H. Burnham & Co., architects. Organ, world's largest at the time, from Louisiana Purchase Exposition, St. Louis 1904, installed 1911. World's largest retail merchandising building when built. John Wanamaker Memorial

Museum on eighth floor includes Wanamaker's office as he left it in 1922. Certified, PHC 1963. 1 ext. photo (1973),* 1 photocopy of ext. photo (c. 1960),* 1 photocopy of int. photo of Grand Court (c. 1960).*

West Arch Street Presbyterian Church (now Arch Street Presbyterian Church). Illustrated

(PA-1536), 1726–32 Arch St., at S.E. corner Eighteenth St. Brick and stone covered with plaster, approx. 85' (six-bay front) × 150', one story with Corinthian tetrastyle portico, low gable roofs with central dome, round-arch windows; three-aisle plan, balconies, Corinthian tetrastyle portico with pediment at chancel.

One of Philadelphia's most elaborate examples of the Italianate Revival. Built 1853–55; Joseph C. Hoxie, architect; Joseph DeNegre, builder; Henry Pasco & Son, plaster work. Original balustrade, urns, bell towers, dome's cupola removed after 1895. Sunday School and lecture room. remodeled 1900; Charles W. Bolton & Co., architects. Certified, PHC 1961; Pennsylvania Register 1970; NR. 5 ext. photos (1974),* 5 int. photos (1976);* 3 data pages (1964).*

Western Saving Fund Society

(PA-1537), 1000–1008 Walnut St., at S.W. corner Tenth St. Brick faced with granite ashlar, 164'-4" (six-bay front) × 120'-2" (originally approx. 88' × 120' with three-bay front), three stories (originally two), flat roof, cyclopean rusticated ground story, great arches on coupled red marble columns, two-story banking room with mezzanine.

Eastern section built c. 1887; James H. Windrim, architect; Allen B. Rorke, contractor. Exterior altered 1902; Furness, Evans & Co., architects. Renovated, western section built c. 1910; John T. Windrim, architect. Third story installed later; demolished 1967. 4 ext. photos (1964),* 1 photocopy of old ext. photo (n.d.);* 8 data pages (including 1889, 1909 insurance surveys).*

White Tower

(PA-1538), 159 N. Broad St., at S.E. corner Race St. Brick faced with white porcelain enamel (not original), approx. 27' (four-bay front) × 25', one story with low squarish tower at northwest corner, flat roof, large single-pane windows with stainless steel sashes, goose-neck lamps at roof line, public dining room, small rear storage and rest rooms.

Fine example of merging symbolism of a firm's advertising image with the modern ahistorical style. Built 1932; Charles J. Johnson (New York), architect. Porcelain enamel facade applied to three walls 1950; Erie Enamel Co. (Erie, Pa.), contractor. One of the oldest fast food chains in the country; founded 1926 in Milwaukee, Wisc. 1 ext. photo (1976).*

Academy of Music. *Cortlandt V. D. Hubbard photo, 1967.*

Bulletin Building. *Jack E. Boucher photo, 1973.*

Girard Trust Corn Exchange Bank. *Jack E. Boucher photo, 1973.*

Harrison Building. *Jack E. Boucher photo, 1962.*

Market Street National Bank Building. *Jack E. Boucher photo, 1973.*

Pennsylvania Academy of the Fine Arts. *Jack E. Boucher photo, 1965.*

Pennsylvania Institution for Deaf and Dumb. *Jack E. Boucher photo, 1973.*

Pennsylvania Railroad, Broad Street Station. *Drawing by Furness, Evans & Co., 1893, courtesy Pennsylvania Railroad.*

Philadelphia and Reading Railroad, Terminal Station, Train Shed. *Photo 1893, courtesy Free Library of Philadelphia.*

Philadelphia City Hall. *Cervin Robinson photo, 1958.*

Philadelphia City Hall, toilet stalls. *Cervin Robinson photo, 1959.*

St. Clement's (Protestant Episcopal) Church. *Jack E. Boucher photo, 1974.*

Scott-Wanamaker House. *Cortlandt V. D. Hubbard photo, 1973.*

Sims-Bilsland House (Franklin Row). *Photo c. 1870, courtesy Historical Society of Pennsylvania.*

West Arch Street Presbyterian Church.
Cortlandt V. D. Hubbard photo, 1976.

Spruce Street Area Study. *Jack E. Boucher photo, 1974.*

South Philadelphia

Moyamensing and Passyunk
Townships

4

Introduction

*L*ONDON'S INFLUENCE on Philadelphia can be seen both in the material culture of the colonial city—its plan, pent-eaved houses in rows, and church spires—and in some of its place names—among these, Southwark. Named after that part of London south of the Thames, Philadelphia's Southwark was once a southern suburb that physically and commercially merged with the city but was politically separated from it until 1854 with South Street, then called Cedar Street, serving as the line of demarcation. Although always outside the city, Southwark emerged as a separate political entity within Philadelphia County in 1762, when it was organized as the District of Southwark.[1]

Other place names reveal an even earlier history for the District. The Lenni Lenape tribe called the area Wicaco, later spelled "Weccacoe." This name is sometimes translated as "pleasant place" or "at the pine tree camp," and was retained by early settlers from Europe.[2] The first of these immigrants were the Dutch, who in 1633 bought a tract along the river they called Schuylkill, meaning "hidden creek."[3] They were followed by the Swedes, who began to plan fur-trading posts in the area after 1638, thereby making it, in terms of recorded history, the oldest part of Philadelphia. The subsequent joint venture of Sweden and the Netherlands was marked more by competition than by cooperation, until in 1655 the Dutch forced the Swedish government out of the area.[4] Yet the mark of Sweden remains in such Southwark street names as Swanson (Andreas Swenson owned an 800-acre tract along the Delaware) and Christian (Christina succeeded Gustavus Adolphus as the Swedish monarch in 1632) and in the presence of the renowned Gloria Dei (illus.), affectionately known as

Old Swedes' Church. Built at the turn of the eighteenth century on the site of an earlier Swedish log church near the Delaware River, it has a Flemish bond with glazed headers and lozenge diaperwork that combine with its steep roof and valuted ceiling to give it a deceptively Swedish flavor. Its form, however, and most, if indeed not all, of its details are English.[5] It served as a mission of Sweden's state chruch until 1845, when it became a part of the Protestant Episcopal communion.[6]

After William Penn was granted his land, the claims of earlier inhabitants were recognized and adjusted, and the entire area south of the City of Philadelphia of the time was divided into Moyamensing Township and Passyunk Township, Indian names that together mean "pigeon droppings" "in the valley."[7] Moyamensing included approximately the eastern half of the territory and Passyunk the western half. Except for the creation of the District of Southwark in 1762, both townships retained their extensive sizes and sparse populations until 1848, when their boundaries were redrawn and Moyamensing was reorganized as a district adjacent to Southwark that extended from river to river.[8]

English colonists began to purchase Southwark land in the 1690's, but housing construction did not accelerate until the 1740's, when the city began to grow. The area quickly developed into a residential community whose people were closely tied to the river and the sea.[9] Representative of these early dwellings is the house at 770 South Front Street, which was built in 1745 by William Finlow as an investment property for two merchants, Maurice and Edmund Nichell. It was a plain house with a brick cornice, simple entrance, and pent eave, but it was comfortable, rising three-and-a-half stories and with a two-story rear ell for sitting room and kitchen.[10] Behind it was a court, one of those deep open spaces that was formed in the interior of large blocks in an effort to carve out more building lots.[11] During the eighteenth and early nineteenth centuries this space was filled with two facing rows of small housing units and was called Clymer's Court after George Clymer, a Philadelphia signer of the Declaration of Independence and United States Constitution. Over the span of two centuries both the house and the rear tenements deteriorated, until in 1969 the house was rehabilitated, its exterior restored, and the tenements rebuilt into middle-income dwellings and the open space landscaped.[12] It stands today as an example of imaginative integration of the old and the new, and a living testimony to the reasonableness of eighteenth-century living patterns.

A block to the north, at the southwest corner of Front and Pemberton streets, is a variation on this plan, known as Workman Place. Instead of the rear tenements being strung out along the length of a single lot, the buildings are positioned roughly at right angles across the width of three lots. The two oldest buildings were built in 1748 for George Mifflin,[13] which accounts for the date and initials laid in glazed headers on the blank end walls facing what is now Pemberton Street. These were separate two-and-a-half-story bandboxes, each divided into two tenements, or apartments, with corner winding stairs.[14] At a right angle to these earliest houses and at the rear of the lots facing Front Street is a row of three two-and-a-half-story bandboxes fronting onto the open courtyard.[15] These tenements probably were built at the same time as the three Front Street houses (numbers 742–46); a fourth house, 740 South Front Street, was demolished when Pemberton Street was cut through before 1849. These commodious three-and-a-half-story dwellings were built along Front Street about 1812 for John Workman, who acquired this extensive piece of real estate in 1810.[16] The compound yielded rental income for the Workman family until 1906, when Lydia S. Clark acquired it as a philanthropic housing project, establishing a pattern that has prevailed ever since. The Octavia Hill Association took over the project in 1942, and acquired additional small dwellings facing the courtyard from Fitzwater Street on the south.[17] By renting the units at rates adjusted to the occupants' incomes and preserving the original setting, the association has retained both the function and environment of Workman Place.

Also representative of Southwark housing is Nathaniel Irish's house (illus.) at 704 South Front Street, built during the 1760's.[18] Its masonry foundation, brick walls, dormer windows, and closets flanking the gable-end chimneys fit Peter Kalm's 1748 description of Philadelphia houses.[19] Other features not mentioned by the Swedish traveler but equally characteristic of colonial Philadelphia houses include the vaulted alleyway that is supported on brick arches in the cellar and cuts through at the street level to the backyard; the London house plan that places the winding stairs in the middle of a side wall in order to best utilize a narrow but not excessively long lot; and the ample use of decorative trim on the interior woodwork of an otherwise modest dwelling.

The characteristics of Philadelphia dwellings related by Peter Kalm and still to be seen in Southwark houses disguise, however, a little known fact: the majority of Southwark houses were not brick but frame, according to the 1811 statistics of Dr. James Mease, physician and amateur historian.[20] In fact, at the turn of the nineteenth century there was a great number of frame dwellings throughout the city, including a deceptively pretentious example on Market Street west of Tenth whose wooden siding was painted to imitate brick.[21] Consequently, George Pearson's (illus.) and Joseph Bird's two-and-a-half-story frame houses at 808–10 and 813–15 South Hancock Street respectively are probably more representative of early Southwark dwellings than was the Irish House. Neglect of the southernmost of Bird's two dwellings a couple of years ago led to the unfortunate decay of the tongue-and-groove weatherboards and to the unexpected discovery of a medieval construction technique evidently carried over from England—the use of brick nogging between the frame members. More representative of late-eighteenth-century frame construction was the house that until 1970 stood at the rear of 33 Catharine Street. Random-width board sheathing was nailed to the frame members and covered with the tongue-and-groove siding. It was much faster to build a house this way than with brick nogging and probably provided about the same degree of insulation.

As Southwark moved into the nineteenth century, it retained its ties to the maritime trade of the adjacent river and the nearby sea. This accounts for the great number of warehouses that stood east of Front Street until they fell to the I-95 expressway in the 1960's. Early in the century they were modestly scaled, as could be seen in John LaTour's 1818 warehouse on Water Street.[22] By the mid-nineteenth century, however, as the city became an industrial center as well as a commercial port, warehouses had become much larger. While LaTour's warehouse measured only 21 feet by 50 feet, for example, not much larger than a dwelling house, Jesse Godley's 1867 warehouse (illus.) at 19–27 Queen Street possessed more than eleven times the square footage. By this time a warehouse's site was determined by railroads as well as by the river, as illustrated by Godley's Argyle Stores, which until 1967 stood next to the Reading Company's tracks at 32–36 Kenilworth Street.

It should not be surprising that the earliest Southwark industry of note was shipbuilding. By the late 1790's Joshua Humphreys' shipyard was located just south of Gloria Dei, where he built a variety of ships, including the sailing frigates United States and Philadelphia. The United States Navy Yard was established nearby in 1801. Its two shiphouses, one rising 103 feet into the air, were built in 1821 and 1822 and remained the city's most conspicuous landmarks until the

Navy Yard was removed farther south to League Island in 1876.[23]
Another famous Southwark landmark is Sparks's Shot Tower. It was
the country's first shot tower when it was built in 1808 by Thomas
Sparks and John Bishop, plumbers. As a conscientious Quaker,
Bishop felt obligated to drop out of the partnership when the U.S.
Army became a customer during the War of 1812. Sparks and his
heirs carried on the business until 1907 when the tower was vacated;
six years later it became part of a neighborhood recreation center,[24]
preserving it both as a local landmark and visual reference point and
as a rare relic of an extinct technology.

   With the emergence of small industries unrelated to maritime
activities, Southwark developed social aspects of a nineteenth-
century industrial town. By the middle of the century low-rent housing was available; extant examples include William Harper's 1847
tenements at 621–23 South American Street and an 1880's rooming
house at 115 Bainbridge Street.[25] Their appearance was followed by
related public and quasi-public institutions such as soup kitchens and
public bath houses. Until 1871 fire protection—and a good deal of
neighborhood camaraderie—were provided by volunteer fire companies,[26] such as Hope Engine Company No. 17, whose exuberant
Italianate hall at 733–35 South Sixth Street was built in 1851 from the
designs of Hoxie and Button, a short-lived but prolific architectural
partnership.[27] The Weccacoe Engine Company No. 9, which moved
to 119 Queen Street about 1843, was one of the few volunteer companies to have had its hall continued as a municipal fire station in
1871.[28] Its present appearance is the result of an extensive remodeling
with the addition of a police patrol house in 1893.[29] The only survivor
of Southwark's public school buildings is the 1869 Ralston School at
the northeast corner of American and Bainbridge streets. Until it was
converted to commercial use in 1928, it was a fine example of the
modified "Philadelphia Plan," a proposal of the noted Philadelphia
architect Samuel Sloan, which provided flexible interior space and
wide, safe front and rear exits.[30]

   Churches increased in number and diversity during the nineteenth century. The Church of St. Philip de Neri at 220 Queen
Street, built in 1840–41, was the first Roman Catholic church in the
area. It survived the "Native American" riots of 1844 and a fire in
1897[31] and stands today opposite Mario Lanza Square as a fine
example of a restrained church design based on Greek Revival principles. Not so fortunate, however, was the 1878 Church of the
Redeemer for Seamen of the Port of Philadelphia at the northwest

corner of Front and Queen streets, which was destroyed by fire in
October, 1974.[32] The destruction of this bold expression of contrast-
ing colors, textures, and forms from the drawing boards of the great
Frank Furness[33] was a loss to both America's architectural history and
Southwark's visual diversity. The church's name described its
primary initial function—to serve as a mission church to seamen—
and is evidence that maritime activity in the neighborhood did not
diminish as manufacturing increased. Eastburn Mariners' Bethel
(Presbyterian) to the north at Front and Delancey streets and the
Baptist Mariners' Bethel (demolished 1966) and the Methodist
Mariners' Bethel to the south at 923 South Front Street and Washington and Moyamesing avenues reinforce the point.

During the eighteenth century and much of the nineteenth the
landscape west and south from Southwark was decidedly rural in appearance. Two of the best known architectural relics to survive from
the period are Bel Air and Stephen Girard's Gentilhommière. Both
are presently located in public parks: Bel Air in Franklin Delano
Roosevelt Park and Gentilhommière in Girard Park. Bel Air has stood
as an architectural mystery, or at least a curiosity, for many decades.
Much of this puzzle stems from its diversion from the standard appearance and plan of the eighteenth-century Philadelphia country
house. It has blank end and rear walls, for example, is only one room
deep, has its chimneys in the rear wall, and has a balcony over its
plain center entrance and projecting from the cove cornice a rounded
canopy over the balcony. These architectural irregularities on a
country house have led to claims of Swedish origin.[34] It is more
likely, however, that it was built about 1720 for former Philadelphia
mayor Samuel Preston, and that its design was determined by city
housing practices and English customs rather than by Swedish influence.[35] Unlike Philadelphia town houses, the country houses on
the city's outskirts did not conform to a pattern. In the openness of
the countryside the gentry seem to have felt free to express their individualism in taste. Gentilhommière, on the other hand, is a much
easier case to explain. When it became the country house of the famous merchant-financier Stephen Girard in December, 1797, only
the eastern eighteenth-century section was standing. Girard added
the finely proportioned center and western sections and the two versatile utility buildings before his death in 1831.[36] The house's interior
has been extensively altered through its use as a community center.

Because of its strategic location along the waterways leading to
Philadelphia, the area south of Southwark became a favorite site for

military installations at the turn of the nineteenth century. In 1798, while under the threat of war with France, the national government began building the Schuylkill Arsenal along the east bank of the Schuylkill River at Washington Avenue, and at the same time it opened a Navy Yard along the Delaware River in Southwark.[37] Also, beginning in the same year, the national government essentially rebuilt and enlarged Fort Mifflin on Mud Island, where American rebels had made a heroic but unsuccessful stand against the British in 1777.[38] A decade later, with the talk of war again facing the nation, the Commonwealth of Pennsylvania built a powder magazine in Passyunk,[39] not far from the fort. Except for Fort Mifflin, whose isolation has allowed it to stand unmolested over the years, these facilities are now gone, victims of modernized defense strategies and establishments. The fort is undergoing restoration as a fine example of early-nineteenth-century fortifications.

By the second quarter of the nineteenth century, Moyamensing began to assume features of a pastoral countryside, a happy compromise between urban congestion and wilderness. This made it a favorite site for social institutions of the day. One of the most handsome and serviceable of these is the United States Naval Asylum (illus.) for retired naval personnel at Gray's Ferry Avenue and Twenty-fourth Street, a short distance from the Schuylkill Arsenal's site. The main building, Biddle Hall (named for the Asylum's first Governor, Commodore James Biddle), was designed by William Strickland and built in 1827–33.[40] It is an exemplary piece of architecture and, in producing its design and plan, William Strickland drew upon his understanding of the history of his profession, his skill as an artist and draughtsman, his awareness of recent technological advances, his knowledge of English and European building precedents, and his sensitivity to the current attitude about social reform. Stuart's and Revett's *Antiquities of Athens* was evidently the source for the broad Ionic octastyle portico that serves as both the focal point and central entrance for the large building. The manner in which the long dormitory wings extend from either side of this block is reminiscent of the European homes for service pensioners. Yet the heaviness that could result from such a large structure is alleviated by placing along the length of these wings piazzas or balconies supported by thin cast-iron columns. This not only solves a design problem but also anticipates a contemporary concern for the healthful effects of fresh air and ventilation. Furthermore, Strickland's early use of expressed cast-iron construction on these balconies illustrates his willingness to experiment with the new materials produced by the Industrial Revolution in an effort to construct as safe, economical, and modern a building as was then possible; in fact, it was reputed to be fireproof with its brick floors, vaulted ceilings, and slate and copper roofs. The beauty and usefulness of Biddle Hall are still evident in spite of the raised roof and dormers that altered its proportions in 1848[41] and raise questions about the soundness of the Navy's decision to vacate the building and neighboring facilities in 1976 in favor of a newly constructed high-rise in Gulfport, Mississippi.

The social theory that a pastoral setting was best for the rehabilitation of criminals helped to determine the site of the Philadelphia County Prison (illus.), or as it was commonly known, Moyamensing Prison or "Old Moko." Designed in 1831 by Thomas U. Walter,[42] it was an excellent representative of what Professor Alan Gowans calls early Victorian eclecticism of taste, making a visual statement through the careful selection of elements from a single historical style.[43] Like its predecessor, Eastern State Penitentiary in the city's Fairmount section, built in 1822–36, Moyamensing Prison was designed in the English Gothic castle style of architecture in order to create the impression of an invulnerable fortress with dismal dungeons within. Its heavy granite walls and octagonal towers served both as visual reminders of the certainty and permanence of punishment and as an effective means of detention and surveillance of prisoners. The adjacent Egyptian Revival Debtors' Prison, also designed by Walter, conveyed essentially the same message in another architectural style.[44] Both buildings were found to be unsuitable for the penal needs of the last half of the twentieth century and were demolished in 1968 without protests from their alumni.

The consolidation of Philadelphia in 1854 led to the division of South Philadelphia into four wards. The first encompassed the former Moyamensing District and Passyunk Township, while the old Southwark District was carved into the second, third, and fourth wards. The area's population grew consistently during the next sixty to eighty years, first westward along South Street and Washington Avenue, then down Broad Street, until by 1914 South Philadelphia included nine wards.[45] Some of these wards were congested and conditions of life there constituted a threat to public health. In 1910, for example, the third ward, between Fitzwater and Christian streets east of Broad Street, had a population density of 129,282 per square mile, a crowded situation as compared to Germantown's density of 14,000 per square mile at the same time.[46] Many of the residents were newcomers from Europe. Irish immigrants continued to flow into the city during the last half of the nineteenth century, some of them into this area. But the turn of the century an increasing number of eastern European Jews moved directly from disembarkation points on Delaware Avenue to the neighborhood below South Street, generally east of Eighth. Thousands of Italian immigrants settled throughout the area, with particularly dense concentrations west of Seventh Street between Bainbridge Street and Snyder Avenue.[47] Perhaps the most exciting living monument of the Italian presence in Philadelphia is the Italian Market, an open-air market extending along Ninth Street from just north of Christian Street to Federal Street. Here hucksters continue Old World merchandising practices that slowly disappeared in other parts of Philadelphia after the Civil War.

By 1920 most of South Philadelphia was filled with corridors of row houses and semidetached dwellings. Since these blocks were usually real estate developments, the houses possessed rigid uniformity. Some developers and builders ritualistically continued the colonial tradition of brick fronts and marble stoops, as can be seen along much of Federal Street. Others carried the classical manner one step beyond its early-nineteenth-century clarity and added hexagonal bays and open-columned porches to red and buff brick fronts. In some cases, architects were utilized to design these developments. One of the better examples is the well-preserved block on South Twentieth Street between Shunk and Porter streets where in 1910 John T. Windrim, working for the Trustees of the Girard Estate,[48] grafted the Colonial Revival style, then much used for suburban houses, onto modest semidetached city dwellings. The visual monotony of these neighborhoods was relieved by handsome churches, such as the Church of St. Charles Borromeo (1868–76) at Twentieth and Christian, St. Gabriel Roman Catholic Church (1902–12) at Twenty-ninth and Dickinson, and Church of St. Thomas Aquinas (1901–13) at Seventeenth and Morris streets.[49] All three were designed by Edwin F. Durang, who was responsible for many of the city's Roman Catholic churches between 1870 and the first World War.

An occasional square or playground provided open space and opportunities for off-street recreation. The most notable of these is Franklin Delano Roosevelt Park, known as League Island Park until 1955. Its development was begun in the early years of this century when the city began reclaiming land on the west side of Broad Street below Pattison Avenue. By 1923 the project was completed and the

365-acre park became the center of the Sesqui-Centennial Celebration in 1926.[50] In addition to stimulating the development of the south end of Broad Street, the Sesqui-Centennial accounted for two permanent structures of note, the American Swedish Historical Museum and the Municipal Stadium. In 1926 the museum, smaller than it is at present, was known as the John Morton Memorial Museum; it was completed in 1935, in time for the three-hundredth anniversary celebration of the arrival of the Swedes in Pennsylvania.[51] The Municipal Stadium on the east side of Broad Street was designed by the Philadelphia architects Simon and Simon to accommodate 73,073 persons, but 120, 757 crowded in to see the first event to be staged there, the famous Dempsey-Tunney fight in 1926.[52] It was renamed the Philadelphia Stadium in 1959 and the John F. Kennedy Stadium in 1964. The biggest athletic event held there in recent years has been the annual Army-Navy football game, yet its presence near the Philadelphia Veterans Stadium, a baseball-football facility completed in 1971, and the Spectrum, a center for basketball, hockey, and boxing that opened in 1967, makes this corner of Philadelphia one of the nation's major sports complexes.

Broad Street has been the main thoroughfare of South Philadelphia for the past century. Many of the area's important social institutions are still to be found along its path: St. Agnes Hospital at Mifflin Street, South Philadelphia High School at Snyder Avenue, Methodist Hospital at Wolf Street, and the United States Naval Hospital at Pattison Avenue. The most impressive of these institutional structures architecturally is the Ridgway Branch of the Library Company of Philadelphia (illus.). The library, a gift from the estate of Dr. James Rush, son of the Signer Dr. Benjamin Rush, was constructed on Broad Street between Carpenter and Christian in anticipation of a strong southward development of the city. The directors of the Library Company, unhappy about the site chosen in 1869, interrupted construction for two years while they went to court in an unsuccessful effort to change the location of the branch library to one less inconvenient for its members.[53] The directors may have been selfishly motivated, but in hindsight they appear to have been prescient, since the presence of the Philadelphia, Wilmington and Baltimore Railroad tracks along Washington Avenue stymied residential development in the area until the advent of the automobile.[54]

Much in the manner of the Pharaohs who used their wealth to build both monuments to their memories and tombs for their bodies, Dr. Rush left a million-dollar trust for the erection of a library building, for the tomes of the Library Company, and tombs for himself and his wife. This may account for the library's morbid and redundantly monumental appearance. In search of overt allusions to erudition and the arts, architect Addison Hutton turned to ancient Greece for inspiration. Yet this building was to be grander than a mere classical temple, so Hutton abandoned the traditional single portico in favor of three: a central Doric octastyle portico and two flanking tetrastyle porticoes arrayed across the 220-foot front. Such pomposity combines blank areas of dark ashlar and heavy columns with exaggerated entasis and uncomely seams to make the building quite uninviting, perhaps a not altogether inappropriate appearance for a private library to have. The cavernous interior offered magnificent spaces and ample book stacks, but its great volume also made it nearly impossible to provide an appropriate environment for the preservation of the library's rare books and manuscripts, to say nothing of its staff and readers. Consequently, in 1966 the Library Company moved its collections, including the remains of Dr. and Mrs. Rush, to new quarters next to the Historical Society of Pennsylvania at Thirteenth and Locust streets,[55] where these two cooperating institutions form one of the nation's greatest private repositories of historical materials. Incidentally, the library's name was one of the conditions of Rush's generosity. Ridgway was the maiden name of his wife Phoebe Ann, the daughter of the prosperous merchant Jacob Ridgway, who was the principal source of James Rush's fortune.

South Philadelphia has shown remarkable resilience over the decades. Its residents take great pride in their neighborhoods. Probably the most disruptive force to attack the area in this century has been I-95, the express highway referred to earlier. Implicit in its construction was the destruction of many dwellings and warehouses, especially many fine old eighteenth-century buildings between Front Street and the Delaware River. In spite of this loss, the old Southwark area is making a comeback as Queen Village.[56] Helped by the success of adjacent Society Hill and by the decision not to construct the Crosstown Expressway along Bainbridge Street, Queen Village, with South Street as its revitalized northern boundary,[57] is a successful community effort at rehabilitation and restoration of many houses and the preservation of others that otherwise might have been lost.

Catalog

Annan, William, House

(PA-1539), 776 S. Front St. Brick with marble trim, approx. 20' (two-bay front) × 31' with two-story rear ell, three-and-a-half stories on raised basement, gable roof, London house plan. Built c. 1800; altered later. Certified, PHC 1957. 1 photocopy of ext. photo (1957);* 3 data pages (1961, including 1822 insurance survey);* HABSI form (1957).*

Arsenal on the Schuylkill (later known as the Schuylkill Arsenal)

(PA-1540), 2620 Gray's Ferry Ave., at N.W. corner Peltz St. at Washington Ave. Begun in 1799 as the nation's third arsenal, Schuylkill Arsenal expanded to include twenty-three buildings on its 8.6 acres when it was demolished in 1962. After 1816 it was used only for manufacturing Army clothing and blankets; during World War II and Korean conflict it became a barracks and recruiting and examination center. Certified, PHC 1957.

Building No. 1, Enlisted Men's Barracks (also known as Luding Hall): Approx. 100' W. of Gray's Ferry Ave. Brick with stone trim, approx. 100' (eleven-bay front) × 42' including wing, three-and-a-half stories, hipped roof, wing open at ground story, 1800 date stone. Built 1800; wing enclosed after 1876; demolished 1962. 4 ext. photos (1958);* 4 data pages (1958).*

Building No. 1A, Executive Officers' Quarters: Approx. 55' N. of Building No. 1. Brick with stone trim, approx. 50' (five-bay front) × 24' with two-story side wing (not original), three-and-a-half stories on raised basement (originally two), hipped roof, one-story entry porch, cornice between second and third stories, third-story round-arch windows. Built 1802; side wing added later; third story, porch added after 1876. 1 ext. photo (1958);* 1 data page (1958).*

Building No. 2, Mifflin Building: Approx. 55' S. of Building No. 1. Brick with stone trim, approx. 150' nine-bay front) × 60', four stories on Peltz St., three stories on courtyard, flat roof (originally hipped), three-story tower with portecochere and octagonal cupola (originally centered on building). Built 1804; alteration after 1876; demolished 1962. 2 ext. photos (1958).*

Building No. 6: Approx. 35' N.W. of Building No. 1. Brick with stone trim, approx. 145' (ten-bay front) × 45', two stories, hipped roof, Tuscan colonnade with balcony along south front, Palladian windows set in blind arches. Built 1804; balcony enclosed, interior altered after 1876; demolished 1962. 2 ext. photos (1958).*

Building No. 8, Powder Magazine: Rear of grounds on Peltz St., approx. 410' E. of Schuylkill River. Coursed granite ashlar ground story, brick upper story, approx. 100' (six-bay front) × 100', two stories, hipped roof, brick groined-arch ceiling on stone posts; surrounded by 10'-wide moat. Built c. 1799; upper story added 1802; demolished 1962. 1 ext. photo (1958),* 1 int. photo (1958).*

Askins-Jones Houses

(PA-1541), 720–24 S. Front St. Brick, Nos. 720–22 each approx. 12' (two-bay front) × 20' with one-story rear ell, No. 724 approx. 15' (two-bay front) × 25', two-and-a-half stories, half-gambrel roofs, London house plans. Built between 1762 and 1769; No. 720 altered to store before 1860, reverted to dwelling after 1931; exterior of No. 722 restored 1970. Certified, PHC 1957. 3 ext. photos (1961, 1967);* 9 data pages (1960);* HABSI form (1957).*

Beck Street Area Study

(PA-1542), 203–15 Beck St., on N. side Beck St., W. of Second St. Court of seven brick row houses, each approx. 16' (two-bay front) × 16' with two-story rear ell, two-and-a-half stories on raised basements (No. 215 three-and-a-half stories), gable roofs, probably city house plans. Built late 18th century. Certified, PHC 1965. 2 ext. photos (1961).*

Bel Air (also known as Belleaire, Singley House, Lasse Cock's Manor House)

(PA-1124), S.W. corner Twentieth St. and Pattison Ave., in Franklin Delano Roosevelt Park (known as League Island Park c. 1922–55). Brick with rubble foundation, approx. 42' (five-bay front) × 21', two-and-a-half stories, gable roof with cove cornice and pent eave around gable end, one-bay balcony over entrance, brick stringcourse, blank end walls; central-hall plan, one room deep, notable interior paneling.

Notable early Georgian country house with distinctive early Philadelphia features. Built between 1714 and 1729, possibly incorporating an earlier building; interior probably remodeled c. 1735; restored 1934–41. Until 1735 country seat of Samuel Preston, Mayor of Philadelphia 1711–14, and Treasurer of the Province of Pennsylvania 1715–43. Maintained by Fairmount Park Commission, City of Philadelphia, since 1929. Certified, PHC 1956. 30 sheets (1940, including plans, elevations, section, details) from Survey, Philadelphia Chapter, A.I.A. (1932); 3 ext. photos (1936), 5 int. photos (1936).

Bellaire. See Bel Air

Bird, Joseph, Houses
(PA-1543), 813–15 S. Hancock St. Wood frame with clapboards, each approx. 14' (two-bay front) × 14' with one-story rear ell, two stories, gable roofs. Rare surviving frame dwellings with original brick nogging. Built after 1786. Certified, PHC 1966. 1 ext. photo (1961);* 3 data pages (1961).*

Bishop-Sparks House
(PA-1544), 948 S. Front St. Brick with marble trim, approx. 18' (two-bay front) × 32' with two-and-a-half-story rear ell, two-and-a-half stories on raised basement, gable roof. Built between 1808 and 1813. John Bishop and Thomas Sparks were initial owners of Sparks's Shot Tower (PA-1621). Certified, PHC 1957. 1 photocopy of ext. photo (1957);* 2 data pages (1961);* HABSI form (1958).

Byrne-Cavenaugh Houses
(PA-1545), 130–32 Queen St., at S.W. corner Hancock St. Brick, each approx. 15' (two-bay front) × 15' with two-story rear frame kitchen wing, three-and-a-half stories, shed roof, round-headed entrance (No. 132 altered to square head), one room to a floor. Built between 1813 and 1819 as tenements at rear of dwelling-tavern at S.E. corner Second and Queen sts. Certified, PHC 1958. 2 ext. photos (1961, 1974);* 6 data pages (1961, including 1835 insurance survey).*

Catharine Street Area Study
(PA-1811), 19–35 Catharine St., N. side Catharine St., in block between Swanson and Front sts. Eight contiguous brick houses and one frame house (No. 35), each approx. 20' (two-bay front) × 22' and

38', three houses with rear ells, two-and-a-half stories (No. 35 two stories), gable roofs, probably originally London house plans. Nos. 19–25 built between 1810 and 1830; Jesse Williamson and David Flickwer, house carpenters. Nos. 27–31 built c. 1791; No. 33 built before 1830; No. 35 built before 1810. Demolished 1967. Certified, PHC 1958. 1 ext. photo (1967).*

Church of the Redeemer for Seamen of the Port of Philadelphia

(PA-1077), 101–7 Queen St., at N.W. corner Front St. Trenton brownstone rubble, brick and frame, approx. 100' (nine-bay front) × 63', one and two stories with squat hipped-roof tower, intersecting gable roofs, Greek-cross plan with two-story school wing, elaborate interior truss work and framing system that eliminates columns between aisles and nave.

Innovative High Victorian Gothic church design. Built by Churchmen's Missionary Association for Seamen of the Port of Philadelphia 1878; Frank Furness, architect; Williams and McNichol, builders; Manly and Cooper, iron railing. Destroyed by fire 1974. Although interdenominational by charter, the church retained close ties with the Protestant Episcopal Church until it merged with the Seamen's Church Institute of Philadelphia and closed its doors 1922. Neighborhood boys' club 1925–68; vacated 1968. Certified, PHC 1957. 5 ext. photos (1961, 1962), 3 int. photos (1961); 5 data pages (1962).

Church of St. Charles Borromeo (Roman Catholic)

(PA-1546), 900 S. Twentieth St., at S.W. corner Christian St. Coursed brownstone ashlar with granite base and copper trim, approx. 75' (five-bay front) × 150', one story with three-stage French-roof tower on southeast corner and five-stage domed tower on northeast corner, gable roof; three-aisle plan, rectangular chancel, balconies.

Fine High Victorian baroque church design. Built 1868–76; Edwin F. Durang, architect; sanctuary and ceiling frescoes by Filippo Costaggini and Lorenzo C. Scataglia. Sandstone cupola with terra cotta dome put on northern tower 1901; Edwin F. Durang, architect; John McShain, contractor. 1 photocopy of old ext. photo (c. 1905),* 1 photocopy of old int. photo (c. 1905),* 1 photocopy of old lithograph (c. 1880).*

Church of St. Philip de Neri (Roman Catholic)

(PA-1547), 220–28 Queen St. Brick covered with coursed stucco, approx. 65' (three-bay front) × 115', two stories, gable roof, four giant

Doric pilasters on front, three-aisle plan. Built 1840–41. Interior extensively renovated after fire 1897; Frank R. Watson, architect. Interior altered 1908; Watson and Huckel, architects. Three-story brick rectory (218 Queen St.) built on east end 1902; Frank R. Watson, architect. Site of Philadelphia Native American riots 1844. Certified, PHC 1966. 2 ext. photos (1974),* 1 photocopy of old ext. lithograph (c. 1880).*

Clifton, John, House

(PA-1548), 852 S. Front St. Brick, approx. 16′ (two-bay front) × 25′ with two-story rear ell, two-and-a-half stories on raised basement, gable roof, side-hall plan. Built between 1745 and 1754; ground story altered to store 1860; altered to dwelling, new front wall added 1938. Certified, PHC 1957. 6 data pages (1961, including 1845, 1887 insurance surveys);* HABSI form (1957).*

Collins, Samuel, House

(PA-1549), 783 S. Front St. Brick with marble trim, approx. 22′ (two-bay front) × 42′, three-and-a-half stories on raised basement, gable roof, round-headed entrance, side-hall plan. Built between 1827 and 1842. Certified, PHC 1957. 1 ext. photo (1961);* 2 data pages (1961);* HABSI form (1958).*

Commandant's Quarters. See United States Naval Base, Quarters "A"

Curtis, John, House

(PA-1550), 785 S. Front St. Brick with marble trim, approx. 19′ (three-bay front) × 30′ with two-story rear ell, two-and-a-half stories, gambrel roof. Built between 1793 and 1840; altered later. Certified, PHC 1957. 1 photocopy of ext. photo (1957);* 3 data pages (1961);* HABSI form (1958).*

Donaldson, William, House

(PA-1551), 14–16 Queen St. Wood frame with clapboard siding, 23′-9½″ (four-bay front) × 17–9″ with one-story rear frame wing, two-and-a-half stories on raised basement, gable roof, originally one room to a floor.

Example of small dwelling of Southwark artisans who made their living on Delaware River. Built c. 1795; rear wing built after 1860; demolished 1967. Certified, PHC 1966. 4 sheets (1966, including plot plan, plans, elevation, section, detail);* 2 ext. photos (1966),* 3 int. photos (1966);* 12 data pages (1964).*

Duché House

(PA-1552), 24 Catharine St. Brick with marble trim, approx. 21' (two-bay front) × 33' with two-story rear ell, two-and-a-half stories on raised basement, gable roof, round-arch entrance with fanlight, side-hall plan. Built before 1827; kitchen added to ell 1855; demolished 1967. Certified, PHC 1963. 2 ext. photos (1966, 1967),* 4 int. photos (1967);* 15 data pages (1966, including 1831–55 insurance surveys).*

Duché-Walker House

(PA-1553), 26 Catharine St. Brick, approx. 19' (two-bay front) × 21' with one- and two-story rear ell, two-and-a-half stories on raised basement, gambrel roof, originally one room to a floor. Built c. 1774; rear ell added before 1831; rear frame kitchen added 1862; demolished 1967. Certified, PHC 1963. 1 sheet (1967, including plot plan, details);* 2 ext. photos (1966, 1967),* 2 int. photos (1967);* 14 data pages (1963, including 1831–71 insurance surveys).*

Eckert-Tarrant House

(PA-1554), 38 Catharine St. Wood frame with clapboard siding, 16'-1³/4" (two-bay front) × 16'-4" with one- and two-story rear ell, two-and-a-half stories, gable roof, originally one room to a floor. Typical small dwelling of Southwark artisans who made their livings on the Delaware River. Built between 1794 and 1802; later rear additions; demolished 1967. Certified, PHC 1963. 3 sheets (1966, including plot plan, plans, elevation, section, details);* 2 ext. photos (1966, 1967),* 2 int. photos (1967);* 9 data pages (1963).*

Elliott, John, House

(PA-1555), 37 Queen St. Brick, approx. 17' (two-bay front) × 18' with two-story rear ell, three-and-a-half stories on raised basement, gable roof, round-arch entrance with fanlight, side-hall plan. Built c. 1813; demolished 1967. Certified, PHC 1957. 3 ext. photos (1961, 1966),* 4 int. photos (1967);* 10 data pages (1964, including 1813 insurance survey).*

Elwell, Henry, House

(PA-1556), 812 S. Front St. Brick with brownstone trim, approx. 18' (two-bay front) × 32' with two- and three-story rear ell, two-and-a-half stories on raised basement, gable roof, round-headed entrance, originally two rooms to a floor with rear stair tower, side-hall plan. Built 1822–23; interior altered later. Certified, PHC 1957. 1 photocopy of ext. photo (1957);* 4 data pages (1961, including 1823, 1850 insurance surveys);* HABSI form (1957).*

Ely-Osbourn Houses and Stores

(PA-1557), 136–38 South St., at S.W. corner Hancock St. Brick with marble trim, approx. 26' (four-bay front) × 60' with three-story wing on corner building and three- and two-story ell on other building (originally approx. 38' × 30' with rear stair tower), three-and-a-half stories, gable roof, rectangular plan with stores on ground story (originally front buildings divided into front shops and rear dwelling rooms with cellar kitchens).

Built as unit of three 1833–34; probably by Ebenezer Osbourn, house carpenter. Corner building altered to tavern, three-story rear wing built 1844; interior Nos. 138–40 altered, two-story kitchen ell built 1844; corner building's front upper stories, rear wing, and storefront replaced 1910; new storefront on No. 138 built 1922; westernmost building (No. 140) demolished between 1956 and 1958; Nos. 136–38 vacated 1967. 1 ext. photo (1967);* 6 data pages (1961, including 1834–44 insurance surveys).*

Finlow-Nichell House

(PA-1158), 770 S. Front St., at N.W. corner Clymer St. Brick, approx. 17' (two-bay front) × 19' with two-and-a-half-story half-gambrel roof rear ell, three-and-a-half stories, gable roof, pent eave between first and second stories, side-hall plan. Built 1745; restored 1969–70. Certified, PHC 1957. 2 ext. photos (1961);* 8 data pages (1961, including 1818, 1834 insurance surveys);* HABSI form (1957).*

First Polish Baptist Church. See Mariners' Bethel (Baptist) Church

Fitzwater Street Area Study

(PA-1559), 11–23 Fitzwater St., on N. side of Fitzwater St. between Swanson and Front sts. Seven contiguous brick houses, each approx. 18' (two-bay front) × 18' with one- or two-story rear ells, two-and-a-half stories on raised basements (No. 17 three stories), originally gable roofs, originally city house plans. Built at various times between 1761 and 1809; demolished 1967. Certified, PHC 1957. 4 ext. photos (1961, 1967);* 38 data pages (1961, 1966, including 1808, 1828 insurance surveys).*

Fitzwater Street Area Study

(PA-1560), 24–32 Fitzwater St., on S. side Fitzwater St. between Water and Front sts. Five brick row houses, each approx. 17' (two-bay front) × 17' with two-story rear ell, three stories, gable roofs, city house plans. Built between 1821 and 1830; demolished 1967. A row of houses built for rental income. Certified, PHC 1957. 2 ext. photos (1967);* 8 data pages (1961, including 1830 insurance survey).*

Flickwer-Williamson Houses

(PA-1561), 809–11 S. Hancock St. Wood frame, each approx. 11' (two-bay front) × 14' with one-story rear ell, two-and-a-half stories, gable roof. Built after 1802; probably by David Flickwer and Jesse Williamson, house carpenters. Certified, PHC 1966. 1 ext. photo (1961);* 3 data pages (1961).*

Franklin Sugar Refinery Warehouse (also known as Harrison, Frazier & Co.)

(PA-1562), 701–15 S. Front St., composing block bounded by Front, Kenilworth, Water, Bainbridge sts. Brick, approx. 122' (eleven-bay front) × 130' (originally approx. 82' × 130'), six stories with mezzanine, three gable roofs, brick-arch vaults under Front and Kenilworth sts. Eastern section built as one-story boiler house c. 1881; raised to six stories and mezzanine for storage purposes and enclosed bridge to buildings in next block built at third story c. 1886; western addition built after 1886, retaining exterior brick wall of adjoining building; bridge later raised three stories. Demolished 1967. 1 ext. photo (1967);* 2 photocopies of plans and views (1882, 1886).*

Fullerton, John, Houses

(PA-1563), 606–8 S. Front St. Two adjoining plastered brick houses, each approx. 14' (two-bay front) × 20' with one-story rear kitchen wing, two-and-a-half stories, half-gambrel roofs, originally one room to a floor. Evidently back buildings built in expectation of front buildings that were never built. Southern house (No. 608) built c. 1748; northern house built c. 1760. Certified, PHC 1957; Pennsylvania Register 1970. 2 ext. photos (1961, 1967);* 8 data pages (1961, including 1765 insurance survey);* HABSI form (1957).*

Gentilhommière (also known as Stephen Girard's Country House)

(PA-140), N.W. corner Twenty-first and Shunk sts. Girard Park. Brick with stucco on lateral wings, 80'-0" (ten-bay front) × 24'-4", two and two-and-a-half stories, gable and hipped roofs, one large room in central section, three-room plan in eastern section, central-hall plan in western section, French influence in casement windows and hardware of eastern section.

Fine example of the evolution of a country house. Eastern section built 1750; center section built c. 1800; western section built after 1825. Country house of Stephen Girard, French-born Philadelphia merchant, financier, and philanthropist 1798–1831. Administered by Department of Recreation, City of Philadelphia. Certified, PHC 1956. 11 sheets (1962, including plot plan, plans, elevations, partial section, details); 12 ext. photos (1940, 1962), 6 int. photos (1940, 1962), 5 photocopies of old ext. photos (c. 1900), 1 photocopy of old watercolor (1891); 20 data pages (1962).

Gentilhommière, Utility Building
(PA-1082), N.W. corner Twenty-first and Shunk sts. Girard Park. Approx. 35' N.E. of Country House. Brick, approx. 17' × 17', two stories with brick-domed cellar, pyramidal roof, upper smoke room, middle cooling room, lower root cellar. One of two identical buildings. Built early 19th century; probably from designs of Stephen Girard. Slightly altered 1901. Certified, PHC 1956. 3 sheets (1962, including plans, elevations, sections, details); 1 ext. photo (1965), 4 int. photos (1962); 4 data pages (1962, 1964).

Glebe House
(PA-139), Old River Road, south of Passyunk Ave. Bridge. Brick, 42'-4" (five-bay front) × 20'-3" with rear kitchen ell, two-and-a-half stories, gable roof with coved cornice, central-hall plan. Built 1698; interior damage by fire and partially rebuilt between 1800 and 1820; demolished before 1956. 14 sheets (1940, including plans, elevations, section, details); 2 ext. photos (1940).

Gloria Dei (Protestant Episcopal) (also known as Old Swedes Church).
Illustrated
(PA-120), 929 S. Water St. Brick, 30'-0¾" (three-bay front) × 79'-1½" with small side wings, one story with one-stage belfry and spire, gable roof with cross-gables, diamond patterns of glazed headers flank entrance; three-aisle plan (earlier two-aisle plan), galleries, semioctagonal chancel.

Notable early church whose English construction and details reflect the acculturation of its congregation. Built 1698–1700; Rev.

Andrew Rudman, builder. Side wings built 1703–5. Interior altered to three-aisle plan 1896; George C. Mason, architect. Wood details from original windows in Architectural Study Collection, Museum Division, Independence National Historical Park. Oldest extant church building in Pennsylvania and one of the oldest in the nation in continuous service. Mission of State Church of Sweden until 1845 when it became part of the Diocese of Pennsylvania of the Protestant Episcopal Church. Certified, PHC 1956; designated National Historic Site 1942. 10 sheets (1936, including plot plan, plans, elevations, sections, details); 10 ext. photos (1937, 1962, including 3 of c. 1830 rectory), 3 int. photos (1937), 4 photocopies of measured drawings (1897);* 2 data pages (1937).

Godley, Jesse, Warehouse. Illustrated
(PA-1564), 19–27 Queen St. Random ashlar with brick front, 109'-5" (nine-bay front) × 102'-2", three stories, flat roof, segmental-head windows, three interior sections with wood columns and stone walls.
    Fine example of urban vernacular architecture. Built c. 1867; demolished 1967. Accessible to river and rail transportation, this warehouse was a major unit in Jesse Godley's storage business. Certified, PHC 1966. 5 sheets (1966, including plot plan, plans, elevation, section, details, hoist machinery);* 3 ext. photos (1961, 1966),* 2 int. photos (1966);* 8 data pages (1966).*

Godley's Argyle Stores
(PA-1565), 32–36 Kenilworth St. Stone with brick front, approx. 63' (seven-bay front) × 68', three stories, flat roof, segmental-arch doors and windows with stone keystones, rectangular plan. Built c. 1868; altered after 1868; demolished 1967. One of Jesse Godley's many warehouses, used for miscellaneous merchandise. 1 ext. photo (1967).*

Harmony Engine Company No. 6
(PA-1566), 730–32 S. Broad St. Brick with granite ashlar front, approx. 34' (three-bay front) × 60', four stories, mansard roof, projecting center semihexagonal bay at second story. Built c. 1868. Site of Franklin Hose Company No. 28 until 1867. Certified, PHC 1971. 2 ext. photos (1963).*

Harper, William, Jr., Houses
(PA-1567), 621–23 S. American St., at N.E. corner Harper's Pl. Two contiguous brick tenements, each approx. 15' (two-bay front) × 90',

three stories, flat roof, rectangular plan with seven dwellings in each building. Example of mid-19th-century low-cost housing. Built c. 1847. 1 ext. photo (1967).*

Harrison, Frazier & Co. See Franklin Sugar Refinery Warehouse

Hart, John, House
(PA-1568), 601 S. Front St., at S.E. corner South St. Brick, approx. 18' (three-bay front) × 36', three-and-a-half stories, gable roof, store in ground-story front room. Built between 1768 and 1771; altered to store before 1860; demolished 1966. Certified, PHC 1957. 1 photocopy of ext. photo (1957);* 6 data pages (1961, including 1771, 1840 insurance surveys);* HABSI form (1957).*

Hart-Patterson House
(PA-1569), 603 S. Front St. Brick, approx. 23' (three-bay front) × 36', three-and-a-half stories on raised basement, gable roof, wooden lintels carved to simulate stone voussoirs, London house plan. Built between 1768 and 1771; altered to store before 1860; demolished 1966. Certified, PHC 1957. 1 ext. photo (1961);* 6 data pages (1961, including 1771, 1840 insurance surveys);* HABSI form (1957).*

Hellings, Benjamin, House
(PA-1570), 931 S. Front St. Brick, approx. 14' (two-bay front) × 17', two-and-a-half stories, gable roof. Built c. 1762; demolished 1961. Certified, PHC 1957. 1 photocopy of ext. photo (1957);* 3 data pages (1961);* HABSI form (1958).*

Henderson, James, House
(PA-1571), 602 S. Front St. Brick, approx. 20' (two-bay front) × 30', three-and-a-half stories, gable roof, London house plan. Built c. 1795; altered to store before 1860; reverted to dwelling later. Certified, PHC 1957; Pennsylvania Register 1970. 3 data pages (1961, including 1795 insurance survey);* HABSI form (1957).*

Hope Engine Company No. 17
(PA-1572), 733 S. Sixth St. Stuccoed brick with cast-iron trim, approx. 17' (three-bay front) × 45', four stories, flat roof, round-arch windows with engaged columns, corbeling at each story, rectangular plan.

Fine example of Italianate volunteer fire hall. Built 1851–52; Hoxie and Button, architects; Edward Storey, builder; William Trout

& Co. (Chester, Pa.), iron work. Ground-story interior and exterior altered 1907. Firehouse 1852–71; fraternal society lodge 1871–1920. Certified, PHC 1971. 2 ext. photos (1963);* 1 data page (1963).*

House

(PA-1573), 30 Catharine St. Brick with stone trim, approx. 16' (two-bay front) × 40' with two-story rear wing, three stories, flat roof. Built after 1871; demolished 1967. 2 ext. photos (1966).*

House

(PA-1069), 33½ Catharine St., at rear of 33 Catharine St. Wood frame with tongue-and-groove random width siding, 15'-4½" (two-bay front) × 14'-5⅞", two-and-a-half stories, gable roof, one room to a floor. Built c. 1790; dismantled above ground story 1962; demolished 1970. Portions of siding preserved at Architectural Study Collection, Museum Division, Independence National Historical Park. 2 sheets (1961, including plot plan, plan, section, isometric detail of exterior corner);* 3 ext. photos (1961),* 2 int. photos (1961);* 4 data pages (1961, 1962).*

House

(PA-1574), 768 S. Front St. Brick, approx. 18' (two-bay front) × 19' with two-story rear ell, two-and-a-half stories, gable roof, originally one room to a floor. Built c. 1744; rehabilitated 1961. Certified, PHC 1957. HABSI form (1957).*

House

(PA-1575), 600 S. Second St., at S.W. corner South St. Brick with marble trim, approx. 20' (two-bay front) × 35', three-and-a-half stories, hipped roof, rectangular plan with ground-story restaurant (originally store). Built early 19th century; new storefront built 1914. Interior remodeled 1970; Venturi and Rauch, architects. 1 ext. photo (1967).*

Houses

(PA-1580), 125–27 Bainbridge St. Two contiguous brick houses, each approx. 20' (three-bay front) × 20' with two-story rear ell, two-and-a-half stories on raised basement, gable roof, arched ground-story alleyway between houses, probably originally city house plan. Modest dwellings of artisans who worked on the Delaware River. Built before 1784; altered later. Certified, PHC 1957. 1 ext. photo (1967);* 4 data pages (1961).*

Houses

(PA-1581), 111–13 Beck St. Brick, each approx. 12' (two-bay front) × 18', two-and-a-half stories, half-gambrel roof. Built mid-18th century. Certified, PHC 1965. 1 ext. photo (1961).*

Houses

(PA-1582), 102–18, 103–15 Clymer St., rear of 770 and 772 S. Front St. Court of seven and nine contiguous brick houses, each approx. 15' (two-bay front) × 15', three stories on raised basement, gable roof. Built early 19th century; rehabilitated 1969. Certified, PHC 1966. 1 ext. photo (1961).*

Houses

(PA-1583), 123–25 League St. Wood frame with clapboards, each approx. 15' (two-bay front) × 18' with rear ell, two-and-a-half stories, gable roof. Built between 1793 and 1806. Certified, PHC 1963. 1 ext. photo (1961).*

Houses

(PA-1584), 39–43 Norfolk St. Three contiguous brick houses, each approx. 14' (two-bay front) × 14', three stories, shed roof, one room to a floor. Built between 1830 and 1833; demolished 1967; 1 ext. photo (1967);* 12 data pages (1964, including 1832, 1833 insurance surveys).*

Houses

(PA-1585), 731–33 S. Front St. Two adjoining brick houses with marble trim, No. 731 approx. 25' (two-bay front) × 40', two stories, flat roof; No. 733 approx. 18' (two-bay front) × 44' with two-story rear wing, three stories on raised basement, flat roof, round-arch entrance. Built after 1860; new storefront added 1911; demolished 1967. 1 ext. photo (1967).*

Hutton, Nathaniel, House

(PA-1586), 814 S. Front St. Brick, approx. 14' (one-bay front) × 20' with three-story rear wing, three-and-a-half stories, half-gambrel roof. Probably a back building built in expectation of a front building. Built between 1785 and 1813; possibly by Nathaniel Hutton, shipwright. Certified, PHC 1964. 1 ext. photo (1961);* 3 data pages (1961).*

Irish, Nathaniel, House. Illustrated

(PA-1013), 704 S. Front St. Brick, 19'-7" (three-bay front) × 33'-10" with two-story rear ell, three-and-a-half stories on raised basement, gable roof, alleyway to rear yard, originally pent eave between first and second stories, originally London house plan, notable original woodwork.

Fine example of 18th-century Philadelphia dwelling. Built between 1762 and 1769; Nathaniel Irish, house carpenter. Interior altered later. Certified, PHC 1957; Pennsylvania Register 1970; NR. 1 sheet (1957, including plans, elevation, section)* courtesy of Philadelphia Historical Commission, 8 sheets (1966, including plot plan, plans, elevations, section, details);* 1 ext. photo (1959), 7 int. photos (1959, 1966*); 8 data pages (1959); HABSI form (1957).*

Keen, James, House

(PA-1587), 946 S. Front St. Brick, approx. 17' (two-bay front) × 30', two-and-a-half stories on raised basement, gable roof, round-headed entrance. Built between 1820 and 1834; wood cornice replaced by brick 1953. Certified, PHC 1957. 1 photocopy of ext. photo (1957);* 2 data pages (1961);* HABSI form (1958).*

Kenilworth Street Area Study

(PA-1588), 109–25 Kenilworth St., on N. side Kenilworth St. between Front and Second sts. Nine contiguous brick houses, each approx. 19' (two-bay front) × 23' with one-, two-, and/or three-story rear ells, two-and-a-half, three, or three-and-a-half stories, originally gable roofs, originally city house and side-hall plans. Built at various times between 1740 and 1800; Nos. 121–23 raised to three stories later. A residential community for waterfront workers for over two centuries. Certified, PHC 1957. 2 ext. photos (1961);* 25 data pages (1960, including 1757–1918 insurance surveys).*

Kitts, Captain John, House

(PA-1589), 609 S. Front St. Brick, approx. 18' (two-bay front) × 20' with two-story rear ell, two-and-a-half stories on raised basement, gambrel roof, city house plan. Built before 1789; altered to store before 1860; demolished 1966. 1 ext. photo (1961);* 2 data pages (including 1789 insurance policy).*

Lamb, Peter, House

(PA-1590), 28 Catharine St. Brick, approx. 16' (two-bay front) × 25' with two-story rear ell, two-and-a-half stories on raised basement, gable roof, round-arch entrance with fanlight. Built after 1821; demolished 1967. Certified, PHC 1963. 1 ext. photo (1966).*

Lasse Cock's Manor House. See Bel Air

Little, John, House
(PA-1591), 618 S. Front St. Brick, approx. 17' (two-bay front) × 30' with two-story rear ell, two-and-a-half stories on raised basement, gable roof, round-arch entrance with fanlight, London house plan. Built 1774; renovated and enlarged c. 1833. Certified, PHC 1957. 1 ext. photo (1966);* 7 data pages (1961, including 1774–1849 insurance surveys);* HABSI form (1957).*

Lyle-Newman Houses
(PA-1592), 905–7 S. Front St. Brick, each approx. 14' (two-bay front) × 28', two-and-a-half stories, gable roof, originally London house plan. Built 1815; demolished 1967. Certified, PHC 1957. 1 photocopy of ext. photo (1957);* 7 data pages (1961, including 1815, 1918 insurance surveys);* 2 HABSI forms (1958).*

McCraig, George, House
(PA-1593), 810 S. Front St. Brick with marble trim, approx. 18' (two-bay front) × 18', three-and-a-half stories, gable roof. Built between 1784 and 1786; two-story rear ell removed after 1887. Certified, PHC 1957. 1 photocopy of ext. photo (1957);* 3 data pages (1961);* HABSI form (1957).*

McDowell-Fritz House
(PA-1594), 808 S. Front St. Brick, approx. 18' (two-bay front) × 22' with two-story rear ell, two-and-a-half stories on raised basement, gable roof, probably London house plan. Built between 1754 and 1761; altered later. Certified, PHC 1957. 2 ext. photos (1961);* 2 data pages (1961);* HABSI form (1957).*

Maloby, Thomas, House and Tavern
(PA-1595), 700 S. Front St., at S.W. corner Bainbridge St. Brick, approx. 17' (two-bay front) × 36', three-and-a-half stories, gable roof, pent eave, originally London house plan. Built c. 1764; altered to store before 1860; one-story rear wing removed later. Interior rehabilitated, exterior partially restored 1972–74. Certified, PHC 1957; Pennsylvania Register 1976; NR. 1 ext. photo (1966);* 10 data pages (1960, including 1774 insurance survey);* HABSI form (1957).*

Mariners' Bethel (Baptist) Church (later known as First Polish Baptist Church and Bethel Christian Center)
(PA-1596), 923 S. Front St. Brick, approx. 40' (three-bay front) × 70' two stories, gable roof. Built 1863; possibly by Stephen D. Button, architect. Small, one-story toilet wing added 1922. Demolished 1966. Certified, PHC 1963. 1 ext. photo (1961).*

Marks-Dunbar House
(PA-1597), 849 S. Front St. Brick, approx. 16' (two-bay front) × 17' with two-story rear ell, two-and-a-half stories on raised basement, gable roof, originally pent eave between first and second stories, originally one room to a floor, later city house plan.
Typical small dwelling of Southwark artisans. Built 1758; rear ell built before 1860, enlarged later; demolished 1967. Certified, PHC 1957. 4 sheets (1967, including plot plan, plans, elevation, section, details);* 2 ext. photos (1966),* 4 int. photos (1966);* 12 data pages (1961, including 1762 insurance survey);*HABSI form (1958).*

Marshall, Joseph, Houses
(PA-1598), 854–56 S. Front St. Brick, each approx. 16' (two-bay front) × 25' with one-story rear addition, two-and-a-half stories, gable roof, probably London house plan. Built c. 1831; altered to store before 1860; rear addition built between 1860 and 1887; new storefront 1930. Certified, PHC 1957. 1 photocopy of ext. photo (1957);* 7 data pages (1961);* 2 HABSI forms (1957).*

Marshall-Morris House
(PA-1599), 774 S. Front St. Brick, approx. 17' (two-bay front) × 34' (originally with two-story rear ell), three-and-a-half stories, gable roof, originally London house plan. Built between 1772 and 1777; altered to store c. 1841; altered to warehouse later. Certified, PHC 1957. 1 photocopy of ext. photo (1957);* 6 data pages (1961, including 1807–41 insurance surveys);* HABSI form (1957).*

Maxfield-Elliott House
(PA-1600), 35 Queen St. Brick, approx. 24' (three-bay front) × 18' with one- and two-story rear ell, three-and-a-half stories on raised basement, gable roof, round-arch entrance with pedimented frontispiece, side-hall plan. Built c. 1802; demolished 1967. Certified, PHC 1957. 2 sheets (1967, including plot plan, fireplace plans and eleva-

tions, details);* 4 ext. photos (1961, 1966, 1967),* 3 int. photos (1967);*
11 data pages (1964, including 1813 insurance survey).*

Mercer, Thomas, Houses
(PA-1601), 2–12 Christian St. Five contiguous brick houses, each approx. 15' (two-bay front) × 30' with three-story rear ell, and one
adjoining brick house approx. 16' (two-bay front) × 18' with two-
story rear ell, three stories on raised basements, gable roofs. Nos. 2–
10 built c. 1833. No. 12 built c. 1844; probably by Alexander McMullin,
bricklayer. Demolished 1962. Certified, PHC 1958. 2 ext. photos
(1961–62),* 2 int. photos (1961–62),* 1 photocopy of ext. photo (1957);*
6 data pages (1961).*

Mikveh Israel Cemetery Gatehouse
(PA-1602), 1114 Federal St. Stuccoed brick, approx. 30' (three-bay
front) × 20', two stories, flat roof, battered recesses flanking door
with corbeled recess above, notable Egyptian Revival details. Built
before 1849; demolished 1963. 2 ext. photos (1963).*

Moffett, Robert, House
(PA-1603), 35 Catharine St. Wood frame with clapboard siding, approx. 20' (three-bay front) × 32', two stories, gable roof. Built
between 1793 and 1810; demolished 1967. Certified, PHC 1963. 1 ext.
photo (1966);* 15 data pages (1963).*

Moore, Captain Thomas, House
(PA-1604), 702 S. Front St. Brick, approx. 18' (three-bay front) × 34'
with two-story rear ell, three-and-a-half stories on raised basement,
gable roof, originally pent eave between first and second stories,
London house plan. Built between 1767 and 1779; possibly by
Nathaniel Irish, house carpenter. Altered to store before 1860;
reverted to dwelling and rear ell later raised to two stories. Certified,
PHC 1957; Pennsylvania Register 1970; NR. 7 data pages (1960);*
HABSI form (1957).*

Moore, John, House
(PA-1605), 734 S. Front St. Brick, approx. 16' (two-bay front) × 33'
with three-story rear ell (originally one story), three-and-a-half
stories (originally two), gable roof, originally London house plan.
Built between 1770 and 1772; house and rear ell raised one story

between 1810 and 1820; ell raised to three stories after 1860. Demolished 1967. Certified, PHC 1957. 1 photocopy of ext. photo (1957);* 4 data pages (1960, including 1772 insurance survey);* HABSI form (1957).*

Moyamensing Prison. See Philadelphia County Prison

Murphy-Johnson House
(PA-1606), 42 Catharine St. Wood frame with clapboard siding, 19'-4¹/₂" (three-bay front) × 17'-4" with two-story rear brick addition and one-story rear ell, two-and-a-half stories, gambrel roof, originally one room to a floor.

Typical small dwelling of Southwark artisans who made their living on Delaware River. Built c. 1795; brick addition later; one-story ell built after 1860; demolished 1967. Certified, PHC 1957. 3 sheets (1966, including plot plan, plans, section);* 1 ext. photo (1967),* 1 int. photo (1967);* 14 data pages (1963).*

Old Swedes Church. See Gloria Dei

Paschall, Jonathan, House
(PA-1607), 36 Christian St. Brick, 16'-0" (two-bay front) × 16'-10" with one- and two-story rear frame ell, two-and-a-half stories, gambrel roof, pedimented frontispiece, originally one room to a floor. One of many small dwellings of Southwark artisans who made their living on Delaware River. Built c. 1749; rear frame ell and addition built later; frontispiece added c. 1800; demolished 1967. Certified, PHC 1963. 4 sheets (1966, including plot plans, elevation, details);* 2 ext. photos (1966);* 6 data pages (1960).*

Pearson, Anthony, House
(PA-1608), 806 S. Front St. Brick, approx. 14' (two-bay front) × 21' with two-story rear ell, two-and-a-half stories, gable roof, city house plan. Built 1754; probably by Anthony Pearson, bricklayer. Certified, PHC 1957. 2 ext. photos (1961);* 3 data pages (1961, including 1764 insurance survey);* HABSI form (1957).*

Pearson, George, Houses. Illustrated
(PA-1609), 808–10 S. Hancock St. Wood frame with clapboards, each approx. 10' (one-bay front) × 12' with one-story rear wing, two-and-a-half stories on raised basement, gable roof. Built late 18th century;

probably by George Pearson, carpenter. Certified, PHC 1966. 1 ext. photo (1961);* 2 data pages (1961).*

Philadelphia County Prison (also known as Moyamensing Prison).
Illustrated

(PA-1096), 1400 S. Tenth St., at S.W. corner Reed St. at Passyunk Ave. Front buildings (offices): Quincy granite ashlar with rubble rear wall, approx. 173' (seventeen-bay front) × 36', two and three stories, flat roofs, battlemented front corner turrets and central two-story octagonal tower on projecting center section, battlemented octagonal tower at front corners of side wings, recessed lancet windows, front part of U-plan.

Two rear buildings (convict quarters): rubble with granite ashlar trim, each approx. 53' × 380', three stories with clerestory skylights, gable roofs, interior center court with three tiers of cells, sides of U-plan. Other rubble two-story buildings for kitchen, workshops in center of U-plan. Prison wall of granite ashlar front and rubble sides and rear; squat battlemented tower at front corners.

Fine example of castellated mode of prison architecture, designed with advanced sanitation and heating system. Built 1832–35; Thomas U. Walter, architect. Ventilation system installed 1849–50; Thomas U. Walter, architect. Workshops built later; rear storage building built 1915; central tower removed later. Vacated 1963; demolished 1968. After 1895 used for detaining prisoners awaiting trial or sentence or serving short sentences. 2 ext. photos (1965),* 7 int. photos (1965),* 1 photocopy of architect's wash drawing (1836);* 2 data pages (1965).*

Philadelphia County Prison, Debtors' Wing

(PA-1097), 1400 S. Tenth St., at S.W. corner Reed St. at Passyunk Ave. Coursed red sandstone ashlar front, sandstone rubble sides and rear, 50'-0¹/₂" (three-bay front) × 86'-2", two stories on raised basement with Egyptian distyle entrance porch, gable roof with front cavetto cornice containing carved winged orb, front windows within battered two-story frames with cavetto cornices, central longitudinal corridor.

Outstanding example of Egyptian Revival design and considered the first archaeologically based Egyptian Revival building in America. Built 1836; Thomas U. Walter, architect. Rear extension built 1837–38; vacated 1963; demolished 1968. Portico preserved at Smithsonian Institution, Washington, D.C. Used for confinement of witnesses and prisoners held for proceedings 1841–68, for confinement of women 1868–1963. 7 sheets (1966, including plot plan, plans, elevations, section, details); 4 ext. photos (1965), 2 int. photos (1965), 1 photocopy of architect's drawing (1836);* 6 data pages (1965).*

Philadelphia Fire Department, Engine Company No. 3, and Patrol House (originally known as Weccacoe Engine Company No. 9)

(PA-1610), 117–21 Queen St. (originally 119 Queen St.). Brick, approx. 56' (four-bay front) × 47' with one-story rear extension (originally 19' × 52', later 37' × 67'), two-and-a-half stories (originally three), gable roof with cross-gambrel (originally flat), brick in herringbone pattern between floors, three equipment doors, rectangular plan.

Fine 19th-century fire station whose development reflects changes in architectural design and municipal administration. Built c. 1843; enlarged c. 1858. New front built, patrol house added to western end 1893; possibly by John T. Windrim, architect; Doyle and Doak, contractors. Interior altered 1904; Philip Johnson, architect. Rear altered and building shortened, interior renovated to dwelling and interior-design studio 1969. Volunteer fire hall c. 1843–71; municipal fire station 1871–1957. Certified, PHC 1963. 1 ext. photo (1973),* 2 photocopies of old ext. photos (1896, 1934).*

Philadelphia, Wilmington and Baltimore Railroad Freight Station Train Shed

(PA-1611), S.E. corner Fifteenth and Carpenter sts., at rear of 1000 S. Broad St. Brick with stone trim, approx. 100' (three-bay front) × 230', one story, gable roof with clerestory skylights, overhanging eaves with iron cinquefoil supports, rectangular plan with exposed iron trusses.

Fine example of vernacular railroad architecture and use of trusses to span large space before introduction of three-hinged arch. Built 1876 as train shed with brick two-story head house fronting on Broad St.; head house demolished 1968. Pennsylvania Railroad acquired control of railroad 1881, changed name to Philadelphia, Baltimore and Washington Railroad 1902, sold buildings 1966. Now Semple Co. Warehouse 5 ext. photos (1969),* 3 int. photos (1969),* 2 photocopies of old ext. photos (1914, 1917).*

Powder Magazine

(PA-124), Magazine Lane. Rubble with brick trim, 53'-7" (two-bay front) × 72'-9", one-and-a-half stories, five buttresses on side walls,

two barrel vaults run the length of the building. Built by Commonwealth of Pennsylvania 1808; vacated 1874; demolished 1940. 4 ext.
photos (1937), 1 ext. photo of two-and-a-half-story, half-gambrel
roof, caretaker's house (1937); 2 data pages (1937).

Purves, Hugh, Store

(PA-1612), 626 S. Second St., at N.W. corner Bainbridge St. Brick, approx. 20′ (three-bay front) × 40′, four stories, flat roof, corner
storefront, decorative iron lintels, rectangular plan. Built c. 1851; demolished c. 1968. Hugh Purves, stove manufacturer, had his store on
the ground story and leased upper-story dwelling rooms c. 1851–96. 1
ext. photo (1967).*

Queen Street Area Study

(PA-1613), 28–38 Queen St., on S. side Queen St. between Swanson
and Front sts. Six contiguous brick houses, each approx. 17′ (two-bay
front) × 17′ with two-story rear ell, two-and-a-half stories on raised
basement, gambrel roof (No. 38 gable roof), probably city house
plans. A residential community of waterfront workers for nearly two
centuries. Built at various times between 1743 and 1793; demolished
1967. Certified, PHC 1957. 2 ext. photos (1966, 1967);* 36 data pages
(1963, 1964, including 1763 insurance survey).*

Ralston School

(PA-1614), 625 S. American St., at N.E. corner Bainbridge St. Brick,
approx. 60′ (six-bay front) × 50′ with two-story lateral addition (east
end), two stories, hipped roof with cross-gables, name and date stone
on south end, ventilator oculus in gable, segmental-arch windows,
rectangular plan with four rooms to a floor and front and rear entrances and stairs.

Fine example of post–Civil War public school house continuing
the earlier modular plan of Samuel Sloan's "Philadelphia Plan." Built
1869; Lewis H. Eshler, architect; George H. Brinkworth, contractor.
Lateral addition built 1928; Francis P. Canavan, architect. Named in
honor of Robert Ralston, prominent mid-19th-century merchant;
school until 1922. Certified, PHC 1973. 1 ext. photo (1967),* 1
photocopy of architect's plan (c. 1869).*

Ridgway Branch of Library Company of Philadelphia. Illustrated

(PA-1616), 900 S. Broad St., composing block bounded by Broad,
Catharine, Thirteenth, Carpenter sts. Brick faced with granite ashlar
and with iron columns and beams in book rooms, approx. 220′

(thirteen-bay front) × 105', two stories on raised basement with central Doric octastyle portico and flanking Doric tetrastyle porticoes with pediments, gable roofs with significant skylight system, central-hall plan with stylized Ionic-colonnade gallery.

Often considered the final expression of the Greek Revival in the United States. Built 1873–78; Addison Hutton, architect. Vacated 1966; altered to recreation center 1973. The oldest circulating library in the United States, the Library Company of Philadelphia was founded by Benjamin Franklin 1731 and the Ridgway Branch was built from funds donated by Dr. James Rush in honor of his wife, Phoebe Ann Ridgway Rush. Certified, PHC 1972. 9 ext. photos (1962),* 9 int. photos (1962),* 1 photo of architect's model (1962),* 4 photocopies of architect's plans and elevations (c. 1873).*

Rooming House

(PA-1617), 115 Bainbridge St. Brick with marble trim, approx. 20' (four-bay front) × 53', four stories, flat roof, molded brick trim between sills and lintels at each floor, rectangular plan. Example of late-19th-century rooming house. Built between 1880 and 1888; ground-story front, interior altered to warehouse 1922. 1 ext. photo (1967);* 3 data pages (1973).*

Schuylkill Arsenal. See Arsenal on the Schuylkill

Siddons, William, House

(PA-1618), 851 S. Front St. Brick, approx. 17' (two-bay front) × 25' with two-story rear ell, two-and-a-half stories on raised basement, gable roof. Built after 1785; ground story altered to store late 19th century; new shopfront built 1920; demolished 1967. Certified, PHC 1957. 1 photocopy of ext. photo (1957);* 2 data pages (1961); * HABSI form (1958).*

Singley House. See Bel Air

South Front Street Area Study

(PA-1812), 600–858 S. Front St., W. side Front St. between South and Catharine sts. Sixty-five contiguous brick houses and one church (Church of the Redeemer), two-and-a-half, three, and three-and-a-half stories, gable and flat roofs. Nearly all of the houses built late 18th and early 19th centuries. 4 ext. view photos (1967).*

Spafford, William, House
(PA-1620), 626 S. Front St., at N.W. corner Bainbridge St. Brick, approx. 20' (three-bay front) × 34' with two-story rear wing, three-and-a-half stories, gambrel roof, brick stringcourses, London house plan, notable oval stairwell. Built before 1762; altered c. 1801; ground story altered to tavern 1907. Certified, PHC 1957. 2 ext. photos (1961, 1966),* 2 int. photos (1966);* 13 data pages (1961, including 1762–1873 insurance surveys);* HABSI form (1957).*

Sparks's Shot Tower
(PA-1621), 129–31 Carpenter St. Brick, approx. 30' diameter at base, six stories (originally seven), conical roof. Built 1808; altered after 1913. The first shot tower erected in the United States and a rare extant example. Used as recreation center since 1913. Certified, PHC 1957. 3 ext. photos (1973, 1976);* HABSI form (1971).*

Stephen Girard's Country House. See Gentilhommière

United States Naval Asylum (now United States Naval Home). Illustrated
(PA-1622), S.W. corner Gray's Ferry Ave. and Twenty-fourth St. at Bainbridge St. Complex of four primary buildings on 9 acres that until 1976 served as the only residence facility for retired naval personnel in the United States. Certified, PHC 1956; Pennsylvania Register 1972; designated National Historic Landmark 1975.

Biddle Hall: Pennsylvania marble ashlar with granite ashlar basement, approx. 385' (thirty-three-bay front) × 135' including domed rear chapel, two-and-a-half stories on raised basement with central Ionic octastyle portico, gable and hipped roofs, lateral dormitory wings with balconies on granite piers and iron columns, symmetrical plan with central axial corridor and central hallways through dormitory wings.

A major Greek Revival building designed to be healthy, efficient, and fireproof. Built 1827–33; William Strickland, architect; John Struthers, marble mason; John O'Neill, carpenter. Cellar added c. 1843; attic altered, windows added 1848. Served as Naval Academy 1840–45 and hospital 1833–68. Name changed to U.S. Naval Home 1889. 6 ext. photos (1964, including detail of iron railing),* 7 int. photos (1964),* 14 photocopies of architectural drawings (including 8 by William Strickland 1826, 5 by W. L. Andrews 1843, 1 by G. W. Mackay 1888),* 2 photocopies of plot plans (c. 1824, 1878),* 1 photocopy of oil painting by Thomas Doughty (1828);* 17 data pages (1965).*

Governor's Residence: Approx. 150' N.E. of Biddle Hall. Brick covered with coursed stucco, approx. 47' (three-bay front) × 100' including two-story rear kitchen wing, two stories with two-story front veranda supported by four pairs slender cast-iron columns, hipped roof, central-hall plan. Built 1844; William Strickland, architect. Kitchen remodeled later. 2 ext. photos (1964),* 3 int. photos (1964),* 1 photocopy of architect's drawing (1844);* 7 data pages (1965).*

Surgeon's Residence: Approx. 150' S.W. of Biddle Hall. Brick covered with coursed stucco, approx. 47' (three-bay front) × 100' including two-story rear kitchen wing, two stories with two-story front veranda supported by four pairs slender cast-iron columns, hipped roof, central-hall plan. Built 1844; William Strickland, architect. Interior altered later. 3 ext. photos (1964, including detail of iron railing);* 8 data pages (1965).*

Laning Hall: Approx. 400' N.W. (rear) of Biddle Hall. Brick with brownstone trim, approx. 325' (twenty-five-bay front) × 85', three stories on raised basement, mansard roof, center section with one-story balustraded porch and Flemish gables, lateral wings, brick quoins, central-hall plan. Built 1868; John McArthur, Jr., architect. Served as hospital 1868–1921; rehabilitated 1938. 2 ext. photos (1964),* 1 int. photo (1964);* 2 photocopies of old ext. photos (1908),* 4 photocopies of old int. photos (1938).*

United States Naval Base, Quarters "A" (also known as Commandant's Quarters)

(PA-1623), Davis Ave., approx. 100' W. of Broad St., near entrance to U. S. Naval Base, League Island. Museum. Brick, 34'-9" (five-bay front) × 38'-3" with two-story rear wing and one-story rear addition, two stories with three-story square corner tower, gable roofs, hipped roof on tower, second-story semi-octagonal bay window on east side, round-arch and segmental-arch windows, bracketed cornice, L-plan with attached wing and tower.

Late example of the Italian villa mode. Built 1874–75; porch added to north and east sides 1901; porch enclosed, porte-cochere and rear one-story brick laundry room built later; interior renovated 1960, 1975. Initially housed civil engineer in charge of developing the base, later Inspector of Ordnance; Commandant's Quarters 1900-1902, dwelling of Captain of the Yard 1903–60; Naval Historical

Museum since 1975. Pennsylvania Register 1975; NR. 1 ext. photo (1976).*

Weccacoe Engine Company No. 9. See Philadelphia Fire Department, Engine Company No. 3, and Patrol House

Wharton, John, Houses
(PA-1624), 919–21 S. Front St. Two brick row houses, each approx. 15' (two-bay front) × 33', two-and-a-half stories on raised basement, gable roof, London house plan, notable round-arch entrance with fanlight and pedimented frontispiece on No. 919. Built c. 1795; exterior of No. 921 renovated mid-19th century; demolished 1967. Certified, PHC 1957. 1 ext. photo (1961);* 7 data pages (1961, including 1812, 1848 insurance surveys);* 2 HABSI forms (1958).*

Wharton-Stewart House
(PA-1185), 27 Christian St. Brick, 13'-6" (one-bay front) × 19'-11", two-and-a-half stories, shed roof, hollow diamond of glazed brick on west exterior wall, originally pent eave along west wall, originally one room to a floor.

Probably a back building built in expectation of a front building. Built between 1745 and 1774; demolished 1967. Certified, PHC 1963. 5 sheets (1966, including plot plan, plans, elevations, section, details); 2 ext. photos (1966); 2 int. photos (c. 1965); 6 data pages (1960).*

Woolfall-Huddell House
(PA-1625), 9 Queen St. Brick with stone trim, approx. 21' (two-bay front) × 22' with two-story rear ell, two-and-a-half stories on raised basement, gable roof, city house plan. Built between 1763 and 1787; demolished 1967. Certified, PHC 1966. 3 ext. photos (1961, 1966);* 11 data pages (1964, including 1791 insurance survey).*

Workman Place
(PA-133), 742–46 S. Front St. and rear, at S.W. corner Pemberton St. Group of ten houses around a central courtyard, including three dwellings on Front St. built 1812–13 for John Workman, lumber merchant, and seven small tenements to the rear built 1748 for George Mifflin, grandfather of the first Governor of Pennsylvania. Variation of interior court plan by which early Philadelphians broke up large blocks into small living complexes. Owned by Workman

family 1812–1906 and by Octavia Hill Association since 1942. Certified, PHC 1957.

House, 742 S. Front St.: Brick with marble trim, approx. 17' (two-bay front) × 38', three-and-a-half stories, gable roof, side-hall London house plan. Built 1812. 6 data pages (1961, including 1812 insurance survey).*

Houses, 744–46 S. Front St.: Brick with marble trim, approx. 20' (three-bay front) × 38' with rear stair tower, three-and-a-half stories, gable roof, side-hall plan. Built 1812–13; No. 746 altered to store 1875, rehabilitated to dwelling after 1900. 2 ext. photos (1961, 1967);* 7 data pages (1960, including 1813–75 insurance surveys).*

Tenements 1–2 and 3–4, 110–12 Pemberton St., at rear of 742 S. Front St.: Brick, each approx. 17' (two-bay front) × 27', two-and-a-half stores, gable roof, originally one room to a floor, originally two tenements to a building, initials "GM" in black headers on north wall of Tenement 1–2, date "1748" in black headers on north wall of Tenement 3–4. Built 1748; Tenement 1–2 altered to single dwelling 1964. 6 ext. photos (1937, 1961*); 5 data pages (1937, 1961,* including 1850 insurance survey).

Tenement 5–6–7, rear of 744–46 S. Front St.: Brick, each approx. 13' (two-bay front) × 17' with two-story rear wing, two-and-a-half stories, half-gambrel roof, one room to a floor. Built 1748. 1 data page (including 1850 insurance survey).*

Gloria Dei. *Jack E. Boucher photo, 1962.*

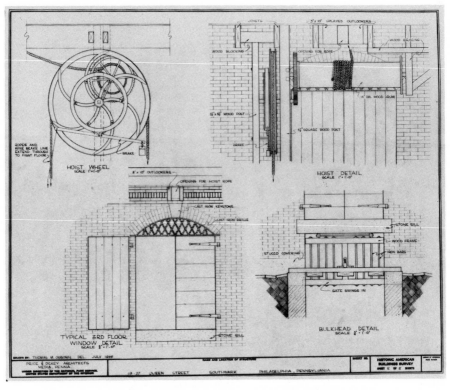

Jesse Godley Warehouse. *Thomas M. Osborn, del., 1966.*

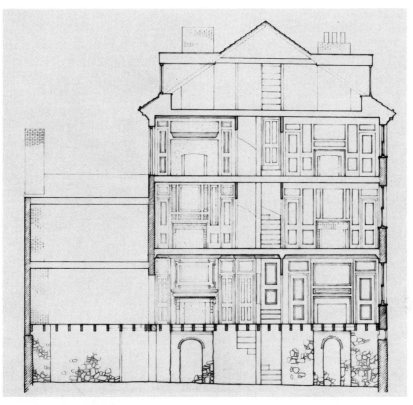

Nathaniel Irish House. *Michael Cunningham, del., 1966.*

Nathaniel Irish House, fireplace detail. *James L. Dillon & Co. photo, 1967.*

George Pearson Houses. *Jack E. Boucher photo, 1961.*

Ridgway Branch of Library Company of Philadelphia. *Jack E. Boucher photo, 1962.*

Philadelphia County Prison. *Oil painting by Thomas U. Walter, c. 1835, courtesy Historical Society of Pennsylvania.*

United States Naval Asylum, Surgeon's Residence. *Jack E. Boucher photo, 1964.*

West Philadelphia

Blockley and Kingsessing
Townships

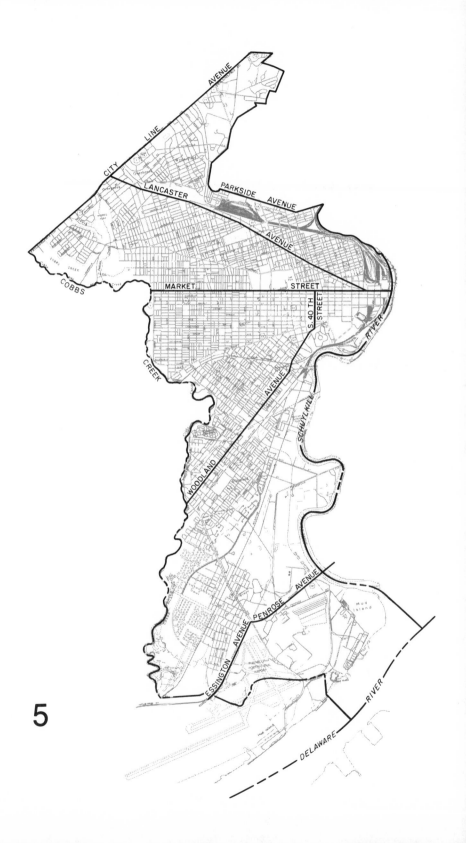

5

Introduction

*K*NOWN AS BLOCKLEY Township and Kingsessing Township in the eighteenth century, West Philadelphia, that part of the city west of the Schuylkill River, was dotted with farms and estates, particularly along the banks of the river. Among the oldest of these buildings is Wynnestay at Fifty-second and Woodbine streets in the city's Wynnefield section.[1] Its two adjoining sections were built in 1689 and 1700 for Dr. Thomas Wynne, an early Quaker convert and missionary, who became William Penn's personal physician and accompanied the Founder to the province in 1682. The first structure was of modest proportions; a pent eave skirted its thick rubble-masonry walls and there was only one room to a floor.[2] With its later additions, however, Wynnestay assumed the appearance of a comfortable early Georgian Pennsylvania country house.[3]

"Of all the Georgian houses in the Philadelphia neighbourhood, none has more striking individuality than Whitby Hall,"[4] which once stood near the present intersection of Fifty-eighth Street and Baltimore Avenue, not far from Cobb's Creek. When Colonel James Coultas, a Philadelphia merchant and man of affairs, acquired his plantation in 1741, a house was apparently already there, but in 1754 he added to it the section commonly known as Whitby Hall.[5] It possessed an unusual combination of architectural elements and building practices. Unlike most country houses, all four of its exterior walls were constructed of coursed ashlar with blank end walls only one room deep.[6] Also, its ridgepole ran the breadth of the house instead of its length so that its gable walls formed the long sides of the building, a practice that was customary for city houses but rarely found in country houses. The stair-and-entry tower was almost cer-

tainly an afterthought. It was awkwardly abutted to the north gable
end, interrupting the pent eave and cove cornice and overlapping
onto a sash window and oculus in the peak. Yet it was a proud addition, pretentiously finished with brick-trimmed tracery windows on
the second story, architrave-framed windows on the third floor, a
dressed-stone stringcourse in between the floors, and a dentil cornice
and three decorative urns at the top.[7] The interior elements were
equally rich and more smoothly integrated. "Enrichment tempered
with dignified reserve" was the conclusion of the English visitor
Nicholas Pickford in 1765.[8] Early in this century the house was dismantled and parts of it were reassembled in nearby Haverford, Pennsylvania, in a manner that belies its former "striking individuality";
most of its interior woodwork was installed at the Detroit Institute of
Art.[9]

Not far from the former site of Whitby Hall stands another
unique country house that expresses even more directly its owner's
personality. The famous naturalist John Bartram built the house in
two stages, in 1730–31 and 1770, facing the Schuylkill near the former
Darby Road (Woodland Avenue). Constructed of massive blocks of
local stone in what a nineteenth-century historian called a "quaint,
old-fashioned style of architecture,"[10] Bartram's house reflects the individuality of a Quaker farmer who bravely left both his personal and
general callings to become a self-styled deist and botanist.[11] From a
distance the house's most striking feature is its elongated Ionic
columns, which with their thick irregularly spaced joints rise two
stories to form a recessed porch; the second floor was apparently enclosed later with beaded tongue-and-groove siding to give the entrance a pinched quality. An additional touch of rustic grandeur is
given by the carved stone crosset-and-volute window surrounds and
the stone panels beneath the heavy sills, all of which are adapted
from contemporary architectural design books.[12] Juxtaposed with this
attempt at worldly sophistication is Bartram's self-conscious procla-
mation of religious independence; in a second-story stone panel is
carved

IT IS GOD ALONE ALMYTY LORD
THE HOLY ONE BY ME ADORD
JOHN BARTRAM 1770.

It was the last word from a man who had lived a running religious debate with his neighbors and had been expelled from the

Darby Friends Meeting for denying the divinity of Jesus Christ.[13] Seven years later he died, leaving behind an impressive contribution to natural history, including the country's first botanical garden, which beginning in the 1920's has been restored by the John Bartram Association.

The grandest eighteenth-century country house of West Philadelphia is Woodlands (illus.). It was the seat of William Hamilton, who inherited the extensive estate along the Schuylkill near Fortieth Street and Woodland Avenue from his grandfather, Andrew Hamilton of Bush Hill, the eminent attorney and designer of Independence Hall. After tactfully sitting out the Revolution and narrowly avoiding the rebels' confiscation of his property, the Anglophile William Hamilton visited Britain and returned to Philadelphia about 1786 to plan an estate befitting the luxurious tastes of a wealthy and hospitable gentleman.[14] His gardens, one of the earliest examples of romantic landscaping in America,[15] have since disappeared, but the house remains as an outstanding American example of the Adamesque-Federal style. In fact, the perambulatory Englishman Nicholas Pickford visited the estate in 1786, and after praising the house's site, design, and appointments, he archly concluded, "I know of no house in England of like Estate where a greater degree of Elegancy and Comfort can be found."[16] In addition to the giant Ionic pilasters that grace the landside terrace and the giant Tuscan portico that faces the river, Hamilton's design incorporated the interplay of surfaces (the Palladian windows recessed within blind arches) and forms (the juxtaposition of the squarish library and study with the oval dining room and parlor, which protrude as exterior bays) that characterized the Adamesque style but was rarely so urbanely executed. After the property became the Woodlands Cemetery in 1843,[17] a caretaker occupied the mansion, and natural growth intruded into the gardens, enhancing the scene in the opinion of contemporaries.[18] Today gravestones and monuments fill the former gardens.

Blockley's rural isolation was broken by the completion of the Lancaster Turnpike in the 1790's and the opening of the Market Street Permanent Bridge on New Year's Day, 1805.[19] The area became increasingly attractive to gentlemen seeking sites for their country estates. They included both newcomers like Italian-born Paul Busti, who moved to his Retreat Farm near Forty-fourth Street and Haverford Avenue about 1799,[20] and established Philadelphians like John Hare Powel, who moved into and renovated Powelton, which had been begun in 1800 near Thirty-second and Race streets.[21] The growth of rural hamlets in the township and small industry along the river, however, began to engulf these estates by mid-century. About 1804 William Hamilton of Woodlands laid out Hamilton Village, which extended north from the Schuylkill to Market Street and west to Forty-first Street.[22] It became an early suburban town and before the Civil War was considered one of the most pleasant villages in the Philadelphia area.[23] About the same time along Haverford Avenue and Spring Garden Street west of the river Judge Richard Peters planned Mantua, whose development was stimulated by the building of the Upper Ferry Bridge in 1812.[24] Thirty-two years later, in 1844, the Philadelphia gentleman George Sidney Fisher visited the neighborhood to view Powelton, which was finally nearing completion. While he praised the house's architecture, he concluded with some regret that "Mr. Powel was foolish to build it. . . . If it were 10 miles from town it would be a delightful residence, but it looks on the coal wharves and mass of brick buildings on the other side of the river & is so near the city that it is constantly liable to trespass & intrusion."[25]

It was during the second quarter of the nineteenth century, while John Hare Powel was overimproving Powelton, that social and charitable institutions began moving into West Philadelphia. The first of these was Blockley Almshouse, designed by William Strickland. Begun in 1830 as a group of four three-story buildings arranged around a large interior courtyard, its grimly gray presence contradicted the optimism and good will that had created it.[26] Beginning in the 1820's the presence of poor people in the midst of a self-consciously democratic and prosperous America posed a challenge, if not an embarrassment, to concerned citizens. Reformers claimed that primary among the roots of poverty were the temptations of a dynamic society in which physical mobility and concomitant urban growth had undermined traditional social sanctions and responsibilities. To remove the indigent from the evils of society and place them in an almshouse where they could be rehabilitated and comforted was thought to be not only humane but also enlightened and practical.[27] Furthermore, if the purpose was to remove the destitute from society, then the farther away they were the better, which accounted for the isolation of Blockley Almshouse on extensive grounds bordering the Schuylkill River south of Thirty-third and Spruce streets.

As the almshouse was being completed, another institution moved to West Philadelphia. Pennsylvania Hospital, which had been caring for the insane since 1794, reflected the enlightened opinion of the times when it was decided in 1831 to build separate quarters for the insane outside the city.[28] The insane asylum of the 1830's and 1840's was not considered a warehouse for lunatics or a last resort for the mentally ill, but a humane institution where psychiatrists were confident mental and nervous disorders could be cured. This confidence was based on the theory that a turbulent and treacherous society caused insanity, and to cure the disease the victim must be removed to a controlled, disciplined environment. The asylum was to be this curative retreat. The problem, however, was to translate psychiatric theory into architectural form. Medical theory dictated the hospital's hundred-acre site on the city's outskirts, since it was argued that if the afflicted must be removed from the anxieties of daily life to the ordered confines of the asylum, then the asylum, in turn, must be removed from the commotion of the community to the tranquility of the countryside.[29] The orthodox plan of the asylum, however, was not fixed until 1836, when the English-trained Isaac Holden won the architectural competition for the Pennsylvania Hospital for Mental and Nervous Diseases (illus.).[30] A pioneer structure of the nineteenth century, it stood near Forty-fourth and Market streets on Paul Busti's former estate. It was the first American hospital built on the echelon, or pavilion-plus-link, plan, a scheme of sprawling attached wings that afforded the best compromise of isolation and communication with a minimal staff.[31] The technical problems of heating and fire prevention led to the installation of twenty-six steam-boiler furnaces and maximum utilization of iron—for columns, stairs, door frames, and window sashes. It is not clear how instrumental Dr. Thomas Story Kirkbride was in deciding the building's plan and appointments, but he certainly became a dedicated booster of the arrangement after he became the asylum's superintendent on its opening in January, 1841. The plan, known as the Pennsylvania School of Hospital Design, was adopted for both insane asylums and general hospitals throughout the United States until after the Civil War.[32]

Other hospitals were erected in the neighborhood during the third quarter of the nineteenth century, but because of their small beginnings, none followed the Pennsylvania School of Hospital Design. The brownstone structure of Christ Church Hospital, which was established in 1772 as a home for the aged, was built from the designs of John M. Gries in 1856–61 on Belmont Plateau near Belmont Avenue and Forty-ninth Street, where its residents could enjoy cool summer breezes and rural scenery. Presbyterian Hospital began in 1872 at Thirty-ninth Street and Powelton Avenue, using two houses that had been renovated for hospital purposes. Its first hospital building, a modest one-story ward pavilion designed by Joseph M. Wilson, was ready in 1874, the year that the Hospital of the University of Pennsylvania opened its more extensive facilities at Thirty-fourth and Spruce streets.[33] In the last quarter of the nineteenth century a number of institutions were established in West Philadelphia, including the Home for Incurables at Forty-seventh Street and Woodland Avenue, Home for Indigent Widows and Single Women at Chestnut near Thirty-sixth Street, Pennsylvania Working Home for the Blind at Thirty-sixth and Lancaster Avenue, and most notably the Pennsylvania Institution for the Instruction of the Blind (illus.), whose complex was built from the designs of Cope and Stewardson at the end of the century.[34]

During the half-century after the opening of the Market Street Bridge, Blockley's economy shifted from agriculture to manufacturing and commerce, and its population grew from a meager 882 inhabitants in 1790 to 6,214 by 1840. Consequently, in 1840 the Borough of West Philadelphia was carved out of Blockley Township by the Court of Quarter Sessions, but since its legality was questioned, the process was repeated four years later by the General Assembly. As the population of the combined borough and township continued to grow, nearly doubling during the 1840's, the borough's area was approximately tripled in 1852. Kingsessing, on the other hand, remained rural. In 1840 more than three-fourths of its people were engaged in farming, and a decade later it possessed fewer than 1,800 inhabitants. Under the Act of Consolidation in 1854, however, the two townships and borough lost their separate identities, if not their distinctions, and became the Twenty-fourth Ward of the City of Philadelphia.[35]

Shortly after the Act of Consolidation, street railways came to West Philadelphia, and during the next twenty years eleven different companies established a recticular complex of horse-powered railways over large parts of the area,[36] helping to make it one of the city's most fashionable suburbs. The model for the antebellum houses that went up along the horsecar lines was Bartram Hall, Andrew M. Eastwick's nationally known picturesque villa designed by Sloan and Stewart in 1850–51 on the former site of John Bartram's botanical gardens.[37] Part of Bartram Hall's visual impact derived from its great scale and superb site, but part also came from its style, a free adaptation of the Italian villa, whose cubic volumes and projecting towers best complemented the cultivated rusticity of the area.[38] Its medieval details, such as a battlemented bay window and free use of corbel tables, led the architects to call Bartram Hall a "Norman villa." More common to the area, however, were the flat, bracketed towers and cornices of the "Tuscan villa," like that of the noted medieval historian Henry Charles Lea (illus.) at 3903 Spruce Street. These towers were more than picturesque ornaments. The four-story tower of Lea's house, for example, served as the entrance foyer, walk-in closets on the upper floors, and a bathroom on the second floor. The well-furnished bathroom, one of two on the floor, was a mark of the owner's affluence. In addition to a walnut bathtub lined with planished copper and fitted with a shower, each bathroom had a copper-lined bidet, a reservoir water closet, and a marble washbasin with marble skirting.[39] Functional as they may have been, towers were not mandatory for Italian villas. The proportions, details, and comfortable interior spaces of the style could be attained without towers or belvederes, as in the case of Samuel Sloan's 1854 design for Joseph Allison's house at 4207 Walnut Street.[40] Yet towers remained popular and were retained when the mansard mode came into fashion after the Civil War.

Block developments also appeared in the wake of the street railways. Again instrumental in setting the pattern was Samuel Sloan, who in 1854 designed Hamilton Terrace, a group of five villas on Forty-first Street below Baltimore Avenue.[41] More extensive and better preserved is Woodland Terrace (illus.), which was built between Baltimore and Woodland avenues on the eve of the Civil War. These two early terrace developments of large semidetached dwellings helped to establish the hierarchical pattern of speculative row housing that was followed throughout the century. This was a status-conscious plan in which the most expensive and picturesque houses occupied the best sites on corner lots or major thoroughfares, with lesser edifices filling in the middle of the rows or side streets. This plan offered monumentality that was not to be had from a continuous row of identical houses, a more democratic but more monotonous plan. Similar brownstone dwellings continued to be built in the neighborhood into the 1860's, but, since most horsecar lines stopped at Forty-third Street, these developments rarely extended west of Forty-fifth Street during the nineteenth century.[42] Plenty of open land, however, remained in West Philadelphia east of Forty-fifth Street, where as late as the 1890's architects designed many handsome houses, such as Wilson Eyre's chaste facade for William Cochran's house at 3511 Baring Street, the Wilson Brothers' Renaissance Revival dwelling for George W. Childs Drexel at Thirty-ninth and Locust, and Will Decker's rusticated Romanesque pile for William J. Swain at 3925 Chestnut Street.[43]

The advent of the automobile after 1900 and the opening of the Market Street Elevated in the spring of 1907 ignited another wave of speculative developments. Instead of the commodious double houses and distinctive mansions of the old streetcar suburb, the West Philadelphia of the auto and Elevated was to be row upon interminable row of two-story dwellings with bay windows set above columned front porches.[44] Apartment buildings, a relatively new housing alternative, were also built near the El, especially in the older parts of West Philadelphia. Hamilton Court, designed by Wilson Brothers and Company in 1901 at Thirty-ninth and Chestnut streets, was one of the area's earliest apartment buildings, but one of the largest was the twelve-story Garden Court Plaza at Forty-sixth and Spruce streets. Built during the late 1920's from the designs of Ralph B. Bencker, it was an up-to-date complex whose parking garage and special radio antenna anticipated future trends. Its name derived from the garden that covered most of the garage's roof.[45] Recent decades have brought updated versions of the row house, either speculative units such as University Mews at Forty-sixth and Spruce streets, or public housing, of which the best example is Louis Kahn's Mill Creek project on Fairmount Avenue between Forty-fourth and Forty-seventh streets.[46]

West Philadelphia's great event after the Civil War was the celebration of the nation's one-hundredth birthday at the International Exhibition of 1876 in nearby Fairmount Park. In an attempt to capitalize on the event, street railways were built in the Park area for the first time, setting off a small real estate boom in the blocks adjacent to the Park. To accommodate the increased traffic, the Pennsylvania Railroad built a new station at Thirty-second and Market streets, and until the Broad Street Station opened in 1881 it remained a major terminal for trains to New York and the West. Most of the physical evidence of the exhibition's direct impact on the area disappeared long ago. The railroad station burned in 1896, and those horsecar lines that did not collapse after the centennial rush were electrified in the 1890's before being integrated into a single urban transit system in 1902.[47] Only Frank Furness's chunky Centennial National Bank at Thirty-second and Market streets remains as a remainder of those fleeting glorious days.

A more lasting impression has been made by the growth of some of the city's great educational institutions. The oldest of these is the University of Pennsylvania, which moved from Ninth and Market streets to Thirty-sixth Street and Woodland Avenue in 1871.[48] Beginning with Thomas W. Richards, who designed College Hall in 1871, the University of Pennsylvania has employed a Who's Who of collegiate architects. The list includes such nineteenth-century greats as Frank Furness, the Victorian Goth who created the massive Library in 1888; Wilson Eyre, Jr., who, in collaboration with Frank Miles Day and Cope and Stewardson during the 1890's, designed the city's finest Creative Eclectic building, the University Museum (illus.); and Cope and Stewardson, who were already famous for their Collegiate Gothic works at Bryn Mawr and Princeton when in 1895 they employed the Jacobethan to produce "a series of architectural adventures" in the Men's Dormitories (illus.) at Pennsylvania.[49] The campus enjoyed another period of expansion after World War II to make its architecture even more diverse. Among the notable names and works of the later era are Eero Saarinen's fortresslike Hill House, with its festive and spacious interior court (1958–60), Mitchell/Giurgola Associates' monumental poured-concrete parking garage at Thirty-second and Walnut streets (1963), and the great Louis I. Kahn's Alfred Newton Richards Medical Research Building (1957–61), which has been called "one of the greatest buildings of modern times."[50]

Twenty years after the university's arrival in West Philadelphia, Drexel Institute of Art, Science, and Industry became a neighbor. Now called Drexel University, it was founded in 1890 by Anthony J. Drexel, one of the nation's leading bankers, to provide a practical industrial education for young people. It opened its doors at Thirty-second and Chestnut Streets in 1892 after the dedication of its main building, an impressive exercise in terra cotta detailing and interior spatial planning by Wilson Brothers and Company.[51] It remains a *tour de force* of academic architecture that Drexel has not matched in the past eighty-five years. Only two dormitories on Thirty-third Street offer competition: Grant Simon's thirteen-story Sarah Drexel Van Rensselaer Dormitory (1930–31) on Powelton Avenue and Vincent G. Kling's Calhoun Hall (1970–71), with its concave facade on Arch Street.[52]

St. Joseph's College is a relative newcomer to West Philadelphia. It moved to its twenty-three-acre campus west of Fifty-fourth Street and City Line Avenue in suburban Wynnefield during the late 1920's, a time of expansion and migration for a number of the city's colleges.

As a small school with initially only one building (a notable Collegiate
Gothic pile begun in 1925 by F. Ferdinand Durang) on a large
campus, St. Joe's had minimal impact on its residential neighborhood.[53] The college grew rapidly during the flush years of the 1960's,
and most of its open space has since been filled in.

As the city pushed westward along the transit system during the
early twentieth century, the neighborhood surrounding Drexel and
Pennsylvania lost its suburban charm, and with the concomitant
stagnation and blight of the Great Depression evolved into a no-
man's-land. West of Forty-fifth Street stretched the rows of houses
and east of Thirty-second Street stood the encroachments of the commercial city, most notably the Thirtieth Street Station and the new
Bulletin Building.[54] Between the two lay a decaying strip occupied by
poor residents, the academic institutions, and roving gangs of
hoodlum youths. In an effort to reverse the trend and to enable the
area's institutions to execute their planned physical expansion
smoothly, the West Philadelphia Corporation was formed in 1959 to
work with the Redevelopment Authority and the City Planning Commission.[55] Out of this cooperative effort has evolved a learning and
research complex, which includes the two adjacent universities, other
nearby institutions, and the new University City Science Center
along Market Street. It has substantially restructured the neighborhood, as dwellings have been replaced by institutional high-rises.
The area's new name, University City, is visually, socially, and
economically appropriate. Nevertheless, nooks and crannies of
nineteenth-century West Philadelphia have survived. One of them,
north of Lancaster Avenue, is Powelton Village, which is gamely try-
ing to resist Drexel's growth.[56] Another is Spruce Hill, which includes a number of pleasant residential streets like St. Mark's Square.
In these remnants of the affluent post–Civil War suburb many mansions, large semidetached houses, and revival style churches[57] still
stand on tree-lined streets, handsomely illustrating this stage of
Philadelphia's growth.

Catalog

Academic Department of the University of Pennsylvania. See University of Pennsylvania, College Hall

Bartram, John, House
(PA-1132), Bartram Park, S. Fifty-fourth St. between Elmwood and Gibson aves. Historic house museum. Coursed ashlar front and stuccoed rubble sides and rear, approx. 49' (four-bay front) × 27' with one-story lateral wings, two-and-a-half stories, gable roof, full-height Ionic stone columns form recessed porch, carved stone window trim, quoins, center entry room with stairs.

Notable example of Pennsylvania stone house with characteristic exterior door to each room but with unique detailing. Built 1730–31; extended eastward one-room depth c. 1770; wings built later; second story of porch enclosed later; restored 1923–25. House and gardens acquired by Fairmount Park Commission, City of Philadelphia 1891. John Bartram, America's first internationally known botanist, built house and lived here 1731–77. Certified, PHC 1956; designated National Historic Landmark 1960.

8 sheets (1940, including plans, elevation, section) from Survey, Philadelphia Charter, A.I.A. (1932); 4 ext. photos (1938), 7 photocopies of measured drawings (n.d.).*

Bleakley House (also known as Cannon Ball Farm House)
(PA-134), Mud Island, approx. 350' W. of Delaware River, approx. 150' N. of north sallyport of Fort Mifflin (originally on Penrose Ferry Rd., approx. 0.5 mi. S.W. of Schuylkill River, approx. 1.24 mi. N.W. of Delaware River, in Southwest Sewage Treatment Works). Brick,

19'-4" (three-bay front) × 36'-10³/₄", two-and-a-half stories on raised basement, jerkin-head roof with pent eave across gable ends, originally pent eave or porch across rear at second story, unusually large chimney on north end, originally two rooms to a floor with corner fireplaces.

A rare surviving example of Philadelphia's characteristic early architecture. Built between 1714 and 1720; side addition with gable roof built later; interior gutted by fire 1947; house moved to present site, side addition demolished 1975. Site of British batteries during siege of Mud Fort (later Fort Mifflin) 1777, when American cannon ball pierced front and back walls. Acquired by City of Philadelphia 1910, Southwest Sewage Treatment Works built around original site of house 1949–54. Certified, PHC 1956. 10 sheets (1940, including plans, elevations, details); 6 ext. photos (1937, 1940); 1 data page (1937).

Blockley Retreat Farm. See Busti Mansion

Blue Bell Tavern
(PA-131), 7303 Woodland Ave., at N.W. corner Park Rd., approx. 170' E. of Cobbs Creek. Coursed rough-hewn fieldstone front, rubble sides and rear with galleting, approx. 18' (two-bay front) × 26' with one-story rubble rear ell with galleting, two stories, gable roof, date stone with initials "HP" on front, open wooden front and side porch. Built 1766. Three-and-a-half-story, five-bay rubble addition built 1801, demolished 1941. Since tavern served as southern gateway to city, George Washington stopped here often; site of his first welcome to Philadelphia as president-elect and his last farewell to city. Acquired by Fairmount Park Commission, City of Philadelphia 1913. Certified, PHC 1958. 2 ext. photos (1936), 10 photocopies of measured drawings (1916);* 2 data pages (1937).

Burnham, George, House. Illustrated
(PA-1627), 3401 Powelton Ave., at N.W. corner Thirty-fourth St. Random rough-faced granite with carved and dressed sandstone trim, approx. 40' (three-bay front) × 45' with two-story rear wing, two-and-a-half stories on raised basement, hipped and gable roofs with cross-gable, ridge cresting, one rear and three corner porches, balcony above one-story entry porch and vestibule, side vestibule and lateral entry hall to stairhall, notable interior woodwork.

Example of suburban domestic architecture by a popular late-19th-century Philadelphia architect. Built 1886; Theophilus Parsons

Chandler, Jr., architect. Corner porches enclosed 1952. Until 1910 the home of George Burnham, senior partner in the Baldwin Locomotive Works, America's foremost manufacturer of locomotives. Drexel University fraternity house since 1967. Certified, PHC 1959. 2 ext. photos (1973).*

Burnside. See Hamilton-Hoffman House

Busti Mansion (also known as Blockley Retreat Farm, later as Kirkbride Mansion)
(PA-1628), Approx. 300' S. of Haverford Ave. at E. side Forty-Fourth St. on grounds of Institute of the Pennsylvania Hospital. Brick with marble trim, approx. 54' (five-bay front) × 38', two-and-a-half stories, gambrel roof with balustrade, originally open one-story porch on south side and one-bay entrance portico on north side, center-hall plan.
    Fine Federal-style country house that survived urban expansion. Built 1794; two-story wing on north side built before 1908, removed 1960; porch on south side removed 1960. Italian-born Paul Busti, General Agent of Holland Land Co. and founder of Buffalo, N.Y., lived here 1799–1826. Property purchased by Pennsylvania Hospital 1836 and served as residence of Dr. Thomas S. Kirkbride, mental hospital superintendent, 1841–83. Acquired by City of Philadelphia 1957 as recreation building and caretaker's residence. Certified, PHC 1956.
    3 ext. photos (1958),* 5 int. photos (1958),* 1 photocopy of old watercolor (c. 1820);* 11 data pages (including 1806–1908 insurance surveys).*

Cannon Ball Farm House. See Bleakley House

Centennial National Bank
(PA-1095), 3142 Market St., at S.E. corner Thirty-second St. (and originally Woodland Ave., since closed). Brick with sandstone and black brick trim, approx. 48' (angled three-bay front) × 92', two stories on raised basement (originally one-story interior with two-story facade), hipped and gable roofs, projecting angled entrance bay with squat columnettes and pediment at roof line with crockets and trefoil design enclosing clock, pinched-top windows on ground story, pointed-arch windows on second story.
    A landmark work whose three-sided front related to a former three-street intersection and whose design marked a significant

development in Frank Furness's style. Built 1876; Frank Furness, architect. Interior altered 1893. Interior remodeled, rear addition built 1899; Frank Miles Day and Bro., architects. Interior altered, exterior simplified 1956; Budd Ross, architect. Branch bank of First Pennsylvania Banking and Trust Co. since 1956. Certified, PHC 1966; Pennsylvania Register 1970; NR. 3 sheets (1965, including plot plan, elevations);* 1 ext. photo (1960).*

College Hall. See University of Pennsylvania, College Hall

Columbia Engine Company
(PA-1629), 3420 Market St. Brick with marble trim, approx. 25' (three-bay front) × 115', three stories, flat roof, round-arch windows at second story, segmental-arch windows at third story, rectangular plan. Built 1867; demolished c. 1969. Built for West Philadelphia Engine Co. but occupied by Columbia Engine Co. 1868–71. 1 photocopy of old ext. photo (c. 1935).*

Cottage
(PA-138), 5901 Woodland Ave. Wood frame with clapboards (gable ends covered with tar paper), 20'-4" (three-bay front) × 14'-2" with one-story rear ell, one-and-a-half stories, gable roof, open front porch, one room with corner winding stairs. Built probably 18th century; rear ell built later; demolished c. 1956. 4 sheets (1940, including plans, elevations, details); 1 ext. photo (1940).

Drexel Institute, Main Building (now Drexel University)
(PA-1630), N.E. corner Thirty-second and Chestnut sts. Museum. Buff brick with rusticated granite basement and terra cotta trim, approx. 200' (eleven-bay front) × 200', three and four stories on raised basement, flat roof, richly decorated giant arched entrance with terra cotta portrait medallions on intrados and statue of Genius of Knowledge on keystone, giant Ionic pilasters, spacious marble entrance hall and large multilevel arcaded central court open to skylight.
Impressive interior space and exterior ornament on a notable late-19th-century academic building. Built 1890–91; Wilson Brothers & Co., architects; Charles McCaul, contractor; terra cotta by New York Architectural Terra Cotta Works. Three-story buff brick and terra cotta Randell Hall built at east end 1901–2; Wilson Brothers & Co., architects; John R. Wiggins, contractor. Drexel Institute was founded 1890 by prominent banker and philanthropist Anthony J.

Drexel to promote technical and vocational education; later known as Drexel Institute of Technology, name changed to Drexel University 1970. Houses Drexel Museum Collection. 3 ext. photos (1973),* 1 photocopy of architectural rendering (1901).*

Fort Mifflin. Illustrated
(PA-1225), Mud Island, W. bank of Delaware River, 0.76 mi. below confluence of Schuylkill and Delaware rivers. Gneiss and brick walls with earth fill, major axis approx. 530', surrounded by moat, flanked by redoubts and demilune, entered from any of three sallyports, wall 11' to 12' high, topped by earthen parapets and bullnose coping, 4' revetments or interior parapet wall, section includes terre plein and ramparts with central parade; star fort plan-type, most gun emplacements along S.W. and S.E. walls, bombproofs beneath east bastion.

Built c. 1772–c. 1798; portion of original stone wall still standing. Certified, PHC 1956; Pennsylvania Register 1970; designated National Historic Landmark 1970. 4 sheets (1970, 1971, including plan, site and location maps, gun emplacement details, bombproof details, wall sections); 18 aerial photos (1970), 6 general view photos (1969, 1970), 4 photos of walls (1970), 11 photos of bombproofs (1969), 5 photos of gun emplacements (1969, 1970), 2 photos of flagpole base (1969), 1 photo of date stone (1969), 1 photo of gun carriage (1969), 27 photos of old maps and views (1777, 1864–66, early 1900's, USGS map 1967, and several n.d.); 3 data pages (1971).

Fort Mifflin, Smith's Shop (PA-1226), Mud Island, W. bank of Delaware river, 0.76 mi. below confluence of Schuylkill and Delaware Rivers. Brick, 22'-8" × 19'-3", one story, gable roof with cornice returns at ends; rectangular one-room plan with herringbone brick flooring, brick and stone forge at S.W. end wall. Built before 1802; restored 1969. 4 sheets (1969, including plan, roof-framing plan, elevations, section, details); 5 ext. photos (1969, 1970, 1971), 5 int. photos (1969, 1970); 2 data pages (1971).

Fort Mifflin, Soldiers' Barracks (also known as Enlisted Men's Barracks) (PA-1227), Mud Island, W. bank of Delaware River, 0.76 mi. below confluence of Schuylkill and Delaware rivers. Brick, 117'-0" (eight-bay front) × 35'-1", two stories, gable roof (superimposed over gambrel roof) with shed roof over one-story colonnade along facade, three gabled dormers on either face of roof; rectangular plan, massive fireplaces of brick and stone. Built before 1800. 5 sheets (1969, including plans, elevations, section, details); 10 ext. photos (1969, 1970), 2 int. photos (1969, 1970); 2 data pages (1971).

Fort Mifflin, Officers' Quarters (PA-1228), Mud Island, W. bank of Delaware River, 0.76 mi. below confluence of Schuylkill and Delaware rivers. Brick, 80'-10" (six-bay front) × 19'-6$^{1}/_{2}$", two stories, gable roof with shed roof over two-story porch on facade, colonnaded porch with gallery, iron gallery rail in trellis pattern and iron strap supports for second floor of porch; rectangular plan, row of four rooms on each floor. Built c. 1814; later alterations. 6 sheets (1969, including plans, elevations, sections, details); 12 ext. photos (1969, 1970), 2 int. photos (1970); 2 data pages (1971).

Fort Mifflin, Arsenal (PA-1229), Mud Island, W. bank of Delaware River, 0.76 mi. below confluence of Schuylkill and Delaware rivers. Brick, 4' thick walls, 44'-2" (three-bay side) × 24'-2" (one-bay front), one story, gable roof with east and west walls parapeted, three circular openings on east and west sides with recessed panels above on east side, metal grilles within six circular openings; rectangular two-room plan, openings that are circular on exterior become arched with flat sills to receive rectangular window sashes on interior. Built c. 1816; later alterations. 2 sheets (1970, including plan, elevations, section, isometric detail); 2 ext. photos (1969), 2 int. photos (1969); 2 data pages (1971).

Fort Mifflin, Storehouse (also known as Commissary) (PA-1230), Mud Island, W. bank of Delaware River, 0.76 mi. below confluence of Schuylkill and Delaware rivers. Brick, 55'-8$^{1}/_{2}$" (six-bay front) × 20'-4", one story, gable roof, originally one room open to rafters. Built between 1839 and 1851; renovated early 1960's. 2 sheets (1970, including plan, elevations, section); 3 ext. photos (1969, 1970); 2 data pages (1971).

Fort Mifflin, Artillery Shed (PA-1231), Mud Island, W. bank of Delaware River, 0.76 mi. below confluence of Schuylkill and Delaware rivers. Brick on three sides with projecting rubble foundations 1' above grade, original open facade of monolithic gneiss columns, frame addition to facade covered with boards and tar paper, 97'-0" (nine-bay front) × 36'-9", one story with interior deck, hipped roof with shed roof over porch addition, cornice returns on ends of porch; rectangular one-room plan, herringbone brick flooring, wooden deck 7' above floor with hoist. Built c. 1837; later additions and alterations. 5 sheets (1970, including plan, elevations, sections, details); 2 ext. photos (1969, 1970), 5 int. photos (1969); 2 data pages (1971).

Fort Mifflin, Commandant's House (also known as Headquarters) (PA-1232), Mud Island, W. bank of Delaware River, 0.76 mi. below confluence of Schuylkill and Delaware rivers. Stuccoed brick, 70'-2$^{1}/_{2}$" (seven-bay front) × 34'-5", one-and-a-half stories, low hipped roof (superimposed over original steep hipped roof), rectangular pavilion on S.E. side of house, octagonal wood cupola with bell-shaped roof, stuccoed pilasters define bays, heavy entablature of corbeled brick; rectangular plan with two-room pavilion, transverse hall between four-room group with central chimney stack and two-room suite with back-to-back fireplaces. Built c. 1796; later alterations and additions. 8 sheets (1970, including plans, elevations, details); 7 ext. photos (1969), 21 int. photos (1969, 1970); 3 data pages (1971).

Fort Mifflin, Mess House (also known as Hospital) (PA-1233), Mud Island, W. bank of Delaware River, 0.76 mi. below confluence of Schuylkill and Delaware rivers. Brick, 49'-6" (five-bay front) × 27'-4", two stories, gable roof with shed roof over two-story porch on facade, colonnaded porch with gallery, rectangular plan. Built c. 1819; many later alterations. 6 sheets (1970, including plans, elevations, section, details); 3 ext. photos (1969); 2 data pages (1971).

Fort Mifflin, Magazines (PA-1234), Mud Island, W. bank of Delaware River, 0.76 mi. below confluence of Schuylkill and Delaware rivers. Includes East and West magazines inside fort, above and below grade magazines in demilune, and below grade and Mortar Magazines in post–Civil War Redoubts; East and West magazines recorded in detail.

Concrete East Magazine with vault 128'-1" × 16'-0¼", brick West Magazine with valut 43'-2" × 15'-1", both magazines: one story, barrel-vaulted, beneath earthen mounds 20' high, stepped brick retaining walls and heavy semicircular-arched doors at entries, offset passages lead to vaults of rectangular plans. West Magazine built c. 1867; East Magazine built c. 1872–86. 3 sheets for East and West magazines (1970, including plans, elevations, sections); for all magazines, 6 ext. photos (1969, 1970, 1971), 7 int. photos (1969, 1970, 1971); 2 data pages (1971).

Fort Mifflin, Northeast Sallyport (PA-1235), Mud Island, W. bank of Delaware River, 0.76 mi. below confluence of Schuylkill and Delaware rivers. Now principal entrance to fort, passage approx. 10'-9" × 42'-6" including retaining walls, runs through N.E. brick curtain wall beneath parapet, barrel vault 9' high, guard rooms to either side with embrasures to exterior of wall, retaining walls extend beyond face of vault as supports against earth ramparts. Massive iron hardware, 6" thick double doors set in frame within semicircular arch, herringbone brick paving. Built c. 1779–98. 4 sheets (1970, including plan, elevation, section, details); 1 ext. photo (1969), 4 int. photos (1969); 1 data page (1971).

Fort Mifflin, Southeast Sallyport (originally known as Main Gate)

(PA-1236), Mud Island, W. bank of Delaware River, 0.76 mi. below confluence of Schuylkill and Delaware rivers. Stone and brick entrance near east bastion, solid brick wall 9' thick with elliptical arched opening 14'-6" high, splayed on interior side, 18' from parade to top of wall, massive double doors framed into arched opening, exterior faced with chamfered gneiss, walkway over entry raised from level of surrounding wall, retaining wall 7' high and 41'-6" long extending from sallyport toward interior of fort for support of adjacent ramparts. Building begun before 1802, completed 1807. 1 ext. photo (1969), 3 int. photos (1969).

Fort Mifflin, Northwest Sallyport (PA-1237), Mud Island, W. bank of Delaware River, 0.76 mi. below confluence of Schuylkill and Delaware rivers. Brick, T-shaped plan including guard rooms to either side of passage, exterior brick masking, barrel-vaulting, retaining walls extending to interior, passage from exterior of scarp curtain wall to ramparts and beneath parapet, massive double doors set in framing of semicircular arch. Built c. 1779–98. 1 ext. photo (1969), 3 int. photos (1969).

Furness Building. See University of Pennsylvania, Library

George House
(PA-1631), 6099 Drexel Rd., at S.E. corner City Line Ave. Stuccoed rubble, approx. 50' (five-bay front) × 15' with two-and-a-half-story rear ell, two-and-a-half stories, gable roof, open columned front porch. Built in three stages during 18th century; altered later; demolished 1958. HABSI form (1958).*

Hamilton-Hoffman House (also known as Burnside)
(PA-1053), N.E. corner Cobbs Creek Parkway and Sixty-first St. Stuccoed rubble, approx. 40' (five-bay front) × 30' with one- and two-story lateral wings (southwestern and northeastern ends respectively), two-and-a-half stories, gable roof, open front and back porches; center-hall plan, elaborate interior trim, carved Federal-style mantels.

Fine example of early-19th-century country estate. Built between 1791 and 1800; porches built c. 1880; two-story wing added later. Demolished 1960. Living-room mantelpiece given to Philadelphia Museum of Art; other woodwork given to St. Peter's Protestant Episcopal Church, Philadelphia. Gavin Hamilton, Jr., was a prominent snuff merchant; Hamilton family owned house from 1832. 2 ext.

photos (1959), 10 int. photos (1959), 1 photocopy of old ext. photo (c. 1890); 8 data pages (1960).

Institute of the Pennsylvania Hospital. See Pennsylvania Hospital for the Insane, Department for Males

Justi, Henry D., House
(PA-1632), 3401 Baring St., at N.W. corner Thirty-fourth St. Random granite ashlar, approx. 45' (five-bay front) × 35' with three-story stuccoed rear ell and one-story lateral rear wing, two-and-a-half stories on raised basement with square three-story corner tower, gable roof with mansard roof on rear ell, open front porch, and semi-hexagonal front bay window, entry in tower with side stairhall.

Example of suburban villas erected in West Philadelphia after the Civil War. Built c. 1867; rear ell built c. 1875; lateral wing and front porch built c. 1890. Home for displaced children since 1922. Henry D. Justi, a manufacturer of artificial teeth and dental instruments, lived here 1867–1922. 2 ext. photos (1963);* 1 data page (1962).*

Kirkbride Mansion. See Busti Mansion

Kirkbride's. See Pennsylvania Hospital for Mental and Nervous Diseases

Lea, Professor Henry Charles, House. Illustrated
(PA-1633), 3903 Spruce St. Stuccoed brick, approx. 50' (five-bay front) × 55', three stories with square four-story corner tower, flat roof with carved brackets, open balustraded semicircular and rectangular porches, ground-story side bays, quoins; entrance foyer in tower with adjacent stairhall, excellent Italianate interior detailing.

Notable example of Italianate suburban villa. Built c. 1857; demolished 1969. Home of internationally known historian and prominent local publisher, Henry Charles Lea, 1857–69; University of Pennsylvania fraternity house 1924–68. Certified, PHC 1962. 3 ext. photos (1968);* 14 int. photos (1968).*

Mount Moriah Cemetery Gate House. Illustrated
(PA-1634), 6299 Kingsessing Ave., at N.E. corner Cemetery Ln. Connecticut brownstone ashlar, approx. 80' × 30' with one-story stuccoed stone side wings, two buildings connected by triple arcade gateway, each two stories, flat roof (originally hipped) with front

corner turrets on southern building, gable roof with Flemish cross-gables on northern building, battlements and urns.

Early picturesque eclectic design. Built 1855; Stephen D. Button, architect; A. H. Ransom, stone builder. Octagonal battlemented tower on southern building and statuary over gateway removed, side wings built later. 5 ext photos (1975).*

Museum of Science and Art. See University of Pennsylvania, University Museum

Overbrook School for the Blind. See Pennsylvania Institution for the Instruction of the Blind

Pennsylvania Hospital for the Insane, Department for Males (now Institute of the Pennsylvania Hospital)
(PA-1635), 111 N. Forty-ninth St. Stuccoed rubble with granite trim, approx. 470' (thirty-three-bay front) × 225' with two flanking rear one-story E-plan wings, three stories with central dome, gable roofs with cross-gables, Doric tetrastyle portico with flat pediment, two octagonal belvederes each on north and south end buildings, triple E-plan with central hall and intersecting corridors.

Fine example of the "Kirkbride System," a plan to increase individual attention to patients and total operational efficiency; originated by Dr. Thomas S. Kirkbride, hospital superintendent 1841–83, and designed by Samuel Sloan, prominent mid-19th-century architect. Built 1856–59; Samuel Sloan, architect; John Sunderland, superintendent of construction. Enlarged 1929, 1931. Certified, PHC 1957; designated National Historic Landmark 1965. 2 ext. photos (1973),* 12 data pages (including 1861–1908 insurance surveys).*

Pennsylvania Hospital for Mental and Nervous Diseases (also known as Kirkbride's). Illustrated
(PA-1636), N.W. corner Forty-fourth and Market sts. Coursed granite ashlar center building and stuccoed rubble wings and end buildings with stone trim, approx. 443' (forty-one-bay front) × 116', two stories on raised basement, gable roofs with central dome, Doric tetrastyle portico with flat pediment, octagonal belvederes at intersection of wings and end buildings, E-plan with central hall and intersecting corridors.

Pioneer example of mental-hospital architecture using extended wings to permit maximum isolation of wards. Built 1836–40; Isaac

Holden, architect; Samuel Sloan, superintendent of construction. One-story addition to north (men's) lodge 1846–47; addition to south (women's) lodge 1849; north wing (Fisher Ward) enlarged 1872–80, 1906. Demolished 1959; portico and flanking walls preserved on site. Used only for women after 1859. Popularly named for Dr. Thomas S. Kirkbride, hospital superintendent 1841–83 and formulator of advanced methods of treating the mentally ill. Certified, PHC 1957. 4 ext. photos (1958),* 4 int. photos (1958),* 1 photocopy of old ext. engraving (c. 1840);* 14 data pages (including 1840–1908 insurance surveys).*

Pennsylvania Institution for the Instruction of the Blind (now Overbrook School for the Blind). Illustrated

(PA-1637), N.E. corner Sixty-fourth St. and Malvern Ave. Yellow plastered stone with terra cotta trim, approx. 400' (forty-five-bay front) × 200', two and three stories, tiled gable roofs with tiled hipped roofs on corner pavilions and tiled shed roofs on courtyard arcades, square bell towers flank arcaded entry porch to octagonal rotunda with low circular turrets and large circular lantern, round-arch and rectangular windows, rectangular plan with two large inner courtyards.

Rare and exceptional example of the Mission style in eastern United States. Built 1897–1900; Cope and Stewardson, architects; Warner H. Jenkins & Co., contractor. Two-story stone addition to superintendent's house built 1910; Cope and Stewardson, architects. Two-story steel-frame Graduate Cottage with brick curtain walls built 1940–41; Zantzinger and Borie, architects. Auditorium behind rotunda burned 1960, replaced by reinforced-concrete auditorium 1961; Francis, Cauffman, Wilkinson and Pepper, architects. One of the oldest institutions for the blind in the country; founded 1834. Name changed to Overbrook School for the Blind 1948. Certified, PHC 1963. 5 ext. photos (1973).*

Potts, Joseph D., House

(PA-1638), 3905 Spruce St. Brick with tile trim, approx. 46' (five-bay front) × 38'with rectangular rear corner tower and two-story rear ell, three-and-a-half stories, hipped roof, segmental pointed-arch windows, center-hall plan, richly carved interior woodwork and hand-painted interior tile.

Important residential work of Joseph M. Wilson, innovative late-19th-century architect of commercial and railroad structures. Built

1850. Extensively enlarged and remodeled c. 1876; Joseph M. Wilson, architect. Front porch removed 1943. Joseph D. Potts, a wealthy steamship and railroad executive lived here 1876–93. Center for foreign students 1918–59; University of Pennsylvania building since 1960. Certified, PHC 1963. 5 ext. photos (1961, including 1 of stable and carriage house),* 6 int. photos (1961);* HABSI form (1962).*

St. Agatha's Roman Catholic Church
(PA-1639), 3801 Spring Garden St., at N.W. corner Thirty-eighth St. Brownstone ashlar, approx. 70' (three-bay front) × 150', one story on raised basement, gable roof with cross-gables, three-stage center tower with pinnacles and battlements, three-aisle plan, galleries, apsidal chancel.

Fine High Victorian Gothic design by a leading late-19th-century church architect. Built 1874–78; vacated 1976; Edwin F. Durang, architect. Octagonal spire erected 1882–83, dismantled 1909. Chapel built and church interior altered 1901, one-story rear addition built 1912; Edwin F. Durang, architect. Interior extensively remolded 1924–26; Maginnis and Walsh (Boston), architects. Roof and top of tower demolished by fire 1966, replaced 1967. Rectory (3813 Spring Garden St.) built 1892; P. Albert Welsh, architect. 2 ext. photos (1973, including 1 color photo),* 1 photocopy of old ext. lithograph (c. 1880).*

St. Francis De Sales Roman Catholic Church. Illustrated
(PA-1640), 4629–35 Springfield Ave., at N.E. corner Forty-seventh St. Coursed rock-faced limestone ashlar with dressed limestone and marble trim, approx. 90' (three-bay front) × 150', one story with two-story facade, gable roofs, central green, yellow, and white tiled Byzantine dome with arcaded lantern and four small tiled domes at base, arched parapet between corner octagonal minarets with tiled domes, round-arch windows with engaged columns and hoods, large oculus at second story of each facade, sculpture over round-arch entrances, three-aisle plan, rectangular chancel.

One of America's finest examples of Late Victorian Byzantine architecture. Built 1907–8; Henry D. Dagit, architect; Melody and Keating, contractors. Sanctuary elements redesigned to conform with ecumenical reforms 1968; Venturi and Rauch, architects. 4 ext. photos (1973, including 3 color photos),* 1 photocopy of architectural drawing (1907).*

St. James Roman Catholic Church
(PA-1641), 3728 Chestnut St., at S.E. corner Thirty-eighth St. Rough-hewn limestone ashlar with marble trim, approx. 67' (three-bay front) × 157', one story on raised basement, gable roof with cross-gables, twin front four-stage towers with pinnacles, rose window, large superimposed pointed-arch entrance, pointed-arch windows, three-aisle plan, apsidal chancel.

Fine example of High Victorian Gothic church. Built 1881–87; Edwin F. Durang, architect. Spire on northeast corner tower removed after 1900; towers altered 1930. 1 ext. photo (1973),* 1 photocopy of old ext. lithograph (c. 1880, including school),* 1 photocopy of old ext. photo (c. 1900),* 1 photocopy of old int. photo (c. 1900);* 2 data pages (including 1869 insurance survey of school).*

St. Mary's Protestant Episcopal Church, Hamilton Village
(PA-1642), 3916 Locust St. (now closed, known as Locust Walk). Random rough-hewn granite ashlar with limestone trim, approx. 85' × 128' with one-story rough-hewn limestone entrance porch and limestone enclosed cloister leading to rear two-story coursed rough-hewn parish house, one story, polychromatic gable roofs with cross-gables, two-stage square tower with pinnacles and spire at northwest corner, pointed-arch, quatrefoil and trefoil windows, one-story brick addition at southeast corner; three-aisle plan, semihexagonal chancel with notable marble and mosaic high altar, exposed wooden trusses.

Small Gothic Revival suburban church whose architecture and function reflect a century of change in its neighborhood. Built 1872–73; Thomas W. Richards, architect. High altar imported from Rome 1875, exhibited at Centennial Exhibition 1876, installed in chancel 1890. Chancel, western entrance porch built 1890; Charles M. Burns, architect. Parish house built 1897; George Nattress, architect. Tower raised to two stories, spire and terra cotta pinnacles added 1907; Herbert J. Wetherill, architect. Enclosed cloister added, chancel altered c. 1910; parish house renovated after fire damage 1936, renovated 1965; spire removed 1973. Founded 1830 as parent parish of West Philadelphia; Hamilton Village derived its name from Hamilton family of Woodlands and was officially incorporated into borough of West Philadelphia 1844. Certified, PHC 1974. 3 ext. photos (1973).*

Tabernacle Presbyterian Church. Illustrated
(PA-1099), 3700 Chestnut St., at S.W. corner Thirty-seventh St. Granite ashlar with limestone trim, approx. 50' (three-bay

front) × 100′ with rear two-story chapel and two-and-a-half-story manse on raised basement at southeast corner, gable roofs with cross-gables, four-stage corner entry and bell tower with gargoyles and pinnacles, octagonal stair tower between church and chapel, open Gothic-arch arcade along chapel; four-aisle cruciform plan, galleries, notable exposed roof truss with angels carved on hammerbeams.

Fine examples of late-19th-century Gothic Revival architecture. Built 1884–86; Theophilus Parsons Chandler, Jr., architect; Oliver Bradin, superintendent of construction; Michael Scully, master mason. 11 ext. photos (1963),* 3 int. photos (1963),* 1 photocopy of old ext. photo (c. 1886),* 2 photocopies of old int. photos (c. 1890, c. 1930),* 1 photocopy of architect's drawing (1884);* 7 data pages (1965).*

University of Pennsylvania, College Hall (originally known as The Academic Department of the University of Pennsylvania)
(PA-1643), S. side Woodland Ave. (now vacated) between Thirty-fourth and Thirty-sixth sts. Random green serpentine ashlar with Leiperville stone, brownstone and Ohio stone trim, 254′-0″ (twenty-five-bay front) × 102′-4″, three and four stories on raised basement, mansard roofs with cross-gables, Gothic tracery windows at second-story center for original chapel, originally two pinnacles on each facade and five-and-a-half-story towers with steep hipped roofs at east and west ends, Gothic entrance porch, E-plan with center entry hall and longitudinal corridors.

Fine example of Collegiate Gothic architecture and one of the country's most advanced academic structures when built. Built 1870–72; Thomas W. Richards, architect; George Watson, builder. Towers and pinnacles removed 1914, 1929. Certified, PHC 1957. 2 ext. photos (1973),* 1 photocopy of architects drawing (c. 1871);* 15 data pages (including 1876 insurance survey and architect's description of plans).*

University of Pennsylvania, Library (now known as Furness Building)
(PA-1644), W. side Thirty-fourth St. S. of Walnut St. (originally at S.W. corner Thirty-fourth St. and Woodland Ave., since closed). Iron and brick with terra cotta and smooth and pecked redstone trim, approx. 140′ (eight-bay front) × 80′ with three-story lateral book stack and side and rear additions, four stories on raised basement with five-story square battlemented tower, tiled hipped roof with cross-

gable on main section with tiled conical roof on apsidal north end and glass gable and shed roofs on book stack, open entrance porch, crockets on dormers and gables, gargoyles on apsidal north end; entry and stairhall with notable iron staircase lead to two original reading rooms and stacks with glass walkways, north room with iron-beam apsidal vaulting, monumental interior detailing.

A major work of Frank Furness, path-making late-19th-century architect, and a masterpiece of library design and function when built; early example of separating reading rooms from book stacks. Built 1888–91; Furness, Evans & Co., architects; terra cotta by New York Architectural Terra Cotta Works. Five-story brick and redstone Duhring Memorial Stack added to book stack 1914–15; Furness, Evans & Co., architects. Two-story reinforced-concrete and brick Henry Charles Lea Library and Reading Room added to rear (along Thirty-fourth St.) 1923–24; Furness, Evans & Co., architects. One-story brick and granite Horace Howard Furness Memorial Library added to front of book stack 1931; Robert McGoodwin, architect. Floor added within main (south) reading room 1922. Duhring Stack (or Wing) altered c. 1947; Harbeson, Hough, Livingston and Larson, architects. Interior renovated 1963. Library and studios of Graduate School of Fine Arts since 1963; name changed to Furness Building in memory of its architect 1963. Certified, PHC 1961. 7 ext. photos (1964),* 3 int. photos (1964);* HABSI form (1962).*

University of Pennsylvania, Men's Dormitories. Illustrated
(PA-1645), S. side Spruce St. at Thirty-seventh St. and Woodland Ave. (since vacated). Brick with sandstone trim, approx. 560' × 42' on Spruce St., approx. 530' × 40' on Woodland Ave., three-and-a-half stories with six-story quadruple entry towers, gable roofs with cross-gables and Flemish cross-gables, balustrades, oriels, arcades, five courts in semitrapezoidal plan.

Integration of architectural detailings, openings, and courts of various sizes and levels into a notable example of Late Victorian Jacobean style. Built 1895–1902; Cope and Stewardson, architects. Enlarged between 1902 and 1910; Cope and Stewardson, architects. Enlarged between 1912 and 1930; Stewardson and Page, architects. Steel-frame additions with brick curtain walls built 1955, 1959; Trautwein and Howard, architects. Certified, PHC 1959. 7 ext. photos (1973, including 1 color photo).*

University of Pennsylvania, University Museum (originally known as Museum of Science and Art). Illustrated

(PA-1646), 3620 South St., at S.E. corner Thirty-third and Spruce sts. Brick with white and colored marble trim, approx. 475' (thirty-eight-bay front) × 190', one, two, and three stories, tiled hipped and gable roofs, round-arch windows, notable sculpture at entrance, U-plan with side and rear extensions.

Notable cooperative work of architecture integrating many historical styles into a building of various levels and units. West court built 1893–99, rotunda behind west court built 1912, middle section and eastward addition to rotunda built 1926–28; Wilson Eyre, Jr., Frank Miles Day and Bro., Cope and Stewardson, collaborating architects; Alexander Calder, sculptor. Planned complex never completed. Five-story reinforced-concrete addition built at southeast end 1969–71; Mitchell/Giurgola Associates, architects. One of the world's leading archaeological museums. Certified, PHC 1959. 4 ext. photos (1973).*

Woodland Terrace Area Study. Illustrated

(PA-1647), 501–19, 500–520 Woodland Terr., E. and W. sides Woodland Terr. between Baltimore and Woodland aves. Eleven brick semidetached houses alternately faced with random brownstone ashlar and stucco, each approx. 32' (four-bay front) × 58' with one-story rear wing, three stories, flat roof, open front and side porches, four-story square side tower on end houses on E. side, transverse entry-hall plan.

Notable example of mid-19th-century speculative terrace housing in the Italianate mode. Built 1861–62; probably by Samuel Sloan, architect. Porches removed from Nos. 500, 520 later; No. 522 demolished before 1963. Paul M. Cret, eminent early-20th-century architect and teacher, lived in No. 516, 1919–45. Certified, PHC 1963; Pennsylvania Register 1971; NR. 8 ext. photos (1973);* 10 data pages (1963, including 1862–63 insurance surveys).*

Woodlands. Illustrated

(PA-1125), S.W. corner Thirty-ninth St. and Woodland Ave., in Woodlands Cemetery. Rubble and stuccoed rubble with brick trim, approx. 86' (five-bay front) × 45', two stories on raised basement, hipped roof, Tuscan tetrastyle portico with pediment on south (river) facade, central stuccoed pavilion with six pilasters and pediment on north facade, circular bays on east and west ends, Ionic frontispiece with fanlight and arched pediment at north entrance, central circular reception hall with stairs to the side. Rubble two-story stable with hipped roof and flanking one-story shed-roof wings approx. 280' north of house.

Notable Adamesque-Federal country house with river and land fronts and the first giant portico on a Philadelphia building. Built c. 1742; extensively enlarged and remodeled for William Hamilton 1787–90. Country seat of eminent Hamilton family until grounds became cemetery 1843; house now superintendent's quarters. Certified, PHC 1956; designated National Historic Landmark 1967. 18 sheets (1940, including plans, elevations, details, plan and elevations of stable) from Survey, Philadelphia Chapter, A.I.A. (1932); 6 ext. photos (1938, including 2 of stable); 1 data page (including 1811 insurance survey).*

Wynnestay

(PA-1648), 5125 Woodbine Ave., at N.E. corner Fifty-second St. Rubble, approx. 55' (five-bay front) × 22' with two-story rear ell (originally approx. 25' × 22'), two-and-a-half stories (originally two), gable roof, pent eave on eastern part, door hood on western end, oculus in west gable end, cornice across gable ends, originally one room to a floor.

Fine example of early Pennsylvania architecture. Built 1689; lateral western extension built 1700; roof of eastern part raised later; rear ell built early 20th century. Built on property of Dr. Thomas Wynne, William Penn's personal physician and Speaker of Provincial Assembly 1683, probably for Wynne's son Jonathan. Certified, PHC 1957. 1 ext. photo (1973).*

George Burnham House. *Jack E. Boucher photo, 1973.*

Church of St. Francis de Sales. *Jack E. Boucher photo, 1973.*

Fort Mifflin, Sally Port. *Cortlandt V. D. Hubbard photo, 1969.*

Professor Henry Charles Lea House. *James L. Dillon & Co. photo, 1968.*

Mount Moriah Cemetery Gate House. *Cortlandt V. D. Hubbard photo, 1975.*

Pennsylvania Hospital for Mental and Nervous Diseases. *Engraving by W. E. Tucker, c. 1840.*

Pennsylvania Institution for Instruction of the Blind. *Jack E. Boucher photo, 1973.*

University of Pennsylvania, Men's Dormitories. *Jack E. Boucher photo, 1973.*

Tabernacle Presbyterian Church. *Jack E. Boucher photo, 1963.*

University of Pennsylvania, University Museum. *Jack E. Boucher photo, 1973.*

Woodland Terrace, *Jack E. Boucher photo, 1973.*

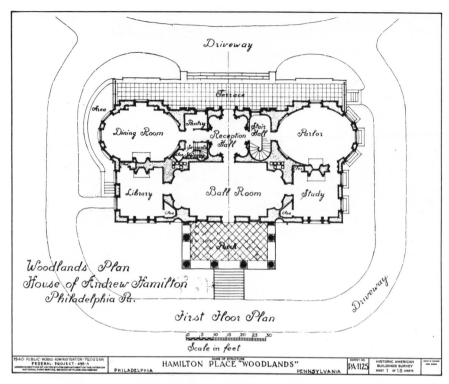

Woodlands. *L. G. Park, del., 1940.*

Fairmount Park

6

Introduction

"I S IT POSSIBLE you have never seen Fairmount Park?
Believe me, then, that it is the most beautiful place of the whole civilized world . . ." claimed the well-traveled writer Lafcadio Hearn in 1889.[1] Surrounded by a city of nearly two million, Fairmount Park offers more than 4,000 acres of open land dotted with recreational and cultural sites. Its borders extend northward from Fairmount along both banks of the Schuylkill River to the Wissahickon Creek and northward along that stream and its tributary, Cresheim Creek, to the city limits at Northwestern Avenue.[2] Its natural history reaches back to pre-Columbian America to include untouched virgin soil strata, rock outcroppings, and ancient trees. In commemoration of the preservation of this natural environment, the Wissahickon Valley was designated a Registered National Natural Landmark in 1964.

Civilization encroached upon the area during the eighteenth and early nineteenth centuries. The falls and rapids of the Wissahickon made it especially suited for mills, while along the Schuylkill gracious manor houses crowned the heights, and gardens and lawns sloped down to the river's edge. These estates served as nuclei for the park's nineteenth-century development and the houses have been seminal projects in historic preservation. Twenty-six of them still stand, joined by three others moved to the park from various sites in Philadelphia.[3] In recognition of this trailblazing role in America's recreation, preservation, and ecological history, Fairmount Park was put on the National Register of Historic Places in 1972.

The genesis of Fairmount Park in 1812 occurred when the city's waterworks were moved from Center Square to the Schuylkill River at Fairmount, a hill on the western outskirts of the original city. The

Fairmount Waterworks (illus.) initially operated in a large two-and-a-half-story steam-engine house, which still stands on the southern edge of the complex. The steam engines proved troublesome and expensive from the beginning of their operation in 1815, and four years later City Councils decided to switch to water power.[4] The new facility, next to the original steam-powered works, required a great deal of work: purchasing the rights to the water power at the Falls of Schuylkill from the Schuylkill Navigation Company, throwing a dam across the river, blasting from the mount a forebay (from which water would power the wheels), and building the machinery and mill house. It was all accomplished within three years, and the system was operating by the end of 1822.[5] The five acres surrounding the initial Fairmount Waterworks had been landscaped into a public garden, making it probably the earliest planned municipal park in the United States.[6] The garden met with such public favor that the landscaped area grew with the waterworks and by 1828 the small park had nearly quintupled in size and had attained a beauty so striking that even the critical Charles Dickens praised it.[7] The present appearance of the waterworks is largely the results of changes made when the waterwheels were replaced by turbines after the Civil War. The pitch roof of the mill house was covered with a flat terrace, and an ornamental peristyle temple structure and flanking wooden huts were placed between the two original end pavilions.[8]

By the mid-nineteenth century a visit to the park generally included a stroll through Laurel Hill Cemetery, located near the river north of the waterworks. Begun in 1836 on the site of the former country seat of Joseph Sims, Laurel Hill was more than a mere burial ground. It was also a romantically landscaped arboretum and sculpture garden highlighted by a handsome Doric gate house (illus.). The young architect John Notman planned twisting roads through the undulating grounds, and the directors planted a wide variety of valuable trees and shrubs and encouraged the erection of tasteful memorials and monuments to make Laurel Hill one of the finest rural cemeteries in the country.[9] These two dissimilar projects, a classically clothed technological wonder and a naturally landscaped graveyard, served as stimuli for Fairmount Park's rapid growth, which began in 1855. In that year Lemon Hill and its approximately forty-five acres were dedicated to public use. Lemon Hill, the graceful Federal-style country house overlooking the waterworks, had been purchased by the city eleven years earlier in order to preserve the purity of the municipal water supply. Two years later Sedgeley and its thirty-three acres were acquired, and shortly thereafter title was obtained to the land between the waterworks and Lemon Hill. Finally in 1867, just about the time that the city was acquiring the Wissahickon Valley and two large estates on the west bank of the river, the Pennsylvania General Assembly set aside a large area bordering the Schuylkill River "forever as an open public place."[10]

By this time the city was clearly concerned as much for park land as for pure water. The river was already a center for racing sculls, a sport that had been inaugurated on its quiet waters in about 1835. Gunwale barges were initially used, followed by outriggers and eventually by sculls,[11] whose gracefulness inspired many of Thomas Eakins's paintings. By the 1850's the growing barge clubs began replacing their shacks with substantial stone buildings,[12] a process that developed into picturesque Boat House Row, which stands in juxtaposition to the classical waterworks and Museum of Art nearby. Perhaps the most noted boathouse is the Undine Barge Club by Furness and Evans, but the bold peaks of Vesper and Malta also contribute to the irregular composition.[13]

A design competition for the park was held in 1858, although the area officially had not yet been set aside for that purpose. The proposal of Sidney & Adams was adopted but the coming of the Civil War seems to have prevented its completion.[14] Later advice was sought from noted landscape architects Robert Morris Copland of Boston and Olmsted and Vaux of New York, but again no plans were executed, although one of Olmsted's proposals survives.[15] It was not until after the Civil War that the park took on a deliberate design. In 1872 the Fairmount Park Commission adopted modest but romantic plans for East Fairmount Park that had been developed by Senior Assistant Engineer H. J. Schwarzmann. Two years later he became the chief engineer and architect of the International Exhibition of 1876, which was located in West Fairmount Park.[16] The site was but a short distance from the Zoological Gardens, which had opened in 1874 as the first zoo in the United States. Schwarzmann also laid out the zoo's grounds,[17] which were filled with imaginatively picturesque structures, such as the gate houses by Furness and Hewitt and the bear pits (illus.) by Theophilus Parsons Chandler, Jr. Both the zoo and the centennial exhibition were accessible to the city by way of the Girard Avenue Bridge (1874) and the Spring Garden Street Bridge (1875).

As one of the great world fairs, the centennial had a major impact on the development of West Fairmount Park. It created circulation

problems that dictated the design of many of the permanent roadways, and it left behind a number of structures, of which only two
remain, Memorial Hall (illus.) and the Ohio House (illus.) It also encouraged the park's sculpture project, which had begun in 1872 with
the founding of the Fairmount Park Art Association, a private group
dedicated to "promote and foster the beautiful in the City of
Philadelphia."[18] The association has given the city notable works by
such greats as Augustus Saint-Gaudens, Daniel Chester French,
Frederic Remington, and Jacques Lipshitz, and while it was not
directly responsible for the grandiose Civil War monument standing
near Memorial Hall—the gift of Richard Smith, a patriotic and
very successful typefounder—it was charged under the benefactor's
will to select and supervise the sculptors of the project.[19] Considered
"one of the most ambitious public monuments erected in Victorian
America,"[20] Smith Memorial possesses works by some of the leading
American sculptors from the turn of the century, and remains a striking landmark.

The park's charms became accessible to people of modest means
in 1896 when the Fairmount Park Transportation Company inaugurated its six-mile trolley run through the park, crossing the river on
the graceful Strawberry Mansion Bridge (illus.). Operations ceased
fifty years later, another casualty of the automobile.[21] The same automobile, however, brought even more people to the park and
further diversified its recreation role. In the summer of 1930 the Robin
Hood Dell presented its first free concert by the Philadelphia
Orchestra,[22] and in 1952 the Playhouse in the Park began its first
season of summer stock under a tent; a permanent structure now
houses the theater. Meanwhile, more tennis courts, swimming pools,
and picnic grounds were being built and promptly filled, reflecting
the increased leisure of the mid-twentieth century and leading some
to fear that even so large a park as Fairmount may be threatened by
overuse.

Catalog

Art Gallery. See International Exhibition of 1876, Memorial Hall

Belmont Mansion. Illustrated
(PA-1649), Belmont Mansion Dr. Brick and rubble with stone trim, approx. 40' (three-bay front) × 35' with attached one- and two-story buildings, three stories (originally two), hipped roof with balustrade, projecting center pavilion with gable, open porch on three sides, center great-hall plan, notable interior details.

Fine example of evolutionary growth of a country house with notable early decorative plaster work. Two-story gambrel-roof rubble cottage west of mansion built before 1742; semioctagonal bay added c. 1744. Two-story gable-roof house at S.W. corner of mansion built 1745. Mansion built c. 1755; rear stair tower, front porticoed porch built c. 1760; ornamental ceiling of great hall installed c. 1762; third story, lattice porch across front built before 1860; porch extended around sides after 1867. Stairway of 1745 house removed 1874; octagonal bay at south end removed 1876. Interiors rehabilitated 1926. One-story brick building at west corner of rubble cottage built after 1931; demolished 1972.

Country house of William Peters, lawyer, jurist, Register of Admiralty 1744–71, and his son Richard Peters, agriculturist, Commissioner of War for Continental Congress 1776–81, member of Pennsylvania Assembly 1788–92, noted federal jurist 1792–1828. Acquired by Fairmount Park Commission, City of Philadelphia 1867. Certified, PHC 1956. 2 photocopies of ext. photos (1931),* 4 photocopies of int. photos (1926);* 4 data pages (1973).*

227

Boat House Row
(PA-1650), E. River Dr. W. of Main Dr. Ten houses of brick or random
rough-hewn stone (one frame with shingles), two, two-and-a-half,
and three stories, gable and hipped roofs with cross-gables and turrets. Rare and picturesque assemblage of river-boat clubhouses. Built
late 19th and early 20th centuries including: No. 2, Fairmount Rowing
Association built early 20th century; No. 3, Quaker City Barge Club
built c. 1860 with later alterations; No. 4, Pennsylvania Barge Club
built 1892 Louis C. Hickman, architect; No. 5, Crescent Boat Club
built c. 1867, alterations and addition 1891, Charles Balderston, architect; No. 6, Bachelors' Barge Club built 1893, Halzehurst and
Huckel, architects; No. 7, University Barge Club built 1870, altered
and enlarged 1893; No. 8, Philadelphia Barge Club built late 19th
century; No. 9, Malta Boat Club built 1870, enlarged 1880, enlarged
and remodeled 1901, G. W. and W. D. Hewitt, architects; No. 10,
Vesper Boat Club built c. 1865, altered and enlarged 1898, Howard
Hagar, architect; No. 11, College Boat Club of University of Pennsylvania built c. 1875; No. 12, Penn Athletic Club Rowing Association
(former West Philadelphia Rowing Club) built 1873 with later alterations; No. 13, Undine Barge Club built 1882–83, Furness and Evans,
architects; No. 14, Philadelphia Girls Rowing Club (former Skating
Club and Humane Society) built 1855; No. 15, Sedgely Club built
1892, Hazlehurst and Huckel, architects; alterations 1902, Arthur H.
Brockie, architect. Boat clubs organized during 1850's, led to formation of Schuylkill Navy 1858 to supervise regattas. 1 ext. photo
(1973).*

British Building. See International Exhibition of 1876, St. George's House

Callowhill Street Bridge. See Spring Garden Street Bridge

Cedar Grove. Illustrated
(PA-1651), Lansdowne Dr. near Black Dr. (originally at S.W. corner
Old Front St. and Harrowgate Ln., Frankford). Historic house
museum. Coursed Wissahickon schist ashlar front and rubble sides
and rear, approx. 38' (six-bay front) × 50' (originally 19' × 34'), two-
and-a-half stories, gambrel roof (orginally gable), open porch on two
sides (originally pent eave), notable interior woodwork of three
periods of 18th-century architecture.
    Fine example of a Philadelphia Quaker farmhouse with furnish-
ings accumulated by five generations of owners. Built c. 1748; rear
one-story kitchen wing raised to two-and-a-half stories 1752;

enlarged one-room width c. 1799; porch replaced pent eave c. 1831. Vacated 1888; acquired by Fairmount Park Commission, City of Philadelphia 1926; moved to present site 1926–27; opened to public 1928; maintained by Philadelphia Museum of Art. In possession of Morris family c. 1793–1926. Certified, PHC 1956. 3 ext. photos (1937, 1973*), 1 photocopy of ext. photo (c. 1900),* 1 photocopy of int. photo (1928);* 13 data pages (1962, including 1791–1868 insurance surveys).*

Centennial Exhibition of 1876. See International Exhibition of 1876, Memorial Hall, Ohio State Building, St. George's House

Centennial Guard Box

(PA-1652), Traffic Triangle, at W. end of Benjamin Franklin Parkway and E. side of Thomas Eakins Plaza, in front of Phialdelphia Museum of Art. (Formerly at N.W. corner E. River Dr. and Fairmount Ave.) Wood frame, octagonal plan with 10'-diameter, one story, concave octagonal roof, large pointed-arch window with engaged column on each side. Notable example of a romantic park structure in the Gothic Revival style. Built c. 1873; restored 1973. Moved to present site 1973. Certified, PHC 1976. 1 ext. photo (1973).*

Chamounix Mansion

(PA-1653), Chamounix Dr. Stuccoed rubble, approx. 45' (three-bay front) × 47', two-and-a-half stories, hipped roof (originally with balustrade), semicircular bay on south facade, ironwork porches on east and north sides (originally columned porticoes), wood porch on west side, center-hall plan.

Example of modest Federal-style country house with complementary addition. Built c. 1802; enlarged, interior altered, three ironwork porches added, Gothic Revival carriage house built 1856. West porch replaced by wood porch c. 1875; interior altered 1885, 1887. Carriage house altered for youth hostel 1972. Acquired by Fairmount Park Commission, City of Philadelphia 1868. Youth hostel since 1964. Certified, PHC 1956; Pennsylvania Register 1972; NR. 3 ext. photos (1973, including 1 of carriage house);* 5 data pages (including 1835–87 insurance surveys).*

Civil War Memorial. See Smith Memorial Arch

The Cliffs

(PA-185), Columbia Ave. Dr. Rubble with scored stucco front and rear, 36'-5" (three-bay front) × 22'-3", two-and-a-half stories, hipped roof with small cross-gables over center bay, pedimented front and rear frontispieces leading directly into living room.

Fine example of Georgian country house. Built 1753; restored 1929; dormers added later; rehabilitated 1976. Acquired by Fairmount Park Commission, City of Philadelphia 1868; vacated 1968. Built for Joshua Fisher, successful merchant, and stayed in possession of this prominent Quaker family until 1868. Certified, PHC 1956; Pennsylvania Register 1971; NR. 14 sheets (n.d., including plans, elevations, details); 1 ext. photo (1973),* 1 photocopy of ext. photo (1931),* 2 photocopies of int. photos (1932);* 2 data pages (including 1854 insurance survey).*

Clunie. See Mount Pleasant

Covered Bridge

(PA-19), Thomas Mill Rd. across Wissahickon Creek. Wooden Howe trough truss with wrought-iron tie rods and cast-iron shoes covered with board and batten and low-pitch gable roof, rubble abutments, 21'-3" × 96'-0".

Employing the most popular of the 19th-century timber truss systems, this remains the last extant covered bridge in Philadelphia. Built c. 1855; reconstructed 1937. Certified, PHC 1976. 1 sheet (1936, including plans, elevations, sections); 3 ext. photos (1973, including 1 color photo).*

Fairmount Waterworks. Illustrated

(PA-1654), Fairmount, at Schuylkill River near Twenty-fifth St. Group of six wooden and plastered stone buildings on granite ashlar bulkheads: wooden Tuscan peristyle temple structure (approx. 40' × 50') flanked by one-story wood frame buildings (approx. 38' × 15') and one-story stone pavilions (approx. 20' × 40') with Tuscan tetrastyle porticoes on top of former stone mill building (approx. 56' × 238'); at south end three-and-a-half-story, gable-roof stuccoed stone house (approx. 72' × 48') for steam engines (known as Greaves Mansion).

Pioneer pumping system of engineering significance when built that now serves as a foil for the Neoclassical Revival Philadelphia Museum of Art behind and above it on site of former reservoir. Steam-engine house built 1812–15; Frederick Graff, engineer and probable architect; Frederick Erdman, carpenter; John Moore, mason. Mill house with end pavilions built 1819–22; Frederick Graff, engineer and probable architect. Engine house altered to saloon, gazebo built at east end of dam 1834. Standpipe with Italianate facade built behind forebay 1852; Frederic Graff (Jr.), architect and engineer. Mill house built on mound at east end of dam 1860–62; H. P. M. Birkinbine, engineer. Level of terrace raised, central peristyle temple structure built, flanking wooden buildings built over mill houses 1867–72; Frederic Graff, architect and engineer. Abandoned as waterworks 1911; standpipe demolished c. 1920. Site of city aquarium 1911–62. Certified, PHC 1956; Pennsylvania Register 1970; designated National Historic Landmark 1976; designated National Engineering Landmark by American Society of Civil Engineers 1975. 4 ext. photos (1973, including 1 color photo);* HABSI form (1970).*

Falls Bridge
(PA-1655), connects E. River Dr. at Calumet St. with W. River Dr. at Neil Dr. across Schuylkill River. Steel with stonemasonry abutments and two stonemasonry piers, approx. 540' × 40', Pratt-type pin-connected trough truss with secondary system of bracing. Built 1894–95; George S. Webster, chief engineer. Designed as double-deck bridge, upper deck never built. 3 photos (1973, including 1 color photo),* 3 photocopies of old photos (1895, 1896),* 1 photocopy of plan (1894).*

Fountain of the Sea Horses (also known as Italian Fountain)
(PA-1656), Aquarium Ln., approx. 310' N.W. of Philadelphia Museum of Art and approx. 200' W. of E. River Dr. Travertine stone with concrete foundation, four sea horses supporting fountain basin with central fountain of circular splayed palmetto forms approx. 10'-8" high, ciruclar pool approx. 155' circumference with inscribed rounded rim.

A modern reproduction that retains a human scale in relation to the nearby Museum of Art. Made c. 1926 as a reproduction of Fontana Dei Cavalli Marini in Borghese Gardens, Rome, by Christoph Untenberger (c. 1770); installed 1928; basin removed later. Gift of Italy to the United States in commemoration of Sesquicentennial of American Independence 1926. 1 photo (1973),* 1 photocopy of old photo (c. 1928).*

Girard Avenue Bridge
(PA-1657), Girard Ave. across Schuylkill River. Wrought-iron quadrilateral Pratt deck truss with four stone piers, approx. 865' × 100', cast-iron railing, lower level for pedestrians, geometric design of structural members.

Considered the widest bridge in the world when it was built,

1872–74; Henry A. Sims and James P. Sims, architects; Clarke, Reeves and Co., engineers; Phoenixville Iron Co., builders. Demolished 1970–71. Parts of iron railing at Smithsonian Institution. 15 photos (1969, 1971),* 1 photocopy of old photo (1906),* 1 photocopy of old stereopair (c. 1880).*

Hatfield House
(PA-1658), N.E. corner Girard Ave. and Thirty-third St. (originally at N.W. corner Hunting Park and Pulaski aves., formerly known as Nicetown Ln. and Clarissa St.). Wood frame with flush board front and random-width beaded-edge clapboard sides and rear, 30'-2" (four-bay front) × 43'-5", two-and-a-half stories with Tuscan five-column portico, hipped roof in front, gable roof in back (original), transverse central-hall plan with entrance on south side, Greek Revival detailing.

Fine example of a small 18th-century house remodeled into a Greek Revival residence. Built c. 1760; enlarged to the front, extensively remodeled c. 1838; open porches built on north and south sides c. 1850. Moved from original site, porches removed 1930. Maintained by Fairmount Park Commission, City of Philadelphia. Certified, PHC 1957; Pennsylvania Register 1970; NR. 6 ext. photos (1938, 1959*), 3 int. photos (1959),* 7 photocopies of old ext. photos (1901, c. 1910, 1929);* 17 data pages (1929, 1958*).

House on Letitia Street. See Letitia Penn House

International Exhibition of 1876, Memorial Hall (also known as Art Gallery). Illustrated
(PA-1659), North Concourse, 0.2 mi. E. of Belmont Ave. Brick with wrought-iron beams and cast-iron columns faced with coursed marble ashlar, approx. 365' (fifteen-bay front) × 210', one story on raised basement with mezzanine at north (rear) end, flat roofs with balustrades and large central square iron and glass dome, on south facade three large round-arch entrances flanked by arcades (originally open), on front corner of main pavilion zinc statues representing Science and Art by A. J. M. Müller (Germany), on top of dome zinc statue of Columbia and at its corners zinc statues of Industry, Commerce, Agriculture, Mining by Müller, originally speed-eagle sculpture on corners of each corner pavilion, handsome Renaissance-style central rotunda with large entry vestibule.

Notable example of Beaux-Arts architecture in United States that influenced many subsequent American and European museum, exposition, and government buildings. Last remaining important Centennial building. Built 1874–76; Hermann J. Schwarzmann, architect; Richard J. Dobbins, contractor. Arcades enclosed, spread-eagle sculpture removed later. Exterior restored, interior rehabilitated 1958–69; Hatfield, Martin, and White, architects. Served as art gallery during Centennial Exhibition 1876; as Pennsylvania Museum of Art (later Philadelphia Museum of Art) 1876–1928; housed some museum collections 1928–1954; used as Fairmount Park Commission offices, park police headquarters, recreational center since 1969. Certified, PHC 1976.

2 ext. photos (1973),* 2 photocopies of old ext. photos (1875, including 1 of dome under construction),* 1 photocopy of old lithograph (1876),* 1 photocopy of plan (1876);* 8 data pages (including 1885 insurance survey).*

International Exhibition of 1876, Ohio State Building (also known as Ohio House). Illustrated
(PA-1660), N.W. corner States Dr. and Belmont Ave. Rusticated sandstone ground story, Dayton stone ashlar second-story pavilion, frame second story with vertical siding and clapboard gables and rear ground story, approx. 40' (five-bay front) × 42', two stories, twin gable roofs with cross-gable, centered gabled pavilion with crockets and one-story entry porch, open side porches, center-hall plan.

Rare example of architectural advertising with names of various Ohio stone dealers, cutters, and quarries carved in stones. Built 1876; Heard and Sons (Cleveland), architects; Aaron Doan and Co. (Philadelphia), builders. Exhibition hall for State of Ohio during Centennial Exhibition 1876; owned by Fairmount Park Commission, City of Philadelphia, and used as dwelling since 1876. Certified, PHC 1963. Rehabilitated 1976. 4 ext. photos (1973, including 2 in color).*

International Exhibition of 1876, St. George's House (also known as British Building)
(PA-1080), W. side States Dr. approx. 0.2 mi. W. of Belmont Ave. Wood frame with applied half-timbering, approx. 90' (eight-bay front) × 65', two stories, gable roofs, picturesque facade, irregular E-plan. Built 1875; Thomas Harris (London, England), architect; J. H. Cundall, superintendent; John Rice, builder. Demolished 1963. 3 ext. photos (1961), 1 int. photo. (1961), 2 photocopies of architectural

drawings (1875), 1 photocopy of ext. lithograph (1876); 4 data pages (1964).

Italian Fountain. See Fountain of the Sea Horses

Laurel Hill (also known as Randolph House)
(PA-13), Edgeley Dr. Brick with marble trim, 31'-6" (three-bay front) × 22'-2" with two-story lateral octagonal addition (north end) and one-story lateral wing (south end), two stories on raised basement, hipped roof with gable roof on lateral wing, Doric frontispiece with pediment, pediment over central pavilion, center entrance with stairhall and lateral hall to octagonal end (originally central-hall plan).

Built c. 1748; one-story wing built 18th century; octagonal addition built early 19th century. Acquired by Fairmount Park Commission, City of Philadelphia 1869. Certified, PHC 1956; Pennsylvania Register 1970. Rehabilitated 1976. 10 sheets (1934, including plans, elevations, details); 1 ext. photo (1938); 2 data pages (n.d.).

Laurel Hill Cemetery Gatehouse. Illustrated
(PA-1743), 3820 Ridge Ave. Stuccoed stone, approx. 65' (six-bay front) × 30' with stuccoed brick lateral wings, two stories, flat roof, flat Doric octastyle portico with barrel-vault entrance. One of the earliest and most outstanding rural cemeteries in 19th-century America. Built 1836; John Notman, architect. Second story inserted in gatehouse, one-story kitchens built behind wings 1885; kitchens raised to two stories, balustrade removed later. 2 ext. photos (1973),* 1 photocopy of unexecuted design (elevation) by William Strickland (c. 1836).*

Lemon Hill
(PA-1010), Lemon Hill Dr. Historic house museum. Stuccoed rubble with granite trim, approx. 60' (three-bay front) × 40', two stories on raised granite ashlar basement, hipped roof, round-headed entrance with large fanlight and double in-curving steps, large central oval pavilion at rear, side verandas with walkway extending across rear, central-hall plan, oval parlor and bedroom.

Notable example of Federal-style country house. Built c. 1799; restored under direction of Fiske Kimball, Director of Philadelphia Museum of Art 1925–26. Between c. 1799 and 1838 country seat of Henry Pratt, merchant whose early work with lemons gave estate its name. Acquired by City of Philadelphia 1844, set aside as a common

1855, forming nucleus of Fairmount Park; under jurisdiction of Fairmount Park Commission since 1867. Headquarters of Colonial Dames of America, Chapter II, since 1957. Certified, PHC 1956.

1 sheet (1962, including gutter detail);* 3 ext. photos (1938, n.d.), 1 photocopy of int. photo (1958).*

Letitia Penn House (also known as House on Letitia Street)
(PA-184), Lansdowne Dr. near W. Girard Ave. (originally at 8 S. Letitia St). Brick, 20'-3" (three-bay front) × 30'-0", two-and-a-half stories, gable roof, notable hood with carved corbels over door, cove cornice, two rooms on ground story with corner fireplaces.

Notable example of small early-18th-century Philadelphia dwelling originally built in a court later opened as a street. Built for Thomas Chalkley between 1713 and 1715; probably by John Smart, house carpenter. Acquired by Fairmount Park Commission, City of Philadelphia, and removed to present site 1883. Restored 1931–32. Once believed to have been built for William Penn on lot of his daughter Letitia. Certified, PHC 1956. 4 sheets (1931, including plans, elevations); 2 ext. photos (1938), 1 photocopy of int. photo (1932);* 1 data page (including 1769 insurance survey).*

Memorial Hall. See International Exhibition of 1876, Memorial Hall

Monastery
(PA-183), Kitchen's Ln. on E. side Wissahickon Creek. Coursed Wissahickon schist front and rubble sides and rear, 33'-1" (four-bay front) × 32'-1½" with one-and-a-half-story lateral kitchen wing, three-and-a-half stories, gable roof, originally pent eave at each story, front and rear open porches, two large rooms on ground story without entrance hall.

Notable example of vernacular domestic architecture of 18th-century Germantown. Built c. 1747; interior altered c. 1800, c. 1850; porch added 20th century. Altered to Children's Museum of Philadelphia 1973. Acquired by Fairmount Park Commission, City of Philadelphia 1897. Received name from association with Brotherhood of Seventh-Day Adventists at Ephrata, whose members sometimes stayed here. Certified, PHC 1956; Pennsylvania Register 1970; NR. 23 sheets (n.d., including plans, elevations, section, details); 1 ext. photo (1973).*

Mount Pleasant (originally known as Clunie)

(PA-1130), Mt. Pleasant Dr. Historic house museum. Scored stuccoed rubble with brick trim, 53'-7" (five-bay front) × 30'-5", two-and-a-half stories on raised basement, hipped roof with balustrade, projecting center pavilion with pedimented Doric frontispiece and Palladian window above, flat-arch lintels with keystones, brick quions and stringcourse, blank end walls; central-hall plan, notable interior detailing. Flanked by two stuccoed brick and stone outbuildings, each approx. 22' square, two stories, concave hipped roof.

Notable example of a Georgian country house with finely carved interior woodwork. Built 1761; restored 1926; rehabilitated 1976. Stone stable in forecourt demolished after 1940. Owned for a short time but never occupied by Gen. Benedict Arnold 1779–80. Acquired by Fairmount Park Commission, City of Philadelphia 1868; maintained by Philadelphia Museum of Art since 1926; gardens maintained by Planter Garden Club. Certified, PHC 1956; designated National Historic Landmark 1960.

31 sheets (1940, including plot plan, plans, elevations, details, plan and elevations of outbuildings and stable) from Survey, Philadelphia Chapter, A.I.A. (1932); 18 ext. photos (1938, 1939, 1960,* 1970,* 1971,* including 2 of outbuilding), 6 int. photos (1971),* 6 photocopies of measured drawings (1897);* 2 data pages (including 1835 insurance survey).*

Ohio House. See International Exhibition of 1876, Ohio State Building

Ormiston

(PA-187), Reservoir Dr. Stuccoed brick, 34'-3" (three-bay front) × 40'-11", two-and-a-half stories, hipped roof with balustrade, open front and rear porches, center-hall plan. Built 1798; porches added before 1855. Acquired by Fairmount Park Commission, City of Philadelphia 1869. Certified, PHC 1956; Pennsylvania Register 1970. 9 sheets (n.d., including plans, elevations); 1 ext. photo (1973);* 2 data pages (including 1855 insurance survey).*

Pennsylvania Museum and School of Industrial Art. See Philadelphia Museum of Art

Philadelphia Museum of Art (originally known as Pennsylvania Museum and School of Industrial Art)

(PA-1661), Fairmount. The Plaza, at N.W. terminus of Benjamin

Franklin Parkway and merging of E. and W. River drives. Steel frame with Minnesota Mankato and Kasota ashlar curtain walls and terra cotta trim, approx. 545' × 325', three and four stories on raised rusticated basement, blue tiled gable roofs with crest-tiles, Corinthian octastyle portico on east end and Corinthian hexastyle portico on west end of central pavilion, Ionic hexastyle porticoes on inner courtyard pavilions with notable polychromatic terra cotta sculpture in tympanum of north pavilion, Ionic tetrastyle porticoes in antis on four outer pavilions, acroteria on pediments, polychromatic palmetto cornice; E-plan, large entrance stairhall with polychromatic Ionic colonnade at upper level.

An outstanding example of early-20th-century eclectic Neoclassical design. Located on the site of the former Fairmount reservoir, the Museum serves as a focal point for Jacques Gréber's Benjamin Franklin Parkway, one of the most successful attempts at baroque urban planning in the United States. Built 1919–28; Horace Trumbauer, Charles C. Zantzinger, and Charles L. Borie, Jr., collaborating architects; C. Paul Jennewein, sculptor of north portico sculpture; John Gregory, sculptor of south portico sculpture (never installed); George A. Fuller Co., contractor. One of America's great art museums. Name changed to Philadelphia Museum of Art 1938. Certified, PHC 1971.

1 ext. photo (1973),* 3 photocopies of ext. photos (1924, 1926, n.d.),* 1 photocopy of int. photo (1927, during construction),* 1 photocopy of aerial photo (1966),* 2 photocopies of architectural drawings (1912, 1915).*

Philadelphia Zoological Gardens, Bear Pits (also known as Bear Dens).
Illustrated
(PA-1662), approx. 200' W. of Thirty-fourth St., approx. 925' S. of entrance pavilions at S.W. corner W. Girard Ave. and Thirty-fourth St. Trenton brownstone rubble, three adjoining semicircular enclosures built into knoll with two terraced semicircular protrusions filling areas between curvatures of enclosures, approx. 86' × 30', front at ground level, rear depressed forming pits; front of each pit has stepped gable with pointed-arch doorway, date stone in center gable. Fine example of 19th-century resolution of romanticism and utility in a new building form. Built 1874; Theophilus Parsons Chandler, Jr., architect. 10 photos (1975);* 1 photocopy of old print (c. 1875).*

Philadelphia Zoological Gardens, Entrance Pavilions
(PA-1663), S.W. corner W. Girard Ave. and Thirty-fourth St. Granite rubble and brick with brick trim, approx. 20' (two-bay front) × 28', two stories, steep hipped roofs with cross-gables and cross-jerkinhead roofs, projecting cornices with bold brackets. Notable example of picturesque design. Built 1875–76; Furness and Hewitt, architects. The Zoological Society of Philadelphia opened this zoo, the nation's oldest, in 1874. 1 ext. photo (1973).*

Randolph House. See Laurel Hill

Ridgeland
(PA-1664), Chamounix Dr. Wissahickon schist rubble, approx. 34' (three-bay front) × 34' with two-story rear frame addition, two-and-a-half stories, gable roof, Tuscan entrance porch, open rear porch, center-hall plan. Built c. 1800; rear addition built late 19th century. Certified, PHC 1956. 3 ext. photos (1959),* 5 int. photos (1959).*

Sedgeley Porter's House (later known as Sedgeley Guard House)
(PA-1665), S.E. corner Sedgeley Dr. and W. Girard Ave. Sandstone rubble with brick trim and local stone rubble additions, approx. 36' (three-bay front) × 18' with flanking lateral extensions, one story, hipped roof, pointed-arch windows and doorway, entrance porch, originally center-hall plan with cellar kitchen.
    Earliest surviving example of the Gothic Revival style in the United States, exhibiting the classical proportions with the pointed-arch details that characterize the style's origins. Built c. 1799 as porter's house in rear of William Cramond's country house Sedgeley; probably by Benjamin Henry Latrobe, architect of Sedgeley. Sedgeley demolished 1857. Porch built mid-19th century; lateral wings, cornice trim added 1897; interior altered later. Acquired by City of Philadelphia 1857; maintained by Fairmount Park Commission since 1867; guard house 1857–1973, maintenance storage since 1973. Certified, PHC 1960. 2 ext. photos (1973);* 1 data page (including 1811 insurance survey);* HABSI form (1971).*

Smith Memorial Arch (also known as Civil War Memorial)
(PA-1666), N.W. and S.W. corners North Concourse and Lansdowne Dr. Two curved forms of smooth rusticated granite with dressed granite trim and bronze statuary, each form approx. 96' (along segment) × 16', each with Palladian arch flanked by rusticated pedestals;

each inner pedestal surmounted by giant Doric columns (7'-6" diameter) with colossal standing statues of Maj. Gen. George Meade by Daniel Chester French, sculptor (south side), and of Maj. Gen. John Reynolds by Charles Grafly, sculptor (north side); outer pedestals surmounted by colossal equestrian statues of Maj. Gen. George B. McClellan by Edward C. Potter, sculptor (south side), and of Maj. Gen. Winfield S. Hancock by J. Q. A. Ward, sculptor (north side); standing statue of Richard Smith, type founder and donor of Memorial, at base of northern inner pedestal, niches filled with eight colossal busts including Union generals, admirals, Pennsylvania's Civil War governor, Memorial's architect, and executor of Smith's will; frieze carved with names of 84 prominent Pennsylvania participants in Civil War. Monumental expression of late-19th-century individualism and patriotism. Built 1897–1912; James H. Windrim and John T. Windrim, architects. Sculpture under auspices of Fairmount Park Art Association. 1 ext. photo (1973).*

Solitude

(PA-1127), S.W. corner W. Girard Ave. and Thirty-fourth St., on Philadelphia Zoological Gardens grounds. Coursed stuccoed rubble, approx. 30' (three-bay front) × 32', two-and-a-half stories, hipped roof, open Ionic porch on east side, Ionic frontispiece with pediment on west side, center-hall plan, notable Adamesque interior plasterwork.

Built 1784–85 for John Penn, grandson of William Penn; rehabilitated 1976. Offices for Philadelphia Zoological Gardens since 1874. Certified, PHC 1956. 9 sheets (1940, including plans, elevations, details) from Survey, Philadelphia Chapter, A.I.A. (1932); 8 ext. photos (1939, 1961,* 1962*), 12 int. photos (1939, 1961*), 1 photocopy of old plan print (n.d.),* 3 photocopies of measured drawings (1897).*

Somerton. See Strawberry Mansion

Spring Garden Street Bridge (also known as Callowhill Street Bridge)

(PA-1667), Spring Garden St. and Callowhill St. across Schuylkill River. Iron with stone abutments, approx. 350' × 48', double-decked roadways with cast-iron arches and stone piers on bridge approach, cast-iron arches and cast- and wrought-iron Whipple truss over river. Built 1874–75; Jacob H. Linville, engineer; Keystone Bridge Co., builders. Arches removed c. 1900; balustrades removed between 1904 and 1910; bridge demolished 1964. 1 photo (1963),* 1 photocopy of old print (1876),* 2 photocopies of old stereopairs (c. 1880),* 6 photocopies of old photos (1874, c. 1890, 1904, 1910).*

Strawberry Mansion (originally known as Summerville, later as Somerton)
(PA-1668), near N.W. corner Dauphin St. Dr. and Edgeley Dr. Historic house museum. Stuccoed rubble and brick, approx. 125' (eleven-bay front) × 30', two-and-a-half and three stories, gable roofs, Tuscan entrance porch, center-hall plan.

Fine example of the evolution of a modest country house into a Classical Revival mansion. Stone center section built 1797–98; three-story brick south wing built 1825; three-story north brick wing built c. 1828; open front and rear porches built 1870, removed c. 1926. Restored 1926–31. Acquired by Fairmount Park Commission, City of Philadelphia 1867; used as restaurant until 1926; since then maintained by Committee of 1926 under direction of Philadelphia Museum of Art. Fine collection of Tucker porcelain, America's first commercially successful high-quality porcelain, in Empire Parlor. Between 1819 and 1842 the country home of Judge Joseph Hemphill, politician and businessman who served six terms in Congress between 1801 and 1831 and was an owner of Tucker porcelain firm c. 1829–35. Supposedly the first strawberries in the United States were successfully grown here after 1820, giving mansion its present name.

1 ext. photo (1938), 2 photocopies of ext. photos (1932),* 3 photocopies of int. photos (1962, 1963);* 7 data pages (including 1873, 1880 insurance surveys).*

Strawberry Mansion Bridge. Illustrated
(PA-1669), connects Ford Dr. in W. Fairmount Park with Strawberry Mansion Dr. in E. Fairmount Park across Schuylkill River. Four wrought-iron arches on three stonemasonry piers with three Warren trusses supporting viaducts on east and west approaches, 1240'-6" × 52'. Built 1894–96; Russell Fair, Jr., engineer; Theodore Cooper, consulting engineer; Phoenix Bridge Co., builders. Built as trolley bridge for Fairmount Park Transportation Co. with vehicular lanes and sidewalk; trolleys stopped running, tracks removed 1946. 2 photos (1973, including 1 color photo),* 1 photocopy of old photo (1908).*

Summerville. See Strawberry Mansion

Sweetbrier

(PA-1670), Lansdowne Dr. Historic house museum. Stuccoed rubble, approx. 53' (five-bay front) × 38', two-and-a-half stories on raised basement, hipped roof, round-arch entrance with fanlight and Tuscan frontispiece, quoins, center-hall plan.

Fine example of a country house in the Federal style. Built 1797; restored by Junior League of Philadelphia 1926. Acquired by Fairmount Park Commission, City of Philadelphia 1868; since 1939 headquarters of The Modern Club, which furnished and maintains the house. Samuel Breck, prominent merchant, public servant, politician, member of Congress 1823–25, lived here 1797–1838. Certified, PHC 1956. Renovated 1976. 2 ext. photos (1973),* 3 photocopies of int. photos (1928).*

Woodford

(PA-1307), S.W. corner Ford Rd. and Greenland Dr. Historic house museum. Brick, approx. 45' (three-bay front) × 25' with two-story rear ell and additions, two stories (originally one) on raised basement, hipped roof with balustrade, projecting central pavilion with pediment, center Palladian window, pent eave on four sides, central-hall plan.

Notable example of a Georgian country house. Built c. 1756; second story added c. 1772; restored 1927–30. Acquired by Fairmount Park Commission, City of Philadelphia 1868; served as Park guard house 1887–1927. Has housed Naomi Wood Collection of household articles under direction of Philadelphia Museum of Art since 1930. Certified, PHC 1956; designated National Historic Landmark 1967. 6 ext. photos (1938, 1960*), 4 int. photos (1960),* 20 photocopies of 1897, 1932 measured drawings;* 2 data pages (1954, including 1769, 1772 insurance surveys).*

Belmont Mansion. *Photo 1926, courtesy Philadelphia Museum of Art.*

Cedar Grove. *Jack E. Boucher photo, 1973.*

Fairmount Waterworks. *Jack E. Boucher photo, 1973.*

International Exhibition of 1876, Memorial Hall. *Photo 1875, courtesy Philadelphia City Archives.*

International Exhibition of 1876, Ohio State Building. *Jack E. Boucher photo, 1973.*

Philadelphia Zoological Gardens, Bear Pits. *Cortlandt V. D. Hubbard photo, 1975.*

Laurel Hill Cemetery Gatehouse. *Jack E. Boucher photo, 1972.*

Strawberry Mansion Bridge. *Jack E. Boucher photo, 1973.*

Northwest Philadelphia

Germantown and Roxborough Townships

7

Introduction

*I*N SEPTEMBER, 1748, the Swedish botanist Peter Kalm rode out of Philadelphia toward the northwest. He passed through a forest, broken by well-tended fields and orchards, until after about a six-mile ride he came upon Germantown. "This settlement," he noted, "has only one street but is nearly two English miles long. It is for the greatest part inhabited by Germans, who come from their country to North America and settle here because they enjoy such privileges as they are not possessed of anywhere else. . . . Most of the houses are built of stone, which is mixed with glimmer and found everywhere around Philadelphia. . . ."[1] Kalm's description has stood the test of time. The oldest, and for many the most fascinating, part of the ancient town remains the double row of stone houses ranging along Germantown Avenue, the former Great Road, residual evidence of Pennsylvania's first linear community.

Germantown has justifiably captured the imaginations of local historians more than any other section of Philadelphia County. It was founded in 1683 by German-speaking settlers as an essentially independent province and during the colonial period developed into a socially separate town, whose material culture has survived to a degree great enough to justify designating its main thoroughfare, Germantown Avenue, a National Historic Landmark.

What people have meant by the term "Germantown" has differed over the years. Initially, it seems to have extended along Germantown Avenue from Roberts Avenue northward to Northwestern Avenue, and stretched from the present Stenton Avenue westward to Wissahickon Avenue. After 1707, when the area became German Township in Philadelphia County, Germantown became known as

the more built-up portion between Wayne Junction and Carpenter
Lane, and the present neighborhood designations gradually came
into use.[2] Mount Airy apparently got its name from Chief Justice
William Allen's country estate, Mount Airy, which was built about
1750 on Germantown Avenue near the present Allen's Lane.[3]
References to Chestnut Hill, the area north of Mount Airy, appear as
early as 1711. The name was self-explanatory, although there is no
evidence that there were any more chestnut trees here than
elsewhere in Philadelphia County. It is a hill, however; in fact, it is
the highest point between Trenton and Bryn Mawr.[4] The more popu-
lous part of the township was organized as Germantown Borough in
1844, and in 1854, when the city and county governments were consolidated, the whole area became the Twenty-second Ward of the
City of Philadelphia, and its boundaries became roughly those of the
original Germantown.[5] Roxborough Township, presently the
Twenty-first Ward, is a natural geographical extension of the former
German Township, and in the early years no distinction was made
between the two places. The two former townships enjoyed a parallel
economic development, and the daily routines of their early inhabitants were often intertwined, particularly in the mills along
Wissahickon Creek. Consequently, Northwest Philadelphia includes
all of the city north and west of Broad Street and Roosevelt Boulevard, with Broad Street its eastern boundary and the Schuylkill River
its southern boundary.

Historians have quibbled about whether the first Germantown
settlers were Dutch, German, or Swiss, but the important point is
that whatever their ethnic origins (most were apparently Dutch), they
were German-speaking and self-consciously called their settlement
Germantown. Yet the population was pluralistic in religion almost
from the beginning, as newcomers arrived and built primarily along
the great road leading to Philadelphia.[6] The earliest settlers were predominantly Quakers, but by 1690 Mennonites were numerous
enough to have their own minister. This group, the first Mennonite
congregation in America, still worships at 6119 Germantown Avenue
in its plain rubble meeting house (illus.) that was built in 1770 on the
site of its 1708 log meeting house. Before 1730 congregations of the
Church of the Brethren (Dunkers) and of the Lutheran and Reformed
churches were established, and some of their early houses of worship
still stand, such as the Church of the Brethren at 6613 Germantown
Avenue, built about 1770 but enlarged and altered in 1897. As
German immigration grew during the eighteenth century, St.

Michael's Lutheran Church became the dominant religious influence in Germantown. Disputes within the congregation led to a host of new churches, such as the more pietistic St. Peter's Lutheran at Barren Hill (1752), the English-speaking Trinity Lutheran (illus.) at Germantown Avenue and Queen Lane (1836), the exclusively German-speaking St. Thomas Lutheran at Herman and Morton streets (1860), and three others in Chestnut Hill and Germantown.[7]

The language disputes at St. Michael's point up a continual tension within the greater Germantown community. Germantown began with a charter assuring political autonomy, but by 1707 an apparently insufficient tax base forced the town into accepting township status and losing political independence.[8] This caused no immediate disruption, since the ethnic distinctiveness of the inhabitants kept the town socially separate from Philadelphia city. By about the middle of the eighteenth century, however, the six-mile trip to Philadelphia no longer assured isolation but only made Germantown far enough removed from the city to be attractive to Philadelphians for summer residences.

Exemplary of these early summer residences is Grumblethorpe (illus.) at 5267 Germantown Avenue, opposite the eastern end of Queen Lane. The German-born Philadelphia merchant John Wister purchased the lot in 1741 and three years later he had his house built,[9] evidently by a local builder, because the structure is a glossary of rural Delaware Valley architectural elements. Like all early Germantown houses, it was built of local Wissahickon schist, a stratified gneiss speckled with mica. The rubble sides and rear and a coursed ashlar front with benches on the stoop were typical of Germantown buildings. Its handling of the pent eave and heavy cornice across the front and one end was repeated in one form or another in a host of Germantown houses, but most explicitly in the Green Tree Tavern, which was built four years later at 6021 Germantown Avenue for Daniel Pastorius, the grandson of the town's founder.[10] Grumblethorpe's balcony, however, has no exact duplicate. The Livesey House (c. 1747) on Wissahickon Creek comes close; it has a balcony but it extends across the entire front, taking the place of a pent eave. Oculi placed in Grumblethorpe's gable near the roof's edge were evidently to provide light and ventilation for the attic. These oculi are probably more evident today in old Pennsylvania barns, but they can be seen in other houses of the period, such as the house at 5434 Germantown Avenue, built c. 1740 and later owned by the noted carriage-maker William Ashmead.[11] The use of two entrances—in the

case of Grumblethorpe one leads to a parlor and the other to a hall—
was apparently a rural German practice and is also found on the
house that was built about 1730 at 6505 Germantown Avenue and
later inhabited by Michael Billmeyer, German printer for the Pennsylvania Assembly.[12] Because of local building traditions, Grumblethorpe stands as an archetype of mid-eighteenth-century Germantown architecture, yet the fact that it was built for a Philadelphian
seeking refuge from the city's summer heat suggests that by the
1740's Germantown was well along the way to becoming an English
town. This point was architecturally reinforced by the town's new
market house, which was erected by 1745 in Market Square and was
nearly identical to its contemporary in Philadelphia, New Market at
Second and Pine streets.[13]

Cultural tensions in Germantown were institutionalized in 1759,
when Germantown Academy was founded. There had been earlier
schools in the township, such as the Beggarstown School at 6669 Germantown Avenue, which provided a rudimentary education for
German-speaking children, but attempts to establish English-speaking public schools were stubbornly resisted by Christopher Sower I,
the renowned German printer and cultural purist. A year after
Sower's death in 1758 a group of men met in Green Tree Tavern to organize what they self-consciously called the Union School of Germantown, and in 1760 ground was purchased on Bensell's Lane,
which has been known as School House Lane ever since the school
building was finished in 1761.[14] It is a large two-story stone building
designed to accommodate both English- and German-speaking
classes, with a small house on each side for the German and English
schoolmasters. It was not long before the school was known as Germantown Academy, a name it retained until 1965. From the beginning, English pupils outnumbered their German counterparts, so it is
not surprising that by 1775 English-speaking residents in the
northern end of Germantown found their numbers increasing fast
enough and the distance to School House Lane great enough to justify building an all-English school on the property of the Upper Burial
Ground at 6309 Germantown Avenue.[15] Concord School (illus.), as it
was euphemistically called, is a rectangular, stuccoed stone structure
with only a central belfry on the roof betraying its original function.
Yet this simple one-room schoolhouse signaled English cultural
dominance in Germantown on the eve of the American Revolution.

About 1750 the newly appointed Chief Justice of the Province of
Pennsylvania, William Allen, built his country house, Mount Airy, on

Germantown Avenue, approximately two miles north of Grumblethorpe. William Allen was not a German; he was an Englishman
and one of the wealthiest and most influential political and business
figures in the colony, in fine, a taste-maker.[16] The fact that such a
prominent man built a summer home in Germantown started a trend.
Spending the summer of 1763 in Allen's Mount Airy, Benjamin
Chew, one of the province's leading attorneys and public servants, so
enjoyed the cool, healthful air that he decided to emulate his friend.
The result was Cliveden (illus.), one of the finest and best documented Georgian country seats in the American colonies.[17] It is a
happy compromise between the English Palladian country house and
the Germantown building tradition. Evidently inspired by country-
house plans to be found in such books as James Gibbs's *A Book of Architecture* (1728), Cliveden consists of a central dwelling flanked to the
rear by two small service dependencies that form a rear court for
formal gardens. The house incorporates some of the finest elements
of Georgian architecture: a dressed sandstone stringcourse, lintels
with keys, a pedimented central pavilion with a Doric frontispiece
and pediment, and dormers with scroll brackets and five ornamental
stone urns distributed across the front of the roof. Yet since Cliveden
was built of local stone by a local master carpenter and mason, it
continues the Germantown tradition of facade architecture (coursed
ashlar front and rubble sides and rear) with gables on the short ends.
Also, since the designer was a provincial gentleman, Benjamin Chew,
the proportions of the central pavilion are rather pinched, and the
handling of the urns and lintels a bit pretentious. The interior is as
grand as the exterior. The entrance hall is the widest room in the
house and nearly half its depth. A screen of four fluted Doric columns
with a finely carved Doric frieze separates the entrance hall from the
rear stairhall, which is flanked by a finely paneled dining room and
parlor. Benjamin Chew succeeded his friend William Allen as the
provincial Chief Justice in 1774, but when the Revolution broke out he
was deported to New Jersey for a year because of his pre-Revolutionary ties to the Tory faction. With his political and economic future
in doubt he sold Cliveden in 1779. Eighteen years later, however, he
regained the estate, and the Chew family resided at Cliveden
continuously until 1972, when it was acquired by the National Trust
for Historic Preservation and opened to the public.[18]

By the time of the Revolution the German era of Germantown
appears to have come to a close. The last significant detached house
to be built in Germantown before the Revolution was the Deshler-

Morris House at 5442 Germantown Avenue, opposite Market Square.
It was built in 1772 for David Deshler, a successful Philadelphia im-
porter of English and West Indian goods and the nephew of John
Wister, the man who helped to begin the summer trek to Germantown more than a generation earlier. Although Deshler was
German-born, his business and charitable activities and architectural
taste suggest that he had been anglicized by the time he approached
retirement in the early 1770's.[19] His summer dwelling is a fine
example of Georgian domestic architecture built on the double-house
plan. Although its ample size was more or less standard for its time,
the plan, in which the center hall extends through to a back door with
stairs off to the side, usually at the rear, was most often reserved for
impressive city and country houses, such as the Reynolds-Morris
House on South Eighth Street and Mount Pleasant in Fairmount
Park. The house is probably best known as Germantown's "White
House." In 1793, when Philadelphia was the nation's capital and in
the throes of the yellow fever epidemic, President George Washington fled the pestilential city to reside here for about a month. He
evidently enjoyed his visit—or perhaps feared a recurrence of the
fever—because he returned the next summer for a six-week stay.
Washington must have found his days here particularly ironic since
the house had served as the headquarters of his victorious opponent,
General Sir William Howe, after the Battle of Germantown
sixteen years earlier.[20] Samuel Morris acquired the house from his
father-in-law, Elliston Perot, in 1836 and for the next 112 years the
Morris family owned the property until it was turned over to the National Park Service in December, 1948, and restored as a historic
house museum.[21] Today it stands as not only America's earliest
"summer White House" and presidential retreat but also as a fine
example of a pre-Revolutionary house and garden.

   The exodus of government officials and private families from the
City of Philadelphia to Germantown during the yellow fever epidemic in 1793 firmly established the town's reputation as a place with
healthful air free from the deadly miasmas of the city. Thereafter the
number of both town and country houses increased, and since they
were executed in the more delicate Federal style, they did much to
change the appearance of Germantown. Across Germantown
Avenue from Benjamin Chew's six-acre estate, John Johnson in 1798
began building a large two-and-a-half-story house onto a rubble
structure that became its rear ell.[22] Later called Upsala (illus.), it is a
fine Federal-style country house with a light, porticoed entrance

porch and finely modeled dentil-and-modillion cornice, but it still retains a Germantown character in its plan and treatment of materials.

To the south, in Vernon Park at 5860 Germantown Avenue, stands a more formal Federal-style country house known as Vernon. As in the case of Upsala and many other Germantown houses, the main part of Vernon was built onto an earlier rubble farmhouse that then became a rear service wing. The front building was built in 1805 for James Matthews, a cane and whip manufacturer, but the first occupants seem to have been John Wister and his family. This John Wister was the grandson of the John Wister who sixty years earlier had had Grumblethorpe built about half a mile down the road.[23] The contrast between the two houses indicates the degree to which the local builders had acculturated. There are no traces of the vernacular rural house in the new section of Vernon; it is a country house of the eighteenth century. The Germantown builders were now incorporated into the Anglo-American world, yet they remained in the backwash of changing taste. The builder of Vernon exhibited his Adamesque-Federal vocabulary by using the proper details, such as fanlights and dentil cornices, in the proper places, but he did not appreciate the subtleties of the language. The house's form, scale, and proportions are those of the Palladian era. It does not possess the interplay of forms and surfaces that can be seen in such paragons of Adamesque-Federal architecture as Woodlands in West Philadelphia. In fact, conceptually and stylistically, Vernon is a continuation of a house type that was first built in the area about 1728 when Stenton was begun at Eighteenth and Courtland streets for James Logan. Since Logan was the leading political figure in early-eighteenth-century Philadelphia, his fine mansion influenced Germantown's architecture even though it stood on the town's geographical fringe. Of course, Vernon is three-quarters of a century later than Stenton, hence stylistically more sophisticated. Its use of a pedimented central pavilion, continuous stringcourse, and projecting frontispiece give it a classical horizontality and three-dimensionality that is lacking in Stenton. Vernon improves in like manner upon the qualities of Cliveden, the other famous local eighteenth-century country house, yet it retains the latter's basic proportions and many of its details, including the shingling across the top of the pediment's cornice.

Improved transportation facilities in the early nineteenth century tied Germantown even more closely to the City of Philadelphia. Notoriously bad roads had greatly impeded communication between the

two places during most of the colonial period, and it was not until
1761 that stage coaches began running between Germantown and
Philadelphia and shortly thereafter to Reading and Bethlehem as
well.[24] The roads remained wretched, however, and travel was at
best uncomfortable until in 1801 the Germantown and Perkiomen
Turnpike Company was chartered and given control of Germantown
Road from Third and Vine streets on the northern edge of
Philadelphia to Perkiomen Bridge in Collegeville, Montgomery
County. This step, however, did not always assure a quick, smooth
trip between Germantown and Philadelphia, and over the decades
conditions failed to improve significantly; in 1874 the town's citizens
went to court and forced the City of Philadelphia to acquire and pave
the road. Meanwhile, leading citizens of Germantown took action to
build a rail connection with both the city and lower Montgomery
County. The Philadelphia, Germantown and Norristown Railroad,
one of the country's earliest, was incorporated in February, 1831, and
construction was begun that summer. By the following fall the road
extended to the Germantown Depot, a tavern at Germantown
Avenue and Price Street.[25] This tavern was replaced by a two-story
station (now the city's oldest railroad station) in 1855, about the same
time that the road was extended into Chestnut Hill by leasing the
newly completed Chestnut Hill Railroad.[26] These rail lines were supplemented in 1859 by the Germantown Passenger Railway Company,
whose horse-drawn streetcars carried an average of 2,500 passengers
daily during its first month of operation along Germantown Avenue
between Phil-Ellena and Eighth streets.[27]

These developments in transportation heralded another era in
Germantown's history, that of garden suburb. The writings of
Andrew Jackson Downing helped to popularize the environment
and architecture of the kinds of residential neighborhoods that Germantown knew during the last half of the nineteenth century. Ivy
Lodge, at 29 East Penn Street, not far from the Penn Street station on
the Philadelphia, Germantown and Norristown Railroad, was built
about 1850, making it one of the town's first suburban villas. Because
of its proximity to Germantown Avenue and the built-up part of the
town, its version of the Italianate is more self-contained than many
Italian villas with their towers, belvederes, wings, and porches. Yet
the static nature of Ivy Lodge's nearly cubic form is relieved by the
use of material (rock-faced local stone ashlar), details (octagonal side
bay, porches, and elongated cornice brackets), and silhouette (arched
dormer set into the cornice). Moving beyond the classical restraint of

Ivy Lodge into the early phases of Picturesque Eclecticism is the
Ebenezer Maxwell House (illus.), built in 1859.[28] Its site at Greene
and Tulpehocken streets, then a sparsely populated neighborhood
with natural plantings, gave the architect the opportunity to develop
a design with a rambling plan and irregular silhouette that was more
suitable for the setting than a classically balanced dwelling. Its prominently shaped and angulated cross-gables indicate that it was
probably meant to be an Elizabethan villa, yet a host of other elements obscure this point. The archivolt treatment of the recessed entrance and the boldly carved porch arcades are reminiscent of the
Romanesque; the porch battlements and truncated lancet window of
the spindly tower are taken from the Gothic; and the mansard roof
suggests French Renaissance. These elements were combined with
random local rubble relieved by dressed ashlar quoins and multiple
diamond-shaped red sandstone and eight different shapes of
windows to create an effect so picturesque that recently
neighbors affectionately called the mansion the "Addams House" in
reference to Charles Addams, the macabre cartoonist whose witches,
ghouls, and monsters often peer from mansarded towers like that of
the Maxwell House. This splendid example of High Victorian
Eclecticism was nearly lost, however, when in the early 1960's efforts
were made to demolish the house to make room first for a service
station and then for a rest home. The Philadelphia Historical Commission interceded to prevent this, allowing enough time for a
dedicated local group to organize and start the present process of
restoring the house into a center for Victorian arts.[29] Next to its preservation, the most pleasant aspect of the Maxwell House is its environment. It is but one star in a galaxy of nineteenth-century suburban
villas. Its immediate neighborhood of Tulpehocken Street, Walnut
Lane, and Wayne Avenue makes up but one constellation; there are
many other picturesque patches to be found bordering such streets as
School House Lane and Upsal Street. The houses in a wide range of
styles remain remarkably little altered, planted firmly on their large
lots along tree-lined byways that are tailor-made for walking tours of
Victorian domestic architecture.

Germantown's second railroad, the Chestnut Hill Line of the
Pennsylvania Railroad (now Penn Central), began operations west of
Germantown Avenue in June, 1884, and almost immediately set off
another rash of residential construction along its path, particularly in
Mount Airy and Chestnut Hill.[30] A fine example of such an affluent
commuter neighborhood is Pelham. Most of its sumptuous Colonial

Revival houses extend along Pelham Street and its tributaries
between Upsal Station, built about 1884 probably by Wilson Brothers
and Company, and the former Pelham Trust Company (now Mount
Airy Branch of the First Pennsylvania Bank), built on Germantown
Avenue in 1907 from the designs of Churchman and Thomas.[31]

Chestnut Hill reflects even more directly the interactions of railroad and suburban development. Henry Howard Houston was the
person most responsible for the orderly development of Chestnut Hill
into what was then called "Philadelphia's prettiest suburb" and is still
acknowledged as "one of the most beautiful and distinctive sections
of the city."[32] Houston was an entrepreneur, one of those
nineteenth-century men who successfully turned their energy and
imagination to business pursuits. His enterprises included wise investments in oil and gold that led to directorships of financial institutions and transportation companies, most notably the Pennsylvania
Railroad, which obligingly built the Chestnut Hill Line through
Houston's extensive land holdings in what was then known as
Wissahickon Heights.[33] In order to give direction to development of
the largest piece of personally owned real estate in the city,[34]
Houston commissioned the noted architects G. W. and W. D. Hewitt
to build two key social centers on Willow Grove Avenue, a short
distance from the former Wissahickon Heights station. One was
Wissahickon Inn, a summer resort hotel that is now Chestnut Hill
Academy, and the other was the Church of St. Martin-in-the-Fields,
for an Episcopal congregation, of course.[35] The neighborhood's environment was at least equally determined by its houses. It was estimated in 1889 that Houston had already erected between eighty and
one hundred dwellings in Wissahickon Heights.[36] All were apparently built for sale, and some, such as the Houston-Sauveur
House (illus.) (1885) at Seminole Avenue and Hartwell Lane, were
initially rented for a few years. The great size and distinctive design of
this robust rendition of the Queen Anne style indicate the extent of
Houston's investment in this real estate venture and help to account
for its stable development. There was no ticky-tacky in Wissahickon
Heights.

Houston's son-in-law, Dr. George Woodward, continued to
contribute to the orderly growth of Chestnut Hill in the early
twentieth century. In addition to freestanding suburban houses,
designed by such practitioners of the Philadelphia suburban style as
Wilson Eyre, Jr., Mantle Fielding, Robert McGoodwin, and Duhring,
Okie and Ziegler, Woodward also built rows, such as the first block of

Benezet Street west of Germantown Avenue, and semidetached dwellings, of which the best examples were built between 1906 and 1928 around Pastorious Park at the end of Lincoln Drive.[37] French Village is a fine example of this conscientious development continuing into the automobile age of the 1920's. This picturesque ensemble of stone dwellings designed by Robert McGoodwin stretches along two twisting lanes on the edge of the Cresheim Creek extension of Fairmount Park to make the most of a triangular piece of land on the corner of McCallum Street and Allen's Lane.[38] Some fine apartment buildings were also erected in Germantown during the 1920's. Alden Park Manor on the former Justus C. Strawbridge estate at Wissahickon Avenue and School House Lane is an outstanding complex that is thought to be the country's first high-rise apartment building planned around open park land in the manner advocated by LeCorbusier only a few years earlier.[39] Notable dwellings are still being built in the area. Three from the early 1960's are Robert Venturi's house for his mother and Louis Kahn's design for Margaret Esherick north of Pastorious Park, and Mrs. Thomas Raeburn White's house by Romaldo Giurgola on the western fringe of Chestnut Hill at 717 West Glengarry Road.[40]

Since the heights of Chestnut Hill, Germantown, and Roxborough offered rapidly moving streams and waterfalls as well as salubrious air, they also possessed attractive sites for mills. Until the advent of steam power in the last half of the nineteenth century, the banks of Wissahickon Creek were dotted with all kinds of mills, including the earliest paper mill in the English colonies. William Rittenhouse, the first minister of Germantown's Mennonites, built his paper mill about 1690 near the present intersection of Lincoln Drive and Rittenhouse Street. The mill and its replacement (the original mill was washed away about a decade after its erection) are now gone, but the millrace and neighboring buildings remain. Among them is a two-story rubble dwelling built in 1707 that is probably best known as the birthplace of William Rittenhouse's great-grandson David Rittenhouse, the noted astronomer, mathematician, and Revolutionary political figure.[41] Gristmills were even more abundant, but like paper mills, they could not compete with the larger steam-powered mills and were abandoned in the mid-nineteenth century, allowed to fall into ruins, and most have since disappeared. Remnants of one of the last of these sites, at the mouth of the Wissahickon, was demolished as recently as 1961. At that time only the house, Shoomac Park, was left, but a gristmill had been erected on the property as early as 1690

by Andrew Robeson, a former Proprietor of the Province of West
Jersey and a member of the Pennsylvania governor's council. The
mill's site along Ridge Road at the point of a ferry across the
Schuylkill could not have been better. By the 1790's, when Robeson's
great-grandson owned the facilities, the mill was grinding nearly
50,000 bushels of wheat a year.[42] The Livezey farm, dating from the
1740's, near the other end of the Wissahickon, has fared better. The
barn is gone now and little more than the foundations of the gristmills remain, but the house, Glen Fern, still stands as the Valley
Green Canoe Club.

Because of its size and flow, the Schuylkill River could not be
turned to manufacturing purposes until a canal was dug. Such a
project was not launched until 1815 when the Schuylkill Navigation
Company was incorporated to develop a waterway along the river's
108 miles between Philadelphia and the anthracite coal mines around
Pottsville. When the project was completed in 1825, it included 51
miles of canals with 114 locks to overcome a total drop of 610 feet.
Forming part of this system was the Flat Rock Canal, which was
opened in early 1819 along the Schuylkill's east bank in the
northwestern end of Philadelphia County.[43] Finding the area's name,
Flat Rock, too prosaic, residents proposed others, and in the mid-
1820's adopted Manayunk, an apparent derivation from an Indian
word.[44] The canal was to make Manayunk what it has been ever
since—a nineteenth-century mill town with a classic ribbon development.[45] The canal carved out an island that provided two additional
banks for water power. By the end of 1822 nine mills stood along the
canal, and by 1830 the number had nearly tripled.[46] By the mid-
nineteenth century the island and the canal's banks were choked with
mills. Among the most extensive operations were the Economy Mills
of Sevil Schofield, remnants of which still stand on the island and
east bank of the canal above Lock Street.[47] The first street back from
the canal, Main Street, developed into the commercial district and the
primary overland route to the Wissahickon and thence into
Philadelphia. Beyond Main Street stand the houses, which have been
forced by Manayunk's topography to cling to the rocky hillside for
which Manayunk is famous. Nearest to the mills is the modest row
housing for workers, and beyond that the more comfortable single
dwellings of the former managerial class, which are usually far
enough removed from the mills to be in Roxborough. Rising above
the rooftops are the spires and towers of the community's churches.
The most imposing of the lot is the brownstone pile of the Roman

Catholic Church of St. John the Baptist at Cresson and Robinson streets.[48]

Industry became so firmly planted in this neighborhood that it easily survived the conversion to steam, which allowed companies to move their mills away from the crowded riverside and up the hill. Few went farther up the hill than Shur's Lane Mills of T. Kenworthy & Bro. at Shur's Lane and Freeland Avenue. Built in two stages, the Worsted Mill in 1876 and the Shoddy Mill in 1882, the complex has had small extensions and offices added over the years,[49] but the two main buildings have barely changed their appearances. The railroad, which turned neighborhoods such as Germantown into residential suburbs, served only to expand industrialism in Manayunk. Even the automobile has not substantially changed the appearance of the town. While Main Street seems never to have recovered from the Great Depression and some old mills have fallen into ruins and disappeared, others are still functioning, and two of the city's largest container manufacturers continue to operate on the canal's island, although the canal itself has not been used since 1916.[50]

Catalog

Armat, Thomas, House

(PA-1671), 5450 Germantown Ave. Coursed Wissahickon schist ashlar front and stuccoed rubble sides and rear, 24'-10" (three-bay front) × 41'-3" with one- and two-story side addition and two-story rear ell (not original), two-and-a-half stories, gable roof, side-hall plan. Built c. 1792; rear brick ell, two-story frame-and-shingle side addition built 19th century; interior altered to shop early 20th century; one-story ashlar addition built in front of side addition 1926; new shopfront, one-story rear concrete-block addition built 1960. Certified, PHC 1957. 2 photocopies of ext. photos (1957, 1960);* HABSI form (1957).*

Ashmead, Albert, House

(PA-1672), 5430 Germantown Ave. Coursed Wissahickon schist ashlar front and rubble sides and rear, 33'-8" (three-bay front) × 32'-1" with two-story rear brick ell and one-story lateral brick addition, two-and-a-half stories, gable roof, side-hall plan. Built before 1796; lateral addition built late 19th century; interior extensively remodeled, frontispiece and triple dormer added 1903. Certified, PHC 1957. 1 photocopy of ext. photo (1957);* HABSI (1957).*

Ashmead, William, House

(PA-1673), 5434 Germantown Ave. Random Wissahickon schist ashlar front and rubble sides and rear, 32'-4" (four-bay front) × 31'-4" with two-story rear ell, two stories, gable roof. Built 1740; rear brick ell built 1906; exterior partially restored, interior altered to bank 1949. Certified, PHC 1957. 1 photocopy of ext. photo (1957);* HABSI form (1957).*

Bank of Germantown. See Clarkson-Watson House

Barron, Commodore James, House
(PA-1674), 5106 Germantown Ave. Stuccoed rubble, 43'-9" (five-bay front) × 22'-5" with two-and-a-half-story rear ell, two-and-a-half stories, gable roof, center-hall plan. Rear ell probably built before 1794; house built between 1794 and 1803; altered to store after 1920. Commodore Barron, commander of the *Chesapeake* when it was attacked by the *Leopard* 1807 and Commandant of Philadelphia Navy Yard 1824–25, 1831–37, owned this house 1839–45. Certified, PHC 1957. 2 photocopies of ext. photos (c. 1900, 1957);* HABSI form (1957).*

Baynton House. See Germantown Historical Society Area Study

Bechtel House. See Germantown Historical Society Area Study

Beggarstown School
(PA-1675), 6669 Germantown Ave., at N.E. corner Springer St. Coursed Wissahickon schist ashlar front and stuccoed rubble sides and rear, 28'-3" (four-bay front) × 18'-3" with one-story rear ell, one story (originally one-and-a-half), gable roof with projecting cornice, one-room plan.
　　Rare example of simple colonial schoolhouse. Built c. 1740; interior altered c. 1840; "restored" 1915; rear ell added later. St. Michael's Lutheran Church parish school, which provided basic education. Certified, PHC 1956; Pennsylvania Register 1970; NR. 1 ext. photo (1973).*

Belfield (also known as Charles Willson Peale House). Illustrated
(PA-1676), 5500 N. Twentieth St., at N.E. corner Clarkson Ave. (not cut through). Stuccoed brick, approx. 55' (five-bay front) × 25' with two- and two-and-a-half-story rear ell (not original), two-and-a-half stories in front, three-and-a-half stories in rear, gambrel roof. Built mid-18th century; enlarged, altered later; porch enclosed, three-story rear corner bay built c. 1923. Charles Willson Peale, versatile and prolific early American painter, lived here 1810–25. Certified, PHC 1956; designated National Historic Landmark 1965. 3 ext. photos (1967),* 3 int. photos (1967);* 3 data pages (including 1923 insurance survey).*

Billmeyer, Michael, House

(PA-1677), 6505–7 Germantown Ave., at N.E. corner Upsal St. Coursed Wissahickon schist ashlar front and rubble sides and rear, approx. 40' (four-bay front) × 33' with two-story rear ells, two-and-a-half stories on raised basement, gable roof, front pent eave, two entrances with side-hall plans.

Fine example of Germantown colonial architecture. Built c. 1730. Acquired by Germantown Historical Society 1961. Michael Billmeyer, newspaper and almanac printer and German printer for Pennsylvania Assembly, lived here 1789–1831. Certified, PHC 1957. 1 ext. photo (1973).*

Blair House

(PA-7-5), 6105 Germantown Ave. Stuccoed rubble, 26'-4" (three-bay front) × 36'-2" with two-and-a-half-story diagonal rear wing, two-and-a-half stories, gable roof, round-arch entrance with fanlight and Tuscan frontispiece, side-hall plan. Rear wing probably built before 1806; house built 1806; demolished c. 1950. 7 sheets (1934, including plans, elevations, sections, details); 4 ext. photos (1934); 2 data pages (1936).

Bringhurst House

(PA-1679), 5448 Germantown Ave. Wissahickon schist rubble front and stuccoed rubble sides and rear, 38'-6" (five-bay front) × 28'-0" with two-story rear brick ell, two stories, gambrel roof (originally gable), center-hall plan. Built mid-18th century; rear ell built mid-19th century; gambrel roof added c. 1875; altered to offices later; exterior partially restored 1960. Certified, PHC 1957. 1 photocopy of ext. photo (1957);* HABSI form (1957).*

Chestnut Hill Academy. See Wissahickon Inn

Chew House. See Cliveden

Church of St. Vincent de Paul (Roman Catholic)

(PA-1680), 101–7 E. Price St., at N.E. corner Lena St. Plastered stone with brownstone trim, approx. 46' × 140', one story, gable roofs with central dome, round-arch windows, giant Tuscan pilasters; Greek-cross plan with extended nave, three aisles, semicircular chancel, notable original interior decorations and paintings.

Fine example of reserved Italianate church design. Built 1849–51; Joseph D. Koecker, architect; William A. and Daniel Ruffner, builders. Dome, transept built 1857; interior renovated 1881, 1901; white marble altar restored c. 1945. Exterior replastered to simulate granite 1945; Earley Studio (Rosslyn, Va.), contractors. First Roman Catholic parish in Germantown. Certified, PHC 1972; Pennsylvania Register 1973. 2 ext. photos (1973),* 1 photocopy of old int. lithograph (c. 1880).*

Clarkson-Watson House (also known as Bank of Germantown)

(PA-1681), 5275–77 Germantown Ave. Museum. Stuccoed rubble, 38'-5" (five-bay front) × 32'-3" with two-story rear wings, two-and-a-half stories on raised basement, gable roof, Tuscan frontispiece with pediment, originally front pent eave, center-hall plan. Built c. 1745; pent eave removed, frontispiece added c. 1775; altered to bank and dwelling c. 1825; altered to stores, rear brick ells built c. 1870, c. 1910. Exterior restored to 1793 appearance 1967–70; Henry J. Magaziner, architect.

Country home 1780–95 of Matthew Clarkson, Mayor of Philadelphia 1792–96. Quarters of Secretary of State Thomas Jefferson and Attorney General Edmund Randolph during yellow fever epidemic of 1793. Home of John Fanning Watson, author of *Annals of Philadelphia* (1830) 1825–47. Bank of Germantown 1825–69. Costume Museum of Germantown Historical Society since 1971. Certified, PHC 1957; Pennsylvania Register 1972; NR. 1 ext. photo (1973).*

Cliveden (also known as Chew House). Illustrated

(PA-1184), 6401 Germantown Ave., in middle of block bounded by Germantown Ave., Johnson, Morton, Cliveden sts. Historic house museum. Coursed Wissahickon schist ashlar front, scored stuccoed rubble sides and rubble rear, 53'-10" (five-bay front) × 44'-4" with two-story rubble rear ell, two-and-a-half stories on raised basement, gable roof with front pediment, stone urns on brick pedestals at front corners and pediment and roof ridge, Doric frontispiece with pediment, quoins, central front-hall and rear-hall plan, notable interior woodwork. Flanked at rear by two-story rubble buildings with stuccoed fronts; northern building connected to house by one-story passage (originally arcade open on south side). Two-story rubble barn and stable at rear of property.

Oustanding example of Georgian domestic architecture with notable interiors. Built 1763–67; Jacob Knor, master carpenter; John

Hesser, mason. Arcade enclosed 1856; rear ell built 1867–68. Country house of Benjamin Chew, Chief Justice of Pennsylvania Supreme Court 1774–76 and Judge of High Court of Errors and Appeals 1791–1808; Chew family residence until 1972 when acquired by National Trust for Historic Preservation. Site of the turning point in the Battle of Germantown 1777. Certified, PHC 1956; designated National Historic Landmark 1961.

17 sheets (1972, including plot plan, plans, elevations, sections, details, landscape plan, plan and elevations of barn-stable);* 40 ext. photos (1967, 1972, including battle-scarred door and barn-stable),* 52 int. photos (1967, 1972, including details of lock, servant bells, floor joists),* 5 photocopies of architectural drawings (c. 1763);* 5 data pages (including 1856–68 insurance surveys).*

Compton (also known as John T. Morris Estate)
(PA-1682), Meadowbrook Ave., near S.W. corner Stroud St. Coursed cyclopean-rusticated granite, approx. 75′ (six-bay front) × 65′ two-and-a-half stories on raised basement with battlemented four-story octagonal tower, hipped, gable, and apsidal roofs with Flemish cross-gables, open porch on two sides, irregular plan with large entrance-stairhall, notable carved interior woodwork. Built 1887–88; Theophilus Parsons Chandler, Jr., architect. Acquired by University of Pennsylvania as part of Morris Arboretum 1932; demolished 1968. 6 ext. photos (1964),* 6 int. photos (1964).*

Concord School. Illustrated
(PA-12), 6309 Germantown Ave. Historic house museum. Stuccoed rubble, approx. 25′ (two-bay front) × 37′, two stories with open square belfry with spire, gable roof, two side entrances to first-floor schoolroom with enclosed stairs in southeast corner, second-floor library (now caretaker's apartment).

Fine example of an 18th-century schoolhouse. Built 1775; probably by Jacob Knor, master carpenter. Established to provide an English education for children too far distant from Germantown Academy. Maintained by Trustees of the Upper Germantown Burying Ground. Certified, PHC 1956. 6 sheets (1934, including plans, elevations, section, plans and section of bell and belfry, details); 1 ext. photo (1973);* 2 data pages (1934).

Conyngham-Hacker House. See Germantown Historical Society Area Study

Deshler-Morris House

(PA-1683), 5442 Germantown Ave., opposite Market Sq. Independence National Historical Park. Historic house museum. Stuccoed rubble, 36'-2" (five-bay front) × 35'-2" with two-story rear ell (not original), two-and-a-half stories on raised basement, gable roof, Tuscan frontispiece with pediment, central-hall plan, notable interior woodwork, adjoining restored 18th-century garden.

Notable Georgian "double-pile" house, whose central-hall plan is unusual for colonial Germantown houses. Built 1772; rear ell built 1838, enlarged 1856, 1868; interior altered 1868, "restored" 1898; lateral addition built on north end 1909; partially restored by National Park Service 1949–50; restoration by National Park Service completed 1975–76. Gen. William Howe's headquarters after Battle of Germantown 1777. Summer residence of President George Washington 1793, 1794. Built for David Deshler, prosperous merchant; in possession of the prominent Morris family 1836–1948; acquired by National Park Service 1948; opened to public 1950. Certified, PHC 1956; Pennsylvania Register 1971; NR. 2 ext. photos (1973),* 4 int. photos (1950, 1960);* 7 data pages (including 1774–1909 insurance surveys).*

Detweiler House

(PA-1684), 8226 Germantown Ave. Coursed Wissahickson schist ashlar front and stuccoed rubble sides and rear, 26'-7" (three-bay front) × 34'-7" with two-story rear ell, two-and-a-half stories, gable roof, open side porch, side-hall plan. Built c. 1760; new front built c. 1800; rear ell built 1902. Certified, PHC 1957. 1 photocopy of ext. photo (1957);* HABSI form (1957).*

Dorfenille House

(PA-1685), 5139 Germantown Ave., at S.E. corner Collom St. Stuccoed rubble, 25'-6" (three-bay front) × 38'-0" with one- and two-story rear wing, two-and-a-half stories, gable roof. Built c. 1795; rear wing built before 1894; ground story altered to store 1924. Certified, PHC 1957. 1 photocopy of ext. photo (1957);* HABSI form (1957).*

Endt, Theobald, House

(PA-1686), 5222–24 Germantown Ave. Stuccoed rubble, 38'-6" (five-bay front) × 41'-3", two-and-a-half stories on raised basement, gable roof, probably originally center-hall plan. Built c. 1795; rear brick addition built later; altered to store and apartments 20th century. Acquired by Germantown Historical Society 1972. Certified, PHC 1957. HABSI form (1957).*

Fielding, Mantle, House
(PA-1687), 28 W. Walnut Ln. Wissahickon schist rubble, approx. 58' (four-bay front) × 30', two-and-a-half stories, gable roof, Tuscan-column porch and balcony across front, center-hall plan. Early example of adaptive use and taking advantage of a vernacular building's rustic qualities for domestic purposes. Built 1796 as barn on adjoining Wyck property. Remodeled into dwelling c. 1891; Mantle Fielding, architect. Porch enclosed later. Mantle Fielding, noted early-20th-century architect and writer, lived here 1892–1939. Certified, PHC 1972. 1 ext. photo (1973).*

First Baptist Church of Germantown
(PA-1688), 36–42 E. Price St. Stuccoed rubble, approx. 50' × 75', one story on raised basement with Corinthian hexastyle portico, gable roof, three-aisle plan. Fine late example of the Greek Revival. Built 1852–53; Samuel Sloan, architect. Steeple added 1862, removed 1887; lecture room remodeled 1892; interior altered 1897. Polite Temple Baptist Church since 1955. Certified, PHC 1965. 2 ext. photos (1973);* 8 data pages (1965, including 1852–53 news clippings).*

Folwell House
(PA-1689), 5281 Germantown Ave. Stuccoed rubble, 24'-1" (three-bay front) × 33'-3" with two-story rear ell, two-and-a-half stories on raised basement, gable roof, side-hall plan. Built late 18th century; brick rear ell built before 1880. Certified, PHC 1957. 1 photocopy of ext. photo (1957);* 6 data pages (including 1881–1916 insurance surveys);* HABSI form (1957).*

Fromberger, John, Houses (now known as Germantown Insurance Co.)
(PA-1690), 5501 Germantown Ave., at N.E. corner Church Ln., on E. side Market Sq. Brick with marble trim, approx. 70' (nine-bay front) × 75' (originally three row houses, probably each approx. 23' × 34'), two-and-a-half stories, gable roof, round-arch entrance with fanlight and pedimented Ionic frontispiece (not original), originally London house plans.
    Built 1795–96; altered many times later; enlarged, altered 1917; renovated after fire 1930. Rehabilitated 1952–54; G. Edwin Brumbaugh, architect. Place of worship for St. Luke's Protestant

Episcopal Church 1813–37; site of (Young) Women's Christian Association of Germantown 1873–1915; Mary Warden Harkness House, boarding home for girls, 1917–41; offices of Germantown Fire Insurance Co. since 1954 whose name changed to Germantown Insurance Co. 1964. Certified, PHC 1956; Pennsylvania Register 1971. 1 ext. photo (1973);* 10 data pages (1965, including 1796 insurance survey);* HABSI form (1957).*

General Wayne Hotel

(PA-1691), 5060 Germantown Ave., at S.W. corner Manheim St. Stuccoed rubble, approx. 32′ (three-bay front) × 22′ with two-and-a-half story rear wing, two-and-a-half stories, gambrel roof with cross-gambrel (originally gable). Built between 1780–85; enlarged, altered, gambrel roof built 1866; front porch removed, ground story altered later. Certified, PHC 1957. 1 photocopy of ext. photo (1957);* HABSI form (1957).*

Germantown Academy (originally known as Germantown Union School)

(PA-7-4), 110 School House Ln., at S.W. corner Greene St. Main building: coursed Wissahickon schist ashlar front and rubble sides and rear, 67′-2″ (five-bay front) × 40′-0″ with two-story rear ell and two-and-a-half-story addition, two-and-a-half stories on raised basement with centered square belfry and spire, gable roof, central-hall plan originally with classroom on each side.

Notable mid-18th-century educational complex. Built 1760–61; Jacob Knor, master carpenter. Rear wing built at northeast end 1879–80; James Kinnier and Sons, architects and builders. Sower Hall built onto rear western end of wing 1904; Robeson Lea Perot, architect. Organized by Pennsylvania Germans as private school 1759; chartered as Public School of Germantown 1784, popularly known as Germantown Academy since 1794. Germantown Lutheran Academy 1965–73; Germantown-Stevens Academy since 1973. Hospital during Battle of Germantown 1777. Certified, PHC 1956; Pennsylvania Register 1971; NR. 13 sheets (1934, including plot plan, plans, elevations, section, details, belfry plan, elevation and section); 7 ext. photos (1934, 1960*); 2 data pages (1936).

German Headmaster's House: S.E. of Main Building. Wissahickon schist rubble, 20′-2″ (two-bay front) × 28′-0″ with rear addition, two stories, gable roof. Built 1760–61; rear addition built later. 3 sheets (1934, including plans, elevations, section, details).

English Headmaster's House: S.W. of Main Building. Wissahickon schist rubble, approx. 20′ (two-bay front) × 28′, two stories, gable roof. Built 1760–61; altered later. 1 sheet (1934, including details); 1 ext. photo (1934).

Germantown Cricket Club (also known as Manheim Club). Illustrated (PA-1693), 5140 Morris St., at S.W. corner Manheim St. Brick with marble trim, approx. 215' (twenty-bay front) × 60', two-and-a-half stories, gable roofs originally with central balustrade, rear Doric colonnade with Tuscan balcony above, U-plan with large central entry hall.

Important Georgian Revival work by a leading late-19th-century architectural firm. Built 1890–91; Charles F. McKim of McKim, Mead and White, architect; A. S. Tourison, contractor. Wings, addition built 1902, 1907; McKim, Mead and White, architects. Interior altered, three-story brick addition built 1911–12; Mantle Fielding, architect. One-story rear brick addition built 1929; Dreher and Churchman, architects. Roof balustrade removed after 1953. Oldest cricket club in the United States; founded 1854. Certified, PHC 1970. 2 ext. photos (1973).*

Germantown Historical Society Area Study
(PA-1694), 5208, 5214, 5218, 5226 Germantown Ave. Three Wissahickon schist rubble houses and one brick house, approx. 25' and 35' (three- and five-bay fronts) × 20' and 30' with two-story rear ells, two-and-a-half stories on raised basements, gable roofs, side-hall and center-hall plans.

Built between 1742 and c. 1790; altered, rear ells added later. Includes Baynton House, No. 5208; Conyngham-Hacker House, No. 5214; Howell House, No. 5218; Bechtel House, No. 5226, which serve as library, headquarters, costume museum, and textile museum respectively of Germantown Historical Society. Certified, PHC 1957; Conyngham-Hacker and Howell Houses on Pennsylvania Register 1971; NR. 5 ext. photos (1976);* 1 data page (including 1827 insurance survey);* 2 HABSI forms (1957).*

Germantown Insurance Company. See Fromberger, John, Houses

Germantown Union School. See Germantown Academy

Glen Fern (also known as Livezey House)
(PA-14), Livezey Ln. at E. side of Wissahickon Creek. Fairmount Park. Wissahickon schist rubble, 34'-0" (four-bay front) × 37'-2" with

one-and-a-half-story lateral ells (originally approx. 34' × 20'), two-and-a-half stories (originály one-and-a-half), gable roof, balcony across south front, two front rooms and large back room (once center-hall plan).

Fine example of 18th-century Delaware Valley farmhouse. Built between 1733 and 1739; enlarged to rear (north) and raised one story c. 1765; ells built before 1853, altered later. Springhouse remains as only extant outbuilding. Purchased 1747 by Thomas Livezey, farmer, miller, justice of peace and provincial commissioner; owned by Livezey family until acquired by Fairmount Park Commission, City of Philadelphia 1869; Valley Green Canoe Club since 1909. Certified, PHC 1956.

12 sheets (1935, including plot plan, plans, elevations, sections, details, barn and springhouse plans, elevations and details); 1 ext. photo (1973);* 8 data pages (1935, including 1833 insurance survey*).

Green Tree Tavern (also known as Pastorius House)

(PA-1695), 6023 Germantown Ave. Wissahickon schist ashlar front and south end, rubble rear and north end, 42'-4" (five-bay front) × 36'-6", two-and-a-half stories, gable end with cornice returns at ends, pent eave on front and south end, 1748 date stone in south gable, center-hall plan.

Notable example of colonial Germantown architecture. Built 1748 for Daniel Pastorius, grandson of Germantown's founder Francis Daniel Pastorius. Moved northward, entrance exterior altered, interior extensively renovated 1930. Germantown Academy organized at meeting here 1759. Offices and parish house of First United Methodist Church of Germantown since 1907. Certified, PHC 1956. 1 ext. photo (1973);* 4 data pages (including 1797–1932 insurance surveys).*

Grumblethorpe (also known as Wister's Big House). Illustrated

(PA-7-1), 5267 Germantown Ave. Historic house museum. Coursed fieldstone ashlar front and rubble sides and rear, 43'-5" (six-bay front) × 34'-4" with three-story rear ell (originally one-and-a-half stories), two stories, gable roof, pent eave with center balcony, center-hall plan with separate door to front parlor.

Initially a country house, Grumblethorpe is a fine example of Delaware Valley vernacular domestic architecture. Built 1744; rear ell enlarged and extended 1750, 1799, 1806, 1819; house altered 1808, restored 1957–67. Headquarters and site of death of British Brig. Gen. James Agnew during Battle of Germantown 1777. Owned and maintained by Philadelphia Society for Preservation of Landmarks since 1940. Certified, PHC 1956; Pennsylvania Register 1971; NR. 10 sheets (1934, including plot plan, plans, elevations, section, details, plan and sections of springhouse and pump); 6 ext. photos (1934, 1973,* including 1 of pump); 9 data pages (1936, including 1764–1887 insurance surveys*).

Grumblethorpe Tenant House. See Wister's Tenant House

Haines House. See Wyck

Hood Cemetery Entrance (originally known as Lower Burying Ground). Illustrated
(PA-1697), 4901 Germantown Ave., at N.E. corner Logan St. Marble, approx. 147' front, horseshoe-arch entry between engaged Corinthian columns and beneath bracketed round arch with cartouche, balustraded front wall. Built 1849; William Johnston, architect; William Struthers, marble mason. Part of original Germantown plan; graves date from 1708. Known as Hood Cemetery since William Hood bequeathed money for entrance in 1850. Certified, PHC 1957. 2 ext. photos (1973).*

House
(PA-1698), 6000–6002 Germantown Ave., at N.W. corner Harvey St. Wissahickon schist rubble, 35'-4" (five-bay front) × 30'-8" with two-story rear ell, two-and-a-half stories, gable roof, originally center-hall plan. Built 18th century; altered to store, one-story brick side addition built later. Certified, PHC 1957. 1 photocopy of ext. photo (1957);* HABSI form (1957).*

House
(PA-1699), 6377 Germantown Ave. Stuccoed rubble, approx. 25' (three-bay front) × 40' with one-story rear ell, two-and-a-half stories on raised basement, gable roof, side-hall plan. Built mid-18th century; interior altered c. 1800; altered to store, rear ell added 20th century. Certified, PHC 1961. 2 ext. photos (1961),* 2 int. photos (1961);* 7 data pages (1961).*

Houston-Sauveur House. Illustrated
(PA-1700), 8205 Seminole Ave., at N.W. corner Hartwell Ln. Random Wissahickon schist ashlar with frame-and-shingle third story, approx. 36' (three-bay front) × 42', three stories, gable and hipped roofs

with cross-gables, enclosed semioctagonal sun porch, open front porch with balcony above, center entry-hall plan. Notable example of Queen Anne style. Built 1885; probably by G. W. and W. D. Hewitt, architects. Initially rental property for Henry H. Houston, steam transportation entrepreneur and developer of Chestnut Hill. 2 ext. photos (1973).*

Howell House. See Germantown Historical Society Area Study

Ivy Lodge
(PA-1701), 29 E. Penn St. Coursed rough-hewn Wissahickon schist, approx. 44' (three-bay front) × 38' with two-story rear wing, two-and-half stories, hipped roof, bracketed cornice interrupted by centered round-arch dormer, front rectangular bays and porch, open porch on east side, semioctagonal bay on west side, center-hall plan. Fine example of Italian villa design. Built c. 1850; Thomas U. Walter, architect. Rear wing built later. Certified, PHC 1970; Pennsylvania Register 1970; NR. 1 ext. photo (1973).*

Jacoby, Wigard, House
(PA-1702), 8327 Germantown Ave. Coursed Wissahickon schist ashlar front and rubble sides and rear, approx. 25' (three-bay front) × 36' with rear one-story concrete-block kitchen ell, two-and-a-half stories, gable roof, round-arch entrance with fanlight and pedimented frontispiece, side-hall plan. Built c. 1794; probably by Wigard Jacoby, carpenter. Two dormers added after 1911; ell built 1948. Certified, PHC 1957. 1 photocopy of ext. photo (1957);* HABSI form (1957).*

Johnson House
(PA-7-7), 6306 Germantown Ave., at N.W. corner Washington Ln. Wissahickon schist ashlar front and rubble sides and rear, 41'-6" (five-bay) × 35'-7¹/₂" with two-story rear ell, two-and-a-half stories, gable roof, pent eave on front and south side, center-hall plan, notable interior paneling.
    Fine example of Germantown vernacular architecture exhibiting German influence in exterior details. Built 1765–68. Owned and maintained by Women's Club of Germantown since 1920. Certified, PHC 1956; Pennsylvania Register 1971; NR. 12 sheets (1934, including plans, elevations, sections, details); 4 ext. photos (1934); 3 data pages (1936).

Keyser Houses
(PA-1704), 5920–26 Germantown Ave., at S.W. corner Haines St.
Two contiguous stuccoed rubble houses, each approx. 19' (two-bay
front) × 22' with one-story rear wing, two stories, gable and flat
roofs. Two stuccoed rubble row houses, each approx. 23' (two- and
three-bay front) × 33' with two-story rear wings, three stories, gable
roofs. Built mid-18th century; altered to stores later. Certified, PHC
1957. HABSI form (1957).*

Keyser, Jacob, House
(PA-11), 6205 Germantown Ave., at N.E. corner Tulpehocken St.
Coursed Wissahickon schist ashlar front and rubble sides and rear,
37'-4" (five-bay front) × 28'-10" with two-story rear ell, two-and-a-
half stories on raised basement, gable roof, pent eave, center-hall
plan. Built after 1744 (rear ell probably part of an earlier house); demolished 1950. 7 sheets (1934, including plot plan, plans, elevations,
cellar sections, details, elevations of barn); 2 data pages (1934).

Livezey House. See Glen Fern

Log House
(PA-143), N.E. corner Courtland and Eighteenth sts. on Stenton
grounds (earlier at Friends' Select School, S.W. corner Race and
Sixteenth sts.). Squared logs with clapboard gable ends, 31'-9" (four-
bay front) × 21'-4" with brick rear ell, two-and-a-half stories, gable
roof, two rooms with central chimney.
    Probably built as barn c. 1790; converted to house c. 1820.
Relocated from Sixteenth St. between Cherry and Race sts. to S.W.
corner Race and Sixteenth sts., rear ell built after 1886; clapboards
removed c. 1925. Moved to Stenton grounds 1969. Maintained by National Society of the Colonial Dames of America since 1969. Certified,
PHC 1957. 4 ext. photos (1940, 1976*).

Loudoun
(PA-1705), 4650 Germantown Ave., at N.W. corner Apsley St. Historic house museum. Stuccoed rubble, 49'-10" (five-bay front) × 46'-
3" including two-story stone lateral and brick rear additions (originally approx. 32' × 30'), two-and-a-half stories with Corinthian
tetrastyle portico, gambrel and hipped roofs, one-story side (west
end) sun porch, center-hall plan.
    Germantown's most imposing Federal-style house. Built 1796–

1801; west section and sun porch built 1829; portico added 1850; rear addition built 1888; rehabilitated 1964. Acquired by Fairmount Park Commission, City of Philadelphia 1939; opened to public 1966. Certified, PHC 1956. 2 ext. photo (1973);* 5 data pages (including 1808–88 insurance surveys);* HABSI form (1957).*

Lower Burying Ground. See Hood Cemetery Entrance

Manheim Club. See Germantown Cricket Club

Maxwell, Ebenezer, House. Illustrated
(PA-1098), 200 W. Tulpehocken St., at S.W. corner Greene St. Museum. Random rough-hewn Wissahickon schist with maroonstone diamond patterns, 78'-3¹/₂" (seven-bay front) × 45'-9¹/₂" including lateral kitchen wings, two-and-a-half stories with square three-story tower, mansard roof with Flemish cross-gables, two open battlemented porches; center-hall plan with lateral passage to kitchen wings.
Fine early example of eclectic suburban villa. Built 1859; probably by Joseph C. Hoxie, architect. Western kitchen wing raised to two stories, western porch added later; porch removed 1970. Restored by Germantown Historical Society as Victorian arts museum 1965– . Ebenezer Maxwell was a successful Philadelphia dry goods merchant. Certified, PHC 1964; Pennsylvania Register 1970; NR.
7 sheets (1966, including plot plan, plans, elevations, details);* 8 ext. photos (1964),* 3 int. photos (1964);* 14 data pages (1964, including 1859 insurance survey).*

Mehl House
(PA-1706), 4821 Germantown Ave. Wissahickon schist rubble, approx. 36' (five-bay front) × 29' with one- and two-story rear ell, two-and-a-half stories on raised basement, gable roof, center-hall plan. Built late 18th century; two-story stone rear ell built later; interior renovated 1906; one-story brick ell built 1933. Certified, PHC 1957. 1 photocopy of ext. photo (1957);* HABSI form (1957).*

Mennonite Meeting House. Illustrated
(PA-15), 6119 Germantown Ave., at N.E. corner Herman St. Wissahickon schist rubble, 30'-1" (three-bay front) × 35'-6" with one-story rear addition, one story, gable roof, two-room rectangular plan. Plain style of a German pietist sect. Built 1770; rear addition built

1907. Established in 1688 as the first Mennonite congregation in America. Certified, PHC 1956; Pennsylvania Register 1972; NR. 7 sheets (1934, 1935, including plot plan, plans, elevations, section, details); 1 ext. photo (1973);* 2 data pages (1934).

Morris, John T., Estate. See Compton

Mutual Fire Insurance Company Building
(PA-1014), 5521 Germantown Ave., at N.E. corner School House Ln. Brick with stone and terra cotta trim, approx. 30' (three-bay front) × 74', two-and-a-half stories, tiled gable roof with cross-gables, general office in front with president's room and directors' room in rear.

Fine example of late-19th-century architectural design with rich use of materials, color, and texture. Built 1884; George T. Pearson, architect; James Kinnier's Sons, builder. Demolished 1959. 3 ext. photos (1959), 2 int. photos (1959), 1 photocopy of old ext. photo (c. 1900); 6 data pages (1959).

Pastorius House. See Green Tree Tavern

Peale, Charles Willson, House. See Belfield

Philadelphia and Reading Railroad, Chestnut Hill Line. See Philadelphia, Germantown and Norristown Railroad, Germantown Depot

Philadelphia, Germantown and Norristown Railroad, Germantown Depot (later Philadelphia and Reading Railroad, Chestnut Hill Line)
(PA-1707), 5731–35 Germantown Ave., at S.E. corner Price St. Brick covered with scored stucco, approx. 50' (three-bay front) × 35' with one-story rear additions (originally one-story train shed approx. 38' × 256'), two stories, gable roof, boldly framed round-arch windows, quoins, rectangular plan.

Fine example of early suburban passenger and freight depot, now Philadelphia's oldest surviving railroad station. Built 1855; probably by Joseph C. Hoxie, architect. Altered to store and offices 1902; Armstrong and Printzenhoff, contractors. Rear train shed replaced by present additions 1934–35; Samuel Rothblatt, contractor. This railroad, Philadelphia's first, began operations 1832; acquired by Philadelphia and Reading Railroad 1870. Certified, PHC 1964. 1 ext. photo (1973).*

Rectory of St. Stephen's Methodist Episcopal Church

(PA-1708), 5213 Germantown Ave. Stuccoed rubble, approx. 26′ (three-bay front) × 40′ with two-story lateral addition, two-and-a-half stories, gable roof, round-arch entrance with fanlight, originally side-hall plan. Built late 18th century; lateral addition, open side and rear porches built mid-19th century. Rectory since 1856. Certified, PHC 1957. 1 photocopy of ext. photo (1957);* HABSI form (1957).*

Rittenhouse House

(PA-16) S.W. corner Lincoln Dr. and Rittenhouse St. Fairmount Park. Whitewashed Wissahickon schist rubble, 44′-0″ (five-bay front) × 18′-4″, two stories, gable roof, originally pent eave on north side, two sections with one room to a floor in each.

Germantown vernacular architecture reflecting strong German influence. Built 1707. Birthplace of David Rittenhouse, early American astronomer and instrument-maker and foreign member of Royal Society. William Rittenhouse, original owner of house, built nearby first paper mill in British colonies 1690. Acquired by Fairmount Park Commission, City of Philadelphia 1890. Certified, PHC 1956. 7 sheets (1935, including plot plan, plans, elevations, details); 3 ext. photos (1925, 1973*); 1 data page (1935).

Robeson House. See Shoomac Park.

Royal House

(PA-1709), 5011 Germantown Ave., at N.E. corner Garfield St. Stuccoed rubble, 26′-3″ (three-bay front) × 36′-0″ with two-story rear ell, two-and-a-half stories, gable roof, side-hall plan. Built probably early 19th century; brick rear ell added later; altered to store and apartments early 20th century; storefront remodeled 1925. Certified, PHC 1957. 1 photocopy of ext. photo (1957);* HABSI form (1957).*

St. Timothy's (Protestant Epsicopal) Church

(PA-1710), 5720 Ridge Ave. Random rough-hewn Wissahickon schist with brick trim, approx. 45′ × 130′ (originally approx. 45′ × 38′) with two-story rear parish house, one story with squat battlemented rear clerestory tower and three-stage square front corner tower, polychromatic gable roof with shed roof over aisles, lancet windows; three-aisle plan with side entrances, semihexagonal chancel, exposed interior wooden trusses.

Fine synthesis of High Victorian Gothic church design. Built

1862–63; Emlen T. Littell, architect. Corner tower added 1871, nave extended, eastern part of parish house built 1874; probably by Charles M. Burns, architect. Nave extended, clerestory tower built, parish house enlarged and connected to church 1885–86; church and parish house renovated 1931; chancel altered later. Certified, PHC 1974. 5 ext. photos (1973).*

St. Timothy's Working Men's Club and Institute
(PA-1711), 5164 Ridge Ave., at intersection of Terrace and Vassar sts. Random Wissahickon schist ashlar with granite trim, approx. 21' (two-bay front) × 46', two-and-a-half stories with three-story front square entry and stair tower, mansard and jerkin-head roofs, Corinthian cast-iron columns at two corners, semihexagonal second-story frame bay on struts, irregular plan.

Built 1877; probably by Charles M. Burns, architect. Founded 1873 by St. Timothy's Protestant Episcopal Church to provide self-improvement and social facilities for residents of working-class neighborhood. Certified, PHC 1974. 2 ext. photos (1973).*

Schaeffer, Harriet D., House
(PA-1712), 433 W. Stafford St. Wissahickon schist rubble with wood frame second story covered with shingles, approx. 55' (six-bay front) × 35', two-and-a-half stories, hipped and jerkin-head roofs, circular porch with apsidal roof, pointed-arch entrance, vestibule leading to open entry hall. Notable example of Shingle style in cottage mode. Built 1888; Wilson Eyre, Jr., architect. Southern gable end altered c. 1920; Wilson, Harris and Richards, architects. 1 ext. photo (1973).*

Shoomac Park (also known as Robeson House)
(PA-1067), S.E. corner Ridge Ave. and Wissahickon Dr. Stuccoed rubble, 55'-4" (six-bay front) × 38'-3" (originally 55'-4" × 22'-9"), two-and-a-half stories, gambrel roof, originally porch at second story along north and west sides, originally one room deep and three rooms wide with hall between north and middle rooms.

Built c. 1759; enlarged c. 1840; large two-story front and side porch built before 1890; demolished 1961. Woodwork sections, paneled chimney breast, corner cupboard in Architectural Study Collection, Museum Division, Independence National Historical Park. Associated with one of the earliest gristmills in Pennsylvania, Certified, PHC 1956. 5 sheets (1961, including plot plan, plans); 3 ext.

photos (1956, 1961, including 1 of carriage house), 4 int. photos (1961); 9 data pages (1961).

Shur's Lane Mills, T. Kenworthy & Bro.
(PA-1713), 428 Shur's Ln., at N.W. corner Freeland Ave. Two Wissahickon schist rubble factories with brick trim at right angles to each other, three stories, gable roofs; Worsted Mill (facing Pechin St.) approx. 55' (five-bay front) × 150' with lateral elevator shaft, Shoddy Mill (facing Shur's Ln.) approx. 72' (seven-bay front) × 149' with lateral fire tower and one- and two-story rear extensions and one-story rear additions; rectangular plans, interior stone walls with iron doors originally separated more combustible processes from finishing work.

Fine example of steam-powered textile mills built in Manayunk and Roxborough after the Civil War. Worsted Mill built 1876; Shoddy Mill built 1882; later minor alterations and additions, including corrugated iron bridge connecting buildings built c. 1921 and rear cinderblock building built 1954. 1 ext. photo (1973),* 1 printed plan and view (1885).*

Sower, Christopher, Jr., House. See Trinity Lutheran Church House

Stenton
(PA-1714), N.E. corner Courtland and Eighteenth sts. Historic house museum. Brick, approx. 50' (six-bay front) × 38', two-and-a-half stories, hipped roof, plain doorway with side lights, asymmetrical side and rear fenestration, center-hall plan; covered porch connects to detached one-and-a-half-story rear service ell.

Significant Georgian "double-pile" house, prototype of Germantown country houses. Built 1728 for James Logan, William Penn's secretary and agent, Chief Justice of the Supreme Court of Pennsylvania, President of Provincial Council. Occupied by Gen. George Washington before Battle of Brandywine and by Gen. William Howe during Battle of Germantown 1777. Acquired by City of Philadelphia 1908; maintained by National Society of Colonial Dames since 1910. Certified, PHC 1956; designated National Historic Landmark 1965. 4 sheets (1976, including plans);* 7 ext. photos (1960, 1967).*

Stenton Barn. Illustrated
(PA-1715), 4685 N. Eighteenth St., at S.E. corner Wyoming Ave. and intersection of Windrim Ave., on Stenton grounds approx.

200' N. of Stenton. Rubble with used brick incorporated into interior wall fabric, approx. 61' × 36', two stories, gable roof, ventilation slits in walls, oculus in peak, inclined approach to large door at upper level, lower stable level, upper level of threshing floor with flanking mows. Fine example of Pennsylvania bank barn without customary forebay. Built between 1787 and 1798; rehabilitated 1974–76. 1 ext. photo (1976).*

Trinity Lutheran Church. Illustrated

(PA-1716), 19 W. Queen Ln. Stuccoed brick, approx. 55' (three-bay front) × 85', two stories with center octagonal steeple and hexagonal spire on wooden rectangular base, gable roof, pedimented enclosed entry porch, Tuscan pilasters between round-arch windows, three-aisle plan, rectangular chancel.

Fine example of a carpenter-planned Italianate church that dominates neighborhood's skyline. Built 1856–57; Jacob and George A. Binder, designers and builders. Clock, Germantown's first community clock, put in steeple 1858; interior frescoed 1870, remodeled 1886. Incorporated 1837 as English Evangelical Lutheran Church, changed name 1860. Certified, PHC 1957. 3 ext. photos (1973).*

Trinity Lutheran Church House (also known as Christopher Sower, Jr., House)

(PA-1717), 5300 Germantown Ave., at N.W. corner Queen Ln. Stuccoed rubble, approx. 40' (five-bay front) × 25' with two-story rear wing and ell, two-and-a-half stories, gable roof, center-hall plan. Rear ell (southwest corner) built c. 1723; front section built between 1755 and 1760; two sections joined 1795; stuccoed brick rear wing (northwest corner) built 1870; interior altered 1917; house renovated 1950. Jacob Bay and Christopher Sower, Jr., operated first permanent American type foundry here 1772–73. Owned by Trinity Lutheran Church since 1836, its church house since 1950. Certified, PHC 1957. 1 photocopy of ext. photo (1957);* HABSI form (1957)*

Upsala. Illustrated

(PA-1718), 6430 Germantown Ave. Historic house museum. Stuccoed rubble sides and rear; coursed Wissahickon schist ashlar front with marble trim, approx. 49' (five-bay front) × 44' with one- and two-story diagonal rear ell, two-and-a-half stories on raised basement, gable roof, Doric entrance porch with pediment, marble stringcourse and lintels with keystones, central-hall plan, notable interior woodwork.

Fine example of Adamesque-Federal style. Rear ell probably built c. 1755; front section built 1798–1801. Restored 1944– . Headquarters of Upsala Foundation since 1944. Certified, PHC 1956; Pennsylvania Register 1971; NR. 3 ext. photos (1973, including 1 of boot-scraper).*

Valley Green Inn

(PA-1719), W. side Upper Wissahickon Dr. near Valley Green Dr. Fairmount Park. Stuccoed rubble, 60'-9" (seven-bay front) × 17'-0" with one-story rear frame addition, two stories, open front and side porches, pent eave on north end, center-hall plan.

Example of an old roadhouse still operating as a restaurant. Built early 19th century; rear additions built later. In the middle of the Wissahickon Valley, designated a National Natural History Landmark 1964. Certified, PHC 1967. 11 sheets (n.d., including plot plan, plans, elevations, section, details);* 1 ext. photo (1973).*

Vernon

(PA-7-2), 5860 Germantown Ave., Vernon Park. Scored stuccoed rubble, 49'-4^1/$_2$" (five-bay front) × 23'-2" with two-story rear ell and two-and-a-half story back building, two-and-a-half stories on raised basement, hipped roof, projecting central pavilion with pediment, Tuscan frontispeice with fanlight, originally central-hall plan.

Fine example of expansion of colonial farmhouse into Federal-style country house. Rear ell built before 1741; main house built 1805; back building added to ell later; interior entensively altered 1898, rehabilitated 1962. Acquired by City of Philadelphia 1892. Germantown Branch of Free Library of Philadelphia 1898–1907; quarters of Site and Relic Society (later known as Germantown Historical Society) 1907–27; now a community center. Certified, PHC 1956. 17 sheets (1934, including plot plan, plans, elevations, details); 7 ext. photos (1934, 1975,*); 4 data pages (1936, including 1808 insurance survey*).

Wissahickon Inn (now Chestnut Hill Academy)

(PA-1720), 500 W. Willow Grove Ave., at S.E. corner Huron St. Random rough-hewn Wissahickon schist ashlar with wood frame third story covered with shingles, approx. 235' (twenty-three-bay front) × 225', three stories on raised basement, hipped roofs with cross-gables, square corner turrets originally with steep pyramidal roofs, large entrance porch and circular turret centered on east facade, U-plan with central-hall corridors.

Fine example of late-19th-century suburban summer resort hotel. Built 1883–84; G. W. and W. D. Hewitt, architects; William C. Mackie, builder. Original porte-cochere and open porches removed 1928. Originally owned by Henry H. Houston, steam transportation entrepreneur and developer of Chestnut Hill. Chestnut Hill Academy since 1898. Certified, PHC 1972. 2 ext. photos (1973).*

Wister's Big House. See Grumblethorpe

Wister's Tenant House (also known as Grumblethorpe Tenant House)
(PA-7-6), 5269 Germantown Ave. Wissahickon schist rubble, 31'-2" (five-bay front) × 30'-11" (originally 18'-4 × 28'-3") two stories (originally one), gable roof, center-hall plan (originally side-hall plan), corner fireplaces in original part.

Example of modest mid-18th-century dwelling as later enlarged. Southeastern section built c. 1745; northwestern section built and southeastern section raised one story early 19th century. Owned and maintained by Philadelphia Society for Preservation of Landmarks since 1940. Certified, PHC 1956; Pennsylvania Register 1971; NR. 9 sheets (1934, including plans, elevations, section, details); 4 ext. photos (1934, 1973*); 1 data page (1934).

Wyck (also known as the Haines House)
(PA-7-3), 6026 Germantown Ave. Historic house museum. Stuccoed stone, 79'-7" (eight-bay front) × 29'-0" (originally 36'-5" × 29'-0") with one-story lateral wing, two-and-a-half stories, gable roof, center-hall plan (in eastern section).

Example of early farmhouse enlarged and remodeled into country house. Western section built c. 1690; eastern section built and open passageway between sections enclosed before 1775. Extensively altered 1824; William Strickland, architect. Lateral western wing built before 1845. Oldest extant house in Germantown; served as British field hospital during Battle of Germantown 1777. Haines family residence until presented to Germantown Historical Society 1973. Certified, PHC 1956; Pennsylvania Register 1970; NR.

22 sheets (1934, including plot plan, plans, elevations, sections, details, plan and elevations of granary, springhouse, smokehouse, and pump); 8 ext. photos (1934, 1973*); 5 data pages (1936, including 1845, 1894 insurance surveys*).

Belfield. *Cortlandt V. D. Hubbard photo, 1967.*

Cliveden. *Jack E. Boucher photo, 1972.*

Concord School. *Jack E. Boucher photo, 1972.*

Germantown Cricket Club. *Jack E. Boucher photo, 1972.*

Grumblethorpe. *Jack E. Boucher photo, 1973.*

Grumblethorpe, water pump. *E. F. Hoffman photo, 1934.*

Hood Cemetery Entrance. *Jack E. Boucher photo, 1973.*

Houston-Sauveur House. *Jack E. Boucher photo, 1972.*

Ebenezer Maxwell House. *Jack E. Boucher photo, 1964.*

Stenton Barn. *Jack E. Boucher photo, 1976.*

Mennonite Meeting House. *Jack E. Boucher photo, 1972.*

Trinity Lutheran Church. *Jack E. Boucher photo, 1972.*

Upsala. *Jack E. Boucher photo, 1973.*

North Philadelphia

8

Introduction

$\mathcal{N}$ ORTH PHILADELPHIA is an area without precise historical, political, or social boundaries, partly because its social and economic configuration has been in a more agitated state of flux than has that of almost any other part of the city. Generally, North Philadelphia is considered that part of the city flanking Broad Street between Girard and Lehigh avenues. For the purposes of this catalogue, however, North Philadelphia is considered that area north of the original city, west of Northern Liberties, and east of Fairmount Park and Germantown, and including the former Spring Garden and Penn districts. Its precise boundaries are Vine Street on the south, Sixth Street on the east, and Cheltenham Avenue on the north. Its western boundary is more irregular: Fairmount Park and Laurel Hill Cemetery north to Allegheny Avenue, Hunting Park Avenue between Allegheny Avenue and Broad Street, and thence northward to the city limits at Cheltenham Avenue.

Bush Hill was probably the most prominent of the few widely scattered eighteenth-century country seats in the area. Built in 1740 at what is now Buttonwood Street between Sixteenth and Seventeenth streets, Bush Hill was better known for its celebrated occupants than for its architecture. The first resident was the famous defender of New York printer John Peter Zenger, lawyer Andrew Hamilton, who built the house and was succeeded in it by his son Pennsylvania Governor James Hamilton. The last occupant was Vice-President John Adams, who lived in the house for two or three years in the early 1790's before it became a hospital during the yellow fever epidemic of 1793. Bush Hill as a name and place, however, remained in the neighborhood for another eighty years. After 1808 the fire-gutted relic was converted into an oil-cloth factory, which it remained until 1871.[1]

Another case in which a name survived an estate is Cherry Hill, the country seat of Benjamin Warner. In 1821 the Commonwealth of Pennsylvania purchased the eleven-acre property for the site of the proposed Eastern State Penitentiary (illus.),[2] known locally as Cherry Hill Prison. At that time most of the region north of the city was open farmland. Some small villages had begun to grow along the thoroughfares, as Francisville had on Ridge Road west of Broad Street, but the greater growth was in the District of Spring Garden. By 1809 Spring Garden was sufficiently built up to be afforded the same legal protection from hazardous manufactories as the city enjoyed, and four years later it was organized as a district from Sixth Street west to Broad and north from Vine to Poplar Street. The district was not extended westward to the Schuylkill until 1827, by which time Eastern State Penitentiary was nearing completion at Twenty-first Street and Fairmount Avenue.[3]

Eastern State Penitentiary was established on the Pennsylvania System, a Quaker-inspired theory of penal reform that had good intentions but hellish results. The reformers believed that if a prisoner were removed from the bad influences of compatriots and put by himself to read the Bible, think penitent thoughts, and engage in light handicrafts, he would leave the prison a humanely reformed person.[4] The Commonwealth commissioned the young, English-trained John Haviland to give the controversial philosophy concrete form. Haviland's proposal had two distinctive and widely copied features: the castellated Gothic facade and the radial plan. Both fit well the philosophy of the institution. The lugubrious dark granite walls with the heavy iron portcullis stood as an impregnable fortress of justice whose dungeons held the wrongdoers of society, and the trauma of entering its gates was heightened by a hood being placed over the head of the prisoner until he reached his isolated cell to begin his period of repentance.[5] The seven cell blocks that radiated outward from the central rotunda like spokes from the hub of a wheel afforded maximum isolation and surveillance of the inmates, who lived every hour of their sentences in their eight-by-ten-foot cells and adjoining high-walled private exercise yards. Although the Pennsylvania System was not officially abandoned until 1915, the maddening condition was ameliorated before the Civil War by a more traditional penal condition—overcrowding.[6] By the mid-twentieth century a growing prison population exceeded the limited space and obsolete facilities of the penitentiary, and a neighborhood landmark of 150 years was vacated in 1970 to become an architectural dinosaur—a curious creature of disproven theory, a massive monster from the past in desperate need of imaginative adaptive use.[7]

The same rural remoteness that was suitable for a prison attracted other institutions as well. Two of the more notable ones were Girard College (illus.) at Girard Avenue and Corinthian Street and Preston Retreat (illus.) at Twentieth and Hamilton streets, both bequeathed by private patrons and designed by the same architect, Thomas U. Walter. Girard College was the earlier, larger, and by far the more famous of the two and, as time has shown, the more enduring. Preston Retreat was demolished in 1963, while Girard College still stands and performs the same function it did when it opened on New Year's Day, 1848—the education of fatherless boys.[8]

Girard College was one of the great institutions of its time. It was an unprecedented expression of philanthropy—two million dollars for the education of the underprivileged. Stephen Girard was a French immigrant who lived a legendary rags-to-riches life to become one of the nation's earliest and most eccentric millionaires.[9] He was the school's benefactor, but it was Nicholas Biddle, a controversial financier and tastemaker, who successfully circumvented the founder's intentions and saw the plans of the talented young Walter carried into execution. In spite of Girard's wishes for a plain, utilitarian main building, Biddle, chairman of the college's board of directors, was determined to use Girard's millions to erect the most imposing and archaeologically correct Greek temple in America.[10] The result is Founder's Hall, the peristyle centerpiece of the forty-one-acre campus. Its architectural significance rests not only in its graceful design and overpowering scale, but also on its ingenious construction. Since Girard's restrictions on interior dimensions and construction were not adaptable to the classical style, Walter was forced to design within the temple form a series of pendentive domes, groin vaults, and cantilevered stairs.[11] Walter hurdled all the obstacles and produced one of the great pieces of American architecture, the climax of the Greek Revival style, but his success was qualified. A structure that met the demands of scale, style, and interior arrangement that circumstances made of it could not also provide the acoustics, lighting, and heat efficiency necessary for a good school building, and it has not been used for classroom purposes since 1916.[12] Yet the building's form fit its function, which was primarily not to be utilitarian but to serve as a monument to a magnanimous man, as its name suggests. And like a Greek temple housing the spirit and effigy of its honored god, Founder's Hall encloses both the body and a statue of Stephen Girard.

Preston Retreat, on the other hand, was much more utilitarian and decidedly less symbolic. It was incorporated in 1836 under the will of Dr. Jonas Preston, a Quaker physician who saw the need for a maternity hospital for indigent women.[13] Walter won the architectural competition with a compact design that took advantage of the Retreat's special function. A Doric tetrastyle portico and flanking side wings gave an austere, imposing presence to the marble front, but the building's virtues lay in its plan. Wards and rooms were organized around two intersecting halls with a central nurses' station and opened onto a veranda that extended around the rear half of the building and gave access to water closets at the rear corners.[14] The longevity of the design—it was never essentially altered—bore testimony to its functionalism. The extensive renovations of 1908 introduced elements of twentieth-century technology (electricity, elevators, fire escapes, refrigeration, and larger furnaces), and the building's end was caused less by the obsoleteness of the physical plant than by the inefficiency of a charitable hospital in a welfare state with centralized institutions.[15] In spite of its modest intentions, Preston Retreat did not open its doors as a hospital for thirty years after its founding. Unlike Girard's investments, Preston's had been limited in both scope and extent, and the quarter-million-dollar estate was crippled the year after the benefactor's death by the suspension of specie payment and the failure of a single company. During most of the twenty-six years that the estate's managers were rebuilding the fund the structure was used by the Foster Home Association, giving children rather than mothers and infants the advantages of the fresh country air.[16]

The environmental advantages that these outlying institutions enjoyed were soon reduced when their presence stimulated improved transportation in their direction. A year after Girard College opened, for example, Ridge Road was paved out to its gates, making the area west of Broad Street less inaccessible.[17] Even before this, the region north of the College, near the Falls of the Schuylkill, had emerged as a suburb with sufficient population and wealth to erect one of the most important Gothic Revival churches in America, the Church of St. James the Less (illus.) at Clearfield and Thirty-second streets. In the early 1840's Robert Ralston, a successful merchant and High Church enthusiast with a country home in the

area, acquired from the Ecclesiological Society, an English group
dedicated to enriching Anglican liturgy and architecture through
study of the medieval past, measured drawings of a thirteenth-
century Anglican country church. Construction began in the fall of
1846, making St. James the Less the first American church built under
the auspices of the ecclesiologists.[18] Most of the development during
this decade, however, remained in the Spring Garden District east of
Broad, especially along Spring Garden Street itself, which in 1847
was flanked by stores west to Tenth Street.[19] The neighborhood's sophistication was illustrated by the diversity of buildings designed by
noted architects. By 1852 the single block of Spring Garden Street
west of Thirteenth included the Commissioners' Hall, probably by
William L. Johnston, St. Mark's Lutheran Church by John McArthur,
Jr., Spring Garden Institute by Hoxie and Button, and the Odd
Fellows' Hall by Samuel Sloan.[20]

On the other side of Broad Street Matthias Baldwin's locomotive
works had been expanding since 1835, helping to make large parts of
western Spring Garden an industrial park.[21] Four noted firms had
plants in the early 1850's along the tracks of the Philadelphia and
Reading Railroad. The Industrial Works of Bement and Dougherty
began in 1851 at Twentieth and Callowhill streets; in 1852 Hoopes and
Townsend established their nuts-and-bolts plant on Buttonwood
Street east of Broad; the next year Bancroft and Sellers (known as
William Sellers and Co. after 1855) built their extensive machine
shops on Pennsylvania Avenue between Sixteenth and Seventeenth
streets; in 1854 at the western end of Callowhill Street near the
Schuylkill, McKeone, Van Haagen and Co. opened the Pennsylvania
Soap-Works which by the 1880's had become the largest soap
manufactory in the state.[22]

Houses of all types were going up in the area during these two
decades. In addition to row houses, whose speculative advantages
were well known, there were cottages, such as the Edgar Allan Poe
House, along Brandywine Street west of Seventh. Built for families of
modest incomes, many of these cottages, including the Poe House,
lost their separate identities when larger houses were attached to
them along the street fronts, making the former cottages back buildings.[23] Terrace houses, a new development promoted particularly by
the noted architect Samuel Sloan, were popular among the middle
class. The main feature of these semidetached houses was the small
grass plot in front of each dwelling. Woodland Terrace 1861–62) in
West Philadelphia possesses the city's best known and best preserved

example of this housing type, but in 1853, Sloan and Stewart had designed twenty-two terrace houses along Green Street between Sixteenth and Eighteenth streets.[24]

Many members of the industrial city's managerial and entrepreneurial class wanted something more opulent than row or terrace houses. It was for them that the North Philadelphia mansions, large stylish dwellings, were built. They appeared in large numbers in the 1850's, particularly on Broad Street, one of Philadelphia's best addresses for more than fifty years, where the mansions of each decade were more grandiose than those of the preceding ones. One of the first of these buildings, Howard Tilden's home at Broad and Brown streets, set the pattern. The *Public Ledger* considered Tilden's house "a magnificent residence . . . of the richest style of Corinthian architecture," further "evidence of the taste of the architects [Hoxie and Button], and the growing desire of our wealthy citizens to surpass one another in the construction of splendid mansions."[25] Neither the architects' taste nor the client's ambition could be called avant-garde, but both illustrated the resolution of the tensions between tradition and change to give a new direction to urban domestic architecture. The house was large—its forty-five-foot front was double that of a colonial town house—but its basic plan of front living quarters and adjoining back service buildings was traditional. Its style was a successful marriage of old forms and new materials, and might be called Georgian Brownstone. Its Corinthian frontispiece, Palladian window, and balustrade were elements long familiar to Philadelphians, but the brownstone walls, enriched window dressings, and skylighted grand staircase were sufficiently Italian to be "modern." Tilden also was operating within a tradition, but one that was increasingly expressed in material terms, that of individual achievement.[26] His business success brought him money, and he used it to buy one of the costliest houses in the city.[27]

Tilden's mansion was but one of many houses that went up in the area during its building boom after 1850. More than 160 houses were constructed between Poplar and Berks streets during the summer of 1863 alone.[28] Some were finished in granite, like the block of ten houses that Sloan and Stewart designed in "the New York style,"[29] but most of the houses were brick, particularly on the lesser streets. Brownstone dwellings, however, were predominant on Broad Street and to a smaller degree on Green Street, where many leading Philadelphians displayed their preference for this somber material for nearly half a century.[30] Brownstone fronts were used on both the 1854 Italian villa of William Gaul, a wealthy brewer, at Broad and Master streets,[31] and the 1864 mansarded "Romanesque" mansion of Matthew Baird, a partner in the Baldwin Locomotive Works, at 814 North Broad Street.[32] The flamboyant French Renaissance mansion (illus.) of traction magnate Peter A. B. Widener at Broad and Girard was finished in 1887 in a highly wrought brownstone, and the stately 1890 mansion (illus.) of financier William H. Kemble at Twenty-second and Green was built of brownstone, as were many of its neighbors.[33]

Banks, stores, market houses, assembly halls, and a rail terminal went up in Spring Garden during the 1850's and 1860's, until by the time of the centennial celebration the former district and the old city were an economic continuum. As a developing fringe of the city, however, the area initially lacked the capital and predictability necessary to assure good architectural design for its commercial and public buildings. Since emphasis was on utility and versatility, most of the commercial structures were architecturally uninteresting, or at best undistinguished. Charles Lewars's 1852 building at the southeast corner of Ninth and Spring Garden streets illustrates this point well. With stores at the street level and assembly rooms and living quarters on the upper floors, the property evidently proved to be a lucrative one for grocer Lewars and his successors, but it was devoid of architectural pretensions; the *Public Ledger* euphemistically spoke of its "new and chaste pattern."[34]

The area's post–Civil War affluence supported better architecture, such as the Northern Saving Fund, Safe Deposit and Trust Company (illus.) at the southwest corner of Sixth and Spring Garden streets. An early and little known work of Furness and Hewitt, it contains many elements that Frank Furness would later develop more fully as his distinctive style matured.[35] Because of its early date, the bank does not assume the massive presence of the architect's later commissions, and for it, unlike nearly all of Furness's later banks, he did not exploit the corner site by putting the entrance on a bevel. Neither did the architects clutter the doorway of the Northern Saving Fund with the columns, corbels, and overhangs that gave a beleaguered appearance to so many of Furness's banks. Instead, the entrance is centered on the building's long Spring Garden Street axis, and the simple round-arch entrance appears inviting rather than intimidating. Other aspects of the design, however, suggest Furness's future histrionics. The projecting central salient, which is corbeled outward above the second story and then slides backward into the crowning tympanum, anticipates the exaggerated examples of Furness's Provident Life and Trust Company (1876–79) and the Kensington National Bank (1877); and the use of contrasting materials, here dressed and pecked granite, would later be handled more vividly. Flat triglyph corbels were also used by Furness's contemporaries,[36] but in the customary function of supporting cornices. Furness, on the other hand, used them more daringly to support projecting upper parts of facades, a practice that is suggested here. Other elements, such as the pinched-top windows and exaggerated keystones and imposts, would appear in very similar forms on Furness's later buildings. The bank has been a local landmark for more than a century, and although it has been enlarged on three different occasions, its additions have been stylistically synonymous with the original structure.[37]

North Broad Street reached its peak in the decade before World War I. Although the extravagantly rich, like Widener and his close associate William Elkins, moved out of the city to palatial estates in suburban counties, splendid mansions were still erected along the great thoroughfare and its adjacent streets into the twentieth century. Brownstone finally lost favor and was replaced by marble and limestone as architects turned to more restrained historical styles: the Second Renaissance Revival, Beaux-Arts Classicism, and the Neoclassical Revival. Alfred E. Burk's house at Broad and Jefferson may be too austere to be the most representative example of the Second Renaissance Revival, but it is distinguished enough to be included in that long list of buildings wishfully attributed to the great Stanford White of the New York firm McKim, Mead and White. In actual fact, however, the house was designed by Simon and Bassett, a fine Philadelphia firm, a year after White's death.[38]

Burk, a millionaire leather manufacturer, was continuing the sixty-year-old drift of self-made men toward Broad Street that stimulated a social rivalry between North Philadelphia and Rittenhouse Square, the more traditionally fashionable neighborhood. Oscar Hammerstein, probably the foremost impresario of his day, came to Philadelphia in 1907 and with great bombast exploited the competition by announcing his plans to build an opera house on North Broad Street to compete with the city's only other center of opera, the venerable Academy of Music.[39] After commissioning the nation's leading theater architect, William H. McElfatrick of New York, to design the massive structure at Broad and Poplar, Hammerstein made good his promise that the Philadelphia Opera House would be finished in a year's time, and brazenly planned its grand opening to coincide with the beginning of the Academy's opera season.[40] Opening night was a social and cultural extravaganza, and the newspapers covered the event in the spirit of a political election or an athletic contest. The result was a draw, for both houses were filled. Perhaps Hammerstein won a moral victory, since, contrary to predictions, "social Philadelphia" did go uptown.[41] Pride had its price, however, and five years later the city's largest stage was dark. Faced with a hopeless $400,000 mortgage, Hammerstein sold out to E. T. Stotesbury, who brought New York's Metropolitan Opera Company to the Opera House for three years, and ever since the building has been known as "the Met." Opened for special events for a few years, the Renaissance Revival theater has been a tomb for memories since 1920. Except for a couple of theatrical spectaculars in the mid-1920's and in 1931, the once ornate structure is used primarily for evangelical purposes,[42] its splendor fading as its cornices crumble and its paint peels.

The Opera House's decline anticipated, if it did not parallel, that of its neighborhood. The beginning of the Broad Street Subway in 1924 reinforced the northward trend of the commercial city. Two of the more notable office buildings erected in the southern limits of the area were the Elverson Building for the *Philadelphia Inquirer* and the Terminal Commerce Building for the Reading Company. These two quite different structures face each other across Broad Street above Callowhill. The *Inquirer's* building, a 1924 application of the Italian Renaissance style to a tall commercial structure, is faced with white limestone and terra cotta and ascends in a series of setbacks to a sleek central tower crowned with a lantern and dome.[43] Although it was designed five years later "to harmonize architecturally with the Elverson Building,"[44] the Reading Company's massive edifice appears squat on its site, although its buff brick walls with their colorful terra cotta ornament rise twelve stories above the street.

The subway also brought changes to the educational institutions that lay along its path. Temple University had been growing slowly since its inception in 1884 as a night school, housed in the basement of Russell Conwell's Grace Baptist Church at Broad and Berks streets. Empowered to grant degrees in 1891, Temple immediately began plans for its own building. This was completed in 1894 from the designs of Thomas Lonsdale.[45] More accessible to its commuting student population, Temple enjoyed a growth during the 1920's and early 1930's exceeded only by that of the 1960's, when it burgeoned

into one of the largest and finest universities in the state. [46] Whereas
the presence of the subway encouraged Temple to expand on its old
site, it induced another Broad Street school, LaSalle College, to move
farther north. Limited to an old mansion at Broad and Stiles streets,
LaSalle saw in the cheap rapid-transit system an opportunity to build
a new campus on the undeveloped land surrounding Twentieth
Street and Olney Avenue. The first building was completed in 1929,[47]
and although the campus is outside our arbitrary boundaries for
North Philadelphia, it reflected events within that area and has
played an important role in the cultural life of the Philadelphia neighborhoods of Logan and Fern Rock.

Spurred on by the inexorable urban expansion, by citizens'
hunger for a suburban environment, and by the building of the
subway, housing construction boomed in the northern reaches of the
city. The region north of Columbia Avenue had been developing
progressively during the 1880's, and by 1890 commodious row
houses were being built as far north as Erie Avenue.[48] The area beyond that, near York Road and Rockland Street where J. B. Lippincott, Jr., built his large stone house in 1891, remained a pleasant
suburb, [49] while commuter railroads helped to transform large parts
of Oak Lane into a district suitable for the development of single
homes and twin houses.[50] Yet great open spaces ranged west of
Broad Street at the turn of the century. When Shibe Park (illus.) was
opened at Twenty-first Street and Lehigh Avenue in the spring of
1909 it stood virtually alone, and some critics feared its distant site
might exceed the loyalty of the city's baseball fans. The fears were un-
founded. Rising attendance forced an extension of the stadium within
two years, and before long it was engulfed by open-porched row
houses.[51] By the time the Great Depression put an end to this boom,
North Philadelphia's frontier had been closed. Houses stood in close
order drill, with the domes and steeples of the area's churches punc-
tuating the low skyline.[52] Aglow with gas logs in their fireplaces and
with electric lights to brighten the dark winter nights, with colorful
flowers decorating their grass plots in the bright summer sun, these
modest dwellings were—and remain—material expressions of the
American Dream.

Catalog

Burk, Alfred E., House
(PA-1722), 1500 N. Broad St., at N.W. corner Jefferson St. Brick faced with coursed Green River limestone ashlar, 42'–10" (three-bay front) × 71'–0" with three-story lateral addition, three stories on raised basement, low hipped tile roof with terra cotta cornice, smooth rusticated ground story, large Corinthian frontispiece with balustrade above, two semicircular bays on south facade; central-hall plan with hall open to roof, notable interior details; detached rear limestone conservatory.

Fine example of Late Victorian Italian Renaissance style. Built 1907–9; Simon and Bassett, architects; John N. Gill & Co., contractors. Rear conservatory altered, enlarged for offices 1949–50; Louis A. Manfredi, architect. Three-story steel-frame northern addition with limestone ashlar curtain walls built 1953–54; Louis A. Manfredi, architect. Alfred E. Burk, leather manufacturer, lived here 1909–21; headquarters of Upholsterers' International Union 1945–70; Temple University School of Social Administration since 1971. Certified, PHC 1971. 2 ext. photos (1973);* 2 data pages (including 1908–53 building permits).*

Cast-Iron Sidewalk. Illustrated
(PA-1723), 1907 N. Seventh St. Cast iron, approx. 11' × 53', constructed of hexagonal plates of approx. 1'-diameter. Rare surviving example of mid-19th-century Ferromania. Built c. 1861; Charles Carnell (machinist and iron founder), probable designer and builder. 8 photos (1972);* HABSI form (1970).*

Cherry Hill Penitentiary. See Eastern State Penitentiary

Church of the Gesu (Roman Catholic). Illustrated
(PA-1724), S.E. corner Eighteenth and Thompson sts. (originally at
N.E. corner Eighteenth and Stiles sts., since closed). Brick with
granite trim, approx. 122' (five-bay front) × 252', one and five stories
with two-story facade, gable roof, square towers with low hipped
roofs flank cyma-curved center gable, coupled Doric colonnade at
ground story, coupled Ionic colonnade at second story, curved pedimented doors; three-aisle plan with four chapels along each side,
balconies, apsidal chancel with two side altars, large flattened barrel
vault, notable decorations.

Notable example of a High Victorian Baroque church with
the country's widest unobstructed nave when built. Part of an ecclesiastical complex that included parish school, college (St. Joseph's),
and nearby hospital (St. Joseph's). Masterpiece of an important local
church architect. Built 1879–88; Edwin F. Durang, architect. Towers
finished after 1895. Interior decorated 1918; Br. Francis C. Schroen,
S.J., decorator. Lower half of interior redecorated, renovated 1952–
56; Brs. Frederick E. Barth, S.J., and George M. Bambrick, S.J.,
decorators. Five-story Jesuit Community House built at rear corner
1911; Fr. Charles Lyons, S.J., designer; Thomas Reilly, builder. 3 ext.
photos (1973).*

Church of St. James the Less (Protestant Episcopal). Illustrated
(PA-1725), 3200 W. Clearfield St., at S.W. corner Thirty-second St.
and intersection of Hunting Park Ave. Random granite ashlar, approx. 36' × 62' with one-story rear vestry, one story, gable roof,
pointed-arch windows, open belfry at west end, three-aisle plan, rectangular chancel.

First and influential example of the Ecclesiological Society's (formerly Cambridge Camden Society) providing plans of a Gothic
church to an American congregation, leading to a fine re-creation of a
13th-century English country parish church and churchyard. Built
1846–50; G. G. Place (England), architect; John E. Carver,
superintendent of construction; three windows by Henry Gerente
(Paris). Chancel remodeled 1878; Charles M. Burns, architect. Open
belfry added c. 1885. Rodman Wanamaker Bell Tower built at edge of
churchyard 1908; John T. Windrim, architect. Certified, PHC 1965;
Pennsylvania Register 1974; NR. 3 ext. photos (1973, including 1 of
doorway sculpture).*

Community College of Philadelphia, Campus II. See United States Mint

Connie Mack Stadium. See Shibe Park

Diamond Street Area Study
(PA-1726), 1601–43 W. Diamond St., N. side Diamond St. between Sixteenth and Seventeenth sts. Twenty-two brick row houses with rock-faced brownstone trim, each approx. 18' (two-bay front) × 26' with three-story rear ell (except corner buildings, 26' × 72' without ells), three stories on raised basements, flat roofs, rock-faced brownstone basements and stringcourses, side-hall plans.

Middle-class row house development designed by a leading late-19th-century architectural firm. Built 1887; Furness, Evans and Co., architects; W. D. Huston, builder; Page Brothers, developers. No. 1601 enlarged, raised one story later; Nos. 1629, 1631, 1637 altered in rear later. Certified, PHC 1974. 2 ext. photos (1973).*

Eagle Hotel
(PA-1727), 601–7 W. Girard Ave., at N.W. corner Sixth St. Brick, approx. 84' (twelve-bay front) × 110', four stories, flat roof, interior courtyard for coaches. Built c. 1858; demolished c. 1958. 4 ext. photos (1957).*

Eakins, Thomas, House
(PA-1728), 1729 Mt. Vernon St. Museum. Brick, approx. 20' (three-bay front) × 75', four stories, mansard roof. Built c. 1854; mansard added 1902; restored 1969–70. Thomas Eakins, famous portraitist and predecessor of American modern painting, lived here from the age of thirteen until his death in 1916. Now an Eakins museum and neighborhood art center administered by the Philadelphia Museum of Art. Certified, PHC 1964; Pennsylvania Register 1970; designated National Historic Landmark 1967. 2 ext. photos (1967);* 2 data pages (1964).*

Eastern State Penitentiary (also known as Cherry Hill Penitentiary). Illustrated
(PA-1729), block bounded by Fairmount Ave., Corinthian Ave., Brown, Twenty-second sts. Coursed granite ashlar wall, 35' height, encloses 11.7 acres and eleven granite ashlar cell-block buildings (originally seven), approx. 80' × 250' and approx. 65' × 335', one and two stories, gable roofs, radial plan around central rotunda; short battlemented towers flank pointed-arch entrance topped with hexagonal

battlemented tower, lancet windows in front wall, battlemented hexagonal towers at front corners.

One of the nation's oldest prisons designed by a foremost prison architect. Its plan is a masterful architectural expression of the Pennsylvania System, a penal philosophy emphasizing solitary confinement, and its design was an early expression of the Gothic Revival in the United States. Built 1823–36; John Haviland, architect; Jacob Souder, master mason. Three radial wings added 1877, one wing 1894, one wing 1911. Acquired by City of Philadelphia 1970. Certified, PHC 1958; Pennsylvania Register 1970; designated National Historic Landmark 1965. 6 ext. photos (1967),* 5 int. photos (1967),* 1 photocopy of isometric view (n.d.).*

Gaul-Forrest House

(PA-1730), 1346 N. Broad St., at S.W. corner Master St. Brick with brownstone ashlar front and painted galvanized iron side trim, approx. 53' (five-bay front) × 50'with one-story brownstone lateral addition and three-story brick L-plan rear addition (originally two-story rear ell with veranda), three stories on raised basement, flat roof, projecting lintels on consoles, projecting center second-story window, center-hall plan.

Fine example of Italianate town house and its later remodeling. Built 1853–54; probably by Stephen D. Button, architect. Veranda enclosed, interior rear ell altered 1860; rear ell raised to three stories, bay window added, one-story lateral addition built 1863; ground story altered to restaurant 1876, used for church services 1877–79. Interior extensively altered, one-story addition enlarged, rear ell remodeled, rear addition built 1880; James H. Windrim, architect; B. Ketcham and Son, contractors.

Built for William Gaul, prominent brewer. Edwin Forrest, America's first outstanding actor, noted for his Shakespearean roles, lived here 1855–72. Site of Philadelphia's School of Design for Women (now Moore College of Art), the country's pioneer school of industrial art for women, 1880–1959. Now Heritage House, neighborhood art and music center. Pennsylvania Register 1971; NR. 1 ext. photo (1973);* 15 data pages (1973, including 1854–76 insurance surveys).*

Girard College, Founder's Hall. Illustrated

(PA-1731), Girard Ave. and Corinthian St. Museum. Coursed marble ashlar, approx. 111' × 169', three stories, gable roof, Corinthian peristyle colonnade with octastyle porticoes, four square vaulted rooms to a floor, stairways in corners. Low segmental groined vaults in lower rooms and penditive domes under roof provide structural soundness and maximum space.

Considered one of the most impressive and monumental Greek Revival buildings in the United States, its innovative internal structure makes it a remarkable blend of structural engineering and architectural design. Built 1833–47; Thomas U. Walter, architect; Findley Highlands, marble mason. Central feature of a school for fatherless boys endowed 1831 by Stephen Girard, America's first multimillionaire philanthropist. Houses sarcophagus of Stephen Girard and the Stephen Girard Collection, probably the best-documented group of furnishings owned by an American of the late-18th and early-19th centuries. Certified, PHC 1956; Pennsylvania Register 1970; designated National Historic Landmark 1969. 7 ext. photos (1973, including 1 color photo),* 1 photocopy of architect's perspective drawing (1835),* 1 photocopy of unexecuted design (plan) by Alexander Jackson Davis (n.d.);* HABSI form (1970).*

Green Street Area Study. Illustrated
(PA-1732), 2201–31 Green St., on N. side Green St. between Twenty-second and Twenty-third sts. Ten large brownstone and brick houses, some faced with granite and brownstone fronts, three, four, and five stories, flat and mansard roofs. The dwellings are architecturally unrelated to one another but as a group they serve as an example of a wealthy late-19th-century neighborhood. Built at various times during the last third of the 19th century. Nos. 2219, 2221 demolished 1963. No. 2201–5 (Kemble-Bergdol House) and No. 2213 (Burnham House) certified, PHC 1967. 15 ext. photos (1963, 1973);* 3 data pages (including 1866 insurance survey).*

Met. See Philadelphia Opera House

Northern Saving Fund, Safe Deposit and Trust Company. Illustrated
(PA-1733), 600 Spring Garden St., at S.W. corner Sixth St. Brick with iron joists, faced with coursed Richmond granite ashlar with rusticated and pecked granite trim, approx. 135' (twelve-bay front) × 76' (originally 97' × 22'), one and two stories, flat roofs, central pedimented salient corbeled out over round-headed entrance, ground-story pinched-top windows, second-story round-headed windows, pecked and rusticated quoins and pinnacles, dark rough-hewn water table, banking room in one story on east end.

Early bank design by two leading 19th-century architects who later developed some of its elements into notable picturesque buildings. Built 1872; Furness and Hewitt, architects. Southern extension (516 N. Sixth St.) built c. 1888; probably by Furness, Evans & Co., architects. Western addition built 1903; George T. Pearson, architect. Another southern extension (514 N. Sixth St.) built 1913–14; Stearns and Castor, architects. Interior altered, entrance lowered to street level, marquee added 1960. Name changed to Northern Trust Co. 1902; acquired by Provident National Bank 1948. 1 ext. photo (1974).*

Philadelphia Opera House (later known as the Met)

(PA-1734), 1400–1418 Poplar St., at S.W. corner Broad St. Steel frame with white brick curtain walls and marble and terra cotta trim, approx. 240' (fourteen-bay front) × 160', four and five stories, flat roof, pedimented central pavilion with open balustraded arcades, simulated smooth rustication on lower stories, originally domes at either end, originally highly lavish interior.

A formerly elegant opera house by a leading theater architect. Built 1908; William H. McElfatrick (Brooklyn), architect; John Morrow and Harry Weichmann, superintendents of construction; Phoenix Iron Co., steel contractors. Interior destroyed by fire, altered 1948; frieze, cornice, pediments removed later. Built for Oscar Hammerstein, an internationally famous early-20th-century impresario. Largest stage in city when built. Opera performances ended 1920; used for various entertainment purposes until 1955 when it became Philadelphia Evangelistic Center. Certified, PHC 1970; Pennsylvania Register 1972. 3 ext. photos (1974),* 1 photocopy of old ext. photo (1925).*

Poe, Edgar Allan, House

(PA-1735), rear of 530 N. Seventh St., at N.W. corner Brandywine St. Museum. Brick, approx. 32' (four-bay front) × 15', three stories, shed roof, center entry with enclosed stairs. Built c. 1841; probably by John Evans, Jr., house carpenter. Partially restored by Richard Gimbel Foundation 1964–66. Back building built before front section (1849). The famous author and editor Edgar Allan Poe lived here 1842–44. Certified, PHC 1962; Pennsylvania Register 1970; designated National Historic Landmark 1962. 2 ext. photos (1967),* 4 int. photos (1967);* 4 data pages (1962).*

Preston Retreat. Illustrated

(PA-1736), 500–518 N. Twentieth St., at N.W. corner Hamilton St. Brick with marble ashlar front and smooth rusticated marble basement, approx. 106' (nine-bay front) × 88', three stories on raised basement (originally two-story facade) with Tuscan tetrastyle portico, hipped roof with large octagonal cupola, T-plan with central hall and transverse corridor.

Important Greek Revival work by Thomas U. Walter, noted architect of classical styles. Built 1837–40; Thomas U. Walter, architect. Used by Foster Home Association 1846–65; operated as maternity hospital for the indigent 1866–1960 as originally intended. Third story raised, pediment added to portico, cupola restored, interior extensively remodeled 1908–9; Edgar V. Seeler, architect. Demolished 1963; portico elements preserved for later construction. Certified, PHC 1956.

3 ext. photos (1963, including 1 of iron gate),* 24 photocopies of architectural competition drawings (1837, including 8 by Thomas U. Walter, 3 by Isaac Holden, 3 by John D. Jones, 5 by William Kelly, 2 by Thomas S. Stewart, 5 by unidentified architect);* 16 data pages (1963, including 1840–1909 insurance surveys).*

Ridge Avenue Farmers' Market Company. Illustrated
(PA-1737), 1810 Ridge Ave., at S.W. corner Ginnodo St. Brick with sandstone trim and cast-iron piers, approx. 93' (five-bay front) × 262', one story, jerkin-head roof with shed roofs along sides, segmental pointed-arch windows. A rare surviving example of a building type that increased in number as open market sheds declined after 1860. Built 1875; Davis E. Supplee, architect. Renovated 1968. Certified, PHC 1968. 4 ext. photos (1973);* HABSI form (1970).*

Shibe Park (later known as Connie Mack Stadium). Illustrated
(PA-1738), 2701 N. Twenty-first St., composing block bounded by Lehigh Ave., Twenty-first, Somerset, Twentieth sts. Reinforced-concrete walls faced with brick and terra cotta trim, wooden roofs on steel trusses, approx. 480' on Lehigh ave. × 525' on Twenty-first St., three stories (originally two-and-a-half) with domed four-story circular corner tower, flat roofs (originally mansard), ground-story terra cotta bands and bas-reliefs and second-story Ionic arcade on grandstand, originally L-plan of corner grandstand and flanking left and right field stands.

First reinforced-concrete baseball park in the country. With a 25,000 seating capacity it was considered a baseball showplace when built and was the oldest park in the major leagues when it was vacated in 1970. Built 1908–9; William Steele and Sons Co., architects and builders. Left and right field stands covered and bleachers built along Somerset St. (north end) 1910, grandstand covered 1912–13; William Steele and Sons Co., architects and builders. Second deck added to left and right field stands 1925; grandstand roof raised, press box and 3500 seats installed 1929. Structure reinforced 1930; Percival M. Sax, contractor. East wall (right field) raised 1935. Lights installed for night baseball 1939; Westinghouse Electric and Manufacturing Co., architects and engineers. Left and right field stands rebuilt 1949; Erny and Nolan, contractor. Electrical scoreboard erected atop right field wall (east end) 1956; infield box seats added 1960. Severely damaged by fire 1971; demolished 1976. Renamed Connie Mack Stadium 1953 in memory of the man who helped to organize Philadelphia Athletics 1901 and managed the team 1901–50, a major league record. Home of American League's Athletics 1909–54, of National League's Phillies 1938–70.

9 ext. photos (1973),* 1 photocopy of old ext. photo (c. 1909),* 1 photocopy of int. photo (1963).* 1 photocopy of aerial photo (1970).*

Spring Garden Institute
(PA-1739), 523–25 N. Broad St., at N.E. corner Spring Garden St. Brick faced with mastic on upper stories (originally rusticated ashlar) and iron sheets simulating rusticated ashlar on ground story, approx. 100' (eleven-bay front) × 57', three stories with domed semicircular bay (not original) above entrance, hipped roof, plain piers rise two stories with arcade at top story, top floor suspended from roof rafters.

Example of institutional application of Italianate mode. Built 1851–52; Hoxie and Button, architects. Northern section, at S.E. corner Broad and Brandywine sts., built as First New Jerusalem Society of Philadelphia 1854. Ground-story stores altered to library 1879; domed bay added 1891; upper stories covered with mastic c. 1891. Four-story steel-skeleton east section with buff brick and terra cotta curtain walls built 1898; Thomas P. Lonsdale, architect. North section occupied by Apprentices' Library, oldest free circulating library in America, 1897–1946; part of Spring Garden Institute 1946–1969. Buildings vacated 1969, demolished 1972. 3 ext. photos (1971);* HABSI form (1970).*

Tanner, Henry Ossawa, House
(PA-1740), 2908 W. Diamond St. Brick with marble trim, approx. 16' (two-bay front) × 28' with two-story rear ell, three stories, flat roof, semi-hexagonal bay at second and third stories covered with alu-

minum siding. Built c. 1871; Daniel H. Bry, builder. Exterior and interior altered later. Henry Ossawa Tanner, America's leading black painter of the late 19th and early 20th centuries, lived here 1872–88. Property held by Tanner family and descendants until 1950. Pennsylvania Register 1976; designated National Historic Landmark 1976. 1 ext. photo (1976).*

United States Mint (now Community College of Philadelphia, Campus II)
(PA-1741), 1600–1644 Spring Garden St., composing block bounded by Spring Garden, Sixteenth, Buttonwood, Seventeenth sts. Coursed gray granite ashlar with smooth rusticated ground story, approx. 318' (nineteen-bay front) × 200', two and three stories, flat roof with balustrade, central flat Roman Ionic tetrastyle portico in antis, projecting second-story lintels with corbels, gold tile and white marble entrance lobby originally with Tiffany Favrille glass mosaics by William B. Van Ingen, arcaded white marble stairhall leading to central red marble rotunda, originally first floor coining operations with visitors' mezzanine, rectangular plan with ivory-white brick inner courtyards.

Fine example of turn-of-century Beaux-Arts government architecture. Built 1898–1901; James Knox Taylor, architect; Edward A. Crane, superintendent of construction; Allen B. Rorke, foundation contractor; Charles McCaul, terrace-level contractor. Altered 1934, 1941. Interior of front half altered 1972–73; Geddes, Brecher, Qualls, Cunningham, architects. Vacated 1969; acquired by Community College of Philadelphia 1971, opened for classes 1973. Tiffany glass mosaics moved to new U.S. Mint, Independence Mall 1971. The third mint in Philadelphia and finest in the country when built. Certified, PHC 1974. 1 ext. photo (1974),* 1 photocopy of architectural rendering (c. 1900).*

Widener, Peter A. B., House. Illustrated
(PA-1742), 1200 N. Broad St., at N.W. corner Girard Ave. Brownstone ashlar and brick, approx. 53' (six-bay front) × 144', four-and-a-half stories on raised basement, hipped roof with four Flemish cross-gables, ogee conical roofs on corner bays, curvilinear walls, curved double-entry stairway, rusticated basement and first stories, carved stone decoration, central-hall plan, notable interior.

Notable example of picturesque eclectic design. Built 1887; Willis G. Hale, architect; George Herzog, decorator. Two-story brick and iron conservatory built at rear 1892; interior altered 1899 when house became H. Josephine Widener Memorial Library 1900–1946. Widener was a developer of street railway syndicates in Philadelphia and Chicago and later a philanthropist. Since 1970 the quarters of The Council of Black Clergy's Institute for Black Ministries at the Conwell School of Theology. Certified, PHC 1963; Pennsylvania Register 1970. 2 ext. photos (1973);* 4 data pages (1964).*

Cast-Iron Sidewalk. *Jack E. Boucher photo, 1972.*

Church of St. James the Less. *Jack E. Boucher photo, 1972.*

Church of the Gesu. *Jack E. Boucher photo, 1973.*

Eastern State Penitentiary. *Cortlandt V. D. Hubbard photo, 1967.*

Girard College, Founder's Hall. *Jack E. Boucher photo, 1973.*

Green Street Area Study (Kemble-Bergdol House). *Jack E. Boucher photo, 1973.*

Northern Saving Fund, Safe Deposit and Trust Company. *Jack E. Boucher photo, 1973.*

FRONT ELEVATION

Preston Retreat. *Drawing by Thomas U. Walter, 1837.*

Ridge Avenue Farmers' Market Company. *Jack E. Boucher photo, 1973.*

Shibe Park. *Jack E. Boucher photo, 1973.*

Peter A. B. Widener House. *Jack E. Boucher photo, 1973.*

Delaware River Corridor

Northern Liberties, Kensington, Richmond, Frankford, Northeast Philadelphia

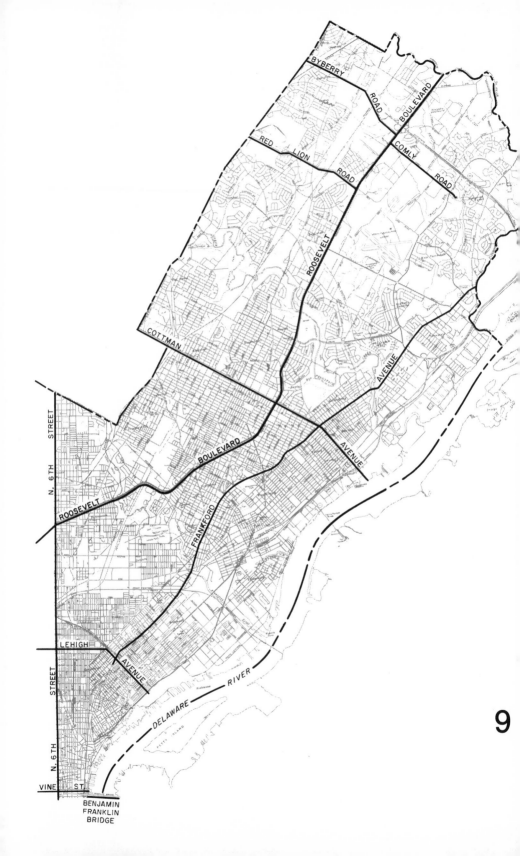

BYBERRY ROAD

BOULEVARD

RED LION ROAD

COMLY ROAD

ROOSEVELT

COTTMAN AVENUE

AVENUE

BOULEVARD

AVENUE

ROOSEVELT

FRANKFORD

N. 6TH STREET

LEHIGH AVENUE

N. 6TH STREET

DELAWARE RIVER

VINE ST.

BENJAMIN FRANKLIN BRIDGE

9

Introduction

SIX LARGE TOWNSHIPS existed north of the City of Philadelphia in William Penn's time: Northern Liberties, Oxford, Bristol, Lower Dublin, Moreland, and Byberry. Rural during the colonial period, they were developed during the nineteenth century as activity radiated outward from the city's commercial center.[1] By 1854, when the outlying hegemonies of the county were consolidated with the city, the area north of Vine Street between the two rivers (excluding Germantown and Roxborough) included seventeen separate governmental entities. Nine of them had been carved from the original Northern Liberties Township, which formed the city's northern border, while the two northernmost and hence most remote and most rural of the townships, Byberry and Moreland, had never been subdivided. Although few people today would recognize the names of all the original subdivisions, some of the boroughs and districts retain their old identities and still inspire neighborhood loyalties. Frankford, organized in 1800, is the oldest of these, but Northern Liberties District (1803), Kensington (1820), Richmond (1847), and Bridesburg (1848) are equally well-known and traditional neighborhoods.[2] The Delaware River with its tributaries is the common tie to all of them and significantly influenced the development of the area, hence its designation here as the Delaware River Corridor. Arbitrary boundaries have been assigned to the corridor: the river on the east, Vine Street on the south, Sixth Street on the west, and the city limits on the north.

Open land combined with the Delaware River to make the area attractive for farms and country houses during the colonial period. One could enjoy the virtues of country living without being uncomfortably remote from the center of the colony's political and economic activity, easily accessible by means of waterways, which offered vastly quicker and more comfortable transport than overland routes. William Penn had his country estate Pennsbury, for example, built outside Philadelphia above the town of Bristol. It was a very long horseback ride from the city's commercial center, but Penn traveled in a private river barge and probably made the trip faster and certainly more comfortably than can modern commuters traveling on expressways in high-powered autos.

Two of the finest country houses along the corridor were Chalkley Hall and Port Royal, standing on either side of Frankford Creek. Thomas Chalkley, the sea-going Quaker missionary-merchant, built Chalkley Hall in 1723 on the south side of the creek "in order to be more retired and for health's sake."[3] He appears to have been too restless to enjoy either, however, and died eighteen years later on the island of Tortola during one of his self-appointed business-missionary expeditions. Chalkley Hall assumed the appearance of an eighteenth-century double-pile country seat at the time of the American Revolution, when Chalkley's son-in-law built the main section of cream-colored sandstone, reportedly brought from England as ballast.[4] Port Royal (1761) was named after the Bermuda birthplace of Edward Stiles, its first owner, and was perhaps even more sumptuous than Chalkley Hall.[5] Its classical proportions, rich interiors (some of which are now installed at Winterthur Museum), and the knowledgeable handling of Georgian details made it one of Philadelphia's finest country houses. The advantages of such rural retreats, which Chalkley anticipated in the early eighteenth century, were still evident a century later and poetically proclaimed by Whittier:

> How bland and sweet the greetings of this breeze,
> To him who flees
> From crowded street and red wall's weary gleam;
> Till far behind like a hideous dream
> The close dark city lies![6]

The rural expanse and many streams were attractive to farmers and millers as well as to country gentlemen, of course, and farmhouses once sprinkled the area. Two of the more historic ones, Benjamin Rush's birthplace and Stephen Decatur's farmhouse, disappeared as late as 1969, the former by a perverse combination of overzealous bulldozing and administrative error and the latter for expansion of the North Philadelphia Airport. Similar structures can still be found in some remote parts of the area, particularly along Red Lion Road, but it goes without saying that Philadelphia's farmhouses are becoming increasingly rare and her colonial mills are apparently nonexistent.

Churches to serve the spiritual needs of the inhabitants sprang up almost at once. Quakers began meeting in the Frankford area as early as 1682, but they did not have a meeting house for another two years; and the present Flemish-bond meeting house at Waln and Unity streets was not built until 1775.[7] Baptists organized in 1685 farther to the north at the Pennepack Baptist Church, the oldest Baptist congregation in Pennsylvania and the eighth oldest in the country.[8] The oldest church building in the area, however, is Trinity Church, Oxford (illus.), which was built for Anglicans in 1711.[9] A number of nineteenth-century additions have not disturbed the original part, which is distinguished by its Flemish bonding with vitrified dark headers relieved by bold lozenge-shaped diapering. Architecturally comparable to Gloria Dei, the former Swedish mission church in Southwark, Trinity Church and its peaceful churchyard stand in bold relief to the neighborhood's twentieth-century dwellings.

Rapid changes followed in the wake of the Revolution. The Penn family was replaced by a new, indigenous government, which saw in the removal of Penns' control over the lands north of the city possibilities for both public and private gain. In 1781 the Commonwealth's need for quick revenue led to the sale of 1,400 lots in Northern Liberties.[10] Those with money and able to distinguish between chance and opportunity invested in Northern Liberties real estate. In the next twenty years rows of dwellings went up throughout the area directly north of Vine Street. Built sometimes in pairs and sometimes in rows, they were generally small, squarish houses, often only two-and-a-half stories high with one room to a floor. Understandably called bandboxes, they were usually crowded onto small side streets or kept out of sight in courts and in those small alleys that cut into the large blocks from through streets.[11]

Phillips Row (later New Market Street), six two-and-a-half-story dwellings built between 1808 and 1811 by its merchant-developer William Phillips, is a good example of such a row. All but one of the houses measured 17 feet by 16 feet; the exception had a twenty-foot front. Except for one carpenter, all the original purchasers were

porters.[12] Similar bandboxes appearing on Olive Street were
not only small, 16 feet by 16 feet with a kitchen ell, but also cheaply
constructed, without basements or brick party walls between the
houses at the garret level. They were "finished in the plainest
manner"; simple winding stairs led to the second story, which was
floored with white pine rather than the more durable heart pine.
Olive Street was initially called Equality Court, but a rear alley of
rental dwellings by any other name remains a rear alley. Patriotic
associations notwithstanding, its developer fell victim to a sheriff's
sale, and in 1811 the alley became known as York Court.[13]

These dwellings may appear quaint and cozy to our eyes, but one
former inhabitant of a New Market Street bandbox has confessed that
"it is no use pretending there was any glamour attached to life in the
bandbox. One would not be truthful to assume that he has a nostalgia
for 'those good old days.' . . . The chief trouble with the bandbox
was that you had no privacy. You not only knew what your neighbor
had for dinner, but you heard all his conversation and knew all his
business," and he yours, it might be added.[14] Cramped as these
houses may have been, they were generally soundly constructed, and
laissez-faire attitudes toward housing for the poor allowed them to
survive into the last half of the twentieth century. In the end they fell
neither to decay nor to enlightened housing policies, but to the
automobile, or more precisely to the Delaware Expressway.

Post-Revolutionary real estate development of the Delaware
River Corridor did not reach beyond Cohocksink Creek, leaving rural
the extensive region to the north. Even Frankford Borough, formed in
1800, had large open spaces and was more an agrarian village than a
commercial town for several decades. Residents, for example, did not
begin to pave and curb their sidewalks until 1849.[15] The Religious
Society of Friends considered this natural quietness ideal for their
proposed psychiatric hospital, and in 1813 construction began on the
Asylum for the Relief of Persons Deprived of the Use of Their Reason,
about two miles west of Frankford.[16] The plan of the plastered stone
asylum, opened in 1817, was very similar to that of the Pennsylvania
Hospital, whose elegant center section was finished in 1805
completing a project begun fifty years earlier.[17] Since Friends
Asylum, as it was named in 1888, was the work of a Quaker carpenter
(Israel Maule), it does not possess the architectural finesse of the
Pennsylvania Hospital. It was severely plain and only a small classical
porch marked its entrance. Many later additions, including an 1871
mansard that replaced the original gable roofs, disguise its initial

simplicity.[18] Yet as the country's first private psychiatric hospital, the building's plan and the staff's practices served as precedents for the development of the famous Kirkbride System, and its architectural expression is to be found in the Pennsylvania Hospital for Mental and Nervous Diseases (1836–40) and the Pennsylvania Hospital for the Insane (1856–59). Friends Hospital, as it is known today, still stands at Roosevelt Boulevard and Adams Avenue.[19] Now a complex of buildings, its spacious landscaped grounds serve as a pleasant greensward in the residential neighborhood.

Open land was still abundant on the western fringes of Kensington as late as the 1850's, when the Protestant Episcopal Church of Philadelphia accepted the gift of the Leamy estate at the southeast corner of Front Street and Lehigh Avenue as the site of their proposed hospital.[20] Many argued that this location was too peripheral to be of service to the city, an argument not lightly to be dismissed since Episcopal was only the third general hospital to be built in a city of over 400,000. In the end, however, the force of environmentalism and faith in urban expansion prevailed, and in May, 1860, the cornerstone for the new hospital (illus.) was laid.[21] Its architect, Samuel Sloan, already nationally known for his hospitals for the insane, was to add to his reputation with this work. Although the hospital's construction was interrupted by the Civil War and the final part was not completed until 1875, the nation knew of the hospital before then as a result of a profusely illustrated fund-raising volume published in 1869 and a long article the following spring in Sloan's *Architectural Review.*[22] Because of Sloan's attention to detail and his full execution of the 1778 recommendations of the French Academy, Episcopal Hospital was considered the finest general hospital in the country during the 1860's and 1870's. Its "pavilion" system, which articulated in the building's form any of its interior functions and was borrowed from Sloan's insane asylums, particularly influenced later hospital design. Although its Trenton sandstone walls and irregular silhouette seem lugubrious today in contrast to the sleek walls and clean lines of the complex's more recent buildings, they projected the sweet-melancholy images that Victorians associated with hospitals.[23]

The same waterways and open land that made the Delaware River Corridor so convenient for country gentlemen also offered militarily strategic sites. In 1816 the United States Army purchased more than twenty acres at the mouth of the navigable Frankford Creek as the site for what has since been known as the Frankford

Arsenal. The choice of the site was not sudden and arbitrary. Even before the four years of internal deliberations leading to the selection of the site, Army ordnance and muskets had been stored there in a stone barn, whose walls were incorporated into a building (now Quarters 2 and 3) erected prior to 1812 when the area served as the Cantonment on Frankford Creek.[24] In the seven years following the arsenal's founding, seven major structures were built in an institutional classical style, and thereafter the arsenal's expansion has paralleled periods of military activity. Two large tracts of land were added in 1849 and 1917, the latter giving the arsenal crucial frontage on the Delaware River when the creek was no longer navigable for twentieth-century vessels. During the Civil War many large manufacturing buildings with bold Victorian details were added to the complex, and after 1895, with the beginning of the American empire, the arsenal gradually expanded, with peaks of activity during the two world wars. Initially a repair, storage, and distribution depot, the arsenal shifted to the manufacture of small arms and artillery ammunition after the Mexican War.[25] Its days are evidently numbered, unfortunately. The post–World War II symbiotic relationship between the military and private industry in weaponry research and development appears to have made the Frankford Arsenal expendable, and it was scheduled to close in mid-1976.[26]

Almost from the beginning of the European presence in the area, the many creeks tributary to the Delaware River were seen as sources of water power, and this eventually became the factor that determined the configuration of the area between the Old City and Frankford Creek. Although popular tradition places one Swede's Mill on Frankford Creek as early as the 1660's, milling was subsidiary to the area's agriculture during the eighteenth century.[27] With the beginning of American industrialism about 1800, a host of mills related to textiles opened along these natural waterways. Globe Mills, Philadelphia's first extensive cotton mill, began operations in 1804 in Kensington, then an unincorporated village extending along the Delaware between Cohocksink Creek and Gunner's Run.[28] Established by Seth Craige, its first contracts were for girth-web to be sold in his saddlery and harness shop at 320 Market Street.[29] Another early mill, the first in Frankford Borough, was begun in 1809 by English-born Samuel Martin, who manufactured woven woolen blankets until the mill burned after the War of 1812.[30] The factory system was firmly ensconced in Kensington and Frankford in the years after that war, mostly by English immigrants who applied in a land of opportunity skills learned in the crucible of the Industrial Revolution.[31]

Because these mills were being built in essentially rural settings there was insufficient housing nearby for the workers.[32] Manufacturers like Thomas Holloway met this need by erecting rows of bandboxes (illus.) near his factory. Holloway, Philadelphia's first marine-engine builder, had a row of ten compact dwellings built along Ellen Street on ground that he had acquired in three large purchases between 1829 and 1832.[33] Although the houses were plain, they suggest Holloway's concern for his workers' well-being. Their pleasant classical proportions, paired round-headed doorways, and arched dormers presented a rhythmic streetscape. The interiors were built on the London house plan, a popular one for nineteenth-century speculative builders. This provided two rooms to a floor with the stairs between the rooms and an entrance hall at one side. While this allowed only a ten-by-twelve-foot front room, the houses were still more commodious and more attractive than the tight bandboxes nearby at Hope and Gray streets with but half the depth of Holloway's and without the small backyards.[34] Holloway's own dwelling, a comfortable three-story brick house, was at the end of the row on the northeast corner of New Market Street.[35] Its proximity to his workers' dwellings and his factory is less an indication of paternalism than of the absence of a transit system.

Although Holloway's row has been demolished for the Delaware Expressway, similar dwellings can still be found along Adams Avenue, particularly on the 1500 block where stone and frame dwellings stand on opposite sides of the street. Those across from the Globe Dye and Bleach Works on Worth Street east of Kinsey exhibit the full range of Philadelphia housing types for workers—singles, twins, and short rows.[36] Not all of the mid-nineteenth-century row houses in Kensington and Northern Liberties were built by local industrialists, however. The area's steady growth attracted real estate entrepreneurs such as Joseph Harrison, who was responsible for the construction of two long rows of houses, one with stores below, in the vicinity of Fourth and Jefferson streets.[37]

About the same time that Holloway was building his workers' houses, the Philadelphia and Trenton Railroad was incorporated, and with the almost simultaneous arrival of both factories and railways the destiny of much of the Delaware River Corridor was sealed. By November, 1834, the new railroad was operating between Kensington and Morrisville, spelling the end to the New York-

Philadelphia stages and drastically reducing the influence of the old inns along Frankford Road.[38] No longer merely a passageway between two cities, the area between Northern Liberties and Frankford was to become Philadelphia's industrial heartland as new waves of European immigrants flowed into the city, and the new railroads merged with the old port facilities.

Since the eighteenth century, shipbuilding had been a lucrative trade for independent shipwrights along the Delaware. With the introduction of steam, shipbuilding became more sophisticated, creating opportunities for those with capital and technical specialties, and in 1830 William Cramp started what was destined to become one of the country's foremost private shipyards. Shortly thereafter, one of Holloway's apprentices headed the nation's foremost propeller-engine manufactory, Reany, Neafie & Company at Beach south of Palmer Street. The Kensington waterfront was teeming with activity, but instead of many small shipyards, six major establishments had consolidated the business.[39]

The Philadelphia and Reading Railroad completed its line from the Falls of Schuylkill to Richmond in 1842 in order to carry its prize cargo from the great anthracite coalfields of Pennsylvania to wharves extending 300 feet into the Delaware River.[40] Out of this developed Port Richmond, today the largest privately owned railroad tidewater terminal in the world, with 85 miles of track on its 225 acres.[41]

The variety of factories in the area reflected the city's diversified economy. By the middle of the nineteenth century Northern Liberties could boast of the country's two leading saw manufacturers: Rowland's Saws on Beach Street, the country's oldest, and Henry Disston's Keystone Saw Works (illus.) on Laurel Street, the country's largest.[42] The neighborhood was already an industrial center when Disston began his saw works in 1846. The works grew steadily until it covered eight acres by the time of the centennial. So persistent was its expansion that two fires, in 1849 and 1864, were less business reverses than opportunities to update facilities.[43] The building at the corner of Front and Laurel streets, erected in 1872 after the second fire, illustrated the functional versatility of these industrial structures, although its functions were not readily visible from the street. Its shell was an example of incongruous architectural splendor. Behind arcaded walls with galvanized iron pilasters and window heads and a richly ornamental cornice were coal storage areas, packing and display rooms, and an office. A passage through the middle of the building carried freight cars to the millyard and coal storage facilities on the ground story; the carved oak paneled office was on the second floor.[44]

In addition to these large manufacturing complexes there were a host of smaller factories in the corridor, such as Isaac Pugh's paper-hangings factory at New Market and Pegg streets. Built about 1842 and enlarged after the Civil War, it was a plain brick structure bereft of architectural ornament; standing in stark contrast to the pulchritude of Disston's later building. It appears that Pugh rented the building after 1849, and although he never became the Disston of paper hangings, he did help to establish one of the city's leading firms of the 1850's, Hart, Montgomery & Company.[45]

The water-powered mills and steam-powered railroads quickly transformed the sleepy villages along the Delaware into factory towns that were economic and social extensions of the city. Industrialists' need for wholesale outlets and storage facilities led to the development of a warehouse district between the river and Third Street that physically and commercially linked Northern Liberties to Philadelphia well before their official consolidation in 1854. The warehouses erected during the 1850's set the pattern for the next generation. Generally of four stories, these brick buildings had iron storefronts at the ground floor and often at the second story as well.[46] Later buildings, such as the Massey and Janney Leather Warehouse at Third and Callowhill streets, employed a wider range of architectural ornament and looked less like a warehouse than some of its older neighbors on Leather Row, but it did not differ from them in construction and organization. Designed by Hutton and Ord in 1878, the broad expanse of its side wall was relieved by bands of white and yellow brick, and a second-story arcade graced the front. Slender incised iron piers marked the ground story, and the interior offered the customary warehouse plan of one large room to a floor with a counting room partitioned off on the first floor and a hoist and hatchways at the rear.[47]

As transportation improved and residents of outlying neighborhoods like Kensington and Frankford were drawn toward the vortex of the city, their shopping districts, churches, and cultural and municipal institutions imitated those of Philadelphia. Some institutions, such as the Frankford Lyceum, were more significant for their existence than for their accomplishments. Formed in 1836 and patterned after the Franklin Institute, the Lyceum was intended to provide a technical education for aspiring workingmen, and in 1840 its building was begun. Interest languished rather quickly, however.

The town was apparently too small to support the institution yet too
big to be without some expression of cultural pride. In 1866 the
Lyceum building was acquired by the Frankford Mutual Insurance
Company, an important local firm whose expansion from a small
office in the building in 1842 to possession of the whole structure
stands as another example of the town's commercial growth.[48]

The Odd Fellows Hall at Third and Brown streets in Northern
Liberties, on the other hand, enjoyed a longer life. It was one of many
such lodge buildings erected during the mid-nineteenth century, the
heyday of private voluntary associations. "Americans form
associations for the smallest undertakings," Tocqueville insisted,
attributing this phenomenon to the nature of democratic societies in
which independent citizens organize in order to overcome their
individual sense of incapacity.[49] Modern scholars pinpoint this sense
of alienation and see these organizations as attempts to counter the
socially disruptive forces of the burgeoning industrial city.[50] When it
was completed in 1847, Odd Fellows Hall served as the meeting place
for at least eighteen lodges in the northern part of the county. Its
Egyptian Revival facade, matched by Egyptian scenery painted on
some of the interior walls, was meant to evoke exotic awe from the
uninitiated and to hint at those mysteries of the order that were to
cement its bonds of brotherhood.[51]

Socially similar to fraternal lodges were the volunteer fire
companies. The erection of new fire halls in the 1840's, such as the
United States Hose Company No. 14 and the Northern Liberty Hose
Company No. 4, further testify to the neighborhood's complexity.[52]
As grass-roots institutions that reflected the ethnic, economic, and
political composition of their neighborhoods, these voluntary
companies "successfully competed with local pubs and workshops
for their members' attention and devotion."[53] They also performed a
crucial civic service, and in recognition of this they began receiving
public appropriations in 1811.[54] As physical expressions of strong
neighborhood associations with quasi-municipal functions, their fire
halls were often self-consciously finished in a builder-designed high
style. Sometimes the company's pride was restrained, as in the case
of the large incised name stone and delicately carved door panels of
the Northern Liberty Company; sometimes it led to exuberant
designs like the awkward and dated Greek Revival of the United
States Company. The companies' competitive spirit, however,
exceeded socially acceptable limits during the 1840's and 1850's, and
after a series of violent confrontations the city established its paid fire

department in 1871. Some of the halls served briefly as municipal fire stations before becoming warehouses and community halls.[55]

Tocqueville included religious groups among America's ubiquitous associations, and appropriately so. America's pluralistic, democratic society has historically encouraged the growth of a wide variety of sects, and while their doctrines complemented the secular faith of democracy in the mid-nineteenth century, their congregations generally reflected narrow social, ethnic, and economic biases.[56] If "the creative soil for religion is social anxiety,"[57] then the rapid growth of the southern part of the Corridor provided not only the bodies and souls to make up the congregations but also the uncertain environment that would draw them to the churches. By the 1850's many congregations had grown sufficiently large and wealthy to enable them to commission noted architects to design their new churches. In Northern Liberties, for example, the Fourth Baptist Church had Joseph C. Hoxie design its well-sited Italianate building (1853–55, demolished c. 1970); in Kensington Samuel Sloan supplied the plans for the First Presbyterian Church (1857–59), which initially had a high steeple where the incongruous ogee dome now rests; in Frankford John McArthur, Jr., furnished the awkwardly eclectic design of the First Presbyterian Church (1859–60); in Holmesburg St. Luke's Episcopal Church hired the renowned Richard Upjohn & Son to produce one of their tasteful Gothic Revival works (1860).[58]

In addition to the imposing Protestant churches there appeared a spattering of modest Roman Catholic churches, beginning in 1833–34 with St. Michael's in Kensington. A decade after its completion St. Michael's was attacked by "Native American" rioters who left it in flames and moved southward to deliver the same fate to St. Augustine's in Old City.[59] Undaunted by the hostile atmosphere and strengthened by the growing number of Catholic immigrants who found work in nearby factories, the area's Roman Catholics not only rebuilt St. Michael's in 1846, but a year earlier engaged Napoleon LeBrun to design the German Catholic Church of St. Peter's in Kensington and the year after the riots two additional churches, St. Ann's in Richmond and St. Joachim's in Frankford, were organized.[60] After the Civil War a burgeoning Catholic population demanded more and larger churches, which with their attendant rectories, convents, and schools were built and rebuilt into the twentieth century.[61] Protestant congregations, on the other hand, saw their membership decline. As the Presbyterian Church of Cohocksink and the First Presbyterian Church of Northern Liberties were being dissolved on the eve of World War I, the Second Presbyterian Church altered the old Shunk School on New Market Street into a community center for Russian, Polish, and Ukranian residents. The trend could not be reversed, however, and the number of Protestant churches continued to decline, until by the 1960's Northern Liberties, once the corridor neighborhood with the largest Presbyterian congregations, did not have a single active Presbyterian church.[62]

In 1854 the Delaware River Corridor was united politically with the City of Philadelphia, and the opening of the North Pennsylvania Railroad in 1855 and the Frankford and Southwark Passenger Railway Company in 1858 helped to make the union a physical and economic fact.[63] As mills and factories multiplied so did the banks that catered to their financial needs. Beginning with the sober brownstone Consolidation Bank on North Third Street, chartered in 1855 and built two years later, new financial institutions were formed and old ones moved to new quarters. During the 1860's and 1870's five banks and trust companies were founded in the area, and the two banks antedating the city's consolidation had new buildings erected.[64] The oldest bank in the corridor, the Bank of Northern Liberties, acquired the building of the Manufacturers and Mechanics Bank about 1870 and had it enlarged and extensively rebuilt from the designs of James H. Windrim, then a young architect who had just made his mark as the winner of the Philadelphia Masonic Hall competition.[65] Windrim's concentration of superimposed pilasters, pediments, arched hoods, and carved ornament at the entrance and the window above lent a touch of baroque exuberance to the facade, but the sobriety of the rusticated quoins and granite balustrade prevailed, keeping the building within the Renaissance tradition of bank design. The Kensington National Bank, on the other hand, stood in vivid contrast to this convention, as might be expected of a bank designed by Frank Furness in the 1870's. Since it repeated the form and decoration of some of his earlier banks, its granite walls of contrasting shades and finishes, rugged details, and colorful interior made it one of the area's foremost architectural attractions.[66]

Supporting these banks and in turn supported by them was a burgeoning group of industrial concerns, as old firms expanded and new ones were established in the last half of the nineteenth century. Frankford Dye Works, established in 1820 by Jeremiah Horrocks, was the oldest such establishment in Philadelphia and on the eve of the Civil War the largest dye works outside New England.[67] Although its expansion continued after the Civil War, its primacy was challenged during the 1870's by new concerns like the Globe Dye Works at Kinsey and Worth streets.[68] The Globe operation began in 1867 with five buildings, which unlike many of the Frankford Works structures were brick rather than frame. In less than seven years seven new buildings were added, the two original Dye Houses were replaced, and the Bleaching House was enlarged.[69] By 1876 it had expanded across Tacony Creek. Further alterations and additions during the 1880's and 1890's made it one of the city's most extensive dye and bleach concerns; it still flourishes at its original site.[70]

One of the greatest industrial establishments in the corridor was John Bromley & Sons carpet mill (illus.). Nationally celebrated for their fine rugs and carpets since before the Civil War, their two large manufacturing factories in Kensington constituted the largest rug and carpet complex in the United States by the turn of the century.[71] Bromley's Lehigh Mills at 201–63 East Lehigh Avenue are especially noteworthy and stand as one of the handsomest nineteenth-century industrial buildings to be found anywhere. A massive brick structure with interior iron columns and beams, it is a U-plan that encompasses a large city block. The west side was built in 1886, probably from the designs of Walter H. Geissinger, the city's leading industrial architect and the architect of the company's other nineteenth-century elements.[72] Projecting brick piers rise out of a plain ground story to frame the windows and reduce the heaviness of the walls, while the dull repetition of this pattern is relieved by corbeled brick parapets flanked by molded minarets at the middle and corners of each long side. The focal point, however, is the 1891 office building, which is firmly planted in the middle of the Lehigh Avenue facade. From the bold granite arches at the street level through the cylindrical minarets protruding from the upper stories to the bulbous French roof, it remains the visual embodiment of vibrant power.

Changing tastes, technologies, and transportation and lower suburban and southern tax rates have combined to undermine the corridor's industrial prominence since World War II. Some industrial giants have closed their doors and quit; others have moved to the cheaper land and limited-access highways of the suburbs. Their old factories are either left vacant or let to a variety of tenants who usually are collectively less productive than the original owners, and because of age and neglect the buildings are vulnerable to an age-old threat, fire.[73] Yet these massive, grimy red brick buildings are monuments to industrial capitalism. Often consuming an entire block and surrounded by rail spurs and small row houses, they are mute reminders of an age when work meant insecurity and hard and dangerous activity relieved by periodic and uncompensated layoffs, while ownership offered wealth, power, and luxury.

Monumental and lucrative as these factories were, they and the railroad ravaged the idyllic surroundings along Frankford and Tacony creeks during the last half of the nineteenth century. The Georgian mansions were abandoned and fell into ruin, and their stripped and rotting hulks were finally torn down in the 1940's and 1950's.[74] (One of Port Royal's dependencies managed to survive because it could be converted to low-rent housing.) This nineteenth-century industrial pattern was so well entrenched in the lower part of the corridor that the twentieth century has brought no significant change. The present physical culture of the upper region, on the other hand, is more the product of the automobile than that of any other part of the city, since large parts of it lay undeveloped until after World War II.

Inland from the waterways a few nineteenth-century suburban villas remain as testimony to its former status as a haven for the city's business elite. Burholme (illus.), a picturesque Italianate villa with a matching carriage house, is one of the finest examples of this building type, which was first popularized by Andrew Jackson Downing in the 1840's. Its crowning glory is an overwhelming belvedere with stained-glass windows and attenuated dome that combines with the arched cross-gable of the hipped roof to give the mansion a bold silhouette. Built about 1859 for Joseph W. Ryerss, president of the Tioga Railroad, Burholme stands on high ground, affording a sweeping view of its forty-eight landscaped acres.[75] The house and grounds were transferred to the city in 1905 by Ryers's daughter-in-law, and today they are part of the extensive Fairmount Park system, which is helping to preserve a splendid survivor of a dying species.[76]

Knowlton falls into the next generation of rural retreats. It was built in 1879–81 for William H. Rhawn, president of the National Bank of the Republic, and is even more unusual and even rarer in some respects than Burholme.[77] While few Italianate villas had the distinctive qualities of the Ryerss mansion, it was still one of many such suburban villas of generally similar design built at the time. Knowlton, on the other hand, is unique. It is the only Philadelphia survivor of a small number of country houses designed by Frank Furness, one of the nation's most innovative nineteenth-century architects. Like Burholme, Knowlton has a matching carriage house and both project lively picturesque silhouettes, but that is the end of their similarities. Juxtaposition of living volumes and the bold

massing of richly textured exterior elements set Knowlton apart from
the classical balance and order of earlier country houses and qualify it
as one of Philadelphia's prized "architectural aberrations."[78]

Improved public transportation after 1890 stimulated real estate
development of the middle part of the corridor, especially the
Frankford and Wissinoming areas. Trolleys were introduced in
Frankford in 1893 and the next year the Reading Railroad completed a
commuter line to the former borough. An even more significant event
in the corridor's history was the opening of the Frankford Elevated
late in 1922.[79] With residents able to travel quickly and cheaply into
the city's commercial center, its army of clerks, secretaries, and
shopkeepers could now heed Whittier and flee to the northeast "from
crowded street and red wall's weary gleam."[80] Open land was
subdivided and parceled out in a manner not unlike that practiced in
Whittier's "close dark city." Single dwellings with front lawns, pale
imitations of an earlier Georgian grandeur, were outnumbered by
row houses or semidetached houses built and organized on the row
house pattern. Housing construction was stifled by the Depression
and World War II, but after 1945 a new generation of consumers
hungered for homes. Geared to the automobile, encouraged by cheap
fuel, and helped by expressway construction, these urban commuters
made possible the explosive development of the great northeast.
Weak attempts at architectural individuality were made, the foul and
crowded courts of the older areas were not repeated, and streets were
laid out in sweeping curves instead of on a rectangular grid, but the
nineteenth-century speculative pattern persisted, because it had
proved so profitable. Called airlites, these compact, low-maintenance
row houses varied little in plan or appearance over two decades.[81]

Among the wave upon wave of mid-brow housing, however,
can be found some imaginative attempts at architectural excellence.
Named for a young hosiery worker killed during an incident
surrounding a 1930 strike,[82] the Carl Mackley Apartments (illus.), on
Castor Avenue facing Juniata Park, were in the vanguard on two
points when they were built in 1933–34. Their construction was
sponsored by the American Federation of Hosiery Workers with the
proceeds of a loan of more than a million dollars from the Public
Works Administration, making it one of the first low-rent
developments built with federal aid.[83] The four buildings in the
group are the work of European-trained Oskar Stonorov in
association with Alfred Kastner. Although they now look like just
another apartment complex (only better), their design was considered

"strikingly modern" in the 1930's, and they mark one of the earliest adaptations of the International style to American domestic architecture. [84] Crisp lines, total absence of applied ornament, and conscious arrangement of the buildings in reference to their grounds are clearly European features, but the residents' organization of life within the development—by way of a credit union, cooperative market, and tenants' steering committee—has historical roots in American utopian communities of an earlier, more romantic era. [85]

Catalog

Bank of Northern Liberties. See National Bank of Northern Liberties

Beth Eden House. See Shunk School

Birely House
(PA-1743), 313 Richmond St. Brick with stone trim, approx. 20' (two-bay front) × 28' with three-story rear ell, four stories on raised basement, mansard roof, round-arch entrance with fanlight and carved stone frontispiece, corbel cornice, side-hall plan. Built c. 1859; mansard probably added later; demolished c. 1969. John Birely (occupant) and Theodore Birely (owner) were local shipbuilders. 1 ext. photo (1967); * 6 data pages (1967).*

Blue Dick, The. See United States Hose Company No. 14

Bromley, John, & Sons, Lehigh Mills. Illustrated
(PA-1744), 201–63 E. Lehigh Ave., composing block bounded by Lehigh Ave., A., E. Somerset, E. Gurney, B sts. Brick with granite trim and interior iron beams and columns, front building approx. 500' (fifty-eight-bay front) × 55', western rear wing approx. 66' × 500', eastern rear wing approx. 66' × 300', one, four, five, and six stories, gable roofs with clerestory skylights, bulbous mansard roof on center office building, projecting piers, turrets, and rock-faced round-headed entrance on center building, U-plan with buildings in courtyard.

Notable late-19th century industrial architecture with massive membering. West wing and six buildings in courtyard built 1886–87;

probably by Walter H. Geissinger, architect. East wing and one building in courtyard built 1889–90; Geissinger and Hales, architects.
Center office building and one building in courtyard built 1891; Geissinger and Hales, architects. Northern end of west wing burned and
replaced, front building connecting wings and office building built
1892; Walter H. Geissinger, architect. Boiler house built 1894; Walter
H. Geissinger, architect. Later additions and alterations to buildings
in courtyard.

John Bromley & Sons was one of the country's largest manufacturers of carpets and rugs; now used for various manufacturing and
wholesale purposes. 2 ext. photos (1973),* 4 photocopies of plans and
printed views (1887, 1892),* 1 photocopy of architectural rendering
(1901).*

Brown, Anna S., House and Shop

(PA-1745), 408 Richmond St. Brick with stone trim, approx. 18' (two-
bay front) × 20' with two-story frame ell, two stories, flat roof,
ground-story shopfront with paneled piers and bracketed cornice.
Example of modest nieghborhood shop-dwelling of mid-19th
century. Built c. 1872; frame ell probably built before 1856; demolished c. 1969. Anna S. Brown, widow, confectioner, had her shop
and residence here 1873–78. 1 ext. photo (1967);* 12 data pages
(1967).*

Browne, John C., Houses

(PA-1746), 427–29 Richmond St. Brick with brownstone ashlar front,
each approx. 18' (two-bay front) × 30' with two-story rear ell, three
stories on raised basement, mansard roofs, side-hall plans. Mid-19th-
century brownstone residences built for investment purposes on
inherited land. Built before 1860; cornice, mansard roof added later;
demolished c. 1969. 1 ext. photo (1967);* 7 data pages (1967).*

Burholme (also known as Ryerss Mansion). Illustrated

(PA-186), Burholme Park, at N.W. corner Cottman and Central aves.
Museum. Stuccoed stone, 74'-9" (five-bay front) × 52'-3" with two-
story rear wings, three stories, hipped roof with large arched cross-
gable, central one-story tower with intersecting barrel-vault roof and
tall dome (not original), veranda on three sides, originally center-hall
plan.

A splendid suburban villa whose picturesque Italian style blends
with the rolling landscape of its 48-acre estate. Built c. 1859; two-story

rear brick kitchen built 1888; tower (or belvedere) added after 1888; interior altered c. 1908–10. Acquired by Fairmount Park Commission, City of Philadelphia 1905; opened to public as Robert W. Ryerss Library and Museum 1910. Built for Joseph W. Ryerss, railroad entrepreneur, and willed to city by son of Robert W. Ryerss. Named for ancestral estate in England. Certified, PHC 1957.

10 sheets (n.d., including plans, elevations, section, details, plan and section of tower); 4 ext. photos (1973, including 1 of carriage house, 1 in color);* 8 data pages (including 1872–88 insurance surveys).*

Butler, Anthony, Houses

(PA-1747), 133, 132–34 Olive St. Three brick houses (Nos. 132–34 contiguous), each approx. 16' (two-bay front) × 16' with two-story rear ell, two-and-a-half stories, gable roofs, ground-story alleyways to backyards, city house plans.

Houses were part of a speculative row of eight dwellings built for Anthony Butler, merchant, on each side of a court cut through from Front St. to New Market St. Built c. 1791; demolished c. 1969. Street known as Equality Court before 1810, as York Court 1811–58, as Onas St. 1858–97. 1 sheet (1967, including plot plan, elevation)* courtesy of Pennsylvania Historical Salvage Council; 2 ext. photos (1967);* 16 data pages (1968).*

Carmalt, Thomas, Houses

(PA-1748), 129–31 Brown St. Two contiguous brick houses, each appprox. 16' (two-bay front) × 20', two-and-a-half stories, gable roof, brick stringcourse, probably originally London house plan. Built between 1795 and 1802; altered later; demolished 1970. 1 sheet (1967, including plot plan, elevation);* 1 ext. photo (1967);* 11 data pages (1968).*

Chalkley Hall

(PA-110), 3869 Sepviva St., at S.W. corner Wheatsheaf Ln. Brick faced with coursed sandstone ashlar and granite trim, 57'-8" (five-bay front) × 46'-11" with two-and-a-half-story stuccoed brick and rubble lateral wing with granite trim, two and three stories, hipped and flat roofs, central pavilion with giant Tuscan pilasters and pediment, Tuscan frontispiece, central-hall plan.

Lateral wing built 1723; main section built 1776; two-story rear wing enlarged to width of house later. Vacated before 1907; demolished 1954. Entrance door installed in American Wing, Metropolitan Museum of Art, New York City. Interior elements at William Penn Memorial Museum, Harrisburg, Pa. Earliest part built for Thomas Chalkey, merchant and itinerant Quaker proselytizer; main section built for Abel James, merchant and Chalkley's son-in-law.

24 sheets (1937, including plans, elevations, section, details); 10 ext. photos (1936, 1937), 1 int. photo (1937), 1 photocopy of old ext. photo (n.d.); 11 data pages (1937, including 1833–59 insurance surveys*).

Clarke, Samuel, House

(PA-1749), 536 New Market St., at S.W. corner Nectarine St. Brick, approx. 14' (two-bay front) × 28', two-and-a-half stories, gable roof, round-arch entrance with fanlight, molded brick cornice. One of five row houses built for rental income for Samuel Clarke, retired merchant and naval officer. Built after 1824; demolished 1969. 1 ext. photo (1967);* 5 data pages (1968).*

Commercial Building

(PA-1750), 214–16 Callowhill St., at S.E. corner American St. Brick with stone trim, 39'-7" (five-bay front) × 33'-4", three stories (No. 214 originally two-and-a-half stories), flat roof (No. 214 originally gable roof), window and side-lights set in blind arch in second story of No. 214, rectangular plan. No. 214 built as dwelling between 1814 and 1825; No. 216 built as store-dwelling between 1851 and 1859; two buildings combined into one, ground story altered later; demolished 1969. 1 ext. photo (1967);* 6 data pages (1968).*

Commercial Building

(PA-1751), 305–7 N. Second St. Brick with cast-iron storefront and trim, approx. 35' (six-bay front) × 70', four stories, flat roof, molded cast-iron Corinthian piers on first and second stories, heavy granite lintels on upper stories, rectangular plan. Built before 1859. Two-story brick addition with steel columns and girders at rear of No. 307 built 1927; J. S. Wilds, engineer. Demolished 1973. 1 ext. photo (1967).*

Commercial Buildings

(PA-1752), 405–7 N. Second St. Two adjoining brick buildings with cast-iron storefronts and stone trim, each approx. 22' (three-bay front) × 42' with one-story rear wing, two stories (originally No. 405

three-and-a-half stories, No. 407 four stories), flat roofs (originally No. 405 probably gable roof), rectangular plan. No. 405 built early 19th century; No. 407 built after 1845; ground stories altered to stores before 1859; upper stories altered to commercial use 1909; lowered to two stories 1951–52; demolished c. 1969. 1 ext. photo (1967).*

Commercial Buildings
(PA-1753), 401–3 Richmond St., at N.E. corner E. Columbia Ave. Brick with stone trim, each approx. 18' (two-bay front) × 70' (No. 403 approx. 30' deep), three stories, flat roofs, shopfronts, rectangular plan. Built c. 1850; shopfronts altered later; demolished c. 1969. 1 ext. photo (1967);* 11 data pages (1967).*

Commercial Buildings
(PA-1754), 123–27 Vine St. Three adjoining brick stores with stone trim, each approx. 17' (two-bay front) × 35' with three- and two-story rear ells, four stories, flat roofs, stores on ground story. Built after 1859; No. 123 lowered to two stories 1957. Demolished 1968. 1 ext. photo (1967).*

Consolidation Bank
(PA-1755), 329–33 N. Third St. Brick with brownstone ashlar front, 41'-6" (five-bay front) × 115'-3" with one-story rear addition, three stories, flat roof, round-arch windows, curved pediment over round-arch entrance, projecting piers and recessed spandrels on upper stories, corbel table beneath cornice, rectangular plan.
Fine example of a mid-19th-century bank in Italianate mode. Built 1857; possibly by Stephen D. Button, architect; George Watson, builder. Parapet date stone removed, interior altered c. 1908. Rear brick addition built 1912; Ralph White, architect. Demolished 1972. Became Consolidation National Bank 1865; ceased banking operations 1908. 3 sheets (1968, including plot plan, plans, elevation)* courtesy Pennsylvania Historical Salvage Council; 1 ext. photo (1967);* 5 data pages (1968).*

Day, Michael, House
(PA-1756), 400 Richmond St., at S.E. corner E. Columbia Ave. Brick with stone trim, approx. 18' (two-bay front) × 65', three stories, flat roof, second-story balcony on rear wing. Built before 1866; altered to saloon later; demolished 1968. 1 ext. photo (1967);* 3 data pages (1967).*

Disston, Henry, & Sons, Keystone Saw, Tool and Steel Works. Illustrated (PA-1757), 1001–5 N. Front St., at N.E. corner Laurel St. Brick with cast-iron posts and galvanized iron trim, approx. 66' (nine-bay front) × 166', two stories, flat roof, round-arch windows with bold lintels, clerestory skylights, rectangular plan with passage through ground story to millyard.

One building in a complex that composed the largest saw and file works in 19th-century America. Built 1872; altered c. 1907; demolished 1968. Served as office and packing and show rooms of firm until 1907 when it became a paper box factory.

7 sheets (1967, including plot plan, plans, elevations, sections, details)* courtesy of Pennsylvania Historical Salvage Council; 2 ext. photos (1967),* 4 int. photos (1967),* 1 photocopy of printed view of Works (1873),* 1 photocopy of Works' plan (1873);* 3 data pages (1967).*

Episcopal Hospital. See Hospital of the Protestant Episcopal Church of Philadelphia

Frankford Town Hall
(PA-1758), 4255 Frankford Ave., at S.E. corner Ruan St. Brick, approx. 30' (three-bay front) × 55', two stories, gable roof. Built 1848; demolished 1960. 1 ext. photo (1959).*

Friendship Engine Company No. 15
(PA-1759), 2200–2204 E. Norris St., at S.E. corner Sepviva St. Brick with stone trim and cast-iron front at ground story, 37'-10" (four-bay front) × 84'-0", four stories (originally three), flat roof, quoins, rectangular plan. Built c. 1860; fourth story installed later by dividing tall second story. Served as Hook and Ladder Company C 1871–78. 1 photocopy of old ext. photo (c. 1935);* 1 data page (1963).*

Goodwin Building
(PA-1760), 317–19 N. Second St. Brick with cast-iron storefront and iron and granite trim, approx. 34' (four-bay front) × 58' with one-story rear wing, four stories, flat roof, molded cast-iron piers on ground story, engaged clustered cast-iron piers and low-pitch segmental pointed arches on second story, originally skylights in rear wing, rectangular plan. Example of mid-19th-century wholesale store. Built between 1854 and 1859. Demolished c. 1973. 1 ext. photo (1967).*

Griffith-Peale House

(PA-1761), 8100 Frankford Ave., at N.W. corner Welsh Rd. (originally known as Bristol Turnpike and Mill St.). Stuccoed rubble, approx. 24' (three-bay front) × 36' with two-story rear ell, two-and-a-half stories, gable roof, round-arch entrance with Tuscan frontispiece, transverse center-hall plan.

Modest Federal-style town house. Built between 1807 and 1810; interior altered, ell built later. Built for Thomas Hockley Griffith, gentleman, who sold the house in 1811. Home of Joseph Temple, merchant and patron of arts 1845–57, Col. John Clark, railroad contractor and politician 1857–72, Dr. James Burd Peale, physician and grandson of painter Charles Willson Peale 1877–80. Owned by Peale's descendants until 1944. Certified, PHC 1960. 3 ext. photos (1959);* 6 data pages (1960, 1962, including 1813 insurance survey).*

Harrison House

(PA-1458), Point No Point, or Port Richmond (street address unknown). Brick, approx. 40' (five-bay front) × 25' with one-story rear wing and frame side addition, two-and-a-half stories, gambrel roof, open porch across the front (probably replacing earlier pent eave), one interior end chimney, probably center-hall plan. Built late 18th century; rear and side additions, front porch built later; demolished after 1936. 2 ext. photos (1936).

Helffenstein, Dr. Abraham, House

(PA-1762), 1008 Shackamaxon St. Brick with brownstone and marble trim, approx. 20' (two-bay front) × 27' with two-story rear ell, four stories, flat roof, smooth rusticated brownstone ground story, wrought-iron balcony on carved stone consoles at second story, alleyway through ground story to backyard, side-hall plan. Example of mid-19th-century town house. Built c. 1849; demolished 1968. 1 ext. photo (1967);* 9 data pages (1967).*

Holloway, Thomas, Houses. Illustrated

(PA-1763), 125–31 Ellen St. Four brick row houses, each approx. 16' (two-bay front) × 28', two-and-a-half stories, gable roof, ground-story alleyway between alternate houses, side-hall London house plan. Part of a row of ten small tenement houses built for a steam-engine and machine manufacturer whose home and factory were nearby. Built c. 1832; demolished 1968. 2 sheets (1967, including

elevation, plans)* courtesy of Pennsylvania Historical Salvage
Council; 1 ext. photo (1967);* 6 data pages (1967).*

Hospital of the Protestant Episcopal Church in Philadelphia (now Episcopal Hospital). Illustrated

(PA-1764), S.E. corner Front St. and Lehigh Ave. Random Trenton
sandstone ashlar, approx. 258' (thirty-one-bay front) × 256' (center
building and chapel) and 200' (wings), two-and-a-half and three-and-
a-half stories on raised basements (chapel one story), gable roofs
(later altered to gambrel roofs on south end of wings), two four-story
towers with concave roofs on wings, minarets, chapel on south end
of center building, originally E-plan with central longitudinal halls
and transverse corridor.

Considered the best general hospital design in the United States
when built and had a great effect on subsequent American hospital
architecture. Center building, chapel, west wing built 1860–62;
Samuel Sloan, architect; James H. Windrim, superintendent of
construction. East wing built 1873–75; Samuel Sloan, architect;
William A. Armstrong, builder. Iron fire escapes added to south end
of wings 1885; wings vacated 1967.

Three-story sandstone Harrison Memorial House of Incurable
Patients built at east end 1889–91; G. W. and W. D. Hewitt, architects; George F. Payne & Co., contractors. Two-and-a-half-story
sandstone Ingersoll Receiving Ward built at west end 1892–94; G. W.
and W. D. Hewitt, architects; Thomas Little & Son, contractors.
North end of center building demolished 1933, replaced by ten-story
steel-frame Tower Building with brick curtain walls 1933–35; Day and
Zimmerman, architects. Harrison Memorial House demolished 1965,
replaced by six-story brick and reinforced-concrete Potter-Morris
Building 1964–66; Vincent G. Kling and Associates, architects. Name
changed to Episcopal Hospital 1964. Certified, PHC 1957.

4 ext. photos (1973),* 3 photocopies of old ext. photos (1869),* 2
photocopies of old int. photos (1869),* 2 photocopies of old prints
(1859),* 2 photocopies of plan (1859, 1869),* 1 photocopy of architectural rendering of Receiving Ward (1893).*

House

(PA-1765), 705 New Market St. Wood frame with clapboards, approx.
18' (two-bay front) × 25' with one-story rear ell, two-and-a-half
stories, gable roof, two rooms to a floor. Built after 1858; demolished
1969. 1 sheet (1967, including plot plan, plans, elevation)* courtesy of
Pennsylvania Historical Salvage Council; 1 ext. photo (1967).*

House

(PA-1766), 720 New Market St., at S.W. corner Olive St. Brick, approx. 17' (three-bay front) × 35' with two-story rear ell, two stories (originally three-and-a-half stories), flat roof (originally probably gable). Built late 18th century. Lowered to three stories and converted to garage 1919; Isadore W. Levin, architect. Lowered to two stories 1942; demolished 1969. 1 ext. photo (1967).*

House

(PA-1767), 106 Produce St. Brick, approx. 20' (three-bay front) × 18' with three-story rear ell (originally two-story rear ell), three-and-a-half stories, gable roof, brick stringcourse at second and third stories, originally ground-story arched alleyway to backyard, probably originally city house plan. Built mid-18th century; rear ell raised one story late 19th century; altered to garage-workshop 20th century; demolished 1969. 1 ext. photo (1967).*

House on Old Turnpike

(PA-1429), Frankford Ave. (street number unknown) Scored stuccoed rubble with brick trim, approx. 22' (three-bay front) × 20' with two-story scored stucco side wing, two-and-a-half stories, gable roof, recessed and paneled entrance with simplified Adamesque frontispiece, brick quoins, brick string-course on wing, probably side-hall plan. Built early 19th century; demolished after 1936. 2 ext. photos (1936).

Houses

(PA-1768), 1035–37 Frankford Ave. Two frame houses with wood sheathing and asbestos shingles, each approx. 14' (two-bay front) × 30' with rear ell, two-and-a-half stories, gable roof. Built after 1828; demolished 1969. 1 ext. photo (1967).*

Houses

(PA-1769), 142–50 Nectarine St. Five adjoining brick houses, each approx. 16' (two-bay front) × 16' with two-story rear ell (three westernmost houses with more attenuated proportions), three stories, gable roofs, alleyways through ground story of alternate houses.

Example of sidestreet dwellings of 19th-century artisans. Nos. 142–44 built between 1823 and 1830; Nos. 146–50 built between 1792 and 1803. Demolished 1969. Street known as Artillery Lane until c. 1785, as Duke St. c. 1785–1858, as Dana St. 1858–97. 1 ext. photo (1967);* 26 data pages (1968).*

Houses and Livery Stable

(PA-1770), 149–55 Noble St. Adjoining brick houses (Nos. 149–51) approx. 16' and 20' (two- and three-bay fronts) × 18' with two-story rear ells, three-and-a-half stories, gable roofs, ground-story alleys to backyards, side-hall plans; brick livery stable (Nos. 153–55) approx. 30' (three-bay front) × 65', two stories, flat roof, rectangular plan. Houses built after 1807; No. 149 by Moses Lancaster, house carpenter. Livery stable built after 1837. Demolished 1969. 1 ext. photo (1967);* 14 data pages (1968).*

Independent Order of Odd Fellows Hall

(PA-1771), 800 N. Third St., at N.W. corner Brown St. Brick covered with plaster, approx. 90' (seven-bay front) × 50', four-and-a-half stories, flat roof, windows with battered surrounds continue over three stories surmounted by corbel-arch windows, polychromatic cavetto cornice, discontinuous floor plans and stairways, third-story truss studding carries second-story auditorium ceiling.

Fine surviving example of Egyptian Revival architecture. Built 1846; possibly by John Haviland, architect. Interior altered 1891; Mettler Brothers, contractors. Destroyed by fire 1976 Lodge hall until 1932; later used for light manufacturing. Certified PHC 1966; Pennsylvania Register 1970; NR. 3 ext. photos (1961),* 1 int. photo (1961);* 8 data pages (1963, including 1849 news article).*

Keen, James, House

(PA-1772), 1015 Shackamaxon St., at N.E. corner Emery St. Brick with stone trim, approx. 20' (two-bay front) × 30' with two-story rear ell, two-and-a-half stories, gable roof, fanlight in gable curtain wall between twin chimneys, richly carved pent eave (not original). Built after 1815; ground story, cornice altered c. 1900; demolished 1968. James Keen, a Kensington ship-joiner and later steamboat inspector, lived here until 1860. 1 ext. photo (1967);* 7 data pages (1967).*

Kensington National Bank

(PA-1773), 2–8 W. Girard Ave., at S.W. corner Frankford Ave. Rock-faced granite with dressed granite trim, approx. 85' (eight-bay front) × 43' with one-story lateral wing, two stories on raised basement, flat roof (originally with crenelation), entrance on bevel with second-story balcony and pinched-top window, boldly carved stone details, decorative wrought-iron window grilles.

Early bank design by an innovative late-19th-century architect. Built 1877; Frank Furness, architect. Western addition built 1920;

Frank R. Watson, architect. One-story southern addition built 1941–
42; Charles H. Ingle, engineer. Interior extensively altered 1954;
Howell Lewis Shay, architect. Kensington Office, First Pennsylvania
Banking and Trust Co. since 1947. 1 ext. photo (1974);* HABSI form
(1962).*

Keyser-Wainwright Building

(PA-1774), 336 Richmond St., at S.W. corner E. Columbia Ave. Brick
with stone trim, approx. 40' (six-bay front) × 20', two stories, flat
roof, bracketed cornice over corner shopfront, rounded corner, rectangular plan. Example of mid-19th-century neighborhood retail
store. Built between 1861 and 1867; ground-story shopfront altered
later; demolished c. 1969. 1 ext. photo (1967);* 7 data pages (1968).*

Knight Building

(PA-1775), 301–3 N. Second St., at N.E. corner Vine St. Brick with
cast-iron storefront and trim, approx. 30' (four-bay front) × 92', four
stories, flat roof, cast-iron balcony at second-story front, ornamented
cast-iron lintels, rectangular plan. Built between 1851 and 1859; demolished 1973. 1 ext. photo (1967).*

Kramer-Rommel Houses

(PA-1776), 915–17 N. Hope St., at N.E. corner Gray St. Two
contiguous brick houses, each approx. 15' (two-bay front) × 13',
three stories, flat roof, one room to a floor. Early-19th-century
expression of an earlier tenement-building type. Built between 1828
and 1832; demolished 1968. 1 sheet (1967, including plot plan, plans,
elevation)* courtesy of Pennsylvania Historical Salvage Council; 2 ext.
photos (1967);* 5 data pages (1968).*

Lawrence, Charles, House

(PA-1777), 400 N. Front St., at N.W. corner Callowhill St. Stuccoed
brick, approx. 20' (three-bay front) × 27' with three-story rear wing,
four stories (probably originally three-and-a-half), mansard roof
(originally probably gable), bold bracketed cornice. Built between
1773 and 1781; altered to tavern before 1867; rear wing, cornice, mansard added later. 1 ext. photo (1967);* 6 data pages (1968).*

Lynfield

(PA-132), 4601 Rhawn St., at S.E. corner Ditman St. Coursed fieldstone ashlar and rubble, approx. 38' (five-bay front) × 38' (originally
38' × 23') with two-story side wing, two-and-a-half stories, gable and

hipped roofs, projecting pavilion with pediment and Adamesque frontispiece, open wooden rear porch, center-hall plan.

Fine example of rural house enlarged into a country house. South ashlar section built c. 1760; enlarged c. 1800 with new north rubble section becoming the front; side wing added mid-19th century; demolished 1942. Entrance in storage, parlor installed as Chippendale Room, William Penn Memorial Museum, Harrisburg, Pa. 3 ext. photos (1937); 1 data page (1937).

McGee, Ann, Houses

(PA-1778), 128–30 Brown St. Two contiguous brick houses, each approx. 11' (two-bay front) × 29', three stories, flat roof, round-arch entrances with fanlights, ground-story alleyway between houses, London house plan without entrance hall. Built between 1830 and 1837; demolished 1969. 1 sheet (1967, including plot plan, plans, elevation);* 1 ext. photo (1967);* 10 data pages (1968).*

Mackley, Carl, Apartments. Illustrated

(PA-1779), block bounded by Castor Ave., Bristol, M, Cayuga sts., on S. side Juniata Park. Four parallel concrete-block units faced with tan tile, each approx. 480' (thirty-four-bay front) × 33', three stories, flat roofs with pavilions, recessed central pavilions, recessed balconies on upper stories; contain 272 apartments.

Notable early example of International style employed on American domestic architecture. Built 1933–34; Oskar Stonorov and Alfred Kastner, architects; W. Pope Barney, superintendent of construction. Named in honor of Philadelphia hosiery worker killed during 1930 strike. 2 ext. photos (1976).*

Maguire-Howell House

(PA-1780), 976 N. Front St. Brick, approx. 18' (two-bay front) × 28' with three- and two-story rear ell, flat roof, round-arch entrance, alleyway through ground story to backyard, side-hall plan with rear stair tower. Built between 1831 and 1844; demolished 1968. 1 sheet (1967, including plot plan, elevation)* courtesy of Pennsylvania Historical Salvage Council; 6 data pages (1968).*

Manderson, Andrew, Houses

(PA-1781), 117–19 Fairmount Ave. Brick with stone trim, each 17'-6" (three-bay front) × 18'-6" with one- and two-story rear ell, three-

and-a-half stories, gable roof, brick stringcourse above first and second stories, dentil cornice, round-arch doorway with fanlight, round-arch alleyway between houses to rear yard, city house plan.

Built between 1801 and 1828; rear ells extended 1833. Certified, PHC 1971. 2 sheets (1967, including elevation, plans)* courtesy of Pennsylvania Historical Salvage Council; 1 ext. photo (1967);* 7 data pages (1967, including 1828–33 insurance surveys).*

Massey & Janney Warehouse
(PA-1782), 355 N. Third St., at S.E. corner Callowhill St. Brick with bands of white and black brick, approx. 20' (three-bay front) × 115', five stories, flat roof, rectangular plan with rear office. Notable example of late-19th-century warehouse. Built 1878; Hutton and Ord, architects. Demolished 1973. Massey & Janney, prominent leather merchants, were here until 1896. 3 ext. photos (1971);* 15 data pages (including architects' specifications and 1878 insurance survey).*

Mount Sinai Cemetery Chapel
(PA-1783), N.E. corner Bridge and Cottage sts. Rock-faced granite and brick with pecked and carved red terra cotta trim, approx. 42' (three-bay front) × 60', one story, intersecting hipped and gable red tile roofs with hipped cross-gables and copper cornice, horseshoe-arch windows, semihexagonal side bay; three-aisle plan, exposed wooden trusses, rear vault.

Richardsonian window and roof motifs mark this as one of the first and best designs by Frank Furness during his declining years. Built 1892; Furness, Evans & Co., architects. Certified, PHC 1974. 5 ext. photos (1973).*

National Bank of Northern Liberties
(PA-1784), 300–302 N. Third St., at N.W. corner Vine St. Brick faced with coursed granite ashlar, approx. 35' (three-bay front) × 94' with two-story lateral addition, two stories, flat roof, pedimented entrance, round-arch windows, quoins, balustrade. Built c. 1835. Enlarged, extensively rebuilt c. 1870; James H. Windrim, architect; George Watson, builder. Addition built c. 1887; probably by James H. Windrim, architect; Allen B. Rorke, contractor. Chartered as Bank of Northern Liberties 1814, it became a national bank 1864. 3 ext. photos (1963),* 1 photocopy of old ext. photo (n.d.);* 2 data pages (1970).*

Northern Liberty Hose Company No. 4 (also known as The Snappers)
(PA-1785), 714 New Market St. Brick with granite trim, approx. 20'
(two-bay front) × 55', three stories, flat roof, carved and paneled apparatus doors, date and name stone in middle of facade, rectangular plan. Fine example of mid-19th-century volunteer fire hall. Built 1846; demolished 1969. Name stone at Philadelphia Fire Museum. Served as Philadelphia Fire Department, Engine Company No. 21 1871–75. Certified, PHC 1964. 2 ext. photos (1963, 1967);* 8 data pages (1964, 1967).*

Pennypack Creek Bridge
(PA-1786), 8300 block Frankford Ave. across Pennypack Creek. Random local stone ashlar, approx. 38' (originally 18') × 110', three round-arches with buttresses. Built 1697–98; eastern (upstream) addition built 1893. Oldest bridge in continuous use in United States. Certified, PHC 1970; designated National Historic Civil Engineering Landmark by American Society of Civil Engineers 1970. 3 photos (1973);* HABSI form (1971).*

Philadelphia Fire Department, Engine Company No. 21
(PA-1787), 826–28 New Market St. Brick with sandstone trim, approx. 27' (four-bay front) × 85', two stories, flat roof, originally elliptical-arch engine doors capped by straight and segmental-arch dentil cornice, tall round-arch windows, heavy straight and segmental-arch iron cornice, parallelogram plan.
    Fine example of late-19th-century municipal fire station. Built 1875; ground-story front replaced mid-20th century. 2 sheets (1967, including plot plan, plans, elevation)* courtesy of Pennsylvania Historical Salvage Council; 1 ext. photo (1967),* 1 photocopy of old ext. photo (1896);* 1 data page (1967).*

Phillips Row
(PA-1788), 903–13 New Market St. Six contiguous brick houses, each approx. 17' (three-bay front) × 16' with two-story rear ell, two-and-a-half stories, gable roofs, ground-story alleyway through alternate houses, city house plan. Row of artisans' homes built on a plan by William Phillips, merchant. Built c. 1810; altered later; demolished 1968. 1 sheet (1967, including plot plan, plans, elevation, section)* courtesy of Pennsylvania Historical Salvage Council; 1 ext. photo (1967);* 24 data pages (1967).*

Port Royal
(PA-111), S. side Tacony St. between Church and Duncan sts. Brick with stone trim, approx. 54' (five-bay front) × 31' with two-story rear ell, two-and-a-half stories on raised basement, hipped roof, central pavilion with pediment, round-arch entrance with fanlight and Doric frontispiece, stone quoins and stringcourse, open rear porches, central-hall plan. Notable Georgian country house. Built 1761–62; demolished c. 1962. Main entrance, doorway, windows, exterior details, plaster ceiling, and entrance hall, parlor, dining-room woodwork installed in H. F. duPont Winterthur Museum, Delaware. 7 sheets (1937, including plans, elevations, details); 5 ext. photos (c. 1900, 1937, including 1 of barn); 5 data pages (1937, including 1770 insurance survey*).

Potts, Horace T., & Co. Warehouse
(PA-1789), 316–20 N. Third St., at N.W. corner Wood St. Steel frame, brick curtain walls with terra cotta trim, approx. 64' (three-bay front) × 176', four stories, flat roof with projecting cornice, pointed-arch windows, glazed brick diaperwork at top story, rectangular plan. Notable example of Venetian Gothic design for a utilitarian building. Built 1896–97; Frank Miles Day & Bro., architects; George Watson & Son, contractors. Interior altered 1917–19; Addison H. Savery, architect. Destroyed by fire 1976. 3 ext. photos (1963, 1974).*

Pratt, Henry Houses
(PA-1790), 229–33 Wood St. Three contiguous brick houses with stuccoed fronts, each approx. 15' (two-bay front) × 17' with two-story rear wing, three-and-a-half stories, gable roofs, probably originally London house plans. Built c. 1800; demolished c. 1970. Investment properties of merchant Henry Pratt, whose country seat was Lemon Hill. 1 ext. photo (1967);* 8 data pages (1968).*

Preisendanz, Christian A., Wagon Works
(PA-1791), 520–26 New Market St. Brick with stone trim, 51'-10½" (six-bay front) × 110'-6", three stories, flat roof, segmental-arch windows, projecting brick piers with corbel-table frame end bays, canted brick string-course on third-story end bays, U-plan with alley through ground story. Built c. 1888; demolished 1969. 1 ext. photo (1967);* 2 data pages (1968).*

Pugh, Isaac, Paper Hangings Factory
(PA-1792), 440 New Market St., at N.W. corner Pegg St. Brick, approx. 42' (four-bay front) × 95' with wing on north side (originally approx. 42' × 73' with one- and two-story front additions), three stories, flat roof, segmental-arch windows, trapezoidal plan.
Example of mid-19th-century factory. Built c. 1842; extended 22' to front between 1880 and 1886; south side (Pegg St.) ground-story windows altered, iron lintels inserted later; demolished 1970. 1 ext. photo (1967);* 5 data pages (1968).

Reybold-Paul House
(PA-1793), 617 New Market St. Brick with marble trim, approx. 20' (three-bay front) × 20' with two-story rear ell, three-and-a-half stories, gable roof, round-arch entrance with fanlight and Adamesque frontispiece, brick stringcourse at second and third stories, sidehall plan. Built between 1803 and 1810; demolished 1969. 1 ext. photo (1967),* 1 photocopy of old ext. photo (1931); * 6 data pages (1968).*

Rich, Comly, House
(PA-1794), 4276 Orchard St. Wood frame with tongue-and-groove siding, approx. 12' (three-bay front) × 15' with one-story rear addition, two stories, gable roof, one room to a floor. Example of a Frankford artisan's dwelling on a deep lot. Built c. 1826; rear wing built later. First house in the United States financed through a savings and loan association 1831. Certified, PHC 1960. 3 photocopies of measured drawings (1974),* 1 photocopy of ext. photo (1960).*

Rotan, William, Houses
(PA-1795), 325–27 Richmond St. Brick with stone trim, each approx. 19' (two-bay front) × 32' with two-story rear ell, two-and-a-half stories, gable roofs, alleyway through ground story to backyard. Example of dwellings for families whose livelihoods were associated with river commerce. Built between 1833 and 1836; demolished c. 1969. William Rotan, Kensington shipwright, rented one house and lived in the other (No. 325) until his death 1875. 1 ext. photo (1967);* 8 data pages (1967).*

Rush, Benjamin, House (also known as Benjamin Rush Homestead)
(PA-1796), N. side Red Lion Rd. between Academy and Knight rds. (later at S.E. corner Keswick and Rayland sts. when streets cut

through). Coursed fieldstone ashlar, approx. 30' (two-bay front) × 22' with two-and-a-half-story ashlar and three-story clapboarded frame lateral additions, two-and-a-half stories, gable roof, originally pent eaves on front and back, open front porch (not original), center-hall plan (originally probably one room to a floor).

South end (left) built c. 1690; center part built probably before 1765; roofs rebuilt, porch added early 19th century; north frame addition built mid-19th century. Demolished 1969. To be rebuilt as museum and community center on Philadelphia State Hospital grounds, Byberry. Birthplace of Benjamin Rush, "father of American psychiatry," signer of Declaration of Independence, Physician General of Continental Army, first professor of chemistry in America's first medical school (University of Pennsylvania), cofounder of America's first abolitionist society, founder of Dickinson College and Franklin and Marshall College. Certified, PHC 1958. 2 ext. photos (1959, including 1 of springhouse),* 1 int. photo (1959, including 1 of mortise-and-tenon joint);* 1 data page (1959).*

Ryerss Mansion. See Burholme

Schaible, Charles, Store and House
(PA-1798), 144 Fairmount Ave. Brick with marble trim, approx. 16' (two-bay front) × 25' with three- and two-story rear ell, three stories, flat roof, wooden storefront with sash doors and bulk window on ground story, arched alley to rear. Example of mid-19th-century store-dwelling. Built c. 1865; two-story addition built early 20th century. Demolished 1969. 1 sheet (1967, including plot plan, elevation)* courtesy of Pennsylvania Historical Salvage Council; 1 ext. photo (1967);* 4 data pages (1968).*

Searle-Bond Houses
(PA-1799), 123 Laurel St. and 1002–4 N. Hope St., at N.W. corner Hope and Laurel sts. Three contiguous brick row houses, each approx. 13' (two-bay front) × 20', two-and-a-half stories, hipped and gambrel roofs, one room to a floor, cellar of corner houses extends under Hope St. sidewalk. Built between 1815 and 1819; demolished c. 1974. Representative of early-19th-century dwellings rented to artisans. 2 sheets (1967, including plot plan, plans, elevations)* courtesy of Pennsylvania Historical Salvage Council; 1 ext. photo (1967)* 6 data pages (1967).*

Shunk School (later known as Beth Eden House)

(PA-1800), 807 New Market St. Brick with marble trim, approx. 40′ (three-bay front) × 64′, two stories, gable roof, projecting lintel on plain corbels over entrance (originally window) and center second-story window, name and date stone in center of gable, rectangular plan with four rooms to a floor.

Fine example of a mid-19th-century public school. Built 1852; Samuel Sloan, architect; Jacob Jones, contractor. Interior altered, reinforced 1916; Louis H. Rush, architect. Front window converted to door, interior altered 1923; Louis H. Rush, architect. Demolished 1969. Named in honor of Francis R. Shunk, educator and public official, who served as Governor of Pennsylvania 1845–48. Beth Eden House, settlement house of Second Presbyterian Church in Philadelphia, 1916–68. Certified, PHC 1967.

3 sheets (1967, including plot plan, plans, elevations)* courtesy of Pennsylvania Historical Salvage Council; 2 ext. photos (1967);* 9 data pages (1967).*

Singerly, Joseph, Houses

(PA-1801), 120–22 Fairmount Ave. Brick with marble trim, each approx. 13′ (two-bay front) × 33′ with one-, two-, and three-story rear ells, three-and-a-half stories, gable roof, round-headed doorway, side-hall plan with rear stair tower. Example of perseverance of 18th-century town house plan. Built between 1831 and 1836; Joseph Singerly, house carpenter. Demolished 1969. 2 sheets (1967, including plot plan, plans, elevation, basement fireplace elevation and details)* courtesy of Pennsylvania Historical Salvage Council; 1 ext. photo (1967);* 6 data pages (1968).*

Smith, John B., House

(PA-1802), 1026 Shackamaxon St. Brick with brownstone trim, approx. 19′ (two-bay front) × 25′ with two-story rear ell, two stories on raised basement, flat roof, rusticated brownstone watercourse, molded galvanized metal cornice with small end minarets, ground-story alleyway, side-hall plan. Built c. 1898; demolished 1968. 1 ext. photo (1967);* 5 data pages (1967).*

Snappers, The. See Northern Liberty Hose Company No. 4

Swope-Weissman Houses

(PA-1803), 527–29 N. Second St. Two contiguous brick houses with stone trim, each approx. 20′ (three-bay front) × 30′ with two-story rear ell, three-and-a-half stories, gable roofs, probably originally side-hall plan. Built c. 1819; ground stories altered to stores before 1859; ground story of No. 529 bricked up 20th century; demolished c. 1969. Rear ell of No. 529 originally three tenements. 2 ext. photos (1967);* 7 data pages (1968).*

Trinity Church, Oxford (Protestant Episcopal). Illustrated
(PA-17), 6900–02 Oxford Ave., at S.W. corner Disston St. (formerly Church Ln.) and Oxford Ave. Brick, 34′-11″ × 75′-0¹/₄″ (originally 25′-2¹/₂″ × 36′-4³/₄″), one story, gable roof, three-stage corner entry and bell tower with octagonal roof, round-arch windows, diamond patterns of glazed headers on original walls, one-aisle cruciform plan with rectangular chancel, originally probably one-aisle rectangular plan.

Example of an early Anglican church incorporating Delaware Valley vernacular building techniques. Built 1711; pews installed 1759; flooring completed, door moved from north side to west end 1807; transepts added 1833; tower (now chancel) built 1839; tower lowered to one story, present corner tower built 1875. Second oldest congregation in Philadelphia and oldest known Episcopal church in Pennsylvania in which continuous services have been held. Certified, PHC 1956.

6 sheets (1934, 1935, 1936, including plot plan, plan, elevations, elevation rendering from old drawing, details, 1709 gravestone rubbing); 3 ext. photos (1973, including 1 color photo),* 2 photocopies of ext. sketch and watercolor (n.d.); 4 data pages (1936).

United States Hose Company No. 14 (also known as the Blue Dick)
(PA-1804), 423 Buttonwood St. Brick covered with coursed stucco with marble trim, approx. 20′ (three-bay front) × 50′, three stories (originally with square frame observation tower), flat roof, recessed second-story balcony with marble Ionic colonnade in antis and ornamental iron fence containing name "United States," marble-pier ground story, originally sculpture of water pump in third-story center niche.

Example of Greek Revival volunteer firehouse. Built 1845; William L. Grubb, carpenter; Charles Twaddell, bricklayer. Parapet, sculpture removed after 1870; demolished 1953. 1 photocopy of old ext. photo (c. 1935).*

Weaver, John, House

(PA-1805), 134 Fairmount Ave., at S.W. corner New Market St. Brick, approx. 20' (two-bay front) × 43', two-and-a-half stories, gable roof, brick stringcourse above first and second stories. Built after 1802; altered to store before 1859; altered to tavern before 1867; altered to store again late 19th century. Demolished 1969. 1 ext. photo (1967);* 6 data pages (1968).*

Whitton, Abednego J., Houses

(PA-1806), 1018–20 Crease St., at S.W. corner Salmon St. Two contiguous frame houses with asbestos siding, each approx. 16' (two-bay front) × 16' with two-story rear ell, two-and-a-half stories, mansard roof, ground-story alleyway between houses, side-hall plan. Built between 1825 and 1830; mansard roof added later; demolished c. 1969. 1 ext. photo (1967);* 7 data pages (1967).*

Wiedersum, George, House

(PA-1807), 405 Richmond St. Brick with marble trim, approx. 18' (two-bay front) × 30', three stories, flat roof, probably originally side-hall plan. Built c. 1848; ground story altered to store later; demolished c. 1969. Southernmost house in a row of eight contiguous houses. 1 ext. photo (1967);* 8 data pages (1967).*

Wildes-Sonder Houses

(PA-1808), 455–57 N. Second St. Two contiguous brick houses with stone trim, each approx. 20' (three-bay front) × 34' with one- and two-story rear wings, three-and-a-half stories, gable roofs, stone stringcourse at third story.

Buildings' history reflects neighborhood's change over two centuries. Built between 1797 and 1799; No. 455 probably by Joseph Wildes, bricklayer; No. 457 probably by Jacob Sonder, bricklayer. Altered to stores before 1859; storefront of No. 455 remodeled 1923; interior of No. 457 altered 1926. Rear wings added before 1886, enlarged between 1922 and 1931. Demolished c. 1969. 1 ext. photo (1967);* 12 data pages (1968).*

John Bromley & Sons, Lehigh Mills. *Jack E. Boucher photo, 1973.*

Burholme. *Jack E. Boucher photo, 1973.*

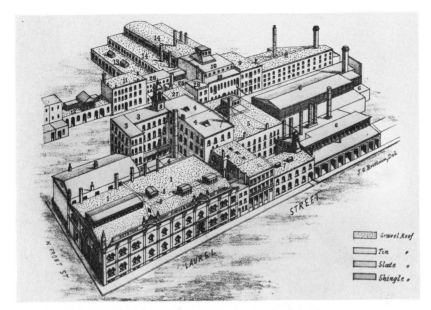

Henry Disston & Sons Keystone Saw, Tool, and Steel Works. *Lithograph 1873, Hexamer General Surveys, courtesy Free Library of Philadelphia.*

Thomas Holloway Houses. *E. McFarlin, del., 1967, courtesy Pennsylvania Historical Salvage Council.*

Hospital of the Protestant Episcopal Church in Philadelphia. *Jack E. Boucher photo, 1973.*

Carl Mackley Apartments. *Jack E. Boucher photo, 1976.*

Trinity Church, Oxford. *Jack E. Boucher photo, 1972.*

Notes

Abbreviations used in notes:
AABN *American Architect and Building News*
Building Permit City of Philadelphia, Department of Licenses and
  Inspection, Construction Unit
Deed Book City of Philadelphia, Department of Records
FFI Franklin Fire Insurance Company
FLP Free Library of Philadelphia
HABS Historic American Buildings Survey
Historic Philadelphia *Historic Philadelphia from the Founding until the*
  Early Nineteenth Century, Transactions of the
  American Philosophical Society 43, Part 1
  (1953)
HSP Historical Society of Pennsylvania
INA Insurance Company of North America
JSAH *Journal of the Society of Architectural Historians*
LCP Library Company of Philadelphia
MA Mutual Assurance Company
PC Philadelphia Contributionship
PHC Philadelphia Historical Commission
PL *Public Ledger*
PMHB *Pennsylvania Magazine of History and Biography*
PRERBG *Philadelphia Real Estate Record and Builders' Guide*
Registry Unit City of Philadelphia, Department of Records

Penn's City: Society Hill

1. Frances Wright d'Arusmont, *Views of Society and Manners in America* (London: Longman, Hurst, Rees, Orme, and Brown, 1821), pp. 83–84.

2. John F. Watson, *Annals of Philadelphia* (enlarged, rev. ed.; Philadelphia: Leary, Stuart Co., 1927), 1:12; J. Thomas Scharf and Thompson Westcott, *History of Philadelphia, 1609–1884* (Philadelphia: L. H. Everts & Co., 1884), 1: 87, 89, 202.

3. Edwin Iwanicki, Brief of Title, 117 Lombard Street, Philadelphia (1961), PHC; PC 918 (March 24, 1764), photocopies at HABS, PHC. For a fuller presentation of the houses and institutions of Society Hill, see Elizabeth B. McCall, *Old Philadelphia Houses on Society Hill, 1750–1840* (New York: The Architectural Book Publishing Co., 1966).

4. Richard S. Fuller, Brief of Title, 220 Spruce Street, Philadelphia (1960), PHC; Iwanicki, Brief of Title for 117 Lombard Street, PHC; rehabilitation plans for 220–22 Spruce Street in PHC files.

5. Brief of Title, 217 Delancey Street, Philadelphia (c. 1957), PHC; Gregory B. Keen, "The Descendants of Jöran Kyn, the Founder of Upland," *PMHB* 5 (1881): 96–97; Henry D. Biddle, "Colonial Mayors of Philadelphia: Samuel Rhoads, 1774," *PMHB* 19 (1895): 64–71.

6. The story of this detailed map, the Clarkson-Biddle Map of Philadelphia, is related by Martin P. Snyder in *City of Independence: Views of Philadelphia Before 1800* (New York: Praeger Publishers, 1975), pp. 62–64, 73–74.

7. Harold Donaldson Eberlein, ed., "Further Passages from the Diary of Nicholas Pickford Esquire, Relating to His Travels in Pennsylvania in 1786," *The Architectural Review* 48 (July 1920): 28.

8. Carl and Jessica Bridenbaugh, *Rebels and Gentlemen: Philadelphia in the Age of Franklin* (New York: Oxford University Press, 1962), p. 109. The Powel House was built in 1765 for Charles Stedman, a merchant who suffered business reverses and sold the property to Powel in 1769. Professor George B. Tatum has written the definitive study of this house and some of its contemporaries: *Philadelphia Georgian: The City House of Samuel Powel and Some of Its Eighteenth-Century Neighbors* (Middletown, Conn.: Wesleyan University Press, 1976). For information of its early ownership, see pp. 4–16, 25–30.

9. Fiske Kimball, "Interior from the Powel House," *Pennsylvania Museum Bulletin* 21 (January 1926): 68; Kimball, "Drawing Room From the Powel House, Philadelphia, 1768," *Pennsylvania Museum Bulletin* 23 (March 1928): 29.

10. This housing form is discussed in William John Murtagh, "The Philadelphia Row House," *JSAH* 16 (December 1957): 9.

11. Brief of Title, 236 Delancey Street, Philadelphia (c. 1957), PHC; MA 2234–35 (June 1806); PC 4745 (March 12, 1830), photocopies at HABS, PHC.

12. Physical evidence shows that an earlier house, built in 1769, was incorporated into the house. PC 1328 (August 24, 1769), photocopy at PHC; Brief of Title, 321 South Fourth Street, Philadelphia (c. 1958), PHC. Hill's biographical sketch appears in William H. Egle, "The Constitutional Convention of 1776: Biographical Sketches of Its Members," *PMHB* 3 (1879): 441–42.

13. Robert E. Cooper, "John Clement Stocker House," HABS Report, PA-1068.

14. Margaret B. Tinkcom, "The New Market in Second Street," *PMHB* 82 (October 1958): 379–96.

15. Edward B. Krumbharr, "The Pennsylvania Hospital," *Historic Philadelphia*, pp. 237–43; George B. Tatum, *Penn's Great Town* (Philadelphia: University of Pennsylvania Press, 1961), pp. 51, 77, 165–66; pl. 35. The most thorough work on the institution's

early years is William Henry Williams, *America's First Hospital: The Pennsylvania Hospital, 1751–1841* (Newark, Del.: University of Delaware Press, 1976).

16. For a brief discussion of the importance of the church to nineteenth- and twentieth-century blacks, see Theodore Hershberg, "Free Blacks in Antebellum Philadelphia," in *The Peoples of Philadelphia,* ed. Allen F. Davis and Mark F. Haller (Philadelphia: Temple University Press, 1973), pp. 120–21.

17. "Richard Allen and Mother Bethel African Methodist Episcopal Church" (Philadelphia: Bethel African Methodist Episcopal Church, 1972), n.p. Allen's autobiography accompanies his biography in R. R. Wright, Jr., *The Bishops of the African Methodist Episcopal Church* (Nashville, Tenn.: The A.M.E. Sunday School Union, 1963), pp. 46–76.

18. *PL,* October 27, 1890.

19. Richard S. Fuller, Brief of Title, 240 South Fourth Street, Philadelphia (1959), PHC; MA 4871 (June 1830), photocopies at PHC; Harold Donaldson Eberlein and Cortlandt Van Dyke Hubbard, *Portrait of a Colonial City* (Philadelphia: J. B. Lippincott Co., 1939), p. 301.

20. Wright d'Arusmont, *Views of Society and Manners in America,* p. 83.

21. *PRERBG* 15 (October 24, 1900): 681; Registry Unit 3S7–23, 2S11–70.

22. A favorable discussion of this process is in Hugh Scott, "Society Hill: Renewal's Mudder Hits the Home Stretch," *Today: The Inquirer Magazine,* March 30, 1969, pp. 6–10, 12, 14–18. A more scholarly examination of the plan and its execution is in Valerie Halverson Pace, "Society Hill, Philadelphia: Historic Preservation and Urban Renewal in Washington Square East" (Ph.D. dissertation, University of Minnesota, 1976).

23. Edmund N. Bacon, the man responsible for the "design structure" of Society Hill, discusses its development in *Design of Cities* (New York: Viking Press, 1967), pp. 243–49.

Penn's City: Old City

1. *The National Register of Historic Places, 1974 Supplement* (Washington, D.C.: U.S. Government Printing Office, 1974), pp. 452–57. The historic districts are Old City Historic District and Elfreth's Alley Historic District. The other National Historic Landmarks are Carpenters' Hall, Christ Church, Elfreth's Alley, Musical Fund Society Hall, Philosophical Hall, Reynolds-Morris House, and Walnut Street Theatre. The major buildings in Independence National Historical Park were designated National Historic Sites in 1948 and earlier. *U.S.S. Olympia* was moved downstream to Penn's Landing Project in June 1976.

2. The Old Court House is similar to that of Amersham, Buckinghamshire, England, 1682. See Horace Field and Michael Bunney, *English Domestic Architecture of the XVII and XVIII Centuries* (London: G. Bell and Sons, 1905), pl. 48. The early political history of the city and county is in J. Thomas Scharf and Thompson Westcott, *History of Philadelphia, 1609–1884* (Philadelphia: L. H. Everts & Co., 1884), 1:157–90. The effort to remove the market houses is related in *Journal of the Common Council of the Consolidated City of Philadelphia* (Philadelphia: W. H. Sickels, 1859), pp. 372–74, 392.

3. Evidence of the building's honored place in America's iconography is the great number of copies or near-copies that have been built. See John Maass, "Architecture and Americanism, or Pastiches of Independence Hall," *Charette* 49 (September 1969): 4–7. The most thorough historical and architectural studies of Independence Hall were

done by the National Park Service during the 1960's; they are called Historic Structures Reports: Independence Hall; and may be obtained from Independence National Historical Park headquarters, 313 Walnut Street, Philadelphia. Its architecture is also evaluated by Hugh Morrison, *Early American Architecture* (New York: Oxford University Press, 1952), pp. 532–37; and George B. Tatum, *Penn's Great Town* (Philadelphia: University of Pennsylvania Press, 1961), pp. 32–33.

4. Jonn Maass, "Philadelphia City Hall: Monster or Masterpiece?" *Journal of American Institute of Architects* 43 (February 1965): 24.

5. For twentieth-century commentary on the violation of Penn's original plan, see John W. Reps, *The Making of Urban America* (Princeton: Princeton University Press, 1965), pp. 167–69; Joseph Jackson, *America's Most Historic Highway: Market Street, Philadelphia* (Philadelphia: John Wanamaker, 1926), pp. 4–5. Nineteenth-century observers were aware of this change in Penn's plan, as noted in John F. Watson, *Annals of Philadelphia* (Philadelphia: Edwin S. Stuart, 1884), 1:225; "Journal of John Mair," *American Historical Review* 12 (October 1906): 80; Charles William Jansom, *The Stranger in America, 1793–1806* (New York: The Press of the Pioneers, 1935), p. 183; Frances Wright d'Arusmont, *Views of Society and Manners in America* (London: Longman, Hurst, Rees, Orme, and Brown, 1821), p. 84. Although eighteenth-century descriptions are rare, late-eighteenth-century and early-nineteenth-century descriptions with references to the waterfront's earlier condition can be found in the following books: J. P. Brissot de Warville, *New Travels in the United States of America, 1788*, trans. Mara Soceanu Vamos and Durand Echeverria (Cambridge, Mass.: Belknap Press of Harvard University Press, 1964), pp. 254–55; John M. Duncan, *Travels through Parts of the United States and Canada in 1818 and 1819* (New York: W. B. Gilley, 1823), 1:188; Wright d'Arusmont, *Views of Society and Manners in America*, p. 84; Jansom, *The Stranger in America, 1793–1806*, p. 183. Reps also discusses the development of the city's plan into the nineteenth century; see *The Making of Urban America*, pp. 160–74. For a thorough, cross-cultural analysis of the evolution of the plan of Philadelphia during the colonial period and the democratic era following the American Revolution, see Anthony N. B. Garvan, "Proprietary Philadelphia as Artifact, " *The Historian and the City*, ed. Oscar Handlin and John Burchard (Cambridge, Mass.: MIT Press, 1963), pp. 177–201. A brief discussion of the extended control of First Purchasers and a reproduction of Surveyor General William Parsons's "Plan of Philadelphia" (c. 1748) showing early ownership of city lots can be found in Nicholas Wainwright, "Plan of Philadelphia," *PMHB* 80 (April 1956): 164–226.

6. Mrs. Charles J. Maurer, Executive Assistant to the Chairman of the Philadelphia Historical Commission, to Stanhope S. Browne, Esq., Chairman of the Committee to Preserve Philadelphia's Historic Gateway, February 9, 1966; Stanhope S. Browne, Esq., to Mrs. Charles J. Maurer, March 24, 1966, copies at PHC.

7. MA 2105 (December 1805), photocopies at PHC and HABS.

8. MA 5889 (December 3, 1870), photocopy at PHC; Morrison, *Early American Architecture*, p. 37.

9. PC 2803–2804 (February 10, 1797), photocopy at PHC; HABS drawings of Warehouse, 329 S. Water Street (July 1966), sheet 2, Frank Sanchis, del.; PC 5228 (April 18, 1837); HABS drawings of Jesse Godley Warehouse, 19–27 Queen Street (July 1966), sheet 2, Thomas M. Osborn, del., copies at HABS, PHC. These buildings are also cited by Margaret B. Tinkcom, "Southwark, A River Community: Its Shape and Substance," *Proceedings of the American Philosophical Society* 114 (August 1970): 339. Brick partition walls also were not uncommon in early-nineteenth-century factories, such as Bazel Graves's soap and candle manufactory on Spruce Street between Front and Second, MA unnumbered (no policy issued) (February 1823), photocopy at PHC.

10. MA 2105 (December 1805).

11. Girard's house is described in John Bach McMaster, *The Life and Times of Stephen Girard* (Philadelphia: J. B. Lippincott Co., 1918), 1: 278–79. The Dowers-Okill House is described in PC 1229 (March 3, 1767). These two dwellings were not exceptions. Of the nine buildings on the east side of Water Street between Arch and Race streets that were surveyed by the Philadelphia Contributionship between 1757 and 1783, four were dwellings, one a dwelling with an adjoining store, and two with stores below, one a tavern with rented rooms above, and one a granary and bake house. PC 377 (June 7, 1757); PC 541 (October 7, 1760); PC 626 (February 16, 1761); PC 1229 (March 3, 1767); PC 2063, 2064, 2065 (September 2, 1783); PC 624 (February 16, 1761); PC 2066, 2067 (September 2, 1783); PC 625 (February 16, 1761), photocopies at PHC.

12. One lintel was displayed at "See What They Sawed, An Exhibition on the Occasion of the 200th Anniversary of Building Carpenters' Hall," Visitors' Center (then First Bank of the United States), Independence National Historical Park, 1970–73.

13. Thompson Westcott, *The Historic Mansions and Buildings of Philadelphia* (rev. ed.; Philadelphia: Walter H. Barr, 1895), pp. 68–78; PC 179 (1763); PC 1767–69 (November 2, 1773), photocopies at PHC.

14. Printer and bookseller Eleazer Oswald kept his shop here until July, 1789, when James Stokes began renting it for his fancy hardware store. Francis White, *The Philadelphia Directory* (Philadelphia: Young, Stewart and McCullogh, 1785), p. 54; James Stokes' Ledger, 1783–1801, pp. 25, 79, Paumgarten Collection, MSS, HSP. Joseph Jackson, *Encyclopedia of Philadelphia* (Harrisburg, Pa.: National Historical Association, 1933), 4: 955–57. By 1854 there were two shops on the ground story, a cooper's ware room in the cellar, and storage on the upper floors of the former coffee house. An old photo (c. 1855) in City Archives shows it to have fallen from its former high fashion as described in Watson's *Annals*, 1:222; PC 179 (Re-Surveys 1842, 1845), photocopy at PHC.

15. Sam Bass Warner, Jr., *The Private City* (Philadelphia: University of Pennsylvania Press, 1968), pp. 49–58.

16. The most thorough discussion of the building's architecture and use is by Agnes Addison Gilchrist, "The Philadelphia Exchange: William Strickland, Architect," *Historic Philadelphia*, pp. 86–95. Updating her research are insurance surveys: PC 11807 (March 1873, August 1903), and FFI 22234 (March 1855), photocopies at HABS, PHC; PRERBG 6 (October 7, 1891): 639; 16 (November 20, 1901): 764.

17. Demolished in the 1850's, the City Tavern has been reconstructed by the National Park Service. PC 1767–69 (November 2, 1773) (Re-Survey November 20, 1834); Mark Forest, "City Tavern to Be Rebuilt," *Today: The Inquirer Magazine*, March 1, 1959, pp. 7–8; *Philadelphia Inquirer*, February 23, 1976.

18. Gilchrist, "Philadelphia Exchange," pp. 86–95; PRERBG 16 (November 20, 1901): 764; PC 11807 (August 1903).

19. Scharf and Westcott, *History of Philadelphia*, 3:2344–45; PRERBG 7 (April 20, 1892): 1099; *PL*, March 25, 1896, March 8, 1901, April 29, 1901, May 2, 1901; *Philadelphia Record*, April 27, 1901.

20. The building of the Bourse was one of the feature architectural and business events of the day. It was begun in 1893 and finished two years later; G. W. and W. D. Hewitt were the architects. A thorough discussion and architectural rendering of the proposed structure appears in *PL*, May 10, 1893. See also PC 10851 (October 1933), photocopy at PHC.

21. In 1849 John Riddell, architect, designed a store for Faust, Winebrenner and Co. on North Third Street and a five-story store at the southwest corner of Third and

Spring streets in this manner. *PL*, April 17, 1849; *McElroy's Philadelphia City Directory for 1858* (Philadelphia: Edward C. and John Biddle, 1858), p. 238.

22. PC 4669 (April 1829); PC 4752 (June 1830); PC 4956 (August 1833); FFI 1654 (September 1836); PC 6060 (November 1844), photocopies at HABS, PHC.

23. PC 4956 (August 1833).

24. The four-story warehouses were at 110–12 Church Alley. PC 4865 (January 1832), PC 4872 (January 1832), photocopies at PHC. Construction of the Naval Asylum began in 1827, was suspended between 1829 and 1832, and completed in 1833. Agnes Addison Gilchrist, *William Strickland: Architect and Engineer, 1788–1854* (Philadelphia: University of Pennsylvania Press, 1950), pp. 7–8; William B. Bassett, "United States Naval Asylum, Biddle Hall," HABS Report, PA–1094, p.l.

25. Richard G. Carrott, "The Architect of the Pennsylvania Fire Insurance Building," *JSAH* 20 (October 1961): 138–39; Harold N. Cooledge, Jr., "A Sloan Checklist, 1849–1884," *JSAH* 19 (March 1960): 36; Samuel Sloan, *City and Suburban Architecture* (Philadelphia: J. B. Lippincott Co., 1859), pl. 56; *PL*, August 28, 1854.

26. Henry-Russell Hitchcock, *Early Victorian Architecture* (New Haven: Yale University Press, 1954), 1:347.

27. Winston Weisman, "Philadelphia Functionalism and Sullivan," *JSAH* 20 (March 1961): 3–19.

28. Agnes Addison Gilchrist, "Jayne Building, Philadelphia, Pa.," HABS Report (PA–188), p. 2. Another source may have been New England, specifically Boston and Providence. Weisman ("Philadelphia Functionalism," p. 6) points out that "structural design involving skeleton stone construction originated in the New England area in the early 1820's" as seen in Alexander Parris's Quincy Market Stores, Boston, and William Holden Greene's Granite Block, Providence, both built in 1824. Since Joseph C. Hoxie and Stephen D. Button, two architects who figure prominently in the use of the utilitarian Italianate, were born and raised in Rhode Island and Connecticut respectively, and Button was married to a Massachusetts woman, it is possible that they brought the new mode to Philadelphia independently. Sixth Census, 1840, Hudson County, New Jersey, MSS, New Jersey State Library; "Biographical Sketch of Mr. S. D. Button, Architect, Philadelphia," *AABN* 37 (July 16, 1892): 37–38.

29. *PL*, July 10, 1852, February 9, 1854, January 22, 1856; Casper Souder, "A History of Chestnut Street" (n.d.), pp. 40–41, MSS, HSP; Registry Unit 1S2–213.

30. M. Field, *City Architecture* (New York: G. P. Putnam & Co., 1853), p. 34.

31. *Ibid.*, p. 17.

32. *PL*, February 9, 1854; January 22, 1856.

33. PC 9439 (c. 1854); PC 4956 (August 1833). Skylights were used on one-story rear buildings without exposed walls during the 1830's, but skylight systems such as that of the Leland Building were not employed until about 1850, when some became quite special, such as the one designed by G. Runge for Morris L. Hallowell & Co., which admitted only "pure northern light" (*PL*, December 20, 1853). An early one-story skylight was at 52 South Second Street (MA 5340, February 1834, photocopy at PHC). Also, iron sliders were used as early as 1833 on the Trotter warehouse, but iron shutters were much more common before the 1850's. Among two contemporary stores on Chestnut Street, numbers 217 and 223, and six on North Third Street between Market and Arch, another wholesale dry goods district, the average dimensions of fireproofs were 7 feet by 6 ¹/₂ feet, and five of them were one story and three were two stories. MA 6350 (May 11, 1853); PC 9956 (December 4, 1857); PC 8688 (September 20, 1851); PC 8683 (November 3, 1851); PC 5570 (February 21, 1854); PC 9336 (February 28,

1854); PC 8146 (August 17, 1850); PC 10505 (January 2, 1861), photocopies at PHC.

34. *McElroy's Philadelphia Directory for 1855* (Philadelphia: Edward C. & John Biddle, 1855), pp. 774–75.

35. PC 2381 (December 1, 1789), photocopy of PHC; Deed Books AWM 48–106, AWM 59–453; *McElroy's Directory 1855*, p. 264.

36. Deed Book TH 61–470; *PL*, January 16, 1854. For a history of this famous firm, see Scharf and Westcott, *History of Philadelphia*, 3:2101–2, and Burton Alva Konkle, "Francis Martin Drexel and Philadelphia Leadership in International Finance," *The Girard Letter* 14 (December 1933): 1–2.

37. *PL*, January 12, 1855; July 31, 1855; *McElroy's Directory 1858*, pp. 774–75.

38. MA 6415 (June 6, 1854), photocopy at HABS, PHC.

39. MA 6499 (July 15, 1855); PC 9571 (August 9, 1854), photocopies at PHC.

40. MA 6499 (July 15, 1855); PC 9571 (August 9, 1854); MA 6415 (June 6, 1854), photocopies at PHC.

41. The early history of the Bank of North America is reviewed by Scharf and Westcott, *History of Philadelphia*, 3:2089–93. The most thorough history of the First Bank of the United States is James O. Wettereau, "The Oldest Bank Building in the United States," *Historic Philadelphia*, pp. 70–79.

42. The Second Bank's history is given by Bray Hammond, "The Second Bank of the United States," *Historic Philadelphia*, pp. 80–85. Its architecture is evaluated by Talbot Hamlin, *Greek Revival Architecture in America* (New York: Oxford University Press, 1944), pp. 75–77, and Tatum, *Penn's Great Town*, pp. 65–66.

43. Gilchrist, *William Strickland*, p. 4.

44. *PL*, April 12, 1849.

45. Tatum, *Penn's Great Town*, pp. 194–95, pl. 119, illustrates and gives the basic historical data for Bank Row.

46. Quoted from FFI 22148 (February 17, 1855), photocopy at HABS, PHC. Construction began in 1854. Insurance Policy 252 (February 11, 1854), MSS, Archives Department, INA.

47. FFI 22148 (February 17, 1855).

48. Ibid; Poulson's Scrap-books 7:159, MSS, LCP. For later alterations see *PRERBG* 18 (January 7, 1903): 1; Building Permits 4910 (1917), 5679 (1919).

49. For the history of this banking house, see Nicholas Wainwright, *The Philadelphia National Bank* (Philadelphia: Philadelphia National Bank, 1953).

50. Harold B. Dickson, *A Hundred Pennsylvania Buildings* (State College, Pa.: Bald Eagle Press, 1954), pl. 67.

51. Poulson's Scrap-books 9:34–35, MSS, LCP. For later alterations see *PRERBG* 7 (June 8, 1892): 1212; 7 (November 9, 1892): 1587; 18 (May 6, 1903): 278.

52. George E. Thomas and James F. O'Gorman, "Catalogue of Selected Buildings," in James F. O'Gorman, *The Architecture of Frank Furness* (Philadelphia: Philadelphia Museum of Art, 1973), pp. 110–15.

53. Tatum, *Penn's Great Town*, pp. 194–95.

54. *PL*, July 22, 1872; William A. Armstrong, Account Book, 1867–75, MSS, HSP; Elizabeth Biddle Yarnall, *Addison Hutton: Quaker Architect, 1834–1916* (Philadelphia: Art Alliance Press, 1974), p. 70; *AABN* 8 (September 18, 1880): 144. For the bank's nineteenth-century history, see Harrison S. Morris, *A Sketch of the Pennsylvania Company for Insurances on Lives and Granting Annuities* (Philadelphia: J. B. Lippincott Co., 1896).

55. Included with the Provident Trust and Guarantee Trust would be the National

Bank of the Republic, 313 Chestnut Street, by Furness and Evans in 1884. Thomas and O'Gorman, "Catalogue," *Architecture of Frank Furness*, pp. 150–51.

56. *PL*, August 10, 1893; *PRERBG* 8 (August 16, 1893): 521; *PC* 14103 (June 1894), photocopy at HABS, PHC.

57. Personnel Office, Minute Book of the Building Committee, MSS, First Pennsylvania Banking and Trust Company.

58. See Nicholas Wainwright, *A Philadelphia Story* (Philadelphia: Philadelphia Contributionship, 1952), for the company's history; pp. 96–128 deal with the erection of the present building.

59. Thomas U. Walter to Directors of Philadelphia Contributionship, July 4, 1835, MSS, PC, photocopies at HABS, PHC.

60. Minutes of the Board of Directors of Philadelphia Contributionship, From 1859, May 18th to 1866, December 16th, pp. 125, 128, MSS, PC, photocopies at HABS, PHC; MA 8908 (December 1898), photocopies at HABS, PHC.

61. MA 8908 (December 1898); *PL*, June 22, 1898.

62. James C. Massey, "Frank Furness in the 1880's," *Charette* 43 (October 1963): 25; Spring Garden Insurance Company, see *AABN* 8 (September 18, 1880): 144.

63. Building Permit 4731 (1911).

64. Building Permit 6442 (1928).

65. Building Specifications, July 24, 1835, MSS, PC, photocopies at HABS, PHC.

66. Bill of Thomas U. Walter, January 3, 1837, MSS, PC; Building Permit 4731 (1911).

67. For praise of this imaginative piece of architecture, see Thomas Hine in *Philadelphia Inquirer*, February 28, 1975; Nessa Forman in *Sunday Bulletin*, August 18, 1974; see also "Penn Mutual Building, Philadelphia," *The Architectural Record* 149 (April 1971): 42.

68. DeWitt C. Baxter, *The Baxter Panoramic Business Directory* (Philadelphia: DeWitt C. Baxter, 1879), n.p.; Francis James Dallett, *An Architectural View of Washington Square* (Philadelphia: By the author, 1968), pp. 28–31.

69. *PL*, April 14, 1852; January 24, 1853; "Horstmann Factory," *Gleason's Pictorial Drawing-Room Companion* 6 (January 21, 1854): 48; Building Permits 1506, 2908 (1934).

70. *Illustrated Philadelphia* (New York: American Publishing Co., 1889), p. 152; *Gopsill's Philadelphia City Directory for 1872* (Philadelphia: James Gopsill, 1872), p. 1164.

71. *PL*, March 26, 1857; August 31, 1857; *Philadelphia Shopping Guide and Housekeeper's Companion for 1859* (Philadelphia: S. E. Cohen, 1859), p. 51; Poulson's Scrapbooks 10:9, MSS, LCP; Nicholas Wainwright, *Philadelphia in the Romantic Age of Lithography* (Philadelphia: Historical Society of Pennsylvania, 1958), p. 153.

72. Jackson, *Market Street*, pp. 220–21; Campbell Scrapbook, vol. 54, MSS. HSP; *PL*, July 27, 1872.

73. *PL*, January 9, 1899, relates the store's early development.

74. *PL*, January 11, 1895, January 6, 1907, March 22, 1907; *PRERBG* 8 (February 8, 1893): 79; 9 (December 26, 1894): 643; 11 (August 12, 1896): 657; 12 (July 7, 1897): 423; Building Permit 1279 (1907); PC 13860 (October 17, 1907) MSS, HSP. The North Seventh Street extension was built in 1917–18, Simon and Bassett, architects; Building Permit 3720 (1917).

The older stores, 709–27 Market, that stand between the end pavilions by Collins and Autenreith were incorporated into the department store during this time with Collins and Autenreith doing the design work. *PL*, September 30, 1898; *PRERBG* 13 (November 16, 1898): 745; 13 (November 23, 1898): 763; 15 (January 3, 1900): 1; 15 (January 17, 1900): 37.

The report that has led to the unquestioned acceptance of Lit Brothers as a cast-iron building appears to have originated in Theo B. White et al., *Philadelphia Architecture in the Nineteenth Century* (Philadelphia: University of Pennsylvania Press, 1953), pl. 72.

75. Hannah Benner Roach, "Elfreth's Alley" (Philadelphia, Elfreth's Alley Association pamphlet, 1972); PC 578 (February 1761); PC 611, (May 1761); PC 1107 (September 3, 1766); PC 2147 (September 1784); PC 2148 (September 1784); PC 2765 (June 30, 1796); PC 2783 (December 2, 1796); MA 1660 (September 1803); PC 6118 (March 7, 1845); PC 6192 (June 4, 1845); PC 6450 (January 8, 1846); PC 7048 (August 30, 1847); PC 475 (Re-Survey April 1811); Fire Association Insurance policy 8862 (April 4, 1844), photocopies at HABS, PHC. Briefs of title of each property by Hannah Benner Roach at PHC.

76. Roach, "Elfreth's Alley."

77. Nos. 113–15 and 121 were new in 1811. MA 3197, 3198 (December 1811); PC 3427 (October 1811).

78. MA 175 (February 1788). The school was built in 1761. Drawing of Old German School-House by D. J. Kennedy (1860), Society Collection, MSS, HSP.

79. PC 1269 (February 1769); PC 1459 (December 1770); Deed Books I1–421 (1763), I10–158 (1771), G8–426 (1747), D12–137 (1762), AWM 73–100 (1848).

Although pent eaves were common during the colonial period, they were built later, as in the case of 131 North Fourth Street, whose pent eave between the second and third floors must have been put up c. 1800 when the third story was added. See PC unnumbered (no policy issued, May 1819), photocopy at PHC. It describes the house with a third story and a pent (neither of which was mentioned in the 1770 survey) and a roof two-thirds worn. Since cedar shingles lasted about thirty years, the pent and third floor were added c. 1800.

80. Census of Philadelphia, 1790, MSS, National Archives, microfilm copy at FLP; County Tax Assessment Ledger, Lower Delaware Ward, 1795, MSS, Philadelphia City Archives.

81. Tatum, *Penn's Great Town*, p. 164, pl. 33; Talbot Hamlin, *Benjamin Henry Latrobe* (New York: Oxford University Press, 1955), p. 152; William John Murtagh, "The Philadelphia Row House," *JSAH* 16 (December 1957): 8.

82. H. M. Pierce Gallagher, *Robert Mills, Architect of the Washington Monument, 1781–1855* (New York: Columbia University Press, 1935), p. 3.

83. Kenneth Ames, "Robert Mills and the Philadelphia Row House," *JSAH* 27 (May 1968): 140–42.

84. MA 191 (November 20, 1795), photocopies at HABS, PHC; Arnold Nicholson, "The Church's White House," *The Episcopalian* 132 (December 1967): 17–19.

85. Murtagh, "Philadelphia Row House," p. 12.

86. PC 4744 (March 9, 1830), photocopies at HABS, PHC.

87. PC 8474 (November 1879; Re-Survey January 1901).

88. It is interesting to note that these patriots evidently dated the nation's birth from the first session of the Continental Congress and the outbreak of hostilities in 1775, not from the signing of the Declaration of Independence. For a discussion of early Quaker meeting houses, see Edwin B. Bronner, "Quaker Landmarks in Early Philadelphia," *Historic Philadelphia*, pp. 210–16. The history of the Free Quakers is presented by Charles Wetherill, *History of the Religious Society of Friends Called by Some the Free Quakers* (Philadelphia: By the author, 1894). Excellent architectural histories of the two buildings under discussion are Lee H. Nelson and Penelope Hartshorne Batcheler, "An Architectural Study of Arch Street Meeting House," (report for

Philadelphia Yearly Meeting of the Religious Society of Friends, 1968), and Charles E. Peterson, F.A.I.A., "Notes on the Free Quaker Meeting House," (report for Harbeson, Hough, Livingston and Larson, AIA, 1966).

89. Donald Richard Friary, "The Architecture of the Anglican Church in the Northern American Colonies: A Study of Religious, Social, and Cultural Expression" (Ph.D. dissertation, University of Pennsylvania, 1971), 1:339, 368; Morrison, *Early American Architecture*, pp. 537–38; Tatum, *Penn's Great Town*, pp. 28–30; Robert W. Shoemaker, "Christ Church, St. Peter's, and St. Paul's," *Historic Philadelphia*, 187–90.

90. Alexander Mackie, "The Presbyterian Churches of Old Philadelphia," *Historic Philadelphia*, pp. 217–22; Tatum, *Penn's Great Town*, pp. 27, 39–40, 69, 160–61, 174 and pls. 23, 61; Theodore Sizer, "Philadelphia's First Presbyterian Church by 'Mr. Trumbul,' " *JSAH* 9 (October 1950): 20–22. For descriptions of the interior appointments of the first two Second Presbyterian churches, see PC 901–3 (June 7, 1763), MA 5516 (September 2, 1837), photocopies at PHC.

91. Tatum, *Penn's Great Town*, pp. 30–32, pl. 13; Scharf and Westcott, *History of Philadelphia*, 2: 1418–21.

92. David Van Horne, *A History of the Reformed Church in Philadelphia* (Philadelphia, 1876); Daniel Miller, *Early History of the Reformed Church in Pennsylvania* (Reading, Pa.: By the author, 1906), pp. 29–31; John T. Faris, *Old Churches and Meeting Houses in and around Philadelphia* (Philadelphia: J. B. Lippincott Co., 1926), pp. 67–75.

93. Fred Pierce Corson, "St. George's Church: The Cradle of American Methodism," *Historic Philadelphia*, pp. 230–36.

94. The erection of Mikveh Israel synagogue and the discriminations the group quietly endured are described by Edwin Wolf 2nd and Maxwell Whiteman, *The History of the Jews of Philadelphia from Colonial Times to the Age of Jackson* (Philadelphia: The Jewish Publication Society of America, 1956), pp. 117–21. For St. Joseph's and St. Augustine's, see Arthur J. Ennis, *Old St. Augustine's Catholic Church in Philadelphia* (Philadelphia: St. Augustine's Catholic Church, 1965), pp. 9–15; Dennis C. Kurjack, "St. Joseph's and St. Mary's Churches," *Historic Philadelphia*, pp. 199–203. The most thorough analysis of the destructive "Native American" riots is Michael Feldberg, *The Philadelphia Riots of 1844: A Study of Ethnic Conflict* (Westport, Conn.: Greenwood Press, 1975). Scharf and Westcott, *History of Philadelphia*, 1: 664–74 describe the riots; and Sam Bass Warner, Jr., *The Private City: Philadelphia in Three Periods of Its Growth* (Philadelphia: University of Pennsylvania Press, 1968), pp. 125–27 discusses the causes of a series of riots between 1834 and 1849.

95. Probably the best account of the early years of the American Philosophical Society is Brooke Hindle, *The Pursuit of Science in Revolutionary America* (Chapel Hill: University of North Carolina Press, 1956), pp. 127–45. For a history of the hall, see William E. Lingelbach, "Philosophical Hall: The Home of the American Philosophical Society," *Historic Philadelphia*, pp. 43–69. *PRERBG* 5 (June 11, 1890): 337; *PL*, August 16, 1890.

96. Charles E. Peterson, "Library Hall: Home of the Library Company of Philadelphia, 1790–1880," *Historic Philadelphia*, pp. 96–128; PC 2414, 2415 (November 1790), photocopy at PHC.

97. The most recent and thorough history of the institute is Bruce Sinclair, *Philadelphia's Philosopher Mechanics* (Baltimore: Johns Hopkins Press, 1975). Its architecture is discussed by Tatum, *Penn's Great Town*, p. 70, and Matthew Baigell, "John Haviland in Philadelphia, 1818–1826," *JSAH* 25 (October 1966): 206.

98. The early history of the institution is given by Arthur M. Kennedy, "The Athe-

naeum: Some Account of Its History from 1814 to 1850," *Historic Philadelphia*, pp. 260–65. For its architecture, see Robert C. Smith, *John Notman and the Athenaeum Building* (Philadelphia: Athenaeum of Philadelphia, 1951); Tatum, *Penn's Great Town*, pp. 92, 98, 184; Dallett, *Architectural View of Washington Square*, p. 20; a copy of Notman's printed specifications (1845) are at HABS; MA 7394 (1868), photocopies at HABS, PHC.

99. Arthur Hobson Quinn, "The Theatre and the Drama in Old Philadelphia," *Historic Philadelphia*, pp. 313–17; Harold Kirker, "The New Theatre, Philadelphia, 1791–92," *JSAH* 22 (March 1963): 36–37; George B. Tatum, "The Early Theatres of Philadelphia," *Bulletin of the Philadelphia Old Town Historical Society* 1 (October 1971): 11–12.

100. Tatum, "Early Theatres," pp. 11–16; Derek Naabe, "Philadelphia's Theatre Heritage: The Arch Street Theatre," *Germantowne Crier* 18 (December 1966): 107–10; Derek Naabe, "Philadelphia's Theatre Heritage: The Walnut Street Theatre," *Germantowne Crier* 22 (Winter 1970): 14–18; Stuart B. Smith, "The Walnut Street Theater, 1809–1834" (M.A. thesis, University of Delaware, 1960), pp. v–vi, 1, 5–7, 16–18; Gilchrist, *William Strickland*, p. 65; Tatum, *Penn's Great Town*, p. 187; Miscellaneous Early Papers, Musical Fund Society Collection, MSS, HSP; *PRERBG* 6 (May 13, 1891): 289; *Philadelphia Record*, August 4, 1893; Building Permit 342 (1946); F. J. Burghart, "Formal Concert Hall," *Today: The Inquirer Magazine*, June 7, 1970, 11.

101. MA 2307 (September 1806, May 1848, June 1881, December 1937); *Biographies of Successful Philadelphia Merchants* (Philadelphia: James K. Simon, 1864), pp. 40–41; *PL*, August 28, 1850, January 21, 1851; PC 2299, 2300 (December 1781), photocopies at HABS, PHC. Briefs of Title for 112–16 Cuthbert Street at PHC. For a study of the area's cast-iron architecture, see Antoinette J. Lee, "The Rise of the Cast Iron District in Philadelphia" (Ph.D. dissertation, George Washington University, 1975).

102. The mall project has been controversial. For a generally favorable evaluation, see Roul Tunley, "Comeback of a Shabby City," *Saturday Evening Post*, December 5, 1959, pp. 32–35, 75–78. Lewis Mumford has praised the decision of Charles E. Peterson, the federal government's chief architect on the project during the 1950's, to preserve more than colonial buildings in Independence National Historical Park, but he has been critical of the mall for imposing on Independence Hall "an aesthetic burden that only a vast palace . . . could hope to carry off." See Mumford's "The Skyline: Historic Philadelphia," *The New Yorker*, November 17, 1956, pp. 138–40, 143–48, February 9, 1957, pp. 100–106, April 6, 1957, pp. 132–41, April 13, 1957, pp. 155–62. The first of the large office structures has proven the best: the Rohm and Haas Building, designed by Pietro Belluschi in collaboration with George M. Ewing Company. "Independence Mall Development," *The Architectural Record* 134 (July 1963): 12; James Van Trump, "Elegance on the Mall: The Rohm and Haas Building in Philadelphia," *Charette* 46 (April 1966): 6–7.

103. The Penn's Landing Project is described in *Third Annual Report to the Board of Directors* (Philadelphia: OPDC Penn's Landing Corporation, 1974).

Penn's City: Center Square

1. John Ord succeeded McArthur, but he apparently had a dispute with some members of the Commission and was replaced in 1894 by W. Bleddyn Powell, who, early in his career, had been associated with Thomas U. Walter. Minutes of the Commissioners for the Erection of the Public Buildings, 1887–1897, MSS, Philadelphia City

Archives; Letter Book of the City Architect, 1901–1906, pp. 14, 54, 65 in ibid. John Maass has written the building's architectural history: "Philadelphia City Hall," *Charette* 44 (January 1964): 23–26; "Philadelphia City Hall: Monster or Masterpiece?" *Journal of the American Institute of Architects* 43 (February 1965): 23–30. For an investigation of the Byzantine politics surrounding the building's erection, see Howard Gillette, Jr., "Philadelphia's City Hall: Monument to a New Political Machine," *PMHB* 97 (April 1973): 233–49. The building's statistics are in Frederick Faust, *The City Hall, Philadelphia* (Philadelphia: F. Faust, 1897), pp. 1–2, 5.

2. City Hall's sculpture is discussed by George Gurney in *Sculpture of a City*, ed. Nicholas B. Wainwright (New York: Walker Publishing Co., 1974), pp. 94–109. Each statue, bas-relief, ornament, and railing is illustrated in *Sculptures and Ornamental Details in Bronze and Iron of the New City Hall*, 5 vols. (Philadelphia, 1883).

3. A Friends meeting house stood on the southwest corner of the square as early as 1687, when it was assumed that the city's plan would be developed more or less uniformly. When this expectation failed to materialize the building was dismantled c. 1700. George B. Tatum, *Penn's Great Town* (Philadelphia: University of Pennsylvania Press, 1961), p. 25.

4. Ibid., pp. 57–58; Talbot Hamlin, *Benjamin Henry Latrobe* (New York: Oxford University Press, 1955), pp. 157–67.

5. Kenneth Ames, "Robert Mills and the Philadelphia Row House," *JSAH* 27 (May 1968): 140–42.

6. Robert Ennis, "Thomas Ustick Walter," Lecture to Philadelphia Chapter, Society of Architectural Historians, December 13, 1973; Letter Book of the City Architect, 1901–1906, p. 54, MSS, Philadelphia City Archives.

7. Tatum, *Penn's Great Town*, p. 71, pl. 64; William John Murtagh, "The Philadelphia Row House," *JSAH* 16 (December 1957): 13.

8. Tatum, *Penn's Great Town*, p. 94.

9. See briefs of title in files of PHC. FFI 24376–79 (July 7, 1856), photocopies at HABS, PHC.

10. See briefs of title in files of PHC.

11. Elisabeth B. Walton, "The Building Art of Wilson Eyre: A Study of 'Queen Anne' Motifs in American Architecture" (M.A. thesis, Pennsylvania State University, 1962), p. 71; Building Permit 8964 (1923).

12. Registry Unit 4S1–162–68, 4S1–173.

13. Tatum, *Penn's Great Town*, pp. 93–94, 185, pl. 94; Harold N. Cooledge, Jr., "A Sloan Check List, 1849–1884," *JSAH* 19 (March 1960): 36.

14. Edward Teitelman and Richard W. Longstreth, *Architecture in Philadelphia: A Guide* (Cambridge, Mass.: MIT Press, 1974), p. 92; Julian Millard, "The Work of Wilson Eyre," *The Architectural Record* 14 (October 1903): 285, 311–12.

15. James C. Massey, "Frank Furness in the 1870's," *Charette* 43 (January 1963): 16; George E. Thomas and James F. O'Gorman, "Catalogue of Selected Buildings," in James F. O'Gorman, *The Architecture of Frank Furness* (Philadelphia: Philadelphia Museum of Art, 1973), p. 99.

16. *Philadelphia and Its Environs* (Philadelphia, 1875), p. 70; Registry Unit 2S23–8, 2S23–233; MA 8545, (September 22, 1883), PC 14047 (March 29, 1894), photocopies at HABS, PHC.

17. *PRERBG* 11 (June 10, 1896): 463; 11 (July 29, 1896): 619; Moses King, *Philadelphia and Notable Philadelphians* (New York: Moses King, 1902), p. 62.

18. Will Book 182 (1895), p. 358, City of Philadelphia, Department of Records; In-

ventory Book 23 (1895), p. 295 in ibid; Orphans' Court, Marriage License No. 99814 (1898); *Gopsill's Philadelphia City Directory for 1894* (Philadelphia: James Gopsill's Sons, 1894), p. 216.

19. *Sunday Bulletin,* April 28, 1974, August 18, 1974; *Philadelphia Inquirer,* October 5, 1974.

20. Norman J. Johnston, "The Caste and Class of the Urban Form of Historic Philadelphia," *Journal of the American Institute of Planners* 32 (November 1966): 334–38.

21. Agnes Addison Gilchrist, *William Strickland: Architect and Engineer, 1788–1854* (Philadelphia: University of Pennsylvania Press, 1950), pp. 34, 62.

22. Phoebe B. Stanton, *The Gothic Revival and American Church Architecture* (Baltimore: Johns Hopkins Press, 1968), pp. 115–25; Jonathan Fairbanks, "John Notman, Church Architect" (M.A. thesis, University of Delaware, 1960), pp. 74–79, 111–33.

23. Alfred Mortimer, *S. Mark's Church, Philadelphia, and Its Lady Chapel* (Philadelphia: By the author, 1909), pp. 22–33.

24. Both churches are documented and discussed by Fairbanks, "John Notman," pp. 88–89, 171–204.

25. May Lilly, *The Story of St. Clement's Church, Philadelphia* (Philadelphia: St. Clement's Church, 1964), p. 5.

26. *Evening Bulletin,* September 10, 1968, September 11, 1968, September 17, 1968, October 10, 1968; *Philadelphia Inquirer,* September 8, 1968, September 15, 1968, September 19, 1968; *Preservation News* 8 (October 1968): 1.

27. *PL,* August 4, 1853, October 7, 1853, November 18, 1870. The First Baptist Church is given as an illustration of the early Romanesque Revival by Carroll L. V. Meeks, "Romanesque before Richardson in the United States," *The Art Bulletin* 35 (March 1953): 17–33. The Arch Street Methodist Church measures up well in comparison with other noted Gothic Revival churches in Philadelphia. See James D. Van Trump, "The Gothic Fane," *Charette* 43 (September 1963): 61. The Arch Street Church is shown in Elizabeth Biddle Yarnall, *Addison Hutton: Quaker Architect, 1834–1916* (Philadelphia: The Art Alliance Press, 1974), pls. 9–10.

28. *PL,* March 9, 1898, April 20, 1898, May 16, 1898, June 23, 1898, December 3, 1898.

29. *PL,* March 10, 1853; *The Presbyterian,* May 28, 1853; Teitelman and Longstreth, *Architecture in Philadelphia,* pp. 94, 102.

30. George Frisbie Whicher, *This Was a Poet: A Critical Biography of Emily Dickinson* (New York: Charles Scribner's Sons, 1936), pp. 99–112. This theory is severely questioned by David Higgins, in *Portrait of Emily Dickinson* (New Brunswick, N.J.: Rutgers University Press, 1967), pp. 79–84.

31. "The Cash Book of the Treasurer of the Building Committee" (1852–1856), MSS, Arch Street Presbyterian Church.

32. *PL,* October 12, 1855.

33. Theo B. White et al., *Philadelphia Architecture in the Nineteenth Century* (Philadelphia: University of Pennsylvania Press, 1953), pp. 27–28, pls. 38, 39; Tatum, *Penn's Great Town,* pp. 94, 186.

34. Haviland's biographer sees the old Pump House, which still stood on Center Square during the 1820's, as the probable source for the institution's elements and composition, which Haviland integrated into a flat planar facade. Matthew Baigell, "John Haviland in Philadelphia," *JSAH* 25 (October 1966): 205–6.

35. The motivations for founding the Academy of Music and the details of the competition are explained by James V. Kavanaugh, "Three American Opera Houses:

The Boston Theatre, The New York Academy of Music, The Philadelphia American Academy of Music" (M.A. thesis, University of Delaware, 1967), pp. 52–60, and Tatum, *Penn's Great Town*, pp. 186–87. The hall's acoustics are still praised; the acoustician Leo L. Beranek considers the Academy "unquestionably the finest opera house in the United States" (*Music, Acoustics and Architecture* [New York: John Wiley & Sons, 1962], p. 165).

36. Tatum, *Penn's Great Town*, p. 95.

37. O'Gorman, *Architecture of Frank Furness*, pp. 32, 38.

38. Thompson Westcott, *The Official Guide Book to Philadelphia* (Philadelphia: Porter and Coates, 1875), p. 260; *PRERBG* 11 (July 1, 1896): 523.

39. The Philadelphia Club is the oldest city club not only in Philadelphia but also in the United States; started in 1834, it is two years older than New York's Union Club. See *The Philadelphia Club, 1834–1934* (Philadelphia, 1934). An engaging discussion of Philadelphia clubs is found in Nathaniel Burt, *The Perennial Philadelphians* (Boston: Little, Brown and Co., 1963), pp. 256–70.

40. Because of its size and function, the building has had only minimal changes during the club's tenure. MA 5520 (October 12, 1837; November 24, 1850; November 15, 1858; March 13, 1889), photocopies at PHC; brief of title in PHC files.

41. That distinction is now history; in late 1975 the club agreed to admit Democrats to membership. *Philadelphia Inquirer*, December 12, 1975; Westcott, *Official Guide*, pp. 247–49; *Sunday Bulletin*, February 11, 1962. For the club's history see Maxwell Whiteman, *Gentlemen in Crisis: The First Century of the Union League of Philadelphia, 1862–1962* (Phildelphia: Union League of Philadelphia, 1975).

42. Tatum, *Penn's Great Town*, p. 107.

43. Alfred Bendiner, *Bendiner's Philadelphia* (New York: A. S. Barnes and Co., 1964), p. 16.

44. *PRERBG* 23 (September 16, 1908): 597; 24 (February 10, 1909): 84; 25 (April 27, 1910): 256; 25 (November 9, 1910): 755; Building Permit 60 (1911).

45. A thorough study of the Masonic Temple's architectural competition is John C. Poppeliers, "The 1867 Philadelphia Masonic Temple Competition," *JSAH* 26 (December 1967): 279–84. For a layman's tour, see Jeannie Autret, "The Sumptuous Clubhouse," *Today: The Inquirer Magazine*, April 28, 1974, pp. 32–33, 35. A contemporary description is "The New Masonic Temple, Philadelphia," *The Architectural Review and American Builders' Journal* 1 (July 1868): 71–73.

46. Tatum, *Penn's Great Town*, p. 117; PC Surveyors' Reports Book 3 (February 25, 1888), p. 57.

47. *Evening Public Ledger*, October 24, 1940; *Philadelphia Record*, March 24, 1946; *Keystone Motorist* 38 (March 1946): 1.

48. *Philadelphia Inquirer*, December 17, 1975; *The New Paper*, January 3, 1976.

49. Richard J. Webster, "Stephen D. Button: Italianate Stylist" (M.A. thesis, University of Delaware, 1963), p. 87.

50. Thomas Hine paid tribute to these stores in *Philadelphia Inquirer*, August 18, 1974.

51. Westcott, *Official Guide*, p. 164; *PL*, August 24, 1875.

52. *PRERBG* 6 (August 6, 1890): 466; *PL*, August 9, 1890; DeWitt C. Baxter, *Baxter's Panoramic Business Directory of Philadelphia* (1880); King, *Philadelphia and Notable Philadelphians*, p. 23.

53. Joseph W. Appel, *The Business Biography of John Wanamaker* (New York: The Macmillan Co., 1930), pp. 74–91. Herbert Adams Gibbons, *John Wanamaker* (New York: Harper and Brothers, 1926), 1:212–39.

54. These two stations are discussed by Carroll L. V. Meeks in *The Railroad Station: An Architectural History* (New Haven: Yale University Press, 1956), pp. 103–4.

55. Tatum, *Penn's Great Town*, pp. 112, 196, pl. 121.

56. *PRERBG* 9 (April 4, 1894): 157; 9 (June 13, 1894): 277; 9 (August 1, 1894): 373; 9 (August 15, 1894): 399; *PL*, January 5, 1895, April 2, 1902.

57. *PRERBG* 8 (July 5, 1893): 413; 8 (August 2, 1893): 489; 8 (October 25, 1893): 653; *Philadelphia Record*, October 7, 1893.

58. Exhaustive descriptions of the Reading Terminal can be found in *PL*, October 14, 1893; *PRERBG* 7 (March 16, 1892): 1020; Walter G. Borg, *Buildings and Structures of American Railroads* (New York: John Wiley & Sons, 1893); and Joseph M. Wilson's eloquent and thorough description in *Transactions of the American Society of Civil Engineers*, 34 (August 1895): 115–84. The balloon shed's technology is explained by Carl W. Condit in *American Building Art: The Nineteenth Century* (New York: Oxford University Press, 1960), pp. 215–18, 329.

59. *PL*, June 9, 1897, May 3, 1898.

60. Carl W. Condit, *The Chicago School of Architecture* (Chicago: University of Chicago Press, 1964), p. 113, pl. 68; Thomas S. Hines, *Burnham of Chicago* (New York: Oxford University Press, 1974), pp. 288, 291.

61. King, *Philadelphia and Notable Philadelphians*, p. 14.

62. Tatum, *Penn's Great Town*, pp. 123, 200, pl. 131. *Golden Book of The Wanamaker Stores* (Philadelphia: John Wanamaker, 1911) contains many illustrations of the new store. A discussion of how its design and technology overcame engineering problems is in "A Modern Department Store: The Construction and Equipment of the Philadelphia Wanamaker Building," *The Architectural Record* 29 (March 1911): 277–88.

63. Tatum, *Penn's Great Town*, p. 200; *Philadelphia Inquirer*, January 21, 1972.

64. George E. Thomas claims that Frank Furness was responsible for the design and that McKim, Mead and White did only the details: "An Essay in Commercial Taste," Lecture at Philadelphia Art Alliance, February 12, 1976. The building is favorably evaluated in relation to its time and place by Tatum, *Penn's Great Town*, pp. 124, 200–201. Excellent photos and drawings of the bank are in *A Monograph of the Work of McKim, Mead & White, 1879–1915* (New York: The Architectural Book Publishing Co., [1915]), 4: pls. 329–31. It is praised in relation to contemporary banks in "Recent Bank Buildings of the United States," *The Architectural Record* 25 (January 1909): 8, 20–21.

65. This unintentional architectural irony does not alter the fact that the Crash and the subsequent Great Depression were disastrous for all associated with the construction industry. The number of issued building permits, which were required for additions, alterations, and demolitions as well as for new construction, plummeted from a high of 11,006 in 1927 to a low of 4,672 in 1933. After that the number climbed to 9,023 in 1937, but fell during the next year's economic downturn and continued to decline during World War II to only 2,945 building permits issued in 1944. It was not until the mid-1950's that the number of permits again reached that of 1927.

66. Tatum, *Penn's Great Town*, pp. 201–2; Agnes Addison Gilchrist, "The Classic Essence of Paul Cret," *Charette* 44 (January 1964): 40–43.

67. Theo B. White, *Paul Philippe Cret* (Philadelphia: The Art Alliance Press, 1973), pp. 21–31.

68. Paul Cret, "The New Building for the Federal Reserve Bank of Philadelphia," *The 3-C Book* 14 (September 1932): 35–36; Cret, "The New Federal Reserve Bank Building," *The 3-C Book* 15 (February 1934): 168–69; Cret, "Sculptural Decoration on Our New Building," *The 3-C Book* 16 (March 1935): 209–10.

69. For two of the most complete discussions of the building, see Robert A. M.

Stern, *George Howe: Toward a Modern American Architecture* (New Haven: Yale University Press, 1975), pp. 108–30; and William H. Jordy, *American Buildings and Their Architects: The Impact of European Modernism in the Mid-Twentieth Century* (New York: Doubleday & Co., 1972), pp. 87–164.

70. Building Permit 1084 (1932).

71. Tatum, *Penn's Great Town*, pp. 125–27, 201, pl. 133. The stages of the Parkway's development are illustrated in *The Fairmount Parkway* (Philadelphia: Fairmount Park Art Association, 1919).

72. A critique of the parkway by Michelle Osborne appeared in the *Evening Bulletin*, October 29, 1965; November 5, 1965; November 12, 1965. Lewis Mumford's acerbic commentary is in "The Sky Line: Philadelphia-I," *The New Yorker*, April 28, 1956; pp. 122–23.

73. James Reichley, "Philadelphia Does It: The Battle for Penn Center," *Harper's Magazine*, February 1957, pp. 49–56. "Philadelphia's Hour of Decision," *Architectural Forum* 96 (June 1952): 118–25.

74. *Penn Center: Redevelopment Area Plan* (rev. ed.; Philadelphia; Philadelphia City Planning Commission, 1952), p. 5.

75. *Progressive Architecture's* First Design Award of 1962 was presented "to the architect Vincent G. Kling *and* his client," the City of Philadelphia, for the MSB, as the Municipal Services Building is known: "Ninth Annual Design Awards," *Progressive Architecture* 43 (January 1962): 114–20. For an illustrated presentation of the building's plan and function, particularly the serviceable space of its lobby and concourse, see "M.S.B.: A Diamond in Philadelphia's Center City?" *Progressive Architecture* 46 (December 1965): 109–16.

76. Mayor's Committee on the Efficient and Appropriate Housing of City Functions, Survey of Municipal Facilities in Center City Area (October 1957).

South Philadelphia

1. For an excellent discussion of the relationship between the material culture of London and Philadelphia, see Margaret B. Tinkcom, "Urban Reflections in a Trans-Atlantic Mirror," *PMHB* 100 (July 1976): 287–313. For the early history of Southwark see J. Thomas Scharf and Thompson Westcott, *History of Philadelphia, 1609–1884* (Philadelphia: L. H. Everts & Co., 1884), 1:256; 3:1175.

2. Henry D. Paxson, *Where Pennsylvania History Began* (Philadelphia: By the author, 1926), p. 30.

3. Scharf and Westcott, *History of Philadelphia*, 1:7.

4. Amandus Johnson, *The Swedish Settlements on the Delaware* (Philadelphia: University of Pennsylvania Press, 1911), 1:117, 182–86; 2:581–616.

5. Hugh Morrison, *Early American Architecture* (New York: Oxford University Press, 1952), pp. 506, 509–10.

6. Paxson, *Where Pennsylvania History Began*, p. 55.

7. Joseph Jackson, *Encyclopedia of Philadelphia* (Harrisburg, Pa.: The National Historical Association, 1932), 3:911, 4:967.

8. Scharf and Westcott, *History of Philadelphia*, 1:684.

9. Before the oldest part of Southwark was demolished in the late 1960's for Interstate–95, an excellent study of the area was made through the cooperation of the American Philosophical Society, Historic American Buildings Survey, and the Philadelphia Historical Commission. See Margaret B. Tinkcom, "Southwark, A River

Community: Its Shape and Substance," in *Proceedings of the American Philosophical Society* 114 (August 1970): 327–42. Hereafter cited as Tinkcom, "Southwark."

10. Edwin Iwanicki, Brief of Title, 770 South Front Street, Philadelphia (1961), PHC; PC 4030 (July 28, 1818), photocopies at HABS, PHC.

11. William John Murtagh, "The Philadelphia Row House," *JSAH* 16 (December 1957): 9.

12. *Philadelphia Inquirer*, June 10, 1969.

13. Will No. 70 (May 16, 1758), Inventory of Estate of George Mifflin, (1759), MSS, City of Philadelphia, Register of Wills.

14. MA 6183 (June 26, 1850), photocopies at HABS, PHC.

15. MA 6184 (June 26, 1850), photocopies at HABS, PHC.

16. MA 3272 (June 1812); MA 3425 (April 1813); MA 3426 (April 1813), photocopies at HABS, PHC; Richard S. Fuller, Brief of Title, Workman Place (1961), PHC.

17. Octavia Hill Association, Inc. leaflet (1961), copy at PHC.

18. Beatrice H. Kirkbride, "Nathaniel Irish House," HABS Report (PA–1013), p. 1.

19. Peter Kalm, *The America of 1750: Peter Kalm's Travels in North America. The English Version of 1770 revised from the Original Swedish with a Translation of New Material from Kalm's Diary Notes*, ed. Adolph B. Benson (New York: Dover Publications, 1966), 1: 99–100.

20. James Mease, *The Picture of Philadelphia* (Philadelphia: B. & T. Kite, 1811), p. 33. He lists 1,466 frame dwellings as compared to 834 brick houses.

21. Survey 2729 (February 13, 1803), MSS, Archives of INA.

22. The LaTour Warehouse was a block north of Southwark in Society Hill, but it was representative of early-nineteenth-century storehouses in the maritime neighborhood. James C. Massey, James Glen Carr and Osmund R. Overby, "John LaTour Warehouse," HABS Report (PA–1056), p. 1.

23. Scharf and Westcott, *History of Philadelphia*, 1:490, 3:2339–40.

24. *Aurora*, October 20, 1808; Stephen N. Winslow, *Biographies of Successful Philadelphia Merchants* (Philadelphia: James K. Simon, 1864), p. 142; M. Antonia Lynch, "Southwark," *Philadelphia History* 1 (1917): 104; Registry Unit 6S12–45. Description, plan, and view of the Shot Tower are in *Hexamer General Surveys* 16 (1880): pl. 1491.

25. Registry Unit 4S22–81, 315, 350; Building Permit 707 (1922).

26. The need for a professional municipal fire department by 1871 is illustrated by the turbulent history of the fierce competition among the volunteer companies. See Andrew J. Neilly, "The Violent Volunteers: A History of the Volunteer Fire Department of Philadelphia, 1736–1871" (Ph.D. dissertation, University of Pennsylvania, 1959), pp. 70–84, 179–82. For a discussion of the social and political roles of Southwark's fire companies, see Bruce Laurie, "Fire Companies and Gangs in Southwark: The 1840's," in *The Peoples of Philadelphia*, ed. Allen F. Davis and Mark H. Haller (Philadelphia: Temple University Press, 1973), pp. 71–88.

27. Hope Engine Company Minutes, 1847–1857, Fire Company Records, MSS, HSP. A brief discussion of the Hoxie and Button partnership and a list of their works are in Richard J. Webster, "Stephen D. Button: Italianate Stylist" (M.A. thesis, University of Delaware, 1963), pp. 54–61, 84–86, 95–97.

28. Registry Unit 5S15–249; *First Annual Message of William S. Stokley, Mayor of the City of Philadelphia, with the Accompanying Documents, June 26, 1873*, pp. 554–56, 663–707. Hereafter mayors' annual messages cited by name and year.

29. PL, May 15, 1893; *PRERBG* 8 (September 6, 1893): 569. Two early views of the fire house are cited in Nicholas B. Wainwright, *Philadelphia in the Romantic Age of Li-*

thography (Philadelphia: Historical Society of Pennsylvania, 1958), pp. 108, 220. A lithograph showing both the 1840's fire house and the enlarged building of the late 1850's is at the Insurance Company of North America Museum. Unfortunately, the 1890's sashes were replaced by simplified single-pane or one-over-one sashes during the building's rehabilitation in 1969.

30. Franklin Davenport Edmunds, *The Public School Buildings of the City of Philadelphia from 1868 to 1874* (Philadelphia: By the author, 1925), pp. 21–24; Building Permit 1502 (1928); Harold N. Cooledge, "Samuel Sloan and the "Philadelphia Plan,' " *JSAH* 23 (October 1964): 151–54.

31. *Historical Sketches of the Catholic Churches and Institutions of Philadelphia* (Philadelphia: Daniel H. Mahoney, 1895), p. 67; Scharf and Westcott, *History of Philadelphia,* 1: 669–71; 2: 1392; *PRERBG* 12 (October 20, 1897): 671; 12; (December 8, 1897): 795; *PL,* October 28, 1897.

32. *Philadelphia Inquirer,* October 1, 1974.

33. James C. Massey, "Frank Furness in the 1870's," *Charette* 43 (January 1963): 16.

34. *Evening Bulletin,* January 10, 1964.

35. These conclusions are based on a study of the house made by Professor Herbert A. Richardson, Glassboro State College, Glassboro, N.J.

36. William B. Bassett, "Stephen Girard's Country House," HABS Report (PA–140), pp. 1–2.

37. Scharf and Westcoff, *History of Philadelphia,* 2:1014–15, 3:2339.

38. Ibid., 1:361–64.

39. It should be noted that the motivation for building the magazine in South Philadelphia was not solely, or even primarily, war fears but that people living near the old magazine at Twenty-third and Walnut streets were apprehensive about its safety. Paxson, *Where Pennsylvania History Began,* p. 62.

40. Agnes Addison Gilchrist, *William Strickland: Architect and Engineer, 1788–1854* (Philadelphia: University of Pennsylvania Press, 1950), p. 73; Edward Shippen, "Some Account of the Origin of the Naval Asylum at Philadelphia," *PMHB* 7 (1883): 129.

41. William B. Bassett, "United States Naval Asylum," HABS Report (PA–1094), p. 3.

42. George B. Tatum, *Penn's Great Town* (Philadelphia: University of Pennsylvania Press, 1961), p. 80.

43. Alan Gowans, *Images of American Living: Four Centuries of Architecture and Furniture as Cultural Express* (Philadelphia: J. B. Lippincott Co., 1964), pp. 291–301.

44. For discussions of the source and symbolism of the Egyptian Revival, see Frank J. Roos, Jr., "The Egyptian Style: Notes on Early American Taste," *Magazine of Art* 33 (April 1940): 218–33, 255; Richard G. Carrott, "The Neo-Egyptian Style in American Architecture," *Antiques* 90 (October 1966): 482–88.

45. John Daly and Allen Weinberg, *Genealogy of Philadelphia County Subdivisions* (2d ed.; Philadelphia: Department of Records, 1966), pp. 64, 84.

46. George Morgan, *City of Firsts* (Philadelphia: Historical Publication Society in Philadelphia, 1926), p. 308.

47. *Philadelphia: A Guide to the Nation's Birthplace* (Philadelphia: William Penn Association of Philadelphia, 1937), pp. 98–106.

48. Thomas Nolan, "Some Recent Philadelphia Architecture," *The Architectural Record* 29 (March 1911): 235.

49. *PL,* May 8, 1876; September 6, 1901; March 10, 1902; April 5, 1902; April 26, 1902; *PRERBG* 17 (April 16, 1902): 241; 27 (May 8, 1912): 303; James D. Van Trump,

"The Column and the Cross in Philadelphia: Three Victorian Classical Churches by Edwin F. Durang," *Charette* 48 (January 1967): 9–12.

50. The city acquired the properties between 1898 and 1900, but a plan was not chosen until 1911. Work was interrupted in 1912, however, when the courts declared unconstitutional the issuance of the supplemental contracts. That matter was resolved by 1917, and the project was begun again only to be delayed by World War I. The public bathing beaches were opened in the summer of 1922 and the final plantings were made and pathways were paved the next year; land for the golf course was purchased between 1927 and 1930. Deeds to City Property, Box No. 93–M, MSS, Philadelphia City Archives; *First Annual Message of Rudolph Blankenburg, 1911,* p. 286; *Second Annual Message of Rudolph Blankenburg, 1912,* p. 14; *Second Annual Message of Thomas B. Smith, 1917,* 2:132–34; *Third Annual Message of J. Hampton Moore, 1922,* p. 119; *Fourth Annual Message of J. Hampton Moore, 1923,* p. 227; *[Second] Annual Message of W. Freeland Kendrick, 1925,* pp. 13–19.

51. *Philadelphia: A Guide to the Nation's Birthplace,* p. 449.

52. Ibid., p. 447.

53. Ellis Paxson Oberholtzer, *Philadelphia: A History of the City and Its People* (Philadelphia: S. J. Clarke Publishing Co., 1911), 2:407–9; Theo B. White, *Philadelphia Architecture in the Nineteenth Century* (Philadelphia: University of Pennsylvania Press, 1953), pp. 31–32.

54. The erection of the railroad's passenger station at Broad Street and Washington Avenue in 1857 seemed to promise economic vitality for the neighborhood, but in time the tracks running east to the Delaware River proved to be more of a barrier than a gateway. *Nineteenth Annual Report of the President and Directors to the Stockholders of the Philadelphia, Wilmington and Baltimore Railroad Company* (Philadelphia, 1857), p. 28.

55. A brief history of the Library Company's stay in its Broad Street quarters and the arguments for its removal are presented in *Why the Library Company Should Move* (Philadelphia: Library Company of Philadelphia, 1960).

56. For a readable account of the social changes being wrought by the renaissance of Southwark, see Carol Byrne, "Queen Village Comes Out of the Closet," *Philadelphia Magazine,* July 1974, pp. 86–91, 158–59, 163–69.

57. Some observers fear, however, that success has spoiled South Street, that during the mid-1970's rising rents east of Sixth Street forced out of the neighborhood many colorful and unconventional shopkeepers that gave the street a bohemian atmosphere. See Kit Konolige, "The Saviors of South Street," *Today: The Inquirer Magazine,* January 27, 1974, pp. 8–14; *Philadelphia Inquirer,* March 7, 1976.

West Philadelphia

1. The first house reputed to have been built in West Philadelphia was John Warner's mansion on or near the present site of Sweetbrier in Fairmount Park. It was dismantled in the 1790's and its timbers were used in building Colonel Edward Heston's house near Fifty-second Street and Lancaster Avenue; the latter dwelling was demolished in 1901 during expansion of the Pennsylvania Railroad. M. Laffitte Viera, *West Philadelphia Illustrated* (Philadelphia: Avil Printing Co., 1903), p. 103.

2. Harold Donaldson Eberlein and Cortlandt Van Dyke Hubbard, *Portrait of a Colonial City: Philadelphia, 1670–1838* (Philadelphia: J. B. Lippincott Co., 1939), pp. 81–84.

3. Hugh Morrison, *Early American Architecture* (New York: Oxford University

Press, 1952), p. 519. A photograph of the house before restoration appears in Viera, *West Philadelphia Illustrated*, p. 92.

4. Eberlein and Hubbard, *Portrait of a Colonial City*, p. 245.

5. Robert Patterson Robins, "Colonel James Coultas, High Sheriff of Philadelphia, 1755–1758," *PMHB* 11 (1887): 50–57.

6. Blank end walls are rare but not unique in Philadelphia country houses. Mount Pleasant (1761) in Fairmount Park has blank end walls two rooms deep, and Bel Air (c. 1714–29) in South Philadelphia has blank end walls one room deep.

7. The only other Philadelphia dwelling known to have urns on it is Cliveden in Germantown. See Margaret B. Tinkcom, "Cliveden: The Building of a Country Seat, 1763–1767," *PMHB* 88 (January 1964): 10. Both the exterior and interior of Whitby Hall are illustrated in Harold Donaldson Eberlein, "Three Types of Georgian Architecture: The Evolution of the Style in Philadelphia," *The Architectural Record* 34 (July 1913): 73–77.

8. Harold Donaldson Eberlein, ed., "Passages from the Diary of Nicholas Pickford Esquire, Relating to His Travels in Pennsylvania in 1765," *The Architectural Review* 47 (June 1920): 145–46.

9. Eberlein and Hubbard, *Portrait of a Colonial City*, pp. 246, 249. Charles E. Peterson, F.A.I.A., tracked down the location of Whitby Hall's woodwork in the fall of 1975.

10. Thompson Westcott, *The Historic Mansions and Buildings of Philadelphia* (rev. ed.; Philadelphia: Walter H. Barr, 1895), p. 182.

11. Ernest Earnest, *John and William Bartram: Botanists and Explorers* (Philadelphia: University of Pennsylvania Press, 1940), pp. 28–33, 60–67; Brooke Hindle, *The Pursuit of Science in Revolutionary America* (Chapel Hill: University of North Carolina Press, 1956), pp. 20–27.

12. For example, see James Gibbs, *A Book of Architecture* (London, 1728), pl. 91; Batty Langley, *The City and Country Builder's and Workman's Treasury of Designs* (London, 1750), pls. 50, 80, 82, 85, 89; Abraham Swan, *The British Architect* (London, 1758), pls. 51, 53, and perhaps the most logical source, *The Rules of Work of the Carpenters' Company of the City and County of Philadelphia, 1786*, annotated by Charles E. Peterson (Princeton: Pyne Press, 1971), pl. 27.

13. Earnest, *John and William Bartram*, pp. 66–67.

14. Eberlein and Hubbard, *Portrait of a Colonial City*, pp. 448–50; Westcott, *Historic Mansions*, pp. 416–22.

15. Sarah P. Stetson, "William Hamilton and His 'Woodlands,'" *PMHB* 73 (January 1949): 26–27.

16. Harold Donaldson Eberlein, ed., "Further Passages from the Diary of Nicholas Pickford Esquire, Relating to His Travels in Pennsylvania in 1786," *The Architectural Review* 48 (July 1920): 30–31.

17. Deed Book RLL 37–139. The cemetery company acquired additional Hamilton property in 1853 and 1867. Deed Books TH 173–546, JTO 32–446.

18. "Hamilton Mansion," *Gleason's Pictorial Drawing-Room Companion*, April 15, 1854, p. 232.

19. The bridge was called permanent because it replaced a form of pontoon bridge, not because of its expected longevity. In fact, the builder, Timothy Palmer, estimated its life expectancy as only thirty or forty years, yet it survived until 1875, when a fire ended its service. David B. Steinman and Sara Ruth Watson, *Bridges and Their Builders* (New York: Dover Publications, 1941), p. 118; Carl W. Condit, *American Build-*

ing Art: The Nineteenth Century (New York: Oxford University Press, 1960), pp. 80–82; Joseph Jackson, *Market Street, Philadelphia: America's Most Historic Highway* (new ed.; Philadelphia: John Wanamaker, 1926), p. 33.

20. Busti was general agent for the Holland Land Company, which received about five million acres of land in western Pennsylvania and New York from the federal government in lieu of a money payment for a loan to the Continental Congress during the Revolutionary War. "Notes and Queries," *PMHB* 7 (1883): 107–8. Although Busti called his estate a farm, the interior elements as well as the columned southern veranda (or piazza) and northern porch were those of a late-eighteenth-century country house rather than a farmhouse. MA 2149 (March 15, 1806), photocopies at HABS, PHC.

21. John Hare Powel, born John Powel Hare, was the nephew and heir of the eminent merchant Samuel Powel, who served as the city's mayor under both the Crown and the new republic. For the nephew's biography, see Henry Simpson, *The Lives of Eminent Philadelphians* (Philadelphia: William Brotherhead, 1859), pp. 808–19. An evaluation of him as a proper Philadelphian who "did nothing, and did it well" is in Nathaniel Burt, *The Perennial Philadelphians* (Boston: Little, Brown and Co., 1963), pp. 156–57. The main section of Powelton was built for Samuel Powel's widow in 1800–1802; John Hare Powel was responsible for the notable Doric pediment, designed by William Strickland in 1825, and the wings, which were finished in the mid-1840's. Charles B. Wood III, "Powelton: An Unrecorded Building by William Strickland," *PMHB* 91 (April 1967): 147–48, 151–55.

22. Townsend Ward, "A Walk to Darby," *PMHB* 3 (1879): 165–66.

23. Jackson, *Market Street*, p. 332.

24. Joseph Jackson, *Encyclopedia of Philadelphia* (Harrisburg, Pa.: The National Historical Association, 1932), 3:868. J. C. Sidney, *Map of the City of Philadelphia* (Philadelphia: Smith and Wistar, 1849).

25. Nicholas B. Wainwright, ed., *A Philadelphia Perspective: The Diary of Sidney George Fisher Covering the Years 1834–1871* (Philadelphia: Historical Society of Pennsylvania, 1967), p. 169.

26. The hospital of the almshouse received patients during the cholera epidemic of 1832, but the complex was not finished until 1835. There were a number of later additions, beginning with a workshop in 1860. Agnes Addison Gilchrist, "Additions to *William Strickland: Architect and Engineer, 1788–1854*," *JSAH* 13 (October 1954): Supplement, 4; Charles Lawrence, comp., *History of the Philadelphia Almshouses and Hospitals from the Beginning of the Eighteenth to the Ending of the Nineteenth Centuries* (Philadelphia: By the author, 1905), pp. 110–12, 116–18, 134–36, 256; *Philadelphia Hospital Reports* (Philadelphia, 1890), 1:91–106. For a glimpse of life in the institution, see Alexander Ames Bliss, *Blockley Days: Memories and Impressions of a Resident Physician, 1883–1884* (Philadelphia: By the author, 1916).

27. David J. Rothman, *The Discovery of the Asylum: Social Order and Disorder in the New Republic* (Boston: Little, Brown and Co., 1971), pp. 155–89.

28. J. Thomas Scharf and Thompson Westcott, *History of Philadelphia, 1609–1884* (Philadelphia: L. H. Everts & Co., 1884), 2:1671.

29. Rothman, *Discovery of the Asylum*, pp. 130–38, 141.

30. Harold N. Cooledge, Jr., has made a case for Samuel Sloan as the architect responsible for the plan and form of the nineteenth-century insane asylum, but Sloan was evidently emulating Holden, since Sloan's 1856 Pennsylvania Hospital for the Insane at Forty-ninth and Market streets was nearly identical in plan, design, and appointments to Holden's hospital at Forty-fourth and Market streets, where the young

Sloan superintended the construction. Furthermore, Holden's building possessed in the 1840's those features that the Association of Medical Superintendents considered fundamental in 1851, two years before Sloan's first independent commission for an insane asylum. Sloan was certainly the propagator of the style, as the thirty-two insane asylums from his designs and a number of articles from his pen testify, but Holden appears to have been the author. Holden lost his opportunity to become America's hospital architect when he returned to England in 1838. Harold N. Cooledge, Jr., "Samuel Sloan and the Philadelphia School of Hospital Design," *Charette* 44 (June 1964): 6–7, 18; Thomas G. Morton, *The History of the Pennsylvania Hospital, 1751–1895* (Philadelphia: Times Printing House, 1895), pp. 165, 174–76; PC 5602 (October 16, 1840), PC 10331–33 (March 6, 1861), photocopies at HABS, PHC; Rothman, *Discovery of the Asylum*, p. 135; Wainwright, ed., *Philadelphia Perspective*, pp. 440–41.

31. Robert C. Smith, "Two Centuries of Philadelphia Architecture, 1700–1900," *Historic Philadelphia*, 301.

32. PC 5602 (October 16, 1840), photocopies at HABS, PHC; Cooledge, "Samuel Sloan and the Pennsylvania School," pp. 6–7. Rothman, *Discovery of the Asylum*, p. 134.

33. James C. Massey, ed., *Two Centuries of Philadelphia Architectural Drawings* (Philadelphia: Philadelphia Museum of Art, 1964), pp. 64–65; Thompson Westcott, *The Official Guide Book of Philadelphia* (Philadelphia: Porter and Coates, 1875), pp. 217–18, 224–25; Scharf and Westcott, *History of Philadelphia*, 2:1676, 1682; *First Annual Report of the Presbyterian Hospital in Philadelphia* (Philadelphia, 1872), pp. 5–9; *Third Annual Report of the Presbyterian Hospital* (Philadelphia, 1874), pp. 52–56.

34. Scharf and Westcott, *History of Philadelphia*, 2:1684; *PL*, April 21, 1886; April 28, 1886; August 16, 1890; April 2, 1897; July 3, 1897; *Philadelphia Record*, August 2, 1893.

35. The contrasting nature of the two townships is illustrated by the fact that in 1840, when 61.1 per cent of the people of Blockley were engaged in manufacturing and commerce, 77.3 per cent of the people of Kingsessing were farming. Scharf and Westcott, *History of Philadelphia*, 1:676, 697, 713; *Heads of Families at the First Census of the United States Taken in the Year 1790: Pennsylvania* (Washington, D.C.: Government Printing Office, 1908), p. 10; *Sixth Census or Enumeration of the Inhabitants of the United States, as Corrected at the Department of State, in 1840* (Washington, D.C.: Blair and Rives, 1841), p. 151; *The Seventh Census of the United States: 1850* (Washington, D.C.: Robert Armstrong, Public Printer, 1853), pp. 178–79; John Daly and Allen Weinberg, *Genealogy of Philadelphia County Subdivisions* (2d ed.; Philadelphia: City of Philadelphia, 1966), p. 64.

36. Harold E. Cox, "Horse Passenger Railways of Philadelphia," *Pennsylvania Traction*, May, 1963, pp. 1–18.

37. Harold Norman Cooledge, Jr., "Samuel Sloan, Architect" (Ph.D. dissertation, University of Pennsylvania, 1963), pp. 50–55.

38. For a discussion of the innovative and picturesque qualities of the Italian villa, see Alan Gowans, *Images of American Living: Four Centuries of Architecture and Furniture as Cultural Expression* (Philadelphia: J. B. Lippincott Co., 1964), pp. 316–27; and George B. Tatum, *Penn's Great Town* (Philadelphia: University of Pennsylvania Press, 1961), pp. 92–94.

39. PC 12021 (July 20, 1874), photocopies at HABS, PHC; Registry Unit 20S9–4; Edward Scully Bradley, *Henry Charles Lea: A Biography* (Philadelphia: University of Pennsylvania Press, 1931), pp. 79–80.

40. *PL*, January 2, 1854; Robert A. Corrigan, Brief of Title, 4207 Walnut Street, Philadelphia (1959), MSS, PHC.

41. Cooledge, "Samuel Sloan, Architect," p. 65.

42. *PL*, March 14, 1863; Cox, "Horse Passenger Railways," pp. 9–11.

43. *PRERBG*, 6 (June 3, 1891): 338; 6 (August 5, 1891): 495; *PL*, May 30, 1893; Moses King, *Philadelphia and Notable Philadelphians* (New York: Moses King, 1902), pp. 68–69; Julian Millard, "The Work of Wilson Eyre," *The Architectural Record* 14 (October 1904): 307–8.

44. Housing construction began west of Forty-fifth Street shortly after the twentieth century began. For example, in the spring of 1901, 222 houses, both twins and singles, were begun on the block bounded by Fifty-first, Fifty-second, Brown, and Parrish streets, and in the same year Wilson Brothers and Company designed twenty-eight row houses for the neighborhood of Fifty-eighth and Arch streets. *PL*, March 19, 1901; May 21, 1901.

Building activity in the area significantly increased after work on the El began in 1903 (see *PL*, April 7, 1903). A month after the El's inauguration, developers were running half-page illustrated ads announcing that their row houses were within two-and-a-half blocks from the Elevated and only ten minutes away from City Hall (see *PL*, April 6, 1907). The Market Street Elevated opened, March 4, 1907, running between Sixty-ninth and Fifteenth streets. A year and a half later it was extended to Second Street (*PL*, March 4, 1907; March 5, 1907; August 3, 1908).

45. *PL*, August 24, 1901; *Building* 9 (July 1929): 7–8. Philadelphia's first apartment hotel, the Gladstone, was erected at Eleventh and Pine streets, 1889–90. George Morgan, *The City of Firsts* (Philadelphia: The Historical Publication Society in Philadelphia, 1926), p. 309; *PRERBG* 4 (November 20, 1889): 549; 5 (January 29, 1890): 37.

46. Vincent Scully, Jr., *Louis I. Kahn* (New York: George Braziller, 1962), pp. 22, 31–32.

47. Cox, "Horse Passenger Railways," pp. 7–8; *PL*, April 19, 1896.

48. Edward Potts Cheyney, *History of the University of Pennsylvania, 1740–1940* (Philadelphia: University of Pennsylvania Press, 1940), pp. 261–63.

49. Annie Gardiner Richards, "Thomas W. Richards and College Hall," *The General Magazine and Historical Chronicle* 53 (Summer 1951): 193–98; Tatum, *Penn's Great Town*, pp. 118–22; George E. Thomas and James F. O'Gorman, "Catalogue of Selected Buildings," in James F. O'Gorman, *The Architecture of Frank Furness* (Philadelphia: Philadelphia Museum of Art, 1973), pp. 165–71; *PL*, May 11, 1897; July 3, 1897. For a generally favorable evaluation of the university's architecture by America's foremost architectural critic of the early twentieth century, see Montgomery Schuyler, "Architecture of American Colleges, V: University of Pennsylvania, Girard, Haverford, Lehigh and Bryn Mawr," *The Architectural Record* 28 (September 1910): 183–200.

50. Quote from Scully, *Louis I. Kahn*, p. 26. Except for Horace Trumbauer's Irvine Auditorium (c. 1926) at Thirty-fourth and Spruce streets, the important architectural works on the Pennsylvania campus are identified in Edward Teitelman and Richard W. Longstreth, *Architecture in Philadelphia: A Guide* (Cambridge, Mass.: MIT Press, 1974), pp. 188–93. Hill House, the women's dormitory at Thirty-fourth and Walnut streets, is not one of Saarinen's greatest works, but it does rate brief commentary: "Court for Coeds at the University of Pennsylvania," *Architectural Forum* 114 (February 1961): 120–21; Allan Temko, *Eero Saarinen* (New York: George Braziller, 1962), p. 117; Rupert Spade, *Eero Saarinen* (London: Thames and Hudson, 1971), p. 123.

Mitchell/Giurgola Associates completed a second poured-concrete parking garage for the university at South Street and Civic Center Boulevard in 1969. The interior organization and treatment of the two structures are similar, and they almost make driving into a parking garage pleasurable, but because of their sites their exterior appearances differ. The isolation of the earlier garage allows an independent boldness, but the proximity of Franklin Field to the latter apparently pressured the architects to employ a symmetry and monumentality of scale that leaves its design less imposing than its predecessor. "University Parking Garage for University of Pennsylvania," *Progressive Architecture* 45 (December 1964): 146–51; "Parking Garage de l'Université de Pennsylvanie, Philadelphie," *Architecture d'Aujourd'hui* 35 (March 1965): 46–47; *Evening Bulletin*, November 26, 1965; May 23, 1969.

The Richards Building is acclaimed for its functional expression and economical use of prestressed reinforced concrete. The blank brick service towers, which project above the flat roofline and contrast with the slender concrete columns and cantilevered spandrels, give the building its dynamic quality, actually a nineteenth-century picturesqueness. For a discussion of the evolution of the building's design and a bibliography of the building, see Scully, *Louis I. Kahn*, pp. 27–31, 124–25.

51. After the death of his father Francis W. Drexel in 1863, Anthony J. Drexel (1826–1893), who had been a partner in Drexel and Company since 1847, continued the family's esteemed name in investment banking in Philadelphia, while his brother Joseph W. formed Drexel, Harjes and Company in Paris, and Francis A. moved to New York where he made J. P. Morgan a junior partner in Drexel, Morgan and Company. Scharf and Westcott, *History of Philadelphia*, 3:2101–2; Burton Alva Konkle, "Francis Martin Drexel and Philadelphia Leadership in International Finance," *The Girard Letter* 14 (December 1933): 1–2. For the construction of the school's main building, see Edward D. McDonald and Edward M. Hinton, *Drexel Institute of Technology, 1891–1941* (Philadelphia: Drexel Institute of Technology, 1942), pp. 18–26, 43–45; *PL*, April 19, 1890; April 23, 1890; January 13, 1892.

52. Building Permit 5079 (1930).

53. St. Joseph's College began in 1851 on Willing's Alley in an addition to the parish house of St. Joseph's Roman Catholic Church. After a brief tenure on Center Square, at the northeast corner of Filbert and Juniper streets, the college in 1868 became part of the notable Jesuit complex around Eighteenth and Stiles streets in North Philadelphia, and it was from that site that it moved to Wynnefield. Francis X. Talbot, *Jesuit Education in Philadelphia: St. Joseph's College, 1851–1926* (Philadelphia: St. Joseph's College, 1927), pp. 36–38, 51–52, 73–77, 128–41.

54. Grandiose, if not grand, the Thirtieth Street Station was built in 1927–34 from the designs of Graham, Anderson, Probst, and White, but its washed-out classicism, as Carroll L. V. Meeks noted, "was old-fashioned before it left the drafting board." *The Railroad Station: An Architectural History* (New Haven: Yale University Press, 1956), p.103. The Bulletin Building was designed in 1953 by George Howe in collaboration with Robert Montgomery Brown. Although its appearance—a gray brick block, with strips of windows, that rests on a glass and stainless-steel loggia—is indebted to Mies van der Rohe, its critical reception has been mixed. Some Philadelphians simply don't like it (Burt, *Perennial Philadelphians*, p. 364), but Lewis Mumford, who found a lot of little things wrong with it, concluded that "it vies with William Strickland's old Custom House as one of the best examples of the classic tradition in Philadelphia" ("The Skyline: Philadelphia—II," *The New Yorker*, May 26, 1956, p. 127).

55. Leon S. Rosenthal, *A History of Philadelphia's University City* (Philadelphia: West Philadelphia Corporation, 1963), pp. 73–80.

56. See Thomas Hine's "Surroundings" column in *Philadelphia Inquirer*, February 8, 1976, for a statement of the problem; for its resolution, see Jan Schaffer's article in *Philadelphia Inquirer*, May 29, 1976.

57. Churches were as much a part of Victorian residential neighborhoods as front porches and grassy plots, and occasionally were part of the development plan. When William Hamilton, for example, laid out Hamilton Village he reserved a lot on Locust Street west of Thirty-ninth Street for what was to become St. Mary's Episcopal Church, which has been in a state of almost continuous evolution since its second structure was begun in 1872. The erosion of residential West Philadelphia and the growth of University City threaten the survival of such architectural gems as Edwin F. Durang's St. Agatha's Roman Catholic Church (1874–78) at Thirty-eighth and Spring Garden Theophilus Parsons Chandler's Tabernacle Presbyterian Church (illus.) (1884–86) at Thirty-seventh and Chestnut streets, Charles M. Burns's Church of the Saviour (1889, rebuilt after 1902 fire) at Thirty-eighth and Ludlow streets, Frank R. Watson's Tabernacle Baptist Church (1896–97) near Fortieth and Chestnut streets, and Carrère's and Hastings's (with T. E. Blake) First Church of Christ Scientist (1909–10) at Fortieth and Walnut streets. The Mary Elizabeth Patterson Memorial Church at Sixty-third and Vine Streets, an imaginative interplay of screened spaces and robust forms by Chandler (1885, 1895–97), got a new lease on life in 1975 when the Church of Christ congregation took over the structure. Still abandoned and facing an uncertain future in 1976 is the Philadelphia Divinity School chapel, a chaste twentieth-century Gothic design by Zantzinger, Borie and Medary (1925–26) that takes advantage of its elevated site on Spruce Street west of Forty-second.

Fairmount Park

1. Quoted in John Maass, *The Glorious Enterpise* (Watkins Glen, N.Y.: American Life Foundation, 1973), p. 24.

2. The most comprehensive work on the park is Esther M. Klein, *Fairmount Park: A History and a Guidebook* (Bryn Mawr, Pa.: Harcum Junior College Press, 1974). Also useful is Theo B. White, *Fairmount Park* (Philadelphia: Philadelphia Art Alliance, 1975).

3. *Antiques* magazine devoted its November 1962 issue to the Fairmount Park houses. There are fully illustrated, well-researched articles on the history and architecture of six of the more notable houses: Cedar Grove, Lemon Hill, Mount Pleasant, Strawberry Mansion, Sweetbrier, and Woodford. There are vignettes about the other houses, as well as articles on some owners of park houses and their portraits and the origins of the park. A scholarly investigation of the interior woodwork of some of these houses is Martin E. Weil, "Interior Architectural Details in Eighteenth-Century Architectural Books and Philadelphia County Houses" (M.A. thesis, University of Delaware, 1967).

4. Nelson Manfred Blake, *Water for the Cities* (Syracuse: Syracuse University Press, 1956), pp. 36–39, 78–79; George B. Tatum, "The Origins of Fairmount Park," *Antiques* 82 (November 1962): 502–3.

5. *Report of the Watering Committee, Read January 9, 1823* (Philadelphia, 1823), p.7; *Annual Report of the Watering Committee, for the Year 1852* (Philadelphia, 1852), pp. 28–29; Thomas Wilson, *Picture of Philadelphia, in 1824* (Philadelphia: T. Town, 1823), pp. 26–34; Harold Donaldson Eberlein, "The Fairmount Waterworks, Philadelphia," *The Architectural Record* 62 (July 1927):65.

6. J. Thomas Scharf and Thompson Westcott, *History of Philadelphia, 1609–1884* (Philadelphia: L. H. Everts & Co., 1884), 3:1853; George B. Tatum, *Penn's Great Town* (Philadelphia: University of Pennsylvania Press, 1961), pp. 63–64.

7. Scharf and Westcott, *History of Philadelphia*, 3:1854; Tatum, "The Origins of Fairmount Park," pp. 503–4; Tatum, *Penn's Great Town*, p. 64, pl. 52; Charles Dickens, *American Notes* (London: Oxford University Press, 1957), p. 98.

8. Eberlein, "The Fairmount Waterworks, Philadelphia," pp. 65–66; *Annual Report of the Chief Engineer of the Water Department of the City of Philadelphia, 1868* (Philadelphia, 1868), p. 8. These reports hereafter cited as *Annual Report*. *Annual Report*, 1869, pp. 5–7; *Annual Report, 1873*, p. 6; *PL*, July 25, 1872.

9. Laurel Hill ranked with Cambridge's Mount Auburn Cemetery and Brooklyn's Greenwood Cemetery, and was particularly praised for its plantings and river view. The concept of rural cemeteries originated abroad with the Père-Lachaise near Paris and the General Cemetery Company at Kensal Green near London; Notman's plan was based on the latter. For a discussion of the cemetery's sculpture, see George Thomas, "The Statue in the Garden," in *Sculpture of a City*, ed. Nicholas B. Wainwright (New York: Walker Publishing Co., 1973), pp. 36–44; Tatum, *Penn's Great Town*, p. 87, pl. 88; *Guide to Laurel Hill Cemetery near Philadelphia* (Philadelphia: Laurel Hill Cemetery Co., 1844), p. 53; R. A. Smith, *Smith's Illustrated Guide to and through Laurel Hill Cemetery* (Philadelphia: W. P. Hazard, 1852); *Rules and Regulations of Laurel Hill Cemetery of Philadelphia* (Philadelphia: Laurel Hill Cemetery Co. 1892), pp. 8–10; Theo B. White, *Philadelphia Architecture of the Nineteenth Century* (Philadelphia: University of Pennsylvania Press, 1953), pls. 31, 32.

10. Tatum, *Penn's Great Town*, p. 88; Scharf and Westcott, *History of Philadelphia*, 3:1855–58.

11. Klein, *Fairmount Park*, p. 56.

12. Scharf and Westcott, *History of Philadelphia*, 3:1871.

13. George E. Thomas and James F. O'Gorman, "Catalogue of Selected Buildings," in James F. O'Gorman, *The Architecture of Frank Furness* (Philadelphia: Philadelphia Museum of Art, 1973), pp. 143–45. G. W. and W. D. Hewitt were responsible for the present appearance of the Vesper and Malta boathouse. *PL*, March 2, 1901; April 1, 1901.

14. Tatum, "The Origins of Fairmount Park," p. 507. One plan from the competition, that of civil engineer Andrew Palles, has survived (Tatum, *Penn's Great Town*, pl. 88). It included both sides of the river between the waterworks and the Philadelphia and Columbia Railroad Bridge. Visual evidence—comparison with a more extensive 1871 plan by Frederick Law Olmsted—suggests that the east part might have been executed (George F. Chadwick, *The Park and the Town* [New York: Frederick A. Praeger, 1966], p. 203). Olmsted's plan is unlike Palles's except for the lower east park, at which point the two plans are almost identical, suggesting that Olmsted was restricted in this older section of the park to what was already there.

15. Maass, *The Glorious Enterprise*, pp. 18, 21; Chadwick, *The Park and the Town*, p, 203; Frederick Law Olmsted, Jr., and Theodore Kimball, *Frederick Law Olmsted: Landscape Architect, 1822–1903* (New York: Benjamin Blom, 1970), p. 15.

16. Maass, *The Glorious Enterprise*, pp. 22–23.

17. Tatum, "The Origins of Fairmount Park," p. 507; Thompson Westcott, *The Official Guide to Philadelphia* (Philadelphia: Porter and Coates, 1875), pp. 185–90; Maass, *The Glorious Enterprise*, p. 22.

18. Wainwright, *Sculpture of a City*, p. 1.

19. Lewis Sharp, "The Smith Memorial," in *Sculpture of a City*, ed. Wainwright, pp. 168–79; George Morgan, *City of Firsts* (Philadelphia: Historical Publication Society, 1926), p. 200; *AABN* 89 (March 31, 1906): 115.

20. Sharp, "The Smith Memorial," p. 179.

21. *Sunday Bulletin*, September 1, 1968.

22. A new indoor-outdoor concert facility, Robin Hood Dell West, opened in June 1976 on George's Hill near Fifty-second Street and Parkside Avenue. Free concerts are still offered, but the best fidelity is enjoyed by the paying customers under the copper-alloy canopy. The older concert shell, Robin Hood Dell East, is still used, but not by the Philadelphia Orchestra. *Philadelphia Inquirer*, June 13, 1976; June 15, 1976; June 16, 1976.

Northwest Philadelphia

1. Peter Kalm, *The America of 1750: Peter Kalm's Travels in North America. The English Version of 1770 Revised from the Original Swedish with a Translation of New Material From Kalm's Diary Notes*, ed. Adolph B. Benson (New York: Dover Publications, 1966), 1:48–49.

2. Edward W. Hocker, *Germantown, 1683–1933* (Philadelphia: By the author, 1933), pp. 21, 29, 51.

3. S. F. Hotchkin, *Ancient and Modern Germantown, Mount Airy, and Chestnut Hill* (Philadelphia: P. W. Ziegler & Co., 1889), p. 349.

4. John J. MacFarlane, *History of Early Chestnut Hill* (Philadelphia: City History Society of Philadelphia, 1914), p. 21.

5. John Daly and Allen Weinberg, *Genealogy of Philadelphia County Subdivisions* (2d ed.; Philadelphia: City of Philadelphia, 1966), pp. 26, 64.

6. Harry M. Tinkcom, Margaret B. Tinkcom, and Grant Miles Simon, *Historic Germantown* (Philadelphia: American Philosophical Society, 1955), pp. 5–6.

7. Ibid., pp. 91, 112; Hocker, *Germantown*, pp. 26–28, 61–67.

8. Hocker, *Germantown*, pp. 51–53.

9. E. Perot Bissell, "Wister's Big House," HABS Report (PA–7–1), p. 2.

10. Tinkcom, Tinkcom, and Simon, *Historic Germantown*, p. 84.

11. Naamen H. Keyser, C. Henry Kain, John Palmer Garber, and Horace F. McCann, *History of Old Germantown* (Philadelphia: Horace F. McCann, 1907), p. 302; John F. Watson, *Annals of Philadelphia* (rev., enlarged ed.; Philadelphia: Leary, Stuart Co., 1927), 2:65–66.

12. Tinkcom, Tinkcom, and Simon, *Historic Germantown*, p. 110; Hocker, *Germantown*, p. 124.

13. For a thorough discussion of Germantown's Market Square and its market house, see Margaret B. Tinkcom, "Market Square," *Germantowne Crier* 19 (September 1967): 69–75.

14. Joseph Jackson, *A History of the Germantown Academy* (Philadelphia: Germantown Academy, 1910), pp. 1–2, 35–40, 43–47;

15. Keyser et al., *History of Old Germantown*, pp. 82–84.

16. An engaging account of William Allen, "the great giant," is given by Carl and Jessica Bridenbaugh, *Rebels and Gentlemen: Philadelphia in the Age of Franklin* (New York: Oxford University Press, 1962), pp. 184–91.

17. Margaret B. Tinkcom, "Cliveden: The Building of a Philadelphia Countryseat, 1763–1767," *PMHB* 88 (January 1964): 3–4. Based on Benjamin Chew's papers at Cliveden, this article is an exceptional discussion of the construction of an eighteenth-

century country house, tracing the evolution of the owner's ideas to the finished structure. For discussions of the house's furnishings, see Alice Winchester, "Living with Antiques: Cliveden, the Germantown home of Mr. and Mrs. Samuel Chew," *Antiques* 76 (December 1959): 532–36; Nicholas B. Wainwright, "Cliveden and Its Furniture," *The University Hospital Antiques Show Catalogue, 1970* (Philadelphia, 1970), pp. 66–69; Raymond V. Shepherd, Jr., "Cliveden," *Historic Preservation* 24 (July–September 1972): 4–11.

18. Burton Alva Konkle, *Benjamin Chew, 1722–1810* (Philadelphia: University of Pennsylvania Press, 1932), pp. 133, 145–51, 189–91, 264; "National Trust Acquires Cliveden in Germantown," *Preservation News* 12 (March 1972): 1, 6.

19. Townsend Ward, "Germantown Road and Its Associations: Part Sixth," *PMHB* 6 (1882): 141–42; Karl Theis Weger, "The Deshler Story," *Germantowne Crier* 7 (September 1955): 7–9, 27.

20. Charles F. Jenkins, *Washington Visits Germantown* (Philadelphia: Germantown Historical Society, 1932), pp. 53–59, 66–68.

21. Brief of Title, 5442 Germantown Avenue, Philadelphia (n.d.), PHC; "Morris House Presentation: May 16, 1950," *Germantowne Crier* 2 (June 1950): 12.

22. Tinkcom, Tinkcom, and Simon, *Historic Germantown*, pp. 104–5.

23. E. Perot Bissell, "Vernon," HABS Report (PA–7–2), p. 1; MA 2742 (October 1808), photocopies at HABS, PHC; Harold Donaldson Eberlein and Cortlandt Van Dyke Hubbard, *Portrait of a Colonial City: Philadelphia, 1670–1838* (Philadelphia: J. B. Lippincott Co. 1939), pp. 520–21.

24. Watson, *Annals of Philadelphia*, 2:28. J. Thomas Scharf and Thompson Westcott, *History of Philadelphia, 1609–1884* (Philadelphia: L. H. Everts & Co., 1884), 1:496.

25. Hocker, *Germantown*, pp. 137–39, 163–73, 253.

26. *Report of the President and Managers of the Philadelphia, Germantown, and Norristown Rail Road Company, to the Stock & Loan Holders for the Year Ending 31st Oct., 1854* (Philadelphia: John Duross, 1854), p. 5. Hereafter cited as *Report of the Philadelphia, Germantown, and Norristown Rail Road. Report of the Philadelphia, Germantown, and Norristown Rail Road, 1855*, p. 7.

27. Hocker, *Germantown*, pp. 223–25.

28. Richard S. Fuller, "200 West Tulpehocken Street," *Germantowne Crier* 16 (December 1964): 111–12.

29. *Philadelphia Inquirer*, November 12, 1964; *Evening Bulletin*, November 19, 1964; January 28, 1965; *New York Times*, January 31, 1965.

30. *Chestnut Hill: An Architectural History* (Philadelphia: Willard S. Detweiler, Jr., 1973), p. 23.

31. Campbell Scrapbooks, vol. 38, MSS, HSP; *PL*, January 19, 1907; February 28, 1907.

32. *PL*, August 30, 1887; *Chestnut Hill: An Architectural History*, p. 22; John Francis Marion, *Bicentennial City: Walking Tours of Historic Philadelphia* (Princeton: Pyne Press, 1974), p. xii.

33. E. Digby Baltzell, *An American Business Aristocracy* (New York: Collier Books, 1962), pp. 141, 234.

34. Frank C. Roberts, Jr., *The Houstons of Philadelphia* (New York: The Newcomen Society in North America, 1954), pp. 11–13.

35. Hotchkin, *Ancient and Modern Germantown*, pp. 403, 425.

36. Ibid., p. 471.

37. Herman L. Duhring, Jr., R. Brognard Okie, and Charles A. Ziegler, "A Practical Housing Development: The Evolution of the 'Quadruple House' Idea," *The*

Architectural Record 34 (July 1913): 47–50; C. Matlack Price, "Architecture and the Housing Problem," *The Architectural Record* 34 (September 1913): 241–47; George Woodward, "Another Aspect of the Quadruple House," *The Architectural Record* 34 (September 1913): 51–55; Harold Donaldson Eberlein, "Pastorius Park, Philadelphia, and Its Residential Development," *The Architectural Record* 39 (January 1916): 24–39.

38. Edward Teitelman and Richard W. Longstreth, *Architecture in Philadelphia: A Guide* (Cambridge, Mass.: MIT Press, 1974), p. 240; Marshall B. Davidson, *The American Heritage History of Notable American Houses* (New York: American Heritage Publishing Co., 1971), pp. 344–45.

39. Edwyn Rorke was the architect and K. M. deVos & Co. the contractor of these twelve-story buildings whose reinforced-concrete construction is sheathed in red brick and terra cotta. The Strawbridge mansion, built in 1885 from the designs of Addison Hutton, serves as a restaurant in the midst of the group. Building Permit 8242 (1926); Lewis Mumford, "The Skyline: Philadelphia-I," *The New Yorker*, April 28, 1956, pp. 121–22; Elizabeth Biddle Yarnall, *Addison Hutton: Quaker Architect, 1834–1916* (Philadelphia: The Art Alliance Press, 1974), p. 56.

40. *Chestnut Hill: An Architectural History*, p. 94.

41. For Rittenhouse's paper mill, see Horatio Gates Jones, "Historical Sketch of the Rittenhouse Paper Mill: The First Erected in America, A.D. 1690." *PMHB* 20 (1896): 315–33. Brooke Hindle has written the latest and probably the best biography of the famous astronomer, *David Rittenhouse* (Princeton: Princeton University Press, 1964). In addition to the colonies' earliest paper mill, Germantown could also boast of Pennsylvania's first gristmill; Scharf and Westcott, *History of Philadelphia*, 3:2295–96. For Mention of late-eighteenth-and early nineteenth-century Germantown mills, see Townsend Ward, "Germantown Road and Its Associations: Part Eighth," *PMHB* 5 (1881): 365–67; "Notes and Queries," *PMHB* 6 (1882): 128; Harrold E. Gillingham, "Calico and Linen Printing in Philadelphia," *PMHB* 52 (1928): 109–10.

42. Susan Stroud Robeson and Kate Hamilton Osbourne, eds., *Historical and Genealogical Account of Andrew Robeson of Scotland, New Jersey and Pennsylvania and of His Descendants from 1653 to 1916* (Philadelphia: J. B. Lippincott Co., 1916), pp. 6–10; Duc de LaRochefoucauld-Liancourt, *Travels Through the United States of North America, the Country of the Iroquois, and Upper Canada, in the Years 1795, 1796, and 1797* (London: R. Phillips, 1799), 1:7; John Frederick Lewis, *The History of an Old Philadelphia Land Title: 208 South Fourth Street* (Philadelphia: By the author, 1934), pp. 86–87.

43. Joseph Starne Miles and William H. Cooper, *A Historical Sketch of Roxborough, Manayunk, Wissahickon* (Philadelphia: George Fein & Co., 1940), pp. 105, 107.

44. Charles V. Hagner, *Early History of the Falls of Schuylkill, Manayunk, Schuylkill and Lehigh Navigation Companies, Fairmount Waterworks, etc.* (Philadelphia: Claxton, Remsen, and Haffelfinger, 1869), p. 81.

45. John Coolidge, *Mill and Mansion: A Study of Architecture and Society in Lowell, Massachusetts, 1820–1865* (New York: Columbia University Press, 1942), p. 23.

46. Hagner, *Early History of the Falls of Schuylkill*, pp. 75–79.

47. Plans, views, and construction and operations data of many of these mills were recorded by E. Hexamer and Son in the *Hexamer General Surveys*. For Schofield's Economy Mills see 8 (1873): pls. 711–12; 12 (1877): pls. 1131–32; 19 (1883): pls. 1809–10.

48. P. C. Keely of Brooklyn designed the church in 1886; it was dedicated in 1894. Eugene Murphy, *The Parish of St. John The Baptist: The First One Hundred Years* (Philadelphia, 1931), pp. 224, 235–36; Miles and Cooper, *Historical Sketch of Roxborough, Manayunk*, p. 177.

49. *Hexamer General Survey*, 20 (1885): pl. 1948.

50. During the mid-1970's what might be called a comeback has been mustered along Manayunk's Main Street, and the neighborhood's convenient location and low-cost housing have attracted young newcomers. Old-time residents, however, are reluctant to see this activity bloom into a renaissance that could drastically change the area's character. See Mary Walton, "Manayunk: New Settlers, Old Ethnics," *Today: The Inquirer Magazine*, March 7, 1976, pp. 10–12, 16, 18. The closing of the canal is cited in Miles and Cooper, *Historical Sketch of Roxborough, Manayunk, p. 112*.

North Philadelphia

1. Thompson Westcott, *The Historic Mansions and Buildings of Philadelphia* (rev. ed.; Philadelphia: Walter H. Barr, 1895), pp. 415–23.

2. Deed Book IH2–279; Thomas B. McElwee, *A Concise History of the Eastern Penitentiary of Pennsylvania* (Philadelphia: Neall & Massey, 1835), pp. 5–6.

3. Other villages would include Rising Sun, Jacksonville, and Nicetown along Germantown Road east of Broad Street. Townsend Ward, "The Germantown Road and Its Associations," PMHB 5 (1881): 18; J. C. Sidney, *Map of the City of Philadelphia* (Philadelphia: Smith and Wistar, 1849); J. Thomas Scharf and Thompson Westcott, *History of Philadelphia, 1609–1884* (Philadelphia: L. H. Everts & Co., 1884), 1:872–73.

4. McElwee, *A Concise History of the Eastern Penitentiary*, pp. 5–15, 41, 51–56.

5. Norman B. Johnston, "John Haviland, Jailor to the World," *JSAH* 23 (May 1964): 101–5; Negley K. Teeters and John D. Shearer, *The Prison at Philadelphia: Cherry Hill* (New York: Columbia University Press, 1957), pp. 56–59; Matthew Eli Baigell, "John Haviland" (Ph.D. dissertation, University of Pennsylvania, 1965), pp. 92–100.

6. Thompson Westcott, *The Official Guide Book of Philadelphia* (Philadelphia: Porter and Coates, 1875), p. 110; Robert W. Kotzbauer, "Echo from the Cells," *Discover: The Sunday Bulletin Magazine*, October 20, 1974, p. 16.

7. *Sunday Bulletin*, January 25, 1970.

8. The best history of Girard College is Cheesman A. Herrick, *History of Girard College* (Philadelphia: Girard College, 1927). Founder's Hall is thoroughly and statistically described in Henry W. Arey, *The Girard College and Its Founder* (Philadelphia: C. Sherman, 1872), pp. 44–56. Preston Retreat was demolished in the spring of 1963 after a long struggle to save it. *Philadelphia Inquirer*, April 28, 1963; May 10, 1963.

9. The most complete history of the man is John Bach McMaster, *The Life and Times of Stephen Girard* (2 vols.; Philadelphia: J. B. Lippincott Co., 1918).

10. Biddle's influence on the design of Founder's Hall is presented by Agnes Addison Gilchrist, "Girard College: An Example of the Layman's Influence on Architecture," *JSAH* 16 (May 1957): 22–25.

11. For an explanation of this great work of masonry that "goes far beyond anything the Greeks attempted," see Carl W. Condit, *American Building: Materials and Techniques from the Beginning of the Colonial Settlements to the Present* (Chicago: University of Chicago Press, 1968), p. 68.

12. Herrick, *History of Girard College*, p. 113.

13. Westcott, *Official Guide*, p. 211; Scharf and Westcott, *History of Philadelphia*, 1:647–48.

14. See HABS photocopies of Walter's drawings and Fire Association 6361 (July 1840), INA 669 (July 1840), PC 5423 (July 16, 1840), copies at HABS, PHC.

15. *PL*, February 10, 1908; *PRERBG*, 23 (July 15, 1908): 455; MA 7150 (June 14, 1909); *Sunday Bulletin*, May 29, 1960.

16. Westcott, *Official Guide*, p. 211; Scharf and Westcott, *History of Philadelphia*, 2:1617, 1680; *Philadelphia Inquirer*, June 11, 1957.

17. *PL*, March 14, 1849.

18. Phoebe B. Stanton, *The Gothic Revival and American Church Architecture* (Baltimore: Johns Hopkins Press, 1968), pp. 3–29.

19. *PL*, March 3, 1847.

20. Samuel Laird, "St. Mark's Evangelical Church: Historical Sketch" (pamphlet, n.d.), MSS, Lutheran Theological Seminary; *PL*, June 12, 1851; July 8, 1851; November 15, 1852; *First Annual Report of the Board of Managers of the Spring Garden Institute* (Philadelphia, 1852), p. 4. This block is illustrated in Nicholas B. Wainwright, *Philadelphia in the Romantic Age of Lithography* (Philadelphia: Historical Society of Pennsylvania, 1958), p. 175.

21. In less than fifty years the Baldwin Locomotive Works extended along Broad Street between Spring Garden and Hamilton streets and westward to Eighteenth Street. The company moved to Eddystone in 1925 but it was another twelve years before the buildings were demolished. Scharf and Westcott, *History of Philadelphia*, 3:2255–57; *Philadelphia: A Guide to the Nation's Birthplace* (Philadelphia: William Penn Association of Philadelphia, Inc., 1937), p. 460.

22. Scharf and Westcott, *History of Philadelphia*, 3:2263–67; *PL*, May 3, 1853; August 31, 1854; Edwin T. Freedley, *Leading Pursuits and Leading Men: A Treatise of the Principal Trades and Manufactures of the United States* (Philadelphia: E. Young, 1856), p. 11; *Royal Road to Wealth* (Philadelphia: Samuel Loag, 1869), pp. 5–20.

23. Transfers of property indicate that the Poe cottage was built between September 1840 and August 1842, and biographers have established that Poe lived there in 1842. Deed Books GS20–415, GS44–194; Hervey Allen, *Israfel: The Life and Times of Edgar Allan Poe* (New York: Farrar & Rinehart, 1934), pp. 428–31. The first mention of the front building was in a title transfer, May 1849. It evidently was not built before then, which explains why this property was assessed in 1847 for $2,000 less than the adjoining lot with its three-story house (532 North Seventh Street). Deed Book GWC53–231; Guardians of the Poor, Poor Tax Register, 1847, p. 30, MSS, Philadelphia City Archives.

24. Poulson's Scrap-books, 7:137, MSS, LCP.

25. *PL*, June 7, 1850.

26. Robin Williams, *American Society* (2d ed., rev.; New York: Alfred A. Knopf, 1960), pp. 417–21.

27. *PL*, June 2, 1853.

28. Ibid., June 12, 1863; July 29, 1863; October 19, 1863.

29. Raised basements and bay windows with balconies above seem to have made these houses on Broad Street near Columbia Avenue in "the New York style." *PL*, July 4, 1853.

30. *PL*, March 11, 1852; July 4, 1853; May 1, 1863; June 13, 1863; September 26, 1863.

31. Gaul's house is better known for its second owner, Edwin Forrest, the great tragedian, who acquired the partially completed house in 1855. James Rees, *The Life of Edwin Forrest* (Philadelphia: T. B. Peterson & Brothers, 1874), pp. 379–81; Richard Moody, *Edwin Forrest* (New York: Alfred A. Knopf, 1960), p. 331. The house is described in FFI 20072 (January 31, 1854); PC 9546 (September 18, 1854), photocopies at HABS, PHC.

32. FFI 31634 (October 5, 1864), MSS, HSP; *PL*, June 2, 1863; October 3, 1864.

33. For a description of the eclectic Widener House, see *Philadelphia Record*, November 28, 1899; August 22, 1943; and *Philadelphia Telegraph*, November 30, 1899. Since the Bergdols, a family of prominent brewers, moved into the Kemble house in 1907, it is generally known as the Kemble-Bergdol House; James H. Windrim was the architect. *PRERBG* 4 (July 3, 1889): 306; 5 (April 19, 1890): 193; 5 (August 13, 1890): 481; *Gopsill's Philadelphia City Directory, 1907* (Philadelphia: James Gopsill's Sons, 1907), p. 204.

34. *PL*, August 13, 1852. Lewars had his grocery in the corner store, evidently rented meeting rooms above it, and lived over the adjacent shop. *McElroy's Philadelphia Directory for 1853* (Philadelphia: Edward C. & John Biddle, 1853), p. 236; *Cohen's Philadelphia City Directory for 1860* (Philadelphia: Hamelin & Co., 1860), p. 546; *McElroy's Philadelphia City Directory for 1865* (Philadelphia: A. McElroy, 1865), p. 401; *Gopsill's Philadelphia City and Business Directory for 1868–69* (Philadelphia: James Gopsill, 1868), p. 962.

35. For a discussion of the development of Furness's mature style, especially in relation to banks, see James F. O'Gorman, *The Architecture of Frank Furness* (Philadelphia: Philadelphia Museum of Art, 1973), pp. 35–45. The building was erected in 1872, but since the bank did not acquire the property until December 11, 1871, it is unlikely that the architects started the design before early 1872. *PL*, July 18, 1872; Registry Unit 4N8–52.

36. See Addison Hutton's Provident Life and Trust Company Building on Fourth Street below Market (1872) and James H. Windrim's Philadelphia Trust, Safe Deposit and Insurance Company Building at 415 Chestnut Street (1873–74). Both are illustrated in O'Gorman, *The Architecture of Frank Furness*, p. 44.

37. Registry Unit 4N8–55; *PRERBG* 6 (November 4, 1891): 704; 18 (July 1, 1903): 407; 18 (July 29, 1903): 483; 28 (August 20, 1913): 544; 28 (December 17, 1913): 816; Castner's Scrapbooks, 8:10, MSS, FLP.

38. The misattribution of buildings to socialite-architect Stanford White is a popular sport, as Brendan Gill illustrates in "True White from False," *The New York Times Magazine*, October 19, 1975, pp. 14–16, 66. Documentation and contemporary description of the Burk House are found in the sources that follow: *PL*, May 3, 1907; *Philadelphia Record*, September 19, 1909; Building Permit 4548 (1908); Thomas Nolan, "Some Recent Philadelphia Architecture," *The Architectural Record* 29 (1911): 227–29.

39. *PL*, March 5, 1907; March 10, 1907; September 8, 1907; March 28, 1908; March 29, 1908.

40. Building Permit 3806 (1908); *PL*, September 1907; *PRERBG* 23 (April 1, 1908): 209.

41. *PL*, November 18, 1908.

42. *Philadelphia: A Guide to the Nation's Birthplace*, p. 463.

43. Rankin, Kellogg, and Crane were the architects. Ibid., p. 458; Building Permit 12502 (1923).

44. The Reading Company's building is interesting in many respects. With a million square feet of floor space, it was the largest commercial structure in the country at the time. It was uniquely sensitive to both the automobile and railroad. Erected on stilts over the Reading's tracks, it became the second "air right" property in Philadelphia and the only one in the world with a public freight station beneath it. At the same time a vehicle concourse extending the width of the building gave access to the city's first underground parking garage. William Steele and Sons were the architects and builders. *Building* 9 (January 9, 1929): 9–10; Building Permits 4246 (1929), 868 (1930).

45. Temple became a university in 1907. Castner's Scrapbooks, 27A:94, MSS, FLP; *PRERBG* 6 (October 21, 1891): 672; 6 (December 30, 1891): 843; 8 (April 26, 1893): 253; *PL*, August 20, 1893; May 4, 1894.

46. *Philadelphia: A Guide to the Nation's Birthplace,* pp. 466–67.

47. *Building* 9 (February 1929): 22, 36.

48. J. W. Anshutz, for example, designed forty-seven two- and three-story brick houses with brownstone trim and turrets at Broad Street and Erie Avenue in 1890. *PL*, April 28, 1890.

49. *Philadelphia Inquirer,* May 9, 1891.

50. T. Henry Ashbury, president of the Enterprise Manufacturing Company, was Oak Lane's primary developer; Amos J. Boyden was his architect. S. F. Hotchkin, *The York Road, Old and New* (Philadelphia: Binder & Kelly, 1892), pp. 71–73.

51. *PRERBG* 23 (February 23, 1908): 131; *PL*, February 26, 1908; April 2, 1908; April 4, 1908; Letter from James Smart, *Evening and Sunday Bulletin*, to Richard J. Webster, July 25, 1972. Some of the houses were begun on the eve of the stadium's construction. For example, 202 two-story row houses were built on the square west of Shibe Park in 1907. *PL*, March 28, 1907.

52. There are many architecturally significant houses of worship in North Philadelphia. A list of the more important ones would include North Broad Street Presbyterian Church, Broad and Green streets, probably by Samuel Sloan (1862–64); (Roman Catholic) Church of the Gesu, Eighteenth and Stiles streets, by Edwin F. Durang (1879–88); (Episcopal) Church of the Advocate, Eighteenth and Diamond streets, by Charles M. Burns (1890–92); St. Francis Xavier Roman Catholic Church, Twenty-fourth and Green streets, by Edwin F. Durang (1893–95, rebuilt after fire 1906–8); (Roman Catholic) Church of St. Edward the Confessor, Eighth and York streets, by Henry D. Dagit (1908); Rodeph Shalom Synagogue, Broad and Mt. Vernon streets, by Simon and Simon (1928); (Roman Catholic) Church of the Holy Child, Broad and Duncannon streets, by George I. Lovatt (1929); Ukranian Catholic Cathedral of the Immaculate Conception, Franklin and Parrish streets, by Julian J. Jastremsky (1964–66); Zion Baptist Church, Broad and Venango streets, by Walter R. Livingston, Jr. (1973).

Delaware River Corridor

1. Stuart M. Blumin, "Residential Mobility within the Nineteenth-Century City," *The Peoples of Philadelphia,* ed. Allen F. Davis and Mark H. Haller (Philadelphia: Temple University Press, 1973), p. 43; Anthony N. B. Garvan, "Proprietary Philadelphia as Artifact," *The Historian and the City,* ed. Oscar Handlin and John Burchard (Cambridge, Mass.: MIT Press, 1963), pp. 197–98.

2. John Daly and Allen Weinberg, *Genealogy of Philadelphia County Subdivisions* (2d ed.; Philadelphia: City of Philadelphia, 1966), pp. 6–10, 51; Ellis Paxson Oberholtzer, *Philadelphia: A History of the City and Its People* (Philadelphia: J. S. Clark, 1911), 2:308–9.

3. Harold Donaldson Eberlein and Cortlandt Van Dyke Hubbard, *Portrait of a Colonial City* (Philadelphia: J. B. Lippincott Co., 1939), p. 152.

4. Ibid.

5. Guernsey A. Hallowell, "Edward Stiles," in *Papers Read before the Historical Society of Frankford* (1909), 2:12.

6. In the summer of 1838, during a temporary stay in Philadelphia, Whittier made many visits to the Frankford area and the former home of Thomas Chalkley. It moved him to write "Chalkley Hall," whose first verse is quoted from *The Complete Poetical Works of John Greenleaf Whittier* (Boston: Houghton Mifflin, 1894), pp. 220, 636.

7. John F. Watson, *Annals of Philadelphia* (rev., enlarged ed.; Philadelphia: Leary, Stuart Co., 1909), 1:140–41; Thomas Creighton, *Frankford: A Good Place to Live in* (Philadelphia, 1909), p. 3; Caroline W. Smedley, "Historical Sketch of Frankford Meeting," in *Papers Read before the Historical Society of Frankford* 2 (1916): 221.

8. The church building at Meeting House and Krewstown roads was rebuilt in 1805. Robert T. Tumbleston, *Old Pennepack Baptist Church* (Philadelphia, 1938), p. 15.

9. E. Perot Bissell, "Trinity Church, Oxford," HABS Report (PA-17), p. 1; Edward Y. Buchanan, *Two Discourses Relating to the Early History of Trinity Church, Oxford, Philadelphia* (Philadelphia: Porter and Coates, 1885), pp. 16, 57.

10. Garvan, "Proprietary Philadelphia," pp. 197–98.

11. William John Murtagh, "The Philadelphia Row House," *JSAH* 16 (December 1957): 9.

12. Wendy H. Robbins, Briefs of Title for 903–13 New Market Street (1967), PHC.

13. Fire Association 1857 (October 3, 1827), photocopy at PHC. Wendy H. Robbins, Briefs of Title for 132–34, 133 Olive Street (1968), PHC; *Philadelphia Directory for 1811* (Philadelphia: James Robinson, 1811), pp. xxxvii, xlvii.

14. Albert Mordell, "Men and Things: Life in a Band Box a Generation Ago," *Evening Bulletin*, July 21, 1937.

15. Josephine Hedges Ewalt, *As It Was in the Beginning* (Chicago: United States Savings and Loan League, 1956), p. 40; *PL*, March 12, 1849.

16. PC 3748–49 (March 1, 1817), photocopy at PHC; *153rd Annual Report, Friends Hospital* (Philadelphia, 1970), p. 11. The building's architectural history can be traced in resurveys of PC 3748 (June 3, 1829; March 30, 1835) and PC 8905 (March 17, 1852; June 2, 1873; February 3, 1880), photocopies at PHC. A brief description of additions to the main building and the erection of additional structures on the grounds is chronologically presented in *Friends Asylum for the Insane, 1813–1913* (Philadelphia, 1913). The lodestone of information on the institution's architecture is the "Minute Book of the Contributors of the Asylum for Persons Deprived of the Use of Their Reason," Friends Collection, Haverford College, Haverford, Pa. This invaluable volume was uncovered at Friends Hospital by Edward Teitelman, M.D., who has done a great deal of research on the institution's architecture.

17. *Friends Asylum for the Insane, 1813–1913*, pp. 65, 95; George B. Tatum, *Penn's Great Town* (Philadelphia: University of Pennsylvania Press, 1961), p. 165, pl. 35; Theo B. White et al, *Philadelphia Architecture in the Nineteenth Century* (Philadelphia: University of Pennsylvania Press, 1953), p. 22, pl. 1. See HABS Drawings of Pennsylvania Hospital, PA–1123, Library of Congress.

18. *153rd Annual Report, Friends Hospital*, p. 11; *Friends Asylum for the Insane, 1813–1913*, pp. 17–18; PC 8905 (June 2, 1873), photocopy at PHC.

19. *153rd Annual Report, Friends Hospital*, p. 1; Harold N. Cooledge, Jr., "Samuel Sloan and the Philadelphia School of Hospital Design, 1850–1880," *Charette* 44 (June 1964):6–7, 18; Edward Teitelman and Richard W. Longstreth, *Architecture in Philadelphia: A Guide* (Cambridge, Mass.: MIT Press, 1974), p. 165.

20. The Hospital of the Protestant Episcopal Church in Philadelphia was chartered on July 18, 1851; it accepted the generous gift of Miss Ann Leamy and her sister Mrs. Elizabeth H. L. Stout in December 1852. Edward F. Leiper, "The Episcopal Hospital," *Medical and Surgical Reports of the Episcopal Hospital* (Philadelphia: William J. Dornan, 1913), 1:19–20.

The legally defined western border of Kensington District in 1820 was Sixth Street; enlarged in 1848, the district spilled above Lehigh Avenue and westward to Germantown Avenue, which ran diagonally into Sixth Street. Although the area has been officially in Philadelphia since 1854, residents still consider it Kensington. Nowadays, however, Kensington's borders are socially rather than legally determined, leading many to consider Front Street the present western boundary. Laws of Pennsylvania, Chapter 52, March 6, 1820; J. Thomas Scharf and Thompson Westcott, History of Philadelphia, 1609–1884 (Philadelphia: L. H. Everts & Co., 1884), 1:689; Rod Townley, "Pride and Prejudice in Kensington," Today: The Inquirer Magazine, June 16, 1974, pp. 27, 31; Peter Binzen, Whitetown, U.S.A. (New York: Random House, 1970), p. 86.

21. Beginning in 1853 patients were treated in the Leamy House, the country seat that had been altered to hospital purposes. By 1860 there was a daily average of twenty-eight patients in the house. Besides Episcopal Hospital there were in the early 1850's only the venerable Pennsylvania Hospital and Philadelphia Hospital, and the latter was essentially a public almshouse. There were also four dispensaries and the Wills Hospital for the blind and lame, but they were small and specialized. Leiper, "The Episcopal Hospital," pp. 19–21.

22. Cooledge, "Philadelphia School of Hospital Design," pp. 6, 18; The Hospital of the Protestant Episcopal Church in Philadelphia: Its Origins, Progress, Work and Want (Philadelphia: J. B. Lippincott Co., 1869); The Architectural Review and American Builders Journal 2 (March 1870): 514–23; Report of the Board of Managers of the Hospital of the Protestant Episcopal Church in Philadelphia 24 (January 1876): 10–11.

23. Cooledge, "Philadelphia School of Hospital Design," p. 18.

24. Frankford Arsenal: Historical Highlights (Philadelphia, [c. 1971]), p. 1; Gwen D. Hollahan, Frankford Arsenal's 150 Year Heritage (Philadelphia, 1966), pp. 3, 6; Odus C. Horney, Frankford Arsenal, 1816–1926 (Philadelphia, 1926), p. 1.

25. Hollahan, pp. 8–9, 11–12; Horney, pp. 2–3; Joseph M. Colby, Frankford Arsenal, 1916–1956 (Philadelphia, 1956), p. 3; Frankford Arsenal: Historical Highlights (Philadelphia, [c. 1971]), p. 4; Scharf and Westcott, History of Philadelphia, 2:1015.

26. Philadelphia Inquirer, June 18, 1969; April 11, 1975.

27. Creighton, Frankford, p. 3. There were some industries unrelated to agriculture, such as a glass works that was established in 1771 at the mouth of Gunner's Run. It apparently was not a very large or profitable enterprise, however, until it was taken over by Dr. Thomas W. Dyott in 1833. Charles H. Cramp, "Old Richmond District and Its Memories," PL, January 10, 1909.

28. Kensington was laid out and named in 1730 by Anthony Palmer, president of Council in the 1740's. Initially it extended southward along the river from Hanover street to Gunner's Run and westward to near Frankford Road. Formed as a district in 1820, it was extended northward in 1848 to Lehigh and Germantown avenues. Laws of Pennsylvania, Chapter 52; Scharf and Westcott, History of Philadelphia, 1:689.

29. Scharf and Westcott, History of Philadelphia, 1:522; Joseph Jackson, Market Street, Philadelphia (Philadelphia, 1918), p. 207; The Philadelphia Directory for 1810 (Philadelphia: James Robinson, 1810), p. 83; The Philadelphia Directory for 1817 (Philadelphia: Edward Dawes, 1817), p. 127.

30. George Castor Martin, "Samuel Martin, Proprietor of the First Textile Mill in Frankford," in Papers Read before the Historical Society of Frankford 2 (1916):243–44.

31. Of the first six mills begun in Frankford between 1816 and 1820, five were established by English-born entrepreneurs. The activities included cotton mills, a block

printing plant, dyeing and bleaching works, and a pottery. William B. Dixon, "Frankford's Early Industrial Development," *The Pamphlet of the Historical Society of Frankford for 1911* (1912), pp. 50–59.

32. As late as the 1860's mills along Frankford Creek were surrounded by farmland and meadows. See, for example, the Frogmore Mills at Frankford Creek and Powder Mill Road, Frankford, in *Hexamer General Survey*, 3, (1867–71): pl. 242.

33. Wendy H. Robbins, Brief of Title for 125–31 Ellen Street (1967), PHC; Edwin T. Freedley, *Philadelphia and Its Manufactures* (Philadelphia: Edward Young, 1859), pp. 317–18.

34. E. McFarlin, Pennsylvania Historical Salvage Council Drawings (B–613), 125–31 Ellen Street (September 1967); (B–514), 915–17 Hope Street (November 1967), copies at HABS, PHC.

35. *McElroy's Philadelphia Directory for 1846* (Philadelphia: John & Edward C. Biddle, 1846), p. 164. Will Book 30, Will No. 92, March 31, 1853, MSS, City of Philadelphia, Register of Wills. Ernest Hexamer and William Locher, *Maps of the City of Philadelphia* (Philadelphia: Hexamer and Locher, 1859), 4:pl. 52. Holloway's house measured 27 feet 6 inches on New Market by 52 feet on Ellen Street.

36. Some of these houses date before 1867, and all were completed by 1874. *Hexamer General Survey*, 4 (1867–71):pl. 334; 6 (1874): pl. 504; 12 (1876): pls. 1065–66; 17 (1881): pls. 1598–99.

37. The houses included eleven dwellings with stores below on Fourth Street north of Jefferson in 1849 and ten dwellings on Lawrence south of George, and twelve on Poplar Street east of Lawrence in 1848. J. and G. A. Binder, popular local builders, built the structures. *PL*, March 30, 1849.

38. The Philadelphia and Trenton Railroad was incorporated in February 1832. The Philadelphia-New York stages stopped running in 1836. Scharf and Westcott, *History of Philadelphia*, 3:2183; Guernsey A. Hallowell, "Transportation," *Frankford: Direction of a Greater Philadelphia* (Philadelphia, 1922), pp. 54, 64.

39. David Budlong Tyler, *The Bay and River Delaware: A Pictorial History* (Cambridge, Md.: Cornell Maritime Press, 1955), p. 132; Freedley, *Philadelphia and Its Manufactures*, pp. 317–18; *PL*, May 7, 1849, April 25, 1851.

40. Scharf and Westcott, *History of Philadelphia*, 3:2185. The rail and wharves complex between Huntington and William streets east of Richmond Street (then Point Road) was already well developed by 1849. See J. C. Sidney, *Maps of the City of Philadelphia* (Philadelphia: Smith & Wistar, 1849).

41. *Port Richmond, Philadelphia* (Philadelphia: Reading Company, 1956), p. 1.

42. Freedley, *Philadelphia and Its Manufactures*, pp. 329–30; *McElroy's Philadelphia Directory for 1856* (Philadelphia: John & Edward C. Biddle, 1856), p. 694; Scharf and Westcott, *History of Philadelphia*, 3:2267.

43. *Philadelphia and Popular Philadelphians* (Philadelphia, 1891), p. 133. Disston began his business earlier on a very small scale near Second and Arch streets. Scharf and Westcott, *History of Philadelphia*, 3:2267.

44. *PL*, July 30, 1872; *Hexamer General Survey*, 8 (June 18, 1873): pls. 693–94; Gray Kirk and E. McFarlin, Pennsylvania Historical Salvage Council Drawings (B–611), 1001 North Front Street (1967). This passage was partly filled in after the building became a paper box factory in 1907. *Boyd's City Directory, 1907* (Philadelphia: C. E. Howe Company, 1907), p. 204; George W. and Walter S. Bromley, *Atlas of the City of Philadelphia* (Philadelphia: G. W. Bromley & Co., 1910), pl. 13.

45. Wendy H. Robbins, Brief of Title for 440 New Market Street (1968), PHC;

McElroy's Philadelphia Directory for 1849 (Philadelphia: John & Edward C. Biddle, 1849), p. 158; *McElroy's Philadelphia Directory for 1859* (Philadelphia: John & Edward C. Biddle, 1859), p. 298; Freedley, *Philadelphia and Its Manufactures,* p. 372; *PL,* February 27, 1851; Nicholas B. Wainwright, *Philadelphia in the Romantic Age of Lithography* (Philadelphia: The Historical Society of Pennsylvania, 1958), p. 146.

46. Three well-preserved examples from the 1850's recorded by the Pennsylvania Historical Salvage Council were the Knight Building at the northeast corner of Second and Vine streets, its neighbor at 305–7 North Second, and the Goodwin Building, 317–19 North Second Street. Registry Unit, 2N18–63, 64; 2N18–71, 149; 2N18–111, 112.

47. FFI 56618 (June 24, 1878). The drawings are at the Historical Society of Pennsylvania, copies at PHC.

48. Robert T. Corson, "The Frankford Lyceum," in *Papers Read before the Historical Society of Frankford* 2 (1910): 90–94.

49. Alexis de Tocqueville, *Democracy in America,* trans. Henry Reeve (New York: Schocken Books, 1961), 2:129–30.

50. The "combined effects of Philadelphia's rapid growth—the endless grid streets, the scattering of churches, stations, and factories, and flood of immigrants, the novelty, the sheer size, and pace of the big city—" are cited by Sam Bass Warner, Jr., *The Private City: Philadelphia in Three Periods of Its Growth* (Philadelphia: University of Pennsylvania Press, 1968), p. 61. An added factor, high residential mobility, has been noted by Stuart M. Blumin, "Residential Mobility within the Nineteenth-Century City," pp. 37–52.

51. The erratic organization of the interior—irregularly shaped rooms, discontinuous floor levels and stairs, gigantic false doors, and one room without windows—leaves no doubt about its ceremonial function, and the presence of seven meeting halls suggests that more than one lodge used the building at the same time. The Odd Fellows vacated the hall in November 1932. *Public Ledger and Daily Transcript,* November 5, 1847; Denis B. Myers, "Independent Order of Odd Fellows Hall," HABS Architectural Report (1963), pp. 1–6; Fred R. Korman, Grand Secretary of Grand Lodge of Pennsylvania, Independent Order of Odd Fellows, to Richard J. Webster, June 18, 1974.

52. Deed Book GS 17–369 (August 22, 1840). A limestone plaque in the middle of the facade of the Northern Liberty Hose Company contained the date of 1846. United States Hose Company, Minutes, 1837–1845, 1845–1848, MSS, HSP; Wainwright, *Philadelphia Lithography,* pp. 88, 234.

53. Bruce Laurie describes changes in the social composition and roles of Southwark fire companies, and although he deals with a section south of the city, his conclusions can be related to Northern Liberties, a section just north of the city and with a similar population: "Fire Companies and Gangs in Southwark: The 1840's," in *The Peoples of Philadelphia,* p. 77.

54. Scharf and Westcott, *History of Philadelphia,* 1:551.

55. Northern Liberty Hose Company served as Philadelphia Fire Department, Engine Company No. 21 until 1875 when a new fire station was built at 826–28 New Market Street. Friendship Fire Company No. 15 at Norris and Sepviva streets was rented by the city for Hook and Ladder Company C until 1878. For a time after that it was known as Friendship Hall. Scharf and Westcott, *History of Philadelphia,* 3:710, 836; *Mayor's Annual Report. Annual Report of Board of Fire Commission* (1875), (1878); George W. and Walter S. Bromley, *Atlas of the City of Philadelphia* (Philadelphia: G. W. Bromley & Co., 1887), 4: pl. K.

56. Kurt B. Mayer, *Class and Society* (New York: Random House, 1955), p. 44; Liston Pope, "Religion and the Class Structure," *Annals of the American Academy of Political and Social Science* 256 (March 1948): 84–91; Tocqueville, *Democracy in America*, 2: 128–29; Max Lerner, *America as a Civilization* (New York: Simon and Schuster, 1957), pp. 711–12; Ralph Henry Gabriel, *The Course of American Democratic Thought* (2d ed.; New York: The Ronald Press, 1956), pp. 26–39; Norman J. Johnston, "The Caste and Class of the Urban Form of Historic Philadelphia," *Journal of the American Institute of Planners*, November 1966, pp. 334–38.

57. Lerner, *America as a Civilization*, p. 710.

58. *PL*, April 20, 1853; Poulson's Scrap-books, 4:16, MSS, LCP; Ernest N. Feind, *A History of the First Presbyterian Church of Kensington, 1814–1939* (Philadelphia, 1939), p. 10; Kathryn I. Welliver, *A History of the First Presbyterian Church of Kensington in Philadelphia* (Philadelphia, 1964), p. 5; Trustees' Records, 1853–1873, pp. 60–62, MSS, Frankford Presbyterian Church. A drawing of the First Presbyterian Church of Frankford, signed by McArthur, is in the possession of the church. For an interior description see Fire Association 43232 (December 17, 1866), photocopy at PHC; Teitelman and Longstreth, *Architecture in Philadelphia*, p. 169.

59. *Historical Sketches of the Catholic Churches and Institutions of Philadelphia* (Philadelphia: Daniel H. Mahoney, 1895), pp. 56–59; Scharf and Westcott, *History of Philadelphia*, 1:666.

60. Joseph Jackson, *Early Philadelphia Architects and Engineers* (Philadelphia, 1923), pp. 236–37; *Historical Sketches of Catholic Churches*, p. 81; *St. Joachim's Church: A History* (Philadelphia, [c. 1974]), n.p.

61. St. Peter's, for example, was rebuilt as a large stone church in 1895–97; St. Michael's was enlarged with a new facade and towers at the turn of the century and a new rectory and a school were finished in 1892; a new St. Ann's was built in 1866–69 and its interior was renovated in 1909, and a new St. Joachim's was built in 1874–80, with its convent and school built in 1866 and 1885 respectively. (The present convent dates from 1921.) *Historical Sketches of Catholic Churches*, pp. 56–59, 81; *History of St. Ann's Church* (Philadelphia, 1920), n.p.; *St. Joachim's Church*, n.p.

62. Robert F. Scott, "Historic Presbyterian Churches of Kensington and the Northern Liberties," Address at 14th Annual Meeting of Friends of Old Pine Street Church, April 28, 1964; Franklin Davenport Edmunds, *The Public School Buildings of the City of Philadelphia from 1845 to 1852* (Philadelphia: By the author, 1915), p. 169; Building Permits 6112 (1916), 7224 (1923).

63. *PL*, February 2, 1854; *Evening Bulletin*, June 21, 1921; Scharf and Westcott, *History of Philadelphia*, 3:2187, 2202.

64. The new institutions were Second National Bank, Frankford (1863), Eighth National Bank (1863) and Shackamaxon Bank (1873) in Kensington, and Security National Bank (1870) and Northern Saving Fund, Safe Deposit and Trust Company (1871) on the boundary of the former Spring Garden District. Thompson Westcott, *The Official Guide Book to Philadelphia* (Philadelphia: Porter and Coates, 1875), pp. 155–61.

65. The Bank of Northern Liberties became a national bank in 1864. Westcott, *Official Guide Book*, p. 155; *Philadelphia and Popular Philadelphians*, pp. 9, 226; Scharf and Westcott, *History of Philadelphia*, 3:2111.

66. George E. Thomas and James F. O'Gorman, "Catalogue of Selected Buildings," in James F. O'Gorman, *The Architecture of Frank Furness* (Philadelphia: Philadelphia Museum of Art, 1973), pp. 95, 117.

67. William B. Dixon, "Frankford's Early Industrial Development," *The Pamphlet of*

the Historical Society of Frankford for 1911 (1912), p. 53; Freedley, *Philadelphia and Its Manufactures,* p. 250; *Hexamer General Survey,* 4 (1867–71): pl. 311; 18 (December 23, 1882): pls. 1729–30.

68. Between 1871 and 1882 the following changes were made: an oil house was added to the Drug Room, the office was enlarged, an extension was built onto the Blue Dyehouse, the Main Building and adjoining buildings were altered on the interior, a boiler house was added to the Engine Room, and the Cylinder Drying Room was altered. *Hexamer General Survey,* 4 (1867–71): pl. 311; 18 (December 23, 1882): pls. 1729–30.

69. Only one building was left unchanged from 1867, and that may have been altered on the inside since an office appears to have been added after 1867. *Hexamer General Survey,* 4 (1867–71): pl. 334; 6 (February 8, 1874): pl. 504.

70. *Hexamer General Survey,* 12 (November 1876): pls. 1065–66; 17 (October 10, 1881): pls. 1598–99; 29 (February 4, 1895): pls. 2843–44.

71. Freedley, *Philadelphia and Its Manufactures,* p. 239; Moses King, *Philadelphia and Notable Philadelphians* (New York: Moses King, 1902), p. 44. The mills' operation is described in *Philadelphia: A Guide to the Nation's Birthplace* (Philadelphia: William Penn Association of Philadelphia, 1937), pp. 518–20.

72. *PL,* April 7, 1886; April 17, 1886; September 12, 1894; *PRERBG* 1 (April 19, 1886):174; 1 (May 31, 1886):244; 4 (August 7, 1889):367; 6 (March 11, 1891):145; 6 (April 15, 1891):226; 7 (April 6, 1892):1067; 7 (April 20, 1892):1099; *Philadelphia Record,* September 12, 1894. For twentieth-century additions and alterations, see Building Permits 970 (1907), 8983 (1909), 582 (1910), 6722 (1912), 9965 (1923), 494 (1924), 921 (1939), 2019 (1941), 2221 (1941).

73. When one of these buildings ignites, it often sets off a multiple alarm fire. In 1975 the former North American Lace Company (1903) at Glenwood and Allegheny avenues fell victim to a nine-alarm fire, and the former Philadelphia Tapestry Mills (c. 1900 and later) at Allegheny Avenue and Mascher Street was consumed in a seven-alarm fire. *Philadelphia Inquirer,* March 1, 1975; August 15, 1975.

74. Carl W. Gatter, a local resident who has done a great deal in researching and documenting Frankford area's historic structures, took photos of Lynfield (c. 1760, c. 1800) during its demolition in 1942, and of Chalkley Hall (1723, 1776) during its demolition in 1954. Port Royal (1761) was demolished c. 1962.´

75. Joseph W. Ryerss acquired the property in January 1859 with a messuage on it, but it was evidently a farm building and not the mansion, which nineteenth-century writers say he built. It was Ryerss's country home, probably used only in the summer; his year-round residence was at 922 Walnut Street. Deed Book ADB 48–82 (January 1, 1859); S. F. Hotchkin, *The York Road, Old and New* (Philadelphia: Binder & Kelly, 1892), p. 424; S. F. Hotchkin, *The Bristol Pike* (Philadelphia: George W. Jacobs & Co., 1893), p. 11; *McElroy's Philadelphia City Directory, 1867,* p. 788;

76. *PL,* October 17, 1907.

77. Rhawn did not become president of the bank until 1884, after the house was completed. Hotchkin, *York Road,* p. 416; O'Gorman, *Architecture of Frank Furness,* pp. 50–51, 125.

78. O'Gorman, *Architecture of Frank Furness,* p. 50, quotes this label, which was evidently first used in the *Architectural Record* at the turn of the century.

79. The Frankford Elevated was begun in 1915 with a scheduled completion date of 1918. It opened in November 1922, its construction costs more than twice the original estimate; both the delay of construction and rise in cost were attributable to World War

I. The El proved popular, and it drew enough business from the railroad to force the closing of the Frankford Station of the Reading Railroad, June 30, 1928. Hallowell, *Frankford*, pp. 3, 68; Walter M. Benner, *Frankford Goes to Town* (Philadelphia: Third Federal Savings and Loan Association, [c. 1960]), pp. 10–11.

80. Whittier's poem was quoted in part earlier. Promotions for new houses in Frankford can be found in Hallowell, *Frankford*.

81. Measuring approximately 16 by 36 feet on lots about three times as deep, these two-story brick houses were modern fusions of the eighteenth-century city house plan and mid-nineteenth-century terrace houses. Henry Berg is said to have designed many of them but so did J. W. Lewin and G. Harold Murphy. Yet it mattered little who the architects were; the result was always similar, uninteresting facades, as seen on the 3100 block (by Murphy in 1946) and the 2200 block (by Lewin in 1950) of Friendship Street. *Philadelphia Inquirer,* September 28, 1975; Murtagh, "The Philadelphia Row House," pp. 11–12; Building Permits 4364–67 (1946), 1961–65, 1970–73 (1950).

82. Mackley, 22, was shot in the head by workers of the H. C. Aberle Company while he was waiting for a green light at the intersection of Rising Sun Avenue and Roosevelt Boulevard. The fatal gun shot climaxed an argument between fellow strikers in Mackley's car and nonstriking workers in an adjacent car. *Philadelphia Inquirer,* March 7, 1930, p. 8. Plaques in memory of five other men and women who died in hosiery strikes during the 1930's are placed on the exterior walls of the buildings.

83. The complex fills a city block, bounded by Castor Avenue, Cayuga, M, and Bristol streets, affording ample interior space for lawns, gardens, and even a swimming pool. Construction began in January 1934 and tenants started moving in a year later. It is appropriate that the houses were built for textile workers, because at the time approximately one-fourth of the city's industrial workers were employed in textile factories that were concentrated in this area. *Philadelphia: A Guide to the Nation's Birthplace,* pp. 515, 525–26.

84. William Pope Barney directed the erection of the buildings. Tatum, *Penn's Great Town,* p. 203; Teitelman and Longstreth, *Architecture in Philadelphia,* p. 164; *Philadelphia: A Guide to the Nation's Birthplace,* p. 526.

85. The community's organization in the 1930's is briefly described in *Philadelphia: A Guide to the Nation's Birthplace,* pp. 526–27.

Index

Abercrombie, Capt. James, House (now Perelman Antique Toy Museum), xxxvii, 4, 9
Aberle, H. C., Co., 382(n.82)
Academic Department of the University of Pennsylvania. *See* University of Pennsylvania, College Hall
Academy of Music (American Academy of Music), 111–12, 121, 290, 355(n.35)
Academy of Notre Dame, 121
Acculturation, 173, 247, 248, 249–50, 251
Acoustics, 112, 121, 355(n.35)
Adamesque-Federal style, 195, 219, 250–51, 277–78. *See also* Federal style
Adams, Clyde S., architect, 125
Adams Ave., 311
Advertising and architecture, 93–94, 117, 118, 233, 293
Air conditioning, central, 141
Airlite houses, 382(n.81)
"Air-right" property, 374(n.44)
Alden Park apartments, 255, 371(n.39)
Allen, Richard, 7, 10, 96
Allen, William, 246, 249
Allison, Joseph, 199
Almshouses. *See* Blockley Almshouse
American Architect and Building News, The, xxvi
American Federation of Hosiery Workers, 319
American Fire Insurance Company, building, 50, 59
American Institute of Architects, xxvi, xxix, xxx, xxxi, xxxii
American Life Insurance Company (Manhattan Building), 59
American Philosophical Society (Philosophical Hall), 56–57, 92
American Society of Civil Engineers, 231, 334
American Swedish Historical Museum (John Morton Memorial Museum), 162

Anderson, W. Nelson, architect, 19, 23, 26, 31
Andrews, W. L., architectural drawings, 187
Annan, William, House, 165
Anshutz, J. W., designs row houses, 375(n.48)
Apartment buildings, 8, 9, 131, 142, 200, 255, 332, 365(n.45)
Apprentices Library, 74, 300
Aquarium, site of City, 231
Arbour, William, House, 4, 10. *See also* Winemore, Phillip, House
Arcade Building (Commercial Trust Building), 115, 122, 138
Arch Street, houses, 76
Arch Street Area Study, 60
Arch Street Friends Meeting, 54, 60
Arch Street Methodist Episcopal Church, 110–11, 122, 355(n.27)
Arch Street Opera-House (Continental Theatre, the Trocadero, the Troc), 112, 122–23
Arch Street Presbyterian Church (Tenth and Arch Sts.), 111
Arch Street Presbyterian Church (Eighteenth and Arch Sts.). *See* West Arch Street Presbyterian Church
Arch Street Theatre, 58
Architects, student training, xxii, xxxiv–xxxv, xxxviii, xxxix
Architectural books, xxiii, xxv–xxviii, xli(n.14), 249, 362(n.12)
Architectural drawings, 187; competition, 24, 61, 112, 121, 197, 225, 299, 316, 368(n.14); early, xxiii, xxvi–xxviii; HABS, xxxii, xxxiv–xxxvi; OPS, xxix; student, xxxiv–xxxv, xlv(n.58)
Architectural records, HABS, xxi
Architectural Iron Works, store fronts by, 66
Armat, Thomas, House, 259

Armory of the National Guard. *See* National Guard's Hall
Armstrong, John, carpenter, 67
Armstrong, William A., builder, 87, 91, 137, 328
Armstrong and Printzenhoff, contractors, 273
Arsenal on the Schuylkill (Schuylkill Arsenal), 159, 165–66
Arsenals. *See also* Frankford Arsenal; Powder Magazine
Art Deco architecture, 61, 118, 133–34
Artillery Lane (now Nectarine St.), 329
Ashbury, T. Henry, developer, 375(n.50)
Ashmead, Albert, House, 259
Ashmead, William, House, 247, 259
Askins-Jones House, 166
Asylum for the Relief of Persons Deprived of the Use of Their Reason. *See* Friends Hospital
Athenaeum of Philadelphia, 57, 60–61, 101; privy, 61
Atwater Kent Museum. *See* Franklin Institute
Augusta Place (Brooks Court). *See* Spring St.
Autenreith, Charles M., architect, 83. *See also* Collins and Autenreith
Ayer, N. W., & Son, Inc., Building (N. W. Ayer & Co.), 61, 101

Badger, Daniel D., 66
Bailey, Joel J. & Co., 51, 83
Baily, William L., architect, 129
Baily and Bassett, architects, 129
Baird, Matthew, 289
Bake House and Oven, 10
Baker and Dallett, architects, 127
Balderston, Charles, architect, 228
Baldwin Locomotive Works, 136, 205, 373(n.21)
Baltimore and Ohio Railroad Station, 123
Bambrick, Br. George M., S. J., decorator, Church of the Gesu, 294
Bancroft and Sellers. *See* Sellers, William, and Co.
Bandboxes, 5, 307–308, 311, 331, 337. *See also* Airlites; Rowhouses
Bank of North America, 42, 47, 48, 61–62
Bank of Northern Liberties. *See* National Bank of Northern Liberties
Bank of Pennsylvania (Philadelphia Bank), 48, 62
Bank of the United States, First (Stephen Girard's Bank), 42, 47, 65, 72

Bank of the United States, Second, xxxiv, 42, 47, 97–98, 128, 366(n.54)
Bank Row, Chestnut St., xxxix, 47–48, 71, 87–88
Banking rooms, design of, 46, 47–48, 87, 88, 91, 92, 97
Banks: Center Square, 117–18, 128–29, 130, 133–34, 141, 148, 357(n.64); Germantown, 254, 262; North Philadelphia, 289–90, 297–98; Northern Liberties, 316, 325, 330–31, 333, 380(n.64); Old City, 46, 47–48, 61–62, 66, 69, 71, 72, 73, 84, 87, 88, 90–91, 92, 97–98; West Philadelphia, 205–6
Baptist Mariners' Bethel, 158
Barney, William Pope, architect, 332, 382(n.84)
Barns, 262, 265, 271, 276–77. *See also* Carriage houses; Stables
Barron, Commodore James, House, 260
Barth, Br. Frederick E., S.J., decorator, Church of the Gesu, 294
Bartram, John, House, xxviii, 194–95, 198, 203
Bartram, John, Hotel. *See* Hotel Walton
Bartram Hall, 198
Baseball Parks. *See* Shibe Park
Batcheler, George D., Jr., architect, 61
Bathrooms, fittings for, 199
Baugh & Sons Co., 62
Bay, Jacob, 277
Bayliss, Dudley C., xlv(n.48)
Baynton House. *See* Germantown Historical Society Area Study
Beaux-Arts Classicism, 108–9, 117–18, 128–29, 135–36, 232, 290, 301
Bechtel House. *See* Germantown Historical Society Area Study
Beck, Paul, Jr., 40, 62
Beck Street Area Study, 166
Beck Warehouse (Baugh Warehouse), 40, 62
Beggarstown School, 248, 260
Behrman, Walter, interior designer, 141
Bel Air (Belleaire, Singley House, Lasse Cock's Manor House), 158, 166–67, 362(n.6)
Belfield (Charles Willson Peale House), 260
Belleaire. *See* Bel Air
Belmont Mansion, 227
Bement and Dougherty, Industrial Works, 287
Bencker, Ralph B., architect, 61, 147, 200
Bendiner, Alfred, 113

Benezet, Anthony, house, xxiv, xli(n.8)
Benjamin Franklin Parkway, 118–19, 124, 237
Bennett and Company, Tower Hall store, 44
Bensell's Lane. *See* School House Lane.
Beranek, Leo L., acoustician, 355(n.35)
Berg, Henry, airlite designer, 382(n.81)
Berger, Carl P., architect, 94
Bernhart, Max A., architect, 136
Berry, Peter L., house carpenter, 63
Berry-Coxe House, 63
Beth Eden House. *See* Shunk School
Bethel African Methodist Episcopal Church (Mother Bethel), 7, 10
Biddle, Nicholas, 285, 372(n.8)
Biddle, Owen, 60
Billmeyer, Michael, House, 248, 261
Bilsland, Ann, 145
Binder, Jacob and George A. (Binder, J. and G. A.), builders, 277, 378(n.37)
Bingham, William, xxiii
Birch, William Russell, *Views*, xxiv
Bird, Joseph, House, 156, 167
Birely, John, shipbuilder, 321
Birely, Theodore, shipbuilder, 321
Birely House, 321
Birkinbine, H. P. M., engineer, 231
Bishop, John, 157, 167
Bishop-Sparks House, 167
Bissell, E. Perot, xxxi, xxxii
Bitter, Karl, sculptor, 138
Blackwell, Reverend Robert, House (St. Peter's Church House), 10–11
Bladen's Court. *See* Elfreth's Alley Area Study
Blair, Reverend Samuel, House (415 Locust St.), 63
Blair House (6105 Germantown Ave.), xxxi, 261
Blake, T. E., architect, 367(n.57)
Bleakley House (Cannon Ball Farm House), xxxii, 203–204
Blight Warehouse, 63
Blockley Almshouse, 196, 363(n.26)
Blockley Township, 193, 195, 198, 364(n.35)
Blodgett, Samuel, Jr., architect, 47, 72
Bloor, A. J., xlii(n.20)
Blue Bell Tavern, xxviii, 204
Blue Dick, The. *See* United States Hose Company, No. 14
Boat House Row, 225, 228
Boittau, Alfred, sculptor, 117, 129
Bolton, Charles L., architect, 122

Bolton, Charles W., & Co., architects, 148
Bonsall, Edward, house carpenter, 17
Bonsall, John, House, 63–64, 101
Booth, Edwin, 46
Borie, Charles L., Jr., architect, 237
Borie Brothers Bank, 66
Boucher, Jack E., xxxvi
Bower, John A., architect, 21
Boyd, D. Knickerbacker, xxix
Boyden, Amos J., architect, 375(n.50)
Bradin, Oliver, 216
Breck, Samuel, 241
Breitnall, David, xli(n.8)
Breton, William L., sketches of Philadelphia, xxiv, xxv
Bridesburg, 305
Bridges, Robert, House, 11
Bridges, 195, 196, 362(n.19); covered, 230; in Fairmount Park, 225, 226, 230, 231–32, 239–40; pedestrian, 122, 138, 172. *See also listings by name*
Bridges-LaTour House, 11
Bridport, George, landscaping design of, 101
Bringhurst House, 261
Brinkworth, George H., contractor, 185
Bristol Township, 305
British Building. *See* International Exhibition of 1876, St. George's House
Broad Street, 162, 288
Broad Street Station. *See* Pennsylvania Railroad, Broad Street Station
Brock, John, Sons & Co., Warehouse, 58, 64
Brockie, Arthur H., architect, 228
Brockie and Hastings, architects, 93
Brokers' Row (S. Third St. below Market), 45–47
Bromley, John, & Sons, Lehigh Avenue Mills, 317, 321–22
Brooks Court (Augusta Place). *See* Spring Street
Brown, Anna S., House and Shop, 332
Brown, Charles T., 30
Brown, Frank Chouteau, xl(n.1)
Brown, Robert Montgomery, architect, 366(n.54)
Brown and Allison, carpenters, 98
Browne, John C., Houses, 332
Brownstone dwellings. *See* Georgian Brownstone
Brumbaugh, G. Edwin, architect, 265
Brumidi, Constantino, decorator, Cathedral of Sts. Peter and Paul, 124
Brutalism, New, 120

Bry, Daniel H., builder, 301
Buell, W. Duncan, architect, 34
Bugbee, Jonathan, architect, 125
Building technology, changes in, 50
Bulfinch, Charles, xxiii
Bulletin Building (Penn Square Building), 123–24
Bulletin Building (30th and Market Sts.), 202, 366(n.54)
Burczynski, Stephen, architect, 125
Burgin, Dr. George H., House, 11, 29
Burholme, 318, 322–23
Burk, Alfred E., House, 290, 293, 374(n.38)
Burnham, D. H., and Company, architects, 115, 132–33, 147–48
Burnham, George, House (3401 Powelton Ave.), 204–5
Burnham House (2213 Green St.). See Green Street Area Study
Burns, Charles M., architect, 127, 147, 215, 275, 294, 367(n.57), 375(n.52)
Burnside. See Hamilton-Hoffman House
Bush Hill, 195, 283
Bussey-Poulson House, 12
Busti, Paul, 195, 197, 205, 363(n.20)
Busti Mansion (Blockley Retreat Farm, Kirkbride Mansion), 195, 205
Butcher & Brother, Warehouse, 64
Butler, Anthony, Houses, 323
Butler, Thomas, House. See Philadelphia Club
Button, Stephen D., architect, 348(n.28); banks, 325; cemetery gatehouse, 212; churches, 14, 30, 74, 110, 180; commercial buildings, 44, 46, 82, 114; competition drawings, 121; houses, 296. See also Hoxie and Button
Byberry Township, 305
Byrne-Cavanaugh House, 167
Byzantine style, 111

Calder, Alexander Milne, sculptor, 106, 140, 218
Callowhill Street Bridge. See Spring Garden Street Bridge
Cambridge Camden Society. See Ecclesiological Society
Campbell, William M., xxix
Canals, Manayunk, 256. See also listings by name
Canavan, Francis P., architect, 185
Cannon Ball Farm House. See Bleakley House
Cantonment on Frankford Creek. See Frankford Arsenal

Carl Schurz Memorial Foundation, xxxiv
Carmalt, Thomas, Houses, 323
Carnell, Charles, 293
Carpenter, Joshua, House, 64
Carpenters' Company, 65
Carpenters' Court, 65
Carpenters' Hall, xxv, xxviii, xli(n.15), 65
Carrère and Hastings, architects, 367 (n.57)
Carriage houses, 66, 134, 229, 318
Carstairs, Thomas, architect, 53
Carver, John E., 294
Casket, The, xxiv, xxv
Cassatt, Alexander, J., 143
Cast-Iron Sidewalk (1907 N. Seventh St.), 293. See also Iron
Catanach, A., carpenter, 32
Catharine Street Area Study, 167–68
Cathedral of SS. Peter and Paul (Roman Catholic), 111, 124
Cedar Grove, 228–29, 367(n.3)
Cemeteries, 181, 195, 211–12, 219, 224, 234, 333, 362(n.17), 368(n.9). See also listings by name
Centennial Exhibition of 1876. See International Exhibition of 1876
Centennial Guard Box, 229
Centennial National Bank, 200, 205–6
Center Square. See Penn's City, Center Square
Center Square Waterworks, 106, 223
Central heating, 24, 197
Central Philadelphia Meeting of Friends, Meeting House. See Twelfth Street Meeting House
Centre Square Towers, 120
Chalkley, Thomas, 235, 306, 324
Chalkley Hall, xxxiii, xlv(n.52), 306, 323–24, 375(n.6), 381(n.74)
Chamber of Commerce. See Commercial Exchange
Chamounix Mansion, 229
Chandler, Theophilus Parsons, Jr., architect, 62, 147; addition, Pennsylvania Fire Insurance Co., 88; Commercial Union Assurance Co., 68; Compton, 263; George Burnham house, 204; Gladstone Hotel, 131; Scott-Wanamaker House, 108, 145; Spring Garden Insurance Co., 99; Tabernacle Presbyterian Church, 216, 367(n.57); Zoo bear pits, 225, 227
Cherry Hill, Benjamin Warner's country house, 284
Cherry Hill Penitentiary (Cherry Hill Prison). See Eastern State Penitentiary

Chestnut Hill, 246, 252, 253–55
Chestnut Hill Academy. *See* Wissahickon Inn
Chestnut Hill Railroad, 252. *See also* Philadelphia, Germantown and Norristown Railroad
Chestnut Street, 44–45, 47–48, 114
Chestnut Street Area Study, 66
Chestnut Street Bridge, 124–25
Chestnut Street Theatre, 57, 58
Chevia B'nai Aviohome MiRussi Synagogue, 7
Chew, Benjamin, 249, 262
Chew House. *See* Cliveden
Children's Museum of Philadelphia. *See* Monastery
Childs, C. G., xli(n.9)
Chinatown, HABS buildings in, 125, 128, 132
Chinatown YMCA (Chinese Cultural and Community Center), 125
Chinese Wall, Broad Street Station, 119
Choragic Monument of Lysicrates, 42
Choragic Monument of Thrasyllus, 57
Christ Church, xxviii, 55, 67
Christ Church Hospital, 197
Christ Church Rectory, 23
Church of St. Charles Borromeo, 161, 168
Church of St. James the Less (Protestant Episcopal), 286–87, 294
Church of St. Luke and the Epiphany (Protestant Episcopal, originally, Church of St. Luke), 126
Church of St. Thomas Aquinas, 161
Church of St. Vincent De Paul (Roman Catholic), 261–62
Church of the Brethren (Dunkers), 246
Church of the Epiphany. *See* Church of St. Luke and the Epiphany
Church of the Gesu (Roman Catholic), 294
Church of the Holy Trinity (Protestant Episcopal), 110, 125–26
Church of the Redeemer for Seamen of the Port of Philadelphia, 157–58, 168, 186
Church of the Saviour, 367(n.57)
Churches: Center Square area, 109–11, 122, 124, 125, 126, 129–30, 143–44, 148, 355(n.27); Delaware River Corridor, 307, 315–16, 339, 380(n.58); Germantown, 246–47, 254, 261–62, 265, 268, 272–73, 274, 277; North Philadelphia, 375(n.52); Society Hill, 7, 10, 13–14, 16, 27–28, 32; South Philadelphia, 154, 157–58, 161; West Philadelphia, 214–16, 367(n.57). *See also listings by name and* Meetinghouses
Churchman and Thomas, architects, 254
City Hall. *See* Philadelphia City Hall
City histories, xxv
City planning. *See* Philadelphia, City Planning
City Tavern, 41, 42
Civil War Memorial. *See* Smith Memorial Arch
Civil Works Administration, xxx, xliv (nn.42, 45)
Civilian Conservation Corps. xxxiii
Clark, John, builder, 73
Clarke, Reeves and Co., engineers, 232
Clarke, Samuel, House, 324
Clarkson, Matthew, 262
Clarkson-Watson House (Bank of Germantown), 262
Class, Robert A., architect, 121
Classical Revival style, 32, 130, 240
Cliffs, The, 229–30
Clifton, John, House, 169
Cliveden (Chew House), 249, 251, 262–63, 362(n.7)
Clubs: Center Square area, 112–14, 134, 135, 139, 140, 146–47, 356(nn.39–41); cricket, 267; rowing, 225, 228; workingmen's, 275
Clunie. *See* Mount Pleasant
Clymer's Court, 154, 177
Cochran, William, house, 200
Colonial Dames of America, Chapter II, 235
Colleges. *See* Universities and Colleges
Collegiate Gothic style, 201, 216
Collins and Autenreith, architects, 49, 51, 65, 75, 83, 89, 350(n.74)
Colonial Revival style, xxvi, xxvii, 161, 253–54
Colonnade Row, 53, 107
Columbia Engine Company, 206
Columbian Exposition (1893), 116, 119, 130
Commercial buildings: Center Square area, 114–16, 122, 123–24, 131–35 *passim*, 142, 147; Delaware River Corridor, 324–25, 326, 331; Old City, 44–51 *passim*, 59–60, 61, 66–71 *passim*, 75, 77, 80–89 *passim*, 93, 94, 97–100 *passim*; North Philadelphia, 289, 291, 374(n.44). *See also* Banks; Warehouses; *and listings by name*
Commercial Exchange (Chamber of Commerce), 42, 67–68
Commercial style, Chicago, 115, 132–33

Commercial Trust Building. *See* Arcade Building
Commercial Union Assurance Co. Building, 68
Committee of 1926, maintains Strawberry Mansion, 240
Community College of Philadelphia, Campus, II. *See* United States Mint
Compton, 263
Concord School, 248, 263
Concrete construction, 120; block, 332; poured, 201, 365(n.50); reinforced, 142, 213, 217, 299, 328, 371(n.39)
Congress Hall (Philadelphia County Court House), xxviii, 79–80. *See also* Independence National Historical Park
Connally, Ernest A., xxxviii
Connie Mack Stadium. *See* Shibe Park
Conrad, Charles, 121
Conservatories, 293, 302
Consolidation Bank (Consolidation National Bank), 316, 325
Continental Theatre. *See* Arch Street Opera-House
Conwell, Russell, 291
Conwell School of Theology, Council of Black Clergy's Institute for Black Ministries, 302
Conyngham-Hacker House. *See* Germantown Historical Society Area Study
Coombe, Griffith, house carpenter, 32
Coombe's Alley. *See* Cuthbert Street
Cooper, Jacob, House, 68
Cooper, Peter, "View," xl
Cooper, Theodore, engineer, 240
Cope, Caleb, 44
Cope, Caleb, & Co., Store (Goldberg's Army-Navy Store), 68
Cope, Edward Drinker, houses, 126
Cope and Lippincott, architects, 60
Cope and Stewardson, architects, 125, 127; Harrison Building, 115, 131; Lady Chapel, St. Mark's, 110, 143–44; Monarch Building, 134–35; Pennsylvania Institution for Instruction of the Blind, 198, 213; University Museum, 201, 218; University of Pennsylvania Men's Dormitories, 217
Copland, Robert Morris, landscape architect, 225
Costaggini, Filippo (Philip), frescoes, 95, 168
Coultas, Col. James, 193
Counting rooms, in commercial buildings, 41, 43, 45

Country houses, 193; Delaware River Corridor, 306–7, 318–19, 331–32, 335, 381(n.75); Fairmount Park, 223–241 *passim*, 362(n.6), 367(n.3); Germantown, 247, 248–49, 250–51, 268–69, 276–77, 278, 279, 362(n.7); North Philadelphia, 283–84; South Philadelphia, 158, 166–67, 172–73, 362 (n.6); West Philadelphia, 193–96, 205, 210, 218–19, 363(n.20)
Court houses (Old Court House or Town Hall), 39
Courts (houses built on a court), 5, 13, 19–20, 95, 154, 155, 177, 189–90
Courts, interior, 43, 89, 183
Courtyards: City Hall, 105; inn, 295
Craige, Seth, 310
Cramond, William, 238
Cramp, William, 312
Cramp and Company, contractors, 80, 147
Crane, Edward A., 301
Creative Eclecticism, 201
Creely, George, bricklayer, 98
Cresson & Co., iron work, 64
Cret, Paul Phillipe, architect, 117, 118, 119, 128
Crump, John, architect, 67, 147
Cummings, G. P., architect, 87
Cundall, J. H., 233
Currie, Dr. William, House, 12
Curtis, Mrs. Cyrus H. K., xxix
Cuthbert Street (Coombe's Alley), 58, 76

Dagit, Henry D., architect, 124, 132, 214, 375(n.52)
Dana Street (now Nectarine St.), 329
Davis, Alexander Jackson, 297
Davis, James, house carpenter, 12
Davis, Seymour, 68
Davis-Lenox House, 12
Day, Frank Miles, architect, 79, 90, 113–14, 139, 201, 218
Day, Frank Miles, and Bro., architects, 206, 218, 335
Day, Michael, House, 325
Day and Klauder, architects, 127, 139
Day and Zimmerman, architects, 137, 328
Decatur, Stephen, 306
Decker, Will, architect, 200
Delancey Street Area Studies, 107–8, 126–27
Delaware Expressway (I-95), 163, 308, 358(n.9)
Delaware River Corridor: boundaries,

305; growth of, 305–20; HABS buildings in, 321–40
Demchick, Israel, 75
DeNegre, Joseph, builder, 148
Department stores, 82–83, 114, 116–17, 147–48
Dependencies, 173, 249, 262, 276–77, 279, 318. *See also* Barns; Springhouses; Stables
Depression, Great, xxviii, xxx–xxxiv, 113, 202, 257, 292, 375(n.65)
Deshler, David, 250, 264
Deshler-Morris House, 250, 264
Design competitions. *See* Architectural drawings, competition
Detroit Institute of Arts, xlii(n.22), 194
Detweiler House, 264
deVos, K. M., & Co., contractors, 371(n.39)
Dewey, Admiral George, 39
Diamond Street Area Study, 295
Dickey, John, architect, 61
Dickey, Weissman, Chandler and Holt, architects, 101
Dickinson, John, 64
Dilworth, Jonathan, house carpenter, 68
Dilworth-Todd-Moylan House (Dolley Madison house), 68–69
Disston, Henry & Sons (Keystone Saw, Tool and Steel Works), 312, 313, 326, 378(n.43)
Doak, James G., & Co., contractors, 146–47
Doan, Aaron, and Co., builders, 233
Dobbins, Richard J., contractor, 233
Dock Street Sewer, 69
Donaldson, William, House, 169–70
Dorfenille House, 264
Doughty, Thomas, 188
Dowers-Okill House, 41, 69
Downing, Andrew Jackson, 252, 318
Doyle and Company, contractors, 80, 124, 127
Doyle and Doak, contractors, 72, 88, 124, 132–33, 184
Dreher and Churchman, architects, 267
Drexel, Anthony J., 201, 206, 366(n.51)
Drexel, Francis M., 46, 69
Drexel, George W. Childs, house, 200
Drexel & Company Building (135–43 S. Fifteenth St.), 127–28
Drexel & Co., Building (34 S. Third St.), 46, 69
Drexel Building (Fifth and Chestnut Sts.), 57

Drexel Institute (now Drexel University), 201, 206–7
Drinker, John, House (241 Pine St.), 5, 12
Drinker, John, House (Krider Gun Shop), 69–70
Drinker's Court (Deimling Place, Bodine St.), 5, 13
Duché House, 170
Duché-Walker House, 170
Dugan, Thomas, builder, 127
Duhring, H. Louis, xliii(n.30), 25
Duhring, Okie and Ziegler, architects, 96, 254
Duke Street (now Nectarine St.), 329
Duncan, John, contractor, 89
Duncannon Iron Company Warehouse, 70
Dunlap-Eyre House, 128
Dunn, William Allen, xxviii
duPont, Henry Francis, xlv(n.52)
Durang, Edwin F., church architect, 95, 121; Gesu, 294, 375(n.52); St. Agatha's, 214, 367(n.57); St. Charles Borromeo, 161, 168; St. Francis Xavier, 375(n.52); St. Gabriel, 161; St. James, 215; St. Thomas Aquinas, 161
Durang, F. Ferdinand, architect, 202
Dyott, Dr. Thomas W., 377(n.27)

Eagle Hotel, 295
Eakins, Thomas, 225
Eakins, Thomas, House, 295
Eastburn Mariners' Bethel (Mariners' Bethel), 13–14
Eastern Office of Design and Construction, National Park Service, xxxv, xl
Eastern State Penitentiary (Cherry Hill Prison), 284–85, 295–96
Eastwick, Andrew M., 198
Ecclesiological Society (Cambridge Camden Society), 287, 294
Eckert-Tarrant House, 170
Eclecticism, Victorian Picturesque, 105
Economy Mills, 256
Edmunds, Franklin D., xliii(n.30)
Edwards and Green, architects, 84
Eggers and Higgins, architects, 124
Egyptian Revival style, 44, 46, 50, 83, 88, 160, 181, 183, 314, 330
Eighth National Bank, Kensington, 380(n.64)
Elfreth, Jeremiah, House, 70
Elfreth's Alley, 52, 53; Area Study, 70–71
Elfreth's Alley Association, 52, 70
Ellen Street, 311

Elliott, John, House, 170
Elliott Building, 45, 66
Ellis, William, and N. W., contractors, 64
Ellison, John B., & Sons, Building, 71
Elverson Building *(Philadelphia Inquirer)*, 291
Elwell, Henry, House, 170–71
Ely-Osbourn Houses and Stores, 171
Endt, Theobald, House, 264–65
English Evangelical Lutheran Church. *See* Trinity Lutheran Church
Enterprise Manufacturing Company, 375(n.50)
Episcopal Community Services of the Diocese of Pennsylvania. *See* St. Paul's
Episcopal Hospital. *See* Hospital of the Protestant Episcopal Church in Philadelphia
Equality Court. *See* Olive Street
Erdman, Frederick, carpenter, 230
Erie Enamel Co., contractor, 149
Erny and Nolan, contractors, 300
Errickson, Michael, carpenter, 89
Esherick, Margaret, 254
Eshler, Lewis H., architect, 185
Estlack, Thomas, House, 14
Evans, Allen, 117, 130
Evans, B. R., drawings, xli(n.12)
Evans, David, master carpenter, 79
Evans, David, Jr., architect, 24
Evans, John Jr., house carpenter, 298
Eyre, Wilson, Jr., architect, 94, 107, 108, 127, 128, 134, 254; Borie Bros. Bank, 66; William Cochran house, 200; Dr. Joseph Leidy house, 133; Clarence B. Moore House, 135; University Museum, 201, 218; Harriet D. Schaeffer house, 275
Eyre and McIlvaine, architects, 107, 127

Factories. *See* Industrial buildings
Fair, Russell, Jr., engineer, 240
Fairmount Park, 255, 367(n.3); Centennial Exhibition in, 225–26; design competition for, 225, 368(n.14); development, 223–26; HABS buildings, 227–41; on National Register of Historic Places, 223; recreational facilities, 226; trolleys, 226, 240
Fairmount Park Art Association, 226, 239
Fairmount Park Commission, 225–241 *passim*, 268, 272, 274
Fairmount Park Transportation Company, 226, 240

Fairmount Waterworks, 106, 224, 230–31; National Engineering Landmark, 231
Falls Bridge, 231
Far East Chinese Restaurant, 128
Farm Journal, 50
Farmers' and Mechanics' Bank, 48, 71
Farmhouses: Delaware River Corridor, 306, 332, 337–38; Fairmount Park, 228; Germantown, 251, 256, 267–68, 278, 279
Faust, Winebrenner and Co., 347(n.21)
Federal Building (Seventh and Arch Sts.), 118
Federal Reserve Bank of Philadelphia (Tenth and Chestnut Sts.), 117–18, 128–29
Federal style: Delaware River Corridor, 327; Fairmount Park houses, 224, 229, 234, 239, 241; Germantown, 251, 271–72, 278; Old City buildings, 54, 79, 92, 102; Society Hill, 5–6, 11, 15, 23, 24, 31, 34, 35; West Philadelphia, 205
Fell–Van Rensselaer House (Penn Athletic Club), 108–9, 129
Fellowship Engine Co., No. 29, 21–22
Fernbach, Henry, architect, 114, 135–36
Few, Joseph, carpenter, 17
Field, J. & S., 64
Field, M., 45
Fielding, Mantle, architect, 127, 254, 267; House, 265
Fife, Matthew, House, 71–72
Finlow, William, 154
Finlow-Nichell House, 171
Fire Association Building. *See* Irvin Building
Fire companies, volunteer: Delaware River Corridor, 314, 334, 339, 379 (nn.52, 53); Society Hill, 21–22; South Philadelphia, 157, 359(n.26). *See also* Philadelphia Fire Department
Firehouses: Center Square, 144; Delaware River Corridor, 314, 326, 334, 339, 379(nn.52, 53, 55); Old City, 86, 88, 90, 101; Society Hill, 21–22; South Philadelphia, 174, 184, 359(n.29); West Philadelphia, 206
Fire Insurance Companies. *See* Insurance Companies
Fire proofing, xli, 41, 187, 197
Fire proofs, 43, 45, 348(n.33)
First Baptist Church, 110, 355(n.27)
First Baptist Church of Germantown, 265
First Church of Christ Scientist, 367(n.57)

First German Reformed Church Area Study, 72–73
First National Bank, 73
First New Jerusalem Society of Philadelphia. *See* Spring Garden Institute
First Pennsylvania Banking and Trust Co., 73. *See also* Centennial Bank; Pennsylvania Company for Banking and Trusts; Pennsylvania Company for Insurances on Lives and Granting Annuities
First Polish Baptist Church. *See* Mariners' Bethel
First Presbyterian Church (Frankford), 315, 380(n.58)
First Presbyterian Church (Kensington), 315
First Presbyterian Church (Washington Square Presbyterian Church), 55, 73, 101
First Presbyterian Church of the Northern Liberties, 315
First Unitarian Church, 129–30
First United Methodist Church of Germantown. *See* Green Tree Tavern
Fisher, Joshua, 230
Fisher, Sidney George, 196
Fisher Building, Chicago, 116
Fitzgerald, Thomas, House, 14
Fitzwater Street Area Study, 171, 172
Flat Rock Canal, 256
Flickwer, David, house carpenter, 168, 172
Flickwer-Williamson Houses, 172
Folsom, Stanton and Graham, architects, 95
Forrest, Edwin. *See* Gaul-Forrest House
Fort Mifflin, xxxii, 204, 207–10
Foster Home Association, 286
Fountain of the Sea Horses (Italian Fountain), 231
Fountains, 33, 98, 231
Fourth Baptist Church, Northern Liberties, 315
Frame construction. *See* Houses, frame and log
Francis, Cauffman, Wilkinson and Pepper, architects, 30, 213
Francisville, 284
Frankford and Southward Passenger Railroad, 316
Frankford Arsenal, 309, 310
Frankford Creek, 310, 378(n.32)
Frankford Dye Works, 316

Frankford Elevated, 319, 381(n.79)
Frankford Lyceum, 313–14
Frankford Mutual Insurance Company, 314
Frankford Road, 312
Frankford Town Hall, 326
Franklin, Benjamin, 6, 24, 49, 56, 74, 89, 92
Franklin Court, 58
Franklin Delano Roosevelt Park (League Island Park), 158, 161, 166
Franklin Institute (Atwater Kent Museum), 57, 65, 74
Franklin Row, 53. *See also* Sims-Bilsland House
Franklin Sugar Refinery Warehouse (Harrison, Frazier & Co.), 172
Frantz, P. Richard, architect, 73
Fraser, John, architect, 32, 44, 113, 147
Fraternal organizations, 113, 140–41, 314
Frazer, Furness and Hewitt, architects, 61
Free Library of Philadelphia, xxix
Free Quakers Meeting House, 54–55, 74–75
French, Daniel Chester, sculptor, 226, 239
French Renaissance style, 105–6, 114, 119, 131, 289, 301–2. *See also* Picturesque Eclectic
French Village, 255
Friends Hospital (Friends Asylum), 308–9, 376(ns.16–19)
Friends Institute, 146
Friendship Engine Company No. 15 (later Friendship Hall), 326, 379(n.55)
Frogmore Mills, 378(n.32)
Fromberger, John, Houses (Germantown Insurance Co.), 265–66
Front Street, North, Area Study, 86
Front Street, South, Area Study, 186
Fuller, George A., Co., contractors, 141, 237
Fullerton, John, Houses, 172
Furness, Frank, architect: bank designs, 289–90, 374(n.35), Centennial National, 200, 206, Girard Trust, 130, 357(n.64), Kensington National, 316, 330, Provident Life and Trust, xxxix, 48, 92; chapels and churches, Church of the Redeemer, 158, 168, Mt. Sinai Chapel, 333, St. Stephen's, addition, 144; houses, 108, 131, 318; libraries, 61, 201, 217; Pennsylvania Academy of the Fine Arts, 112, 137; railroad stations, 123. *See also* Frazer, Furness and Hewitt; Fur-

Furness, Frank *(Cont.)*
  ness and Evans; Furness, Evans & Co;
  Furness and Hewitt.
Furness and Evans, architects, 59, 225,
  228, 349–50(n.55)
Furness and Hewitt, architects, 112, 137,
  225, 238, 289, 298
Furness Building. *See* University of Penn-
  sylvania Library
Furness, Evans & Co., architects, 49, 59,
  127, 138; Arcade Building, 115, 122;
  Diamond Street houses, 295; Girard
  Bank, 117; Reliance Insurance Co.
  building, 93; remodeling or additions to
  Philadelphia Club, 140; PSFS, 7th &
  Walnut Sts, 91; Western Saving Fund,
  148.

Garages, parking, 200, 201, 365(n.50),
  374(n.44)
Garden, C. H., & Co., Building, 75
Garden Court Plaza, 200
Gardens, 5, 129, 195, 198, 203, 224, 236,
  250, 264
Gas, illuminating, 46
Gatehouses, cemetery, 181, 211–12, 234
Gatter, Carl W., 381(n.74)
Gaul, William. *See* Gaul-Forrest House
Gaul-Forrest House, 289, 296, 373(n.31)
Geddes, Brecher, Qualls, Cunningham,
  architects, 301
Geissinger, Walter H., industrial ar-
  chitect, 317, 332
Geissinger and Hales, 332
Gendell Building, 66
General Wayne Hotel, 266
Gentilhommière (Stephen Girard's
  Country House), 158, 172–73; Utility
  Building, 173
George, Henry, House, 130
George House, 210
Georgian Brownstone style, 288
Georgian Period, xxvii–xxviii
Georgian Revival style, 107, 133, 267
Georgian style: churches, 28, 55, 66–67;
  clubs, 140; country houses, 167, 193–
  94, 230, 236, 241, 249, 250, 262–63, 276,
  306, 323–24; houses, 4–5, 9, 21, 22, 25,
  28, 34, 68–69; Independence Hall, 77–
  78
Gerentee, Henry, 294
German Catholic Church of St. Peter's,
  Kensington, 315, 380(n.61)
Germantown: growth of, 245–57; HABS

buildings, 259–79; Kalm's description,
  245
Germantown Academy (Germantown
  Union School, Germantown Lutheran
  Academy, Germantown-Stevens
  Academy), xxxi, 248, 266
Germantown Avenue (Germantown
  Road, the Great Road), 245, 252
Germantown, Bank of. *See* Clarkson-
  Watson House
Germantown Cricket Club (Manheim
  Club), 267
Germantown Historical Society, 262, 272,
  279; Area Study, 267
Germantown Insurance Co. (German-
  town Fire Insurance Co.). *See* From-
  berger, John, Houses
Germantown Passenger Railway Com-
  pany, 252
Germantown and Perkiomen Turnpike
  Company, 252
Gesu, Church of the. *See* Church of the
  Gesu
Gibbs, James, 249
Gilbert, John, 52
Gilbert, Joseph, builder, 127
Gilbert's Alley. *See* Elfreth's Alley
Gilchrist, Agnes Addison, xxxvi
Gill, John N., and Company, contractors,
  293
Girard, Stephen, 14, 41, 72, 285; Collec-
  tion, 297; Estate, 69, 86, 161
Girard Avenue Bridge, 225, 231–32
Girard Bank. *See* Girard Trust Corn Ex-
  change Bank
Girard College, Founder's Hall, 285–86,
  296–97, 372(n.10)
Girard National Bank, 72
Girard Park, 158
Girard Row, 14
Girard Trust Corn Exchange Bank (Girard
  Bank), 117, 130, 357(n.64)
Giurgola, Romaldo, architect, 254
Gladstone Hotel (Greystone Apart-
  ments), 131, 365(n.45)
Glass, Tiffany Favrille mosaics, 301
Glass works, 29, 377(n.27)
Glebe House, 173
Glen Fern (Livezey House), 247, 256, 267–
  68
Globe Mills (Globe Dye and Bleach
  Works), 310, 311, 317, 381(nn.68, 69)
Gloria Dei (Old Swedes), xxviii, xxxii,
  154, 173–74, 307

Godley, Jesse, Warehouse, 40, 75, 156, 174

Godley's Argyle Stores, 156, 174

Godley's Stores. *See* Granite Street Vaults

Goforth, William Davenport, xxvii, xlii (n.21)

Goldberg's Army-Navy Store. *See* Cope, Caleb, & Co., Store

Goode, Ned, xxxvii

Goodwin Building, 326, 379(n.46)

Gordon, George, Building, 75

Gothic Revival style: bridges, 124; churches, 109–10, 122, 143–44, 168, 215, 274, 286, 294, 315, 355(n.27), 367(n.57); colleges, 216; commercial buildings, 44, 335; guard houses, 229, 238; houses, 135; prisons, 160, 183, 284

Grace Baptist Church (Temple University), 291

Graff, Frederic, Jr., architect and engineer, 231

Graff, Frederick, architect and engineer, 230, 231

Grafly, Charles, sculptor, 239

Graham, Anderson, Probst and White, architects, 366(n.54)

Grand Depot Store. *See* Wanamaker, John, Store

Granite Block, Providence, R. I., 348 (n.28)

Granite Street Vaults (Godley's Stores), 75

Gray Street, 311

Grèber, Jacques, 118, 237

Greek Revival style, xlvi(n.62), 109; banks, 47, 84, 90–91, 97–98; churches, 27, 55, 73, 126, 157, 265; commercial buildings, 49, 89–90; firehouses, 314, 339; Franklin Institute, 57, 74; hospitals and asylums, 111, 137–38, 187, 286, 298–99; houses, 20, 35, 82, 92–93; libraries, 186; theaters, 100–101; waterworks, 106

Green Street Area Study, 297

Green Tree Tavern, 247, 268

Greene, William Holden, 348(n.28)

Greenwood Cemetery, Brooklyn, N.Y., 368(n.9)

Gregory, John, sculptor, 237

Greystone Apartments. *See* Gladstone Hotel

Gries, John M., architect, 47, 62, 71, 197

Griffith, Thomas Hockley, 327

Griffith-Peale House, 327

Grist mills. *See* Mills

Gropius, Walter, architect, 118

Groves, Daniel, master mason, 85, 144

Grubb, William L., carpenter, 339

Grumblethorpe (Wister's Big House), xxxi, 247, 268–69

Grumblethorpe Tenant House. *See* Wister's Tenant House

Guarantee Trust and Safe Deposit Company, 48, 349(n.55)

Guard houses, 229, 238, 241

Gunner's Run, 377(n.27)

HABS (Historic American Buildings Survey), history, xxi–xl

HABSI (Historic American Buildings Survey Inventory), xlvi(n.63)

Hagar, Howard, architect, 228

Haines House. *See* Wyck

Hale, Willis G., architect, 101, 301–2

Hall, John, House, 14–15

Hall-Wister House, 15

Hallowell, Morris L., & Co., 348(n.33)

Hamilton, Andrew, 195, 283

Hamilton, Gavin, Jr., 210

Hamilton, William, 195, 196, 219, 367(n.57)

Hamilton Court, 200

Hamilton-Hoffman House (Burnside), 210–11

Hamilton Terrace, 199

Hamilton Village, 196, 215, 367(n.57)

Hammerstein, Oscar, 290–91, 298

Hancock, Joseph M., builder, 127

Hancock, Gen. Winfield Scott, 239

Hansell, John, House, 75

Harbeson, Hough, Livingston and Larson, architects, 129, 137, 217

Harlan, Edward, carpenter, 130

Harmony Engine Company No. 6 (Franklin Hose Company No. 28), 174

Harper, William Jr., Houses, 157, 174–75

Harris, Thomas, architect, 233

Harrison, Charles C., 108

Harrison, David, 67

Harrison, Henry, Houses, 58, 76

Harrison, John, 67

Harrison, Joseph, 108, 311

Harrison Building, Fifteenth & Market Sts., 115, 120, 131

Harrison Building, Tenth and Filbert Sts. *See* Monarch Building

Harrison, Frazier & Co. *See* Franklin Sugar Refinery Warehouse

Harrison House (Point No Point), 327
Harrison Row, 107
Hart, John, House, 175
Hart, Montgomery & Company, 313
Hart-Patterson House, 175
Hart, Samuel, & Sons, contractors, 122
Hatfield, James S., architect, 133
Hatfield House, 232
Hatfield, Martin and White, architects, 233
Hauf, Leopold, Jr., 91
Havens, Louis D., contractor, 131
Haviland, John, architect, xxvii, 24, 27, 61; Colonnade Row, 53, 107; Eastern State Penitentiary, 284, 295–96; First Presbyterian Church, 55, 73; Franklin Institute, 74; IOOF Hall, 330; Pennsylvania Fire Insurance Co., 44, 88; Pennsylvania Institution for Deaf and Dumb, 111, 137, 355(n.34); Walnut St. Theatre, 58, 101
Hazell, Alexander T. C., architect, 141
Hazelhurst and Huckel, architects, 10, 94, 143, 146, 228
Heap, George, "East Prospect," xxiii
Heard and Sons, architects, 233
Hearn, Lafcadio, quoted, Fairmount Park, 223
Helffenstein, Dr. Abraham, House, 327
Hellings, Benjamin, House, 175
Hemphill, Joseph, 240
Henderson, James, House, 175
Henle, Fritz, xxxvi
Hensel, Colladay & Co., 76
Heritage House. See Gaul-Forrest House
Herzog, George, decorator, 106, 140, 141, 147, 301
Hesser, John, mason, 263
Heston, Col. Edward, 361(n.1)
Hewitt, G. W., architect, 67, 112, 125
Hewitt, G. W. and W. D., architects, 125, 228, 270, 328, 368(n.13); Bourse, 89, 347(n.20); Wissahickon Inn, 254, 278–79
Hickman, Louis C., architect, 90, 228
Highlands, Findley, marble mason, 297
Hill, David, House, 15
Hill, Henry, 15
Hill, Howard Carter, architect, 85
Hillman, Charles, L., xxviii, xlii(n.25), xliii(n.30)
Hilyard, Eber, House, 16
Historic buildings, recording of, xxxviii–xxxix

Historic Mansions and Buildings of Philadelphia, The, xxvi
Historic Sites and Buildings Act (1935), xxxv–xxxvi
Historical Society of Pennsylvania, architectural drawings in, xxviii
Hockley, Thomas, House, 108, 131
Holden, Isaac, architect, 197, 212–13, 299, 363(n.30)
Holland, Leicester B., xxx–xxxi, xl(n.1), xliii(n.30), xliv(nn.40, 43)
Holloway, Thomas, 311, 327–28, 378 (n.35)
Holme, Thomas, 39
Holy Redeemer Chinese Catholic Church and School, 132
Holy Trinity, Church of the. See Church of the Holy Trinity
Holy Trinity (German) Roman Catholic Church, 16
Home for Incurables, Forty-seventh & Woodland, 198
Home for Indigent Widows and Single Women, Chestnut St. near Thirty-sixth, 198
Hood, William, 269
Hood Cemetery Entrance (Lower Burying Ground), 269
Hoopes and Townsend, 287
Hope Engine Company No. 17, 157, 175–76
Hope Hose Company, No. 6, 21
Hope Street, 311
Hopkins, Harry L., xxx
Hopkinson House, Washington Square, 101
Horrocks, Jeremiah, 316
Horstmann Factory, 51
Hospital of the Protestant Episcopal Church in Philadelphia (Episcopal Hospital), 309, 328, 376(n.20), 377(n.21)
Hospital of the University of Pennsylvania, 198
Hospitals, 6, 162, 196–98, 286, 298–99, 309, 328, 363(n.26); psychiatric, 24, 197, 212–13, 308–309, 363(n.30)
Hotel Metropole. See Hotel Walton
Hotel Walton (Hotel Metropole, John Bartram Hotel), 132
Hotels, 132, 254, 266, 278–79, 295
House on Letitia Street. See Letitia Penn House
Houses, 16, 17, 176–77, 269; artisans', 169, 170, 176, 180, 182, 329, 336, 337;

cottages, 206, 287, 298, 373(n.23); development, 161, 254, 255, 319; frame, 156, 167, 169, 170, 172, 176, 177, 181, 182, 328, 329, 336, 340; frame, brick nogging, 156, 167; log, 271; mill-workers', 311, 327; "New York Style," 288, 373(n.29); North Philadelphia mansions, 288–89, 290; offices and shops in, 21, 22, 171, 185, 311, 322, 324, 337, 374(n.34), 378(n.37); rooming house, 186; semi-detached, 161; tene-ments, 167; terrace, 199, 218, 287–88. *See also* Airlites, Bandboxes; Country houses; Row houses; Villas

Housing: low rent, 155, 157, 174–75, 190, 319–20, 332, 382(nn.82, 83); public, 200; for rental income, 14, 34, 70, 172; speculative, 106, 107, 141–42, 145, 199, 218

Houston, Henry Howard, 254, 270, 279

Houston-Sauveur House, 254, 269–70

Howe, George, architect, 118, 124, 366 (n.54)

Howe and Lescaze, architects, 91, 118, 141

Howell & Brothers Building, 77

Howell House. *See* Germantown Historical Society Area Study

Howell, J., and Company, 67

Hoxie, Joseph C., architect, 45, 51, 66, 87, 101, 315, 348(n.28); Caleb Cope & Co., Store, 68; Maxwell House, 272; Phil-adelphia, Germantown and Nor-ristown Railway station, 273; West Arch St. Presbyterian Church, 111, 148

Hoxie and Button, architects, 288; Brock & Sons warehouse, 64; Hope Engine Co, No. 17, 157, 175, 359(n.27); Spring Garden Institute, 287, 300

Hoxsie, Solomon K., 68, 82

Hubbard, Cortlandt Van Dyke, xxxvi

Hudders, John R., carpenter, 82

Huston, W. D., builder, 295

Hutchinson & Co., 46, 83

Hutton, Addison, architect, 24, 85, 99, 371(n.39), 374(n.36); Arch St. Meth-odist Church, chapel, 122; Lippincott house, 133; Pennsylvania Co. for Insur-ing Lives & Granting Annuities, 48, 87; PSFS, 700 Walnut St., 91; Ridgway Branch Library, 163, 186

Hutton and Ord, architects, 313

Hutton, Nathaniel, House, 177

IBM Building, 120

Ickes, Harold, xxx, xliv(n.43)

Independence Hall (State House of Penn-sylvania), 39–40, 77–79; drawings of, xxviii; restorations, xxviii

Independence Mall, 58

Independence National Historical Park: architectural study collection, 13, 19, 31, 34, 60, 63, 65, 66, 69, 72, 73, 79, 81, 85, 174, 176, 275; buildings in, 44, 57, 58, 80; development of, xxxiv–xxxv

Independent Order of Odd Fellows, halls, 113, 287, 314, 330, 379(n.51)

Industrial buildings: Delaware River Cor-ridor, 310, 312–13, 316–17, 321–22, 326, 335, 336, 381(n.73); North Philadelphia, 287; Old City, 51, 94; South Phil-adelphia, 156–57, 187

Industrial parks, 287

Ingle, Charles H., engineer, 331

Institute of the Pennsylvania Hospital. *See* Pennsylvania Hospital for the Insane, Department for Males

Insurance companies, Center Square, 114, 132–33, 135–36; Delaware River Corridor, 314; Germantown, 266, 273; Old City, 48–50, 59, 68, 80, 81, 87, 88, 89, 91, 93, 94, 99; Society Hill, 22, 29

Insurance Patrol, 60, 80

Insurance Row, 48–50

International Exhibition of 1876 (Centen-nial Exhibition), xxvi, 200, 215, 225–26; Memorial Hall, 226, 232–33; Ohio State Building, 226, 233; St. George's House (British Building), 233

International style, 118, 320, 332

Irish, Nathaniel, house carpenter, 181

Irish, Nathaniel, House, 155, 156, 177–78

Iron (cast iron, wrought iron, ironwork): bridge, 276; decorative uses, 48, 71, 95, 144, 175, 184, 326, 330, 339; fire escape, 328; fronts (facades and shop fronts), 51, 58, 64, 66, 75, 76, 83, 86, 87, 94, 95, 97, 98, 100, 101, 313, 324, 326, 350(n.74); protective devices, 41, 45, 348(n.33); structural use, 43, 48, 70, 80, 84, 95, 137, 197, 297, 299, 313, 321, 324, 326

Irvin Building (Fire Association Building), 50, 80

Irwin and Leighton, contractors, 129

Italian Fountain. *See* Fountain of the Sea Horses

Italian Market, 161

Italianate style (Italian Revival): bank design, 46, 48, 69, 71, 73, 83, 91, 127, 316, 325; churches, 95, 111, 148, 262, 315; commercial buildings, 44, 46, 69, 98, 114, 291; fire halls, 157, 175; houses, 108, 288, 289, 293, 296; institutions, 57, 60–61, 113, 121, 300; villas, 188, 199, 211, 218, 252, 270, 318, 322
Iungerich Warehouse, 80
Ivy Lodge, 252, 270

Jacobean Revival style, 108, 144
Jacobethan style, 201
Jacksonville, N. Philadelphia village, 372(n.3)
Jacoby, Wigard, House, 270
Jastremsky, Julian J., 375(n.52)
Jayne, Dr. David, 81
Jayne Building, 44, 80–81
Jefferson, Joseph, House, 17
Jefferson Fire Insurance Company Building, 81
Jenkins, Warner H., & Co., contractors, 213
Jennewein, C. Paul, sculptor, 237
Jewelers' Row, 50
John Bartram Association, 195
John Bartram Hotel. See Hotel Walton
John F. Kennedy Stadium (Municipal Stadium, Philadelphia Stadium), 162
Johnson, Charles, carpenter, 27
Johnson, Charles J., architect, 149
Johnson, John, 250
Johnson, Philip, architect, 184
Johnson House, 270
Johnston, William L., architect, 65, 81, 269, 287
Jones, John D., contractor, 121, 299
Jordon-Stoddart House, 17
Justi, Henry D., House, 211
Junior League of Philadelphia, Inc., 74

Kaighn, Robert, 127
Kahn, Louis I., architect, 200, 201, 255
Kalm, Peter, xxiii, xl(n.3), 155, 245
Kastner, Alfred, architect, 319
Kearsley, Dr. John, 67
Keely, P. C., architect, 371(n.48)
Keen, James, House, 946 S. Front St., 178
Keen, James, House, 1015 Shackamaxon St., 330
Keen Building, 66
Keen and Mead, architects, 147
Kelly, J. Wallace, sculptor, 61
Kelly, William, 299

Kemble, William H., 289, 374(n.33). See also Green Street Area Study
Kenilworth Street Area Study, 178
Kennedy, David J., xli(n.12)
Kennedy, Robert G., architect, 147
Kensington, 305, 310, 312, 377(nn.20, 28)
Kensington National Bank, 290, 316, 330–31
Kenworthy, T., and Bro., mills, 257
Kessler, George, contractor, 68
Ketcham, B., and Son, contractors, 296
Ketcham and McQuade, contractors, 61
Ketchum, Benjamin, builder, 68
Keyser, Jacob, House, 271
Keyser Houses, 271
Keyser-Wainwright Building, 331
Keystone Automobile Club. See Philadelphia Art Club
Keystone Bridge Co., builders, 239
Keystone Telephone Co., 68
Kid-Chandler House, 81
Kid-Physick House, 82
Kilgore, John, carpenter, 82
Kimball, Fiske, 234
Kimball, Francis H., architect, 139
Kimball, L. P., architect, 123
Kimball and Thompson, architects, 139
Kingsessing Township, 193, 198, 364 (n.35)
Kinnier, James, and Sons, architects and builders, 266
Kinnier's, James, Sons, builder, 273
Kirkbride, Dr. Thomas Story, 197, 205, 212
Kirkbride Mansion. See Busti Mansion
Kirkbride System. See Pennsylvania Hospital for Mental and Nervous Diseases
Kirkbride's. See Pennsylvania Hospital for Mental and Nervous Diseases
Kitts, Captain John, House, 178
Klauder, Charles Z., architect, 128
Kline, Herman H., architect, 26
Kling, Vincent G., architect, 120, 201, 358(n.75)
Kling, Vincent G., and Associates, 133, 328
Knauer Foundation for Historic Preservation, 19
Kneass, Samuel, 24
Kneass, Strickland, architect and engineer, 125
Knight, Daniel, builder, 89
Knight, Robert, builder, 89
Knight Building, 331, 379(n.46)
Knor, Jacob, carpenter, 262, 266

Knowlton, 318–19
Koch, Richard, xl(n.1)
Kosciuszko, Gen. Thaddeus, House, 17
Kramer-Rommel Houses, 331
Krider Gun Shop. *See* Drinker, John, House
Krumbhaar, David, xlv(n.57)

Lamb, Peter, House, 178
Lancaster, Moses, house carpenter, 330
Land Title Building, 115, 132–33
Landreth Building, 66
Langenheim, Frederick, xxv, xlii(n.17)
Langenheim, William, xxv, xlii(n.17)
La Salle College, 292
Lasse Cock's Manor House. *See* Bel Air
LaTour Warehouse, 18, 156, 359(n.22)
Latrobe, Benjamin Henry, architect, 53, 97, 98, 106, 238
Laub, Sidney, 101
Laurel Hill (Randolph House), 234
Laurel Hill Cemetery, 224, 234, 368(n.9)
Lawrence, Charles, House, 331
Lea, Professor Henry Charles, House, 199, 211
Lea and Febiger, 50
Leamy House. *See* Hospital of the Protestant Episcopal Church in Philadelphia
Leather Row, 313
LeBrun, Napoleon, architect, 85, 95, 112, 121, 124, 315
Le Corbusier, 118, 255
Lee, Robert M., House, 82
LeGrand, Claudius, sculptor, 72
Leidy, Dr. Joseph, Jr., House (Poor Richard Club), 133
Leland, Charles, 46
Leland Building, 44, 45, 66, 82, 348(n.33)
Lemon Hill, 224, 234–35, 335, 367(n.3)
Lescaze, William, architect, 118. *See also* Howe and Lescaze
Letitia Penn House (House on Letitia Street), xxiv–xxv, 235
Levering and Garrigues, steel contractors, 131
Levin, Isadore W., architect, 329
Levy, L. J., Company, 51
Lewars, Charles, building, 289, 374(n.34)
Lewin, J. W., designs airlites, 382(n.81)
Lewis Building, 44, 46, 66
Library Company of Philadelphia, 57, 163, 185–86
Library of Congress, HABS records in, xxi, xxx–xxxi, xxxii

Linville, Jacob H., engineer, 239
Lippincott, J. B., 292
Lippincott, J. B., & Co., 50, 51, 83
Lippincott, Joshua B., House, 133
Lipshitz, Jacques, sculptor, 226
Lit Brothers, 51, 82–83, 350(n.74)
Littell, Emlen T., architect, 275
Little, John, House, 179
Little, Thomas, & Son, contractors, 328
Livery stable, 330
Livezey House. *See* Glen Fern
Livingston, Walter R., Jr., architect, 375(n.52)
Lloyd, John, and Associates, architects, 9
Log houses, xl(n.3), 271
Logan, James, 251, 276
Logan Square, 119
Lombard Street Area Study, 18
London Coffee House, 41, 347(n.14)
Long, M. A., architect, 123
Lonsdale, Thomas P., architect, 291, 300
Lossing, Benson J., xxv, xli(n.15)
Loudoun, 271–72
Louisiana Purchase Exposition, 116, 147
Lovatt, George I., architect, 16, 139, 375(n.52)
Loving, F. Bryan, architect, 101
Lower Burying Ground. *See* Hood Cemetery Entrance
Lower Dublin Township, 305
Lyle-Newman House, 179
Lynfield, 331–32, 381(n.74)
Lyons, Fr. Charles, S.J., designer, Community House, Church of the Gesu, 294

McArthur, John, contractor, 24, 287
McArthur, John, Jr., architect, 24, 97, 111, 315, 380(n.58); First National Bank, 48, 73; Laning Hall, U.S. Naval Asylum, 188; Philadelphia City Hall, 105, 139–40, 353(n.1)
McAuley, William J., xxvii, xlii(n.21)
McCaul, Charles, contractor, 92, 133, 139, 206, 301
McClare, Samuel, 83
McClare-Hutchinson Building, 83
McClellan, Gen. George B., 239
McCloskey and Co., contractors, 132
McCraig, George, House, 179
McCrea, James, House, 84
McCrea, John, developer, 107, 127
McDonald, Malcolm, House, 18
McDowell-Fritz House, 179

McElfatrick, William H., architect, 290, 298
McGee, Ann, Houses, 332
McGoodwin, Robert, architect, 217, 254, 255
Mackay, G. W., 187
McKean, Thomas, Jr., House, 18–19
McKee, Harley J., xxxviii
McKeone, Van Haagen and Company, 287
Mackie, William C., builder, 279
McKim, Charles F., architect, xxvi, 267
McKim, Mead and White, architects, 117, 130, 267, 290, 357(n.64)
Mackley, Carl, Apartments, 319–20, 332, 382(nn.82, 83)
McMullin, Alexander, bricklayer, 181
McMullin, Robert, House, 19
McPherson, William C., and Son, contractors, 135, 140
McShain, John, contractor, 168
Magaziner, Henry J., architect, 30, 100, 262
Maginnis and Walsh, architects, 214
Maguire-Howell House, 332
Mahoney, John T., architect, 111, 124
Maller, Rev. Mariano, 111, 124
Maloby, Thomas, House and Tavern, 179
Malta Boat Club, 225, 228, 368(n.13)
Man Full of Trouble Tavern (Cove Cornice House; Stafford's Tavern), xxxiii, 19. See also Paschall, Benjamin, House.
Manayunk (Flat Rock), 256–57, 276, 372(n.50)
Manderson, Andrew, House, 332–33
Manfredi, Louis A., architect, 293
Manhattan Building. See American Life Insurance Company Building.
Manheim Club. See Germantown Cricket Club
Manly and Cooper, 168
Manning Street Area Study. See Marshall's Court Area Study
Mantua, 196
Manufacturers Club, 113
Manufacturers and Mechanics Bank, 316
Maps, Philadelphia, xxiii, xxix, xliii(n.32)
Mariners' Bethel. See Eastburn Mariners' Bethel
Mariners' Bethel (Baptist) Church (First Polish Baptist Church and Bethel Christian Center), 158, 180
Maris, J. M., & Co., 51, 83

Market Street Area Study, 84
Market Street Elevated, 200, 365(n.44)
Market Street Meeting House, 146
Market Street National Bank Building (One East Penn Square), 118, 133–34
Market Street Permanent Bridge, 196, 198, 362(n.19)
Markets, 42: Germantown, 248; High Street, 39, 42, 55; Italian, 161; "New Market," 6, 21–22, 248; Reading Terminal, 139; Ridge Avenue Farmers', 299
Marks-Dunbar House, 180
Marshall, Joseph, Houses, 180
Marshall-Morris House, 180
Marshall's Court Area Study (Manning Street), 19–20
Martin, Samuel, 310
Martin, Sydney E., architect, xxix, xliii(n.31), 92
Martin, Stewart, Noble and Class, architects, 121
Mary Elizabeth Patterson Memorial Church, 367(n.57)
Mary Warden Harkness House. See Fromberger, John, Houses.
Mask and Wig Club, 134
Mason, George C., xxvi, xxvii, xlii(nn.23, 24), 79, 174
Mason, George C., & Son, architects, 144
Mason, James S., & Co., Store, 84
Masonic Temple. See Philadelphia Masonic Temple
Massara, Carl, architect, 35
Massara, Carl, and Associates, architects, 21
Massey, James C., xl
Massey and Janney Leather Warehouse, 313, 333
Matlack, Timothy, 74
Matthews, James, 251
Matthewson, Ernest J., architect, 80
Maule, Israel, carpenter, 308
Maxfield-Elliott House, 180–81
Maxwell, Ebenezer, House, 253, 272
Meade, Gen. George G., 239
Meany, John, developer, 53, 106, 145
Mears-Heaton House, 20
Mease, Dr. James, 156
Mechanics Bank (Norwegian Seamen's Church), 84
Meeks, Carroll L. V., 366(n.54)
Meem, John Gaw, xl(n.1)
Meetinghouses, 60, 74, 146, 246, 272–73, 307, 354(n.3)

Megargee Building, 85
Mehl House, 272
Mellon, Thomas, House, 20
Mellor and Meigs (Mellor, Meigs and Howe), architects, 134
Mellor, Meigs and Howe, architects, 91
Melody and Keating, contractors, 214
Memorial Hall. *See* International Exhibition of 1876
Mennonite Meeting House, 246, 272–73
Mercer, Thomas, Houses, 181
Merchants' Exchange. *See* Philadelphia Exchange Company
Merchants' Hotel, 85
Meredith House. *See* Stone-Penrose House
Met, The. *See* Philadelphia Opera House
Methodist Mariners' Bethel, 158
Metropolitan Museum of Art, New York, 5, 25
Mettler Brothers, contractors, 330
Mifflin, George, houses. *See* Workman Place
Mifflin, John, Houses, 20
Mikveh Israel Cemetery Gatehouse, 181
Mikveh Israel Synagogue, 56, 352(n.94)
Military installations, 159, 165, 184–85, 188, 207–10, 309, 310
Mill Creek Project, 200
Miller, M. Luther, xxviii
Mills, Robert, architect, 10–11, 20, 53, 78, 106, 145
Mills, 223, 254; Frankford, 310, 377(n.31), 378(n.32); Manayunk, 256–57; steam-powered, 257, 276; water-powered, 255–56; Wissahickon, 255–56, 274, 371(n.41)
Milltowns, 256
Milner, John D., architect, xxxvii, 30
Mission 66, National Park Service, xxxv–xxxvii
Mission style, 213
Mitchell, Thomas, House, 20
Mitchell/Giurgola Associates, architects, 50, 88, 201, 218, 365(n.50)
Modern Club, The, 241
Moffett, Robert, House, 181
Monaghan, C. T., muralist, 142
Monarch Building (Harrison Building), 115, 134–35
Monastery, the, 235
Moore, Clarence B., House, 135
Moore, John, mason, 230
Moore, John, House, 181–82
Moore, Captain Thomas, House, 181

Moore College of Art (Philadelphia School of Design for Women). *See* Gaul-Forrest House
Moorish style, 44
Moreland Township, 305
Moroney, William, 95
Morris, John T., Estate. *See* Compton
Morris, Samuel, 250
Morris Arboretum. *See* Compton
Morris Brewery Vaults, 85
Morris, Tasker & Morris, 24
Morrow, John, 298
Moses, Arnold H., architect, 81
Mother Bethel African Methodist Episcopal Church. *See* Bethel AME Church
Mount Airy, 246, 249
Mount Auburn Cemetery, Cambridge, Mass., 368(n.9)
Mount Moriah Cemetery Gatehouse, 211–12
Mount Pleasant (Clunie), xxviii, 236, 250, 362(n.6), 367(n.3)
Moyamensing Prison. *See* Philadelphia County Prison
Moyamensing Township, 154, 159, 160
Muller, A. J. M., sculptor, 232
Mumford, Lewis, 366(n.54)
Municipal Services Building, 119, 358(n.75)
Murphy, G. Harold, designs airlites, 382(n.81)
Murphy-Johnson House, 182
Museum of Science and Art. *See* University of Pennsylvania, University Museum
Museums, 112, 119, 136–37, 201, 218, 230, 233, 236–37
Musical Fund Society Hall, 58, 85–86
Mutual Assurance Company, 7, 22, 28–29
Mutual Fire Insurance Company Building, 273
Myers, Jacob, builder, 59

Naomi Wood Collection, 241
National Bank of Northern Liberties, 316, 380(n.65)
National Bank of the Republic, 349(n.55)
National Guard's Hall (Armory of the National Guard), 86
National Historic Civil Engineering Landmarks, 231, 334
National Historic Landmarks: Center Square, 126, 137; Fairmount Park, 231, 236, 241; Germantown, 260, 263, 276; North Philadelphia, 295, 296, 297, 298,

National Historic Landmarks *(Cont.)* 301; Old City, 345(n.1); Society Hill, 15–16, 21–22, 24–25, 32; South Philadelphia, 187; West Philadelphia, 203, 207, 212, 219

National Historic Sites, 174, 345(n.1)

National Natural History Landmark, 223, 278

National Park Service (United States Department of the Interior), xxx–xl *passim*, 250, 264. *See also* Independence National Historical Park

National Recovery Act, xxx

National Society of the Colonial Dames of America, 271, 276

National Trust for Historic Preservation, 249, 263

Nattress, George, architect, 215

Neave, Samuel, House and Store, xxxvii, 21

Nectarine Street (Artillery Lane, Duke Street, Dana Street), 329

Neely, R. R., architect, 144

Neoclassical Revival style, 115–16, 117, 119, 230, 237, 290

Nevel, Thomas, House, 21

New Century Club of Philadelphia, 135

"New Market," 21–22

New Market Street. *See* Phillips Row

New York Architectural Terra Cotta Works, 206, 217

New York Mutual Life Insurance Company Building (Victory Building), 114, 135–36

Newhall, George M., Engineering Co. Ltd., 126

Newman and Harris, architects, 139

Nicetown, 372(n.3)

Nichols, Minerva Parker, architect, 135

Norman style, 113

Norris, Joseph P., Jr., 22

Norris-Cadwalader House (Mutual Assurance Company), 7, 22

North American Lace Company, 381 (n.73)

North Pennsylvania Railroad, 316

North Philadelphia: development of, 283–92; HABS buildings in, 293–302

Northern Liberties, 313

Northern Liberty Hose Company No. 4 (The Snappers), 314, 334, 379(nn.52, 55)

Northern Saving Fund, Safe-Deposit and Trust Company (Northern Trust Company, 1902; now Provident National Bank), 289, 298, 380(n.64)

Norwegian Seamen's Church. *See* Mechanics Bank

Notman, John, architect, 111; Athenaeum, 61; Holy Trinity, 125; Laurel Hill Cemetery, 224, 234, 368(n.9); St. Clement's, 143; St. Mark's, 110, 143–44

Nowicki, Joseph, architect, 125

Oak Hall Store, 114

Oak Lane, 292, 375(n.50)

Octavia Hill Association, 155, 190

Odd Fellows Hall. *See* Independent Order of Odd Fellows

O'Gorman, James, xxxvii

Ohio House. *See* International Exhibition of 1876

Old City Historic District, 58

Old Colonial Architectural Details in and around Philadelphia, xxvii

Old Customs House. *See* Second Bank of the United States

Old First Reformed Church, United Church of Christ (First German Reformed Church), 52, 56, 72–73

Old Philadelphia Colonial Details, xxviii

Old Philadelphia Development Corporation, 8

Old Philadelphia Survey, xxix–xxxi, xxxii

Old Pine Street Church. *See* Third Presbyterian Church

Old St. Mary's. *See* St. Mary's Church

Old Swedes Church. *See* Gloria Dei

Olive Street (Equality Court, Onas Street, York Court), 308, 323

Olmsted, Frederick Law, architect, 225, 368(n.14)

Olmsted Brothers and Co., landscape architects, 101

Olmsted and Vaux, landscape architects, 225

Olympia, U.S.S., 39

Onas Street. *See* Olive Street

One East Penn Square. *See* Market Street National Bank Building

O'Neill, John, builder, 85, 187

O'Neill, John P., xlv(n.48)

Opera houses, 112, 122–23, 290–91, 298

Oram, H. C., & Co., 62

Ord, John, architect, 140, 353(n.1)

Organs, church, 96, 116

Ormiston, 236

Osbourn, Ebenezer, house carpenter, 171

Oswald, Eleazer, 347(n.14)
Otis, Harrison Gray, house, Boston, Mass., xxiii
Overbrook School for the Blind. *See* Pennsylvania Institution for the Instruction of the Blind
Oxford Township, 305

Page Brothers, developers, 295
Palladian church architecture, 55
Palladian Revival style, 124
Palles, Andrew, engineer, 368(n.14)
Palmer, Anthony, 377
Palmer, John, mason, 67, 96
Palmer, John, House, 4, 22, 34
Palmer, Timothy, builder, 362(n.19)
Pancake, Philip, House, 23
Pancoast-Lewis-Wharton House, 23
Paper mills, 254, 274
Parker, Minerva. *See* Nichols, Minerva Parker
Parks, 158, 161, 166, 318, 361(n.50). *See also* Fairmount Park *and other listings by name*
Parnum, Edward, architect, 12
Parris, Alexander, 348(n.28)
Parry, Charles T., House, 136
Paschall, Benjamin, House, 23
Paschall, Jonathan, House, 182
Pasco, Henry, & Son, plasterers, 148
Passyunk Township, 160
Pastorius, Daniel, 247, 268
Pastorius House. *See* Green Tree Tavern
Pastorius Park, 254
Payne, George F., & Co., contractors, 328
Peabody and Stearns, architects, 108, 129
Peale, Charles Willson, 78, 260
Peale, Charles Willson, House. *See* Belfield
Pearson, Anthony, House, 182
Pearson, George, House, 156, 182–83
Pearson, George T., architect, 273, 298
Pederson, Erling, architect, 127
Pei, Ieoh Ming, architect, 8
Peking Mandarin Palace style, 125
Pelham, commuter neighborhood, 254
Pelham Trust Company (now First Pennsylvania Bank), 254
Pemberton House, 65
Penal reform, 284
Pencoyd Iron Works, steel contractors, 139
Penn, John, 239
Penn, William, 105, 140, 306

Penn Center, 115, 119–20
Penn District, 283
Penn Mutual Life Insurance Company Building, 129 S. Third, 87; Sixth and Walnut Sts., 50
Penn Square Building. *See* Bulletin Building
Penn Street Station (Philadelphia, Germantown and Norristown Railroad), 252
Pennepack Baptist Church, 307
Pennock, J. E., and A. L., contractors, 68, 131, 135, 136
Penn's City, 3–8, 39–58, 105–20; HABS buildings in, 9–36, 59–102, 121–49
Penn's Landing, 58
Pennsylvania: HABS records, xxxii; hospital design, 197; penal reform system, 284, 296; State House. *See* Independence Hall
Pennsylvania Academy of the Fine Arts, 112, 136–37
Pennsylvania Company for Banking and Trusts, 62
Pennsylvania Company for Insurances on Lives and Granting Annuities: 431 Chestnut St., 48, 71, 87–88; 304 Walnut St., 88
Pennsylvania Fire Insurance Company Building, 44, 50, 88
Pennsylvania Horticultural Society, 81, 82
Pennsylvania Hospital, 6, 23–24, 196–97, 308, 377(n.21)
Pennsylvania Hospital for the Insane, Department for Males (Institute of the Pennsylvania Hospital), 212, 309, 363(n.30)
Pennsylvania Hospital for Mental and Nervous Diseases (Kirkbride's), 197, 212–13, 309, 363(n.30)
Pennsylvania Institution for the Deaf and Dumb (Philadelphia College of Art), 111, 137–38, 355(n.34)
Pennsylvania Institution for the Instruction of the Blind, 198, 213
Pennsylvania Museum of Art. *See* Philadelphia Museum of Art
Pennsylvania Museum and School of Industrial Art. *See* Philadelphia College of Art
Pennsylvania Railroad, 114, 200, 202, 253–54, 366(n.54); Broad Street Station, 115, 119, 122, 138
Pennsylvania Soap Works, 287

Pennsylvania Working Home for the Blind, 198
Pennypack Creek Bridge, 334
Penrose, Thomas, 40
Pent eaves, 351(n.79)
Peoples Bank, 48
Père-Lachaise cemetery, Paris, 368(n.9)
Perelman Antique Toy Museum. *See* Abercrombie, Captain James, House
Perot, Robeson Lea, architect, 266
Perry, William Graves, xl(n.1)
Perseverance Hose Company No. 5, 88
Peters, Richard, 196
Peters, William, 227
Peterson, Charles E., 14
Philadelphia: City Hall (U.S. Supreme Court Building), 40, 79; City Hall (Penn Square), 105–106, 114, 119, 122, 139–40, 353(n.1); city planning, 115, 118–20, 202, 237; City Planning Commission, 8, 202; commercial district, 40, 42–51, 60, 66, 348(n.33); courthouse, xxv; Depression affects, 357(n.65); dry goods district, 348(n.33); early views, xxiii, xxiv; Fire Department, 90, 184, 314–15, 326, 334, 359(n.26), 379(n.55); German community, 52, 72–73; HABS in, xxii; Historic Districts, 345(n.1); landmarks, xxiii; neighborhoods, 7, 196, 201, 202, 283, 284, 292, 305, 372(n.3); retail district, 51; sports complex, 162; theater district, 57–58; westward development, 107, 114, 119, 196, 198, 199–200, 202
Philadelphia and Columbia Railroad Bridge, 368(n.14)
Philadelphia and Reading Railroad, Terminal Station (Reading Terminal), 115, 135, 138–39, 312, 319, 381(n.79). *See also* Reading Company
Philadelphia and Trenton Railroad, 311, 378(n.38)
Philadelphia Art Club (Keystone Automobile Club) 113, 139
Philadelphia, Baltimore and Washington Railroad. *See* Philadelphia, Wilmington and Baltimore Railroad
Philadelphia Bourse, 43, 89, 347(n.20)
Philadelphia Club (Thomas Butler House), 112, 140, 356(nn.39, 40)
Philadelphia College of Art. *See* Pennsylvania Institution for the Deaf and Dumb
Philadelphia Contributionship for Insuring of Houses from Loss by Fire, 49, 50, 89

Philadelphia County Courthouse. *See* Congress Hall
Philadelphia County Prison (Moyamensing Prison), 160, 183; Debtor's Wing, 160, 183–84
Philadelphia Customs House. *See* Second Bank of the United States
Philadelphia Divinity School chapel, 367(n.57)
Philadelphia Evangelistic Center. *See* Philadelphia Opera House
Philadelphia Exchange Company (Merchants' Exchange) 42, 89–90
Philadelphia Functionalism (utilitarian Italianate), 44, 45. *See also* Italianate
Philadelphia, Germantown and Norristown Railroad, 252
Philadelphia High School for Young Ladies, 142
Philadelphia Historical Commission, 253
Philadelphia Hose Company, 86, 90
Philadelphia Hospital, 377(n.21)
Philadelphia Inquirer, 291
Philadelphia Labor Institute, 86
Philadelphia Maritime Museum, 71, 93
Philadelphia Masonic Temple, 113, 140–41
Philadelphia Museum (Peale's Museum), 78
Philadelphia Museum of Art (Pennsylvania Museum and School of Industrial Art; Pennsylvania Museum of Art), 5, 25, 119, 136, 230, 231, 233, 236–37, 295
Philadelphia National Bank, 71, 87
Philadelphia Opera House (later known as the Met), 290–91, 298
Philadelphia Orchestra, 226, 369(n.22)
"Philadelphia Plan," school design, 157, 185
Philadelphia Saving Fund Society: PSFS building, 118, 141; 700 Walnut St., 91, 101; 306 Walnut St., 90–91
Philadelphia Society for the Preservation of Landmarks, 15, 25, 269, 279
Philadelphia Society for Promoting Agriculture, 82
Philadelphia Tapestry Mills, 381(n.73)
Philadelphia Trust, Safe Deposit and Insurance Company, 48, 91, 374(n.36)
Philadelphia Veterans Stadium, 162
Philadelphia, Wilmington and Baltimore Railroad, 162; freight station train shed, 184
Philadelphia Zoological Gardens, 225, 237, 238, 239

Phillips, William, 307, 334
Phillips Row (later New Market St.), 307, 334
Philosophical Hall (American Philosophical Society), 57, 92
Phoenix Bridge Co., builders, 240
Phoenix Iron Company, steel contractors, 298
Phoenixville Iron Co., builders, 232
Photogrammetry, xxxix, xlvi(nn.66–70)
Physick, Dr. Philip Syng, 16, 24
Physick-Conner House, 24
Piazzas, 6
Pickands, Thomas, House, 24–25
Pickford, Nicholas, 194, 195
Picturesque Eclectic Style, 253, 301
Piles, John, House, 25
Place, G. G., architect, 294
Planter Garden Club, 236
Playhouse in the Park, 226
Plowman, George W., architect, 123
Plumbing fixtures, for commercial buildings, 45, 47, 49
Poe, Edgar Allan, House, 287, 298, 373(n.23)
Polite Temple Baptist Church. See First Baptist Church of Germantown
Poor Richard Club. See Leidy, Dr. Joseph, Jr., House
Poppliers, John, xl
Port Folio, The, xxiv
Port Richmond (Point No Point, Point Road), 312, 378(n.40)
Port Royal, xxxiii, xlv(n.52), 306, 318, 335, 381(n.74)
Portico Square, 53, 107, 141–42
Potter, Edward C., sculptor, 239
Potter, W. Woodward, 41
Potts, Horace T., & Co., Warehouse, 335
Potts, Joseph D., House, 213–14
Powder Magazine, 159, 184–85, 360(n.39)
Powel, John Hare, 196, 363(n.21)
Powel, Samuel, House, 5, 25, 344(n.8)
Powell, W. Bleddyn, architect, 105, 140, 353(n.1)
Powelton, 196, 363(n.21)
Powelton Village, 202
Praissman, Joseph A., architect, 26, 94
Pratt, Henry, 234
Pratt, Henry, Houses, 335
Preisendanz, Christian A., Wagon Works, 335
Presbyterian Church of Cohocksink, 315
Presbyterian Hospital, 198
Preston, Dr. Jonas, 286

Preston, Samuel, 158, 167
Preston Retreat, 285, 286, 298–99, 372(n.8)
Price, William D., architect, 83
Price, William L., architect, 142
Price and McLanahan, architects, 142
Prisons, 160, 183–84. See also listings by name
Provident Life and Trust Company Bank, xxxix, 48, 92, 290, 349(n.55)
Pryor, Richard E., xlv(n.57)
Public School of Germantown. See Germantown Academy
Publishers' Square. See Washington Square
Pugh, Isaac, Paper Hangings Factory, 313, 336
Purves, Hugh, Store, 185

Quaker plain style, 54, 60, 74
Quakers, 246, 284
Quarry Street, houses, 77
Queen Anne style, 254
Queen Street Area Study, 185
Queen Village. See South Philadelphia (Southwark)
Quincy Market Stores, Boston, Mass., 348(n.28)

Railroad stations: freight, 184, 361 (n.54), 374(n.44); passenger, 200, 254, 366(n.54), Baltimore and Ohio, 123, Broad Street, 114–15, 119, 138; Philadelphia, Germantown and Norristown, 252, 273; Philadelphia and Reading, 115, 138–39
Railroads, 135, 156, 252, 316, 382(n.79); commuter, 292; and industrial development, 257, 312; and suburban growth, 253–54
Raley, Robert L., xlvi(n.63)
Ralston, Robert, 286–87; House, 60, 92–93
Ralston School, 157, 185
Randall, John, builder, 101
Randolph House. See Laurel Hill
Rankin, Kellogg, and Crane, architects, 374(n.43)
Ransom, A. H., 212
Rath, Frederick L., xlvi(n.63)
Reading Company, the (Philadelphia and Reading Railroad), 273, 287, 291, 374(n.44)
Reading Terminal. See Philadelphia and Reading Railroad, Terminal Station

Real Estate Title Insurance Company. *See* Land Title Bank and Trust Company
Reany, Neafie & Company, 312
Rectory of St. Stephen's Methodist Episcopal Church, 274
Red Lion Road, 307
Redevelopment Authority, 8, 202
Reed, Earl H., xl(n.1), xlvi(n.63)
Reed Houses, 26
Reed's, Jacob, Sons, Store, 142
Reeves, Buck & Co., 64
Reilly, Thomas, builder, 294
Reliance Insurance Company of Philadelphia, Building, 49, 93
Remington, Frederic, sculptor, 226
Renaissance Revival style, 62, 108–9, 115, 116, 139, 147, 200
Reservoirs, 237
Restaurant and apartment building (110 N. Ninth St.), 142
Restaurants, 128, 142, 148–49
Reybold-Paul House, 336
Reynolds, General John F., 239
Reynolds-Morris House, 54, 93, 250
Rhawn, William H., 318, 381(n.77)
Rhoads, Samuel, master carpenter, 4, 24, 26
Rhoads-Barclay House, 4, 26
Rice, John, builder, 71, 73, 75, 233
Rich, Comly, House, 336
Rich-Truman House, 26
Richard Gimbel Foundation, 298
Richards, Thomas W., architect, 201, 216
Richardson, H. H., xxxvi
Richardson, Nathaniel, House, 27
Richmond, district, 305
Ricketts Circus, 57
Riddell, John, 67, 87, 347(n.21)
Ridge Avenue Farmers' Market Company, 299
Ridgeland, 238
Ridgway Branch of Library Company of Philadelphia, 163, 185–86
Riggs & Brother: Navigator Statue, 93–94
Rising Sun, village, 372(n.3)
Rising Sun Tavern. *See* Letitia Penn House
Rittenhouse, William, 254, 274, 371(n.41)
Rittenhouse House, 254, 274
Rittenhouse Square, 107, 108–9, 110, 121, 129, 133, 142–43
Ritter and Shay, architects, 134
Roach, F. Spencer, architect, 74
Roach, Hannah Benner, 52

Robbins, George A., architect, 23
Roberts, Frank C., & Co., engineers, 124
Roberts, George B., architect, 12, 15
Roberts, Spencer, architect, 142, 146
Robeson, Andrew, 256
Robeson House. *See* Shoomac Park
Robin Hood Dell East, 226, 369(n.22)
Robin Hood Dell West, 369(n.22)
Robinson, Cervin, xxxvi
Robinson, William, House, 27
Rogers, T. Mellon, architect, xlii(n.24), 78
Rogers-Cassatt House, 142–43
Romanesque Revival style, 7, 10, 110, 125, 143, 200, 289, 355(n.27)
Romantic Classicism, 47
Roney, John, House, 94
Roos, Frank J., xxxviii
Roos, Philip W., architect, 114, 136
Roos and Brosam, architects, 136
Rorke, Allen B., contractor, 76, 89, 148, 301, 333
Rorke, Edwyn, architect, 371(n.39)
Ross, Budd, architect, 206
Rotan, William, Houses, 336
Roth, Gabriel B., architect, 145
Row houses: Center Square, 106–7, 127, 141–42, 145; Germantown, 254–55, 265, 271; North Philadelphia, 287, 292, 295, 375(nn.48, 50, 51); Northern Liberties, 307–8, 311, 319, 323, 327, 329, 334, 337, 340; Society Hill, 14, 15, 19, 22, 25, 26, 34; South Philadelphia, 161, 166, 172, 189; speculative developments, 53, 97, 323, 334; West Philadelphia, 200, 202, 365(n.44). *See also* Airlites; Bandboxes
Rowland's Saws (Beach Street), 312
Rowley-Pullman House, 94
Roxborough, 246, 256
Royal House, 274
Royal Insurance Company Building, 94
Rubicam, Charles A., builder, 68, 95–96
Rudman, Rev. Andrew, builder, Gloria Dei, 174
Ruffner, William A., and Daniel, builders, 262
Rumpp, C. F., & Sons, Inc., Factory, 94
Runge, G., architect, 46, 69, 112, 121, 348(n.33)
Rush, Benjamin, House, 306, 336–37
Rush, Dr. James, 162–63
Rush, Louis H., architect, 338
Rush, William, sculptor, 27, 96
Russell, Capt. James, House, 95

Russell, Walter, architect, 101
Russell Building, 94–95
Ryan, J. J., & Co., excavation contractors, 139
Ryerss, Joseph W., 318, 323, 381(n.75)
Ryerss Mansion. *See* Burholme

Saarinen, Eero, architect, 201, 365(n.50)
Sabitini, Raphael, sculptor, 61
Sailmakers' lofts, 41
Sailor, Samuel, woodcarver, 93
St. Agatha's Roman Catholic Church, 214, 367(n.57)
St. Andrew's Protestant Episcopal Church (St. George's Greek Orthodox Cathedral), 7, 27
St. Ann's Roman Catholic Church, Richmond, 315, 380(n.61)
St. Augustine's (Roman Catholic) Church, 56, 95, 315
St. Charles Hotel, 95–96
St. Clement's (Protestant Episcopal) Church, 110, 143
St. Francis De Sales Roman Catholic Church, 214
St. Gabriel (Roman Catholic) Church, 161
Saint-Gaudens, Augustus, sculptor, 144, 226
St. George's Greek Orthodox Cathedral. *See* St. Andrew's Church
St. George's Methodist Episcopal Church, 56, 96
St. James the Less. *See* Church of St. James the Less
St. James Roman Catholic Church, 215
St. James Street Area Study, 96
St. Joachim (Roman Catholic) Church, Frankford, 315, 380(n.61)
St. John the Baptist (Roman Catholic) Church, 257
St. Joseph's College, 201–2, 366(n.53)
St. Joseph's Willings Alley, 56, 366(n.53)
St. Luke and the Epiphany, Church of. *See* Church of St. Luke and the Epiphany
St. Luke's Protestant Episcopal Church, Germantown. *See under* Fromberger, John, Houses
St. Luke's Episcopal Church, Holmesburg, 315
St. Mark's Lutheran Church, Spring Garden, 287
St. Mark's (Protestant Episcopal) Church, 109–10, 143–44

St. Martin's-in-the-Fields (Protestant Episcopal) Church, 254
St. Mary's Protestant Episcopal Church, Hamilton Village, 367(n.57)
St. Mary's (Roman Catholic) Church (Old St. Mary's), 27–28
St. Michael's Lutheran Church, Fifth and Appletree Sts., 56
St. Michael's Luthern Church, Germantown, 247
St. Michael's (Roman Catholic) Church, Kensington, 56, 315, 380(n.61)
St. Paul's Protestant Episcopal Church (Old St. Paul's), 96
St. Peter's Kensington. *See* German Catholic Church of St. Peter's
St. Peter's Lutheran Church, Barren Hill, 247
St. Peter's (Protestant Episcopal) Church, Third and Pine Sts., 28, 96, 210
St. Peter's Church House. *See* Blackwell, Reverend Robert, House
St. Philip de Neri (Roman Catholic) Church, 157, 168–69
St. Stephen's Protestant Episcopal Church, 109, 144
St. Thomas Lutheran Church, 247
St. Timothy's (Protestant Episcopal) Church, 274
St. Timothy's Working Men's Club and Institute, 275
St. Vincent De Paul. *See* Church of St. Vincent De Paul
Sansom, William, 53
Sansom, William, House, 97, 101
Sansom's Row, 53, 106
Saunders, W. B., Company, 50
Saur and Hahn, architects, 87
Savage, John, developer, 53, 106, 145
Savery, Addison H., architect, 335
Savery and Sheetz, architects, 130
Sax, Percival M., contractor, 300
Saxton, Joseph, xxv
Scataglia, Lorenzo C., frescoes by, 168
Schaeffer, Harriet D., House, 275
Schaible, Charles, Store and House, 337
Schenck Building, 51, 97
Schively, Henry, House, 28
Schofield, Sevil, 256
School House Lane (Bensell's Lane), 248
School of Industrial Art. *See* Philadelphia Museum of Art
Schoolhouses, 52, 73, 95, 248, 260, 263, 266, 316, 338

Schools, 121, 132, 137, 142, 198, 213, 294, 296, 300; modular design for, 185. *See also* Universities and Colleges *and listings by name*

Schroen, Br. Francis C., S.J., decorator, Church of the Gesu, 294

Schuylkill Arsenal. *See* Arsenal on the Schuylkill

Schuylkill Hose, Hook & Ladder Company No. 24, 144

Schuylkill Navigation Company, 224, 256

Schuylkill Navy, 228

Schwarzmann, Hermann J., architect and engineer, 225, 233

Scott, James P. 108, 145

Scott-Wanamaker House, 108, 144–45

Scully, Michael, mason, 216

Sculpture, 27, 93, 96: for Ayer Building, 61; Broad St. Station, 138; City Hall, 106; in Fairmount Park, 226, 232, 239; Federal Reserve Bank, 117–18, 129; First Bank of the United States, 72; on Philadelphia Museum of Art, 237; St. Stephen's, 144; University Museum, 218

Searle-Bond Houses, 337

Second Empire style, 136, 140, 147

Second National Bank, Frankford, 380(n.64)

Second Presbyterian Church (now First Presbyterian), 111

Second Presbyterian Church, Northern Liberties, 316, 338

Second Renaissance Revival (Late Victorian Italian Renaissance style), 290, 293

Security National Bank, Spring Garden District, 380(n.64)

Sedgeley, porter's house, 224, 238

Seeds, Thomas M., Jr., contractor, 90

Seeler, Edgar V., architect, 49, 80, 111, 123, 127, 299

Sellers, Horace Wells, architect, xxviii, xxix, xliii(n.30), 78, 79, 135, 143

Sellers, William and Co. (Bancroft and Sellers), 287

Semple Co. Warehouse. *See* Philadelphia, Wilmington and Baltimore Railroad Freight Station Train Shed

Sesqui-Centennial Celebration, 162

Seventh Street, North, Area Study, 86

Shackamaxon Bank, Kensington, 380 (n.64)

Sharpless Brothers, 51

Shay, Howell Lewis, architect, 331

Shibe Park (Connie Mack Stadium), 292, 299–300

Shingle style, 275

Shipbuilding, 156, 312

Shippen, William, 29

Shippen-Wistar House (Mutual Assurance Company), 7, 28–29

Shoomac Park (Robeson House), 255, 275

Shop fronts (Old City), 63, 64, 66, 67, 71, 77, 86, 87, 94, 99

Shops. *See* Houses, offices and shops in

Shot Towers, 157, 187

Shunk School (Beth Eden House), 316, 338

Shur's Lane Mills, T. Kenworthy & Bro., 257, 276

Siddons, William, House, 186

Sidney & Adams, landscape architects, 225

Silverman and Levy, architects, 121

Simon, Grant M., architect, 201

Simon and Bassett, architects, 83, 293, 350(n.74)

Simon and Simon, architects, 162, 375 (n.52)

Simons, Albert, xl(n.1)

Sims, Henry, architect, 111, 232

Sims, James P., architect, 232

Sims, Joseph, 145

Sims, Joseph Patterson, xxviii, xxxii

Sims-Bilsland House (2 Franklin Row), 106, 145

Singer, John, Warehouse, 58, 98

Singerly, Joseph, builder, 87

Singerly, Joseph, Houses, 338

Singley House. *See* Bel Air

Sink-Burgin House, 29

Site and Relic Society (Germantown Historical Society), 158

Skylights, 321, 326; banking room design, 45, 47, 87, 89, 92, 348(n.33); clerestory, 183, 184; Ridgway Library, 186

Skyscrapers: Center Square area, 113, 115, 116, 118, 120, 122, 141; early Philadelphia, 59, 80–81; institutional, 202

Slate Roof House, 42

Sloan, Samuel, architect, 213, 309, 363(n.30); Bennett and Co. Tower Hall, 44; Episcopal Hospital, 328; First Baptist Church, Germantown, 265; First Presbyterian Church, Kensington, 315; Joseph Harrison house, 108; Harrison Row, 107; IOOF Hall, Spring Garden, 287; Keen Building, 66; North Broad St. Presbyterian Church, 375 (n.52); "Philadelphia Plan" for school design, 157, 185; Shunk School, 338; terrace houses, 199, 218, 287

Sloan and Hutton, architects, 91

Sloan and Stewart, architects, 198

Smart, John, house carpenter, 235

Smedley, Walter, architect, 146

Smith, Daniel, Jr., House, 29

Smith, John B., House, 338

Smith, John D., carpenter and builder, 146

Smith, J. K., and E. K., 81

Smith, Richard, 239

Smith, Robert, master carpenter, xli(n.6), 25, 28, 65, 67, 96

Smith, R. Morris, architect, 30

Smith Memorial Arch (Civil War Memorial), 226, 238–39

Smithsonian Institution, 25

Smythe Building, 98

Snappers, The. See Northern Liberty Hose Company No. 4

Society Hill, and HABS, xxxvii

Society Hill Synagogue. See Spruce Street Baptist Church

Society Hill Towers, 8

Solitude (John Penn's House), 239

Solomon Brothers, contractors, 9

Somerton. See Strawberry Mansion

Sonder, Jacob, bricklayer, 340

Sons of Temperance Fountain, 98

Souder, Charles F., House, 98

Souder, Jacob, master mason, 296

South Philadelphia: development, 153–63; HABS buildings in, 165–90

Southeast Square. See Washington Square

Southern Loan Company of Philadelphia (Tradesmen's National Bank of Philadelphia), 30

Southwark, 153–54

Southwark Hose Company No. 9, 30

Sower, Christopher, Jr., 277

Spafford, William, House, 187

Sparks, Thomas, 157, 167

Sparks Shot Tower, 157, 187

Spectrum, The, 162

Sports complex (South Philadelphia), 162

Spring Garden District, 284, 287–88, 289, 300

Spring Garden Institute, 287, 300

Spring Garden Insurance Company Building, 49, 99

Spring Garden Street Bridge (Callowhill Street Bridge), 225, 239

Spring Garden Street Bridge (Upper Ferry Bridge), 196

Spring Street (Brooks Court, Augusta Place), 95

Springhouses, 15, 268, 269, 279

Spruce Hill, 202

Spruce Street Area Study, 145–46

Spruce Street Baptist Church (Society Hill Synagogue), 7, 30

Stables, 134. See also Barns

Stafford's Tavern. See Man Full of Trouble Tavern

Stage coaches, 311–12, 378(n.38)

Stearns and Castor, architects, 83, 298

Steel: stainless, 141, 366(n.54); structural use, 43, 50, 61, 80, 116, 129, 136, 139, 147, 213, 217, 293, 298, 299, 300, 328

Steele, William and Sons Company, architects and builders, 71, 134, 299, 374 (n.44)

Stenciled decorations, 144

Stenton, 251, 276–77

Stephen, Thomas, 100

Stewardson and Page, architects, 217

Stewart, Thomas, House, 31

Stewart, Thomas S., architect, 126, 299

Stewart, Noble, Class & Partners, architects, 88

Stiles, Edward, country house. See Port Royal

Stiles, Edward, House, 128 N. Front St., 99

Stiles, William, House, 31

Stocker, John Clement, House, 6, 31

Stokes, James, 41, 347(n.14)

Stone-Penrose House (Meredith House), 99–100, 101

Stonorov, Oskar, architect, 319, 332

Storey, Edward, builder, 175

Stortz, John, & Son, Store, 100

Stotesbury, E. T., 291

Stotz, Charles Morse, xl(n.1)

Strawberry Mansion (Summerville, Somerton), 240, 367(n.3)

Strawberry Mansion Bridge, 226, 240

Strawbridge, Justus C., 255, 371(n.39)

Streetcars, horse drawn, 198, 199, 200, 252. See also Transportation, public

Strickland, William, architect, 58, 96, 196, 363(n.21), 366(n.54); drawings by,

Strickland, William *(Cont.)*
  xxiv, 24, 234; Independence Hall Tower, xxvii, 78; Mechanics Bank, 84; Merchants' Hotel, 85; Musical Fund Society Hall, 85–86; Philadelphia Merchants' Exchange, 42, 89–90; St. Peter's tower, 28; St. Stephen's Church, 109, 137, 140, 144; 2nd Bank of U.S., 47, 97–98; U.S. Naval Asylum, 43, 159, 187–88; Wyck, remodeling, 279
Stride-Madison House, 31–32
Struthers, J., and Son, marble masons, 89, 91
Struthers, John, marble mason, 73, 84, 90, 97, 144, 187
Struthers, William, marble mason, 14, 71, 269
Struthers & Sons, marble contractors, 140
Sturges, John H., xlii(n.19)
Suburbs, 198–200, 252–53, 286, 292
Subways, 291–92. *See also* Transportation, public
Sully, Thomas, House, 32
Sunderland, John, 212
Summer Street (Swarthmore Place), 77
Summers-Worrell House, 32
Summerville. *See* Strawberry Mansion
Supplee, Davis E., architect, 299
Swain, William J., 199
Swede's Mill (Frankford Creek), 310
Sweetbrier, 240, 367(n.3)
Swope-Weissman Houses, 338–39

Tabernacle Baptist Church, 367(n.57)
Tabernacle Presbyterian Church, 367 (n.57)
Talbot, Fox, xxv
Tanner, Henry Ossawa, House, 300–301
Taverns, xxiv, 19, 41, 179, 204, 248, 268
Taylor, James Knox, architect, 301
Teitelman, Dr. Edward, 376(n.16)
Temple University, 291, 293, 375(n.45)
Tenth Presbyterian Church. *See* West Spruce Street Presbyterian Church
Terminal Commerce Building, the (Reading Co.), 291, 374(n.44)
Thalheimer and Weitz, architects, 89, 139
Theaters, 100, 112, 122–23, 226
Third Presbyterian Church (Old Pine; Corporate name: Third, Scots, and Mariners' Church), 32
Third Street, North, Area Study, 87
Thomas, Walter H., xliii(n.30)

Thomas, Churchman and Molitor, architects, 127
Thomas Jefferson University, 106
Thornton, Dr. William, 57
Tiffany and Bottom, 98
Tilden, Howard, 288
Tingley, W. B., and Company, 46
Tioga Railroad, 318
Titus, Lloyd, architect, 127
Tornatore, Rev. John B., 111, 124
Tourison, A. S., contractor, 267
Town houses, 54, 93, 102. *See also* Houses
Tradesmen's National Bank of Philadelphia, 88
Transportation, public, 198, 200, 202, 365(n.44). *See also* Frankford Elevated; Railroads; Stage coaches; Streetcars; Subways; Trolleys
Transportation Building, 120
Trautwein and Howard, architects, 217
Trautwine, John C., 24
Trinity Church, Oxford (Protestant Episcopal), xxxii, 307, 339
Trinity Lutheran Church, 247, 277
Trinity Lutheran Church House (Christopher Sower, Jr., House), 277
Trocadero, The. *See* Arch Street Opera-House
Trolleys, 7, 226, 240, 319
Trotter, Nathan and Company, 43, 45, 348(n.33)
Trout, William & Co., 175
Trumbauer, Horace, architect, 113, 133, 147, 237, 365(n.50)
Trump, Robert T., 15, 26
Turnpikes, 251–52
Tuscan style. *See* Italianate
Tuthill, Louisa C., xxv, xli(n.14)
Tutleman Brothers & Faggen Building, 100
Twaddell, Charles, bricklayer, 339
Twelfth Street (Friends) Meeting House (Meeting House of Central Philadelphia Meeting of Friends), 146

UGI Building, 110
Undine Barge Club, 225, 228
Union League of Philadelphia, 112–13, 146–47, 356(n.41)
Union School of Germantown. *See* Germantown Academy
United States Bonded Warehouse, 33
United States Circuit Court, 79

United States District Court, 79
United States Hose Company No. 14 (Blue Dick), 314, 339
United States Marine Corps Museum, 66
United States Mint (Community College of Philadelphia, Campus II), 301
United States Mint, Independence Mall, 301
United States Naval Asylum (United States Naval Home), 43, 159–60, 187–88, 348(n.24)
United States Naval Base, Quarters "A" (Commandant's Quarters), 188–89
United States Navy Yard, 156, 159
United States Supreme Court Building. See Independence Hall, Philadelphia City Hall
Universities and Colleges, 201–2, 216–18, 291–92, 365(n.50), 366(nn.51, 53)
University City, 202, 367(n.57); Science Center, 202
University Mews, 200
University Museum. See University of Pennsylvania
University of Pennsylvania, 201, 365 (n.50); College Hall, 201, 216; Library, 201, 216–17; Men's Dormitories, 201, 217; University Museum, 201, 218
Untenberger, Christoph, sculptor, 231
Upjohn, Richard, xxvi
Upjohn, Richard & Son, 315
Upper Burying Ground, Germantown, 263
Upper Ferry Bridge. See Spring Garden Street Bridge
Upsala, 251, 277–78; Foundation, 278
Urban renewal, xxix, xxxvii. See also Redevelopment Authority

Valley Green Canoe Club, 256, 268
Valley Green Inn, 278
van der Rohe, Mies, architect, 118, 366 (n.54)
Van Ingen, William B., mosaics by, 301
Van Ravenswaay, Charles, xlii(n.17)
Van Rensselaer, Sarah Drexel Fell (Mrs. Alexander Van Rensselaer), house. See Fell–Van Rensselaer House
Vaughan, Samuel, architect, 92
Vaults: brewery, 85; brick, Franklin Sugar Refining Co., 172; storage, 55, 75, 100
Vaux, George, 101
Vaux, Roberts, xxiv

Ventilation systems, 159, 183, 185
Venturi & Ranch, architects, 176, 214
Venturi, Robert, architect, 255
Vernacular architecture (Delaware Valley): church, 339; domestic, 235, 247–48, 249, 251, 261, 268–69, 270, 274; railroad stations, 184; schoolhouse, 260; warehouse, 174
Vernon, xxxi, 251, 278
Vesper Boat Club, 225, 228, 367(n.3)
Victorian (High) Baroque, 168
Victorian Byzantine style, 214
Victorian Eclectic Classicism, 80
Victorian Gothic style, 108, 112, 127, 130, 214
Victorian Jacobean style, 217. See also Jacobethan
Victorian Society of America, Philadelphia Chapter, 51
Victory Building. See New York Mutual Life Insurance Company Building
Villas, suburban, 198–99, 211, 252–53, 270, 272, 289, 296, 318, 322–23
Vint, Thomas C., xl(n.1), xliv(n.45)

Wade, Angus S., architect, 132
Wagon works, 335
Wallace, Philip B., xxviii
Walnut Street Jail, xli
Walnut Street Theatre, 58, 100–101
Walter, Joseph S., builder, 89
Walter, Thomas U., architect, 24, 27, 28, 67, 81, 96, 128; City Hall, 105, 140, 353(n.1); Girard College, Founder's Hall, 285, 297; Ivy Lodge, 270; Moyamensing Prison and Debtor's Wing, 160, 183; Philadelphia Contributionship, 49, 89; PSFS, 91; Portico Square, 53, 106–7, 142; Preston Retreat, 285, 286, 299; Spruce St. Baptist Church, 30; Winder houses attributed to, 35
Wanamaker, John, 108. See also Scott-Wanamaker House
Wanamaker, John, Store, 114, 116–17, 147–48
Ward, J. Q. A., sculptor, 239
Ware, William Rotch, xxvii–xxviii
Warehouses: design of Philadelphia, 40–41, 43; on Heap's "East Prospect," xxiii; in Northern Liberties, 313, 333, 335; Society Hill, 18, 33; Southwark, 156, 172, 174. See also listings by name

Warner, Benjamin, 284

Warner, John, 361(n.1)

Washington Hose Company No. 10, 101

Washington Square, 50; Area Study, 101–2

Washington Square Presbyterian Church. *See* First Presbyterian Church

Waterman, Thomas T., xxxii–xxxiii, xlv (nn.48, 52)

Water power: creeks supply, 223, 246, 255–56, 310; for Fairmount Waterworks, 224; for Manayunk mills, 256

Water Street, North, houses, 76, 77

Water Street, South, Area Studies, 29–30, 99

Water Trough and Fountain (S. Ninth St.), 33

Waterworks, Center Square, 106, 355 (n.34). *See also* Fairmount Waterworks

Watson, Frank R., architect, 169, 331, 367(n.57)

Watson, George, builder, 216, 325, 333

Watson, George and Son, contractor, 62, 335

Watson, George, & Sons, contractors, 129

Watson, John Fanning, xxiv–xxv, xli (nn.11, 12)

Watson and Huckel, architects, 169

Watson, Huckel and Co., architects, 139

Weaver, John, House, 340

Webster, George S., engineer, 231

Weccacoe Engine Company No. 9, 157

Weichmann, Harry, 298

Welsh, Albert, architect, 214

West Arch Street Presbyterian Church (Arch Street Presbyterian Church), 111, 148

West Philadelphia: development, 193–202, 361(n.1), 365(n.44); HABS buildings, 203–19

West Philadelphia Engine Co. *See* Columbia Engine Co.

West Spruce Street Presbyterian Church (Tenth Presbyterian), 111

Westcott, Thompson, xxvi

Western Saving Fund Society, 148

Westinghouse Electric and Manufacturing Co., 300

Wetherill, Herbert Jr., architect, 215

Wetherill, Joseph, House, 34

Wetherill, Samuel, 74

Wharton, Isaac, House, 34

Wharton, John, Houses, 189

Wharton, Joseph, House, 4, 34

Wharton-Stewart House, 189

Whitby Hall, xlii(n.22), 193–94

White, Ralph, architect, 325

White, Stanford, architect, 290, 374(n.38)

White, Mrs. Thomas Raeburn, 254

White, Bishop William, House, 54, 102

White Tower, building, 148–49

Whitehead, Russell F., xxviii, xlii(n.26)

Whittier, John Greenleaf, on Chalkley Hall, 306, 319, 375(n.6)

Whitton, Abednego J., Houses, 340

Wiccaco (Weccacoe), 153

Widener, Peter A. B., House, 289, 290, 301–2

Wiedersum, George, House, 340

Wiggins, John R., contractor, 74, 94, 206

Wildes, Joseph, bricklayer, 340

Wildes-Sonder Houses, 340

Wilds, J. S., engineer, 324

Williams, A. H., Sons, contractors, 134

Williams-Hopkinson House, 34–35

Williams-Mathurin House, 35

Williams and McNichol, builders, 168

Williamsburg, Va., architects endorse HABS, xxx, xliii–iv(n.38)

Williamson, Jesse, house carpenter, 168, 172

Willing, Charles, xxviii

Wilson, C. Henry, 71

Wilson, Joseph M., architect, 198, 213

Wilson Brothers and Company, architects, 92, 143, 200; Drexel Institute, 201, 206–7; North section, Broad St. Station, 138; Reading Terminal train shed, 115, 139; row houses, 365(n.44); Upsal station, 254

Wilson, Harris and Richards, architects, 275

Winder, Clarence E., 68

Winder, William H., Houses, 35

Windrim, James H., architect, 72, 84, 147, 296, 328; Bank of North America, 48, 62; Bank of Northern Liberties, 316, 333; Commercial Exchange, 42, 68; Dock St. sewer, 69; Kemble-Bergdol House, 274(n.33); Masonic Temple, 113, 140–41; Peoples Bank, 48; Philadelphia Trust, 48, 91, 374(n.36); Smith Memorial Arch, 239; Western Saving Fund, 148

Windrim, John T., architect, 48, 62, 91, 148, 161, 184, 239, 294

Winemore, Phillip, House, 35

Wissahickon Creek, mills along, 223, 246, 255–56

Wissahickon Heights, 254

Wissahickson Inn (Chestnut Hill Academy), 254, 278–79
Wissahickon Valley, National Natural History Landmark, 223, 224
Wister, Frances A., xxix
Wister, John, 247, 250
Wister's Big House. *See* Grumblethorpe
Wister's Tenant House (Grumblethorpe Tenant House), 279
Women's Club of Germantown, 270
Wood, George, House, 35–36
Wood, Robert, 64, 71, 121
Wood, William Halsey, architect, 143
Woodford, xxviii, 241, 367(n.3)
Woodland Terrace, 199, 218, 287
Woodlands, The, 195, 218–19, 251
Woodlands Cemetery, 195, 219, 362(n.17)
Woods, Captain John, House, 36
Woodward, Dr. George, 254–55
Wooley, Edmund, master carpenter, 77
Woolfall-Huddell House, 189
Woolmington, C. H., 80
Workman, John, 155, 189

Workman Place, xxxii, 155, 189–90
Works Progress Administration, 97
Worth Street, 311
Wren, Sir Christopher, 55
Wright, Frances, on Philadelphia, 3
Wyck (Haines House), xxv, xxxi, 279
Wynne, Dr. Thomas, 193, 219
Wynnefield, 193, 201, 366(n.53)
Wynnestay, 193, 219

Yang, C. C., architect, 125
York Court. *See* Olive Street

Zantzinger, Charles C., architect, 237
Zantzinger and Borie, architects, 213
Zantzinger, Borie and Medary, architects, 367(n.57)
Zimmers, Hugh M., architect, 15, 25
Zion (German) Lutheran Church, 52, 56
Zoological Gardens. *See* Philadelphia Zoological Gardens
Zoological Society, 238